Plant Stress Mitigators

AF739059

Anukool Vaishnav • S.S. Arya • D K Choudhary
Editors

Plant Stress Mitigators

Action and Application

Springer

Editors
Anukool Vaishnav
Department of Plant
and Microbial Biology
University of Zurich
Zurich, Switzerland

S.S. Arya
Department of Botany
Maharshi Dayanand University
Rohtak, India

D K Choudhary
Amity Institute of Microbial Technology
Amity University
Noida, India

ISBN 978-981-16-7761-8 ISBN 978-981-16-7759-5 (eBook)
https://doi.org/10.1007/978-981-16-7759-5

© The Editor(s) (if applicable) and The Author(s), under exclusive license to Springer Nature Singapore Pte Ltd. 2022
This work is subject to copyright. All rights are solely and exclusively licensed by the Publisher, whether the whole or part of the material is concerned, specifically the rights of translation, reprinting, reuse of illustrations, recitation, broadcasting, reproduction on microfilms or in any other physical way, and transmission or information storage and retrieval, electronic adaptation, computer software, or by similar or dissimilar methodology now known or hereafter developed.
The use of general descriptive names, registered names, trademarks, service marks, etc. in this publication does not imply, even in the absence of a specific statement, that such names are exempt from the relevant protective laws and regulations and therefore free for general use.
The publisher, the authors, and the editors are safe to assume that the advice and information in this book are believed to be true and accurate at the date of publication. Neither the publisher nor the authors or the editors give a warranty, expressed or implied, with respect to the material contained herein or for any errors or omissions that may have been made. The publisher remains neutral with regard to jurisdictional claims in published maps and institutional affiliations.

This Springer imprint is published by the registered company Springer Nature Singapore Pte Ltd.
The registered company address is: 152 Beach Road, #21-01/04 Gateway East, Singapore 189721, Singapore

Preface

Climate change and continuously increasing population are adding enormous pressure on the agricultural sector to provide consumers good quality food. In order to meet the qualitative and quantitative needs of the consumer, agricultural scientists must focus on plant stress physiology and different types of mitigators to alleviate the negative effects of stresses on plant productivity. Plant stress mitigators including beneficial microbes and other organic compounds (green chemicals, botanical extracts, microbial-derived products, etc.) help plants to cope with different types of biotic and abiotic stresses. The inclusion of such eco-friendly approaches in agricultural practices will help to achieve our goal for sustainable agriculture production. Over the past decade, several studies have been conducted on plant stress mitigators to find out new bioactive compounds and beneficial microorganisms to reveal the metabolism involved in remarkable crop performance and quality under different stress conditions. These mitigators have gained maximum attention of farmers and industrialists in incorporating these products in organic farming practices to quantitative and qualitative foods. Hence, these plant stress mitigators have a huge global market. The plant stress mitigators follow different action mechanisms for enhancing plant growth and stress tolerance capacity including nutrient solubilizing and mobilizing, biocontrol activity against plant pathogens, phytohormone production, soil conditioners, and many more unrevealed mechanisms.

This book discusses the stress alleviation action of different plant stress mitigators on crops grown under optimal and suboptimal growing conditions (abiotic and biotic stresses). The area of interest also includes potential contributions regarding the effect as well as the molecular and physiological mechanisms of plant stress mitigators on nutrient efficiency, product quality, and the modulation of the microbial population quantitatively and qualitatively. The content of this book is divided into three parts, viz. (1) Climate change impacts on plant and soil health, (2) Microbe mediated plant stress mitigation, and (3) Advances in plant stress mitigation. We strongly believe that this compilation of high-standard scientific chapters on the principles and practices of plant stress mitigators will foster knowledge transfer among scientific communities, industries, and agronomists and will enable a better understanding of the mode of action and application procedure of these mitigators in different cropping systems.

The editors would like to acknowledge all the contributors for their efforts in making this book worthy to disseminate complete knowledge for scholarly involvement around the globe. We are grateful to our family members for their constant support and blessings. Also, the editors would like to thank the handling editors and Springer team for the opportunity to publish this book.

Zurich, Switzerland — Anukool Vaishnav
Rohtak, India — S.S. Arya
Noida, India — D K Choudhary

Contents

About the Editors

Anukool Vaishnav is working as a Swiss Excellence Fellow in the University of Zurich, Switzerland, and also associate with GLA University, Mathura, India as an Assistant Professor.

S.S. Arya is working as an Assistant Professor in the Department of Botany, Maharshi Dayanand University (MDU), Rohtak, India.

D K Choudhary is a Member of National Academy of Sciences (MNASc) and is currently working as an Associate Professor in Microbiology at Amity University, Noida.

Part I
Climate Change Impacts on Plant and Soil Health

Climate Change Impact on Alteration of Plant Traits

Pooja Ahlawat, Babita Rani, Anita Kumari, Seema Sangwan, Pooja, Sarita Devi, and Vinod Goyal

Abstract One of the main drivers of organism growth and species distributions is climate; thus, a drastic change in climate has many consequences for plants, be it droughts, heat waves, or increased flooding. Besides these cumulative effects of global warming, rising concentration of carbon dioxide and rapid increase in temperature affect the persistence, growth, and reproduction of plants. These impacts would have devastating consequences on natural vegetation, agriculture, and humankind too and are very difficult to predict. A large number of studies reflected the importance of climate change, investigating how plants respond to rising CO_2 concentration and temperature as well as interaction with other environmental factors. Consequences of these impacts were already felt, as the 0.5° warming of Earth's temperature in the last 50 years shifted the distribution of plant and animal species and crop yields have decreased. In this chapter, we will focus on the effects of increasing atmospheric CO_2 concentration, high temperature, drought stress, and their interaction with plant developmental processes in roots and leaves, plant species distribution, phenology, and reproductive structures of plants. This compendium of research provides an important means for predicting shift in forests, ecosystems, and crop patterns in the coming decades and for finding ways to protect and adapt plants in order to avoid the harmful effects of global climate change.

Keywords High CO_2 · High temperature · Plant species · Plant phenotypes

P. Ahlawat · B. Rani · A. Kumari (✉) · S. Devi · V. Goyal
Department of Botany and Plant Physiology, CCS Haryana Agricultural University, Hisar, Haryana, India

S. Sangwan
Department of Microbiology, CCS Haryana Agricultural University, Hisar, Haryana, India

Pooja
ICAR—Sugarcane Breeding Institute, Regional Center, Karnal, Haryana, India

© The Author(s), under exclusive license to Springer Nature Singapore Pte Ltd. 2022
A. Vaishnav et al. (eds.), *Plant Stress Mitigators*,
https://doi.org/10.1007/978-981-16-7759-5_1

Introduction

Climate is a long atmospheric pattern in a particular region. The weather can change from time to time, daily, monthly, or yearly. However, climate is a weather pattern over a period of more than 30 years. The desert area can experience a rainy week but in the long run has little rain fall. The desert area has a dry climate and the climate is almost constant, so organisms can adapt to it. The enormous diversity of life on Earth occupies most of the various climates that exist, but the climate changes. The change in climate due to natural causes is extremely slow. As the climate changes, the organisms that live in the area are at risk of adaptation, relocation, or extinction.

The climate of the Earth has changed many times. Changes in atmospheric greenhouse gases also affect the climate. Many anthropogenic activities result in huge emission of greenhouse gases. Gases such as carbon dioxide trap solar heat in the atmosphere, raising the temperature of Earth's surface. These activities are dramatic and dangerously changing the global climate. Average global temperature has been rising since about 1880 (Fig. 1). Due to greenhouse gases, Earth's atmosphere temperature is increasing gradually. Severe climate change can lead to more severe weather patterns: more tornadoes, typhoons, and hurricanes. Where there is less rainfall, there is more rainfall elsewhere. When the climate changes, the habitat of living things changes. In the history of the Earth, the climate has changed multiple times, but the changes have occurred slowly over thousands of years. But due to human activities, the pace of climate change has accelerated which led to severe and prolonged changes in atmospheric composition (such as rise in CO_2 concentration),

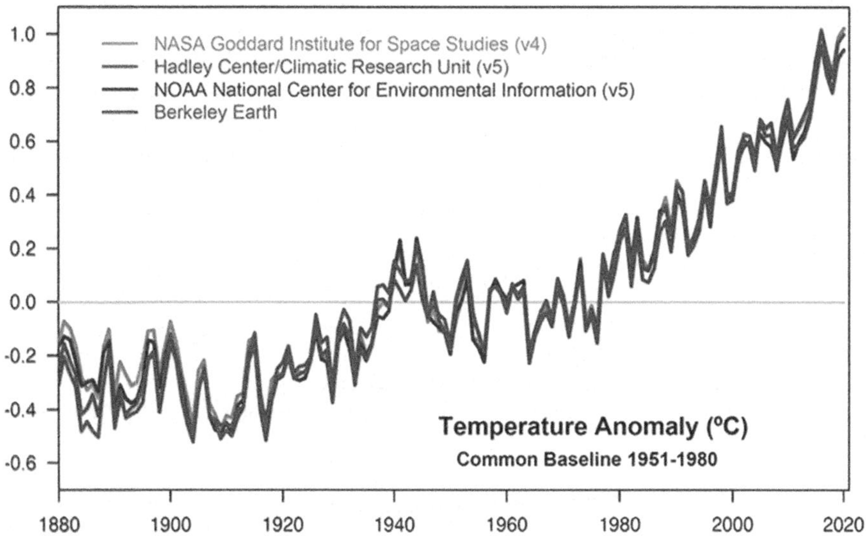

Fig. 1 The data shows continuous rise in temperature in the past few decades. The highest temperatures were recorded in 2016 and 2020. Source: NASA's Goddard Institute for Space Studies

temperature, precipitation, and hydrological cycles. These changes in turn affect the flora and fauna of the Earth.

The climate change which causes the increase in global temperature greatly affects plants by modifying their developmental, physiological, and metabolic activities. The impact of climate change is observed on crop yield and quality. These effects might be positive or negative, but the net effect is likely to be negative which poses great menaces to the global food safety. Plants are affected directly and indirectly by climate change. Direct impact of climate change includes the following.

Variations in Plant Species Distribution

Climate change factors, such as changes in local temperature and precipitation that exceed the permissible range of phenotypic plasticity of the species, lead to change in its distribution, which is unavoidable. Evidences support that change in the regional climate results in shifting of plant species in their distribution in altitude and latitude. However, it is difficult to predict how the range of species changes with climate and to differentiate these changes from all other anthropogenic environmental changes such as eutrophication, acid rain, and habitat destruction. The rapid change in the climate has great potential to alter plant species distribution as well as render many species out of adaptable climates. Information on how species can adjust and support quick changes is still relatively limited.

Variations in Life Cycles (Phenology) of Plants

Climate greatly influences the phenology of plants as they are intricately linked to the climatic factors such as photoperiod and temperature which acts as an important stimulus for the flowering. Along with flowering and fruiting (Xia and Wan 2013), other processes like germination (De Frenne et al. 2012) and leaf emergence (Jeong et al. 2011) are influenced by changing climate. An interesting example of how warming of climate influences fertilization is of the early spider orchid (*Ophrys sphegodes*). It depends upon deception to reproduce. A vital step in its pollination is to trick the lonely male bees into thinking that the plant is its female partner by secreting pheromone (Robbirt et al. 2010). This phenomenon is called pseudocopulation. The orchid's body is bulbous and crimson which looks like an insect. This works because the blooming time of orchid coincides with the emergence of male bees from hibernation but before female bees appear. But as the temperature is rising and period of growing season is prolonged leading to earlier onset of spring, female bees are emerging earlier than orchid bloom. With each degree Celsius of rise of temperature, there are decreases in the period of 6.6 days in the emergence of male and female bees which provides orchid less chances to reproduce (Robbirt et al. 2010; Robbirt et al. 2014).

Indirect Impacts of Climate Change

Climate change is happening right now; there is a dire need of change to bring down discharge of greenhouse gases which cause heating of the atmosphere. Climate change is absolutely linked to the greenhouse gases which have a great effect on the crop productivity (IPCC 2014). The elevation in CO_2 concentrations increases the net photosynthesis as well as improves the water use efficiency of plants (Deryng et al. 2016). As photorespiration is reduced due to elevated CO_2, it results in higher photosynthesis, but this effect is observed only in C_3 plants like wheat, rice, and soybean. If this rise in CO_2 remains constant, it will negatively affect the climate although crop yield will increase (Senapati et al. 2019; Wei et al. 2019). Ozone is one of the more reactive oxidants which also leads to severe damage to plant tissue and leaf injuries as well increases the pace of senescence and finally cell death (Vandermeiren et al. 2009). Changes in O_3 will have a considerable detrimental impact on key agricultural crop yields.

Climate change has led to shift in pattern of rainfall. Severity of rainfall has increased the incidences of flood, and intense drought spells and offseason precipitations are expected. A significant loss of plant is observed if the crop faces offseason rainfall during its critical stages (Lobell and Burke 2008). An intense rainfall during winter and autumn increases the pest population in oilseed rape which might lead to occurrence of diseases (Sharif et al. 2017). Unseasonable damage in the young plants of soybean and maize is observed due to more pronounced rainfall during spring (Urban et al. 2015). Severe rainfall is associated with extreme floods. Countries like Bangladesh and China have much of their harvest area within flood-threatened regions. Food safety is at risk in such countries where floods either delay crops or destroy the cultivable land (Fig. 2).

Flooding causes low O_2 and CO_2 levels due to anoxia; in addition to that, high salt content of seawater will lead to ionic and osmotic stress in plants. It has been observed that if *Brassica napus* is exposed to seawater flooding conditions, it will face lower seed mass and lower number of siliques which will contribute to reduction in crop productivity (Hanley et al. 2019).

With the rise in temperature and decrease in rainfall, certain regions face frequent and prolonged drought events. Drought affects the growing season negatively and causes a severe drop in crop production, a major effect on crop productivity (Saadi et al. 2015; Lesk et al. 2016; Zipper et al. 2016). The most critical stage to be affected by drought is the reproductive stage which greatly leads to reduction in crop yield. Drought faced at this stage leads to ovary abortion in *Zea mays*, and pollens are rendered infertile in barley and wheat, and all these factors result in a decrease in the number of kernels and overall biomass of the plant (Boyer and Westgate 2004). In generalized terms drought leads to water deficiency in plants which sends signal to

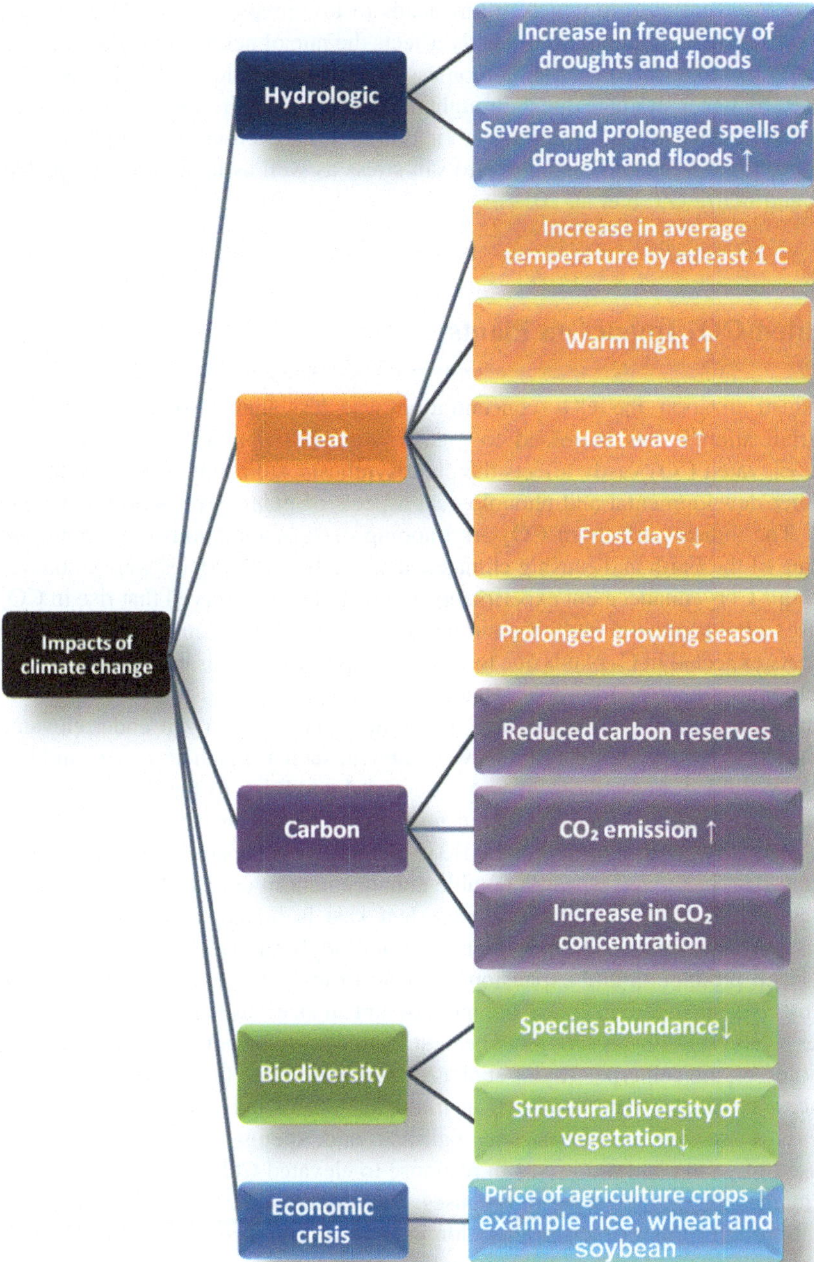

Fig. 2 Impacts of climate change (up arrow corresponds to increase while down arrow to decrease)

plant for stomatal closure which in turn leads to low intake of CO_2. Decreases in CO2 concentration within plants directly affects the rate of net photosynthesis which is directly linked to the reduction in biomass of plant (Garofalo et al. 2019). Salinity is caused by water shortage due to drought conditions. Osmotic stress due to salinity reduces the plant growth significantly. Yield is reduced even in tolerant plants such as barley, sugar beet, and cotton due to salt stress as well as in sensitive crops like wheat and maize (Zörb et al. 2019).

Elevated CO_2 Impact on Plants

Pre-industrialization the CO_2 concentration was less than 280 ppm, but post-industrialization it has increased to greater than 400 ppm today (Meehl et al. 2007). Elevated CO_2 vastly affects the photosynthesis, stomatal gaseous exchange, and other developmental and phonological aspects of plants (Ainsworth and Long 2005). The indirect effects of CO_2 are trapping of solar radiations contributing to warming of the Earth and causing climate change. About 40 species were studied at 12 free-air CO_2 enhancement experiments in which it was observed that rise in CO_2 increases photosynthetic carbon assimilation rates by 31% (Ainsworth and Long 2005). As expected in C_3 plants due to increase in photosynthesis in response to high CO2, biomass increased by 20%. Reich et al. (2014) reported a 33% increase in biomass in a grassland provided with optimum water and nitrogen under elevated CO_2; however, when other conditions were altered, such as much less water and low nitrogen provided, lower biomass was observed. In addition to shoot biomass, root biomass also increased significantly in response to high concentration of CO2 in many plants (Madhu and Hatfield 2013). In many crop species such as peanut, soybean, rice, bean, and wheat, a notable enhancement in crop yield was observed (Hatfield et al. 2011). Increased average leaf size is linked to the aboveground biomass by increasing CO_2 in poplar trees and soybean (Dermody et al. 2006). Higher cell production and expansion can lead to increased leaf size, and both of these mechanisms seem to contribute increase in leaf size, with enhanced CO_2 in the species where these processes have been studied. These responses vary quite a bit depending on the cell type. Taylor et al. (2008) reported that increased level of CO_2 enhances the size of epidermal cell in immature leaves but not in mature hybrid *Populus* × *euramericana*, whereas palisade and spongy mesophyll cell size increased in young and old leaves in response to elevated CO_2.

In many species, increasing CO_2 causes a considerable increase in root biomass. More thorough studies describe alteration in architecture of root as well as changes in the structure of cell to shed light on how this rise in biomass can occur. However, no comprehensive study has looked at how all of these responses are conserved between and within different plant species. Mini-rhizotron tests in soybean showed that increased CO_2 enhance the length of root, especially at shallow and intermediate soil depth, and that enhanced CO_2 associated with lower rainfall enhances the

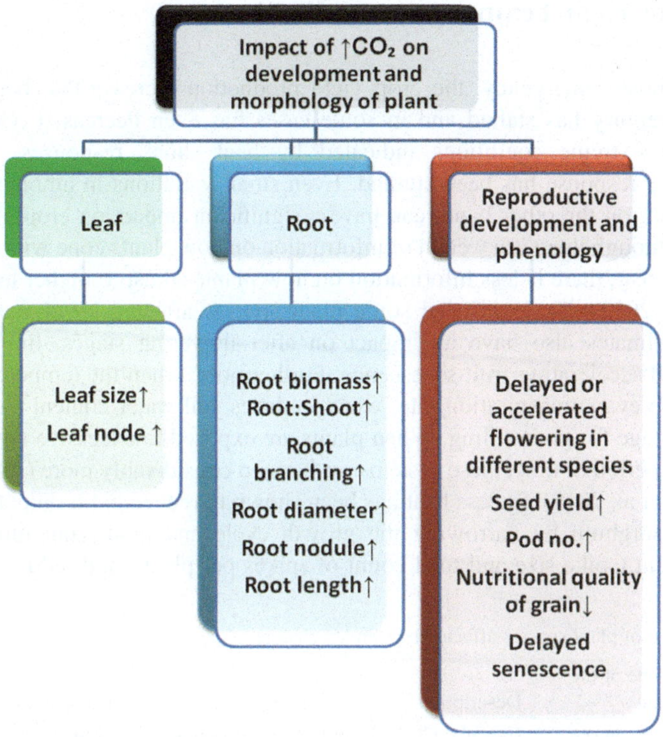

Fig. 3 Effect of high concentration of CO_2 on development and morphology of plant (up arrow corresponds to increase while down arrow to decrease)

quantity and root nodule density which is inhabited by nitrogen-fixing (Bradyrhizobia) bacteria (Gray et al. 2013, 2016) (Fig. 3).

Although increased CO_2 boosts seed yield in a variety of crop plants, the grain nutritional quality suffers as a result of variation in ion profiles, particularly lower iron and zinc concentration (Loladze 2014). Increased CO_2 also lowers the nitrogen and protein content of legume crop seeds (Myers et al. 2014). The increased CO_2 enhanced seed output by an average of 9% across 18 soybean genotypes across repeated growing seasons, but the fractionating coefficient dropped by 11% (Bishop et al. 2014). The increased amount of seed yield caused by high CO_2 varied with climate, with seed yield decreasing to zero in dry and hot conditions (Gray et al. 2016).

Impact of High Temperature on Plants

Due to climate emergency, the crop yield production increase that began in the previous century has stalled and in some cases has even decreased (Quint et al. 2016). At extreme conditions indicated by heat shock responses, the high-temperature response has been studied. Even small variations in ambient growing temperature, on the other hand, can have a significant impact on crop growth and output. Although there is a wealth of information on how plants cope with extremely damaging heat, there is less information on how plants adjust to higher temperature (Vu et al. 2019). Table 1 listed some plant species affected by climate change. Warmer climates also have an impact on after flowering stages, limiting grain growth and accelerating fruit senescence. Furthermore, when the temperature rises, so does the evapotranspiration rate, which reduces soil water content and leads to water shortage for grain filling. When plants are exposed to excessive temperatures for a short period of time, these systems are harmed considerably more (Asseng et al. 2019; Liu et al. 2019). Excess heat has been shown to reduce grain output in wheat, rice, and sorghum by narrowing the growth cycle and modifying inflorescence development (spike size and total count of spikes per plant), and reduce grain size

Table 1 List of plant species affected by climate change

S. No.	Name of the plant	Description	References
1.	Soybean (*Glycine max*)	Elevated CO_2—Root length and number of nodules increased Drought—In shallow depth soil root length decreased, no. of pods decreased, and accelerates maturity	Gray et al. (2013), Bishop et al. (2014), Gray et al. (2016)
2.	Rice (*Oryza sativa*)	High temperature leads to spikelet sterility and delayed flowering Drought leads to reduced plant height, biomass and tillers, overall reduction in yield	Jagadish et al. (2007), Pandey and Shukla (2015)
3.	Wheat (*Triticum aestivum*)	Drought and high temperature lead to early maturation and yield reduction as well as reduced pollen viability	Naresh Kumar et al. (2014), Daloz et al. (2021), Trnka et al. (2019)
4.	Barley (*Hordeum vulgare*)	Reduced grain filling which results in yield loss due to drought and high temperature	Daničić et al. (2019), Olesen et al. (2011)
5.	Maize (*Zea mays*)	High temperature and water deficiency lead to significant reduction in yield	Adhikari et al. (2015)
6.	Cotton (*Gossypium* spp.)	Slight increase in temperature has positive impact on cotton plant growth and lengthens the cotton growing season. Elevated CO_2 could favor cotton yields	Li et al. (2020), Bange (2007)
7.	Pearl millet (*Pennisetum glaucum*)	Drought at flowering stage decreases grain filling which results in unstable yields	Gloria (2013), Azare et al. (2020)

and grain per spike (Lesk et al. 2016; Fahad et al. 2017). Furthermore, yield losses in *Brassica juncea*, *Brassica rapa*, and oilseed rape are due to a decrease in the number of silique per plant and number of seed per silique, as well as abnormal pod development (Angadi et al. 2000; Morrison and Stewart 2002). Extreme heat also changes the seed content, resulting in a decline in crop quality. High temperature lowers the oil, protein, and carbohydrate content of seeds in cereals and oilseed crops (Jagadish et al. 2015; Fahad et al. 2017). Elevated temperatures in wheat have been demonstrated to diminish the quantities of important protein while causing the buildup of proline and soluble carbs (Qaseem et al. 2019).

Plants require more water in tropical climates owing to enhanced evaporation and transpiration and reduced water absorption by the root, resulting in an overall drought (Heckathorn et al. 2013). Various responses of plasmatic membrane fluidity and aquaporins to enhance temperature in roots have been seen in studies using a variety of crops. Water intake in warmer soil appears to be positively correlated with aquaporin functioning in pepper and wheat (Maurel et al. 2015). Variations in climate also affect nutrient content. The influence of temperature on nutrient uptake varies depending on the crops, just as it does with water. Relatively warm soils inhibit root growth and reduce nutrient uptake, resulting in lower macro- and micronutrient levels in tomatoes (Giri et al. 2017). Supra-optimal root temperature causes a reduction in root and branch growth in *Andropogon gerardii*, a fodder plant. Higher temperatures have a mild effect on nitrogen intake, but their efficient utilization is significantly hampered. Warm temperatures, on the other hand, had no effect on maize potassium, phosphorus, and nitrogen uptake, but increased temperatures appear to modestly decrease potassium and phosphorus uptake (Hussain et al. 2019). Rising temperature responsible for the negative root reactions jeopardizes water and nutrient uptake, resulting in a significant drop in crop output.

Increased temperature can affect reproductive growth by changing the schedule of reproductive stages or damaging reproductive structures with heat. When a plant is sown at a higher temperature, reproductive developmental events tend to occur sooner. When *Arabidopsis* is cultivated at a high temperature, for example, an early transition of flowering has been studied extensively (Balasubramanian et al. 2006; van Zanten et al. 2013). As the temperature goes up, many crop species said to advance more quickly through vegetative and reproductive development, up to a species-specific ideal, and then after, growth and development decline and finally stop (Hatfield et al. 2011). Flowering is accelerated in high-temperature environments, which may limit the plant's ability to collect the resources needed for successful gamete production (Zinn et al. 2010). Burghardt et al. (2016) revealed that in numerous *Arabidopsis* accessions, variable high temperatures induced flowering to commence even sooner than steady high temperatures, implying that range of temperatures, in addition to mean temperature, regulates flowering time. By changing the cold season chilling circumstances, higher temperatures may impact the timeline and effectiveness of reproductive development.

Conclusion

Changing climate will influence plant growth in such a way that will have a profound impact on agricultural plant and natural ecosystem function. Temperatures will rise in the future, as will the prevalence of extreme occurrences such as heat waves and droughts, as well as changes in the atmospheric composition (IPCC 2014). In the future, due to climate change plant developmental flexibly will be crucial for ecosystem function and crop productivity. The amount of trials undertaken in physiologically relevant environmental stresses now limits our capacity to recognize and anticipate plant growth response to climate changes. For instance, we know more about the molecular mechanism underlying plant responses to high drought conditions than we do about the mechanisms underlying mild drought conditions (Clauw et al. 2015). Merging of data from different levels of biological study, particularly molecular investigations of growth process at better spatial resolution, is required for a more comprehensive knowledge of plant responses to the aspects of climate change. Molecular studies should be conducted in various species at the cell type and tissue level in controlled and realistic situations to examine the effect of developmental and molecular alterations on entire plant morphology and yield. Ultimately, plants will not be unaffected by climate change elements in isolation: increasing temperatures, altering hydrological cycle, and an increase in the frequency of extreme weather events will all occur as concentration of greenhouse gases rise in the future (IPCC 2014). It will be crucial to improve our mechanistic understanding of plant developmental responses to various, interacting aspects of climate change in order to predict implications on agricultural and natural systems.

References

Adhikari U, Nejadhashemi AP, Woznicki SA (2015) Climate change and eastern Africa: a review of impact on major crops. Food Energy Sec 4:110–132

Ainsworth EA, Long SP (2005) What have we learned from 15 years of free-air CO_2 enrichment (FACE)? A meta-analytic review of the responses of photosynthesis, canopy properties and plant production to rising CO_2. New Phytol 165:351–372

Angadi SV, Cutforth HW, Miller PR, McConkey BG, Entz MH, Brandt SA (2000) Response of three brassica species to high temperature stress during reproductive growth. Can J Plant Sci 80: 693–701

Asseng S, Martre P, Maiorano A, Rötter RP, O'Leary GJ, Fitzgerald GJ (2019) Climate change impact and adaptation for wheat protein. Glob Chang Biol 25:155–173

Azare I, Dantata I, Abdullahi M, Adebayo A, Aliyu M (2020) Effects of climate change on pearl millet (*Pennisetum glaucum* [L. R. Br.]) production in Nigeria. J Appl Sci Envl Manag 24(1): 157

Balasubramanian S, Sureshkumar S, Lempe J, Weigel D (2006) Potent induction of Arabidopsis thaliana flowering by elevated growth temperature. PLoS Genet 2:0980–0989

Bange M (2007) Effects of climate change on cotton growth and development. Aust Cotton Grower 28(3):41–45

Bishop KA, Betzelberger AM, Long SP, Ainsworth EA (2014) Is there potential to adapt soybean (*Glycine max Merr.*) to future [CO_2]? An analysis of the yield response of 18 genotypes in free-air CO_2 enrichment. Plant Cell Environ 38:1765–1774

Boyer JS, Westgate M (2004) Grain yields with limited water. J Exp Bot 55:2385–2394

Burghardt LT, Runcie DE, Wilczek AM, Cooper MD, Roe JL, Welch SM, Schmitt J (2016) Fluctuating, warm temperatures decrease the effect of a key floral repressor on flowering time in Arabidopsis thaliana. New Phytol 210:564–576

Clauw P, Coppens F, DeBeuf K, Dhondt SV, Daele T, Maleux K, Storme V, Clement L, Gonzalez N, Inzé D (2015) Leaf responses to mild drought stress in natural variants of Arabidopsis. Plant Physiol 167:800–816

Daloz A, Rydsaa J, Hodnebrog Ø, Sillmann J, van Oort B, Mohr C, Agrawal M, Emberson L, Stordal F, Zhang T (2021) Direct and indirect impacts of climate change on wheat yield in the indo-Gangetic plain in India. J Agri Food Res 4:100132

Daničić M, Zekić V, Mirosavljević M, Lalić B, Putnik-Delić M, Maksimović I, Dalla Marta A (2019) The response of spring barley (*Hordeum vulgare* L.) to climate change in northern Serbia. Atmos 10(1):14

De Frenne P, Graae BJ, Brunet J (2012) The response of forest plant regeneration to temperature variation along a latitudinal gradient. Ann Bot 109:1037–1046

Dermody O, Long SP, DeLucia EH (2006) How does elevated CO_2 or ozone affect the leaf-area index of soybean when applied independently? New Phytol 169:145–155

Deryng D, Elliott J, Folberth C, Müller C, Pugh TAM, Boote KJ (2016) Regional disparities in the beneficial effects of rising CO_2 concentrations on crop water productivity. Nat Clim Chang 6: 786–790

Fahad S, Bajwa AA, Nazir U, Anjum SA, Farooq A, Zohaib A (2017) Crop production under drought and heat stress: plant responses and management options. Front Plant Sci 8:1147

Garofalo P, Ventrella D, Kersebaum KC, Gobin A, Trnka M, Giglio L (2019) Water footprint of winter wheat under climate change: trends and uncertainties associated to the ensemble ofcrop models. Sci Total Environ 658:1186–1208

Giri A, Heckathorn S, Mishra S, Krause C (2017) Heat stress decreases levels of nutrient-uptake and -assimilation proteins in tomato roots. Plan Theory 6:6

Gloria C (2013) Genetic characterization of pearl millet (Pennisetum Glaucum (L.) R. Br.) genotypes in Zambia. Unpublished MSc. Dissertation submitted to the University of Zambia. Pp. 1–103

Gray SB, Strellner RS, Puthuval KK, Shulman R, Siebers M, Rogers A, Leakey ADB (2013) Nodulation of field grown soybean is enhanced by free-air CO_2 enrichment only when combined with drought stress. Funct Plant Biol 40(2):137–147

Gray SB, Siebers M, Locke AM, Rosenthal DR, Strellner RS, Paul RE, Klein SP, Ruiz Vera UM, Mc Grath J, Dermody O, Ainsworth EA, Bernacchi CJ, Long SP, Ort DR, Leakey ADB (2016) Intensifying drought eliminates the expected benefits of elevated [CO_2] for soybean. Nat Plants 2, 16132

Hanley ME, Hartley FC, Hayes L, Franco M (2019) Simulated seawater flooding reduces oilseed rape growth, yield and progeny performance. Ann Bot 125:247–254

Hatfield JL, Boote KJ, Kimball BA, Ziska LH, Izaurralde RC, Ort D, Thomson AM, Wolfe D (2011) Climate impacts on agriculture: implications for crop production. Agron J 103:351–370

Heckathorn SA, Giri A, Mishra S, Bista D (2013) Heat stress and roots. In: Tuteja N, Gill SS (eds) Climate change and plant abiotic stress tolerance. Weinheim: Wiley-VCH Verlag GmbH & Co., Weinheim, pp 109–136

Hussain HA, Men S, Hussain S, Chen Y, Ali S, Zhang S (2019) Interactive effects of drought and heat stresses on morpho-physiological attributes, yield nutrient uptake and oxidative status in maize hybrids. Sci Rep 9:7

IPCC (2014) Climate change 2014: synthesis report. In: Pachauri RK, Meyer LA (eds) Contribution of working groups I, II and III to the fifth assessment report of the intergovernmental panel on climate change. IPCC, Geneva, p 151

Jagadish S, Crauford PQ, Wheeler TR (2007) High temperature stress and spikelet fertility in rice (Oryza sativa L.). J Exp Bot 58:1627–1635

Jagadish SVK, Murty MVR, Quick WP (2015) Rice responses to rising temperatures - challenges, perspectives and future directions. Plant Cell Environ 38:1686–1698

Jeong SJ, Ho CH, Gim HJ, Brown ME (2011) Phenology shifts at start vs end of growing season in temperate vegetation over the northern hemisphere for the period 1982–2008. Glob Chang Biol 17:2385–2399

Lesk C, Rowhani P, Ramankutty N (2016) Influence of extreme weather disasters on global crop production. Nature 529:84–87

Li N, Lin H, Wang T, Li Y, Liu Y, Chen X, Hu X (2020) Impact of climate change on cotton growth and yields in Xinjiang, China. Field Crops Res 247:107590

Liu B, Martre P, Ewert F, Porter JR, Challinor AJ, Müller C (2019) Global wheat production with 1.5 and 2.0_C above pre-industrial warming. Glob Chang Biol 25:1428–1444

Lobell DB, Burke MB (2008) Why are agricultural impacts of climate change so uncertain? The importance of temperature relative to precipitation. Environ Res Lett 3:7

Loladze I (2014) Hidden shift of the ionome of plant sex posed to elevated CO_2 depletes mineral salt the base of human nutrition. elite:1–29

Madhu M, Hatfield JL (2013) Dynamics of plant root growth under increased atmospheric carbon dioxide. Agron J 105:657–669

Maurel C, Boursiac Y, Luu DT, Santoni V, Shahzad Z, Verdoucq L (2015) Aquaporins in plants. Physiol Rev 95:1321–1358

Meehl GA, Stocker TF, Collins WD, Friedlingstein P, Gaye AT, Gregory JM, Kitoh A, Knutti R, Murphy JM, Noda A, SCB R, Watterson IG, Weaver AJ, Zhao ZC (2007) Global climate projections. In: Solomon S, Qin D, Manning M, Chen Z, Marquis M, Averyt KB, Tignor M, Miller HL (eds) Climate change 2007: the physical science basis. Contribution of Working Group I to the Fourth Assessment Report of the Intergovernmental Panel on Climate Change. Cambridge University Press, Cambridge/New York, NY

Morrison MJ, Stewart DW (2002) Heat stress during flowering in summer. Brassica Crop Sci 42: 797–803

Myers SS, Zanobetti A, Kloog I, Huybers P, Leakey ADB, Bloom AJ, Carlisle E, Dietterich LH, Fitzgerald G, Hasegawa T, Holbrook NM, Nelson RL, Ott-man MJ, Raboy V, Sakai H, Sartor KA, Schwartz J, Seneweera S, Tausz M, Usui Y (2014) Increasing CO_2 threatens human nutrition. Nature 510:139–142

Naresh Kumar S, Aggarwal P, Swaroopa Rani D, Saxena R, Chauhan N, Jain S (2014) Vulnerability of wheat production to climate change in India. Clim Res 59(3):173–187

Olesen JE, Trnka M, Kersebaum KC, Skjelvag AO, Seguin B, Peltonen-Sainio P, Rossi P, Kozyra J, Micale F (2011) Impacts and adaptation of European crop production systems to climate change. Eur J Agron 34:96–112

Pandey V, Shukla A (2015) Acclimation and tolerance strategies of rice under drought stress. Rice Sci 22:147–161

Qaseem MF, Qureshi R, Shaheen H (2019) Effects of pre-anthesis drought, heat and their combination on the growth, yield and physiology of diverse wheat (*Triticum aestivum* L.) genotypes varying in sensitivity to heat and drought stress. Sci Rep 9(477):43477-z. https://doi.org/10.1038/s41598-019-

Quint M, Delker C, Franklin KA, Wigge PA, Halliday KJ, van Zanten M (2016) Molecular and genetic control of plant thermo morphogenesis. Nat Plants 2:15190

Reich PB, Hobbie SE, Lee TD (2014) Plant growth enhancement by elevated CO_2 eliminated by joint water and nitrogen limitation. Nat Geo Sci 7:920–924

Robbirt KM, Davy AJ, Hutchings MJ, Roberts DL (2010) Validation of biological collections as a source of phenological data for use in climate change studies: a case study with the orchid Ophrys sphegodes. J Ecol 99(1):235–241

Robbirt K, Roberts D, Hutchings M, Davy A (2014) Potential disruption of pollination in a sexually deceptive orchid by climatic change. Curr Biol 24(23):2845–2849

Saadi S, Todorovic M, Tanasijevic L, Pereira LS, Pizzigalli C, Lionello P (2015) Climate change and Mediterranean agriculture: impacts on winter wheat and tomato crop evapotranspiration, irrigation requirements and yield. Agric Water Manag 147:103–115

Senapati N, Brown HE, Semenov MA (2019) Raising genetic yield potential in high productive countries: designing wheat ideotypes under climate change. Agric For Meteorol 271:33–45

Sharif B, Makowski D, Plauborg F, Olesen JE (2017) Comparison of regression techniques to predict response of oilseed rape yield to variation in climatic conditions in Denmark. Eur J Agron 82:11–20

Taylor G, Tallis MJ, Giardina CP, Percy KE, Miglietta F, Gupta PS, Gioli B, Calfapietra C, Gielen B, Kubiske ME, Scarascia-Mugnozza GE, Ket K, Long SP, Karnosky DF (2008) Future atmospheric CO_2 leads to delayed autumnal senescence. Glob Chang Biol 14:264–275

Trnka M, Feng S, Semenov MA, Olesen JE, Kersebaum KC, Rötter RP, Semerádová D, Klem K, Huang W, Ruiz-Ramos M, Hlavinka P, Meitner J, Balek J, Havlík P, Büntgen U (2019) Mitigation efforts will not fully alleviate the increase in water scarcity occurrence probability in wheat-producing areas. Sci Adv 5:eaau2406. https://doi.org/10.1126/sciadv.aau2406

Urban DW, Roberts MJ, Schlenker W, Lobell DB (2015) The effects of extremely wet planting conditions on maize and soybean yields. Clim Chang 130:247–260

Van Zanten M, Bours R, Pons TL, Proveniers MCG (2013) Plant acclimation and adaptation to warm environments. In: Franklin/temperature and plant development. John Wiley & Sons, Inc., Oxford, pp 49–78

Vandermeiren K, Harmens H, Mills G, De Temmerman L (2009) Impacts of ground-level ozone on crop production in a changing climate. In: Singh SN (ed) Climate change and crops. Environmental science and engineering. Springer, Berlin, pp 213–243

Vu LD, Xu X, Gevaert K, De Smet I (2019) Developmental plasticity at high temperature. Plant Physiol 181:399–411

Wei PS, Chiu HH, Hsieh YC, Yen DL, Lee C, Tsai YC (2019) Absorption coefficient of water vapor across atmospheric troposphere layer. Heliyon 5:e01145. https://doi.org/10.1016/j.heliyon.2019.e01145

Xia JY, Wan SQ (2013) Independent effects of warming and nitrogen addition on plant phenology in the inner Mongolian steppe. Ann Bot 111:1207–1217

Zinn KE, Tunc-Ozdemir M, Harper JF (2010) Temperature stress and plant sexual reproduction: uncovering the weakest links. J Exp Bot 61:1959–1968

Zipper SC, Qiu J, Kucharik CJ (2016) Drought effects on US maize and soybean production: spatiotemporal patterns and historical changes. Environ Res Lett 11:94021

Zörb C, Geilfus CM, Dietz KJ (2019) Salinity and crop yield. Plant Biol 21:31–38

Climate Change: A Key Factor for Regulating Microbial Interaction with Plants

Awadhesh Kumar Shukla, Vinay Kumar Singh, and Sadanand Maurya

Abstract In the current scenario, global climate alteration is a serious concern influencing every organism on the earth. The predicted variation in temperature and precipitation in the environment is may be due to changing climatic conditions. This change may promote difficulty and ambiguity in agricultural practices and generally threaten sustainable management. The beneficial microbes associated with plants may stimulate plant growth promotion and also enhance disease resistance activity against a variety of abiotic and biotic stresses. Moreover, climate change also directly influences the crop productivity and the structural dynamics of the relationships among insect pests, diseases, and crops. In this chapter, the regulation of microbial interaction with plants in relation to climate change is critically discussed.

Keywords Climate · Heavy metals · Greenhouse gas · Stress · Microbial interaction

Introduction

Every organism is being affected either directly or indirectly by continuously changing environmental conditions. Due to improper activities for the last few decades by anthropogenic activities for the last few decades, the earth's environment is changing quickly, and the impacts are being seen in unicellular and multicellular organisms. It is prudently predicted by the researchers that climate is changing in its pace in the coming centuries, and several parameters would be directly affected in the environment (Houghton et al. 2001). Scientists have predicted that climate change may directly influence the soil biological activity over a longer period of time (Conant et al. 2011). It is demonstrated earlier that temperature and

Vinay Kumar Singh and Sadanand Maurya contributed equally with all other contributors.

A. K. Shukla (✉) · V. K. Singh · S. Maurya
Department of Botany, K.S. Saket P.G. College, Ayodhya, Uttar Pradesh, India

© The Author(s), under exclusive license to Springer Nature Singapore Pte Ltd. 2022
A. Vaishnav et al. (eds.), *Plant Stress Mitigators*,
https://doi.org/10.1007/978-981-16-7759-5_2

precipitation are the two major influencing factors and consequently affect the activity of microbes present on the earth crust (Cheng et al. 2017). Microbes are ubiquitous and cosmopolitan in nature; however, in soil the microorganisms play a pivotal role in biogeochemical cycling and ecosystem functions and services (Conant et al. 2011). It is determined by earlier researchers that temperature may potentially involve and influence the metabolic activity and development of various microorganisms (Bradford et al. 2010). Precipitation is one of the environmental factor, important for enhancing the microbial activity in dried soils (Austin et al. 2004; Li et al. 2018). Although there is scanty information on the effects of water availability on microbial activity, the effects of different climatic conditions in the environment directly or indirectly are correlated with the functionality of microbes. According to the reports of Intergovernmental Panel on Climate Change (IPCC), the atmospheric CO_2 concentration is continuously increasing in the environment (IPCC Climate Change, 2007). Due to continuous increasing temperature, it is possible that the moisture content of water is expected to decrease in some areas and consequently create drought situations in different areas of the world (Le Houerou 1996). The continuous changing environmental conditions are affecting by and large the plant and microbe interactions in the soil ecosystem adversely.

Owing to the changing environmental condition, the physiological activities of the plant are also little bit changed. They possess inherent potential to choose different pathways in order to complete their life cycle for their better metabolic activity and survival. It is demonstrated that, under warming condition, the plants are able to sprout out and show early stage of flowering in the growing season (Cleland et al. 2007; Wolkovich et al. 2012). It is assumed that change in climatic condition may alter root phenology and plant-rhizosphere interactions.

The plant growth-promoting microorganisms have a potential to colonize inside and nearby areas of the phyllosphere and rhizosphere. The soil directly attached with the root surface is impacted by root exudates released by microorganisms along with population density of the microorganisms (Bent 2006; Lugtenberg and Kamilova 2009). It is reported that some microorganisms survive and penetrate through the root and rhizosphere of the host plants and promote the metabolic activity as an endophyte. Plant growth-promoting arbuscular fungi such as arbuscular mycorrhizae, ectomycorrhizae, endophytic fungi, and plant growth-promoting rhizobacteria are the group of microorganisms that promote and help in growth of crop plant for sustainable agriculture (Kloepper and Schroth 1978; Das and Verma 2009). These microbes are exploited as biocontrol agents against a variety of phytopathogens, as potential biofertilizers, as phytostimulators in agriculture, and phytoremediators for decontamination of contaminated environment (Lugtenberg and Kamilova 2009). These important mechanisms adopted by the microorganisms in soil microbiota are adversely affected by altered environmental conditions.

Plants and microbes play an important role in maintaining life on earth. In this case, they interact with each other mutually. But in nature, it is not so easy because they are under a variety of environmental pressures. Several greenhouse gases like carbon dioxide (CO_2), methane (CH_4), ozone (O_3), nitrous oxide (N_2O), chlorofluorocarbon (CFC), etc. and heavy metals such as Cd, Pb, and Hg are factors which interfere the interaction between plants and microbes. This chapter mainly focuses on the impact of environmental pressure on the interaction of microbes with the plant.

Impact of Greenhouse Gases (GHGs) on Microbial Interaction

Greenhouse gases enhance the temperature of the planet Earth by trapping the light energy radiating from the sun's rays and reflecting it back to the atmosphere. The global climate change is a serious concern for every, organism residing on earth and it is predicted that the radically transformed in coming century and would be affected due to variation in environmental conditions (Houghton et al. 2001). It is evident that atmospheric carbon dioxide concentration is also increasing continuously and altering the environmental conditions (IPCC 2007). Due to the warming condition of the environment, the soil water content in the crop plant is expected to decrease in some areas, and consequently the plant-microbe interaction is largely affected (Le Houérou 1996). It is reported earlier that the global warming potential may occur due to increased influx of greenhouse gases, and after temperature enhancement, ozone levels may influence the structural composition of microbial communities and its functional dynamics, which directly or indirectly influence the further coevolution of plants and their pathogens (Garrett et al. 2006; Eastburn et al. 2011; Singh et al. 2019). Moisture contents in soil may largely influence the dynamics of microbial community composition of the soil. It is predicted that small changes in soil moisture, there is seen that fungal community may shows shifting from one dominant member to another while no change in bacterial communities were observed (Kaisermann et al. 2015). After successful colonization of microbial community in the rhizosphere, the endophytic bacteria may alleviate temperature or drought stress on plants by inducing a response (Yang et al. 2009; Aroca and Ruiz-Lozano 2009).

Stress and Heavy Metal Pollution

Heavy metal pollution is usually found in the area of huge mining and agricultural activities. Heavy metal stress can alter to produce a variety of microbial community patterns. It is reported earlier that changes in physico-chemical and biological properties of the soil do not necessarily change the indigenous soil microbes (Peréz-de-Mora et al. 2006). The areas of soil, where the organic carbon content of soil is high due to heavy metal pollution, may lower the efficacy of the microbial population in mineralization of organic compounds. This is a plausible indication of

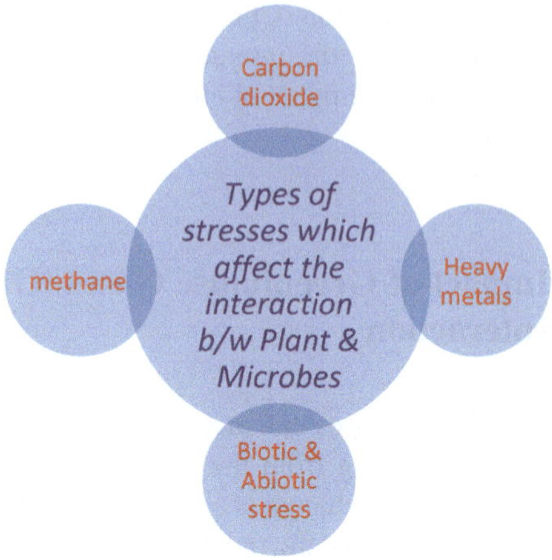

Fig. 1 Effect of stresses on microbes

the heavy metal pollution on soil microbial communities (Kozdroj and Jdvan 2001). Not only microbes but plants are also affected by heavy metals. Stress is a condition in which plants adopt a new type of mechanism for survival from external difficult situations. Stress can be two types; it is either abiotic or biotic. High soil salinity, cold, drought, and heavy metal toxicity come under the category abiotic stress. In plants, biotic stress is due to living organisms, specifically viruses, bacteria, fungi, insects, and weeds. Arbuscular mycorrhizal fungi (AMF) possess a potential to respond with higher temperature and promote plant colonization for the majority of strains for growth and development (Fitter et al. 2000). During drought conditions, there is a significant impact seen on the ectomycorrhizal fungi, its colonization, and community structures (Shi et al. 2002; Swaty et al. 2004). Figure 1 and Table 1 described the microbial interaction that could be altered by a variety of stresses such as heavy metal stress, abiotic and biotic stresses, etc.

The number of biotic stresses also affects the interaction between plants and microorganisms. It is established that fungi can act as parasite either in the form of necrotrophic fungi that kill the host cell by secretion of toxin or biotrophic fungi which directly feed on living host cells. They are responsible for inducing a variety of disease symptoms such as leaf spots and cankers in plants (Laluk and Mengiste 2010; Doughari 2015; Sobiczewski et al. 2017). Other microbes, such as nematodes, directly feed on the plant parts and primarily cause soilborne diseases leading to nutrient deficiency, stunted growth, and wilting (Lambert and Bekal 2002; Osman et al. 2020).

Table 1 Effects of stresses on microbial community

Sr. no.	Types of abiotic and biotic stresses	Inferences	References
1.	CO_2	Under higher CO_2 conditions, the nutrients like N act as a limiting factor for enhanced fertilizer input in agriculture	Compant et al. (2010)
2.	CFC_s	Effects of halocarbons such as emission of CFC_s, HFC_s, and others lead to the depletion of ozone, which is responsible for adverse effect on plants, humans, and environment	Aggarwal et al. (2013)
3.	Heavy metals	It contaminates the soil and water and creates toxicity; it has become one important constraint to crop productivity and quality	Singh et al. (2016)
3. (a)	Copper	Copper (cu) is accumulated in plant tissues and is difficult to be scavenged	Petrovic and Krivokapic (2020)
3. (b)	Zinc	Excessive zinc (Zn) in plant cells causes alteration in physiological processes of the plants	Liang and Yang (2019)
3. (c)	Lead	Lead (Pb) accumulation in plants causes physiological problems, such as DNA damage and destroying root and shoot systems	Gichner et al. (2008)
3. (d)	Cadmium	The ecological effect of cadmium (cd) ions was investigated on plants *Lactuca sativa* seeds. The results revealed that cd inhibited the microbial growth	Vardhan et al. (2019)
4.	Biotic stress	Biotic stress is negatively influenced by other living organisms such as insects and plant parasitic nematodes	O'shaughnessy and Rush (2014)
4. (a)	Nematodes	It feeds on the plant parts and primarily causes soilborne diseases	Bernard et al. (2017), Osman et al. (2020)
4. (b)	Virus	Causes local and systemic damage resulting in chlorosis and stunting	Pallas and García (2011)
4. (c)	Fungal parasites	They cause several diseases like leaf spots, vascular wilts ET in crop plants	Laluk and Mengiste (2010), Doughari (2015), Sobiczewski et al. (2017)

Conclusion and Future Perspective

Nowadays, climate change is a very serious matter. Every organism including unicellular and multicellular residing on planet Earth is directly or indirectly influenced by changing environmental conditions. So, for the betterment of our future, we should be aware and take the right decision properly. Most of the microbes have been investigated that are useful and can be exploited for improvement of sustainable agriculture. It is well known that a variety of microbes are associated with plant, and they can withstand to cope up with biotic and abiotic stresses. It is known that some of the indigenous microbial communities play a role for maintaining the plant health. Hence, it is vital to exploit and promote beneficial microbial communities. Generally, high-throughput molecular technologies for the analyses of DNA and RNA directly from the microbes using metagenomic, proteomic, and transcriptomic approaches would be helpful for revealing the hidden knowledge of microbial community dynamics in soil and plants or other environments. It is warranted that research is needed on the effects of climate change on microbial communities through experimental studies using pyrosequencing approach.

References

Aggarwal A, Kumari R, Mehla N, Singh RP, Bhatnagar S, Sharma K, Sharma K, Amit V, Rathi B (2013) Depletion of the ozone layer and its consequences: a review. Am J Plant Sci 4(10): 1990–1997. https://doi.org/10.4236/ajps.2013.410247

Aroca R, Ruiz-Lozano JM (2009) Induction of plant tolerance to semi-arid environments by beneficial soil microorganisms – a review. In: Lichtfouse E (ed) Climate change, intercropping, pest control and beneficial microorganisms, sustainable agriculture reviews, vol 2. Springer, Dordrecht, pp 121–135

Austin AT, Yahdjian L, Stark JM, Belnap J, Porporato A, Norton U, Ravetta DA, Schaeffer SM (2004) Water pulses and biogeochemical cycles in arid and semiarid ecosystems. Oecologia 141(2):22135. https://doi.org/10.1007/s00442-004-1519-1

Bent E (2006) Induced systemic resistance mediated by plant growth-promoting rhizobacteria (PGPR) and fungi (PGPF). In: Tuzun S, Bent E (eds) Multigenic and induced systemic resistance in plants. Springer, New York, pp 225–259

Bernard GC, Egnin M, Bonsi C (2017) The impact of plant-parasitic nematodes on agriculture and methods of control. In: Shah MM, Mahamood M (eds) Nematology-concepts, diagnosis and control. IntechOpen, London. https://doi.org/10.5772/intechopen.68958

Bradford MA, Watts BW, Davies CA (2010) Thermal adaptation of heterotrophic soil respiration in laboratory microcosms. Glob Chang Biol 16:1576–1588. https://doi.org/10.1111/j.1365-2486.2009.02040.x

Cheng L, Zhang N, Yuan M, Xiao J, Qin Y, Deng Y et al (2017) Warming enhances old organic carbon decomposition through altering functional microbial communities. ISME J 11(8): 1825–1835. https://doi.org/10.1038/ismej.2017.48. https://escholarship.org/uc/item/8mp28182

Cleland EE, Chuine I, Menzel A, Mooney HA, Schwartz MD (2007) Shifting plant phenology in response to global change. Trends Ecol Evol 22:357–365

Compant S, Clément C, Sessitsch A (2010) Plant growth-promoting bacteria in the rhizo-and endosphere of plants: their role, colonization, mechanisms involved and prospects for utilization. Soil Biol Biochem 42(5):669–678

Conant RT, Ryan MG, Ågren GI, Birge HE, Davidson EA, Eliasson PE, Evans SE, Frey SD, Giardina CP, Hopkins FM, Hyvönen R, Kirschbaum MUF, Lavallee JM, Leifeld J, Parton WJ, Megan Steinweg J, Wallenstein MD, Martin Wetterstedt JÅ, Bradford MA (2011) Temperature and soil organic matter decomposition rates – synthesis of current knowledge and a way forward. Glob Change Biol 17:3392–3404. https://doi.org/10.1111/j.1365-2486.2011.02496.x

Das A, Varma A (2009) Symbiosis: the art of living. In: Varma A, Kharkwal AC (eds) *Symbiotic fungi principles and practice*. Springer), Berlin, pp 1–28. https://doi.org/10.1007/978-3-540-95894-9

Doughari J (2015) An overview of plant immunity. J Plant Pathol Microbiol 6:104172. https://doi.org/10.4172/2157-7471.1000322

Eastburn D, Mcelrone A, Bilgin D (2011) Influence of atmospheric and climate change on plant-pathogen interactions. Plant Pathol 60:54–69. https://doi.org/10.1111/j.1365-3059.2010.02402.x

Fitter AH, Heinemeyer A, Staddon PL (2000) The impact of elevated CO2 and global climate change on arbuscular my-corrhizas: a mycocentric approach. New Phytol 147:179–187

Garrett KA, Dendy S, Frank E, Rouse M, Travers S (2006) Climate change effects on plant disease: genomes to ecosystems. Annu Rev Phytopathol 44:489–509. https://doi.org/10.1146/annurev.phyto.44.070505.143420

Gichner T, Žnidar I, Száková J (2008) Evaluation of DNA damage and mutagenicity induced by lead in tobacco plants. Mutat Res 652:186–190

Houghton JT, Ding Y, Griggs DJ, Noguer M, van der Linden PJ, Dai X, Maskell K, Johnson CA (eds) (2001) Climate change 2001: the scientific basis. The Press Syndicate of the University of Cambridge, Cambridge

IPCC (2007) Climate change 2007: the physical science basis. Cambridge University Press, Cambridge

Kaisermann A, Maron P-A, Beaumelle L, Lata J-C (2015) Fungal communities are more sensitive indicators to non-extreme soil moisture variations than bacterial communities. Appl Soil Ecol 86. https://doi.org/10.1016/j.apsoil.2014.10.009

Kloepper J, Schroth MN (1978) Plant growth-promoting rhizobacteria on radishes. In: IV international conference on plant pathogenic bacteria. France 2:879–882

Kozdroj J, Jdvan E (2001) Structural diversity of microbial communities in arable soils of a heavily industrialized area determined by PCR-DGGE fingerprinting and FAME profiling. Appl Soil Ecol 17(1):31–42

Laluk K, Mengiste T (2010) Necrotroph attacks on plants: wanton destruction or covert extortion? Am Soc Plant Biol 8:e0136. https://doi.org/10.1199/tab.0136

Lambert K, Bekal S (2002) Introduction to plant-parasitic nematodes. Plant Health Instruct 10: 1094–1218. https://doi.org/10.1094/PHI-I-2002-1218-01

Le Houérou HN (1996) Climate change, drought and desertification. J Arid Environ 34(2): 133–185ISSN 0140-1963. https://doi.org/10.1006/jare.1996.0099

Li G, Kim S, Han S, Chang H, Du D, Son Y (2018) Precipitation affects soil microbial and extracellular enzymatic responses to warming. Soil Biol Biochem 120. https://doi.org/10.1016/j.soilbio.2018.02.014

Liang J, Yang W (2019) Effects of zinc and copper stress on antioxidant system of olive leaves. IOP Conf Ser Earth Environ Sci 300:52058

Lugtenberg B, Kamilova F (2009) Plant-growth-promoting rhizobacteria. Annu Rev Microbiol 63: 541–556

O'Shaughnessy SA, Rush C (2014) Precision agriculture: irrigation. In: Van Alfen NK (ed) Encyclopedia of agriculture and food systems. Academic Press, pp 521–535, ISBN 9780080931395. https://doi.org/10.1016/B978-0-444-52512-3.00235-7

Osman HA, Ameen HH, Mohamed M, Elkelany US (2020) Efficacy of integrated microorganisms in controlling root-knot nematode Meloidogyne javanica infecting peanut plants under field conditions. Bull Natl Res Centre 44:1–10. https://doi.org/10.1186/s42269-020-00366-0

Pallas V, García JA (2011) How do plant viruses induce disease? Interactions and interference with host components. J Gen Virol 92:2691–2705. https://doi.org/10.1099/vir.0.034603-0

Peréz-de-Mora A, Burgos P, Madejón E, Cabrera F, Jaeckel P, Schloter M (2006) Microbial community structure and function in a soil contaminated by heavy metals: effects of plant growth and different amendments. Soil Biol Biochem 38:327–341. https://doi.org/10.1016/j.soilbio.2005.05.010

Petrovic D, Krivokapic S (2020) The effect of cu, Zn, cd, and Pb accumulation on biochemical parameters (proline, chlorophyll) in the water caltrop (Trapa natans L.), Lake Skadar, Montenegro. Plan Theory 9:1287

Shi L, Guttenberger M, Kottke I, Hampp R (2002) The effect of drought on mycorrhizas of beech (Fagus sylvatica L.): changes in community structure, and the content of carbohydrates and nitrogen storage bodies of the fungi. Mycorrhiza 12(6):303–311

Singh S, Parihar P, Singh R, Singh VP, Prasad SM (2016) Heavy metal tolerance in plants: role of transcriptomics, proteomics, metabolomics, and ionomics. Front Plant Sci 6:1143. https://doi.org/10.3389/fpls.2015.01143

Singh VK, Shukla AK, Singh AK (2019) Impact of climate change on plant–microbe interactions under agroecosystems. In: Chaudhary KK, Kumar A, Singh AK (eds) Climate change and agricultural ecosystems. Woodhead Publishing, Cambridge, pp 153–179. https://doi.org/10.1016/B978-0-12-816483-9.00007-4

Sobiczewski P, Iakimova E, Mikiciński A, Wêgrzynowicz-Lesiak E, Dyki B (2017) Necrotrophic behaviour of Erwinia amylovora in apple and tobacco leaf tissue. Plant Pathol 66:842–855. https://doi.org/10.1111/ppa.12631

Swaty R, Deckert R, Whitham T, Gehring C (2004) Ectomycorrhizal abundance and community composition shifts with drought: predictions from tree rings. Ecology 85:1072–1084. https://doi.org/10.1890/03-0224

Vardhan KH, Kumar PS, Panda RC (2019) A review on heavy metal pollution, toxicity and remedial measures: current trends and future perspectives. J Mol Liq 290:111197

Wolkovich EM, Cook BI, Allen JM, Crimmins TM, Betancourt JL, Travers SE, Pau S, Regetz J, Davies TJ, Kraft NJ, Ault TR, Bolmgren K, Mazer SJ, McCabe GJ, McGill BJ, Parmesan C, Salamin N, Schwartz MD, Cleland EE (2012) Warming experiments underpredict plant phenological responses to climate change. Nature 485(7399):494–497. https://doi.org/10.1038/nature11014

Yang J, Kloepper JW, Ryu CM (2009) Rhizosphere bacteria help plants tolerate abiotic stress. Trends Plant Sci 14:1–4

Climate Change and Soil Fertility

Preeti Singh, Asha Sharma, and Jyotika Dhankhar

Abstract Soil is essential for carrying out variety of ecosystem services and managing world food productivity. Climate, along with few others is the utmost dominating factor which determines and gives rise to formation of variety of individual soils in any given climatic area. More developed soil is found in hotter and wetter climate, whereas cooler and dried climates have less developed soil. Climate change pertains to remarkable modifications in various abiotic factors like global temperature, patterns of wind, precipitation, greenhouse gases (GHGs) and other climatic factors which take many decades or even more longer time periods to occur. Biological, physical and chemical functions of soil are hugely impacted by climatic change directly or indirectly. Precipitation, temperature, wind patterns, greenhouse gases and moisture regimes impose direct impact, whereas indirect impacts are induced by human adaptations such as crop rotation changes, irrigation, tillage practices etc. Soil is the mode of expression of plant responses to climatic conditions of a particular area. Change in climatic conditions such as rainfall, temperature etc. severely affect crop production in agricultural systems, as capability of crops to attain maturity is halted. Land degradation in the form of soil erosion, salinization, desertification etc. is a consequence of drastic climatic change. Hydrological cycle gets disturbed significantly following change in climate which ultimately degrades soil health and fertility. Elevated temperature, altered precipitation, atmospheric nitrogen deposition, rising atmospheric carbon dioxide (CO_2) levels and imbalance of other greenhouse gases (GHGs) ultimately lead to global warming and impose high impacts on functions performed by soil. Hence all these factors highly influence future usage of the soil as well. Essentially, climatic change and soil fertility are deeply and inextricably connected together.

Keywords Soil fertility · Climate change · Soil properties · Soil health · Climate change drivers · Carbon sequestration

P. Singh (✉) · A. Sharma
Deptartment of Botany, Maharshi Dayanand University, Rohtak, Haryana, India

J. Dhankhar
Department of Food and Technology, Maharshi Dayanand University, Rohtak, Haryana, India

© The Author(s), under exclusive license to Springer Nature Singapore Pte Ltd. 2022

A. Vaishnav et al. (eds.), *Plant Stress Mitigators*,
https://doi.org/10.1007/978-981-16-7759-5_3

Introduction

The phenomenon of climate change is global and taking place continuously since the existence of earth. Since last decade, it has become a key subject of concern among scientists and politicians. Its history is full of extreme hot and cold cycles which have been noticed comparatively faster in the last two centuries (Fauchereau et al. 2003). As the earth's history is proceeding, the climatic parameters such as precipitation, temperature, CO_2 concentration etc. have changed globally as well as regionally. Apart from such extraterrestrial causes, there are certain terrestrial causes as well like forest fires, volcanoes, global ice alterations, snow covers etc. which have brought considerable changes. Average global temperature has increased by more than 15 °C in the last hundred years, which is not only because of natural causes but also due to anthropogenic causes. Less water evaporation from agricultural land than natural forests, carbon dioxide and heat emission from industrial areas and nitrous oxide (N_2O) and methane (CH_4) emission from agricultural lands are some major and significant causes. It is estimated that the mean global temperature will rise in twenty-first century by 2–3 °C more, majorly due to high usage of fossil fuels and intensely traditional agriculture practices (Karmakar et al. 2016). Climate has a big role in altering soil fertility and limiting crop yield. Only half or even less of the crop genetic and climatic potential yield is achieved by farmers for a given cultivar choice, site and sowing date. How it affects soil health in variety of ways has been studied by many scientists, agriculturalists and economists. Crop cultivar and climatic change are the only factors which limit the potential crop yield when other factors are optimal (Sinha and Swaminathan 1991; Saseendran et al. 2000; Aggarwal and Sinha 1993; Rao and Sinha 1994). The implications of variety of combinations of irrigation input, nitrogen amount and climatic variability chiefly at high and low irrigation inputs have been studied, and it was reported that when moisture supply is adequate such as in north Indian regions like Haryana and Punjab, yield benefits are attained at application of higher amount of nitrogen. However, in regions where moisture supply is limited or moderate, the rising trends in crop yields are not up to comparatively lesser values of nitrogen. It is hard to decide the most appropriate levels of nitrogen fertilizers to maximize the crop yield returns in view of ambiguous nitrogen responses which are related strongly to the sufficient rainfalls after monsoons during the growing period of crops (Kalra and Aggarwal 1994).

Soil temperature increases when air temperatures are higher, which in turn increase the reaction rate of solutions and other reactions controlled by diffusion (Buol et al. 1990). Solubility of gaseous and solid components of soil keeps on increasing and decreasing, but it may take years for the consequences to come out. In addition, decomposition rate of soil organic matter accelerates with higher temperatures which results in decreasing carbon to nitrogen (C-N) ratio as carbon is released into the atmosphere. However, such effects are counterbalanced to some extent by higher biomass of roots and crop residues (litter) from plants as a result of plants' response to higher atmospheric carbon dioxide. As soil organic matter decomposition is influenced by soil temperature, it results into the release and uptake of nutrients and other metabolic processes in plants. High temperature also influences soil chemical reactions which affect soil organic matter and minerals. Soil microbes which control major processes such as nutrient flow and soil productivity are also affected by climate change especially by changing atmospheric CO_2 concentrations. The major functions served by soil microorganisms are (a) sequestration of carbon and various minerals mainly nitrogen, sulfur and phosphorus in their living biomass itself and so act as a major reservoir of available nutrients for plants and (b) nutrient transportation and transformation. Altered carbon to nitrogen (C-N) ratio of litter impacts metabolic and other physiological processes of soil micro-flora which also affects trace gas production in soil (Patil and Lamnganbi 2018).

Soil erosion is also one of the major causes of reducing soil fertility. Soil erosion rates can be increased by climate change such as increasing rainfall, high wind pressure etc. which in turn greatly reduces the productivity of soil. A high rate of sedimentation is found in rivers, streams and reservoirs due to increased soil erosion. Drought/decreased rainfall can also elevate soil erosion through winds which take away residues from upper fertile layers of soil during dry seasons (Parry et al. 1999). If rate of soil erosion is not checked, farmers will eventually be forced for the abandonment of their lands due to constant impoverishment of soil and hence soil productivity. Thus it won't be wrong to say that soil erosion is among the key threats to the food production in dry and warmer regions where rainfall is equal to zero. Other factors responsible for degradation of soil such as soil sodicity (Patil and Lamnganbi 2018), increasing soil salinity, soil acidification, water logging, structural decline etc. are emerging due to various agricultural practices and other anthropogenic causes (Carter et al. 1997; Slavich 2001).

As soil plays the most essential role in providing almost all the micro- and macronutrients to the crops grown in it, it becomes very important to study the change related to its chemical, physical and biological properties in respect to climatic change. Impact of a wide range of global climatic change such as rising global temperature and carbon dioxide concentration, altering precipitation, increasing nitrogen deposition etc. should be considered while defining soil fertility in relevance to climate change. The relationship between climate change and soil properties in response to temperature, rainfall and carbon dioxide is having a better clarity in today's time as variety of studies have been conducted by many scientists, climatologists and agriculturists (Tao et al. 2003; Arias et al. 2005; Moebius et al. 2007; Reynolds et al. 2009). So the aim of this chapter is to brief the effect of climatic changes on soil fertility and other chemical, physical and biological properties. Before going deep into the main topic, here is a small introduction of what is soil, its major types, soil fertility, climate change and its causes etc.

Climate Change

Climate is a constant dynamic entity rather than a static one. These are not just atmospheric fluctuations which contribute to climate change, but the atmosphere as a whole including hydrosphere, lithosphere, cryosphere and other extraterrestrial influences is responsible for climate change. Interaction and variations among various above mentioned factors (temperature, rainfall, GHGs concentration, wind patterns etc.) decide climate of a particular area in a given time. According to Intergovernmental Panel on Climate change (IPCC) of the United Nations (UN), climate change refers to any sort of change whether caused by natural alterations or anthropogenic factors over a period of time. According to the Framework Convention on Climate Change (FCCC), climate change pertains to change of climate that resulted directly or indirectly from human activities which alter atmosphere composition globally, which is in addition to natural alterations of climate occurred over equivalent time periods. As per IPCC studies (Fig. 1), an increase has been noticed in global atmospheric concentrations of methane (CH_4), carbon dioxide (CO_2), and nitrous oxide (N_2O) from the pre-industrial period to year 2005 (IPCC 2007).

Change in crucial climatic parameters during 1973–2005 have been studied by a few scientists. They found a significant decline in snow covers and mountain in both hemispheres. Melting of icecaps and glaciers resulted in rise of sea level. An average rise of 1.8 mm of sea level per year was noticed during 1961–2005 (Rahmstorf et al. 2007). During 1993–2007, rate of melting of icecaps became much faster and rise in sea level has almost doubled to 3.1 mm per year on an average. Since 1960, the westerly winds of mid-latitude in both south and north hemispheres become much more strengthened. Since 1970 longer and much intense droughts have been reported mainly in tropical and subtropical regions of the world. Decreased

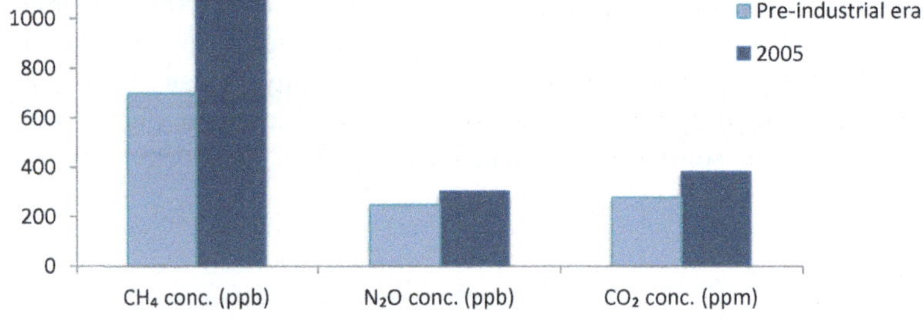

Fig. 1 Rising of atmospheric concentration of methane (CH_4), nitrous oxide (N_2O), and carbon dioxide (CO_2) from pre-industrial era to 2005 (Karmakar et al. 2016)

precipitation and increased temperature resulted in drought conditions. Changed wind patterns, sea surface temperatures (SSTs) and decreased snow covers and icecaps additionally contributed to drought.

Various anthropogenic activities also contributed to a great extent to climate change. Land use alterations and intensifications also caused land degradation and desertification (IPCC 2020). Constant emission of GHGs from developed and mechanized countries is causing seasonal unpredictability, rise in sea levels and hydro- meteorological events etc. (Adnan et al. 2011; Yuksel 2014). The annual concentrations of carbon dioxide growth rate were greater in the last 10 years (during 1995–2005, 1.9 ppm/year), and variability has been recorded in growth rates every year and an average increase rate was found to be around 1.4 ppm from 1960 to 2005 (IPCC 2007). Industrial progressions and increased population contributed largely in emission of carbon dioxide through industrial processes and fossil fuel burning globally, and it has been estimated that from 1970 to 2010 such processes alone have contributed around 78% of the sum of GHG emission (EPA 2010; IPCC 2014). So climate change reflects alterations within the earth's atmosphere, lithosphere and hydrosphere including ecosystems like forests, ocean, snow covers, glaciers icecaps etc.

Major Causes of Climate Change

Climate change is an actual and critical challenge which is affecting population and environment globally (Braman et al. 2010). According to various studies till date, the main causes of climate change are natural as well as manmade (Fig. 2). The International Panel on Climate Change (IPCC 2007) has discussed about natural and anthropogenic drivers of climate change. Alterations in land surface or soil properties, solar radiations, imbalance of greenhouse gases (GHGs) and aerosols are responsible for causing energy imbalance. Heat escape is prevented by imbalance of greenhouse gases, which ultimately cause global warming. Few gases which are quite long lived stay in the atmosphere almost permanently without responding chemically or physically to any alterations in temperature. This is called forcing climate change. Gases like water vapors respond in both ways, physically and chemically to temperature changes, known as feedbacks. Discussed below are natural and anthropogenic causes of climate change (IPCC 2007).

Natural Causes

For climate change the root drivers are anthropogenic/human activities, but there are a few major natural factors also which highly influence climate (EPA 2010; IPCC 2013). Earth's climate is not static; rather it has dynamicity and it keeps constantly changing through the natural cycles in environment. The climate change has been studied by scientists and climatologists worldwide, and evidences were found from

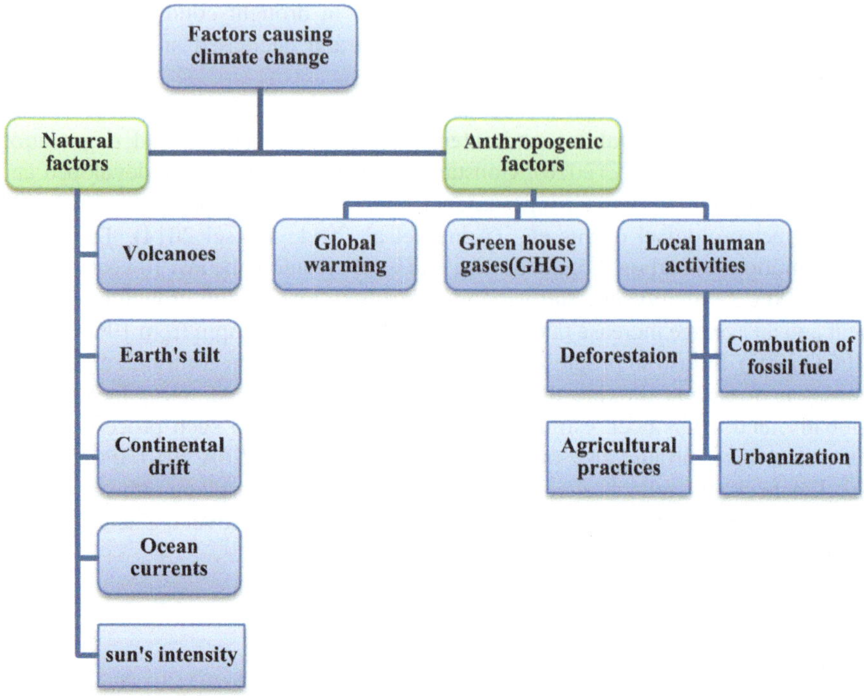

Fig. 2 Major causes of climate change

pollen, tree rings, ocean sediments, snow cores etc. Out of the several natural factors which are responsible for climate change, volcanoes, earth's tilt, continental drift, ocean currents, meteorites and comets, sun's intensity etc. are more prominent. Oceanographers and geologists investigated the influence of ocean currents on climate change and found that modifications in ocean currents significantly affect earth's climate (Cunningham 2005; Tierney et al. 2013). A constant directed movement of ocean water produced by several forces such as wind, temperature, salinity differences, breaking waves etc. which act upon the water forms an ocean current (England et al. 2014). Circulation of oceanic currents plays a vital role in regulation of global climate and maintenance of net primary productivity (NPP) of marine ecosystem (Duteil et al. 2014). Records of ice core from Greenland suggested that sudden shifts in the circulation strength prompted remarkable fluctuations in temperature during the glacial periods (Jayne and Marotzke 2001; Fischer and Knutti 2015). Earth's tilt angle, its eccentricity, relative position in space etc. also influence climate change (Ruddiman 2007). Ice sheet and glaciers melting cause expansion of sea levels (Peterson et al. 2013). As oceans get warmer due to high temperature, they tend to expand as warmer seas take more space (Fischer and Knutti 2015; Savage et al. 2015). The sun's intensity causes either hitting or cooling of the earth's surface

(IPCC 2018) which can be a major cause of drought in high-intensity summers in many areas worldwide.

Anthropogenic Causes

Global warming caused chiefly by human activities is one of the major cause of climate change. There are many anthropogenic factors such as deforestation, industrialization, fossil fuel combustion, crop cultivation, pollution etc. which are greatly responsible for increasing greenhouse gases (GHGs) in the atmosphere. Methane (CH_4), carbon dioxide (CO_2), nitrous oxide (N_2O) and water vapors are major greenhouse gases which do not allow heat to escape from earth's surface; hence average global temperature increases. Carbon dioxide released through human activities is far more than it can be absorbed by oceans and plants. Such gases stay in the environment for several years, therefore even if emission of such gases were stopped today, global warming will not be stopped instantly. The temperature varies in different parts of world, but since almost the last 25 or more years, it is increasing everywhere (Yuksel 2014; IPCC 2018). Greenhouse gases are responsible for making planet earth habitable for all living organisms by trapping long-wave (thermal) radiations which are emitted from the surface of the earth, maintaining earth's surface mean temperature around 15 °C, which would have been −18 °C in absence of these gases in the atmosphere (Rakshit et al. 2009). This phenomenon is called greenhouse effect. Water vapor is one of the most crucial greenhouse gase and a part from it, carbon dioxide contributes substantially, while methane, ozone and nitrous oxide have smaller contributions. In recent years, there has been a constant remarkable increase in concentrations of carbon dioxide, nitrous oxide and methane. Chlorofluorocarbon (CFC), one of the greenhouse gase, has also been added to the atmosphere in significant amounts through human activities. Increased concentrations of abovementioned gases have been estimated to increase earth's mean temperature by 0.5 °C since 1860, and in the next coming 40 years, it is estimated to increase by 1.5 °C. (Mitchell 1989). Hence, imbalanced concentrations of greenhouse gases are a bigger cause of global warming (Fig. 3, structure of green house gases, GHGs).

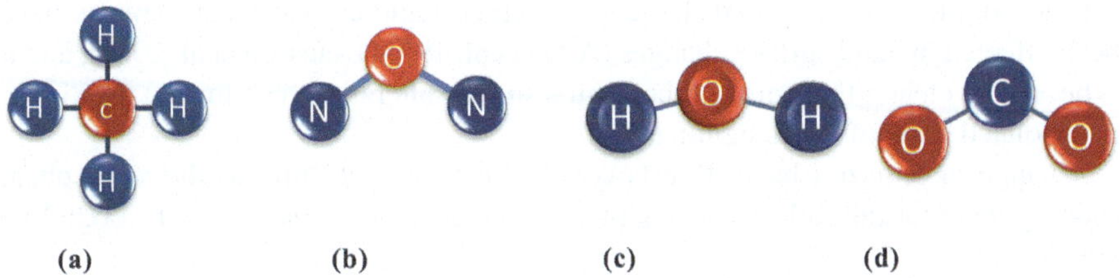

Fig. 3 Gases which contribute to greenhouse effect: (**a**) methane (CH_4), (**b**) nitrous oxide (N_2O), (**c**) water vapor (H_2O), (**d**) carbon dioxide (CO_2)

- *Water vapor:* It is the most abundantly present greenhouse gas in the atmosphere and plays a feedback role to the climate. When earth's temperature is warmer, the amount of water vapor increases and so as the probability of precipitation and cloud formation. So it makes the most crucial feedback mechanism in the atmosphere. It is a crucial infrared radiation absorber (Held and Souden 2000). Clouds and water vapors are accountable for 25 and 49% of thermal absorption respectively (Schmidt et al. 2010).
- *Carbon dioxide (CO_2)*: It is the most crucial but a minor greenhouse gas of the atmosphere which accounts for 20% of thermal absorption (Schmidt et al. 2010). It is produced mainly by natural phenomenon such as volcanic eruptions, respiration processes and through human causes such as fossil fuel burning, land uses and deforestation. Since the last two centuries, atmospheric carbon dioxide concentration has dramatically climbed up (Mittler and Blumwald 2010; IPCC 2014). It persists for several years in the atmosphere.
- *Methane (CH_4)*: It is a hydrocarbon gas which is produced both from natural sources and anthropogenic sources or human activities, involving waste decompositions in agricultural lands mainly rice cultivation land, decay and decomposition of plant materials in various wetlands, ruminant digestion, gas/oil production and management of wastes. Production of food and human excreta alone produces half of the total amount of methane present in the atmosphere. Methane by far is more active than carbon dioxide, but it is present in very less amount in the atmosphere. It produces ozone (O_3) and carbon monoxides (CO), which controls OH concentrations in the troposphere (Wuebbles and Hayhoe 2002).
- *Nitrous oxide (N_2O)*: The nitrous and nitric oxide (N_2O and NO, respectively) are strong GHGs emitted by soil. Their concentration is constantly increasing in the atmosphere as a result of human interventions (IPCC 2014). Microbial activities in nitrogen-rich soil cause high emission of N_2O (Hall et al. 2008). Emission through human activities such as agriculture and fossil fuel burning and emission through soil are the major factors for NO in the atmosphere (Medinets et al. 2015).

Local Activities

Various studies have exposed the level to which regional and local climates have been affected by land surface changes (Allan et al. 1995; Claussen et al. 2001), and it is becoming clearer that some of the earth's surface alterations can impact the distant and isolated parts of earth significantly.

Change in Amazon basin forests cover affect moisture flux to the atmosphere causing regional rainfall due to regional convection (Roy and Avissar 2002; Xue 1997).

Soil and Soil Fertility

Soil is one of the most crucial natural resource which support varied life forms on earth. It is a mixture of various macro- and micro-nutrients, liquids, gases, organic matter and organisms which support life together. Its formation takes thousands of years of interactions of various forces including parent materials, climate, varied forms of living organisms and microbes etc. (Karmakar et al. 2016). The layer of the earth which is made up of soil is called the pedosphere which serves many functions such as production of food, fuels and other biomasses; medium for water storage, purification and supply; production of raw materials; platform for cultural and physical heritage; habitat for plants, animals and other living organisms; platform for interactions among various ecosystems and modifier of atmosphere etc.

Soil fertility is the state of a soil which allows it to supply adequate amount of nutrients, minerals and water in an appropriate balance for specified plant growth when other biotic and abiotic growth factors such as soil micro-flora, temperature, moisture, light and other physical conditions of soil like porosity, bulk density, pH, texture etc. are favorable. (Fig. 4, showing major factors which impact soil fertility).

Management of soil fertility plays a significant role in agricultural productivity and crop yield. For the sustainment of soil fertility, the amount of soil nutrients extracted from soil must be replenished equally. Chemical, physical and biological properties of soil such as soil texture and structure, soil porosity, bulk density, water-holding capacity, soil pH, electrical conductivity, nutrient pool etc. are greatly impacted by climatic conditions. These soil properties are majorly responsible for proper functioning of soil.

Some Major Functions of Soil

- *Source of food, fiber, fuels, and other biomass*: Plant roots are anchored in the soil to gain all the supply of essential nutrients and water. They also interact with soil microbes such as various bacteria, fungi, etc. to form beneficiary associations. The soil also provides appropriate temperature to roots for their normal functioning by controlling temperature fluctuations. So by providing all the suitable and essential conditions, soil supports plant growth for food, fiber and other biomass production.
- *Climate regulator*: Soil with a good health is the most important storehouse of terrestrial carbon. Sustainable management of the soil can help in decreasing greenhouse gas (GHGs) emissions into the atmosphere and thereby helps in carbon sequestration. Soil is the major sink for carbon sequestration, and contains around double of the amount possessed by atmosphere and triple of the carbon as possessed by forests (IPCC 2007). In contrast, poor management of soil can lead to increased levels of CO_2 into the atmosphere. Hence by managing soil properly, humans can help a little in reducing greenhouse gas (GHGs) emissions hence

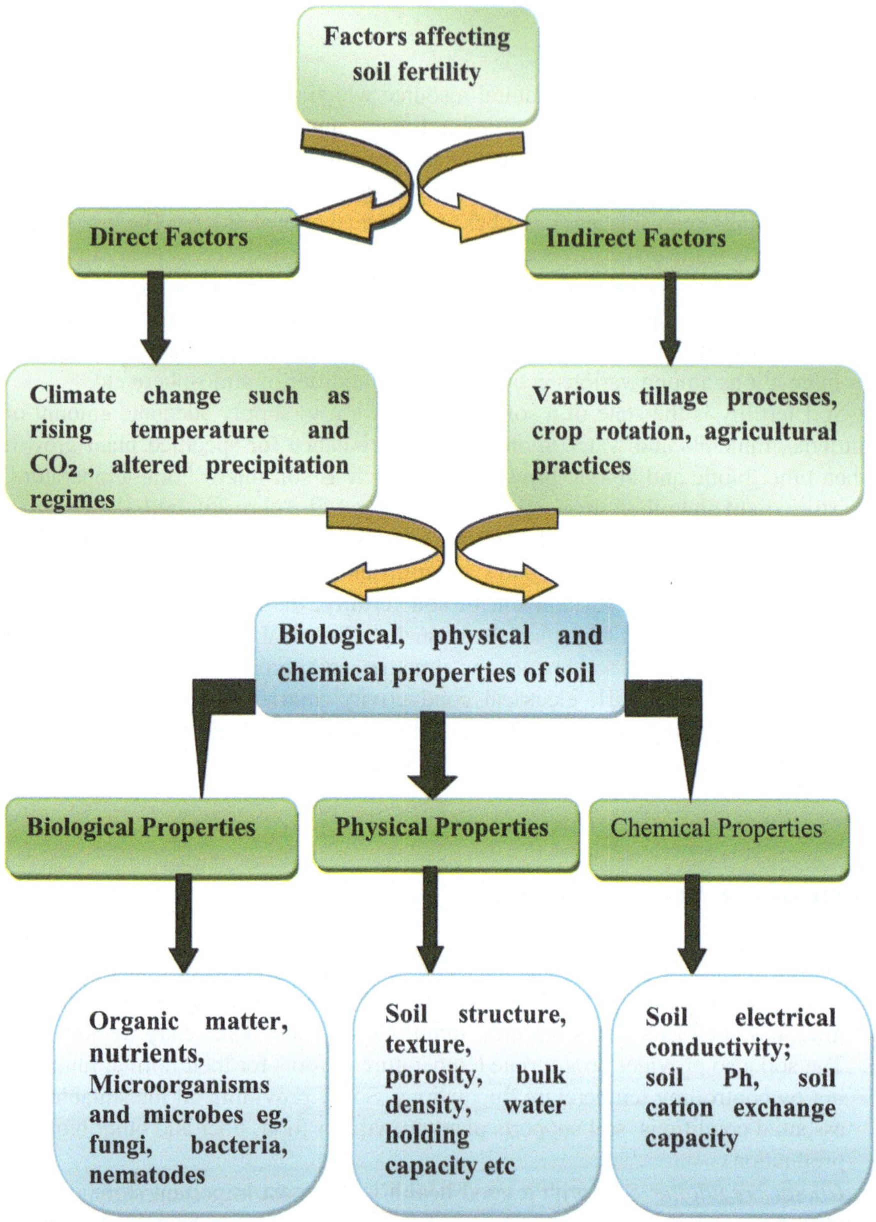

Fig. 4 Major factors affecting soil fertility:- climate change imposes direct impact on soil fertility, whereas various agricultural practices such as tillage, crop rotation, irrigation etc. impose indirect impact on soil fertility

preventing global warming. The carbon pool of soil has become narrower now because of a number of human activities such as agriculture and several other disturbances and hence soil has lost big amount of carbon into the atmosphere (Houghton 1999; Schimel 1995; Lal 1995).

- *Interactions among different ecosystems*: Soil can absorb large amounts of carbon dioxide in the form of soil organic matter, thus decreasing the total atmospheric carbon concentrations. It also helps in decaying and decomposing waste materials and transforms those to new ones through interactions with various microbes present in soil itself. Soil also regulates supply and purification of water by filtering and capturing contaminants between its particles and provides cleaner water in rivers, aquifers etc.
- *Habitat for living organisms and gene pool reserve*: Soil indeed is a habitat for varied life forms and a reserve of gene pool as varied forms of living organisms on earth, whether they be plants, animals, microbes etc. born, live and die on earth itself. Most of the plants, algae, fungi and other microbes grow and live in soil, so soil definitely contains a huge gene pool inside it. It provides all the food, fiber and essential nutrients for the survival of living beings. Soil supports all life forms by decomposing dead and decaying material and recycling the same for future usage.
- *Source of raw materials*: Soil provides all the necessary raw materials for the production of varieties of processed food and fiber for human, fodder for cattle, fuel for various human activities, construction material for buildings, wood, paper and rubber etc.
- *Platform for cultural and physical heritage*: Soil is a source of preserved and protected physical artifacts of our past history which can help us in better understanding cultural heritage. Soil also gives an idea about people's migration from place to place and their settlements in ancient times.
- *Basis for manmade structures*: Soil provides base for all the manmade structures and materials such as cement, sand, wood, glass etc. for building constructions, roads and highways etc. The firmness, bearing strength, shearing strength and compressibility of soil should be properly tested and considered before constructing anything to ensure a better foundation. Analysis of physical properties of soil gives a better understanding and application of the soil to engineering usage as well.
- *Source of pharmaceuticals*: Soil microorganisms, herbs, shrubs, trees etc. are some of the major sources of pharmaceuticals. Various parts of plants such as leaves, roots, bark, stem, bud, flower, latex etc. have been used in pharmaceutical industries to manufacture various therapeutic drugs from time immemorial. Various bacteria, fungi and other microbes also serve as a good source of pharmaceutical drugs. The first antibiotic penicillin was extracted from a fungus called *Penicillium*.
- *Bioremediation*: It is a process of treating contaminated water, soil, media or any other surface material, by changing environmental conditions to enhance growth of microbes and further degrade the target pollutants. Soil microbes play a crucial role in bioremediation.

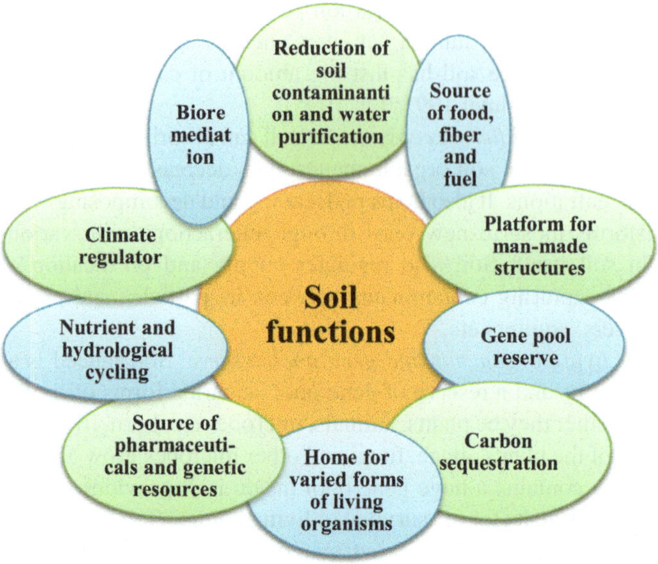

Fig. 5 Various functions that soil performs—soil plays a major role in our ecosystem to sustain life of all living beings. Some major functions of soil have been briefed in this particular diagram given above-

- Hence, soil imposes a major impact on environmental interactions and ecosystem stability for the sustainment of varied life forms on earth (Fig. 5 enlisting major functions of soil).

Major Types of Soils in India (Chauhan and Dahiphale, 2020)

Indian soil has been classified into eight categories by the Indian Council of Agricultural Research (ICAR), New Delhi. These are as follows:

- *Alluvial soil*: It mainly occurs in plains of Ganga, Satluj and Brahmaputra rivers and in valleys of Narmada and Tapi. This soil suits very good to crops like maize, rice, oilseed, sugarcane etc. It is further divided into two categories. The first one is Khadar soil, which is silt enriched and the second one is Bhangar soil which contains nodules of calcium carbonate. It is loamy to clayey textured.
- *Red soil*: Red soil is the second largest group of soil found in India. It is abundant in ferric oxides which gives its characteristic red color. It is best suited to cultivation of tobacco, orchards, pulses, cotton, oilseed, potato, wheat, millet etc.
- *Laterite soil*: These soils are mainly found in hills of eastern and western ghats, Vindhya, Odisha, West Bengal, North Cachar hills etc. This soil is not so rich in organic matter and contains fewer quantities of nitrogen, potassium, lime and

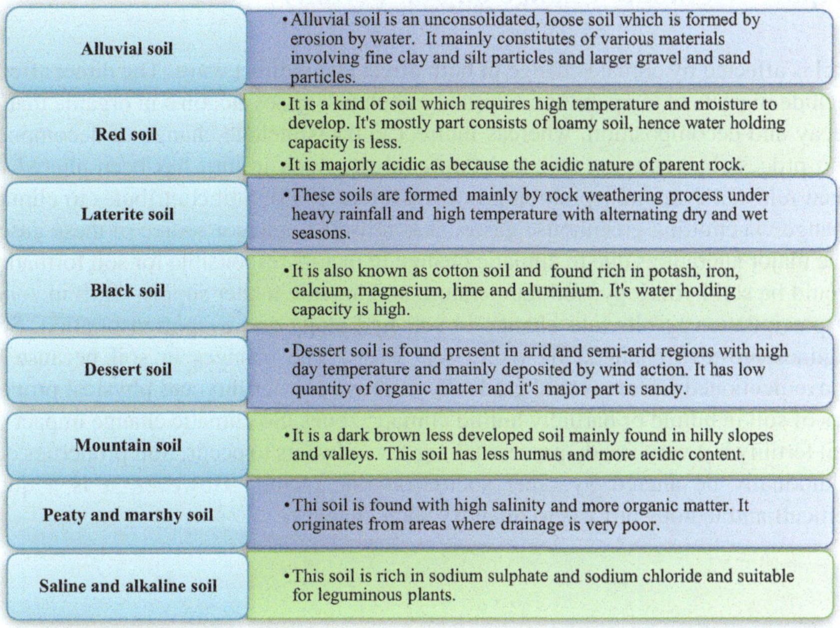

Fig. 6 Major types of soils in India, as classified by the Indian Council of Agricultural Research (ICAR), New Delhi, India

organic matter. It is rich in aluminum and iron and best suited for the cultivation of sugarcane, raghi, rice, cashew etc.

- *Black soil*: Another name of black soil is cotton soil. This soil formation took place from cretaceous lava. In India, it is found in Maharashtra, parts of Gujrat, Western areas of MP, Karnataka, Rajasthan, Tamil Nadu and Jharkhand. Because of having high water retention capacity, it is best suited for tobacco, castor, linseed and citrus fruit cultivation.
- *Desert soil*: It contains low quantity of organic matter and majorly constitutes of sand. It has very less amount of soluble salts and moisture content with less water-retaining capacity. This soil is suitable for crops which require very less amount of water such as bajra, guar, pulses, fodder etc.
- *Mountain soil*: This soil is having very less organic matter and humus content and best suited for crops such as legumes, orchards and fodder crops.
- *Peaty and marshy soil*: This soil is rich in organic matter and saline content. These are majorly found in Sundarbans deltas, Kerala, deltas of Mahanadi etc.
- *Saline and alkaline soil*: This soil is rich in sodium chloride content and best suited for leguminous crops (Fig. 6, briefing major types of soils found in India, classified by ICAR, New Delhi).

Impact of Climate Change on Soil Fertility

Soil is affected by climate change in both direct and indirect ways. The direct affects include impacts of temperature, rainfall, greenhouse gases etc. on soil organic matter decay and decomposition, whereas indirect impacts include change of decomposition process by changing soil moisture content which in turn has been altered by plant relevant evapotranspiration process (Defera 2005). Soil contributes to climate change via emitting greenhouse gases, as soil itself is a major source of these gases. The major alterations due to climatic change in factors responsible for soil formation would be soil hydrology, soil temperature and organic matter supply. Shift in zones of precipitation results into change in soil hydrology and evapotranspiration. Soil regimes are also changed by rising sea levels. The changes in soil because of abovementioned factors would gradually improve soil fertility and physical properties of soil in humid or partially humid climatic zones. So climatic change impact on soil fertility is a very steady process taking several years to occur. Soil properties can additionally be altered by other socioeconomic factors. However, it is a quite difficult and tedious process to quantify such changes.

Direct Impacts of Climatic Change on Soil Fertility

Any change whether it be direct or indirect in the soil function will eventually change the soil fertility. Constant release of carbon dioxide from vegetation and its addition into soil are expected to alter precipitation, temperature and evaporation with a naturally accompanying rise in turnover of organic matter, increased CO_2 losses in organic and mineral soil etc. Other functions of soil are also affected by such losses for e.g. soil structure, water retention capacity, stability, nutrient adequacy and soil erosion. However, elevated nutrient release ultimately resulting into increased soil fertility and plant production can counteract such effects. Enhanced formation of peat and release of methane can be expected in the area with high rainfall, whereas areas having decreased precipitation experience loss of CO_2, fewer peat formation and moisture scarcity for crops (particularly in areas with superficial soils) and forest soils, thus affecting reproductive capacity and survivability of invertebrates found in soil, forest foraging patterns, food chain etc. (Chander 2012). Atmospheric deposition of the nitrous oxide to soil will be enhanced with increased rainfall which may lead to soil disturbances, soil erosion and polluted soil and surface water which further affects soil fertility badly (Defera 2005).

Indirect Impacts of Climate Change on Soil Fertility

The climatic changes as a whole are generally expected to enhance crop yields (e.g. sugar beet, sunflower, and winter wheat) resulted from the integrated effects of long growing seasons, CO_2 fertilization and efficiency of radiation use which mainly applies to plants operating C_3 photosynthetic pathway (Pathak et al. 2012; Mihra and Rakshit 2008) but not C_4 pathways essentially (Allen Jr. et al. 1996). Enhanced CO_2 increases the overall dimension and mass of C_3 plants. Structural components get more photoassimilates during the vegetative phase development to support the light-harvesting complex. Some crops like sunflower, oilseed and potatoes are found to have increased yield when grown in drought conditions having light textured soil. Enhancement in yield of grasses is also expected. Tree growth is found to be increased under high temperature and rising carbon dioxide levels in the atmosphere. Other oblique impacts of climatic change on soil fertility are through elevated CO_2-induced growth rates or water-use efficiencies, through rise in sea level, through elevated or reduced vegetation cover induced by climate change or altered human impacts on soil etc.

Table 1 Relationship between soil properties and affected soil processes (Anjali and Dhananjaya, 2019)

Soil properties	Affected soil processes
Physical properties	
Bulk density	Soil compaction and structural status
Protective cover of soil	Movement of soil nutrients and water, carbon and nitrogen fixation, stabilization of soil
Soil porosity	Plant's available air and water capacity, yield capacity
Structure of the soil	Organic matter turnover and soil aggregate stability
Infiltration capacity of soil	Soil water movement and availability
Water distribution and availability	Soil texture, soil macro-pores, field capacity, and permanent wilting point
Rooting depth	Soil salinity and plant water availability
Chemical properties	
Soil electrical conductivity	Threshold level of plant microbial activity
Soil ion exchange capacity	Soil nutrients and water
Soil pH	Threshold of biological and chemical activities
Biological properties	
Soil organic matter (SOM) Mineralization and macro-organic matter	Soil organism's metabolic activity, organic matter storage decomposition of pant residues, formation of macro-aggregates, Net N flux from mineralization and immobilization
Soil respiration	Activity of microbes
Plant available N, K, P	Soil's available nutrients
Microbial diversity	Availability and cycling of nutrients
Microbial biomass	Microbial activity
Microbial quotient	Substrate use efficiency

Impact of Climate Change on Various Parameters of Soil

Climate change imposes great impact on different soil parameters (Table 1) which in turn affects soil fertility. Given below is a detailed account of various soil parameters and affected soil processes under changing climate.

Impact on Soil Physical Parameters

Most of the physical soil processes are uptakes, losses, movements and transfers of substances like water, minerals, salts, silicates, carbonates, sugars, organic matter etc. from one plant part to another. These processes often include oxidation-reduction reactions. Uptakes are mostly concerned with gain of organic matter, water and oxygen molecules through hydration and oxidation reactions. Losses mainly consist of substances suspended or dissolved in water percolating from the earth's surface through permeable soil. The extremities of precipitation, elevated temperatures, rising carbon dioxide concentration etc. interact with earth's ecosystem and are anticipated to affect a number of soil processes. Factors like availability of moisture content in soil, salt stress, C and N ratios, soil nutrient content, soil structure and texture, porosity, electrical conductivity, soil biodiversity, water (H_2O) movements through soil particles, water congestion and roots penetration inside the soil are some of the alterations in physical attributes of soil which climate change can bring about and hence impose high risk on soil fertility. Given below is a detailed explanation of how climate change affects these soil properties and thus soil fertility.

Structure and Stability of Soil

Soil structure is a very crucial property which gives idea about collective arrangements of soil particles together. The combined association and organization of soil particles into soil mass makes the soil structure. Quantities of air and water found in the soil is controlled by soil structure. Soil structure, its biological and chemical properties along with efficient management practices determine the combined stability and soil resistance against external forces such as higher intensity of rain and aggressive land use in the form of crop cultivation (Dalal and Moloney 2000; Moebius et al. 2007). Soil stability is a functional physical health indicator of soil as it maintains crucial functions of soil ecosystem such as organic carbon accumulation, water movement and storage, interactions between root and soil microflora, soil infiltration capacity, soil erosion etc. Quality and quantity of organic and inorganic matter present in the soil matrixes, methods of cultivation and other physical processes of soil sturdily influence the quality and nature of the soil structure. Decreased levels of organic matter in soil lead to decline in infiltration rates and aggregate stability and increased compaction and erosion susceptibility of soil, thus strongly affecting soil fertility (Bot and Benites 2005; Karmakar et al. 2016).

Soil Water Availability

The water in the soil can be influenced by climate change in a number of ways such as the following: heavy rainfall causes change in the soil water within a few hours; rising temperature results in enhanced evapotranspiration, hence water loss from the soil and last but not least is the kind of land usage. Regimes of soil moisture and soil water balance reflect the control of climate-hydrology-vegetation-land use change (Varallyay 1990a, b; Varallyay and Farkas 2008). For example, increased temperature imposes risk of increased evaporation and transpiration in plant, less rainfall which results in decreased water infiltration and storage in the plants, surface runoff in hilly areas etc. Several processes involved in soil formation such as turnover of organic matter, structure and texture formation, weathering, gleying, podzolisation etc. are majorly affected soil moisture regimes (Varallyay 2010).

Soil Texture

A comparative percentage of silt, clay and sand in a soil define its texture. It is straightly prone to change in climate. Soil texture differentiation is greatly impacted by humid, sub-humid, arid and sub-arid climatic zones. Low content of organic matter, nutrients and moisture, loose soil structure with more sand particles etc. are few of the characteristics of soil in arid zones. Moreover soil is regularly subjected to extreme temperature, irradiance and short precipitation regimes in arid and semi-arid zones which bring about moisture in the soil for a very short period of time; hence soil fertility is greatly reduced. Soil of humid areas contains relatively more amount of clay but is more prone to erosion (Brinkman and Brammer 1990; Scharpenseel et al. 1990).

Soil Porosity

Soil porosity refers to measurement of presence of empty spaces in a soil mass. It is measured in fraction which is the ratio of total volume of nonsolid part to total solid part of soil. Movements and conduction of water, nutrients and air in agricultural production also depend upon soil porosity. Soil water-storing capability and necessary air in root zone for the growth of the plant growth are provided by pore size distribution (Reynolds et al. 2002). Physical qualities of soil, pore volume functions and micro-porosity, bulk density etc. are strongly associated with pore characteristics. There is a direct influence of water release characteristics and soil porosity on a range of soil characteristics such as aeration capacity, field capacity and water availability of soil. Pore size distribution and soil porosity also influence hydraulic properties of soil such as water-holding capacity, hydraulic conductivity, infiltration rates etc. Since soil hydraulic conductivity is a function of pore size distribution, soil with coarse texture has bigger pore size and relatively high saturated hydraulic conductivity than soil with fine texture. It has been reported in various studies that interactive surfaces between solid and liquid, particularly in loamy soil having higher clay content are affected by alterations in temperature regimes. Soil pore size distribution and porosity are closely associated with enzymatic activities and root development in soil. Altering climate scenarios such as rising carbon dioxide (CO_2) concentrations, temperatures and extreme precipitation regimes are likely to affect biological activities of soil such as aggregate stability, soil microbial interactions, root development, plant growth and other soil functions in the future in unexpected ways.

Plant Nutrient Availability

The quality and content of nutrients and organic matter present in the soil is a measure of capacity of the soil to sustain plant growth and on the other hand, it also recognizes and identifies significant and critical values for assessing environmental hazards. Soil organic carbon cycle and nitrogen cycles are intimately and deeply linked together. Hence climatic change factors such as rising temperatures, altered precipitation and deposition of atmospheric nitrogen are likely to affect nitrogen cycle and probably other plant nutrient cycling such as sulfur and phosphorus.

Soil Bulk Density

Soil bulk density is frequently evaluated to properly describe the soil condition and firmness/compactness concerned to soil utilization and management (Hakansson and Lipiec 2000). It is usually found in negative correlation with soil organic carbon or soil organic matter (Weil and Magdoff, 2004). Increase in temperature causes increasing decomposition rate which in turn is responsible for loss of organic carbon from the soil (Hakansson and Lipiec 2000). It may further lead to increased bulk density of soil rendering soil additionally prone to compaction namely (Davidson and Janssens 2006) stress of climate change such as draught, high intensity rainfall, land management activities etc. (Birkas et al. 2009).

Rooting Depth

Plant available water capacity, soil organic carbon content, soil salinity and many other properties of soil are affected by rooting depth which indicates key constraints in soil (Birkas et al. 2009; Dalal and Moloney 2000). In the conditions of extended drought periods, the effects of soil limitations such as salt stress and elevated concentrations of chloride ions are expected to be more on plant water availability and therefore plant productivity (Dang et al. 2008; Rengasamy 2010).

Soil Surface Cover

Soil surface cover serves as a platform for a wide range of crucial ecological functions such as safety guard for soil surface; functions to help in nutrient and water retention on soil surface, carbon fixation and N fixation (Box and Bruce 1996); sustains indigenous seed germination etc. A number of soil conditions related to its structure such as soil formation, soil crust etc. are chiefly related to soil sodicity and are used in characterizing soil health under changing climate. A range of soil processes can be affected by the formation of soil crusts such as oxygen diffusion, water infiltration, surface water evaporation, soil erosion etc.

Soil Temperature Regimes

A close connection has been noticed between temperatures of air and soil, a rise in air temperature results in rise in soil temperature and vice-versa. Soil temperature is largely controlled by solar radiation intensity (Box and Bruce 1996). The gain and loss of solar radiations at earth's surface, evaporation process and heat transmission through soil and convection transfer via water and gas movements mainly govern the soil temperature regimes. Like soil moisture, soil temperature as well is a major factor in most of the soil processes. Higher soil temperature will speed up soil processes such as decomposition rate of soil organic matter, activity of soil microbes and release of nutrients, nitrification rates and mineral weathering process. Nature of plant life existing on the earth's surface also affects soil temperature regimes.

Soil Chemical Parameters

Loss of salt and mineral content, pH imbalance, change in electrical conductivity etc. would be some of the main effects of chemical changes in the soil under the influence of climate change. Process of leaching results in loss of salts and other nutrients, whereas salinization occurs as a result of upward movements of water due to evapotranspiration or decreased precipitation/irrigation (Brinkman and Sombroek 1996). The composition of clay mineral fractions usually modify slightly in centuries except transformations of halloysites, which are produced under perennially moisture circumstances subjected to episodic drying or the steady drying out of goethite to hematite under high temperatures or harsh drying conditions. Modifications in the surface characteristics of the clay are usually slower than movements of salt which takes place more quickly than alterations in bulk composition. These surface modifications have a leading influence on the physical and chemical properties of the soil (Brinkman 1985, 1990).

Change in the bulk composition of clay and clay mineral surface fraction of soil is brought about by a number of processes, enlisted below (Brinkman 1982). Every process may be inhibited or accelerated by changing external environmental conditions because of global changes (Brinkman and Sombroek 1996).

- Enhanced leaching can fasten up hydrolysis by water having CO_2 concentrations which also eliminate basic cations and silica.
- Ferrolysis, a cyclical method of clay transformations along with dissolutions arbitrated by modifying reduction and oxidation processes of iron, which reduces soil capacity of cation exchange by interlayering of aluminum in bulging clay minerals, might take place wherever soils are subjected to leaching and alternative oxidizing and reducing reactions. In future hotter world, it might happen over larger regions than in present scenario particularly in areas of high latitudes and monsoon climate.
- Cheluviation, which liquefy and eliminates iron and particularly aluminum by organic acid chelating, can be fasten up by enhanced leaching rates.
- Amorphous silica and aluminum salts are produced by clay mineral dissolution by mineral acids, e.g. in areas where sulfidic substances are oxidized with an improved drainage system.
- Reverse weathering processes, i.e. forming and transforming clay under neutral to robustly alkaline environment, which might produce, e.g., analcime or palygorskite and montmorillonite; it can initiate in dry regions during global warming and will persist in areas which are presently arid.

Soil Salinization, Acidification, and Sodicity

Rising in temperature levels can be forecasted, but forecasting about precipitation is less certain in almost all parts of world. Leaching and acidification of soil increase with significant rise in precipitation. High precipitation also leads to loss of nutrients because of leaching. The course of change towards increased soil leaching or elevated rate of evaporation will be dependent upon the temperature alterations, rainfall amount and subsequent change in land utilization and its management strategies. In both cases the situation possibly will lead to significant change in soil.

Upsurged alkalization and salinization will occur in regions where rate of evaporation is increased or precipitation decreased (Varallyay 1994). Temporary levels of salinity increase as capillary rise takes over, transporting salts into root zone of sodic soil. Soil leaching during periodic precipitation might be restricted because of surface sealing. Concentrations of salts are increased with excessive drying of soil solution. On the other hand, the rigorousness of salty scalds because of resultant salinization might fade away as levels of groundwater drop in line with decreased precipitation. This growth might have important impact on arid areas. In the areas where salt stress/salinity is a resultant of recharging processes, salinization increases with increasing upstream recharge rainfalls (Peck and Allison 1988). Increase in concentrations of atmospheric CO_2 could decrease salinity impact on vegetation growth (Nicolas et al. 1993). Climate change impacts have been studied in the coastal lowland acid sulfate soil (Bush et al. 2010). Increased proportions of precipitation, warmer temperatures, increased frequency of tropical cyclones, sea level rise etc. are some of the predicted impacts of change in climate. Each and every one of these anticipated impacts has straightforward relevance to the landscapes having coastal acid sulfate soil.

Nutrient Acquisition

In areas having more aridity, crop yield declines (Lal 2000). Deficiency of soil moisture affects the crop productivity and yield by influencing nutrient transport (Gupta 1993). Water is an exclusive medium of transport of nutrients from soil to root; water scarcity thus reduces the diffusion of nutrients to short as well as long distances (Mackay and Barber 1985; Barber 1995). Root hair length and surface increased in response to water deficiency which capture nutrients having less mobility such as sulfur (Lynch and Brown 2001). Nutrient acquisition capacity of root decreases; root function and growth get impaired. Reduced oxygen, carbon and nitrogen in drought inhibit fixation of nitrogen in legumes (Gonzalez et al. 2001; Ladrera et al. 2007; Athar and Ashraf 2008). Soil losses its significant amount of nutrients when extreme rainfall occur in the form of soil leaching (Tang et al. 2008; Zougmore et al. 2009), e.g., nitrate leaching (Sun et al. 2007). Agricultural fields having soil which is inadequately drained or is experiencing regular and intense precipitation might have completely hydrated soils which ultimately cause hypoxic conditions. Functions and growth rate of roots are hampered by reduction of crop yield and formation of phytotoxic solutes due to elemental toxicity such as toxicity of iron, boron, manganese, aluminum etc. when redox status of soil is modified under anoxic condition (Table 2, showing relevance between climate change and consequent mineral stress).

A possible relation between soil processes and mineral stress is given in Table 2. Hypoxia may also lead to deficiency of nutrients as the active transportation of cations and anions into roots is ATP driven, while ATP is synthesized in mitochondrion through electron transport chain which is oxygen dependent (Drew 1988; Atwell and Steer 1990). Considerable amount of nitrogen losses may also occur in conditions of hypoxia through denitrification since nitrate is the alternative electron acceptor in microorganisms when oxygen is not present (Prade and Trolldenier 1990). High soil temperature can result in increased nutrient uptake (almost 300-fold) by expanding the area of root surfaces and therefore enhancing nutrient diffusion rate and influx of water (Ching and Barbers 1979; Mackay and Barber 1984). As transpiration rate is increased in warmer temperatures, plants more readily tend to obtain nutrients which are water soluble such as sulfate, nitrates, Mg, Ca etc. as temperature rises. Increase in rhizosphere temperature may also encourage acquisition of nutrients by enhancing uptake of nutrients via quicker ion diffusing rates

Table 2 Possible relationship between changing climatic factors and mineral stress (Clair and Lynch 2010)

Sn. no.	Soil processes	Changing climate factors	Nutrient stress
1.	Organic carbon status of soil	Precipitation and temperature regimes, CO_2 concentration	All the nutrients
2.	Soil leaching	Heavy rainfall	Ca, Mg, NO_3, and SO_4
3.	Salinity	Temperature and rainfall	Na, K, Ca, and Mg
4.	Plant phenology	Temperature	K, N, P
5.	Biological nitrogen fixation	Soil temperature and water scarcity	N
6.	Nitrogen cycle	Temperature and water scarcity	N
7.	Soil erosion	Precipitation and water scarcity	General loss of SOC, fertilizers, and nutrients
8.	Soil redox condition	Heavy rainfall/flooding	B, Mn, Al, Fe
9.	Mass flow of nutrients and ions	Temperature, drought, CO_2	Ca, Si, mg, NO_3, and SO_4
10.	Mycorrhizae	Carbon dioxide	Zn, P, and N
11.	Root growth	CO_2, soil temperature, and drought	Almost all nutrients

and increased metabolism of roots (Bassirirad 2000). Nevertheless, plenty of soil moisture is a main factor upon which nutrient uptake in warmer temperatures depends. Increased temperature causes vapor pressure deficit under drought conditions which in turn causes closure of stomata and consequently decreases nutrient uptake (Abbate et al. 2004; Cramer et al. 2009).

Soil pH

pH of the soil is significantly influenced by nature of plant life occurring on soil, soil parent material, weathering process and time and other climatic conditions. It is a vital sign of soil health. Soil pH therefore is incorporated in integrative tests of soil health to analyze influences of changes of land use and various agricultural practices. The majority of soils do not suffer quick change in pH due to climate change. Nevertheless, climate change influences on other soil processes such as flow of nutrients, soil water availability, status of soil organic matter, soil structure etc. which in turn affects soil pH and hence soil fertility (Reth et al. 2005).

Electrical Conductivity

Electrical conductivity (EC) of soil is a measure of concentration of salts present in the soil. It is able to notify trends in crop performance, salt concentration, nutrient cycling and other biological activities of the soil. Besides pH measurements, it also acts as a substitutive measure of structural decline in soil, particularly in sodic soils (Arnold et al. 2005). Electrical conductivity functions as a chemical marker to notify biological quality of soil in regard to crop management trends (Gil et al. 2009). Rising temperatures and declining precipitation enhance the electrical conductivity in response to climate change (Smith et al. 2002).

Cation Exchange Capacity and Sorption of Soil

Cation exchange capacity (CEC) and sorption of the soil are very crucial parameters of the soil principally in the retention of main nutrient ions such as K^+, Ca^{2+}, Mg^{2+}, etc. and Al^{3+} and Mn^{3+} mobilization. These soil parameters therefore are helpful in indicating condition of the soil and informing about its capacity to soak up nutrients as well as chemicals and pesticides (Ross et al. 2008). Organic pollutants and heavy metal adsorption are also influenced by CEC. Soil organic matter, pH, and decomposition rate are few of the factors which affect ECE. Loss of soil organic matter can result in enhanced leaching process of the basic cations in reaction to lofty and severe rainfall, therefore transporting alkalinity to waterways from soil (Davidson and Janssens 2006).

Soil Biological Parameters

The organisms which live in the soil are very well adapted to alterations in the environmental conditions. For evaluation of soil health properties in climatic change response, functions of soil health indicators become very crucial.

Soil Organic Matter

Soil organic matter consists of a widespread range of biotic and abiotic components. It is a very crucial component of soil which constantly varies in its functions and properties because of regular environmental fluctuations (Weil and Magdoff 2004). It supports the soil in many ways and serves many functions such as source of nitrogen and sink of carbon, regulates nutrient cycling for e.g. sulfur and phosphorus cycles etc. It is able to form complex with multiple organic compounds and ions and provides habitat for a variety of microbes, plants and animals. It provides water-retention capacity, hydraulic conductivity and aggregate stability to soil (Haynes 2008; Weil and Magdoff 2004). As soil organic matter serves most of the functions attributed to soil, its reduction leads to decline in soil fertility and biodiversity and also loss of structure and texture of soil, declined water retention capacity, high erosion and bulk density causing soil compaction. It helps in absorbing atmospheric CO_2, thereby lessening global warming. It also helps in mitigating floods following heavy precipitation by absorbing and storing large quantities of water. It enhances soil resilience and hence soil fertility (Eamus 1991; Gifford 1992; Gunderson and Wullschleger 1994).

Low-Density Fractions and Macro-Organic Matter

Low-density (or light) fractions and macro-organic constituents of soil organic matter consist majorly of particulate animal and plant residues which are generally free from minerals and easily decomposable and serve as substrate for soil microorganisms. They also serve as a nutrient reservoir (Post and Kwon 2000; Wagai et al. 2009). Such low-density fractions and macro-organic matter are receptive to soil managements and can act as before time indicators for calculating the effectiveness of varying management systems in adaptive climate change response (Gregorich et al. 1994). As temperature increases, labile soil organic carbon rapidly depletes (Brinkman and Sombroek 1996; Knorr et al. 2005).

Soil Carbon and Carbon to Nitrogen (C-N) Ratio

Elevated temperatures and periodic rainfall events stimulate soil microbial activities such as decomposition, mineralization etc. They will direct to decline in accumulations of biomass, reduction of soil carbon content and decreased ratio of carbon and nitrogen (Rosenzweig and Hillel 2000; Anderson 1992; Lal 2004). Elevated concentrations of atmospheric CO_2 improve water-use efficiencies of plants. It will enhance production of biomass per mm of available soil water. In the conditions of water deficiency, rate of decomposition remains superior to net primary production (NPP). This course of action makes the dried conditions favorable for reduction of organic carbon. Biomass loss occurs because of water scarcity and hence annual and perennial vegetation reduces (Kimball 2003).

Soil Flora, Fauna, and Microbial Biomass

Soil microbial biomass is the main biotic component found in soil organic matter. It is a dynamic carbon pool in the soil and susceptible marker of disturbances in soil processes in relation to soil energy and nutrient dynamics together with mediation of transfer among soil organic carbon fractions. Nevertheless, it is receptive to temporary and transient changes in environment. Soil flora and fauna being crucial components of soil play a vital role in nutrient flow and release and influence various soil parameters such as soil structure, porosity, bulk density etc. Vegetation zone migration and ecosystem change significantly affect soil flora and fauna which is very less migratory in response to altered precipitation and enhanced temperature.

An additional noteworthy impact of change in climate on fauna and flora of the soil is through elevated atmospheric CO_2 concentrations which lead to increased growth of plant and carbon allocation underneath the ground rendering the

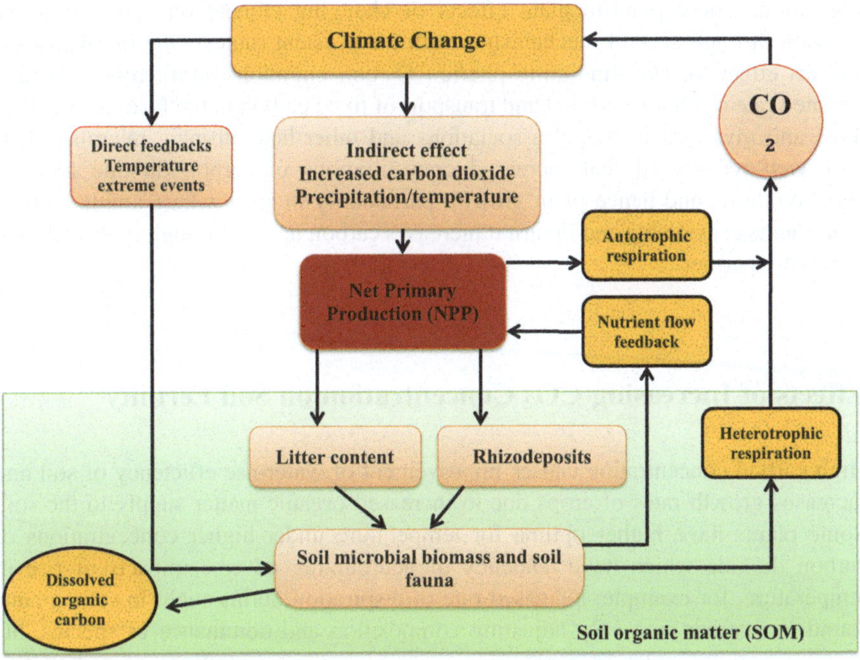

Fig. 7 Soil microbial communities are affected by climate change in both direct and indirect ways. Soil microbial activities direct GHGs to atmosphere through direct and indirect feedbacks (PS—Bardgett et al. 2008). Soil microbe activities feedback GHGs to the environment/atmosphere and contribute significantly to the global warming (Fig. 7). Direct effects on soil micro-flora include the climatic influence on soil microbes, greenhouse gas (GHG) production, altered temperature and precipitation and other tremendous climatic events while indirect effects resulted from climate-driven alterations in plant diversity and productivity which modify soil physicochemical state, the carbon supply to the soils and litter decomposition and carbon release by microbial communities (Bardgett et al. 2008; Jenkinson et al. 1991)

population of microbes to speed up rate of nitrogen fixation, denitrification, nitrogen immobilization, enhanced mycorrhizal alliance, enhanced soil aggregation and last but not least enhanced mineral weathering. Functional activities of soil microbes which feedback GHGs to the environment are affected by change in climate both directly and indirectly. Direct effects include modifications in soil microbial activities, temperature and rainfall regimes and carbon concentrations, whereas indirect effects alter physicochemical properties of soil. Direct climate-microbe feedback is SMO decomposition and the belief that heterotrophic microbial activity will be increased with increased global warming, thereby enhancing the CO_2 efflux to atmosphere and dissolved organic carbon export by hydrologic leaching (Fig. 7). Indirect microbe feedback includes circuitous effects on the communities of soil microbes and their activities and therefore the prospective for microbial response to climatic changes through its impact on growth of plant and composition of vegetation. These plant-mediate effects of changing climate on soil microbiota function through several mechanisms, with inconsistent routes. One mechanism is indirect effects of elevating atmospheric CO_2 concentrations on microbes, through elevated rate of photosynthesis and transport of fixed carbon in the form of sugars to roots and mycorrhizal fungal associations and other heterotrophic microbes. It is now well recognized that increased concentrations of carbon dioxide increase photosynthesis and hence plant growth, particularly in areas where ample amount of nutrients is available and in turn it increases carbon uptake through roots and their symbiotic partner.

Effects of Increasing CO_2 Concentration on Soil Fertility

High carbon concentration causes improvement of water-use efficiency of soil and increases growth rates of crops due to increased organic matter supply to the soil. Some plants have higher optima for temperature under higher concentrations of carbon dioxide which have tendency of neutralizing adverse impacts of rise in temperature, for example, increased rate of respiration during night in some plants found in warmer areas. By adjusting composition and dominance of species, the reduced growth cycles of a species due to higher concentrations of carbon and temperature is compensated. In agro-ecosystems, long-duration selection or change in pattern of cropping could eradicate infertile periods which may take place due to the shorter growth cycles of the main crops.

Good amount of litter, high root biomass, mycorrhizal colonization, soil microbe activities etc. are few factors which lead to high plant productivity. Supply of nitrogen to plants is constructively affected by nitrogen-fixing microbes. High carbon partial pressure and carbon activities in the soil air and soil water, respectively are needed by improved microbial activities to release more nutrients during soil mineral weathering. Likewise phosphate uptake is also improved by soil mycorrhizal activity. All these effects work in a synergistic manner with high nutrient uptake by rigorous root system because of high CO_2 concentrations.

The better microbial activities are likely to boost the amount of plant nutrient cycling through organisms found in soil. High amount of root biomass adds in soil organic matter, which demands the provisional immobilization and flow of soil nutrients. Elevated C-N ratios in the litter, observed under elevated CO_2 concentrations demand slower rate of decomposition and further sluggish re-mobilization of soil nutrient from litter and soil nutrient pool and provide additional time for the absorption into the soil.

Improved microbial activities due to higher carbon and temperature produce larger quantities of polysaccharides and other compounds which act as soil stabilizers. High amount of litter, organic matter litter or crop deposits, root biomasses etc. tends to encourage the activities of soil macro-fauna for better nutrient assimilation.

Effects of Temperature and Precipitation Change in Varied Climates

Increased precipitation in tropical areas and monsoon climates elevated soil leaching in the soils having high infiltration rates and well drainage causing transient flooding and water saturations, thus decreasing the rate of decomposition of organic matter. It might influence a noteworthy percentage of particularly the healthier soils in, for example, Sub-Saharan Africa. They will additionally generate larger and more regular runoff on soil in slanting terrain with sedimentation down slope. Soils having good resilience against such modifications will have sufficient capacity of cation exchange and anion sorption to diminish loss of nutrients during soil leaching and possess a stability in terms of structure and texture for maximizing soil infiltration capacity and quick bypass flow through the soil during precipitation of high intensity.

Higher water-use efficiency and productivity due to higher concentrations of atmospheric carbon would be likely to enlarge the ground neutralizing the impacts of higher temperatures in subtropical, sub-humid and semi-arid regions of the earth's surface. Higher local precipitation and rising inter- and intra-annual variability will possibly lead to the reduction of production of soil dry matter and consequently reduce organic matter content of soil in due course. Episodic soil leaching during intense precipitation with reduced standing plant life possibly will desalinize few of soils of well-drained locations, cause high runoff in other soils and soil salinization in places having depressions or in the areas where water table is high. Soils of such arid and high rainfall regions have more structural stability and strong mixed systems of constant macro-pore and consequently rapid infiltration rates, high water availability and deep water table as well.

In arid zones, elevated temperatures require high evaporation. In the areas where there's adequate soil moisture such as in areas having plenty of irrigation, this might lead to salinization in soil if water management of farm or land or proper irrigation and drainage are insufficient. In contrast, current studies position to enhance crop salt tolerance under increasing atmospheric CO_2 concentrations. Slight increase in precipitation is absorbed by increased rate of evapotranspiration of crops anticipated at high temperatures in temperate climates so that net chemical and hydrological impacts on soils reduce.

The harmful effects of elevated temperature on the soil organic matter are compensated by organic matter supply from robustly growing vegetation having high rate of photosynthesis, evapotranspiration and water-use efficiencies in higher atmospheric CO_2 concentrations (Buol et al. 1990) (Table 3, enlisting different soil characteristics affected by different climatic factors).

Table 3 Relationship between climatic factors and soil properties (Pareek, 2017)

Sn. no.	Climate change	Soil characteristic
1.	High precipitation	• Leaching of nutrients • High soil erosion and surface runoff • Increased arid region's productivity • Increased soil organic matter (SOM) • Increased soil moisture • More reduction of nitrates and Fe • Increased nitrogen loss
2.	Low precipitation	• Reduced availability of soil nutrients • Reduced soil organic matter • Reduced soil salinization
3.	Elevated temperature	• Reduced moisture content • Reduced rate of soil mineralization • Decreased soil organic matter • Increased rate of soil respiration • Loss of soil structure
4.	Increasing CO_2 concentration	• Accelerated flow of nutrients • Increased soil organic matter • Increased soil water-use efficiency • More availability of carbon to soil microorganisms

Mitigation Strategies

The residue management and conservation tillage aid in controlling few of soil properties and mitigating the unfavorable impacts of climate change on soil fertility in the following ways (Sharma 2011).

- *Soil structure and aggregation*: Soil structure and aggregation is a very crucial property of soil. It's aggregation refers to joining of soil particles together into secondary units. Soil aggregates which are water stable assist in maintaining good soil structure and rate of infiltrations and also protect soil from erosion. It's collective binding materials are organic compounds and mineral substances which participate in aggregation. The organic substances are derivatives of soil micro-flora such as earthworms (annelids), various types of fungi e.g., actinomycetes, bacteria etc. Soil aggregation can be directly affected by plants through root exudates, stem, leaves etc. and leaching material from weathering process and dead and decaying plant residues, canopies etc. which defend aggregation against the breakdown effects of scratches by wind forces, raindrop and dispersion caused by water and root actions. Good aeration, greater surface water entry and more capacity of water retention are few characteristics of well-aggregated soils in comparison to poorly aggregated one. Soil aggregation is intimately linked to organic matter content and biological activities of the soil. The viscous materials which bind soil components into aggregates are formed mainly by the presence of diverse living organisms in the soil. Thus, soil aggregation is improved by practices which support biotic constituents of soil profile. Since the substances responsible for binding are themselves susceptible to degradation by microbial mass, there's a need to replenish organic matter frequently to maintain soil aggregation. To preserve the soil aggregates when they are formed, there's a great need to diminish the factors which degrade and devastate them. Surface crusting is resisted by the soil which is well aggregated. The soil which is poorly aggregated is crusted by impact of raindrops by clay particle disbursement on soil surfaces obstructing the pores immediately underneath and closing them instantly before soil dries out rendering following precipitation more susceptible to surface runoff. On the other hand, crusting is resisted by bulge aggregated soil because the aggregates which are water stable are less prone to break away when raindrops hit them. Types of soil management practices which guard the soil from impact of raindrops will reduce soil crusting as well as enhance flow of water into the soils.

- *Soil Fertility and Organic Matter*: Conservation of agricultural/farming practices facilitate soil organic matter improvement through various ways such as accumulating organic wastes and other crop and tree residues by (a) green manure and fertilizer usage, rotational leguminous crops, lesser tillage and supplementary irrigation, (b) sowing seeds into the soil without any soil disturbances and fertilizer addition through tools followed by proper chemical weed management system and (c) managing surface residues by practicing decreased tillage, recycling of residues and including leguminous species in crop rotation. It is also needed to standby few residues for soil applications which in the future will assist in improving soil fertility, tilth and productivity.
- *Soil erosion, Crusting and Hydraulic Conductivity*: Soil crusting, water storage capacity and hydraulic conductivity are affected by tillage process. It has been noticed that the change in silt, clay and sand proportions soil texture etc. occurs as a result of mixing and inversion of soil caused by tillage related factors such as tillage instrumentation and depth, operation mode and soil erosion impact etc. Germination and seedling emergence are severely affected by soil crusting because of cumulative dispersion and rearrangement and resorting of soil particles during heavy precipitation followed by dry season. Soil aggregate dispersion is prevented by conservation tillage process, which also helps in increasing drenched hydraulic conductivity of the soil. Combined effect of increased hydraulic conductivity and infiltration capacity of the soil as a result of conservation tillage oversaturates/drenches the soil. Moreover tillage process also helps in less evaporation and hence increases soil water retaining capacity.
- *Relation Between Soil Porosity and Bulk Density*: An inverse relationship is found between soil porosity and bulk density. Unploughed soil has more tillage layer density than ploughed soil. With the involvement of residues, tilled soil density get decreased. Soil get compacted with automation using advanced machinery which eventually results in decreased soil porosity and enhanced bulk density of soil. Soils having less organic matter are naturally compacted. But the effectiveness of act of practicing preservation tillage to counterbalance the compaction will be only effective if there are sufficient residues, whereas severe tillage practice adversely affects the fauna of the soil, which indirectly affect the soil porosity and bulk density (Patil and Lamnganbi 2018).

Conclusion

The evaluation of forecasted impacts of changing climate on soil fertility is complicated because of the uncertainties in the weather forecasting. Land dreadful conditions and degradation issues are intimately related to unfavorable effects of changing climate. Soil has a great mitigating potential in terms of carbon and many benefits have been provided by the carbon stocks for example soil fertility, water-holding capacity, nutrient flow, workability and other positive attributes of soil. The conservation agriculture has revealed optimistic results in diminishing soil degradation events. In implementation of conservation tillage and residue managing strategies, it is crucial to recognize absolute package of practices based on rigorous research for every agro-ecological area. Nevertheless, the management practices which are site specific for water and soil conservation, crop enhancement and integrated nutrient management are required to be recognized to beat the impact of climate change on chemical, physical and biological properties of soil.

References

Abbate PE, Dardanelli JL, Cantarero MG, Maturano M, Melchiori RJM, Suero EE (2004) Climatic and water availability effects on water-use efficiency in wheat. Crop Sci 44:474–483

Adnan AT, Pierre C, Yves A, Luc N, Bashir A (2011) Modeling snowmelt-runoff under climate scenarios in the Hunza River basin, Karakoram Range, Northern Pakistan. J Hydrol 409(1–2):104–117

Aggarwal PK, Sinha SK (1993) Effect of probable increase in carbon dioxide and temperature on wheat yields in India. J Agric Meteorol 48:811–814

Allan DL, Adriano DC, Bezdicek DF, Cline RG, Coleman DC et al (1995) Soil quality: a conceptual definition. Soil Sci Soc Am Agron News 7:1995

Allen LH Jr, Barker JT, Boote KJ (1996) The CO_2 fertilization effect: higher carbohydrate production and retention as biomass and seed yield. In: Global climate change and agricultural production. Direct and indirect effects of changing hydrological, pedological and plant physiological processes. Food and Agricultural Organization of the United Nations, Rome. ISBN-13: 9789251039878

Anderson JM (1992) Responses of soils to climate change. Adv Ecol Res 22:163–210

Anjali MC, Dhananjaya BC (2019) Effect of climate change on soil chemical and biological properties–a review. Int J Curr Microbiol App Sci 8(2):1502–1512

Arias ME, Gonzalez-Perez JA, Gonzalez-Villa FJ, Ball AS (2005) Soil health-a new challenge for microbiologists and chemists. Int Microbiol 8:13–21

Arnold SL, Doran JW, Schepers J, Wienhold B (2005) Portable probes to measure electrical conductivity and soil quality in the field. Commun Soil Sci Plant Anal 36:2271–2287

Athar HR, Ashraf M (2008) Strategies for crop improvement against salinity and drought stress: an overview. In: Ashraf M, Ozturk MA, Athar HR (eds) Salinity and water stress: improving crop efficiency. Springer, New York, NY, ISBN-13: 9781402090653, pp 1–16

Atwell BJ, Steer BT (1990) The effect of oxygen deficiency on uptake and distribution of nutrients in maize plants. Plant Soil 122:1–8

Barber SA (1995) Soil nutrient bioavailability: a mechanistic approach, 2nd edn. John Wiley and Sons, New York, NY

Bardgett RD, Bowman WD, Kaufmann R, Schmidt SK (2005a) A temporal approach to linking aboveground and belowground ecology. Trends Ecol Evol 20:634–641

Bardgett RD, Freeman C, Ostle NJ (2008) Microbial contributions to climate change through carbon cycle feedbacks. ISME J 2:805–814

Bassirirad H (2000) Kinetics of nutrient uptake by roots: responses to global change. New Phytol 147:155–169

Birkas M, Dexter A, Szemok A (2009) Tillage induced soil compaction as a climate threat increasing stressor. Cereal Res Commun 37:379–382

Bot A, Benites J (2005) The importance of soil organic matter key to drought-resistant soil and sustained food and production. FAO soils bulletin 80. Food and Agriculture Organization of the United Nations, Rome, p 78

Box JE, Bruce RR (1996) The effect of surface cover on infiltration and soil erosion. In: Agassi M (ed) Soil erosion, conservation and rehabilitation. Marcel Dekker, Inc., New York, pp 107–118

Braman L et al (2010) Climate change adaptation: integrating climate science into humanitarian work. Int Rev Red Cross 92(879):693–712

Brinkman R (1982) Clay transformations: aspects of equilibrium and kinetics. In: Bolt GH (ed) Soil chemistry, B. physicochemical models. Developments in soil science 5B, 2nd edn. Elsevier, Amsterdam, pp 433–458

Brinkman R, Brammer H (1990) The influence of a changing climate on soil properties. Proceedings of the Transactions 14th International Congress of Soil Science, August 1990, Kyoto, Japan. pp. 283–288

Brinkman R, Sombroek WG (1996) The effects of global change on soil conditions in relation to plant growth and food production. In: Bazzaz FA, Sombroek WG (eds) Global climate change and agricultural production: direct and indirect effects of changing hydrological, pedological and plant physiological processes. Food and Agriculture Organization, Rome. ISBN-13: 9789251039878

Brinkman, R., 1985. Mineralogy and surface properties of the clay fraction affecting soil behavior and management. In: Soil physics and rice, Woodhead, T. (Ed.). International Rice Research Institute, Los Banos, ISBN-13: 9789711041465, pp: 161–182

Brinkman R (1990) Resilience against climate change? In: Scharpenseel HW, Shomaker M, Ayoub A (eds) Soil minerals, transformations and surface properties: eh, pH. Elsevier, New York, pp 51–60

Buol SW, Sanchez PA, Kimble JM, Weed SB (1990) Predicted impact of climatic warming on soil properties and use. Am Soc Agron 53:71–82

Bush RT, Sullivan LA, Burton ED, Johnston SG, Keene AF, Wong VNL, Mosely L (2010) Climate change impacts on acid sulfate soil landscapes. In: Proceedings of the Earth, Energy and the Environment: 20th annual VM goldschmidt conference, June 13–18, 2010, Geochemical society and European Society for Geochemistry, Knoxville, TN

Carter MR, Gregorich EG, Anderson DW, Doran JW, Janzen HH, Pierce FJ (1997) Concepts of soil quality and their significance. In: Gregorich EG, Carter MR (eds) Soil quality for crop production and ecosystem health, vol 25. Elsevier, Ottawa, ON, pp 1–19

Chander S (2012) Impact of climate change on insects. In: Pathak H, Aggarwal PK (eds) Climate change impact, adaptation and mitigation in agriculture: methodology for assessment and application. Indian Agricultural Research Institute, New Delhi, p 131

Chauhan HS, Dahiphale P (2020) Soils in India: a review. Int J Res Educ Scient Method 8:1

Ching PC, Barbers SA (1979) Evaluation of temperature effects on K uptake by corn. Agron J 71: 1040–1044

Clair SBS, Lynch JP (2010) The opening of Pandora's Box: climate change impacts on soil fertility and crop nutrition in developing countries. Plant Soil 335:101–115

Claussen M, Brovkin V, Ganopolski A (2001) Biogeophysical versus biogeochemical feedbacks of large-scale land cover change. Geophys Res Lett 28:1011–1014

Cramer MD, Hawkins HJ, Verboom GA (2009) The importance of nutritional regulation of plant water flux. Oecologia 161:15–24

Cunningham SA (2005) RRS discovery cruise 279 (04 Apr–10 may 2004): a transatlantic hydrography section at 24.5° N cruise report 54, 1–199 (Southampton oceanography Centre, Southampton, 2005) http://eprints.soton.ac.uk/17527/

Dalal RC, Moloney D (2000) Sustainability indicators of soil health and biodiversity. In: Hale P, Petrie A, Moloney D, Sattler P (eds) Management for sustainable ecosystems. Centre for Conservation Biology, Brisbane, pp 101–108

Dang YP, Dalal RC, Mayer D, McDonald M, Routley R, Schwenke GD et al (2008) High subsoil chloride concentrations reduce soil water extraction and crop yield on vertisol in northeastern Australia. Aust J Agric Res 59:321–330

Davidson EA, Janssens IA (2006) Temperature sensitivity of soil carbon decomposition and feedbacks to climate change. Nature 440:165–173

Defera (2005) Impact of climate change on soil functions. Final Project Report, Research and Development, London. https://doi.org/10.1016/j.soilbio.2014.09.025BoxJE,BruceRR

Drew M (1988) Effects of flooding and oxygen deficiency on the plant mineral nutrition. In: Tinker E, Tinker PB, Lauchli A (eds) Advances in plant nutrition. Greenwood Publishing Group, Westport, CT, ISBN-13: 9780275929442, pp 115–159

Duteil O, Böning CW, Oschlies A (2014) Variability in subtropical-tropical cells drives oxygen levels in the tropical Pacific Ocean. Geophys Res Lett 41(24):8926–8934

Eamus D (1991) The interaction of rising CO_2 and temperatures with water use efficiency. Plant Cell Environ 14:843–852

England MH, Mcgregor S, Spence P, Meehl GA, Timmermann A, Cai W, Sen GA, Mcphaden MJ, Purich A, Santoso A (2014) Recent intensification of wind-driven circulation in the Pacific and the ongoing warming hiatus. Nature Climate Change 4:222–227

Fauchereau N, Trzaska M, Rouault M, Richard Y (2003) Rainfall variability and changes in southern Africa during the 20th century in the global warming context. Nat Hazards 29(2): 139–154

Fischer EM, Knutti R (2015) Anthropogenic contribution to global occurrence of heavy-precipitation and high temperature extremes. Nat Clim Chang 5(6):560–564

Gifford RM (1992) Implications of the globally increasing atmospheric CO_2 concentration and temperature for the Australian terrestrial carbon budget: integration using a simple-model. Aust J Bot 40:527–543

Gil SV, Meriles J, Conforto C, Figoni G, Basanta M, Lovera E et al (2009) Field assessment of soil biological and chemical quality in response to crop management practices. World J Microbiol Biotech 25:439–448

Gonzalez EM, Galvez L, Royuela M, Aparicio-Tejo PM, Arrese-Igor C (2001) Insights into the regulation of nitrogen fixation in pea nodules: lessons from drought, abscisic acid and increased photoassimilate availability. Agronomie 21:607–613

Gregorich EG, Carter MR, Angers DA, Monreal CM, Ellert BH (1994) Towards a minimum data set to assess soil organic matter quality in agricultural soils. Can J Soil Sci 74:367–385

Gunderson CA, Wullschleger SD (1994) Photosynthetic acclimation in trees to rising atmospheric CO_2: a broader perspective. Photosynth Res 39:369–388

Gupta JP (1993) Wind erosion of soil in drought-prone areas. In: Sen A, Kar A (eds) Desertification and its control in the thar, Sahara and Sahel regions. Scientific Publishers, Jodhpur, ISBN-13: 978–8172330446, pp 91–105

Hall S, Huber D, Grimm N (2008) Soil N_2O and NO emissions from an arid, urban ecosystem. J Geophys Res 113:G01016. https://doi.org/10.1029/2007JG000523

Hakansson I, Lipiec J (2000) A review of the usefulness of relative bulk density values in studies of soil structure and compaction. Soil Till Res 53:71–85

Haynes RJ (2008) Soil organic matter quality and the size and activity of the microbial biomass: their significance to the quality of agricultural soils. In: Huang Q, Huang PM, Violante A (eds) Soil mineral-microbe-organic interactions: theories and applications. Springer, Berlin, pp 201–230

Held I, Souden B (2000) Water vapor feedback and global warming. Annu Rev Energy Environ 25: 441–475

Houghton RA (1999) The annual net flux of carbon to the atmosphere from changes in land use 1850 to 1990. Tellus 50B:298–313

Intergovernmental Panel on Climate Change (2013) Climate Change 2013: the physical science basis. In: Stocker TF, Qin D, Plattner G-K, Tignor M, Allen SK, et al (eds) Contribution of working group I to the fifth assessment report of the intergovernmental panel on climate change. Cambridge University Press, Cambridge/New York

Intergovernmental Panel on Climate Change (IPCC) (2014). Climate change 2014: synthesis report. In: Pachauri RK, Meyer LA (eds) Contribution of working groups I, II and III to the fifth assessment report of the intergovernmental panel on climate change. IPCC, Geneva

Intergovernmental Panel on Climate Change (IPCC) (2018) Summary for Policymakers. In: Masson-Delmotte V, Zhai P, Pörtner H-O, Roberts D, Skea J, Shukla PR, Pirani A, Moufouma-Okia W, Péan C, Pidcock R, Connors S, Matthews JBR, Chen Y, Zhou X, Gomis MI, Lonnoy E, Maycock T, Tignor M, Waterfield T (eds) Global warming of 1.5°C. An IPCC special report on the impacts of global warming of 1.5°C above pre-industrial levels and related global greenhouse gas emission pathways, in the context of strengthening the global response to the threat of climate change, sustainable development, and efforts to eradicate poverty. In Press

Intergovernmental Panel on Climate Change (IPCC) (2020) Climate change and land. An IPCC special report on climate change, desertification, land degradation, sustainable land management, food security, and greenhouse gas fluxes in terrestrial ecosystems; Summary for Policymakers.

IPCC (2007) Climate change 2007: mitigation. Contribution of working group III to the fourth assessment report of the intergovernmental panel on climate change. Cambridge University Press, Cambridge/New York

Jayne SR, Marotzke J (2001) The dynamics of ocean heat transport variability. Rev Geophys 39: 385–411

Jenkinson DS, Adams DE, Wild A (1991) Model estimates of CO_2 emissions from soil in response to global warming. Nature 351:304–306

Kalra N, Aggarwal PK (1994) Evaluating water production functions for yield assessment in wheat using crop simulation models. In: ten Berge HFM, Wopereis MCS, Shin JC (eds) Nitrogen economy of irrigated rice: field of simulation studies. SARP Research Proceedings. AB-DLO Publisher, Wageningen, pp 254–266

Karmakar R, Das I, Dutta D, Rakshit A (2016) Potential effects of climate change on soil properties: a review. Sci Int 4:51–73

Kimball BA (2003) Response of plants to elevated atmospheric CO2. Ind J Plant Physio:18–24

Knorr W, Prentice IC, House JI, Holland EA (2005) Long term sensitivity of soil organic carbon turnover to warming. Nature 433:298–301

Ladrera R, Marino D, Larrainzar E, Gonzalez EM, Arrese-Igor C (2007) Reduced carbon availability to bacteroids and elevated ureides in nodules, but not in shoots, are involved in the nitrogen fixation response to early drought in soybean. Plant Physiol 145:539–546

Lal R (1995) Tillage systems in the tropics: management options and sustainability implications. FAO Soils Bull. No. 71. Food and Agriculture Organization of the United Nations, Rome, p 180

Lal, R., 2000. Soil management in the developing countries. Soil Sci., 165: 57–72.

Lal R (2004) Soil carbon sequestration to mitigate climate change. Geoderma 123:1–22

Lynch JP, Brown KM (2001) Topsoil foraging-an architectural adaptation of plants to low phosphorus availability. Plant Soil 237:225–237

Mackay AD, Barber SA (1984) Soil temperature effects on root growth and phosphorus uptake by corn. Soil Sci Soc Am J 48:818–823

Mackay AD, Barber SA (1985) Soil moisture effect on potassium uptake by corn. Agron J 77:524–527

Medinets S, Skiba U, Rennenberg H, Butterbach-Bahl K (2015) A review of soil NO transformation: associated processes and possible physiological significance on organisms. Soil Biol Biochem 80:92–117

Mihra PK, Rakshit A (2008) Consequence of climate change for Indian agricultural productivity and land use. Int J Agric Environ Biotechnol 1:160–162

Mitchell JFB (1989) The greenhouse effect and climate change. Rev Geophys 27:115–139

Mittler R, Blumwald E (2010) Genetic engineering for modern agriculture: challenges and perspectives. Annu Rev Plant Biol 61:443–462

Moebius BN, van Es HM, Schindelbeck RR, Idowu OJ, Clune DJ, Thies JE (2007) Evaluation of laboratory-measured soil properties as indicators of soil physical quality. Soil Sci 172:895–912

Nicolas ME, Munns R, Samarakoon AB, Gifford RM (1993) Elevated CO_2 improves the growth of wheat under salinity. Aust J Plant Physiol 20:349–360

Pareek N (2017) Climate change impact on soils: adaptation and mitigation. MOJ Eco Environ Sci 2 (3):136–139. https://doi.org/10.15406/mojes.2017.02.00026

Parry M, Rosenzweig C, Iglesias A, Fischer G, Livermore M (1999) Climate change and world food security: a new assessment. Glob Env Chan 9:51–67

Pathak H, Aggarwal PK, Singh SD (2012) Climate change impact, adaptation and mitigation in agriculture: methodology for assessment and application. Indian Agricultural Research Institute, New Delhi, pp 1–302

Patil A, Lamnganbi M (2018) Impact of climate change on soil – a review. Int J Chem Stud:1

Peck AJ, Allison GB (1988) Groundwater and salinity response to climate change. In: Pearman GI (ed) Greenhouse: planning for climate change. Brill Archive, Melbourne, pp 238–251. ISBN-13: 9780643048638

Peterson TC, Hoerling MP, Stott PA, Herring SC (2013) Explaining extreme events of 2012 from a climate perspective. Bull Am Meteorol Soc 94(9):S1–S74

Post WM, Kwon KC (2000) Soil carbon sequestration and land-use change: processes and potential. Glob Chan Bio 6:317–327

Prade K, Trolldenier G (1990) Denitrification in the rhizosphere of rice and wheat seedlings as influenced by plant K status, air-filled porosity and substrate organic matter. Soil Biol Biochem 22:769–773

Rahmstorf S, Cazenave A, Church JA, Hansen JE, Keeling RF, Parker DE, Somerville RC (2007) Recent climate observations compared to projections. Science 316:709–709

Rakshit A, Sarkar NC, Pathak H, Maiti RK, Makar AK, Singh PL (2009) Agriculture: a potential source of greenhouse gases and their mitigation strategies. IOP Conf Ser: Earth Environ Sci 6

Rao D, Sinha SK (1994) Impact of climate change on simulated wheat production in India. In: Rosenzweig C, Iglesias A (eds) Implications of climate change for international agriculture: crop modelling study. U.S. Environmental Protection Agency, Washington, DC, pp 1–10

Rengasamy P (2010) Soil processes affecting crop production in salt-affected soils. Func Plant Bio 37:613–620

Reth S, Reichstein M, Falge E (2005) The effect of soil water content, soil temperature, soil pH-value and root mass on soil CO2 efflux-a modified model. Plant Soil 268:21–33

Reynolds WD, Bowman BT, Drury CF, Tan CS, Lu X (2002) Indicators of good soil physical quality: density and storage parameters. Geoderma 110:131–146

Reynolds WD, Drury CF, Tan CS, Fox CA, Yang XM (2009) Use of indicators and pore volume-function characteristics to quantify soil physical quality. Geoderma 152:252–263

Rosenzweig C, Hillel D (2000) Soils and global climate change: challenges and opportunities. Soil Sci 165:47–56

Ross DS, Matschonat G, Skyllberg U (2008) Cation exchange in forest soils: the need for a new perspective. Eur J Soil Sci 59:1141–1159

Roy SB, Avissar R (2002) Impact of land use/land cover change on regional hydrometeorology in Amazonia. J Geophys Res Atmos 107(D20):LBA-4

Saseendran SA, Singh KK, Rathore LS, Singh SV, Sinha SK (2000) Effects of climate change on rice production in the tropical humid climate of Kerala, India. Clim Chang 44:495–514

Savage M, Mujica A, Chiappe F, Ross I (2015) Climate finance and water security: Bangladesh case study. Oxford Policy Management Limited, Oxford

Scharpenseel HW, Schomaker M, Ayoub A (1990) Soils on a warmer earth: effects of expected climate change on soil processes, with emphasis on the tropics and sub-tropics. Elsevier, Amsterdam, p 273

Schimel DS (1995) Terrestrial ecosystems and the carbon-cycle. Glob Chang Biol 1:77–91
Schmidt GA, Ruedy RA, Miller RL, Lacis AA (2010) Attribution of the present-day total greenhouse effect. J Geophys Res 115:D20106
Sharma KL (2011) Adverse effects of climate change. Model training course on 'Impact of Climate Change in Rainfed Agriculture and Adaptation Strategies'. November 22–29, 2011, CRIDA, Hyderabad, India; 68–84
Sinha SK, Swaminathan MS (1991) Deforestation, climate change and sustainable nutrition security: a case study of India. Climate Change 19:201–209
Slavich P (2001) Soil health: a systems approach to soils. In: Soil health: the foundation of sustainable agriculture. Proceedings of a Workshop on the Importance of Soil Health in Agriculture, Wolongbar, NSW, Australia,20–21 June 2001. Ed. R. Lines-Kelly (NSW Agriculture: Wollongbar, NSW) pp. 2–7
Smith JL, Halvorson JJ, Bolton H Jr (2002) Soil properties and microbial activity across a 500 melevation gradient in a semi-arid environment. Soil Biol Biochem 34:1749–1757
Sun Y, Solomon S, Dai AG, Portmann RW (2007) How often will it rain? J Clim 20:4801–4818
Tang JL, Zhang B, Gao C, Zepp H (2008) Hydrological pathway and source area of nutrient losses identified by a multi-scale monitoring in an agricultural catchment. Catena 72:374–385
Tao F, Yokozawa M, Hayashi Y, Lin E (2003) Future climate change, the agricultural water cycle and agricultural production in China. Agric Ecosyst Environ 95:203–215
Tierney JE, Smerdon JE, Anchukaitis KJ, Anchukaitis, and Seager R. (2013) Multidecadal variability in east African hydroclimate controlled by the Indian Ocean. Nature 493(7432):389–392
United States Environmental Protection Agency (EPA) (2010) United States Environmental Protection Agency (EPA). Climate Change Science Facts. www.epa.gov/climatechange
Varallyay G, Farkas CS (2008) Impact of climate change on soils. In: Harnos ZS, Csete L (eds) Climate change: environment-risk-society. Szaktudas Kiado Haz, Budapest, pp 91–129. (In Hungarian)
Varallyay G (1990a) Influence of climatic change on soil moisture regime, texture, structure and erosion. In: Scharpenseel HW, Schomaker M, Ayoub A (eds) Soils on a warmer earth: effects of expected climate change on soil processes, with emphasis on the tropics and sub-tropics, Elsevier, Amsterdam, pp 39–49, ISBN: 9780080869889
Varallyay G (1990b) Potential impact of climate change on soil moisture regime and soil degradation processes. Proceedings of the International Seminar on Future Research Trends in MAB, August 20–29, 1990, Tokyo, Japan, pp. 256–267
Varallyay G (1994) Climate change, soil salinity and alkalinity. In: Rounsevell MDA, Loveland PJ (eds) Soil responses to climate change. Springer-Verlag, Heidelberg, pp 39–54, ISBN: 978-3-642-79220-5
Varallyay G (2010) The impact of climate change on soils and on their water management. Agron Res 8:385–396
Wagai R, Mayer LM, Kitayama K (2009) Nature of the "occluded" low-density fraction in soil organic matter studies: a critical review. Soil Sci Plant Nutr 55:13–25
Weil RR, Magdoff F (2004) Significance of soil organic matter to soil quality and health. In: Magdoff F, Weil RR (eds) Soil organic matter in sustainable agriculture. CRC press, Florida, pp 1–43
Ruddiman WF (2007) Earth's climate: past and future, 2nd edn. Macmillan, New York
Wuebbles DJ, Hayhoe K (2002) Atmospheric methane and global change. Earth Sci Rev 57:177–210
Xue Y (1997) Biosphere feedback on regional climate in tropical North Africa. Q J R Meteorol Soc 123:1483–1515
Yuksel G (2014) Sea surface temperature anomalies at the Mediterranean coast of Turkey (period: 1968–2010). Proc Soc Behav Sci 31:476–486
Zougmore R, Mando A, Stroosnijder L (2009) Soil nutrient and sediment loss as affected by erosion barriers and nutrient source in semi-arid Burkina Faso. Arid Land Res Manage 23:85

Impacts on Plant Growth and Development Under Stress

N. Yadav, Monika, A. Kumar, N. Kumar, Mamta, Heena, S. Kumar, and S.S. Arya

Abstract In environment different forms of stress are present due to which all growth and development processes of plants are disturbed. These stress are mainly classified into two groups, biotic stress and abiotic stress. Biotic stress is caused by living organisms like virus, bacteria, pathogen and fungi. Abiotic stress includes temperature stress (high and low), drought and salinity stress. Some plants change their morphology and physiological activities according to environmental changes and raise their ability to fight against stress. But some plants do not have capability to survive in adverse conditions and die because these stress would disturb their normal routine activities like reduced plant growth and height, reduced photosynthetic pigment, loss in transpiration rate and also disrupted water transport process. Nowadays genetic engineered crops are used against stress. Salinity stress have adverse effect in seed germination, survival rate, plant morphology, development and yield, and downregulation of photosynthesis and respiration rate. Plant oxidative stress develops as a result of overproduction of reactive oxygen species (ROS). NADPH oxidases and peroxidases are responsible for the production of ROS. Plants grown under chilling stress were facing maintenance of cellular membrane structure because it destroys membrane. During reproductive stage chilling stress showed browning, necrosis (appearance of dead tissues), and sterility at the time of anthesis (floral opening). Elevated CO_2 exposed crops like soybean, wheat, rice, peanut and bean have shown increased leaf nodes number and leaf size, shoot biomass, seed yield. Heat stress effects on germination, seedling survival, vegetative growth and reproductive development of plants. In this chapter we will briefly study about how stress affects the plant growth and development.

Keywords Biotic stress · Photosynthesis · Temperature · Yield

---

N. Yadav · Monika · A. Kumar · N. Kumar · Mamta · Heena · S.S. Arya (✉)
Department of Botany, Maharshi Dayanand University, Rohtak, India

S. Kumar
Department of Environmental Sciences, Maharshi Dayanand University, Rohtak, India

© The Author(s), under exclusive license to Springer Nature Singapore Pte Ltd. 2022
A. Vaishnav et al. (eds.), *Plant Stress Mitigators*,
https://doi.org/10.1007/978-981-16-7759-5_4

Introduction

Plants do not always live in their ideal environment, i.e. ideal light, temperature, mineral supply and humidity, water and biotic factors. Moreover, in nature, plants are attacked by various microorganisms like bacteria, fungi, animals and viruses. Because plants are immobile in nature, they cannot escape from stress and thus die. Different types of stress (biotic and abiotic) affect regulation of plant development and cause economic losses through decreased yield of crops and production of seeds and grains for livelihood. Result from water-deficit condition is disturbance in agriculture through this disturbance food production also effected (Chaves et al. 2002). Two types of effects are recorded on plant (direct and momentary) generally under elevated carbon dioxide (CO_2) concentrations in atmosphere. Because of high concentration CO_2 at the site of enzyme, rate of photosynthesis increases. Second, loss of water in transpiration form would be downregulated because of closed nature of stomata. If stomata remains in condition Respiration rate affected it should be decrease and increase depend upon changes in plant chemical composition, leaf morphology and anatomy against stress. Anthropogenic activities like burning of fossil fuels and cutting of trees (deforestation) have great impact on atmospheric greenhouse gases, increase CO_2 concentration from 730 ppm to 1000 ppm till now, these changes alter the climate condition result in warming of climate, Meehl et al. in (2007) and IPCC (2014) also reported through this global average surface temperature 1.0–3.7 °C also increase during this same time; Result of climate changes also included drought condition that are already arid (IPCC 2014). In response to water-deficit condition various initiation development process altered like inhibition of lateral roots (Babé et al. 2012) and alter developing time of species and reduced the size and number of leaves (Balasubramanian et al. 2006; Dermody et al. 2006). Different stress affect plants growth parameters and drought stress is also included. Result of water stress is numerous changes occur in plant which is mainly depends son time duration and severity of water deficit. The salts that give rise to salinity come mainly from aerial deposition of ocean aerosols and weathering of rocks via rain or wind (Rengasamy 2002). The main salt of saline soils is NaCl, but sometimes there are also significant concentrations of Ca^{2+}, Mg^{2+}, $SO_4{}^{2-}$ and $CO_3{}^{2-}$. Soil salinity has a significant impact on food industry in various countries of world, including the USA, Australia, China, India and Pakistan. Hot temperatures can have both effect (reversible, irreversible) on plant growth parameters curling of leaves patches on plant body. Effect of heat stress mainly depend on the intensity of heat and time duration because some plant has capability to adapt and some have showed damaging effect. Threshold temperature routine where in natural environment heat stress begins to plant sciences, whereas very extreme temperatures, which can damage many aspects of plant metabolism, are more relevant to culinary sciences. All over the globe water is one of the most important limiting factors which determine the species distribution and primary production in terrestrial. In crops various water deficiency symptoms occur due to inadequate rainfall and soil bound water (Sekhon et al. 2010; Vadez et al. 2011). Chaves et al. (2002) observed result of water-deficit condition agriculture, through this it affects food production in the world, resulting in famine. Shao et al. (2008a, b) gave some information from his study, for normal plant growth water is necessary component but permanent or temporary water deficit limits the growth and development of natural or cultivated plants. Massad et al. (2012) and Tezara et al. (1999) observed that Water stress affects the ecosystem and agriculture by decrease the growth and photosynthesis process in plant and also disturb human activities. Salinity stress have fallen different effect on crops growth and development because according to Khan and Duke (2001) 23% world's cultivated land is saline. Salinity can also affect plant growth via disbalancing the nutrient amount which were essential for plant growth (Tester and Davenport 2003). Nearly about 70 years ago chilling injury to crop plants was systematically observed by scientist. In chilling stress chloroplasts are the most severely impacted organelle as compared to mitochondria, nuclei and other organelles. During chilling stress thylakoids showed swelling characters in which disappearance of starch granules occur, and a peripheral reticulum which are arising from inner membrane. Chilling stress also affect the organellar development and ontogeny. Heavy metal stress affects the seed germination process because these provide toxicity condition for their growth. Biotic stress is caused by various microorganisms like fungi and bacteria. Biotic stress mainly affects the yield and plant height and weight parameters.

Stress Induced Different Stages in Plant

Selye gave the original stress concept who differentiates stress into three phases. But on the basis of Selye concept Larcher and Lichtenthaler classified it into mainly four different classes and added fourth regeneration phase into this concept (Fig. 1). Plants have optimum conditions (i.e. amount of light, water and mineral supply) for their growth and physiology. First three stress response phases occur when stressors and complex stress events occur and fourth one is that when release stressors called regeneration phase (when stressors have no severe effect on plant, Fig. 2). These four phases are shown in Fig. 1.

Plants growing under stress begin to change their physiological standard condition to cope with stressor, some will activate different defence mechanism. When stressor is removed, after that new standard of physiology can occur depending on duration and intensity of stress.

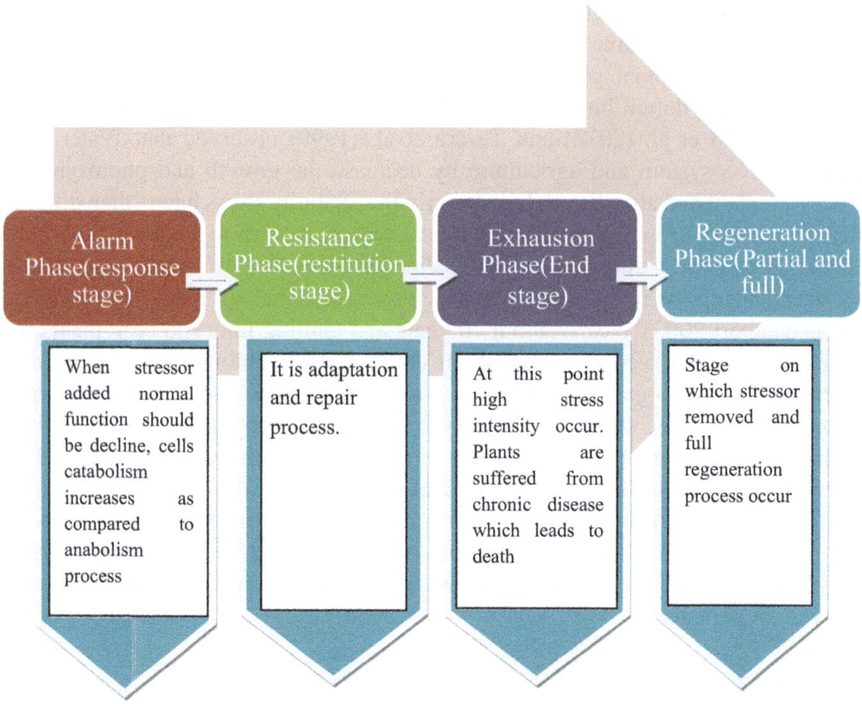

Fig. 1 Larcher and Lichtenthaler classified stress into four different stages: (i) Alarm Phase, (ii) Resistance Phase, (iii) Exhaustion Phase, (iv) Regeneration Phase

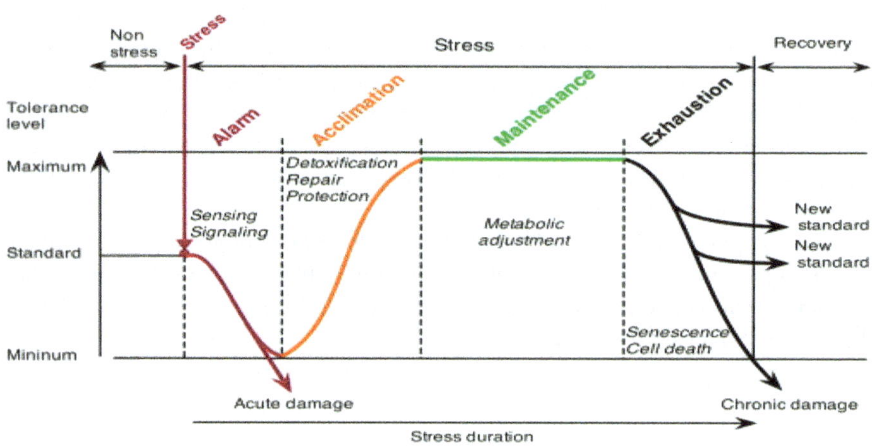

Fig. 2 Graphical representation of different phases of stress and their response induced against stress exposure (Mireille et al. 2012)

Stress Concept in Plant

In plants mainly two types of stress factors are involved. These are natural and anthropogenic stress factors acting on terrestrial vegetation.

I. *Natural Stress Factors*: In natural stress factors mainly abiotic stress is discussed. Stress which is produced by natural environment factors is called abiotic stress. For example, water deficit(drought), wind and saline soil includes. Abiotic stress is not regulated by humankind. Heat, chilling, high radiation like photoinhibition and photooxidation, water desiccation naturally, mineral nutrient deficiency (e.g. nitrogen, potassium) and flood condition are categorized under natural stress.
II. *Anthropogenic Stress Factors*: These are stress factors which are induced by human activities and not by natural environment. This stress includes mainly herbicides, pesticides and fungicides. It also includes various types of air pollutants (e.g. SO_2, NO, NO_2, NOx) produced by industries, increasing ozone (O_3) and photochemical smog by human activities, formation of ROS (O_2, radicals $O_2^{\cdot}-$ and $OH^{\cdot}$, H_2O_2) and peroxyacylnitrates like photooxidants. Acid rain gives negative impact on crops by making soil's pH acidic in nature and inducing mineral deficiency. Human activities are also responsible for making soil nutrient table unstable for nutrient uptake, e.g. overproduction of nitrogen, deposition of dry and wet nitrite, lead and cadmium, and ammonium overproduction during breeding stations. UV-A and UV-B radiation has increased. Change in global climate and greenhouse gases in environment are the results of human activities.

Oxidative Stress (Reactive Oxygen Species)

In plant biology, different definitions of oxidation stress are given by various scientists. First, it is the 'physiological state'; when oxidation exceeds reduction, loss of electrons are more than gain of electrons. Due to oxidative stress it leads to dysfunction of cell enzymes and structural components of cell membrane. In oxidative stress, 'lack of electrons' process occurs due to which long-term imbalance process is produced. Second, it is one of the 'stress factors' (same as to other stress factors salinity, water deficit) which can induce cell injuries, defence reactions and trigger signalling cascades. Both definitions are related to each other and they can also be combined. In most cases activation of O_2 makes this molecule more active for reaction; therefore, it is often called as the stress caused by 'reactive oxygen species,' 'oxygen-derived species,' 'oxygen free radicals,' etc. In oxidative stress another important class is Reactive nitrogen species (RNS)s. The major 'reasons' for oxidative stress are as follows: (i) if there is severe change in cell physiology, unbalancing occurs between ROS generation and detoxification; (ii) ROS which is produced by special enzymes is a constituent part of immunity response and stress-related signalling (used in defence mechanism and adaptation). Activation of cell wall peroxidases and NADPH oxidases in transition metal and ozone or ultraviolet stress factors are directly responsible for the production of deleterious ROS (Apel and Hirt 2004; Rao et al. 1996; Ranieri et al. 2003; Zhang et al. 2010). Wong et al. (2007) observed that FRET technique has clarified the mechanisms of this phenomenon demonstrating that cytosolic Ca_{2+} directly stimulates Rac–Rboh interaction. Being directly activated by cytosolic Ca^{2+}, NADPH 'works in concert' with ROS activated Ca^{2+}- permeable cation channels to generate and amplify stress-induced Ca^{2+} and ROS transients (Demidchik and Maathuis 2007; Demidchik et al. 2009). The more Ca^{2+} appears in the cytosol, the more $O^{2\bullet}-$ is generated and, vice versa, $O^{2\bullet}-$ activates ROS- activated cation channels through which calcium ions influx occur (Demidchik and Maathuis 2007; Demidchik et al. 2009). Reactive oxygen species with proven importance for plant physiology include hydroxyl radical, hydrogen peroxide, singlet oxygen, superoxide radical and nitric oxide (forming peroxynitrine). H_2O_2 in the presence of transition metals (hydroxyl radical) is also responsible for oxidative damage. Both photosystems and mitochondrial complex (I & III) are responsible for ROS generation, ETC of peroxisomal membrane, peroxisomes' matrix xanthine oxidase, plasma membrane's NADPH oxidases and peroxidases that are expressed in all cells. Oxidative stress also changes the permeability of bio membrane. The various studies found on oxidative stress, lipid peroxidation, oxidative stress and antioxidative enzymatic related information in plants (Sharma et al. 2012; Wahid et al. (2007, 2009). Due to its nature of highly reactive, attacking, unsaturated fatty acids and sulphydryl groups, free radicals are toxic. Oxidation of fatty acids will disrupt membrane and alter its permeability. Oxidation of sulphydryl groups on proteins will result in a loss of activity of enzymes, Rubisco. When activity of these enzyme is lost, it will disrupt the ultrastructure of cell and its organelles and also inhibit critical metabolic pathways such as respiration and photosynthesis. Cell membrane, which is made up of PUFA, is highly sensitive to oxidative stress because of stress changes in their membrane permeability, fluidity and functions. Apel and Hirt (2004) found that abiotic factors like drought may disrupt the ROS production and scavenging equilibrium. If ROS is suddenly increased with abnormal rate, it leads to irreversible changes in photosynthesis rate and ultimately plant would die. Some nonenzymatic antioxidants like tocopherol, flavonoids, carotenoids and glutathione (GSH) are present which detoxify the ROS molecules such as singlet oxygen, superoxide and H_2O_2. Plants follow some mechanisms for enzymatic ROS scavenging, which include ascorbate peroxidase (APX), superoxide dismutase (SOD), glutathione peroxidase (GPX) and catalase (CAT) (Vernoux et al. 2002). Verslues et al. (2006) reported many methods for ROS quantification in plants. Pei et al.(2000) gave information about water deficit condition, under water stress ROS may act as positive response(signalling) like stomatal regulation in Arabidopsis and maintenance of root elongation; In maize gravitropism and auxin signalling (Joo et al. 2005). Verma and Mishra (2005) reported that chlorophyll degradation and reduction in membrane fluidity and selectivity is caused by ROS. By measured lipid peroxidation and chlorophyll loss we can calculate the oxidation damage (Del Rio et al. 2005). Carotenoid pigments have several roles in photosynthesis but also follow oxidative defence mechanism (Gill and Tuteja 2010).

Salinity Stress

All over globe a significant part of world's land area is salt effected via saline stress and sodicity. Munns (2005) informed that irrigated land represents only 15% of total cultivated land. Near seashores and estuaries terrestrial plants faced high salt concentration under natural condition. Estuaries are defined as the transition zone between sea water and fresh water. Far inland, geologic marine was deposited naturally and salt seepage can washed into adjoining areas, making them unstable for agriculture. According to Khan and Duke (2001) world's cultivated lands contain 23% of salinity, 37% is sodic. Accumulation of salt from irrigation is big problem for agriculture land. Plant growth and productivity is limited by using poor quality water for irrigation and saline soil because salt is abiotic stress which also effect plants yield. Plants adapt various mechanisms to tolerate and avoid this stress; e.g. changes occur at their cellular structure, physiology and molecular level. In different ways plant is affected by salinity such as specific-ion toxicity and/or nutritional disorders, and osmotic effect (Läuchli 1999). Effect of stress also varies with species to species, its genotype, plant age, ionizing component and the salt solution ratio. During salt stress cells shrink because of loss of water and after that it can gain water from surrounding and come to its original shape and size. After this recovery, the cell division and enlargement rate decrease leading to reduce root and leaf growth. During saline soil condition plant can uptake salt from roots; due to excessive salt uptake plant shows some morphological injuries on their body and leaves. These sign and result of stress were shown with passage of time slowly. After weeks, shoot injury is noticed, and after months overall plants have showed symptoms. Within month a clear difference was observed on overall growth and injury with comparison to control. Munns (2002, 2005) had developed two phase growth response under salinity with temporal differences. The first phase includes quick response against salt stress (within minutes) and therefore rapid growth reduction activity occurs. Quick response occurs because roots are the primarily response regulator to change its physiology according to the surrounding environment. Saline soil can alter their osmotic effects and cell water relations. Therefore, reduced ability to absorb water is similar to water deficit, resulting in differences in genotypes. Munns (2002) observed that after initial decreased growth of leaf few minutes later, there is sudden increase in growth rate until a new stationary state is formed, which totally depends on salinity environment present around the root. The second response is not as fast as first response. It will takes days, weeks or months. During second response plant can accumulate salt in leaves which leads to toxicity. Older leaves accumulate salt first, then the younger leaves accumulate. Salt toxicity leads to reduced total photosynthesis leaf area and sometimes death of plants also happens. Munns (2002) reported on carbon balancing in plants; when photosynthate supply is disrupted, overall carbon balancing is also affected. Munns and Termaat (1986), Munns (2002, 2005) and Munns et al. (2006) reported that if accumulation of salt in leaf occurs, salt load increases in leaf vacuole; it will increase salt concentration up to its toxicity level and leads to leaf death and morphological injury in plants. Through reduction in total photosynthetic leaf area and leaves dying we can determine the survival of the plant. In case whether the ratio of production of young leaves greater than old leaves die which leads to high flower and seed production because sufficient amounts of leaves till remain for photosynthesis process. If whole process would altered i.e. the formation of new leaves slower than the old leaves death, the life span should be decreased and plant will die. In that condition there is no sufficient photosynthate supply occurring during reproductive stage and produce non-viable seeds. Based on this two-phase concept, both salt-sensitive and salt-tolerant plants show reduction in growth parameter during initial stage due to salts present outside the root changes their osmotic effect. Munns et al. (2006) reported that in the second phase, there is difference between a salt tolerance species and sensitive species, because the sensitive species has inability to prevent accumulation of salt during transpiration process. Accumulated salt becomes toxic for plant species. Evaporation and transpiration process makes soil concentrated because of loss of pure water from plant parts (mainly from leaves). During evaporation process, the water is lost from soil and salt concentrates become more in soil. During irrigation process, if concentrated soil presents at that time there is less chance to flash out salts and the salts rapidly reached at their toxic level which is injurious for plant growth and development for salt-sensitive species.

Above 50 Mm salinity level can cause various negative impact on plant growth and development like inhibited root and shoot growth in various mesophytes. In salinity condition stomatal closure activity is also observed which has negative effect on photosynthesis rate, respiration rate and transpiration rate. Sodium chloride (NaCl) reduced all the parameters including root length, plant length and number of leaves. Salinity stress also effect the seed germination and plant survival percentage with concentration of salt (Fig. 3). Therefore, we can easily say that when salt concentration increases it leads to reduction of the plant survival rate and seed germination. With increasing concentration of NaCl accumulation of salt in shoot, it will show a negative effect on shoot growth leading to necrosis morphological injury. Presently because of salinity in many species like cotton, spinach stomatal closing process occurs. During stomatal closed condition, it inhibits respiration and transpiration process. Some species of crops like bean and maize are more sensitive against salinity stress. So, because of salinity the enzyme which is responsible for nitrogen assimilation process is also negatively regulated. Salinity condition can inhibit the activity of enzyme nitrate reductase which helps to reduced nitrate to nitrite. It also effects the uptake and transport of nitrate from shoot and root and thereby causes reduction in growth rates. Salinity stress also effect the ion transport system of plant like it inhibits uptake of potassium and monovalent cations. Bayuelo-Jimenez et al. (2003) observed from his studies that plants like beans and maize salt in-tolerant plant have ability to exclude Na^+ whereas Gorham (1990) observed that bread, wheat and some salt-tolerant species reduced the transport rate

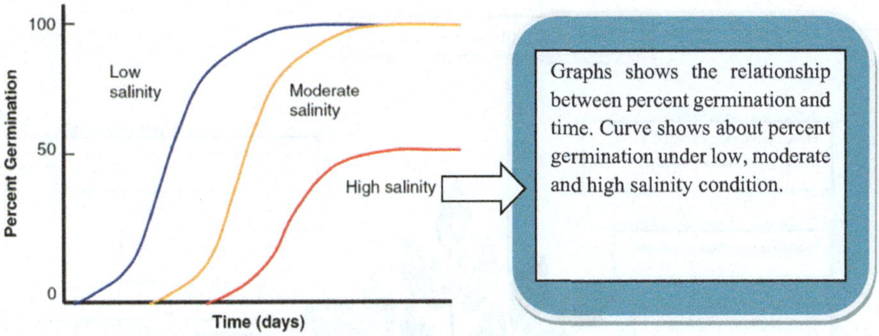

Fig. 3 Graph shows the relationship between percent germination and time. (i) Blue curve shows the percent germination under low salinity condition. (ii) Yellow curve shows about the percent germination under moderate salinity. (iii) Red curve shows about the percent germination under high salinity (Lauchli and Grattan 2007)

of Na^+ to shoots. Reproductive and vegetative stages would be affected under salinity stress which has founded implications based on whether the harvested organ is leaf, fruit, fibre, stem, leaves, shoot and grains and fruit. During saline condition plants life cycle is also affected, because in wheat and rice it will affect the flowering stage and maturity stage. Salinity also reduces more shoot growth as compared with root growth, so number of florets per ear reduced and sterility rate increased (Fig. 4). According to Munns (2002), metabolic and cellular process are also affected similarly like in drought stress condition. New leaves production rate is mainly depending on soil water potential. If water potential is low, then it will uptake more salt concentration from root as compared to water which leads sometimes to toxicity to plant growth and development. In growing tissues, the salts itself do not build up at toxicity level which inhibit their growth (phloem meristematic tissues are faded largely). Salt which is taken up by plant does not directly inhibit the growth of new leaves because through xylem salt is passed to different locations in plant parts and in vacuole it is stored. Salt stress have more negative impact on shoot than root growth; therefore it reduced the flower number and increased sterility rate of plants. Salinity stress mainly effects the seed germination process; sometimes it could delay germination process, although most plants are adapting to the salinity environment by changing their cellular modification and there may be no difference occurring on seed germinating process. At low salinity, during photosynthesis process CO_2 uptake increases as compared to control. But if salinity is higher, it will have negative effect on plant contents like amino acid, cellular proteins, soluble sugar concentration, starch and sucrose. Proline content has shown different results like at lower salinity it will be increased (1–3 mS cm^{-1}) and at higher salinity (8–11 mS cm^{-1}) it will be decreased. At 2–4 mS cm^{-1} salinity glycine betaine and proline both accumulations were high but as compared to proline accumulation of glycine betaine is more. When NaCl concentration increased in the growth medium, it could affect the chlorophyll pigments. Chl a content decreased more as compared

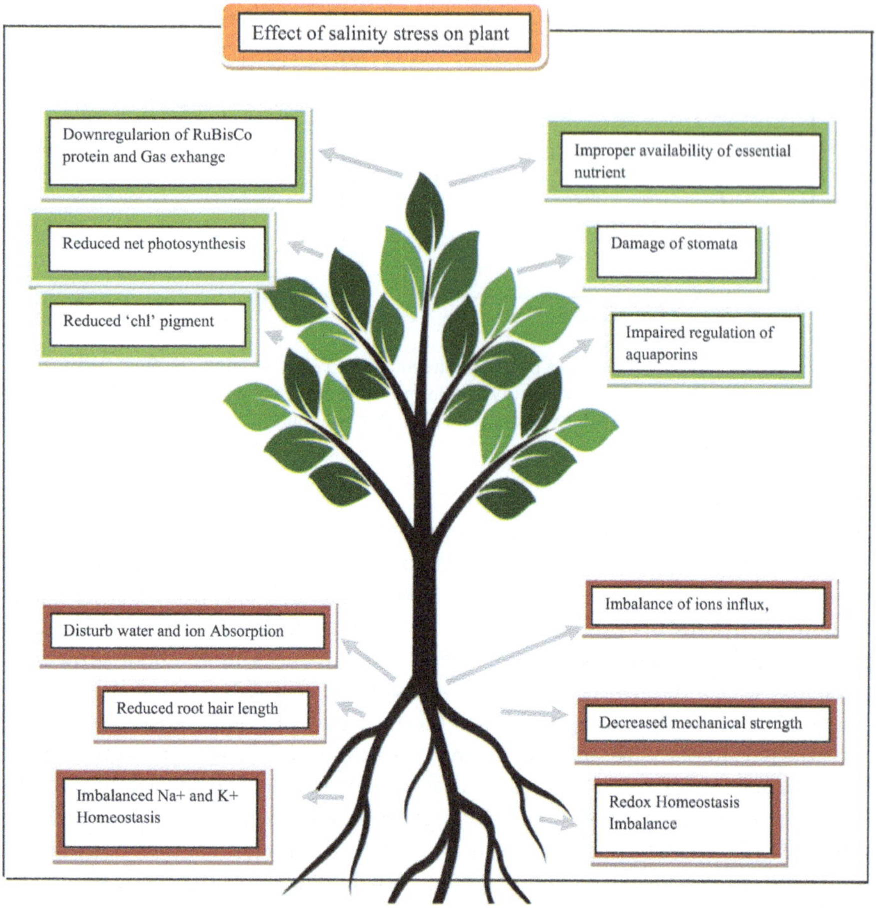

Fig 4 Effect of salinity stress on plant growth and development. Salinity affects the 'chl' pigment, rate of photosynthesis, gaseous exchange and stomatal structure. It can also disturb the basal portion of plant via disturbing their ion and water absorption, homeostasis imbalance

to Chl b content and same result was found in carotenoid condition with respect to salinity.

Water Stress

According to Bohnert and Jensen (1996), in abiotic stress several stress were included but water stress is the major stress faced by the farmers for their crop cultivation. Under water-deficit condition water potential and turgor pressure of plant could be low so that plant lost their ability to do normal physiological

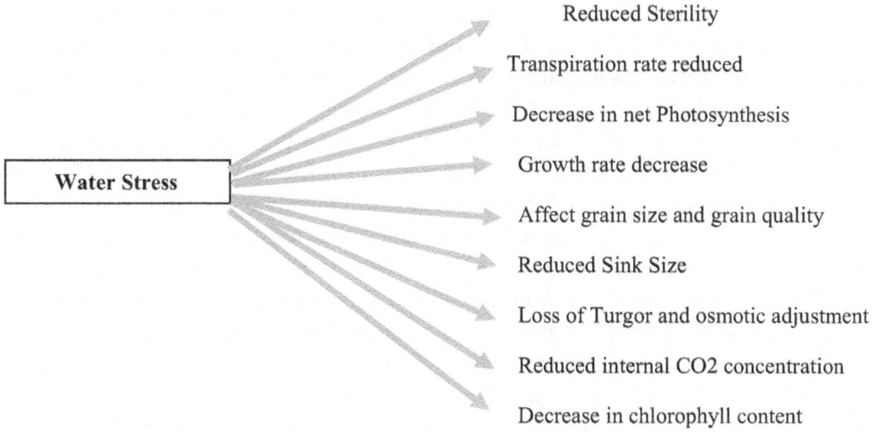

Fig. 5 Water stress have negative impact on plants . It decreases its sterility rate, transpiration process and growth of plant. Qualitative parameters like grain size and grain quality are also affected under water stress condition

functions. Drought occurs in many parts of the world every year, which is frequently experienced in the field grown plants under arid and semi-arid climates. Due to water stress numerous changes have been observed in plants related to its growth parameters mainly. Non-woody plants consist of about 80–90% water and water is the major medium for transporting the minerals and nutrient to all over the plant for doing their physiological functions. Mainly two reasons are responsible for a plant to experience water stress. First is limiting water supply to the root and second one is transpiration rate becomes high. Water stress are mainly caused by saline stress or drought condition. The primary cause of water stress is drought and high soil salinity. During high saline soil flooding, temperature-stressed plant is unable to take water from soil through roots due to 'physiological drought'. Drought, as an abiotic stress, is multidimensional in nature. Drought affects plants according to their level of organization. If drought occurs for a long time, many plants will dehydrate by losing H_2O during the process of transpiration and evaporation and die. Water stress have negative effect on photosynthesis and chloroplast. Water potential of the cell should be reduced and elevate their solutes into extracellular matrix. When leaf (expressed per unit leaf area) is expose to mild water stress, it shows very little response in leaf growth because it decreased the photosynthetic activity in plant (Fig. 5). However, we can say that stomatal conductance and leaf photosynthesis are affected by mild water stress. During early stage of water stress if stomata closure occurs, water use efficiency should be increased because stomatal closure inhibits the transpiration process. If mesophyll plants are exposed to severe stress, then dehydration condition occurs which inhibits the photosynthesis process and disturbs cell metabolism process. Results came from many studies that water stress highly effects stomatal conductance than photosynthesis. There is some relation between stomatal conductance and photosynthesis process with CO_2 concentration. If high

CO_2 is supply to plant, the effect of stress will be decline through which we can find directly the difference between stressed or non-stressed plant. Sometimes question arise in our mind 'Does water stress directly affect translocation?'; in plant phloem is used for translocation process. Transportation through phloem mainly depends on turgor pressure. Under water stress condition the water potential of cell decrease so turgor would be affected. Therefore, if turgor is low, then it will inhibit the translocation and assimilation process. So, water stress indirectly effects the translocation. However, from many experiments it concluded that translocation process is less affected. Under water stress wheat plant growth shows a decrease in leaf water potential and water content. Siddique et al. (2000) and Li et al. (2015) also observed that in wheat plant canopy temperature is responsible for drought during anthesis and vegetative growth. Under water stress reduced photosynthetic rate was observed in wheat (Arora et al. 2001; Boudjabi et al. 2015). It is the hydraulic signal which is responsible to recognize water stress in plant such as water potential, uptake of water and turgidity (Novák and Lipiec 2012). Under water deficit condition,translocation of water inhibited because there is no transport of water occur through xylem, which leads to wilting of leaves.

Another relationship was observed between LWP and plants yield in rice (Jongdee et al. 2002), soya bean (Djekoun and Planchon 1991), maize plant (Cary and Fisher 1971), sorghum (Jones and Turner 1978), wheat (Winter et al. 1988), sunflower (Boyer 1968), cotton (Grimes and Yamada 1982) and other many crops. During field work experiments we can measure LWP by using pressure chamber apparatus. Zhang and Davies in (1989) and Liu et al. in (2003) reported that during water stress condition ABA hormone is responsible for stomata closure. Water stress disrupt the cellular membrane and increase its permeability, which leads to efflux of electrolyte from the cell membrane. Effect of water stress vary from species to species in plants. In wheat plant had been observed more damage than maize plant because loss of chl and H_2O is more in C_3 plants than C_4 plants. It is similar as oxidative damage, because in oxidative stress there is increase in malondialdehyde and hydrogen peroxide content. High amount of enzymatic and non-enzymatic antioxidants is observed in C_4 plants. Nayyar and Gupta (2006) reported that leaves and root of wheat plant have higher catalase activity as compared to maize. Under water stress stomata closure occurs due to which carbon dioxide diffusion is inhibited in leaves and second it can disrupt the metabolic activity of plant. Therefore, CO_2 assimilation should be low in leaves. Flexas et al. (2004) showed that downregulation of photosynthesis process under water deficit inhibits the plant growth, yield and survival rate. Water stress causes a drop in photosynthetic potential, which disrupts the metabolic process. It leads to decreased RWC (relative water content) and assimilation of CO_2. The ultimate result is that the RuBisco activity is also disturbed which leads to decrease in photosynthesis and ATP synthesis (Lawlor 2002). Sign of water stress is observed in rice plants by measuring their chlorophyll pigment and senescence rate. Chlorophyll content is reduced in rice plant under stress. Starch mobilization (in stem) is also affected by water stress, which shortened the grain filling rate. Some enzymes are enhanced by water stress condition, e.g. sucrose phosphate synthase which gave positive response under water.

During water stress condition, ABA accumulation and proline like osmolytes process occur which is responsible for chlorosis and wilting of leaves. Cell division rate would be effected so cell growth, cell enlargement process should be prohibited. Sometimes under water stress reactive ascorbate, ROS and glutathione radical scavenging compounds formed which negatively affects the plant growth and development. When stomatal closure condition occurs, there is no gaseous exchange, transpiration and carbon assimilation process occurring. Mineral uptake and transport activity are also effected, leads to decrease leaf area and also alters the assimilation process in plant organs. It prevents gaseous exchange, lowers transpiration rates, and reduced carbon absorption rates during photosythesis. It would decrease the mineral nutrition and their transport via affecting xylem and phloem pathways. Assimilation of nutritive substances and metabolism process shows slow in response and leads to decreased leaf surface area. Water stress altered the elasticity nature of cell wall and homeostasis maintenance process is also disturbed. So, ion distribution step is also disbalanced under water stress. In soya bean plant photosynthesis rate decreased with mild and severe water stress by 40% and 70%. Water stress also disrupt the electron transport chain mechanism by shifting electron from cytochrome to other alternative pathways. Ribas-Carbo et al. (2005) showed that drought-subjected plants can increase their electron partitioning alternative pathway. Electron partitioning to alternative pathways is seen to increase by 40% in drought-subjected plants. Akir (2004) found that growing maize plants in a water-stressed environment had a significant impact on growth indices such as plant height, leaf area index, and grain yield (100-kernel weight) during harvesting. Under water stress, plants have been found to lose 28–32 percent of their dry weight. Final dry weight of plant also reduced 28–32%. It was totally supported by Baher et al. (2002). Baher also added some more parameters from his study like decrease in total dry and fresh weight of plants. Prolonged water stress also effects carbon assimilation and exchange process of plants. Shoot-related parameters like stem yield, diameter of plant stem, tiller count, plant's height, number of leaves, essential oil rate and plant's biological yield also effected under water stress. Farahanil et al. (2009) reported that there is increase in essential oil percentage in drought. In water-deficit condition Abas (2006) showed decrease in auxillary shoots, length of auxillary shoot, and fresh and dry herb yield. Prado and Maurel (2013), Sadok and Sinclair (2010) and Vadez et al. (2011) reported that aquaporins channels are present on roots and leaves cell membrane which are regulated by ABA during stress. Plants response towards stress mainly depends on field condition, plant species and plant developmental behaviour (Mittler and Blumwald 2010). Vigna catjang showed directly proportional result of proline content and hydrogen peroxide to water stress. Under drought condition auxin and activated cytokinin play important role, auxin level increase in roots as compared to shoots (Pospisilova et al. 2005; Dilrukshi et al. 2015). Havlová et al. (2008) measured the response of hormone by primary root growth. Gupta et al. (2001) and Allahverdiyev et al. (2015) observed information from the experiment of water stress during anthesis stage and showed the reduction in result of test weight, grains number, harvest index and biological yield. A thicker cuticle layer aids in the reduction of water loss through transpiration. Because the cuticle layer accounts for 5–10% of total leaf transpiration, it has been destroyed if the leaf has been exposed to extreme stress.

Temperature Stress

Temperature stress is classified into three categories, namely, heat stress, chilling stress and freezing stress. Temperature stress have negative impact on plant growth, photosynthesis, and germination rate. Severe stress may change their molecular process and results in death of plant. Various studies have been done at macrolevel and microlevel of plants. But still many questions arise from temperature stress. First, how plants sense these changes occurred through change in temperature and translate these signals into particular reactions. From many studies we observed that it is PIF4 protein which is responsible for detection of temperature. During cold, PIF4 becomes less active so plant doesn't follow their normal growth pattern. During chilling stress, PIF4 is less active—in other words, the plant rarely grows and does not follow normal growth pattern. PIF4 is highly active at high temperatures; it acts as growth-promoting genes, so plant grows taller. Till date, PIF4 complete function is unknown.

Chilling Stress

Temperature at which freezing process does not occur is called chilling temperature. Plants are submitted to a chilling stress when exposed to low non-freezing temperatures. Some plants are sensitive to chilling stress. Sensitive plants when exposed under chilling stress did not follow their regular development function properly. These plants will develop chilling stress symptoms morphologically and after some time die. Some species of plant can change their molecular mechanism according to stress and adapt the unfavourable environment. Boyer (1982) gave the definition of chilling temperature, i.e. temperature range 0–20 °C (non-freezing) has been recognized as chilling temperature. In maize, rice, tomato and soya bean chilling stress is having a major impact on plant growth. Temperature is a major determinant on which wild or cultivated crops distribution of species depends mainly (Woodward et al. 2004). Chilling sensitivity mainly varies from species to species and their ecotypes. Chilling stress symptoms were observed at cellular and subcellular level. It also has an impact on species at the molecular level, altering protein folding complexes, disrupting enzyme reaction rates and destabilizing protein structure, lowering photosynthesis rates and accumulating reactive oxygen species, and making cell walls stiff. Lyons (1973) observe that tropical, subtropical and temperate latitude are chilling sensitive. Warm habitat plants are injured when exposed to low,

Table 1 Chilling sensitive of selected species with symptoms

Species type	Common name	Observation
1. Chilling-sensitive		
Cucumis sativum	Cucumber	Chloroplast swelling, thylakoid dilation, randomly tilted grana stacks, formation of peripheral reticulum. In stroma accumulation of lipid droplets and special type of thylakoid like serpentine observed Chloroplasts disintegrate with prolonged chilling. No injury in mitochondria
Glycine max	Soya bean	
Fragaria virginiana	Strawberry	
Lycopersicon esculentum	Tomato	
Nicotiana tabacum	Tobacco	
Zea mays	Maize	
Nicotiana tabacum	Tobacco	
Phaseolus vulgaris	Common Bean	
Paspalum dilatatum	Dallis Grass	
Pisum sativum	Pea	
Sorghum spp.	Sorghum	
2. Extremely chilling sensitive		
Episcia reptans	Flame violet	Chilling-induced injury also observed in mitochondria Rapid chilling injury results in cell lysis
Saintpaulia ionantha	African Violet	
Vigna radiata	Mung Bean	
Ephedra Vulgaris	Ephedra	
Gossypium hirsutum	Cotton	

non-freezing temperature. Plants such as maize *(Zea mays)*, banana (*Musa sp.)*, tomato (*Lycopersicon esculentum)*, cotton (*Gossypium hirsutum* cucumber (*Cucurbita sp.)* and soya bean (*Glycine max))* are particularly sensitive and will exhibit signs of injury when exposed to temperature below 10–25 °C (Table 1). Even some temperate plants such as potato, asparagus and apple experience injury at temperature above freezing (0–5 °C). Chilling injury stress mainly depends on species, age of the plant and duration of the low temperature exposure . Young seedlings typically show injury signs like wilting, chlorosis and reduction in leaf expansion. In extreme cases reproductive stages are also sensitive to chilling temperature; results observed in these conditions are browning and appearance of dead tissue(necrosis) and at the time of anthesis (floral opening) sterile flower. *Lycopersicon esculentum* is strongly affected by a growth temperature of

10–12 °C. On the contrary, the wild tomato *Lycopersicon hirsutum* has mainly ten different ecotypes and each ecotype represents different response against stress. Physiological age is another factor that controls chilling. In orchid *Phalaenopsis*, younger leaves are more sensitive than older leaves. Reproductive phase is one of the main phases which is sensitive against chilling stress. Failure during reproductive stage is an example of chilling injury. Pfannschmidt (2003) said that chilling stress response in chloroplast could be observed by calculating their photosynthesis rate (Fig. 6). Photoreceptor and membrane have ability to sense the change in temperature. Photosynthetic complexes are mainly present in thylakoid membrane. Low temperatures can make the enzymatic photosynthesis activity slow by providing rigidity to the cell membrane. Lipid composition of chloroplast membrane plays an important role in plant photosynthesis. Fujii et al. (2017), Jung et al. (2016) and Legris et al. (2016) observed that the thermosensory molecules are sensed by phototropins and phyB photoreceptor which perceive temperature fluctuation. According to Kodama et al. in 2008 Phototropin 2 is a blue light receptor which optimize the photosynthesis process under cold avoidance response. Calvin cycle is also effected by chilling stress by reducing their enzymatic activity. In chilling-sensitive plants low temperature caused uncoupling in thylakoid reaction and effected site is H-ATPase (Peeler and Naylor 1988; Terashima et al. 1989a, b). Sonoike (1995) observed that low temperature in cucumber leaves destroyed the ETC components like F_X, F chain unit like F_A, F_B and PQ A_1 in the Fe-S centre of PSI. Result came from different studies told low temperature cause irreversibly damage the photosystem (Tjus et al. 1998a, b; Teicher et al. 2000; Kudoh and Sonoike 2002; Zhang and Scheller 2004; Zhang et al. 2014). Lipid composition plays an important role in sensitive and tolerance response under chilling tolerance. In chloroplasts phosphatidylglycerol plays an important role in photosynthetic process. Somerville (1995), Wu et al. (1997) and Routaboul et al. (2000) said that the distribution of unsaturated fatty acid chains is greatly affected under chilling environment. Pribil et al. (2014) and Li and Yu (2018) reported some additional lipids in thylakoid membrane like digalactosyl diacylglycerol (DGDG), monogalactosyldiacylglycerols (MGDGs), sulphoquinovosyl diacylglycerols (SQDGs) and phosphatidylcholine (PC). Novitskaya and Trunova (2000), Degenkolbe et al. (2012) and Skupień et al. (2017) did experiment on membrane fluidity affected under stress and result showed that in some cases chilling would change the membrane composition from compactly arranged into open type which is responsible for membrane fluidity, e.g. cucumber .Cold stress positively effects the wheat plant result as increasing its number and size of chloroplast and length of grana (Venzhik et al. 2016). During chilling stress, for maintenance various RNA-binding proteins are needed, and if they are lost or mutated, then yellow and pale leaves occur (Kusumi et al. 2011; Kupsch et al. 2012; Gong et al. 2014; Song and Zhao 2014; Zhang et al. 2016; Cui et al. 2018). RNA binding protein is also responsible for seed germination process. Under low temperature stress, Gong et al. (2014) and Wu et al. (2016) found that gene alteration at the transcriptional level occurs. Chilling stress had a negative impact on ETC, and genes involved in RNA and protein synthesis were also affected (Liu et al. 2010; Gu et al. 2014; Wang et al.

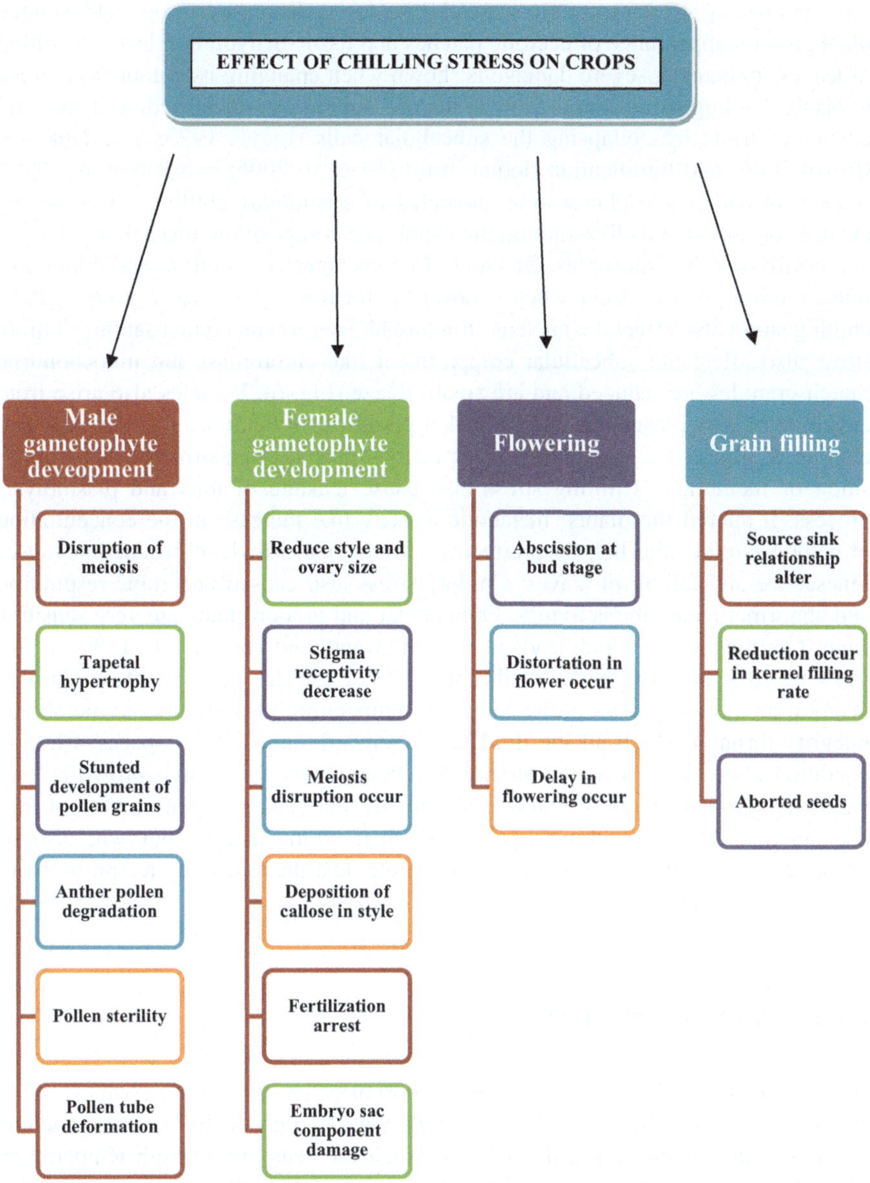

Fig. 6 Effect of chilling stress on male gametophyte, female gametophyte, flowering and grain filling

2016; Morita et al. 2017). Mutations in RNA genes like tcd5 result in disrupting the synthesis of chlorophyll, which leads to chlorosis and inhibition of growth of plant (Liu et al. 2010; Gu et al. 2014; Wang et al. 2016; Morita et al. 2017). In cotton plant seed germination process should be downregulated against chilling stress. Various types of chilling stress symptoms were observed; e.g. decrease in crop yield, stunted plant growth, appearance of necrotic patches and tissue dehydration leads to wilting of leaves. In banana, severe damage is shown when changing its colour from green to black. Chilling stress causes rubbery texture and uneven ripening of tomatoes and cucumber fruits by collapsing the subcellular cells (Lyons 1973; Van Dijk and Brown 2006) and lignification (loquat fruit; Cai et al. 2006; Crisosto et al. 1999). If tobacco and maize plants were subjected to continuous chilling effect, severe changes occur in plants like altering the membrane composition, unstacking of grana and destroying the chloroplast envelope. In some species, swelling and dilation of mitochondria process occur when exposure is for 6 h at 5 C; e.g. *Episcia reptans* chilling stress also effect the nucleus structure like chromatin condensation. Chilling stress also affect the subcellular compartment like chloroplast and mitochondria. Starch granules are reduced and are small in size (Fig. 6). Vesicles also arise from envelope of endoplasmic reticulum called peripheral reticulum in the exposure to chilling. Chilling stress also effect some cellular changes like structure and composition of membrane. Chilling stress also cause leakage of ions and plasmolysis process. It altered the plants' metabolic activity like increase in the concentration of carbon dioxide and help in production of ethylene. Ethylene production causes senescence and falling of leaves. Chilling stress also caused anaerobic respiration and abnormal metabolic activities. Ornamental and indoor plants are very sensitive against chilling stress which leads to stunted growth and may be killed. Plants also show necrotic area on leaves in the presence of low temperature. After chilling necrotic area also appears on leaves. Low temperature also effects the membrane integrity through which all the fluid moves towards intracellular spaces. Chilling condition also affects vascular strands by changing its colour into brown colour, e.g. Avogado. Life cycle of plants could be affected because ripening time of fruit had changed. If ripen timing is changed, then aroma and flavour will also be changed. Sweet potato propagules, for example, lost their capacity to sprout under a cooling environment.

High Temperature Stress

High temperature (HT) stress is one of the most affective abiotic stress which highly effect the plant metabolic activity, its length, yield and productivity. Plants include different types of biochemical activities which are sensitive to high temperature. Plants have both positive and negative aspects towards high temperature; it varies from species to species. Above 45 °C temperature has adverse effect on plant and plants are unable to survive. Because of high temperature plants show dehydration condition. Hydrated and non-growing tissues are less affected and survival rate is more as compared to unhydrated and growing tissues under high temperature. Seeds and pollen grains are good example of dehydrated tissue. Different plant organs had shown different tolerance rate; e.g. dry seeds endure 120 °C, pollen grain can endure 70 °C but some growing tissues are sensitive at 45 °C. Various problems are faced by plant under conditions of high radiations and temperature of a leaf by 5 °C. Plants such as *Tidestromia oblongifolia* and some desert grasses thrive at temperature up to 50 °C. Up to a certain level the vegetative growth like leaf appearance and node were increased with increase in temperature. In several plant species, reproductive development generally has lower tolerance to temperature than vegetative development. Hatfield et al. (2008, 2011) reported that exposure to high temperature during reproductive stage would affect pollen viability and fertilization process. Even after returning to 30 °C following a temperature increase from 30 °C to 33 °C during the endosperm division phase, the size of the kernel and its development were recovered (Ouattar and Crookston 1984). Commuri and Jones in 2001 observed that high temperature (above 30 °C) effect on maize plant, cell division, and size of grain and harvesting yield reduced. It also damaged amyloplast replication. First of all, growth phase mainly affected germination step. High temperature stress also decreased the seed germination rate and it varies from one species to another (Johkan et al. 2011; Kumar et al. 2011). Various steps involved during seed germination were altered like seedlings are abnormal. Radicle and plumule growth would be reduced and seeds vigour also changed during temperature stress. Essemine et al. (2010) also reported that high temperature inhibits seedling production. Above 45 °C in wheat plant the seed germination rate will be reduced and sometimes embryo is damaged and caused cell death (Cheng et al. 2009). Rice cultivar exposure to high temperature reduced plant height, number of tillers and total biomass (Mitra and Bhatia 2008).

Some vascular plants have ability to tolerate and survive the temperature above 50–55 °C, e.g. agave and cacti. Herrero and Johnson (1980), Schoper et al. (1987) and Dupuis and Dumas (1990) reported if temperature is more than 35 °C pollen viability decreased in maize plant. Because vapour pressure plays an important role in pollen viability. Due to high temperature vapour pressure deficit enhanced and pollen viability would be decreased (Fonseca and Westgate 2005). Rice (Orzya sativa L.) shows a similar response(maize) against temperature because it also declines in pollen viability when it exceeds 33 °C (Kim et al. 1996) (Table 2). High temperature most sensitive physiological activity of plant is photosynthesis (Crafts-Brandner and Salvucci 2002). In chloroplast, thylakoid membrane is the primary site of injury at high temperature (Wang et al. 2009; Marchand et al. 2005). Heat temperature changes the organization of thylakoid structure and causes swelling of grana and loss of grana stacking under heat stress. Under HTs the photosystem II (PSII) activity should be decreased and effect the photosynthetic pigments (Fig. 7). The decline in chl pigment also is a result of heat stress observed in sorghum at 40 °C during day and 30 °C at night. In soya bean plant heat stress significantly decreased parameters, e.g. 18% decrease in total chlorophyll content (18%) and chlorophyll a and a/b ratio also reduced. Stomatal conductance and net photosynthesis rate also reduced in soya bean plant under high temperature. Biochemical parameters like total soluble sugar content of leaf and sucrose content also declined in heat condition. If rice plants are exposed to temperature 33° for 5 days, it leads to decreases in photosynthetic rate. Greer and Weedon observed that with increasing temperature from 25 to 45 °C average rates of photosynthesis of Vitis vinifera leaves decreased by 60% due to stomatal closure. If heat stress continued with plant, it would destroy

Table 2 Effect of high temperature stress on plant species

Crops with common name	Heat treatment on various growth stage	Major effect
Capsicum annuum (Chili pepper)	During mature and reproductive stage if exposure to 38/30 °C	Seed per fruit number changed fruit and reduced fruit weight and width
Oryza sativa (Rice)	At heading stage if Exposure to temperature above 33 °C for continue 10 days	Pollen viability decreased and spikelet fertility rate should change
Triticum aestivum (Wheat)	At maturity stage or during grain filled if temperature is 37/28 °C for 20 days	Grain filling steps, maturity will be completed in short time and kernel weight and yield reduced
Hordeum vulgare (Sorghum)	At maturity stage if exposure to 30/40 °C	Thylakoid membrane changed its composition, chl content reduced, antioxidant enzyme activity decreased, increased reactive oxygen species ROS, also disrupt ETC by decreasing photosystem II (PSII) photochemistry, yield component less in number found
Zea mays (Maize)	During reproductive stage for continuing 14 day exposure to temperature 27/35 °C (night/day)	Reduced the supply of photosynthate which decrease the synthesis of cellulose and hemicellulose. Decrease ear expansion and cob length
Glycine max (Soya bean)	On flowering stage if 14 days temperature is 38/28 °C (day/night)	Spongy and palisade tissue layer increased its thickness, decreased stomatal conductance of leaf, destroy plasma membrane structure, cristae, matrix and mitochondrial membrane would be distorted
Nicotiana tabacum (Tobacco)	Early growth stage if 43 °C for 2 h	Net photosynthetic rate decreased, apparent quantum yields also decreased, reduced carbon dioxide efficiency during photosynthesis, antioxidant enzyme activity also negatively affected

all stored carbohydrate reserved food and plant leads to starvation. At reproductive stage exposure to heat stress had shown negative effect on seed and fruit production (Fig. 7). Floral bud and flower would be aborted at high temperature stress. Heat can damage the meiosis process in male and female floral organs, causing pollen tube development to be disrupted, ovule viability to be lost, and stigma and style position to be shifted. The reasons behind increasing sterility is HTs, because of impaired meiosis in female and male organs, disrupted pollen tube, disappearance of ovule viability, stigmatic and style positions had changed, number of pollen grain decreased by stigma, interrupted fertilization process, restrict the endosperm growth,

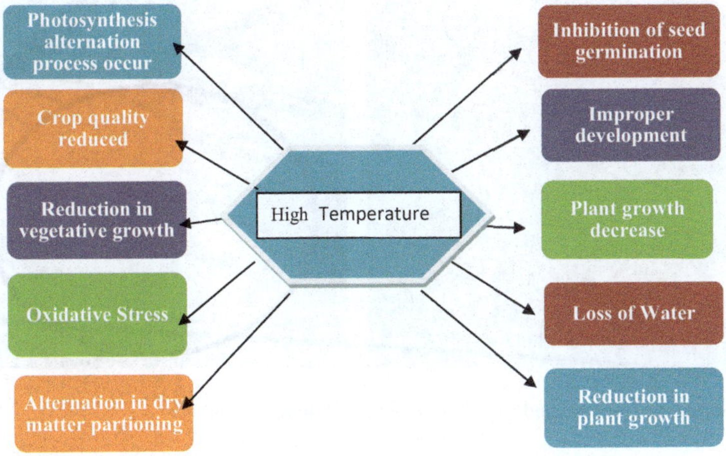

Fig. 7 Diagram shows how high temperature affects plant activity. It can reduce crop quality, decrease plant growth and inhibit germination of seed

due to unfertilized and pro-embryo condition numbers of pollen should be decreased on the stigma which caused the sterility of seed in rice plant (Hurkman et al. 2009; Ahamed et al. 2010). In rice pollen germination decreased up to 36% than control and increased sterility in spikelet up to 61% as compared to control under high temperature. Hay and Porter (2006) and Reynolds et al. (2007) reported that high temperature affects the photosynthesis process because of alternation occurring in membrane stability. Therefore the respiration costs would be enhanced. This statement is supported by Cicchino et al. (2010) and Högy et al. (2013) in wheat and maize crops. If exposure to high temperature occurs during flowering stage, it will show reduction in yield component and grain filling process will also be uncompleted. High temperature 33–40 °C in maize significantly affects the plant biomass and light capturing capacity of plants. In maize, millet and sugarcane reason behind reduced relative growth is reduction in net assimilation rate under HTs. Morphological symptoms arising from heat stress are sunburn patches on leaves and twigs. Ht promotes the leaf senescence and abscission process and it will damage the fruit by producing discoloration effects. Mohammed and Tarpley in Mohammed and Tarpley (2010) showed that in sugarcane plant drying and rolling of leaves occur under heat condition. Heat stress reduced the number of tillers in wheat crops when temperature is 30/25 °C during day and night. In wheat with promoted shoot elongation, number of tillers reduced under heat stress falling on day and night time. HT can alter the total phenological life cycle of plants, duration of grain filling, size of grains, etc. This action causes denaturation and aggregation of proteins in

Fig. 8 Shrinkage of petals and Leaf burn due to high temperature

plants under intense heat stress, and particular cells and tissues follow a programmed cell death cycle (Fig. 8).

High temperature muted the chloroplast omega-3 fatty acid saturase, which is found in the chloroplast membrane of tobacco plants, according to Murakami et al. (2000). PSII is highly sensitive against the heat temperature stress. Spring wheat for their growth needs hot climate, so it shows positive effect under heat stress by increasing its photosynthetic rate per unit leaf area and increasing in kernel number (Reynolds et al. 1994). During anthesis stage kernel number and dry spike weight correlate to each other (Fischer 1985). Heat stress affect the various stages of plant but primarily effect the seed germination process negatively in wheat plant (Johkan et al. 2011; Hossain et al. 2013; Essemine et al. 2010; Yamamoto et al. 2008). High temperature sometimes worked as ethylene hormone by boosting the abscission of leaves via reducing photosynthesis and reducing plant growth and meristem length (Kosova et al. 2011). If night temperature reached up to 25 °C, it will affect the tiller production process in plant and leads to decrease in number of tillers (Rahman et al. 2009). According to Nawaz et al. (2013), Bennett et al. (2012) and Yu et al. (2014), reproductive stage of wheat represents the detrimental stage of heat stress, e.g. temperature rise by one degree from the optimum leads to decrease in yield component in wheat. High temperature will change the protein expression by degenerating mitochondrial membrane. Reduced ATP accumulation in membrane and oxygen uptake in wheat embryo leads to decreased seed quality, seed mass, seed size and seed germination process (Balla et al. 2012; Hampton et al. 2013). Increase in temperature by 1–2 °C from optimum temperature reduces its seed mass by shortening the period of grain filling cycle in wheat (Nahar et al. 2010). Daily minimum night temperatures above 15 °C occur in many subtropical and tropical zones where cowpeas are grown (Nielsen and Hall 1985). In 'Groundnut' if during morning there is high temperature exposure, then fruit set reduced, whereas afternoon no effect was found on fruit set (Vara Prasad et al. 2000). An early morning flowering line was bred and found to flower a few hours earlier and showed less spikelet sterility than the parental cultivar which flowered when temperatures were higher (Ishimaru et al. 2010). In tomato, because of heat stress failure of fruit set occurs (reduced yield) (Peet et al. 1997; Zhang et al. 2011, Zhang et al. 2012). High-temperature stress also affect many physiologic traits, including fresh and dry weight of plant and leaf area in tomato (Shaheen et al. 2016; Zhou et al. 2017). Other vegetative effects include reduced photosynthetic efficiency (Bar-Tsur et al. 1985; Criddle et al. 1997), reduced assimilate translocation, reduced mesophyll resistance, and enhanced disorganization of cellular organs (Chen et al. 1982). The most damaging impact is on fruit yield. The yield reduction is related primarily to reduced fruit set, which may not occur for many reasons, including adverse effects on meiosis of ovules and pollen mother cells, reduced pollen shed resulting from impaired development of the endothecium in the anthers, stigma position (exerted under heat stress), number of pollen grains retained on stigma, pollen germination rate, growth of pollen tube, ovule viability, fertilization–postfertilization processes, and endospermic growth (Driedonks et al. 2016; Peet et al. 1997; Sato et al. 2002; Zhou et al. 2017). Other indirect yield-reducing effects of heat stress include fruit cracking, malformation of fruits (e.g. cat facing), and a malformed blossom-end scar. In sum, heat stress reduces the fruit number, quality, and marketable yield of tomato. In cowpea genotypes during floral bud formation exposure to heat stress degenerates the floral bud. Degenerated floral bud will not be able to produce any type of flower. Floral bud with arrested growth at red (655–665 nm)/far-red (725–735 nm) ratios of 1.3 or 1.6 are able to grow under high temperature but if ratio increased 1.9 then growth will be prohibited (Ahmed et al. 1992).

Elevated CO_2 Affecting Plant Development and Morphology

When the Industrial Revolution started in atmosphere the CO_2 concentration started increasing day by day and now it has reached 400 ppm from 280 ppm (Meehl et al. 2007). CO_2 directly regulates the downstream process in plant development by affecting gas exchange mechanism during photosynthesis (Ainsworth and Long 2004). Because of potent greenhouse nature of CO_2, it has also direct effects which contributes to climate change and global warming (Meehl et al. 2007). In response to elevated CO_2, C_3 species increased biomass on average 20% (Ainsworth and Long 2004; Reich et al. 2014) reported that water, nutrient and CO_2 relationship also play important role in biomass production. Those regions which contain sufficient amount of water, nitrogen and have high CO_2 increased plants biomass above the ground surface up to 33% in grassland as compared to lower nutrient and drier condition. Many species with elevated CO_2 concentration root biomass, shoot biomass and yield of crops also positively increased with CO_2 concentration, e.g. wheat, rice, peanut and beans (Reviewed in Madhu and Hatfield 2013; Reviewed in Hatfield et al. 2011). Dermody et al. (2006) reported from his studies that increase in CO_2 concentration increased leaf size and leaf nodes in soya bean plant and increased underground root region, root length and production of nodule; pod and seed number also increased with exposure to high carbon dioxide (Morgan et al. 2005; Bishop et al. 2014). Elevated carbon dioxide helps to increase cell division process and regulates the expansion process which is responsible for making leaf big in size in poplar and soya bean. Variation in genotype is also induced by carbon dioxide concentration, e.g. hybrid *Populus euramericana*. Young and old leaves would increase in size by changing the spongy and palisade cells size in developing leaves (Taylor et al. 2003). Masle (2000) reported in wheat that increased CO_2 had cell specific in nature in case of leaf anatomy than epidermal anatomy. Evidence of cell wall expansion effect had been shown in poplar, but it is totally dependent on age of leaf (Ranasinghe and Taylor 1996; Ferris et al. 2001; Taylor et al. 2003). Under elevated CO_2 condition mainly three types of growth response occur, species specific, temporally specific and spatially specific. Leaves mostly follow species specific cell type for their response. Woodward and Kelly (1995) observed that in Arabidopsis stomatal index also reduced with elevated CO_2 concentration. Same result was reported by Gray et al. (2000) with reference to stomatal index. Elevated CO_2 exposure results as reduced stomatal density in Arabidopsis (Engineer et al. 2014). Further work was done by Engineer et al. (2014); they observed that in extracellular signalling pathway carbonic anhydrases used for CO_2 contribution over stomatal development. To identify QTL for stomatal response to elevated CO_2, mapping population of *P. trichocarpa* and *P. deltoides* were utilized (Ferris et al. 2002). QTL for stomatal trait responsiveness to elevated CO_2 were identified by authors, but candidate genes of these species have not been identified (Ferris et al. 2002). Shoot architecture would be altered by carbon dioxide concentration because in soya bean plants number of vegetative nodes present on shoot increased (Dermody et al. 2006). Some shoot architecture modification was also noticed in wheat plants, e.g. number of branches and tillers increases and meristem length will also be promoted (Christ and Korner 1995; Nicolas et al. 1993; Slafer and Rawson 1997). Rice plants also positively responded against CO2 elevation by increasing plant height and number of tillers on plant (Jitla et al. 1997). Morita et al. (2015) identified some binding proteins for chlorophyll and starch accumulated responsive regulators present in rice, e.g. CRCT and CONSTANS. CRCT stands for chlorophyll a/b Binding Protein. If CRCT overexpression occurs in leaves, starch accumulation also increased, which is responsible for the production of wider lateral branches by modifying the tillering angle (Jin et al. 2008). Carbon metabolism is also altered by elevated CO_2. Increases in transcription rate, which is responsible for starch, sugar metabolism, glycolysis, TCA cycle, and Electron Transport Chain, carbon assimilation during photosynthetic process, and respiration rates during night were observed in soya bean by Leakey et al. (2009) and Fukayama et al. (2011). Similarly result was given by Markelz et al. (2014) in Arabidopsis plant. Photosynthesis process and growth would increase under elevated CO_2 exposure in some members of Fabaceae family (Legumes) and C_4 plant species (Rogers et al. 2009). Glucose act as signalling molecules for plant growth and development and elevated CO_2 increases the flux of carbohydrate and metabolite which is required for growth and root biomass. Rogers et al. (1997) reported that acquisition of water and mineral will increase root:shoot ratio compared with control experiment. Gray et al. had performed Minirhizotron experiments in soya bean plant at various soil depth (shallow, intermediate) and the result shows that high CO_2 increased root length. Precipitation rate would be low which is used for increasing root density nodule number. Crookshanks et al. (1998) observed that forest and agriculture species root biomass increased because CO_2 enhance their root length, its diameter and root branching, and lateral root number (Madhu and Hatfield 2013). Elevated CO_2 altered the root system architecture, because CO_2 distributed the roots length towards water resources for increasing water accumulation in root. Nutrients are distributed to various depth under the soil region; it makes roots able to gather these nutrients from resources (Lynch 2015). In sorghum, cotton and wheat, elevated CO_2 made modification on shoot region more than root via increasing lateral root number (Pritchard and Rogers 2000). Rogers et al. (1992) did experiment on anatomically features who gave statement that CO_2 positively affect the root diameter, cortex and stele region and root volume. Crookshanks et al. (1998) observed from his experiment that there is increase in cortical cell expansion and extensibility found in Arabidopsis. The CO_2 concentration would affect the inter-specific genetic variation in root density in *Populus deltoides* and *Populus trichocarpa*, with *P. trichocarpa* showing a stronger root magnitude (Rae et al. 2007). Ions profile is significantly affected under CO_2 stress which is responsible for reduction of nutritional quality like iron (Loladze 2014; Myers et al. 2014). Nitrogen and protein content in seed would decline under carbon elevation (Jablonski et al. 2002; Myers et al. 2014). Jablonski et al. in 2002 collected data from 79 species which were growing under elevated CO_2 environment during reproductive stage; he found that it shows positive response in many parameters, e.g. fruits, flowers and seeds number had increased up to 16–19% and seed mass 25% with respect to control. Stimulation of seed yields mainly depend on climate (CO_2) it will be diminishing to zero dry and hot climate (Ruiz-Vera et al. 2013; Bishop et al. 2014). Morgan et al. (2005) observed the result of increased seed yield is just because of increase in their pod number, seeds per pod and mass number. Bishop grew 18 genotypes of soya bean in different growing seasons in 2014 and achieved a 9 percent average rate of seed output, but the partitioning coefficient was reduced to 11 percent (Bishop et al. 2014). Fruit production also varies from place to place, e.g. under elevated CO_2 crop fruit increased on an average 28% while wild species has showed less affect (Jablonski et al. 2002). CO_2 is also responsible for delayed reproductive development in soya bean (Castro et al. 2009). Life cycle of plants is altered by CO_2, e.g. duration from sowing to harvesting in soya bean species would be extended because of the production of new node (Dermody et al. 2006; Castro et al. 2009). But Arabidopsis plants did not show delaying process of flowering (Springer and Ward 2007). But high CO_2 concentration alter the FLC gene expression then delaying flowering process noticed in Arabidopsis genotype (Springer et al. 2008). Elevated carbon dioxide causes delay in senescence in tree species, e.g. poplar (*P. tremuloides and P. euramericana*) shows autumnal senescence. Taylor et al. (2008) and Tallis et al. (2010) reported that anthocyanin pigment process increased due to elevated CO_2 concentration; anthocyanin biosynthesis process positively increases in tree species (*P. euramericana*) which is used for delaying senescence. Gould (2004) studied about anthocyanin's pigment and gave the statement that anthocyanin helps in protection from UV damage, pathogen stress, and scavenging of ROS.

Heavy Metal Stress

In modern world various soils pollutants are present because of anthropogenic activities, industries. Heavy metals are also included in these pollutants like copper, cobalt, nickel cadmium, chromium in soil particle, lead and mercury in air and huge amounts are accumulated in water ecosystem. Soil is most important for plant growth because soil provide all nutrients. For seed germination, growth and survival soil is nonrenewable and valuable source. Seeds are largely affected by heavy metal stress by decreasing their germination rate. Wang et al. (2003), Ahmad and Ashraf (2011) and Pourrut et al. (2011) reported about decreased root, shoot and dry weight of plants. Cell membrane should be disrupted leads to productivity lost and seed toxicity. Li et al. (2015) made the decreasing order of seed germination according to metal Hg > Cd > Pb > Cu. Cadmium negatively effects reserved food material by damaging the cell membrane due to which leakage of nutrient process occur. Food mobilization into cell like glucose, starch, amino acid, and soluble sugar content are disrupted by damaged cell membrane (Rahoui et al. 2010; Sfaxi-Bousbih et al. 2010). The accumulation and over-accumulation of lipid peroxidation products was observed in seeds (Ahsan et al. 2007; Smiri et al. 2011). Sunflower seedlings produce reactive oxygen species under copper stress. ROS produces oxidative stress which decrease the catalase activity by altering the protein structure (Pena et al. 2011). Symptoms of metal toxicity (Cd, Hg) and water deficiency in barley had shown similar effects. The effect of heavy metals in alfalfa produced oxidative stress and depletion in glutathione was reported (Tamás et al. 2010; Hernandez et al. 2012). Presence of heavy metal in soil highly affects seed germination process and seedling growth (Zhang et al. 2002), as stress affects the seed germination in wheat. Reduced radicle and plumule length of plant species would be observed under excess of As concentration in soil, e.g. *Helianthus annuus* L. seedlings. In sessile plants roots are the first sensory organ that encountered heavy metal stress. Heavy metal stress affect the enzyme activity by reacting with protein sulphydryl group which affect the protein metabolism of plant. Plants with damaged chloroplast membranes lose photosynthetic pigment and nutrients, which is an indication of heavy metal stress (Li et al. 2006; Singh et al. 2009; Ahsan et al. 2010). Keunen et al. (2011), Kikui et al. (2005), Panda et al. (2009), Buendía-González et al. (2010), Gangwar et al. (2010, 2011), Gangwar and Singh (2011), Eleftheriou et al. (2012), Hayat et al. (2012), Silva (2012) and Anjum et al. (2014) reported that all plant growth were hampering under heavy metal toxicity effect, plants grown under heavy metal soil had decreased growth and yield of plant. Doncheva et al. (2005), Sundaramoorthy et al. (2010), Hossain et al. (2012a, b) and Thounaojam et al. (2012) studied that in the presence of heavy metals, plants' mitotic activity has been decreased in various plant species, which consequently suppressed the root. Heavy metal affects the cell division cycle, Cr caused delaying the cell cycle process which inhibits cell cycle and results in reduced root growth (Sundaramoorthy et al. 2010). In chlorophyll magnesium (Mg) molecule can be replaced with Ni (Kupper et al. 1996), and Ni damages the thylakoid membrane and increases the degradation of chlorophyll (Molas 2002; Gajewska et al. 2006).

Biotic Stress

Biotic stress is defined as the stress which is caused by living organisms. Bacteria, fungi, nematodes, viruses and insects are some stress causal organisms. These organisms produced disease in plants. Biotic stress reduced the crop production rate, a 37.2% loss of rice, 28.3% yield loss of wheat, 40.4% loss of potatoes, 31.3% loss of maize, 28.8% loss of cotton and 26.3% loss of soya beans (Wang et al. 2013) (Fig. 9). Fungi, virus and bacteria cause different diseases to the plant, and fungi caused disease more than any other. Bacteria, herbivore and other microorganisms can cause wilting of leaves, root rot, seed damage and patches present on leaves. Both the NO and reactive oxygen species (ROS) can be induced upon herbivory. These plant defence molecules may be generated downstream from Ca^{2+} signals (and concomitant membrane depolarization) during plant defence responses to herbivory. Similar to some pathogenic compounds released by fungi or microbes, herbivores also introduce oral secretions into the wounded region of the plant. Many pathologists observed that orally secreted compounds from insects are able to induce Ca^{2+} and membrane depolarization in plant cells (Mithöfer and Boland 2008; Bricchi et al. 2010).

Biotic stress consists of damage to plants through other living organisms. For controlling (fungi, bacteria) these types of stress various types of pests should be used. Genes which are used for pigment formation and electron transport downregulated by biotic stress. By collecting the host plant sample and sampling time we can measure the total damage. Biotic stress can be downregulating the photosystem I and Photosystem II reaction centre activities. ATP synthase activity and light-harvesting complex which is associated with PSII would be prohibited under biotic stress. De Vos et al. (2005), Devoto et al. (2005) and Coram and Pang (2007) reported that not all proteins present in light-harvesting complex would be downregulating but some remains same and not show any significant effect under biotic stress, e.g. exposure to methyl jasmonate (MeJA) had shown rare effect. RAS (root architecture) help to sense these changes and gave response against biotic stress. Higginbotham et al. (2004) observed the RSA role in interaction of pathogen and plants. Plant with high root length will have less fungal infection. In *T. aestivum* causal agents of root rot is *Pythium ultimum* and *Pythium debaryanum*, if the root length is more than its rate of infection chances should be reduced. In contrast, Berta et al. (2005) and Simonetta et al. (2007) gave statement from his studies that in

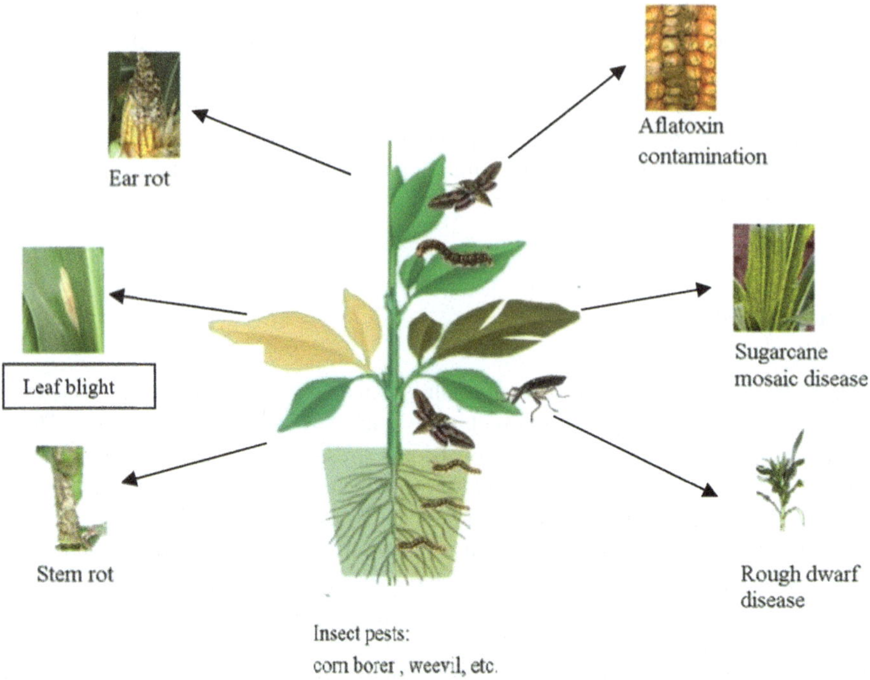

Fig. 9 Pathogen-related biotic stress in plant and disease caused by pathogen

S. lycopersicum infection Rhizoctonia is causal agent which reduced root length, root tip number and magnitude of branching which absorb water from deep soil layers and transport it to shoot for growth process. Therefore, we can say that there is correlation between root length density and extent of pathogen infection. Root length density is inversely proportional to rate of pathogen infection means if RLD increases then rate of infection decreases. Even under sufficient soil moisture environment present surrounding the plant, some pathogens have ability to reduce plant water content in plant. For example, under high water potential condition U. phaseoli infection in P. vulgaris had showed wilting effect not because of water content but because of destroyed xylem. Duniway and Durbin (1971) reported that some toxins were released from U. phaseoli which inhibit stomatal closure. If stomata would remain open, it would not be able to control over water loss through stomata and cuticle layer will be disrupted. Cells reduced its water potential because of stomatal opening. Drought affected shoot water potential, transpiration, and leaf turgidity in the same way (Burman and Lodha (1996). *M. phaseolina*, the causal agent of the diseases charcoal rot and stem blight in *V. vinifera*, exhibits combined drought and biotic stress effects. The *Xylella fastidiosa* induced leaf scorch infection exhibits a significant reduction in stomatal conductance and water potential of the leaf, aggravating scorch symptoms more in drought stressed plants than in well-watered plants (McElrone et al. 2003). Waxy nature of cuticle layer helps in protection against pathogen. Marcell and Beattie (2002) exposed maize crop against Clavibacter michiganensis pathogen which is causal agent of leaf blight. Experiment was done on millet and maize crops with wild-type genotype and (gl4) glossy mutants of *Z. mays*. After experiment they concluded that gl4 mutants contain more bacterial colonies compared with wild type. Cuticle Marcell and Beattie (2002) say it disturbs the waxy production pathway in the cuticle.

References

Abas H (2006) Effect of water deficit stress on growth, yield and essential oil content of draco cephalum moldavica. Iran J Med Aroma Plant 22(33):256–261

Ahamed KU, Nahar K, Fujita M, Hasanuzzaman M (2010) Variation in plant growth, tiller dynamics and yield components of wheat (Triticum aestivum L.) due to high temperature stress. Adv Agric Bot 2:213–224

Ahmad MS, Ashraf M (2011) Essential roles and hazardous effects of nickel in plants. Rev Environ Contam Toxicol 214:125–167

Ahmed FE, Hall AE, DeMason DA (1992) Heat injury during floral development of cowpea (Vigna unguiculata, Fabaceae). Am J Bot 79:784–791

Ahsan N, Lee SH, Lee DG, Lee H, Lee SW, Bahk JD et al (2007) Physiological and protein profiles alternation of germinating rice seedlings exposed to acute cadmium toxicity. C R Biol 330:735–746

Ahsan N, Lee DG, Kim KH, Alam I, Lee SH, Lee KW et al (2010) Analysis of arsenic stress-induced differentially expressed proteins in rice leaves by two-dimensional gel electrophoresis coupled with mass spectrometry. Chemosphere 78:224–231

Ainsworth EA, Long SP (2004) What have we learned from 15 years of free-air CO2 enrichment (FACE)? A meta-analytic review of the responses of photosynthesis, canopy properties and plant production to rising CO2. New Phytol 165:351–372

Akir R (2004) Effect of water stress at different development stages on vegetative and reproductive growth of corn. Field Crop Res 89(1):1–16

Allahverdiyev TI, Talai JM, Huseynova IM, Aliyev JA (2015) Effect of drought stress on some physiological parameters, yield, yield components of durum (Triticum durum desf.) and bread (Triticum aestivum L.) wheat genotypes. Ekin J Crop Breed and Gen 1(1):50–62

Anjum NA, Gill SS, Gill R, Hasanuzzaman M, Duarte AC, Pereira E et al (2014) Metal/metalloid stress tolerance in plants: role of ascorbate, its redox couple, and associated enzymes. Protoplasma 251:1265–1283

Apel K, Hirt H (2004) Reactive oxygen species: metabolism, oxidative stress, and signal transduction. Ann Rev Plant Biol 55:373–399

Arora A, Fosfuri A, Gambardella A (2001) Specialized technology suppliers internationals spill overs and investment: evidence from the chemical industry. J Dev Econ 65(1):31–54

Babé A, Lavigne T, Séverin J-P, Nagel KA, Walter A, Chaumont F, Batoko H, Beeckman T, Draye X (2012) Repression of early lateral root initiation events by transient water deficit in barley and maize. Philos. Trans R Soc Lond B Biol Sci 367:1534–1541

Baher ZF, Mirza M, Ghorbanli M, Rezaii MB (2002) The influence of water stress on plant height, herbal and essential oil yield and composition in Satureja hortensis L. Flav Fragrance J 17(4): 275–277

Balasubramanian S, Sureshkumar S, Lempe J, Weigel D (2006) Potent induction of Arabidopsis thaliana flowering by elevated growth temperature. PLoS Genet 2:0980–0989

Balla K, Karsai I, Bencze S, Veisz O (2012) Germination ability and seedling vigour in the progeny of heat-stressed wheat plants. J Acta Agron Hung 60:299–308

Bar-Tsur A, Rudich J, Bravdo B (1985) Photosynthesis, transpiration and stomatal resistance to gas exchange in tomato plants under high temperatures. J Hortic Sci 60(3):405–410
Bayuelo-Jimenez JS, Debouck DG, Lynch JP (2003) Growth, gas exchange, water relations and ion composition of Phaseolus species grown under saline conditions. Field Crops Res 80:207–222
Bennett D, Izanloo A, Reynolds M, Kuchel H, Langridge P, Schnurbusch T (2012) Genetic dissection of grain yield and physical grain quality in bread wheat (Triticum aestivum L.) under water limited environments. Theor Appl Genet 125:255–271
Berta G, Sampo S, Gamalero E, Massa N, Lemanceau P (2005) Suppression of Rhizoctonia root-rot of tomato by Glomus mossae BEG12 and Pseudomonas fluorescens A6RI is associated with their effect on the pathogen growth and on the root morphogenesis. Eur J Plant Pathol 111:279–288
Bishop KA, Betzelberger AM, Long SP, Ainsworth EA (2014) Is there potential to adapt soybean (Glycine max Merr.) to future (CO2)? An analysis of the yield response of 18 genotypes in free-air CO2 enrichment. Plant Cell Environ 38:1765–1774
Bohnert HJ, Jensen RG (1996) Strategies for engineering water-stress tolerance in plants. Trend Biotechnol 14(3):89–97
Boyer JS (1982) Plant productivity and environment. Science 218:443–448
Boyer JS (1968) Relationship of water potential to growth of leaves. J Plant Physiol 43:1056–1062
Boudjabi S, Kribaa M, Chenchouni H (2015) Growth, physiology and yield of durum wheat (Triticum durum) treated with sewage sludge under water stress conditions. EXCLI J 14:320–334
Bricchi I, Leitner M, Foti M, Mithöfer A, Boland W, Maffei ME (2010) Robotic mechanical wounding (MecWorm) versus herbivore-induced responses: early signaling and volatile emission in Lima bean (Phascolus lunatus L.). Planta 232:719–729
Buendía-González L, Orozco-Villafuerte J, Cruz-Sosa F, Barrera-Díaz CE, Vernon-Carter EJ (2010) Prosopis laevigata a potential chromium (VI) and cadmium (II) hyperaccumulator desert plant. Bioresour Technol 101:5862–5867
Burman U, Lodha S (1996) Macrophomina phaseolina induced changes in plant water relations of resistant and susceptible cowpea genotypes. Indian Phytopathol 49:254–259
Cai C, Chen KS, Xu WP, Zhang WS, Li X, Ferguson I (2006) Effect of 1-MCP on postharvest quality of loqual fruit. Postharvest Biol Technol 40:155–162
Castro JC, Dohleman FG, Bernacchi CJ, Long SP (2009) Elevated CO2 significantly delays reproductive development of soybean under free-air concentration enrichment (FACE). J Exp Bot 60:2945–2951
Chaves MM, Pereira JS, Maroco J, Rodrigues ML, Ricardo CPP, Osorio ML, Carvalho I, Faria T, Pinheiro C (2002) How plants cope with water stress in the field? Photosynth Growth Ann Bot 89(7):906–917
Cheng L, Zou Y, Ding S, Zhang J, Yu X, Cao J, Lu G (2009) Polyamine accumulation in transgenic tomato enhances the tolerance to high temperature stress. J Integr Plant Biol 51:489–499
Chen HH, Shen ZY, Li PH (1982) Adaptability of crop plants to high temperature stress. Crop Sci 12:719–725
Christ RA, Korner C (1995) Responses of shoot and root gas exchange, leaf blade expansion and biomass production to pulses of elevated CO2 in hydroponic wheat. J Exp Bot 46:1661–1667
Cicchino M, Edreira JIR, Uribelarrea M, Otegui ME (2010) Heat stress in field grown maize: response of physiological determinants of grain yield. Crop Sci 50:1438–1448
Commuri PD, Jones RD (2001) High temperatures during endosperm cell division in maize: a genotypic comparison under in vitro and field conditions. Crop Sci 41:1122–1130
Coram TE, Pang ECK (2007) Transcriptional profiling of chickpea genes differentially regulated by salicylic acid, methyl jasmonate and amino cyclopropane carboxylic acid to reveal pathways of defence-related gene regulation. Funct Plant Biol 34:52–64
Crafts-Brandner SJ, Salvucci ME (2002) Sensitivity of photosynthesis in a C4 plant, maize, to heat stress. Plant Physiol 129:1773–1780

Criddle RS, Smith BN, Hansen LD (1997) A respiration based description of plant growth rate responses to temperature. Planta 201(4):441–445
Crookshanks M, Taylor G, Dolan L (1998) A model system to study the effects of elevated CO2 on the developmental physiology of roots: the use of Arabidopsis thaliana. J Exp Bot 49:593–597
Crisosto CH, Mitchell FG, Ju ZG (1999) Susceptibility to chilling injury of peach, nectarine, and plum cultivars grown in California. HortScience 34:1116–1118
Cui X, Wang Y, Wu J, Han X, Gu X, Lu T et al (2018) The RNA editing factor DUA1 is crucial to chloroplast development at low temperature in rice. New Phytol
De Vos M, Van Oosten VR, Van Poecke RM et al (2005) Signal signature and transcriptome changes of Arabidopsis during pathogen and insect attack. Mol Plant-Microbe Interact 18:923–937
Degenkolbe T, Giavalisco P, Zuther E, Seiwert B, Hincha DK, Willmitzer L (2012) Differential remodeling of the lipidome during cold acclimation in natural accessions of Arabidopsis thaliana. Plant J 72:972–982
Del Rio D, Stewart AJ, Pellegrini N (2005) A review of recent studies on malondi aldehyde as toxic molecule and biological marker of oxidative stress. Nutr Metab Cardiovasc 15:316–328
Demidchik V, Maathuis FJM (2007) Physiological roles of nonselective cation channels in plants: from salt stress to signalling and development. Tansley Rev New Phytol 175:387–405
Demidchik V, Shang Z, Shin R, Thompson E, Rubio L, Chivasa S, Slabas AR, Glover BJ, Schachtman DP, Shabala SN, Davies JM (2009) Plant extracellular ATP signaling by plasmamembrane NADPH oxidase and Ca2+ channels. Plant J 58:903–913
Dermody O, Long SP, DeLucia EH (2006) How does elevated CO2 or ozone affect the leaf-area index of soybean when applied independently? New Phytol 169:145–155
Devoto A, Ellis C, Magusin A, Chang HS, Chilcott C, Zhu T, Turner JG (2005) Expression profiling reveals COI1 to be a key regulator of genes involved in wound- and methyl jasmonate-induced secondary metabolism, defence, and hormone interactions. Plant Mol Biol 58:497–513
Dilrukshi SK, Nagahatenna LP, Whitford R (2015) Tetrapyrrole-based drought stress signaling. Plant Biotechnol J 13(4):447–459
Djekoun A, Planchon C (1991) Tolerance to low leaf water potential in soyabean genotypes. Euphytia 55:247–253
Doncheva S, Ameńos M, Poschenrieder C, Barceĺo J (2005) Root cell patterning: a primary target for aluminium toxicity in maize. J Exp Bot 56:1213–1220
Driedonks N, Rieu I, Vriezen WH (2016) Breeding for plant heat tolerance at vegetative and reproductive stages. Plant Reprod 29:67–79
Dupuis L, Dumas C (1990) Influence of temperature stress on in vitro fertilization and heat shock protein synthesis in maize (Zea mays L.) reproductive systems. Plant Physiol 94:665–670
Duniway JM, Durbin RD (1971) Detrimental effect of rust infection on the water relations of bean. Plant Physiol 48:69–72
Eleftheriou EP, Adamakis ID, Melissa P (2012) Effects of hexavalent chromium on microtubule organization, ER distribution and callose deposition in root tip cells of Allium cepa L. Protoplasma 249:401–416
Engineer CB, Ghassemian M, Anderson JC, Peck SC, Hu H, Schroeder JI (2014) Carbonic anhydrases, EPF2 and a novel protease mediate CO2 control of stomatal development. Nature 513:246–250
Essemine J, Ammar S, Bouzid S (2010) Impact of heat stress on germination and growth in higher plants: physiological, biochemical and molecular repercussions and mechanisms of defence. J Biol Sci 10:565–572
Farahanil HA, Valadabadil SA, Daneshianl J, Khalvati MA (2009) Evaluation changing of essential oil of balm (Melissa officinalis L.) under water deficit stress conditions. J Med Plant Res 3(5): 329–333

Ferris R, Sabatti M, Miglietta F, Mills RF, Taylor G (2001) Leaf area is stimulated in Populus by free air CO2 enrichment (POPFACE) through increased cell expansion and production. Plant Cell Environ 24:305–315

Ferris R, Long L, Bunn SM, Robinson KM, Bradshaw HD, Rae AM, Taylor G (2002) Leaf stomatal and epidermal cell development: identification of putative quantitative trait loci in relation to elevated carbon dioxide concentration in poplar. Tree Physiol 22:633–640

Fischer RA (1985) Number of kernels in wheat crops and the influence of solar radiation and temperature. J Agric Sci 105(2):447–461

Flexas J, Bota J, Loreto F, Cornic G, Sharkey TD (2004) Diffusive and metabolic limitations to photosynthesis under drought and salinity in C3 plants. Plant Biol 6:269–279

Fonseca AE, Westgate ME (2005) Relationship between desiccation and viability of maize pollen. Field Crops Res 94:114–125

Fujii Y, Tanaka H, Konno N, Ogasawara Y, Hamashima N, Tamura S et al (2017) Phototropin perceives temperature based on the lifetime of its photoactivated state. Proc Natl Acad Sci U S A 114:9206–9211

Fukayama H, Sugino M, Fukuda T, Masumoto C, Taniguchi Y, Okada M, Sameshima R, Hatanaka T, Misoo S, Hasegawa T, Miyao M (2011) Gene expression profiling of rice grown in free air CO2 enrichment (FACE) and elevated soil temperature. Field Crop Res 121:195–199

Gajewska E, Sklodowska M, Mazur J (2006) Effect of nickel on antioxidative enzyme activities proline and chlorophyll contentsin wheat shoots. Biol Plant 50:653–659

Gangwar S, Singh VP (2011) Indole acetic acid differently changes growth and nitrogen metabolism in Pisum sativum L. seedlings under chromium (VI) phytotoxicity: implication of oxidative stress. Sci Hortic 129:321–328

Gangwar S, Singh VP, Prasad SM, Maurya JN (2010) Modulation of manganese toxicity in Pisum sativum L. seedlings by kinetin. Sci Hortic 126:467–474

Gangwar S, Singh VP, Srivastava PK, Maurya JN (2011) Modification of chromium (VI) phytotoxicity by exogenous gibberellic acid application in Pisum sativum (L.) seedlings. Acta Physiol Plant 33:1385–1397

Gill SS, Tuteja N (2010) Reactive oxygen species and antioxidant machinery in abiotic stress tolerance in crop plants. Plant Physiol Biochem 48:909–930

Gong X, Su Q, Lin D, Jiang Q, Xu J, Zhang J et al (2014) The rice OsV4 encoding a novel pentatricopeptide repeat protein is required for chloroplast development during the early leaf stage under cold stress. J Integr Plant Biol 56:400–410

Gorham J (1990) Salt tolerance in Triticeae: K/Na discrimination in synthetic hexaploid wheats. J Exp Bot 41:623–627

Gould KS (2004) Nature' s Swiss army knife: the diverse protective roles of anthocyanins in leaves. J Biomed Biotechnol 5:314–320

Gray JE, Holroyd GH, van der Lee FM (2000) The HIC signalling pathway links CO2 perception to stomatal development. Nature 408:713–716

Grimes DW, Yamada H (1982) Relation of cotton growth and yield to minimum leaf water potential. Crop Sci 22:134–139

Gu L, Xu T, Lee K, Lee KH, Kang H (2014) A chloroplast-localized DEAD-box RNA helicase AtRH3 is essential for intron splicing and plays an important role in the growth and stress response in Arabidopsis thaliana. Plant Physiol Biochem 82:309–318

Gupta NK, Gupta S, Kumar A (2001) Effect of water stress on physiological attributes and their relationship with growth and yield of wheat cultivars at different stages. J Agro Crop Sci 186(1): 55–62

Hampton JG, Boelt B, Rolston MP, Chastain TG (2013) Effects of elevated CO2 and temperature on seed quality. J Agric Sci 151:154–162

Hatfield JL, Boote KJ, Fay P, Hahn L, Izaurralde RC, Kimball BA, Mader T, Morgan J, Ort D, Polley W, Thomson A, Wolfe D (2008) Agriculture. In: The effects of climate change on agriculture, land resources, water resources, and biodiversity in the United States

Hatfield JL, Boote KJ, Kimball BA, Ziska LH, Izaurralde RC, Ort D, Thomson AM, Wolfe DW (2011) Climate impacts on agriculture: implications for crop production. Agron J 103:351–370

Havlová M, Dobrev PI, Motyka V, Štorchová H, Libus J, Dobrá J, Malbeck J, Gaudinová A, Vanková R (2008) The role of cytokinins in responses to water deficit in tobacco plants over-expressing trans-zeatin O-glucosyltransferase gene under 35S or SAG12 promoters R. Plant Cell Environ 31(3):341–353

Hay RKM, Porter JR (2006) The physiology of crop yield. Blackwell Publishing Ltd., Oxford

Hayat S, Khalique G, Irfan M, Wani AS, Tripathi BN, Ahmad A (2012) Physiological changes induced by chromium stress in plants: an overview. Protoplasma 249:599–611

Hernandez LE, Villasante CO, Montero-Palmero MB, Escobar C, Carpena RO (2012) Heavy metal perception in a microscale environment: a model system using high doses of pollutants. In: Gupta DK, Sandalio LM (eds) Metal toxicity in plants: perception, signaling and remediation. Springer-Verlag, Berlin; Heidelberg, pp 23–37

Herrero MP, Johnson RR (1980) High temperature stress and pollen viability in maize. Crop Sci 20: 796–800

Higginbotham RW, Paulitz TC, Kidwell KK (2004) Virulence of Pythium species isolated from wheat fields in eastern Washington. Plant Dis 88:1021–1026

Högy P, Poll C, Marhan S, Kandeler E, Fangmeier A (2013) Impacts of temperature increase and change in precipitation pattern on crop yield and yield quality of barley. Food Chem 136:1470–1477

Hossain MA, Piyatida P, da Silva JAT, Fujita M (2012a) Molecular mechanism of heavy metal toxicity and tolerance in plants: central role of glutathione in detoxification of reactive oxygen species and methylglyoxal and in heavy metal chelation. Aust J Bot 2012:872–875

Hossain MA, Hossain MD, Rohman MM, da Silva JAT, Fujita M (2012b) Onion major compounds (flavonoids, organosulfurs) and highly expressed glutathione-related enzymes: possible physiological interaction, gene cloning and abiotic stress response. In: Aguirre CB, Jaramillo LM (eds) Onion consumption and health. Nova Science Publishers, New York, pp 49–90

Hossain A, Sarker MAZ, Saifuzzaman M, Teixeira da Silva JA, Lozovskaya MV, Akhter MM (2013) Evaluation of growth, yield, relative performance and heat susceptibility of eight wheat (Triticum aestivum L.) genotypes grown under heat stress. Int J Plant Prod 7:615–636

Hurkman WJ, Vensel WH, Tanaka CK, Whitehand L, Altenbach SB (2009) Effect of high temperature on albumin and globulin accumulation in the endosperm proteome of the developing wheat grain. J Cereal Sci 49:12–23

IPCC (2014) Climate change 2014 synthesis report. In: Core Writing Team, Pachauri RK, Meyer LA (eds) Contribution of working groups I, II and III to the fifth assessment report of the intergovernmental panel on climate change. IPCC, Geneva, p 151

Ishimaru T, Hirabayashi H, Ida M, Takai T, San-Oh YA, Yoshinaga S, Ando I, Ogawa T, Kondo M (2010) A genetic resource for early-morning flowering trait of wild rice Oryza officinalis to mitigate high temperature-induced spikelet sterility at anthesis. Ann Bot 106:515–520

Jablonski LM, Wang X, Curtis P (2002) Plant reproduction under elevated CO2 conditions: a meta-analysis of reports on 79 crop and wild species. New Phytol 156(1):9–26

Jin J, Huang W, Gao JP, Yang J, Shi M, Zhu MZ et al (2008) Genetic control of rice plant architecture under domestication. Nat Genet 40(11):1365–1369

Jitla DS, Rogers GS, Seneweera SP, Basra AS, Oldfield RJ, Conroy JP (1997) Accelerated early growth of rice at elevated CO2 (Is it related to developmental changes in the shoot apex?). Plant Physiol 115(1):15–22

Johkan M, Oda M, Maruo T, Shinohara Y (2011) Crop production and global warming. In: Casalegno S (ed) Global warming impacts-case studies on the economy, human health, and on urban and natural environments, Rijeka, pp 139–152

Jongdee B, Fukai S, Cooper M (2002) Leaf water potential and osmotic adjustment as physiological traits to improve drought tolerance in rice. Field Crop Res 76:153–164

Jones MM, Turner NC (1978) Osmotic adjustment in leaves of Sorghum in response to water deficits. Plant Physiol 61:122–126

Joo JH, Yoo HJ, Hwang I, Lee JS, Nam KH, Bae YS (2005) Auxin-induced reactive oxygen species production requires the activation of phosphatidylinositol 3-kinase. FEBS Lett 579:1243–1248
Jung JH, Domijan M, Klose C, Biswas S, Ezer D, Gao M et al (2016) Phytochromes function as thermosensors in Arabidopsis. Science 354:886–889
Keunen E, Remans T, Bohler S, Vangronsveld J, Cuypers A (2011) Metal-induced oxidative stress and plant mitochondria. Int J Mol Sci 12:6894–6918
Khan MA, Duke NC (2001) Halophytes- A Resource for the Future. Wetl Ecol Manag 6:455–456
Kikui S, Sasaki T, Maekawa M, Miyao A, Hirochika H, Matsumoto H et al (2005) Physiological and genetic analyses of aluminium tolerance in rice, focusing on root growth during germination. J Inorg Biochem 99:1837–1844
Kim HY, Horie T, Nakagawa H, Wada K (1996) Effects of elevated CO2 concentration and high temperature on growth and yield of rice. II. The effect of yield and its component of Akihikari rice. Jpn J Crop Sci 65:644–651
Kodama Y, Tsuboi H, Kagawa T, Wada M (2008) Low temperature-induced chloroplast relocation mediated by a blue light receptor, phototropin 2, in fern gametophytes. J Plant Res 121:441–448
Kosova K, Vitamvas P, Prasil IT, Renaut J (2011) Plant proteome changes under abiotic stress-contribution of proteomics studies to understanding plant stress response. J Proteome 74:1301–1322
Kudoh H, Sonoike K (2002) Irreversible damage to photosystem I by chilling in the light: cause of the degradation of chlorophyll after returning to normal growth temperature. Planta 215:541–548
Kumar S, Kaur R, Kaur N, Bhandhari K, Kaushal N, Gupta K, Bains TS, Nayyar H (2011) Heat-stress induced inhibition in growth and chlorosis in mungbean (Phaseolus aureus Roxb.) is partly mitigated by ascorbic acid application and is related to reduction in oxidative stress. Acta Physiol Plant 33:2091–2101
Kupper H, Kupper F, Spiller M (1996) Environmental relevance of heavy metal-substituted chlorophylls using the example of water plants. J Exp Bot 47:259–266
Kupsch C, Ruwe H, Guusewski S, Tillich M, Small I, Schmitz-Linneweber C (2012) Arabidopsis chloroplast RNA binding proteins CP31A and CP29A associate with large transcript pools and confer cold stress tolerance by influencing multiple chloroplast RNA processing steps. Plant Cell 24(10):4266–4280
Kusumi K, Sakata C, Nakamura T, Kawasaki S, Yoshimura A, Iba K (2011) A plastid protein NUS1 is essential for build-up of the genetic system for early chloroplast development under cold stress conditions. Plant J 68:1039–1050
Läuchli A (1999) Salinity-potassium interactions in crop plants. In: Oosterhuis DM, Berkovitz GA (eds) Frontiers in potassium nutrition. Potash and Phosphate Institute, Norcross, pp 71–76
Lauchli A, Grattan S (2007) Plant growth and development under salinity stress, biology
Lawlor DW (2002) Limitation to photosynthesis in water stressed leaves: Stomata vs. metabolism and the role of ATP. Ann Bot 89(7):871–885
Leakey ADB, Xu F, Gillespie KM, McGrath JM, Ainsworth EA, Ort DR (2009) Genomic basis for stimulated respiration by plants growing under elevated carbon dioxide. Proc Natl Acad Sci U S A 106:3597–3602
Legris M, Klose C, Burgie ES, Rojas CC, Neme M, Hiltbrunner A et al (2016) Phytochrome B integrates light and temperature signals in Arabidopsis. Science 354:897–900
Li HM, Yu CW (2018) Chloroplast galactolipids: the link between photosynthesis, chloroplast shape, jasmonates, phosphate starvation and freezing tolerance. Plant Cell Physiol 59:1128–1134
Li WX, Chen TB, Huang ZC, Lei M, Liao XY (2006) Effect of arsenic on chloroplast ultrastructure and calcium distribution in arsenic hyperaccumulator Pteris vittata L. Chemosphere 62:803–809
Li J, Cang Z, Jiao F, Bai X, Zhang D, Zhai R (2015) Influence of drought stress on photosynthetic characteristics andprotective enzymes of potato at seedling stage. J Saudi SocAgricul Sci

Liu F, Andersen MN, Jensen CR (2003) Loss of pod set caused by drought stress is associated with water status and ABA content of reproductive structures in soybean. Funct Plant Biol 30:271–280

Liu E, Yan C, Mei X, He W, Bing S, Ding L, Liu Q, Liu S, Fan T (2010) Long-term effect of chemical fertilizer, straw, and manure on soil chemical and biological properties in Northwest China. Elsevier 158:173–180

Loladze I (2014) Hidden shift of the ionome of plants exposed to elevated CO2 depletes minerals at the base of human nutrition. elife 3:e02245

Lynch JP (2015) Root phenes that reduce the metabolic costs of soil exploration: opportunities for 21st century agriculture. Plant Cell Environ 38:1775–1784

Lyons JM (1973) Chilling injury in plants. Ann Rev Plant Biol 24:445–466

Madhu M, Hatfield JL (2013) Dynamics of plant root growth under increased atmospheric carbon dioxide. Published in Agron J 105:657–669

Markelz RJC, Lai LX, Vosseler LN, Leakey ADB (2014) Transcriptional reprogramming and stimulation of leaf respiration by elevated CO2 concentration is diminished, but not eliminated, under limiting nitrogen supply. Plant Cell Environ 37:886–898

Marcell LM, Beattie GA (2002) Effect of leaf surface waxes on leaf colonization by Pantoea agglomerans and Clavibacter michiganensis. Mol Plant Microb Interact 15:1236–1244

Marchand FL, Mertens S, Kockelbergh F, Beyens L, Nijs I (2005) Performance of high arctic tundra plants improved during but deteriorated after exposure to a simulated extreme temperature event. Glob Change Biol 11:2078–2089

Massad TJ, Dyer LA, Vega CG (2012) Cost of defense and a test of the carbon-nutrient balance and growth-differentiation balance hypotheses for two co-occurring classes of plant defense. PLoS One 7:e7554

Myers SS, Zanobetti A, Kloog I, Huybers P, Leakey AD, Bloom AJ, Usui Y (2014) Increasing CO2 threatens human nutrition. Nature 510(7503):139–142

McElrone AJ, Sherald JL, Forseth IN (2003) Interactive effects of water stress and xylem limited bacterial infection on the water relations of a host vine. J Exp Bot 54:419–430

Meehl GA, Stocker TF, Collins WD, Friedlingstein P, Gaye AT, Gregory JM, Kitoh A, Knutti R, Murphy JM, Noda A, Raper SCB, Watterson IG, Weaver AJ, Zhao Z-C (2007) Global climate projections. In: Solomon S, Qin D, Manning M, Chen Z, Marquis M, Averyt KB, Tignor M, Miller HL (eds) Climate change: the physical science basis. Contribution of Working Group I to the Fourth Assessment Report of the Intergovernmental Panel on Climate Change. Cambridge University Press, Cambridge, pp 747–845

Mithöfer A, Boland W (2008) Recognition of herbivory-associated molecular patterns. Plant Physiol 146:825–831

Mitra R, Bhatia CR (2008) Bioenergetic cost of heat tolerance in wheat crop. Curr Sci 94:1049–1053

Mittler R, Blumwald E (2010) Genetic engineering for modern agriculture: challenges and perspectives. Ann Rev Plant Biol 61(4):43–462

Mohammed AR, Tarpley L (2010) Effects of high night temperature and spikelet position on yield-related parameters of rice (Oryza sativa L.) plants. Eur J Agron 33:117–123

Molas J (2002) Changes of chloroplast ultrastructure and total chlorophyll concentration in cabbage leaves caused by excess of organic Ni (II) complexes. Environ Exp Bot 47:115–126

Morgan PB, Bollero GA, Nelson RL, Dohleman FG, Long SP (2005) Smaller than predicted increase in aboveground net primary production and yield of field-grown soybean was found when (CO2) is elevated in fully open-air. Glob Chang Biol 11:1856–1865

Morita R, Sqgio M, Hatanaka T, Misoo S, Fukayama H (2015) CO2-Responsive CONSTANS, CONSTANS-Like, and time of chlorophyll a/b binding protein expression 1 protein is a positive regulator of starch synthesis in vegetative organs of rice. Plant Physiol 167(4):1321–1331

Morita R, Nakagawa M, Takehisa H, Hayashi Y, Ichida H, Usuda S et al (2017) Heavy-ion beam mutagenesis identified an essential gene for chloroplast development under cold stress

conditions during both early growth and tillering stages in rice. Biosci Biotechnol Biochem 81: 271–282
Munns R (2002) Comparative physiology of salt and water stress. Plant Cell Environ 25:239–250
Munns R (2005) Genes and salt tolerance: bringing them together. New Phytol 167:645–663
Munns R, Termaat A (1986) Whole plant responses to salinity. Aust J Plant Physiol 13:143–160
Munns R, James RA, Läuchli A (2006) Approaches to increasing the salt tolerance of wheat and other cereals. J Exp Bot 57:1025–1043
Murakami Y, Tsuyama M, Kobayashi Y, Kodama H, Iba K (2000) Trienoic fatty acids and plant tolerance of high temperature. Science 287:476–479
Nahar K, Ahamed KU, Fujita M (2010) Phenological variation and its relation with yield in several wheat (Triticum aestivum L.) cultivars under normal and late sowing mediated heat stress condition. Not Sci Biol 2(3):51–56
Nawaz A, Farooq M, Cheema SA, Wahid A (2013) Differential response of wheat cultivars to terminal heat stress. Int J Agric Biol 15:1354–1358
Nayyar H, Gupta D (2006) Differential sensitivity of C3 and C4 plants to water deficit stress: Association with oxidative stress and antioxidants. Environ Exp Bot 58(1–3):106–113
Nicolas ME, Munns R, Samarakoon AB, Gifford RM (1993) Elevated CO2 improves the growth of wheat under salinity. Funct Plant Biol 20(3):349–360
Nielsen CL, Hall AE (1985) Responses of cowpea (Vigna unguiculata (L.) Walp.) in the field to high night air temperature during flowering. I. Thermal regimes of production regions and field experimental system. Field Crop Res 10:167–179
Novák V, Lipiec J (2012) Water extraction by roots under environmental stresses. In: Halasi-Kun J, Stekauerová V, Fodor I, Nagy V, Sinóros-Szabó B, Lo Pinto R (eds) Pollution and water resources. Columbia University Seminar Proceedings: Impact of Anthropogenic Activity and Climate Changes on the Environment of Central Europe and USA. Slovak Academy of Sciences – Hungarian Academy of Sciences – Columbia University
Novitskaya GV, Trunova TI (2000) Relationship between cold resistance of plants and the lipid content of their chloroplast membranes. Dokl Biochem 371:50–52
Ouattar JRJS, Crookston RK (1984) Thermal environment during endosperm cell division and grain filling in maize: effects on kernel growth and development in vitro. Crop Sci 24:133–137
Panda SK, Baluska F, Matsumoto H (2009) Aluminum stress signaling in plants. Plant Signal Behav 4:592–597
Pei ZM, Murata Y, Benning G et al (2000) Calcium channels activated by hydrogen peroxide mediate abscisic acid signalling in guard cells. Nature 406:731–734
Peeler TC, Naylor AW (1988) A comparison of the effects of chilling on thylakoid electron transfer in Pea (Pisum sativum L.) and cucumber (Cucumis sativus L.). Plant Physiol 86:147–151
Peet MM, Willits DH, Gardner R (1997) Response of ovule development and post-pollen production processes in male-sterile tomatoes to chronic, sub-acute high temperature stress. J Expt Bot 48:101–111
Pena LB, Azpilicueta CE, Gallego SM (2011) Sunflower cotyledons cope with copper stress by inducing catalase subunits less sensitive to oxidation. J Trace Elem Med Biol 25(3):125–129
Pfannschmidt T (2003) Chloroplast redox signals: how photosynthesis controls its own genes. Trends Plant Sci 8:33–41
Pospisilova J, Vagner M, Malbeck J, Travnickova A, Batkova P (2005) Interactions between abscisic acid and cytokinins during water stress and subsequent rehydration. Biol Plant 49(4): 533–540
Pourrut B, Shahid M, Dumat C, Winterton P, Pinelli E (2011) Lead uptake, toxicity, and detoxification in plants. Rev Environ Contam Toxicol 213:113–136
Prado K, Maurel C (2013) Regulation of leaf hydraulics: from molecular to whole plant levels. Front Plant Sci 4:255
Pribil M, Labs M, Leister D (2014) Structure and dynamics of thylakoids in land plants. J Exp Bot 65:1955–1972

Pritchard SG, Rogers HH (2000) Spatial and temporal deployment of crop roots in CO2-enriched environments. New Phytol 147:55–71

Rae AM, Tricker P, Bunn S, Taylor G (2007) Adaptation of tree growth to elevated CO2: quantitative trait loci for biomass in Populus. New Phytol 175:59–69

Rahman MA, Chikushi J, Yoshida S, Karim AJMS (2009) Growth and yield components of wheat genotypes exposed to high temperature stress under control environment. Bangladesh J Agric Res 34:361–372

Rahoui S, Chaoui A, El-Ferjani EJ (2010) Membrane damage and solute leakage from germinating pea seed under cadmium stress. Hazard Mater 178:1128–1131

Ranasinghe S, Taylor G (1996) Mechanism for increased leaf growth in elevated CO. J Exp Bot 47(296):349–358

Ranieri A, Castagna A, Pacini J, Baldan B, Sodi AM, Soldatini GF (2003) Early production andscavenging of hydrogen peroxide in the apoplast of sunflower plants exposed to ozone. J Exp Bot 54:2529–2540

Rao MV, Paliyath C, Ormrod DP (1996) Ultraviolet-B- and ozone-induced biochemical changes inantioxidant enzymes of Arabidopsis thaliana. Plant Physiol 110:125–136

Reich PB, Hobbie SE, Lee TD (2014) Plant growth enhancement by elevated CO2 eliminated by joint water and nitrogen limitation. Nat Geosci 7:920–924

Rengasamy P (2002) Transient salinity and subsoil constraints to dryland farming in Australian sodic soils: an overview. Aust J Exp Agric 42:351–361

Reynolds MP, Balota M, Delgado MIB, Amani I, Fischer RA (1994) Physiological and morphological traits associated with spring wheat yield under hot, irrigated conditions. Funct Plant Biol 21(6):717–730

Reynolds MP, Pierre CS, Saad ASI, Vargas M, Condon AG (2007) Evaluating potential genetic gains in wheat associated with stress-adaptive trait expression in elite genetic resources under drought and heat stress. Crop Sci 47:172–189

Ribas-Carbo M, Taylor NL, Giles L, Busquets S, Finnegan PA, Day DA, Lambers H, Medrano H, Berry JA, Flexas J (2005) Effects of water stress on respiration in soybean leaves. Plant Physiol 139(1):466–473

Rogers HH, Peterson CM, McCrimmon JN, Cure JD (1992) Response of plant roots to elevated atmospheric carbon dioxide. Plant Cell Environ 15:749–752

Rogers HH, Prior SA, Runion GB, Mitchell RJ (1997) Root to shoot ratio of crops as influenced by CO2. Plant Soil 187:229–248

Rogers A, Ainsworth E et al (2009) Will elevated carbon dioxide concentration amplify the benefits of nitrogen fixation in legumes? Plant Physiol 151:1009–1016

Ruiz-Vera UM, Siebers M, Gray SB, Drag DW, Rosenthal DM, Kimball BA, Ort DR, Bernacchi CJ (2013) Global warming can negate the expected CO2 stimulation in photosynthesis and productivity for soybean grown in the midwestern United States. Plant Physiol 162:410–423

Routaboul J-M, Fischer SF, Browse J (2000) Trienoic fatty acids are required to maintain chloroplast function at low temperatures. Plant Physiol 124:1697–1705

Sadok W, Sinclair TR (2010) Transpiration response of "slow- wilting" and commercial soybean (Glycine max, L. Merr.) genotypes to three aquaporin inhibitors. J Exp Bot 61:821–829

Sato S, Peet M, M. Thomas J.F. (2002) Determining critical pre- and post-anthesis periods and physiological processes in Lycopersicon esculentum Mill. exposed to moderately elevated temperatures. J Expt Bot 53:1187–1195

Sekhon HS, Singh G, Sharma P, Bains TS (2010) Water use efficiency under stress environments. In: Yadav SS, McNeil DL, Redden R, Patil SA (eds) Climate change and management of cool season grain legume crops. Springer Press, Dordrecht/Heidelberg/London/New York

Sfaxi-Bousbih A, Chaoui A, El Ferjani E (2010) Cadmium impairs mineral and carbohydrate mobilization during the germination of bean seeds. Ecotoxicol Environ Saf 73:1123–1129

Shaheen MR, Ayyub CM, Amjad M, Waraich EA (2016) Morpho-physiological evaluation of tomato genotypes under high temperature stress conditions. J Sci Food Agr 96:2698–2704

Shao HB, Chu LY, Lu ZH, Kang CM (2008a) Primary antioxidant free radical scavenging and redox signaling pathways in higher plant cells. Int J Biol Sci 4:8–14

Shao HB, Chu LY, Shao MA, Zhao CX (2008b) Advances in functional regulation mechanisms of plant aquaporins: their diversity, gene expression, localization, structure and roles in plant soil-water relations. Mol Membr Biol 25:1–12

Sharma P, Jha AB, Dubey RS, Pessarakli M (2012) Reactive oxygen species, oxidative damage, and antioxidative defense mechanism in plants under stressful conditions. J Bot 1–26:Article ID 217037

Siddique MRB, Hamid A, Islam MS (2000) Drought stress effectson water relations of wheat. Bot Bull Acad Sin 41:35–39

Silva S (2012) Aluminium toxicity targets in plants. J Bot 2012:219462. https://doi.org/10.1155/2012/219462

Simonetta S, Avidano L, Berta G (2007) Morphogenetic effects induced by pathogenic and non-pathogenic Rhizoctonia solani Kühn strains on tomato roots. Caryologia 60:141–145

Singh N, Ma LQ, Vu JC, Raj A (2009) Effects of arsenic on nitrate metabolism in arsenic hyperaccumulating and non-hyperaccumulating ferns. Environ Pollut 157:2300–2305

Skupień J, Wójtowicz J, Kowalewska Ł, Mazur R, Garstka M, Gieczewska K et al (2017) Dark-chilling induces substantial structural changes and modifies galactolipid and carotenoid composition during chloroplast biogenesis in cucumber (Cucumis sativus L.) cotyledons. Plant Physiol Biochem 111:107–118

Slafer GA, Rawson HM (1997) CO2 effects on phasic development, leaf number and rate of leaf appearance in wheat. Ann Bot 79(1):75–81

Smiri M, Chaoui A, Rouhier N, Gelhaye E, Jacquot JP, El Ferjani E (2011) Cadmium affects the glutathione/glutaredoxin system in germinating pea seeds. Biol Trace Elem Res 142:93–105

Somerville C (1995) Direct tests of the role of membrane lipid composition in low-temperature-induced photoinhibition and chilling sensitivity in plants and cyanobacteria. Proc Natl Acad Sci U S A 92(14):6215

Song CL, Zhao XH (2014) The preparation of an oligochitosan-glycosylated and cross-linked caseinate obtained by a microbial transglutaminase and its functional properties. Int J Dairy Technol 67:110–116

Sonoike K (1995) Selective photoinhibition of photosystem I in isolated thylakoid membranes from cucumber and spinach. Plant Cell Physiol 36:825–830

Springer CJ, Orozco RA, Kelly JK, Ward JK (2008) Elevated CO2 influences the expression of floral-initiation genes in Arabidopsis thaliana. New Phytol 178:63–67

Sundaramoorthy P, Chidambaram A, Ganesh KS, Unnikannan P, Baskaran L (2010) Chromium stress in paddy: (i) nutrient status of paddy under chromium stress; (ii) phytoremediation of chromium by aquatic and terrestrial weeds. C R Biol 333:597–607

Tallis MJ, Lin Y, Rogers A, Zhang J, Street NR, Miglietta F, Karnosky DF, De Angelis P, Calfapietra C, Taylor G (2010) The transcriptome of populus in elevated CO2 reveals increased anthocyanin biosynthesis during delayed autumnal senescence. New Phytol:415–428

Tamás L, Mistrík I, Huttová J, Halusková L, Valentovicová K, Zelinová V (2010) Role of reactive oxygen species-generating enzymes and hydrogen peroxide during cadmium, mercury and osmotic stresses in barley root tip. Planta 231:221–231

Taylor G, Tricker PJ, Zhang FZ, Alston VJ, Miglietta F, Kuzminsky E (2003) Spatial and temporal effects of freeair CO_2 enrichment (POPFACE) on leaf growth, cell expansion, and cell production in a closed canopy of poplar. Plant Physiol 131:177–185

Teicher HB, Møller BL, Scheller HV (2000) Photoinhibition of photosystem I in field-grown barley (Hordeum vulgare L.): induction, recovery and acclimation. Photosynth Res 64:53–61

Terashima I, Huang L-K, Osmond CB (1989a) Effects of leaf chilling on thylakoid functions, measured at room temperature, in Cucumis sativus L. and Oryza saliva L. Plant Cell Physiol 30: 841–850

Terashima I, Shen J-R, Katoh S (1989b) Chilling damage in cucumber {Cucumis sativus L.) thylakoids. In: Tazawa M, Katsumi M, Masuda Y, Okamoto H (eds) Plant water relations and growth under stress, pp 470–472

Tester M, Davenport R (2003) Na+ tolerant and Na+ transport in higher plant. Ann Bot 191:503–527

Tezara W, Mitchell VJ, Driscoll SD, Lawlor DW (1999) Water stress inhibits plant photosynthesis by decreasing coupling factor and ATP. Nature 401:914–917

Thounaojam TC, Panda P, Mazumdar P, Kumar D, Sharma GD, Sahoo L et al (2012) Excess copper induced oxidative stress and response of antioxidants in rice. Plant Physiol Biochem 53:33–39

Tjus SE, Møller BL, Scheller HV (1998a) Photosystem I is an early target of photoinhibition in barley illuminated at chilling temperatures. Plant Physiol 116:755–764

Tjus SE, Teicher HB, Møller BL, Scheller HV (1998b) Photoinhibitory damage of barley photosystem I at chilling temperatures induced under controlled illumination is also identified in the field. In: Garab G (ed) Photosynthesis: mechanisms and effects. III. Kluwer Academic Publishers, Dordrecht, pp 2221–2224

Vadez V, Kholova J, Choudhary S, Zindy P, Terrier M, Krishnamurth L, Kumar PR, Turner NC (2011) Whole plant response to drought under climate change. In: Yadav SS, Redden R, Hatfield JL, Lotze- Campen H, Hall AE (eds) Crop adaptation to climate change. Wiley-Blackwell, Chichester

Van Dijk PA, Brown AK (2006) Emotional labour and negative job outcomes: an evaluation of the mediating role of emotional dissonance. J Manag Organ 12(2):101–115

Vara Prasad PV, Craufurd PQ, Summerfield RJ, Wheeler TR (2000) Effects of short episodes ofheat stress on flower production, pod yield and yield components of groundnut (Arachis hypogaea L.). J Exp Bot 51:777–784

Verma S, Mishra SN (2005) Putrescine alleviation of growth in salt stressed Brassica juncea by inducing antioxidative defense system. J Plant Physiol 162:669–677

Vernoux T, Sanchez-Fernandez R, May M (2002) Glutathione biosynthesis in plants. In: Inze D, Montagu MV (eds) Oxidative stress in plants. Taylor and Francis, London, pp 297–311

Venzhik Y, Talanova V, Titov A (2016) The effect of abscisic acid on cold tolerance and chloroplasts ultrastructure in wheat under optimal and cold stress conditions. Acta Physiol Plant 38:63

Verslues PE, Agarwal M, Katiyar-Agarwal S, Zhu J, Zhu JK (2006) Methods and concepts in quantifying resistance to drought, salt and freezing, abiotic stresses that affect plant water status. Plant J 45:523–539

Wahid A, Gelani S, Ashraf M, Foolad MR (2007) Heat tolerance in plants: an overview. Environ Exp Bot 61:199–223

Wahid A, Arshad M, Farooq M (2009) Cadmium phytotoxicity: Responses, mechanisms and mitigation strategies. In: Lichtfouse E (ed) Advances in sustainable agriculture—Book series, vol 1. Springer, Dordrecht, pp 371–403

Wang W, Vinocur B, Altman A (2003) Plant responses to drought, salinity and extreme temperatures: Towards genetic engineering for stress tolerance. Planta 218:1–14

Wang JZ, Cui LJ, Wang Y, Li JL (2009) Growth, lipid peroxidation and photosynthesis in two tall fescue cultivars differing in heat tolerance. Biol Plant 53:247–242

Wang R, Gao F, Guo BG, Huang JC, Wang L, Zhou YJ (2013) Short-term chromium-stress-induced alterations in the maize leaf proteome. Int J Mol Sci 14:11125–11144

Wang Q, Li R, Cai H, Awasthi MK, Zhang Z, Wang JJ, Ali A, Amanullah M (2016) Improving pig manure composting efficiency employing Ca-bentonite. Ecol Eng 87:157–161

Winter SR, Musick JT, Porter KB (1988) Evaluation of screening techniques for breeding drought resistant winter wheat. Crop Sci:512–516

Wong HL, Pinontoan R, Hayashi K, Tabata R, Yaeno T, Hasegawa K, Kojima C, Yoshioka H, Iba K, Kawasaki T, Shimamoto K (2007) Regulation of rice NADPH oxidase by binding of Rac GTPase to its N-terminal extension. Plant Cell 19:4022–4034

Woodward FI, Kelly CK (1995) The influence of CO2 concentration on stomatal density. New Phytol 131(3):311–327

Woodward FI, Lomas MR, Kelly CK (2004) Global climate and the distribution of plant biomes. Philos Trans R Soc Lond Ser B Biol Sci 359:1465–1476

Wu J, Lightner J, Warwick N, Browse J (1997) Low-temperature damage and subsequent recovery of fab1 mutant Arabidopsis exposed to 2°C. Plant Physiol 113:347–356

Wu L, Wu J, Liu Y, Gong X, Xu J, Lin D et al (2016) The rice pentatricopeptide repeat gene TCD10 is needed for chloroplast development under cold stress. Rice 9:67

Yamamoto Y, Aminaka R, Yoshioka M, Khatoon M, Komayama K, Takenaka D, Yamashita A, Nijo N, Inagawa K, Morita N, Sasaki T, Yamamoto Y (2008) Quality control of photosystem II: impact of light and heat stresses. Photosynth Res 98:589–608

Yu Q, Li L, Luo Q, Eamus D, Xu S, Chen C, Wang E, Liu J, Nielsen DC (2014) Year patterns of climate impact on wheat yields. Int J Climatol 34:518–528

Zhang J, Davies WJ (1989) Sequential response of whole plant water relations to prolonged soil drying and the involvement of zylem sap ABA I the regulation of stomatal behaviour of sunflower plants. New Phytol 113:167–174

Zhang S, Scheller HV (2004) Photoinhibition of photosystem I at chilling temperature and subsequent recovery in Arabidopsis thaliana. Plant Cell Physiol 45:1595–1602

Zhang J, Jiang XD, Li TL, Chang TT (2012) Effect of elevated temperature stress on the production and metabolism of photosynthate in tomato (Lycopersicon esculentum L.) leaves. J Hort Sci Biotechnol 87(4):367–373

Zhang Y, Yu Z, Fu X, Liang C (2002) Noc3p, a bHLH protein, plays an integral role in the initiation of DNA replication in budding yeast. Cell 109:849–860

Zhang HX, Zhang FQ, Xia Y, Wang GP, Shen ZG (2010) Excess copper induces production of hydrogen peroxide in the leaf of Elsholtzia haichowensis through apoplastic and symplastic Cu Zn superoxide dismutase. J Hazard Mater 178:834–843

Zhang ZS, Yang C, Gao HY, Zhang LT, Fan XL, Liu MJ (2014) The higher sensitivity of PSI to ROS results in lower chilling-light tolerance of photosystems in young leaves of cucumber. J Photochem Photobiol B 137:127–134

Zhang J, Terrones M, Park CR, Mukherjee R, Monthioux M, Koratkar N, Kim YS, Hurt R, Frackowiak E, Enoki T, Chen Y, Chen Y, Bianco A (2016) Carbon science in 2016: status, challenges and perspectives. Elsevier Carbon 98:708–732

Zhou R, Yu XQ, Ottosen CO, Rosenqvist E, Zhao LP, Wang YL, Yu WG, Zhao TM, Wu Z (2017) Drought stress had a predominant effect over heat stress on three tomato cultivars subjected to combined stress. BMC Plant Biol 17:1-3

Consequences of Climate Change Over Soil Dynamics: An Update

Alok Bharadwaj

Abstract Soil is our motherland that contains various minerals and lots of organic matter in addition to air and water. Due to the strong connection between weather, topology, and floral rate, the soil acts such a habitat in which a variety of microorganisms survive freely. Among all ecosystems, soil has the most diversified ecosystem around the world that possesses bacteria, fungi, protists, virus, archaea, etc. These microbes have been found to perform various functions like increasing soil fertility and crop yield, nutrient recycling for pollutant detoxification, and minimizing the production of greenhouse gases like methane and nitrous oxides. The survival of all these microbial communities depends upon the relation among weather, geology, and vegetation of the particular habitat. As per the data available through Food and Agriculture Organization (FAO), due to soil erosion there must be approximately 20–80% losses occurred in agriculture crop production, and this soil erosion occurred due to various human activities and climate change. In this chapter, the effect of climate change on soil properties has been discussed. The changes in climate adversely affect the microbial diversity in soil; henceforth in the coming future, it may result in decrease in soil carbon level, increase in soil-borne greenhouse gas levels, and change in plant-soil interaction eventually resulting in the decrease in soil fertility.

Keywords Climate change · Soil microflora · Population dynamics · Soil ecosystem and nitrogen fixation

Introduction

It has been observed that change in the climate not only adversely affects the microbial population but also affects the interaction between them (Wookey et al. 2009; van der Putten 2012). Natural microbial population possessed high degree of

A. Bharadwaj (✉)
Department of Biotechnology, GLA University, Mathura, UP, India
e-mail: alok.bhardwaj@gla.ac.in

© The Author(s), under exclusive license to Springer Nature Singapore Pte Ltd. 2022
A. Vaishnav et al. (eds.), *Plant Stress Mitigators*,
https://doi.org/10.1007/978-981-16-7759-5_5

heat resistance, variation in life history behavior, and spreading ability. Among all these microbial communities, there must be some interaction that may be beneficial, harmful, or neutral depending upon the factors associated with climate change (Vandenkoornhuyse et al. 2015). Previous studies revealed that climate change adversely affect the interactions among different microbial species which results in modified biodiversity and various ecosystems (Walther et al. 2002; Gottfried et al. 2012; Langley and Hungate 2014) including some studies focused on soil population (Schimel et al. 2007; de Vries et al. 2012).

It has been found globally that change in climate is a key factor that affects the soil microflora and subsequent processes. Soil microflora is considered to be the most vital ecosystem because it has great diversity. Soil microflora consists of great diversity including bacteria, algae, fungi, and viruses, along with protozoa and nematodes. In the present scenario, change in climate occurs mainly due to the total sum impact of biological disturbances along with the human interventions. Moreover, due to these disturbances, it is very complicated to find out that how these soil microbial communities will react to such situation. For any ecosystem, environmental inconsistency is an essential component as some disturbances are unavoidable like seasonal disturbances, etc.

It has been found in several studies that microbes are the key player in exchanging greenhouse gases between the soil and environment (Conrad 1996; Falkowski et al. 2008), but no literature was available for explaining the role of these soil microorganisms in the evolution of earth's climate. Moreover, soil microbial flora performs a wide range of functions in the ecosystem like control of the amount of soil organic matter, soil carbon sequestration, and release of greenhouse gases, therefore improving the soil physical health that results in better plant growth and also yield. Apart from this soil microflora also helps in the decomposition of the organic matter, secretion of nutrients made available for the growth of plants, and degradation of toxic and harmful chemical substances along with enhanced mineral solubilization that improves the soil health and also structure.

In the present chapter, a discussion was performed over the soil biodiversity and their functions along with the impact of climate change over soil microbiome and key methods of protection of soil biodiversity.

Biodiversity in Soil Microflora and Functions

It has been noticed that there is no adequate information available on the diversity of soil microbial flora as well as the relationship between microbial diversity and soil function (Lavelle et al. 1997; Wolterr 1991). Mostly these microorganisms play a key role in various biochemical processes, e.g., degradation of soil organic matter and increased microbial activity along with control over microbial diversity. Moreover, apart from mentioned activities, there are several mesophilic microbes that are responsible for regeneration and also help in improving the soil organic matter resulting in the recovery of soil health. However, in soil ecosystem some larger organisms (e.g., earthworms and myriapods) are present, and due to their burial action, they have the capability to alter their habitat and thus transform the trophic resource base of microbes which are small in size and have less motility (Lavelle et al. 1997). Due to all these activities, voids are formed along with enhanced water infiltration and development of stable soil aggregates resulting in the development of a humic top horizon in soil.

Bacteria

As far as microorganisms are concerned, like bacteria and fungi, it has been observed that they are major contributor of the energy flow (i.e., more than 90%) in soil ecosystem (Coleman and Crossley 1996; Nannipieri and Badalucco 2003). Apart from this, viruses are also the normal inhabitant in soil and are capable of infecting all the living cells ranging from bacteria to large animals. Among these microorganisms, bacterial population is found to be the maximum on the earth (Torsvik and Ovreas 2002), and it has been calculated that approximately 4–6 $\times$ 10^{30} bacterial cells survive on the earth. Out of that more than 90% reside in the soil and its subsurface (Whitman et al. 1998). It has been calculated that 1 gram of soil approximately contains one billion bacterial cells and more than 10,000 bacterial genomes (Torsvik and Ovreas 2002). Moreover, in the temperate grassland soil, estimated bacterial biomass is about 1–2 t/ha, i.e., corresponding to the weight of one cow (Killham 1994) and estimated 3–5% of total soil organic matter content. Actinobacteria are the filamentous bacteria having branching filaments resembling fungal mycelium which are commonly present in soil and have the capability to decompose organic matter and many toxic pollutants. In soil, certain autotrophic bacteria also survive, i.e., capable of undergoing the photosynthesis and are called cyanobacteria. These cyanobacteria secrete a protein called geosmin that has the characteristic aroma of freshly moistened soil or compost. In soil these bacteria reside in the intracellular spaces between the soil particles and form the aggregations of cells with soil particles (Donlan 2002).

Decomposition of complex organic compounds into simpler ones is a very typical and time-consuming process, and this process continued since ancient times on the earth. Microorganisms are the key players of this process and help in decomposing process and finally convert organic nitrogen back to mineral nitrogen. Denitrifying bacteria have the ability to convert nitrate into atmospheric nitrogen under anaerobic environment (Fig. 1). Now further, free-living or symbiotic bacteria fix this atmospheric nitrogen and are made available for the plants. In the first step, microorganisms (bacteria or fungi) undergo the process of ammonification in which organic nitrogen from decaying animals or plants is converted into ammonium (NH^{4+}). This mineralization step is performed by a variety of microbes. Furthermore, certain plants and microorganisms bypass this mineralization step as they may utilize organic nitrogen (Nannipieri and Paul 2009). In the second step, nitrification process takes place after ammonification, in which ammonia is converted to nitrites (NO^{2-})

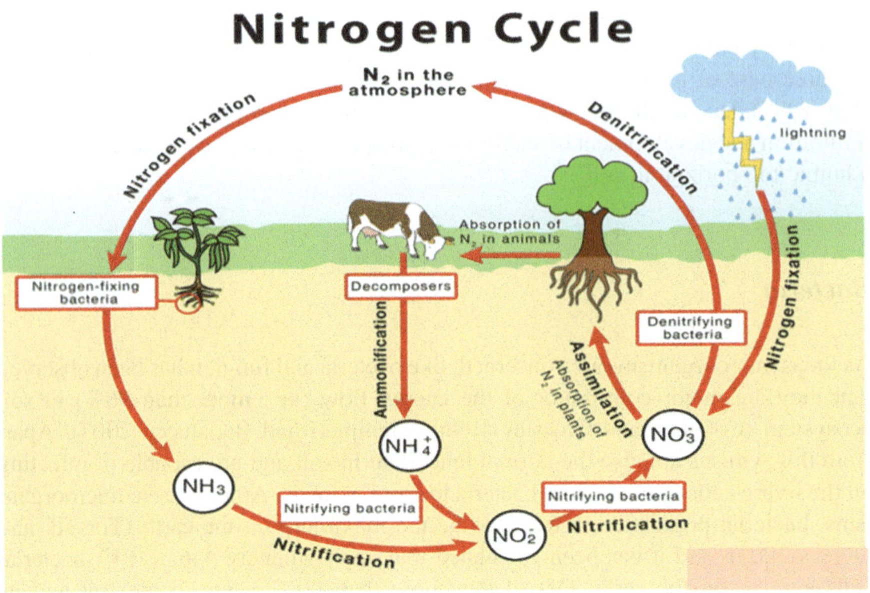

Fig. 1 Nitrogen cycle

that are toxic to plants, with the help of a specialized group of chemotrophic bacteria known as ammonia-oxidizing bacteria (AOB) (Fig. 1). Moreover, there are certain bacteria that are able to oxidize these nitrites into nitrates (NO^{3-}) that pose no harm and are beneficial for plant growth. In the next step, reduction of nitrate or nitrite to nitrous oxide takes place under anoxic environment by certain bacteria. Ammonium ($NH^{4\ +}$) or nitrate (NO^{3-}) ions are readily utilized by the plants either through their root hairs from soil or by the symbiotic interaction with *Rhizobium* bacterium. On the other hand, when nitrate is not utilized by plants, denitrification process takes place. In this process nitrates get reduced and get back to nitrogen gas (N_2) into the atmosphere with the help of certain bacteria under anaerobic environment. These bacteria are of much significance as they utilize nitrogen for respiration in place of oxygen.

Fungi

The second most diversified group among soil microorganisms is fungi. These are the oldest and largest organisms on earth, ranging from microscopic unicellular yeasts to multicellular forms like rhizomorphs and fruiting bodies. Till now more than 80,000 fungal species have been isolated and identified from soil. The extent of total fungal diversity has been projected at about 1.5 million species (Hawksworth 1991). It has been found that approximately one million individual fungi have been isolated from 1 gram of soil, whereas in temperate soil the fungal biomass is around 2–5 t/ha (Killham 1994). Since fungi are heterotrophs, henceforth they depend on organic substrates (like dead remains of plants or animals) to attain carbon for their growth and development. Furthermore, few fungal species survive on complex organic materials for carbon, catabolism of sugars, starches or lignin, and cellulose within the wood.

Certain fungal species are parasitic, i.e., they cause diseases in plants, animals, and other organisms. On the other hand, some fungi show symbiotic interaction with plants, i.e., mycorrhizal associations that result in the improved nutrient supply to the plants. Mostly these mycorrhizal associations transpire with terrestrial plants, while others are host-specific. Moreover, it has been observed that a single plant host may undergo a number of various mycorrhizal fungi around a single rhizosphere (Perotto et al. 1996). In addition to it, certain other organisms like earthworms or large arthropods also have the capability to decompose organic matter (dead plant parts, etc.). Prominently, certain species of fungi secretes a glycoprotein known as "glomalin" that plays a key role in the formation of soil aggregates because of its sticky property (Rillig 2004; Purin and Rillig 2007).

Impact of Climate Change over Soil Microbiome

We have already discussed about the microbial diversity residing in the soil in the previous section. Now here we focused on the impact of various physiological and biochemical changes that occur in the soil microflora due to climate change. Various impacts of environmental changes over soil ecosystem are discussed here.

Elevated CO_2

Based on experimental data available from the previous studies, due to climate change the increased eCO_2 content may pose great impact over soil microbial flora. Due to increased eCO_2 content, changes occur in the functioning of soil microbial flora which can be evaluated by the screening of that gene through metagenomics. One example is BioCON grassland experiment, where it was found that eCO2-stimulated amplification in gene families directly linked with degradation, nitrate reduction, and nitrogen fixation, while decreased abundances of gene families are associated with glutamine synthesis and anaerobic ammonium oxidation (Tu et al. 2017). Moreover, in arid grassland ecosystem, microbial genes showed increased amplification associated with degradation, N_2 fixation, C fixation, methane metabolism, N mineralization, and denitrification (Yu 2018).

In the previous study, employing meta-analysis and modeling clearly concluded that eCO_2 concentration enhanced the photosynthesis process along with amount of carbon in soil. Moreover, it has been observed over a long time period that due to

eCO_2, microbial decomposition of soil organic matter increased (Van Groenigen et al. 2014). It has been noticed that long-lasting impact of eCO_2 over soil carbon stocks is mainly based on the availability of water and other nutrients that directly affects photosynthesis process, microbial degradation, and accumulation of soil carbon.

Increased Temperature

Temperature is an important environmental factor that estimates the growth pattern and yields of pure microorganisms. In response to higher temperature, microorganisms have the capability to alter the lipid proportion of plasma membrane resulting in the decreased membrane fluidity and presentation of heat shock proteins (HSP). As the assessment of the impact of high temperature over soil microflora is a much complex process in situ, nowadays advanced sequencing and functional gene arrays have shown community and functional gene shifts as an indication to enhanced temperature in the surroundings (Melillo et al. 2017).

As per the finding available from the previous studies, its long-term impact results in acclimation of microbial respiration and associated microbial mechanisms in four phases, i.e., speedy carbon loss by respiration process, reorganization of microbial communities, development of more diversified microbial species along with increased soil respiration among heated plots in comparison to controls, and a drop in additional recalcitrant carbon pools with predictable alteration in the organization of microbial population (Melillo et al. 2017). For short-term exposure the acclimation of soil respiration was observed as decreased microbial biomass and thermal adaptation of soil respiration. Finally it has been concluded that decreased carbon availability leads in drop among the fungal and actinomycete population, while on the other hand, population of oligotrophic bacteria increased (DeAngelis et al. 2015).

Permafrost Thaw

In the Arctic, where the temperature is very low, global warming presents very complex situations, i.e., thaw of permafrost soils. Since the permafrost soils are rich in carbon content, the impact of climate change over permafrost thaw is enormous (Turetsky et al. 2019). During the process of permafrost thaws, ice melts; as a result more water is available for enhancing the microbial activity. Due to enhanced microbial activity, there is an increase in the decomposition of soil organic matter along with the emission of gases like CO_2 and CH_4 (Mackelprang et al. 2016). In general permafrost thaw represents the alteration in the soil moisture capacity that has direct impact over activity, e.g., certain thawed permafrost surroundings have the capability to produce methane, a greenhouse gas (Tas et al. 2018). It has become evident from the metagenome sequencing study that the functional potential of microbial population under permafrost soil differs from the microbial population residing in active soil layers (Tas et al. 2018); henceforth it can be concluded that permafrost thaw modifies the microbial flora of such soils. Besides this, the population of *Actinobacteria* increases with depth into the permafrost soil.

Drought

Among mesic grassland ecosystem, drought presents serious issue due to climate change. It has been observed that due to increase in the drought conditions, there is reduction in the microbial activities which is necessary for sustainability of ecosystem. Moreover, due to drought, amount of water present between the soil particles also decreases; as a result decomposition of soil organic carbon also reduces (Schimmel 2018). Among soil microorganisms different mechanisms are available to cope with the drought stress. Some of these mechanisms are dormancy, osmoregulation, and manufacturing of extracellular enzyme. In osmoregulation, microorganisms collect solute (osmolytes) to hold turgor pressure (Schimmel 2018). Though under extreme drying conditions, the osmolytic accumulation of solute might be energetically costly (Boot et al. 2013). Soil microflora has the ability to recover and grow again when the water content is made available to them. Moreover, soil microflora is capable of producing extracellular polymeric substances for retaining water at low matric potentials. Certain bacterial genera like *Bacillus* and *Actinomycetes* can tolerate such drought situation due to their capability to become dormant during dehydrated conditions (Naylor et al. 2017).

Increased Precipitation and Flooding

It has also been observed that climate change affects the precipitation rate leading to excessive rainfall at the cost of ice, leading to the minimized ice pack and enhanced freeze-thaw cycles (Sorensen et al. 2016). Due to increased precipitation, the moisture content increases; henceforth the gap between soil particles becomes water logged, producing anaerobic conditions, therefore creating an environment suitable for methanogenesis and denitrification process leading to the discharge of CH_4 and N_2O. Under situation of extensive flooding, it has been noted that microbial action decreases because of depletion of resources creating a "boom and bust" condition (Sjogaard et al. 2018).

Increased Fire Frequency or Intensity

Long dried fire period along with unsustainable land management procedures, the incidences of fire are increasing worldwide day by day. Every incidence of fire leads to the discharge of huge volume of CO_2 into the atmosphere (Sun et al. 2015). As the upper layer of soil ignites due to fire, it produces more heat. Moreover, the soil with increased moisture content, soil heating might be deferred and destroyed maximum microbial flora through the process of pasteurization. Uncontrolled fire condition may lead to separation of the soil aggregates, resulting in the reduction in soil aeration. In certain cases, post-fire consequences lead to soil degradation and erosion. Due to fire, there is a significant reduction in the carbon and nitrogen content of soil (Hinojosa et al. 2016).

Protection of Soil Biodiversity

As we have already discussed in this paper, soil has a great microbial diversity. We can save the soil microbial diversity by using following methods.

Mulching/Light Soil Sealing

In mulching process, we covered the top soil layer to guard it from soil erosion, in turn increasing its productivity. As we know mulching process is employed before the commencement of crop growing period and may be repeated as per the requirement. Mulching process helps to warm the soil to hold heat and moisture. Materials used for mulching process are mainly the organic remains (like plant remains, hay, bark) as well as manure, compost, sludge, rubber, and plastic films.

Application of Organic Residues (Compost/Manure/Sludge)

Incorporation of animal manure, sewage sludge, or other organic wastes like coffee-berry pulp or compost ameliorates the soil organic carbon content. For best agricultural practices, it is necessary to permit the degradation of organic components for appropriate time prior to employing them into the field. It is necessary as incorporation of organic components immobilizes available nitrogen in the soil because microbial population required both carbon and nitrogen for growth and development.

Fertilizers

As we all know, prolonged incorporation of chemical fertilizers (inorganic nitrogenous) to the microbial population helps in utilizing that nitrogen for their growth and development. This results in enhanced degradation of soil organic matter, which in turn decreases the soil organic matter and finally deteriorates the soil structure along with less water-holding capacity.

Crop Management

Choice of Crop Species

The selection of crop to be cultivated is very crucial as it depends on the type of habitat accessible to soil microbial population. One excellent example is legume crops that have the ability of nitrogen fixation in soil with the help of mutualistic association with *Rhizobium* bacterium.

Crop Rotations

It is a key method employed for retarding the growth and multiplication of pests and pathogens. As we change the crop variety, it will eventually affect the associated microbial population.

Conclusion

In this chapter we have discussed about different microbial communities residing in soil along with the functions they perform. Apart from this we have also focused on the impact of climate change over microbial diversity as well as about the key processes required for the protection of soil microflora. From the above literature, it has become evident that microbial population in soil plays a key function in upholding the soil carbon content which made it available to the plants for their growth and development. With the help of these soil microbial communities, we can preserve the soil for our coming generation. Henceforth, there is an urgent requirement for monitoring the consequences of climate change on soil microflora that carry out various functions needed to sustain the environment. Advanced technology for weather forecasting is required so that we can make the strategies to save these soil microbial communities from the consequences of climate change well in advance.

References

Boot CM, Schaeffer SM, Schimel JP (2013) Static osmolyte concentrations in microbial biomass during seasonal drought in a California grassland. Soil Biol Biochem 57:356–361

Coleman DC, Crossley DA (1996) Fundamentals of soil ecology. Academic Press, Amsterdam

Conrad R (1996) Soil microorganisms as controllers of atmospheric trace gases (H_2, CO, CH_4, OCS, N_2O, and NO). Microbiol Rev 60:609–640

de Vries FT, Liiri ME, Bjornlund L, Bowker MA, Christensen S, Setala HM, Bardgett RD (2012) Land use alters the resistance and resilience of soil food webs to drought. Nat Clim Chang 2: 276–280

DeAngelis KM et al (2015) Long-term forest soil warming alters microbial communities in temperate forest soils. Front Microbiol 6:104

Donlan RM (2002) Biofilms: microbial life on surfaces. Emerg Infect Dis 8:881–890

Falkowski PG, Fenchel T, Delong EF (2008) The microbial engines that drive Earth's biogeochemical cycles. Science 320:1034–1039

Gottfried M, Pauli H, Futschik A, Akhalkatsi M, Baranc˘ok P, Alonso JLB, Coldea G, Dick J, Erschbamer B, Kazakis G (2012) Continent-wide response of mountain vegetation to climate change. Nat Clim Chang 2:111–115

Hawksworth DL (1991) The fungal dimension of biodiversity - magnitude, significance, and conservation. Mycol Res 95:641–655

Hinojosa MB, Parra A, Laudicina VA, Moreno JM (2016) Post-fire soil functionality and microbial community structure in a mediterranean shrubland subjected to experimental drought. Sci Total Environ 573:1178–1189

Killham K (1994) Soil ecology. Cambridge University Press, Cambridge

Langley JA, Hungate BA (2014) Plant community feedbacks and long-term ecosystem responses to multi-factored global change. AoB Plants 6:12

Lavelle P, Bignell D, Lepage M, Wolterr V, Roger P, Ineron P, Heal OW, Dhillion S (1997) Soil functioning in a changing world: the role of invertebrate ecosystem engineers. Eur J Soil Biol 33:159–193

Mackelprang R, Saleska SR, Jacobsen CS, Jansson JK, Taş N (2016) Permafrost meta-omics and climate change. Annu Rev Earth Planet Sci 44:439–462

Melillo JM et al (2017) Long-term pattern and magnitude of soil carbon feedback to the climate system in a warming world. Science 358:101–104

Nannipieri P, Badalucco L (2003) Biological processes. In: Bembi DK, Nieder R (eds) Processes in the soil–plant system: modelling concepts and applications. The Haworth Press, Binghamton

Nannipieri P, Paul EA (2009) The chemical and functional characetrization of organic N and its biotic components. Soil Biol Biochem 41:2357–2369

Naylor D, DeGraaf S, Purdom E, Coleman-Derr D (2017) Drought and host selection influence bacterial community dynamics in the grass root microbiome. ISME J 11:2691–2704

Perotto S, ActisPerino E et al (1996) Molecular diversity of fungi from ericoid mycorrhizal roots. Mol Ecol 5(1):123–131

Purin S, Rillig MC (2007) The arbuscular mycorrhizal fungal protein glomalin: limitations, progress, and a new hypothesis for its function. Pedobiologia 51(2):123–130

Rillig MC (2004) Arbuscular mycorrhizae, glomalin, and soil aggregation. Can J Soil Sci 84(4): 355–363

Schimel J, Balser TC, Wallenstein M (2007) Microbial stress-response physiology and its implications for ecosystem function. Ecology 88:1386–1394

Schimmel JP (2018) Life in dry soils: effects of drought on soil microbial communities and processes. Ann Rev Ecol Evol Syst 49:409–432

Sjogaard KS, Valdemarsen TB, Treusch AH (2018) Responses of an agricultural soil microbiome to flooding with seawater after managed coastal realignment. Microorganisms 6:1–18

Sorensen PO, Templer PH, Finzi AC (2016) Contrasting effects of winter snowpack and soil frost on growing season microbial biomass and enzyme activity in two mixed-hardwood forests. Biogeochemistry 128:141–154
Sun H et al (2015) Fungal community shifts in structure and function across a boreal forest fire chronosequence. Appl Environ Microb 81:7869–7880
Tas N et al (2018) Landscape topography structures the soil microbiome in Arctic polygonal tundra. Nat Commun 9:777
Torsvik V, Ovreas L (2002) Microbial diversity and function in soil: from genes to ecosystems. Curr Opin Microbiol 5(3):240–245
Tu Q et al (2017) Metagenomic reconstruction of nitrogen cycling pathways in a CO_2-enriched grassland ecosystem. Soil Biol Biochem 106:99–108
Turetsky MR et al (2019) Permafrost collapse is accelerating carbon release. Nature 569:32–34
Van der Putten WH (2012) Climate change, above- ground-belowground interactions and species range shifts. Annu Rev Ecol Evol Syst 43:365–383
Van Groenigen KJ, Qi X, Osenberg CW, Luo YQ, Hungate BA (2014) Faster decomposition under increased atmospheric CO2 limits soil carbon storage. Science 344:508–509
Vandenkoornhuyse P, Quaiser A, Duhamel M, Le Van A, Dufresne A (2015) The importance of the microbiome of the plant holobiont. New Phytol 206(4):1996–1206
Walther GR, Post E, Convey P, Menzel A, Parmesan C, Beebee TJC, Fromentin JM, Hoegh-Guldberg O, Bairlein F (2002) Ecological responses to recent climate change. Nature 416:389–395
Whitman WB, Coleman DC et al (1998) Prokaryotes: the unseen majority. Proc Natl Acad Sci U S A 95(12):6578–6583
Wolterr V (1991) Soil invertebrates – effects on nutrient turnover and soil structure – a review. J Plant Nutr Soil Sci 154:389–402
Wookey PA et al (2009) Ecosystem feedbacks and cascade processes: understanding their role in the responses of Arctic and alpine ecosystems to environmental change. Glob Chang Biol 15: 1153–1172
Yu H (2018) Et al. elevated CO2 and warming altered grassland microbial communities in soil top-layers. Front Microbiol 9:1790

Physiological Effects of Drought Stress in Plants

Arash Hemati, Ebrahim Moghiseh, Arian Amirifar, Morteza Mofidi-Chelan, and Behnam Asgari Lajayer

Abstract The most common effect of the drought stress is to reduce the water potential, the turgor pressure in the growing cells, and thus the lack of turgor pressure necessary for their growth. Lack of water accelerates cell differentiation. Under drought stress, root, stem, leaf, and fruit growth decreases. Also, in these conditions, not all plant organs are affected equally. As a rule, due to drought stress, the ratio of leaves to stems decreases. Older leaves and leaves that are exposed to shade usually die sooner, slow down tillering, and increase tiller death in tillering species. Physiologic effects of water stress contain so many cases such as reduction of relative water content (RWC), reduction of intercellular space during wilting, effect of drought stress on photosynthesis, effect of drought stress on respiration, effect of drought stress on photosynthetic derivative distribution, effect of drought stress on metabolism, accumulation of sugars, drought stress and protein breakdown, and the effect of drought stress on ABA hormone, which are among the most important of these cases.

Keywords Water · Stress · Physiology · Plant

A. Hemati · B. Asgari Lajayer (✉)
Department of Soil Science, Faculty of Agriculture, University of Tabriz, Tabriz, Iran
e-mail: h-asgari@tabrizu.ac.ir

E. Moghiseh
Nuclear Agriculture Research School, Nuclear Science and Technology Research Institute, Karaj, Iran

A. Amirifar
Department of Biology, Islamic Azad University of Urmia, Urmia, Iran

M. Mofidi-Chelan
Department of Range and Watershed Management, Faculty of Natural Resources, Urmia University, Urmia, Iran

© The Author(s), under exclusive license to Springer Nature Singapore Pte Ltd. 2022
A. Vaishnav et al. (eds.), *Plant Stress Mitigators*,
https://doi.org/10.1007/978-981-16-7759-5_6

Introduction

Crop susceptibility to drought depends on the economic yield of the plant (Khoshmanzar et al. 2019). Investigation of Firoozabadi et al. (2003) showed that the amount of root yield under normal conditions, mild stress, and severe stress that was applied continuously during the growing season was 6.58, 8.45, and 7.34 tons per acre, respectively. According to Chołuj et al. (2008), drought stress reduced both root yield and glucose yield due to stress time. However, leafy vegetables such as lettuce, which their economic product is fresh leaves, are so sensitive to drought that they will not perform well in seasons and areas where high evaporation is required even if they are irrigated frequently. Crops such as alfalfa, which are cultivated for their dry matter, have a high resistance to drought, so that a balanced lack of water has a little effect on reducing economic yield (Agrawal and Dadlani 1994; Moustafa-Farag et al. 2020). Drought susceptibility in annual plants that are grown for seed or fruit production varies depending on the stage of development in which they are stressed. In these plants, they are generally more drought tolerant in the vegetative state than in the reproductive stages. Plants such as cotton and tomato that have unlimited growth and are able to produce many leaves after flowering are more resistant to drought than plants that have limited growth. In sunflower, the occurrence of a dry period in the vegetative growth stage caused a decrease in grain yield due to lack of leaf growth after re-irrigation, reduced leaf area, and photosynthesis potential (Mao et al. 2020). Most of the annual seed-producing plants are sensitive to drought in the early stage of flowering. If the grains are stressed in the early stages of reproductive growth, their grain yield will be significantly reduced. Corn is highly sensitive to drought during the flowering stage. The main reason for this sensitivity is the delay in the emergence of female organs, which makes the pistil unprepared to accept them when the pollens arrive. Wheat is also sensitive to drought just before pollination. In this case, meiotic division is likely to be impaired, and healthy pollen production is reduced (Levitt 1980). In plants that grow indefinitely, drought has less effect on seed production, because these plants are able to produce more flowers and seeds after drought from rain and irrigation. Drought in the late reproductive period causes the seeds and fruits to shrink. Reduction of grain size due to drought depends on the balance between photosynthetic source and reproductive reservoir and the effect that drought has on these two components (De Micco and Aronne 2012).

In general, the effects of drought at different stages of development on limited and unlimited seed yield can be observed in the model plant. Drought has shrinking effects in grain yield in plants with unlimited growth; in plants with limited growth, drought in the vegetative stage determines the number of seeds in them. As a result, the final grain yield is the number of seeds (which is affected by drought in the vegetative stage and early flowering stage) multiplied by the weight of the seed affected by drought in vegetative stage, which determines the canopy of photosynthetic capacity, and the dryness in the filling stage, which reduces grain size. Seed photosynthesis is the opposite in plants that grow indefinitely because the number of leaves and yield increases cumulatively. The drought of the previous stages will have relatively little effect on the later stages (Agrawal and Dadlani 1994).

Effects of Drought Stress on the Plant

General Effects of Drought Stress

The most common effect of drought stress is to reduce the water potential, the turgor in the growing cells, and thus the lack of expansions necessary for their growth (Ghassemi et al. 2018; Khoshru et al. 2020). The effects of drought stress can be divided into anatomical and physiological changes. Physiological changes include the opening and closing of stomata, different movements of plant organs, structure of proteins and enzymes, amount and action of hormones, carbohydrate metabolism, pattern of accumulation of substances in the cell membrane, photosynthesis, respiration, and rate of action of photosynthesis. Anatomic changes can be seen in reducing the size of cells and the distances between them, thickening of the cell wall, more development of mechanical tissues, and reducing the number of stomata per unit area and changes in organ growth (Levitt 1980; Heshmat et al. 2020).

Drought Effect on Cell and Tissue Growth

Because plant growth is the result of cell growth, it is necessary to consider cell growth in three stages of division, development, and differentiation in relation to drought stress. It is often concluded that cell division is less sensitive to drought stress than its enlargement (leaf development). In support of this result, corn leaf development and early germination were limited to 0.75 MPa, but cell division continued (Vyas et al. 1985).

Similarly, soybean tissue culture showed a decrease in cell size (Sinclair 1985). The growth of sugarcane sprouts (possibly by cell division) was reduced to a potential of 5 atmospheres and stopped completely at 20–30 atmospheres. This difference can be explained by the fact that developing and enlarging cells need several times more water than cell division (Kumar et al. 1994). A 1996 study by Gzik on sugar beet found that drought stress increased proline content, which reduced plant growth and net leaf weight. When the lack of water is present in the plant, cell growth ends earlier than when there is plenty of water, and the structural differentiation of cells begins earlier. In general, a lack of water accelerates cell differentiation. It is found that the final effect on growth will depend on the frequency and duration of stress periods. Drought stress does not affect all plant organs in the same way. As a rule, leaf-to-stem ratio decreases due to drought stress, and older and shaded leaves die sooner (Doorenbos and Kassam 1979; Hayat et al. 2020). Also water scarcity decreases speed of tillering stage and increases tiller death in tillering species. In maize, the ability of the cell wall to expand decreases due to lack of water, while the ability of the root cell to expand increases (Westage and Boyer 1985). With the onset of drought stress, the root growth rate decreases. However, root growth is less affected than the aerial parts of the plant, and in general, the root-to-stem ratio increases. Roots are reduced and grow toward the water in the soil as long as they are close to the water. When the rainfall is light and frequent, only a small part of the potential root area is moistened, and root penetration is limited to the shallow layer. In this case, the plant will be sensitive to dry periods, especially in the later stages of the growth (Black and Ong 2000).

During the drought stress, the growth stops and continues after stress relief. The amount of damage to plants depends on physiologic age, amount of water stress, length of stress period, and species of plants. In general, organs of the plant that grow rapidly during stress are damaged more than other organs (Mao et al. 2020).

Vyas et al. (1985) showed that sesame in drought stress strongly affects the internode length, capsule length, and biomass, but the traits of first capsule height to ground, number of capsules, average capsule thickness and 1000 seed weight, root length, and number of branches whether with capsules or not were less affected by stress.

Physiological Effects of Drought Stress

Reduction of Relative Water Content (RWC)

One of the most important changes due to drought stress is reduction of leaf relative water content (Heshmat et al. 2020). This index can show the ability of plants to tolerate drought stress. Kaiser et al. (1985) categorized the potential effects of increased defoliation as follows: By reducing the relative moisture content of the leaf (RWC) between 70% and 100%, photosynthesis is reduced due to the closure of the stomata, which is rapidly reversible. Upon reaching the relative moisture content between 35% and 70%, photosynthetic capacity is only slowly improved with rehydration. Light inhibition is the main cause of decreased photosynthesis under stress condition. Electron transfer is also one of the restrictive actions, and if the relative water content of the leaf is reduced to less than 30%, photosynthetic capacity is reduced due to membrane damage that it is in chloroplasts leads to death. To calculate the relative water content, the leaves are separated from the plant and weighted. They are then submerged in water to reach the maximum turgor pressure for re-weighing, calculating indicators such as relative water content, water saturation deficit, and relative saturation deficit. Because relative water content and water saturation reduction require accurate determination of leaf dry weight, it is often preferable to use relative saturation reduction. The proposed method for selecting cereal seedlings for drought resistance is based on a relative turgidity index (Kumar et al. 1994). The index of relative water content of leaves to determine the water status is proper than the parameter of water relation, because the relative moisture content of the leaf is related to the cell volume and may better express the balance between leaf water supply and evapotranspiration (Sinclair 1985). In sucrose-transmitting species, sucrose and hexose levels increase as the target level decreases with water stress (Vyas et al. 1985). The increase in sucrose and hexose levels appears to be due to increased starch hydrolysis and sucrose synthesis. Accumulation of sucrose and hexose is performed to play the role of osmotic regulation in these species (Vyas et al. 1985). In drought conditions, stomatal conductance and leaf photosynthesis also decreased (Saghafi et al. 2020). Water stress does not affect all aspects of plant development equally. Some processes are very sensitive to increasing drought stress, while other processes are less affected by water stress (Nayyar and Gupta 2006). Decrease in relative humidity in plants causes morphologic changes in them, such as reducing the number and size of leaves and finally stopping leaf growth and falling of them, reducing the number of nodes.

Reduction of Intercellular Space during Wilting

If the cells are in a state of wilting (in some organs), due to reabsorption of water and increase of cell water content, it increases the intercellular space, and the contact between the cells decreases. Conversely, if the intercellular water content decreases, the released cell walls collapse and exit. Water flows from the intercellular space. Therefore, reducing the amount of tissue water by about 30% in tobacco reduces the gas flow to the tissues (Kirda 2002).

In corn leaves, the decrease in gas volume is more than the decrease in total leaf volume, which coincided with a decrease in tissue water in the scale of 0.6–0.65 of the maximum amount of water (Sharp et al. 1994). Intercellular space in sunflower leaves decreased from 0% to 50% water reduction interval (Sharp et al. 1994). Citrus leaves have strong surfaces that prevent it from shrinking and reducing the intercellular space. Reducing the intercellular space reduces or eliminates the gas flow path in the leaf and also reduces the diffusion flow in the intercellular liquid phase. It is comparable to aperture resistance (Kirda 2002). As the leaf water potential decreases until it reaches a critical or threshold of water potential, the changes in stomatal resistance are small, and at a value below the threshold, the stomatal resistance decreases significantly (Black and Ong 2000). Sharp et al. (1994) showed that the potential of the water critical point for stomatal closure varies in genotypes of plant species and cultivars, but there are less differences between cultivars. Also, the amount of potential of leaf water in which the stomata closes varies according to the position, leaf, shoot, age, and the rate of stress progression (Kaiser et al. 1985).

The Effect of Drought Stress on Photosynthesis

The most important physiologic effect of water stress is to limit photosynthesis. When plants are exposed to water stress, often photosynthesis is not the first reaction to be affected. As drought stress increases, photosynthesis decreases significantly and usually reaches zero at higher stress levels. Water scarcity reduces photosynthesis by reducing leaf area, closing stomata, and reducing carbon fixation efficiency. Decreased leaf area due to drought stress is an important cause of reduced crop yield. The decrease in photosynthesis in stress-affected plants is primarily attributed to the closure of stomata, which simultaneously prevents the fixation of CO_2 due to damage to the photosynthetic system (De Swaef and Steppe 2010). It has also been suggested that carbohydrate accumulation may reduce photosynthesis when growth is confined by drought stress (Gzik 1996). Water deficiency also affects the activity of enzymes that mediate photosynthesis in the dark. A number of researchers have reported that the activity of enzymes that mediate the photosynthesis in the dark is effective.

Some researchers have reported that activity of imported enzymes such as ribulose 1, carboxylase 5 bisphosphate, ribulose 5 phosphate kinase, and phosphopyruvate carboxylase was reduced due to drought stress. Also water potential less than 0.5 MPa reduces chlorophyll production and chlorophyll a to b ratio. Degradation of chlorophyll molecules is accelerated by drought stress. In maize, which is a C4, chlorophyll molecule degradation is much greater in mesophilic cells than in cells or vascular sheaths (Westage and Boyer 1985). Degradation of chlorophyll a and b molecules in chloroplasts and loss of light-receiving lamellae are affected by drought stress. Another study on sugar beets showed an increase in chlorophyll content. This researcher attributed this to the shrinkage of plant cells under drought stress conditions and greater accumulation of chlorophyll in these conditions (Mohammadian et al. 2003). Khafagi and El-Lawendy (1997) also reported an increase in chlorophyll content under drought stress in sugar beet. Due to the high differences between and within plant species, it is difficult to generalize the impact of drought stress on photosynthesis (Khafagi and El-Lawendy 1997).

The Effect of Drought Stress on Respiration

Lack of water, which is enough to close the stomata and reduce photosynthesis, usually reduces dark (true) breathing. But the decrease in dark breathing rate will be less than its amount in photosynthesis. By reducing the leaf water potential of sunflower from −0.4 to −1.8 MPa, the rate of leaf photosynthesis decreased by 70% while the reduction in respiration rate reached 33% (Gray 1984). Drought stress in sugar beet increased dark respiration and breakdown of starch into sucrose, and in general drought stress increased the sugar content in this plant (Ober 2001). In sunflower stalks of maize and soybeans, dark respiration decreased as much as photosynthesis meaning that the lack of water in dark respiration for stems decreases more than its amount in leaves. With the decrease of water potential in soybean, the amount of respiration also decreased, but in the range of −1.6 to −0.4 MPa, no decrease was assumed in respiration. This context demonstrates that enzyme complexes are relatively associated with wilting. Sometimes with reduction of plant's water content, an increase in respiration occurs; therefore it is possible that drought stress causes hydrolysis of starch and its conversion to sugars and provides more substrate for respiration (Ferus and Arkosiova 2001).

Drought stress in sunflower and beans has small effect on respiration, and photorespiration has less effect than gross photosynthesis. Therefore, the ratio of photorespiration to gross photosynthesis increases with drop of leaf water potential. Decreased activity of RuBp carboxylase shows up in parallel with pure photosynthesis, so it can be concluded that the relative increase in photorespiration is associated with the inhibition of the activity of enzymes in the Calvin cycle and reduces the photosynthetic efficiency due to drought stress (Ferus and Arkosiova 2001).

The Effect of Drought Stress on the Distribution of Photosynthetic Derivatives

In some cases, drought stress can change the pattern of photosynthetic material allocation. If drought stress occurs in the filling stage of soybean, by reducing grain formation in the final stages of grain growth and by transferring derivatives, reduction of grain size happens. Also, reducing leaf size and change in root-to-system ratio indicates a change in derivative distribution pattern (Sinclair 1985). The effect of water deficiency on the distribution of photosynthetic materials in different plant organs depends on plants' age, stress intensity, occurrence or non-occurrence of stress in earlier stages of growth, and degree of stress sensitivity in different organs, for example, the dwarf species of lolium temulentum is more sensitive to water deficiency and the pathway of photosynthetic derivatives to young leaves, pods, and roots. While under favorable conditions, the pathway of photosynthetic derivatives is intended to meet the reproductive growth needs of the plant, but in conditions of water scarcity, the current photosynthetic material decreases and the proportion of stored material that is transferred to the seeds increases (Black and Ong 2000).

Mohammadian et al. (2003) announced that in plants under drought stress conditions, the transfer of photosynthetic substances is affected and causes leaf saturation by these substances, which may lead to confinement of photosynthesis.

Drought Stress Effects on Metabolism and Accumulation of Sugars

Drought stress increases respiration and decreases photosynthesis, thereby reducing vegetative growth and reducing the utilization of carbohydrates. Dehydration and high temperatures during the growing season increase sugar in the roots and increase the impurities of sugar beet roots, especially nitrogenous compounds (Kumar et al. 1994). Increased amylase activity in stressed leaves as a result of starch hydrolysis is intensified, and the concentration of soluble sugars is increased. The conversion of starch to sugar is often correlated, but there have been reports of no close association between decreased polysaccharides and increased simple sugars (Hussein et al. 2008). If the plants are exposed to drought stress frequently, the plant will suffer from hunger related to lack of carbohydrates in parallel with stomatal closure. The conversion of structural sugars and polysaccharides to soluble sugars regulates the osmotic pressure and thus counteracts cell water and loss of turgor continuation (Ferus and Arkosiova 2001). Carbohydrates and starch levels in cotton plants during the wilting point were one-half and one-third of plants in optimal conditions. Decreasing the leaf water potential in soybeans increases soluble sugars and decreases insoluble sugars, including soluble sugars which showed an increase in glucose and fructose in leaf water potential of −2.5 MPa; the amount of soluble sugars was doubled compared to the control (Ferus and Arkosiova 2001). Ferus and Arkosiova (2001) experimented the effect of drought stress on sesame and measured various factors and concluded that the amount of soluble sugars in stressed plants increased compared to the control sample.

Drought Stress and Protein Breakdown

The effect of wilting on changes in protein structure has been studied by scientists. In tobacco and sunflower, due to drought stress, in the lower leaves, proteins were converted to asparagine and glutamine and then transferred to younger leaves, in which the synthesis of new proteins was used again. The leaves also accelerate the aging of the leaves. Under the drought stress effects, the plant copes with the stress at cellular level, which results in the consumption of materials and high energy, which is spent on cell construction under optimal conditions. On the other hand, a decrease in intercellular energy, which is mostly used as ATP in the vital activities of the plant, and decreased synthesis of essential proteins are also two of the cases that are affected by environmental stresses. Barlow et al. (1977) reported that a delay for 3 hours caused a 40% reduction in intercellular ATP and increased free amino acids by about 20%. Perhaps this phenomenon could be interpreted as stress causing free amino acid synthesis. Proline is one of the amino acids that accumulate in a number of laboratory and field products in response to drought stress in plant leaves. Under stress, the amount of proline may reach 40 to 100 times the initial level (Aspinall and

Paleg 1981). Aspinall and Paleg (1981) believe that the process of proline accumulation in drought-sensitive cultivars is faster than drought-resistant cultivars, and this can be attributed to the difference in the amount of reduced leaf water potential. Accordingly, Moustafa-Farag et al. (2020) suggested that proline accumulation may be a sign of a stress adaptation mechanism, but there is insufficient evidence to support this theory. Proline accumulation in cells has been discussed to regulate osmotic pressure, reduce cell water loss, maintain turgidity and also solubility of different proteins, conserve albumin, and prevent their breakdown. On the other hand, water stress causes the accumulation of toxins such as ammonium ions and proline and prevents the destructive effects of this toxic substance on metabolism. Most reported cases of proline accumulation are as follows: (1) through its synthesis from other amino acids such as glutamic acid and arginine, (2) reduction of oxidation, and (3) participation of these amino acids in the process of protein synthesis in dry conditions and vice versa. Proline is oxidized to other amino acids such as glutamic acid. Gzik (1996) reported an increase in proline and free amino acid composition in sugar beet. Moustafa-Farag et al. (2020) showed that in rye under drought stress, total amino acids are released to a lesser extent than proteins. With the exception of proline, which is more likely to be synthetized from other amino acids, some amino acids were initially increased in sunflower under drought stress, but proline accumulated only under severe stress. Sharp et al. (1994) showed that in low water potential conditions, in the end of the maize root and in the elongation zone, the amount of proline reaches ten times per unit length compared to their non-growth stress conditions; the increase in proline plays an essential role in osmotic regulation. And it continues to have elongated roots. Vyas et al. (1985) experimented with drought in sesame and reported that stress increased proline. Aspinall and Paleg (1981) stated that a positive correlation between proline accumulation and drought resistance could not be used as an indicator of drought resistance in grain screening in breeding programs. The differences in the results are not only due to the plant species but also due to the amount of wilting, leaf age to be measured, and other conditions. Betaine accumulation, like proline, has been reported in association with drought stress. According to the research done by Hanson et al. (1977), the accumulation of betaine resulting from its new synthesis is from two organic precursors during drought stress. Under drought stress, barley leaves accumulate betaine at a rate of 200 nmol per square decimeter of leaf area per day. Another important effect of drought stress is photic inhibition due to the destruction of photosystem 2 in chloroplasts (Nayyar and Gupta 2006). The photic inhibition obtained by drought stress or other stresses is detected by chlorophyll fluorescence, which we will discuss in detail in the next section.

Drought Stress Effects on Abscisic Acid Hormone

One of the first effects of wilting in mesophytic plants is a decrease in growth inhibitors. Therefore, under the influence of drought stress, the synthesis and accumulation of ABA in the leaves and roots of stressed plants occur (Sharp et al. 1994). Abscisic acid is a growth inhibitor whose wilting conditions increase its concentration and accelerate its synthesis rate when the isolated leaves of cotton, chickpea, and bean are exposed to air, and up to 9% heavier, water is lost and withered (for 4 years at 22 °C). ABA was synthesized and accumulated in the dark (Agrawal and Dadlani 1994). The ability of ABA synthesis in the response of plant organs to wilting in leaves and young organs is higher than old organs. ABA enables the recovery of cell turgor by controlling the opening and closure of stomata, and thus the application of ABA in wheat and barley under stress has increased water consumption efficiency (Peña-Valdivia et al. 2010). De Swaef and Steppe (2010) reported an increase in abscisic acid concentration under drought stress conditions in sugar beet. Some of the effects of drought on plant growth are explained by the effects of abscisic acid concentrations. The general response of plants to high concentrations of abscisic acid is to reduce stem growth, which is more sensitive to increasing concentrations of abscisic acid than to decrease the turgor. The effects of abscisic acid on root responses to water stress are not well understood.

Conclusion

Crop susceptibility to drought depends on the conditions and type of the plant. For instance, most annual plants that produce seeds are sensitive to drought in the early stages of flowering. Also, in plants with unlimited growth, drought has less effect on seed production. With the onset of drought stress, root growth rate decreases; however, root growth is less affected than the growth of the aerial parts of the plant. The amount of damage to plants depends on physiologic age, amount of water stress, length of stress period, and plant species. In general, organs of the plant that grow rapidly in the event of stress are damaged more than other organs. In drought conditions, stomatal conductance and leaf photosynthesis are also reduced. Water stress does not affect all aspects of plant growth equally, and some processes are very sensitive to increased water stress, while other processes are less affected by water stress. The most critical effect of water stress is the limitation of photosynthesis, and drought stress sometimes changes the pattern of photosynthetic material allocation. Drought stress increases respiration and decreases photosynthesis, thereby reducing vegetative growth and reducing the utilization of carbohydrates. Decreased synthesis of essential proteins and accumulation of ABA in the leaves and roots of plants are also some of the cases that are affected by environmental stresses.

References

Agrawal P, Dadlani M (1994) Techniques in seed science and technology. New Delhi J 42:13–20

Aspinall D, Paleg L (1981) Proline accumulation: physiological aspects. Academic Press, Sydney, p 205

Barlow E, Boersma L, Young L (1977) Photosynthesis, transpiration and leaf elongation in corn seedling at suboptimal soil temperatures. Agron J 69(1):95–100

Black C, Ong C (2000) Utilization of light and water in tropical agriculture. Agric For Meteorol 104:25–47

Chołuj D, Karwowska R, Ciszewska A, Jasińska M (2008) Influence of long-term drought stress on osmolyte accumulation in sugar beet (*Beta vulgaris* L.) plants. Acta Physiol Plant 30(5):679

De Micco V, Aronne G (2012) Morpho-anatomical traits for plant adaptation to drought. In: Aroca R (ed) Plant responses to drought stress. Springer, Berlin, Heidelberg. https://doi.org/10.1007/978-3-642-32653-0_2

De Swaef T, Steppe K (2010) Linking stem diameter variations to sap flow, turgor and water potential in tomato. Funct Plant Biol 37:429

Doorenbos J, Kassam A (1979) Yield response to water food and agriculture organization of the united nation No. 33

Ferus P, Arkosiova M (2001) Variability of chlorophyll content under fluctuating environment. Acta Fytotechn Zootechn 4(2):123–125

Firoozabadi M, Abdollahian-Noghabi M, Rahimzadeh F, Moghadam M, Parsaeyan M (2003) Effects of different levels of continuous water stress on the yield quality of three sugar beet lines. Sugar Beet J Iran 19:133–142. (In Persian)

Ghassemi S, Farhangi-Abriz S, Faegi-Analou R, Ghorbanpour M, Asgari Lajayer B (2018) Monitoring cell energy, physiological functions and grain yield in field-grown mung bean exposed to exogenously applied polyamines under drought stress. J Soil Sci Plant Nutr 18(4): 1108–1125

Gray D (1984) The role of fluid drilling in plantestablishment. Asp Appl Biol 7:153–172

Gzik A (1996) Accumulation of proline and patterns of alpha amino acids in sugar beet plant in response to osmotic. Water and salt stress. Environ Exp Bot 36:29–30

Hanson A, Nelson C, Everson E (1977) Evaluation of tree proline accumulation as an index of drought resistance using two contrasting barely cultivars. Crop Sci 17:720–726

Hayat F, Ahmed MA, Zarebanadkouki M, Javaux M, Cai G, Carminati A (2020) Transpiration reduction in maize (*Zea mays* L.) in response to soil drying. Frontiers. Plant Sci 10(1695)

Heshmat K, Asgari Lajayer B, Shakiba MR, Astatkie T (2020) Assessment of physiological traits of common bean cultivars in response to water stress and molybdenum levels. J Plant Nutr. https://doi.org/10.1080/01904167.2020.1822395

Hussein MM, Kassab OM, Ellil AA (2008) Evaluating water stress influence on growth and photosynthetic pigments of two sugar beet varieties. Res J Agric Biol Sci 4(6):936–941

Kaiser K, Guff D, Outlaw W (1985) Sugar content of leaves of desiccation-sensitive and desiccation-tolerant plants. Naturzoissenschaften 72:608–609

Khafagi OMA, El-Lawendy WI (1997) Effect of different irrigation intervals on sugar beet growth, plant water relations and photosynthetic pigments. Ann Agric Sci Moshtohor 35(1):305–319

Khoshmanzar E, Aliasgharzad N, Neyshabouri MR, Khoshru B, Arzanlou M, Asgari Lajayer B (2019) Effects of Trichoderma isolates on tomato growth and inducing its tolerance to water deficit stress. Int J Environ Sci Technol 17:869–878

Khoshru B, Moharramnejad S, Hosseinzadeh Gharajeh N, Asgari Lajayer B, Ghorbanpour M (2020) Plant microbiome and its important in stressful agriculture. In: Varma A, Tripathi S, Prasad R (eds) Plant microbiome paradigm. Springer Nature, Switzerland AG. https://doi.org/10.1007/978-3-030-50395-6_2

Kirda C (2002) Deficit irrigation practices: deficit irrigation scheduling based on plant growth stages showing water stress tolerance. FAO. http://www.fao.org/docrep/004/Y3655E/Y3655E00.htm

Kumar A, Singh DP, Singh P (1994) Influence of water stress on photosynthesis, transpiration, water-use efficiency and yield of *Brassica juncea* L. Field Crop Res 37(2):95–101

Levitt J (1980) Response of plant environmental stress, vol 2. Academic Press, New York, pp 30–42

Mao W, Zhu Y, Wu J, Ye M, Yang J (2020) Modelling the salt accumulation and leaching processes in arid agricultural areas with a new mass balance model. J Hydrol 591:125329

Mohammadian R, Rahimian H, Moghaddam M, Sadeghian (2003) The effects of stress drought on chlorophyll fluorescence and content of chlorophyll. Pak J Biol Sci 6(20):1763–1769

Moustafa-Farag M, Mahmoud A, Arnao MB, Sheteiwy MS, Dafea M, Soltan M, Elkelish A (2020) Hasanuzzaman, M; Ai, S. Melatonin-induced water stress tolerance in plants: recent advances. Antioxidants, 9, 809

Nayyar H, Gupta D (2006) Differential sensitivity of C_3 and C_4 plants to water deficit stress: association with oxidative stress and antioxidants. Environ Exp Bot 58(1–3):106–113

Ober E (2001) The search for drought tolerance in sugar beet. Br Sugar Beet Rev 69(1):40–43

Peña-Valdivia CB, Sánchez-Urdaneta AB, Meza RJ, Juárez Muñoz J, García-Nava R, Celis VR (2010) Anatomical root variations in response to water deficit: wild and domesticated common bean (*Phaseolus vulgaris* L.). Biol Res 43(4):417–427

Saghafi D, Asgari Lajayer B, Ghorbanpour M (2020) Engineering bacterial ACC deaminase for improving plant productivity under stressful conditions. In: Sharma V, Salwan R, Al-Ani LKT (eds) Molecular aspects of plant beneficial microbes in agriculture. Elsevier Inc. https://doi.org/10.1016/B978-0-12-818469-1.00022-5

Sharp R, Wu Y, Voetberg G, Saab I, Lenobel M (1994) Confirmation that ABA accumulation is required for maize primary root elongation at low water potentials. J Exp Bot 45:1743–1751

Sinclair T (1985) Water relations of field grown soybean under drought stress. Crop Sci 26:993–998

Vyas S, Kathuj S, Garg K (1985) Performance and metabolic alternations in sesamum indicum under different intensities of water stress. Ann Bot 56(3):323–331

Westage M, Boyer J (1985) Osmotic adjustment and the inhibition of leaf, root, stem and silk growth at low water potentials in maize. Planta 164:459–540

Bioremediation-Based Microorganisms to Break Down Pollutants Decelerate Due to Climate Change

Savita Mishra, Sandhya Hora, Ritesh Mishra, and Pawan Kumar Kanaujia

Abstract Global climate change is one of the most worrisome political and scientific issues from the last hundred years, which is being aggravated by human-mediated increases in greenhouse gases. The need of the hour is to understand regulation of carbon exchanges between soil, oceans, and atmosphere. Moreover, response of ecosystem through climate-ecosystem feedbacks could amplify or reduce provincial and universal climate change. Bioremediation is a general process to treat contaminated water, soil, oil spills, and subsurface material with the help of microorganisms to degrade the target pollutants. Microorganisms involved play vital roles in the transformation of pollutants to biodegradable substance. Microorganisms use oxidation-reduction process to donate or accept electron from pollutant substance and break down into nontoxic products. In both these approaches, additional nutrients, minerals, vitamins, and pH buffers are used to optimize conditions for the microorganisms. The extreme diverse metabolic activities cause breakdown of these contaminants. However, climate change has significant effect on microbial activity. Pollutants exhibit cellular toxicity that leads to microbial growth inhibition. It has been known that microbial activity is accountable for the cycling of crucial elements in the environment. Impairment in growth of microbe can have negative effects on microbial activity as well as in the bioremediation process. In the present chapter, we have discussed the consequence of pollutants on microbial activity in bioremediation process.

Savita Mishra and Sandhya Hora contributed equally to this work.

S. Mishra
Amity Institute of Biotechnology, Amity University, Noida, India

S. Hora
National Institute of Biologicals, Noida, India

R. Mishra
Institute of Plant Sciences and Genetics in Agriculture, The Hebrew University of Jerusalem, Rehovot, Israel

P. K. Kanaujia (✉)
HIMT Group of Institutions, Greater Noida, India
e-mail: pawan.kumar@himt.ac.in

© The Author(s), under exclusive license to Springer Nature Singapore Pte Ltd. 2022
A. Vaishnav et al. (eds.), *Plant Stress Mitigators*,
https://doi.org/10.1007/978-981-16-7759-5_7

Keywords Soil pollution: Microbial activity · Pesticides · High temperature · Bioremediation

Introduction

The occurrence of harmful substance in our surrounding is termed as "pollutants." These toxic elements, molecules, or particles have adverse affects on humans as well as on our environment (Borah et al. 2020). At present, major environmental processes are constantly experiencing pollution and climate change-related stresses, hence compromising the whole ecosystem. Indeed, chemical waste, poor disposal industrial waste, and accidental leakage have caused severe environmental concerns in the world (Anderson et al. 2018). Plastic usage has resulted in potentially harmful contaminants in the environment, with 20–42% of total global plastics now stored on land and predicted to biodegrade slowly (Pignattelli et al. 2021). Temperature increase has been observed with changing patterns of the seasonal cycle, intensity, and extremes of natural disasters such as drought, flooding, and cyclones, also likely increased from the last century by 2–18 °C. Anthropogenic activity produces greenhouse gases (GHGs) such as CO_2, CH_4, N_2O, and chlorofluorocarbons, which cause the greenhouse effect (CFCs) (Shukla et al. 2019). In the previous three decades, global climate-related disasters have escalated significantly, resulting in significant economic losses of agricultural supplies. From 1980 to 1990, there were 149 disasters, compared to 332 from 2004 to 2014 (Dastagir 2019).

Microorganisms like bacteria, fungi, and plants have major involvement in the maintenance of ecological cycle in the ecosystem (Deveau et al. 2018). Bacteria have ubiquitous capabilities, and when it gets nutrients and favorable conditions for metabolism and cell division, it grows enormously. However, biological response to pollutants varies due to different organisms' different sensitivities toward the same chemical, and their sensitivity is completely dependent on the time and intensity of exposure to the pollutant (D'Souza et al. 2018). Biological effects calculated at or near a contaminated place tend to yield extra accurate and relevant information about impact of pollution. Furthermore, the microorganisms that exist in these places can be used as biomarkers for contamination (Li et al. 2019). Bioremediation, which depend on the utilization of microorganisms to break down contaminants, has the potential to significantly reduce pollution from the ecosystem. Remediation of pollutants is termed as utilization of microorganisms to reduce pollutants to non-hazardous or to less hazardous forms (Kumar and Sharma 2019).

Soil Pollution

Contamination of soil due to unnatural toxic substance is a serious environmental concern. Pollution occurs in soil due to the intentional (waste dumping) or unintentional application of a single chemical or complex mixes of chemicals, such as municipal waste, industrial effluents, sewage sludge, and others (Ndiaye et al. 2000) (Fig. 1). The soil matrix is complex and material characterized by a variety of parameters such as temperature, presence of metallic/nonmetallic ions, pH, granulometric composition, and oxido-reductive compounds. It also consists of several organic substances such as pesticides, PCBs (polychlorinated biphenyls), humic acids, PAHs etc. (León-Santiesteban and Rodríguez-Vázquez 2017). Abovementioned features affect the microorganism's activity and soil enzymes. Moreover, pesticides create a serious environmental issue due to time-consuming degradation. Most of the time, the pesticides devolve chemical interaction with inorganic and organic compounds, which generate cytotoxic effects both on humans and on microorganisms (Verma et al. 2014).

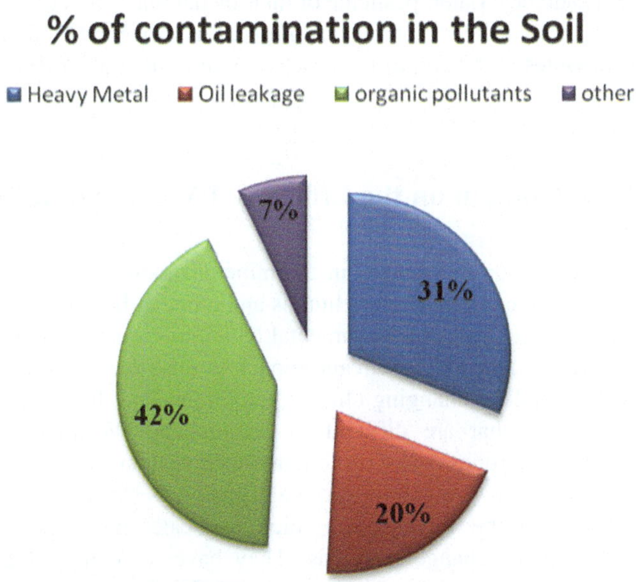

Fig. 1 Types of soil pollutants (Stojić and Prokić 2019)

Bioremediation of Soil Pollution

Soil microorganisms play a key role in the flow of elements from soil to ecosystem. Decomposition of organic matter and inorganic residues through these microbes regulates various nutrient as well as carbon cycles (Mahdi et al. 2017). Besides, microorganisms are responsible for degradation and transformation of synthetic organic compounds and waste materials. Microorganisms also influence the physical properties of soil (Cardelli et al. 2017). Hence, they are good indicator of soil health and give an untimely signal of soil quality as well as early warning of soil degradations (Wang et al. 2020).

Herbicides impair bacterial metabolism, resulting in reduction in soil disruption and enzymatic activity of biological nitrogen binding (Wołejko et al. 2020). Herbicides, such as Roundup, prevent growth of non-symbiotic diazotrophs for up to 30 days after they are introduced into the soil environment. For instance, some bacterial *species* including *Achromobacter*, *Streptomyces*, and *Arthrobacter* use glyphosate to break down products such as carbon, nitrogen, and phosphorus for their activity and growth (Zhan et al. 2018; Sviridov et al. 2015). Microbes are first converted into more water soluble by oxidation, reduction, or hydrolysis during metabolism of pesticides. Later, pesticide or their metabolites are interacting with an amino acid or sugar to make it more water soluble. It is then converted into nontoxic secondary metabolites at the end of the process (Van Eerd et al. 2003) (Table 1).

Effect of Soil Pollution on the Efficacy of Microbial Activity

Different microbes in soil system aid in the mineralization process as well as the improvement in organic matter such as humus and other carbon-related compounds (Bardgett et al. 2008). Various environmental conditions, such as temperature and weather, moisture, etc., have a significant impact on this process. Microbes' nature also changes as a result of changing climate conditions, and they create extremely intricate relationships that are difficult to comprehend (Wołejko et al. 2020; Mukherjee 2019). Microbes adapt to climate-mediated stress by shifting their resource allocation from growth to survival strategies. However, very few are known in relation to the fundamental microbial-scale mechanisms that drive ecosystem-scale climate change reactions. They have to regulate breakdown of organic materials in soil and its release into the atmosphere as carbon dioxide (CO_2) (Schimel et al. 2007; Jansson and Hofmockel 2020). Mostly, utilizing the pesticides can destroy some microorganisms, subsequently creating condition to grow other microorganisms (Sharma et al. 2020). Changes in soil moisture and temperature caused due to climate change can have a significant impact on essential soil processes including organic compound decomposition and nutrient cycling, which are partly or completely dependent on the microbe's soil activity (Burns et al. 2013). Alteration in humidity, temperature, freeze-thaw, and wet-dry cycles

Table 1 List of source of soil pollution, related active pollutants, and bioremediation microbes

Source of soil pollution	Pollutants	Microorganism for bioremediation	Ref.
Petroleum derivatives	Mixture of hydrocarbons and nitrogen, sulfur, and organometallic compounds, oxygenated organic compounds, sulfur gas, and heavy metals	*P. alcaligenes*, *P. mendocina*, *P. putida*, *P. veronii*, *Achromobacter*, *Flavobacterium*, *Acinetobacter*	Gospodarek et al. (2019), Kuppusamy et al. (2020), da Silva et al. (2020), Abatenh et al. (2017)
Urban source	Toxic metals such as aluminum (Al), cadmium (cd), manganese (Mn), copper (cu), nickel (Ni), lead (Pb), and zinc (Zn) and PAHs (polycyclic aromatic hydrocarbons)	*Coprinellus radians*, *Rhizobium*	Aranda et al. (2010), Karigar and Rao (2011), Abdu et al. (2017)
Agrochemical source	Insecticides, herbicides, fungicides, and farm fuel spill	*Streptomyces hygroscopicus* and *Streptomyces viridochromogenes* (*Sorghum bicolor (L.) Moench.*), (*Hordeum vulgare L.*) *Bacillus*, *Pseudomonas*, *Arthrobacter*, and *Micrococcus*	Wołejko et al. (2020), Parte et al. (2017), Malik et al. (2017)
Oil leakage	Crude oil, hexane, toluene, n-hexane, paraffin oil, vegetable oil, motor oil kerosene	*Bacillus*, *Micrococcus*, V*ibrio*, *Moraxella*, *Flavobacterium*, *Achromobacter spanius*, *Pseudomonas putida*, and *Pseudomonas aeruginosa*	Yanxun et al. (2011), Wang et al. (2015), Kora (2018), Stancu (2020)
Defense sites	Chlorinated solvents [perchloroethylene (PCE), trichloroethene (TCE), chloroform, etc.] and metals (Cd, Cu, etc.)	*Mycobacterium aurum*, *Dehalococcoides*, *Dehalobacter*, and *Dehalogenimonas* spp.	Vogel (2017), Ramakrishnan et al. (2011), Baldwin et al. (2017)
Cattle/sheep dip sites	Dichlorodiphenyltrichloroethane (DDT), arsenic, and other insecticides	*Bacillus*, *Achromobacter*, *Brevundimonas*, *Microbacterium*, *Ochrobactrum*, *Pseudomonas*, *Comamonas*, *Stenotrophomonas*	Mesa et al. (2017), Mallick et al. (2014), Ramakrishnan et al. (2011), Steinfeld et al. (2015)

leads to modification in overall development and composition of soil microorganisms. Furthermore, climate induced changes in ecological parameters as well as both the organization and physiology of soil microbes (Alkorta et al. 2017; Zak et al. 2011).

Impact of Pesticides on Microbial Activity

Pesticide's active components destroy the soil environment, impacting microorganisms that live there. As a result, they may impair the proper sequence of metabolic processes in soil biogeochemical cycles. Organochlorides are one of the most dangerous compounds among all pesticides due to their volatile nature as they have the ability to volatilize from soil to atmosphere and be transported from one place to another (Verma et al. 2014; Łozowicka et al. 2016). Due to their movable quality, these types of pollutants create new contamination site and affect microbial activity (Łozowicka et al. 2017). Recent study showed that, in most situations, using fungicide such as chlorothalonil, captan, and benomyl decreases the activity of some fungi in soil, leaving niches for other bacteria to thrive in (Burns et al. 2013). Additionally, pollutants also have a major impact on soil enzymes, which are important catalysts in modifying microorganism habitat in soil (Łozowicka et al. 2017; Jian et al. 2016). Various metabolites have been produced by pesticides due to several biochemical and physiochemical processes. Some are lethal, whereas others have no effect on microorganism. However, most of the times these metabolites obliterate some microbes subsequently creating niches inhabited by other microorganisms.

A research conducted by Pal et al. (2008) elucidates that fungicides like pencycuron, dithianon, or prochloraz can decrease the quantity of nonpathogenic saprophytic soil fungus (Pal et al. 2008; Tejada et al. 2011). Similarly, a significant microbial biomass decreases with the use of tebuconazole and pencycuron which are prominent pesticides in rice crops (Muñoz-Leoz et al. 2011). We have listed some important pesticides and their effect of microbial activity (Table 2).

Impact of Increased Soil Temperature on Microbial Activity

Evaluation of different ecological conditions highlighted that increases in aridity may result in a decline in the genetic potential and stability of the soil microbiome (Jansson and Hofmockel 2020). It has been observed that higher soil temperatures lead to an increase in soil metabolic activity and organic matter decomposition, resulting in a higher CO_2 flow, but no increase in microbial biomass (Melillo et al. 2017). However, some soil fungi has shown increased growth with the boost of temperature. Several field researches have examined the soil microbe response with respect to short- and long-term rise in the temperature (Guhr et al. 2015). Study on

Table 2 Effect of pesticides on soil microbial activity (adapted from Wołejko et al. 2020, Alkorta et al. 2017)

Pesticides	Affected microbes	Influence on microbiological activity
Benzoylurea	Different species of *Pseudomonas* and *Paracoccus*	Dimilin, flufenoxuron, and insect growth regulators (IGR) such as novaluron which inhibits microbes in acidic sand soil
Benzimidazole (glyoxalin)	*Rhizobacteria* species such as *Bacillus pumilus* SE34, different species of *Rhodococcus* like *R. qingshengii*	Methyl-2-benzimidazole carbamate (MBC) reduces soil microbiomes
Neonicotinoid	*Pseudomonas* sp., various species of *Bacillus* like *B. subtilis* FZB24, B. *amyloliquefaciens* IN937a, *B. pumilus* SE34, and some species of *Rhizobia*	The total viable number of bacteria is not affected by thiamethoxam or imidacloprid
Organophosphates	Different species of *Pseudomonas* like *P. diminuta* and *P. aeruginosa*, *Burkholderia cepacia*	• Dimpylate reduces growth of *Proteus vulgaris*, a bacterium that produces urease • Phorate, chlorpyrifos, and azodrin have no effect on the total number of bacterial viability. However, chlorpyrifos alters cell shape, resulting in formation of pleomorphic cells • The usage of methamidophos, commonly known as monitor and Curacron, reduces microbial biomass by 41–83% • Ethion also has a negative impact on soil microbes • Phorate reduces the total number of bacteria as well as nitrogen-fixing bacteria, and malathion increases the amount of denitrifying bacteria. On the other hand, populations of nitrifying bacteria and fungi were unaffected
Carbamates	Some species of *Sphingomonas*, *Achromobacter*, *Arthrobacter*, *Flavobacterium*, and *Pseudomonas* *B. pumilus* SE34, *Stenotrophomonas maltophilia* M1	Carbofuran amplifies the number of bacteria in the soil
Aryloxyphenoxypropionate	*Bacteroidetes*, *Proteobacteria* and *Pseudomonas azotoformans* QDZ-1	It has been seen that microorganism biomass, nitrogen, and carbon levels were considerably higher in fenoxaprop-treated soils than in control soils

(continued)

Table 2 (continued)

Pesticides	Affected microbes	Influence on microbiological activity
Urea	*Jerusalem artichoke*, *Sterigmatocystis nigra* *Sphingomonas* strain SRS2, *Rhizoctonia solani*, *Variovorax paradoxus*, *Rhizopus japonicus* *Pseudomonas* sp., *Cunninghamella elegans*, *Mortierella isabellina*, *Delftia acidovorans*, *Talaromyces wortmanii*, *Aspergillus Niger*, *Bacillus sphaericus*	• *Bradyrhizobium* sp. is negatively affected by isoproturon, which boosts bacterial population but lowers actinomycetes and fungi • 16S rDNA polymerase chain reaction (PCR)-denaturing gradient gel electrophoresis (DGGE) revealed that isoproturon degradation was linked to *Sphingomonas* species proliferation
Triazine	*Raoultella planticola*, *Paenarthrobacter nicotinovorans*, *Phanerochaete chrysosporium* (a white-rot fungi), *Cryptococcus laurentii*	*Bradyrhizobium* species is degraded by Gesaprim Total nitrogenase activity is reduced by Gesatop and prometryn

the Harvard forest highlighted that long-term warming resulted in depletion of soil organic compounds and related reductions in microbial biomass, which leads to long-term negative effects for soil sustainability (Romero-Olivares et al. 2017; Schindlbacher et al. 2011). Another study conducted by Bradford et al. expressed that increased temperature often increased soil respiration, microbial biomass, and organic matter decomposition. However, all effects depleted with long-term worming. They have also hypothesized that increased microbial activity depletes labile soil carbon substrates, and trade-offs happen as microbiomes (i) acclimatize, (ii) alter in composition, (iii) restrict their biomass to respond to changes in environmental circumstances and substrate accessibility (Allison and Martiny 2008; Bradford 2013; Classen et al. 2015). A statistical analysis revealed that by the end of the twenty-first century, there is 95% chance that our earth will have warmed by more than 2 °C; in fact, likely range of global temperature increase is 2.0–4.9 °C with potentially increase in humanity but not in terms of bioremediation performance (Pachauri et al. 2014; Raftery et al. 2017). Some pollutants are proven to have enhanced toxicity with the rise of heat, which leads to degradation of soil processes and ecosystem services, resulting in lower soil quality and less bioremediation. Similarly, increased intensity and frequency of extreme weather events may increase the risk of pollutant mobilization resulting in the increase of the size of the area that requires remediation (Rohr et al. 2013; Landis et al. 2013).

Impact of Altered Soil pH on Microbial Activity

Soil pH has a significant influence on diversity and richness of soil microbiomes as it provides salinity. All factors such as variety of minerals, soil organic compounds, and redox conditions have direct impact on the habitats for microorganism (Fisher 1936; Osuolale et al. 2017). For instance, fire can change the soil pH; thus it can be a major factor for soil as well as microbial diversity. Researchers have observed that there was a significant increase in fungal diversity after fire break in boreal forests, most probably due to increased pH, although it decreased with time (Sun et al. 2015). The long-term functional impacts of these changes in soil microbial community are still unknown. A cross-biome study suggested that bulk soil compositions of desert soil and microbial communities can be separated from microbiomes in other soil habitats. Mostly arid soils have an alkaline pH, which can be the main influencing factor for microbial community composition (Fisher 1936; Fierer et al. 2012). A study conducted by Fierer et al. suggested that dry environment conditions created a distinct cluster when compared to other biomes with high pH (Delgado-Baquerizo et al. 2018). Biodegradation can take place at any pH, although in most aquatic and soil systems, pH ranging from 6.5 to 8.5 are ideal for biodegradation. It has been seen that most of the metabolic activities are carried out by microorganisms susceptible for slight change in pH (Abatenh et al. 2017).

Impact of Changed Soil Moisture Content on Microbial Activity

Soil microorganisms have developed various approaches such as osmoregulation, dormant condition, and synthesis of extracellular enzymes to manage drought-related stress (Barnard et al. 2013). Soil ecosystem is very sensitive for moisture content or water content which can limit microorganism activity and growth, and in addition the diffusion of nutrients and carbon substrates affects organic pollutant biodegradation; on the other hand, high water content might limit microorganism oxygen supply (Howard and Howard 1993). Understanding the microbial characteristics that confer drought tolerance in ecosystems could help forecast and manage ecosystem responses to climate change. Since they have the ability to retain function and become dormant under dry conditions, members of several bacterial species, such as *Actinobacteria*, have been found to concentrate on drought-affected soils (Bouskill et al. 2016; Naylor et al. 2017). The latest report suggested that some fungi contribute in maintenance of carbon and nitrogen cycle in low water content. As soil moisture levels rise, soil pores become water-filled and anaerobic, allowing methanogenesis and denitrification to occur, as well as the emission of N2O and CH4 greenhouse gases (Treseder et al. 2018). Drought ecosystems have moisture-limiting circumstances for microbial activity, resulting in a negative feedback on microbial metabolic activity and loss of carbon in soil due to microbial respiration.

Studies of forest ecosystems show that during dry seasons, litter isoenzyme diversity and phenol oxidase activity, as well as microbial biomass, decrease significantly (Di Nardo et al. 2004; Krivtsov et al. 2006).

Water Pollution

Freshwater, which accounts for only 2.7% of all water, is essential for human survival since it is used for residential, industrial, and agricultural purposes. Sources of water pollution can be both geogenic such as evapotranspiration and water-rock interactions and anthropogenic like industrial, agricultural, and domestic waste (Coyte et al. 2019; Madhav et al. 2020). These sources of water can be categorized into inorganic compounds (e.g. heavy metals), organic compounds andmicroorganims (Coelho et al. 2015). Despite the fact that industrial water consumption is small in contrast to agricultural water consumption, unregulated disposal of industrial waste on land and in surface water bodies renders water resources unsuitable for future purposes (Schweitzer and Noblet 2018). Any healthy water body has gradient of dissolved oxygen which decreases with increasing depth. The dissolved oxygen is needed by aquatic organisms for their cellular respiration. Microorganisms decompose dead organic matter to carbon dioxide (CO_2) and terminal electron acceptor to water (H_2O) and reduce oxygen (O_2) level (Kumar et al. 2011). Heavy metal pollution in the environment as a result of anthropogenic and industrial activity has caused irreversible damage to aquatic ecosystems. Deep-sea ecosystems encompass roughly two thirds of the world's surface (Herring 2001; Thurber et al. 2014). Oil spills are another kind of incident that makes the petroleum (organic) pollution becomes the main contamination in ocean (aquatic) environment. Microorganisms present in environments can degrade and transform petroleum pollutant in the water (April et al. 1998). The oceans can help to moderate the effects of global climate change by functioning as a CO_2 sink and absorbing the heat produced by greenhouse effect (Reid et al. 2009). However, change in pH, temperature, salinity, oxygen content, and patterns of particulate organic carbon (POC) which sink toward the bottom have a variety of consequences for the marine biota (Doney et al. 2012; Nagelkerken and Connell 2015).

Bioremediation of Water Pollution

Treatment of wastewater has evolved into a major social, technological, economic, and political issue. Every country's modern legislation imposes environmental regulations. The nature of contaminants in industrial wastewater, as well as their concentrations, is determined by their source. Industrial wastewater is often hotter than municipal or agricultural wastewater when it exits the units (Caicedo et al. 2019). Due to higher concentration of pollutants in the industrial waste, separation, transformation, and further decomposition must all be accomplished using specific methods. Generation of algae protoplasm in photosynthesis participates with the presence of carbon, phosphorus, and nitrogen which are the critical factors for growth of algae as well as the eutrophication of lake. Photoautotrophic microorganism *Phormidiumbohneri* can remove nitrogen and phosphorus in convinced environment (Sylvestre et al. 1996). The removal of nitrogen by microbes is generally accomplished with nitrification and denitrification, whereas the removal of phosphorus by microbes is through the process of absorbing phosphorus of *Aeromonas* on aerobic or anaerobic conditions. As a result, it is utilized extensively and provides effective way to eliminate the nitrogen and phosphorus in water (Cheng 2014). For the elimination of the majority of pollutants, biological treatment is extremely successful. Biodegradation techniques are essentially slower, do not allow for large degrees of elimination, and are not ideal for substances that are harmful to microorganisms, notwithstanding their success and cost-effectiveness (Khan and Malik 2018). Under microbial process, microorganisms used oil as an organic carbon source during bioremediation, resulting into breakdown of oil components to low molecular weight compounds. This technology accelerates natural biodegradation under optimized conditions such as pH, temperature, oxygen supply, water content, nutrients, and the existence or addition of suitable microbial population (Banat et al. 1991). According to the report, *Planctomycetes*, *Gemmatimonadetes*, *Proteobacteria*, *Bacteroidetes*, *Nitrospirae*, *Acidobacteria*, *Actinomycetes*, *Chloroflexi*, and *Flavobacterium* are among the bacteria found in deep-sea surface sediments (Danovaro et al. 2017). Activated sludges are complex microbial ecosystems made up of Bacteria, Archaea, Eukarya, and viruses. They are the most prevalent biological wastewater treatment application. Implementation of integrated wastewater systems with biological filters appears to be a potential move toward future wastewater treatment. In contrast, few studies on the effects of bio-filters on the microbial community have been published (Lukwambe et al. 2019). The bacterioplankton community plays an important role in aquatic food web by contributing to energy flow, circulating nutrients, and serving as bio-indicators of environmental factors in ecosystems (Labbate et al. 2016).

Influence of Abiotic Factors on Microbial Activity in Bioremediation of Water Pollutants

Temperature, pH, oxygen, carbon, and nutrition availability are all important factors in microorganism growth. The composition of urban and industrial wastewaters is heterogeneous, containing a variety of inorganic and organic contaminants. At certain concentrations, such chemicals can be harmful to microbial growth and limit their culture development. For example, the presence of 400 mg/L phenol can inhibit the growth of the microalgae *Chlorella vulgaris (Chlorophyta)* (Klekner and Kosaric 1992). CO_2 concentrations have been increasing exponentially over the last two decades, causing a 0.3–0.4 unit decrease in sea pH (Minich et al. 2018). Most of the macro-algae, including *Macrocystis pyrifera*, use both bicarbonate (HCO_3-) and CO_2 as inorganic carbon sources; therefore an increase in CO_2 will lower utilization of the less efficient bicarbonate pathway. Similarly, increased temperature causes a risk to macro-algae growth, particularly those which live near their maximum temperature tolerances. It has been seen that spore formation, germination, and recruitment are all negatively affected by temperatures above 18 °C (Minich et al. 2018; Harley et al. 2012; Gaitán-Espitia et al. 2014). Ocean acidification (change in pH) can reduce the thermal tolerance range of aquatic organisms, making them more susceptible to extreme temperatures and impairing their performance (Alma et al. 2020).

Influence of Biotic Factors on Microbial Activity in Bioremediation of Water Pollutants

The generation of hazardous chemicals, enzyme induction, symphony of microbial community, and the quantity of microorganisms can all influence the removal of contaminants from effluents (Pacheco et al. 2020). *Bacillus*, *Rhizobium*, *Flavobacterium*, *Sphingomonas*, and *Hyphomonas* are among the bacteria that have been found to be helpful for *Chlorella vulgaris*. By delivering fixed atmospheric nitrogen, *Bacillus pumilus* ES4 has been demonstrated to stimulate *Chlorella vulgaris* development (Hernandez et al. 2009). Many bacteria, on the other hand, have been reported to have a deleterious impact on algal growth. Algae-associated bacteria from taxa *Rhodobacteraceae*, *Flavobacteriaceae*, and *Saprospiraceae* have been linked to *Delisea pulchra* bleaching. It has been reported that *Microbacterium* sp. LB1 was found to be responsible for algal cell lysis and damage to *Choricystis* minor laboratory cultures, resulting in a dry weight drop up to 34% after 120 h of cultivation (Ivanova et al. 2014) (Table 3).

Conclusion

Organic and inorganic toxins have harmed the earth's ecology through deliberate and unintentional leakage, manufacture, and generation of debris. Microbes naturally offer lots opportunities for cleaning up polluted regions; hence microbial bioremediation of these contaminations has captivated the interest of researchers all over the world. Microorganisms also have a significant role in managing greenhouse gas emissions by participating in heterotrophic respiration (CO_2), denitrification (N_2O), and methanogenesis (CH_4). Human activities directly or indirectly affect the greenhouse gas emissions (mainly CH_4, N_2O, and CO_2), pollution (mainly eutrophication), and agriculture (mainly land use). Microorganisms, on the other

Table 3 Effect of biotic factors on bioremediation of water pollution

Microalga	Bioremediation	Interacting bacteria	Effect	Ref.
Beneficial bacteria				
Chlorella vulgaris	Textile wastewater	*Bacillus pumilus*	In N-free media, cell density increased by 150%	Lim et al. (2010), Xin and Xuefeng (2001)
Chlorella ellipsoidea	Heavy metal such as Zn^{2+} Cd^{2+} Ni^{2+} Cu^{2+}	*Brevundimonas* sp.	Algal cell density increased three times after 7 days	Xin and Xuefeng (2001)
Chlorella sorokiniana IAM	Wastewater and magnesium nickel removal from wastewater	*Microbacterium trichotecenolyticum*	Growth rate increased by 16%	Chen et al. (2020), Watanabe et al. (2005)
Unfavorable bacteria				
Delisea pulchra	Acid rock drainage remediation	*Rhodobacteraceae*, *Saprospiraceae*, and *Flavobacteriaceae*	Caused bleaching of the seaweed	Zozaya-Valdés et al. (2017), Brune and Bayer (2012)
Nannochloropsis oceanica IMET1	Sludge treatment plant of municipality sewage and petroleum	*Gram-negative bacteria: Alteromonas*, *Cytophaga*, *Flavobacterium*, *Pseudomonas*, *Saprospira*, *Vibrio*, and *Pseudoalteromonas*	Responsible for rot symptoms and gall in seaweed	Wang et al. (2012)
Chlorella vulgaris	Textile wastewater	*Pseudobodo* sp. *KD51*	50% decrease in chlorophyll content	Chen et al. (2014)
Dunaliella salina, *Platymonas subcordiformis*, and *Microcystis aeruginosa*	Bioremediation of nonylphenol (NP) water waste	*Pseudobodo* sp. *KD51*	Growth inhibition	Chen et al. (2014), Wang et al. (2019)

hand, provide significant potential for resolving human-caused issues through enhanced agricultural outputs, biofuel generation, and pollution cleanup. Knowledge of parameters influencing biological and physicochemical processes is essential for efficient bioremediation. Molecular biology tools are being developed to determine target-specific bacteria and their chemical degradation pathways, and there is still a potential to learn more about the natural bioremediation process. Future research prospects in microbial bioremediation offer enormous of chances for aspiring scientists. Climate change is expected to disrupt the earth's ecosystem as a whole, as it changes the climate and its accompanying repercussions, which can modify microbial populations. However, research indicates higher microbial activity with elevated CO_2 and temperature. Scientists predict that climate change will alter the earth's ecosystem as a whole. The findings could have implications for future research into climate change in nature. Climate change and microbial processes in nature will certainly be the interesting field of future research.

References

Abatenh E, Gizaw B, Tsegaye Z, Wassie M (2017) The role of microorganisms in bioremediation-A review. Open J Environ Biol 2(1):038–046

Abdu N, Abdullahi AA, Abdulkadir A (2017) Heavy metals and soil microbes. Environ Chem Lett 15(1):65–84

Alkorta I, Epelde L, Garbisu C (2017) Environmental parameters altered by climate change affect the activity of soil microorganisms involved in bioremediation. FEMS Microbiol Lett 364(19)

Allison SD, Martiny JBH (2008) Resistance, resilience, and redundancy in microbial communities. Proc Natl Acad Sci 105(Supplement 1):11512–11519

Alma L, Kram KE, Holtgrieve GW, Barbarino A, Fiamengo CJ, Padilla-Gamiño JL (2020) Ocean acidification and warming effects on the physiology, skeletal properties, and microbiome of the purple-hinge rock scallop. Comp Biochem Physiol A Mol Integr Physiol 240:110579

Anderson CM, Kissel KA, Field CB, Mach KJ (2018) Climate change mitigation, air pollution, and environmental justice in California. Environ Sci Technol 52(18):10829–10838

April TM, Abbott SP, Foght JM, Currah RS (1998) Degradation of hydrocarbons in crude oil by the ascomycete Pseudallescheria boydii (Microascaceae). Can J Microbiol 44(3):270–278

Aranda E, Ullrich R, Hofrichter M (2010) Conversion of polycyclic aromatic hydrocarbons, methyl naphthalenes and dibenzofuran by two fungal peroxygenases. Biodegradation 21(2):267–281

Baldwin BR, Taggart D, Chai Y, Wandor D, Biernacki AL, Sublette K et al (2017) Bioremediation management reduces mass discharge at a chlorinated DNAPL site. Groundwater Monitor Remediat 37(2):58–70

Banat IM, Samarah N, Murad M, Horne R, Banerjee S (1991) Biosurfactant production and use in oil tank clean-up. World J Microbiol Biotechnol 7(1):80–88

Bardgett RD, Freeman C, Ostle NJ (2008) Microbial contributions to climate change through carbon cycle feedbacks. ISME J 2(8):805–814

Barnard RL, Osborne CA, Firestone MK (2013) Responses of soil bacterial and fungal communities to extreme desiccation and rewetting. ISME J 7(11):2229–2241

Borah P, Kumar M, Devi P (2020) Types of inorganic pollutants: metals/metalloids, acids, and organic forms. Inorganic pollutants in water. Elsevier, Cambridge, MA. pp 17–31

Bouskill NJ, Wood TE, Baran R, Ye Z, Bowen BP, Lim H et al (2016) Belowground response to drought in a tropical forest soil. I. Changes in microbial functional potential and metabolism. Front Microbiol 7:525

Bradford MA (2013) Thermal adaptation of decomposer communities in warming soils. Front Microbiol 4:333

Brune KD, Bayer T (2012) Engineering microbial consortia to enhance biomining and bioremediation. Front Microbiol 3:203

Burns RG, DeForest JL, Marxsen J, Sinsabaugh RL, Stromberger ME, Wallenstein MD et al (2013) Soil enzymes in a changing environment: current knowledge and future directions. Soil Biol Biochem 58:216–234

Caicedo C, Rosenwinkel KH, Exner M, Verstraete W, Suchenwirth R, Hartemann P et al (2019) Legionella occurrence in municipal and industrial wastewater treatment plants and risks of reclaimed wastewater reuse. Water Res 149:21–34
Cardelli R, Vanni G, Marchini F, Saviozzi A (2017) Characterization and origin of organic and inorganic pollution in urban soils in Pisa (Tuscany, Italy). Environ Monit Assess 189(11):1–17
Chen Z, Lei X, Zhang B, Yang L, Zhang H, Zhang J et al (2014) First report of Pseudobodo sp, a new pathogen for a potential energy-producing algae: chlorella vulgaris cultures. PLoS One 9(3):e89571
Chen Z, Qiu S, Amadu AA, Shen Y, Wang L, Wu Z et al (2020) Simultaneous improvements on nutrient and mg recoveries of microalgal bioremediation for municipal wastewater and nickel laterite ore wastewater. Bioresour Technol 297:122517
Cheng J (2014) Bioremediation of contaminated water-based on various technologies
Classen AT, Sundqvist MK, Henning JA, Newman GS, Moore JAM, Cregger MA et al (2015) Direct and indirect effects of climate change on soil microbial and soil microbial-plant interactions: what lies ahead? Ecosphere 6(8):1–21
Coelho LM, Rezende HC, Coelho LM, de Sousa PA, Melo DF, Coelho NM (2015) Bioremediation of polluted waters using microorganisms. Adv Bioremed Wastewater Poll Soil 10:60770
Coyte RM, Singh A, Furst KE, Mitch WA, Vengosh A (2019) Co-occurrence of geogenic and anthropogenic contaminants in groundwater from Rajasthan. India Sci Total Environ 688:1216–1227
D'Souza G, Shitut S, Preussger D, Yousif G, Waschina S, Kost C (2018) Ecology and evolution of metabolic cross-feeding interactions in bacteria. Nat Prod Rep 35(5):455–488
da Silva S, Gonçalves I, Gomes de Almeida FC, Padilha da Rocha e Silva NM, Casazza AA, Converti A et al (2020) Soil bioremediation: overview of technologies and trends. Energies 13(18):4664
Danovaro R, Corinaldesi C, Dell'Anno A, Rastelli E (2017) Potential impact of global climate change on benthic deep-sea microbes. FEMS Microbiol Lett 364(23):fnx214
Dastagir MR (2019) Role of microorganisms in managing climate change impacts. Microbial interventions in agriculture and environment. Springer, Singapore. pp 1–16
Delgado-Baquerizo M, Oliverio AM, Brewer TE, Benavent-González A, Eldridge DJ, Bardgett RD et al (2018) A global atlas of the dominant bacteria found in soil. Science 359(6373):320–325
Deveau A, Bonito G, Uehling J, Paoletti M, Becker M, Bindschedler S et al (2018) Bacterial–fungal interactions: ecology, mechanisms and challenges. FEMS Microbiol Rev 42(3):335–352
Di Nardo C, Cinquegrana A, Papa S, Fuggi A, Fioretto A (2004) Laccase and peroxidase isoenzymes during leaf litter decomposition of Quercus ilex in a Mediterranean ecosystem. Soil Biol Biochem 36(10):1539–1544
Doney SC, Ruckelshaus M, Emmett Duffy J, Barry JP, Chan F, English CA et al (2012) Climate change impacts on marine ecosystems. Annu Rev Mar Sci 4:11–37
Fierer N, Leff JW, Adams BJ, Nielsen UN, Bates ST, Lauber CL et al (2012) Cross-biome metagenomic analyses of soil microbial communities and their functional attributes. Proc Natl Acad Sci 109(52):21390–21395
Fisher RA (1936) The use of multiple measurements in taxonomic problems. Ann Eugenics 7(2): 179–188
Gaitán-Espitia JD, Hancock JR, Padilla-Gamiño JL, Rivest EB, Blanchette CA, Reed DC et al (2014) Interactive effects of elevated temperature and pCO2 on early-life-history stages of the giant kelp Macrocystis pyrifera. J Exp Mar Biol Ecol 457:51–58
Gospodarek J, Petryszak P, Kołoczek H, Rusin M (2019) The effect of soil pollution with petroleum-derived substances on Porcellio scaber Latr. (Crustacea, Isopoda). Environ Monit Assess 191(1):38
Guhr A, Borken W, Spohn M, Matzner E (2015) Redistribution of soil water by a saprotrophic fungus enhances carbon mineralization. Proc Natl Acad Sci 112(47):14647–14651
Harley CDG, Anderson KM, Demes KW, Jorve JP, Kordas RL, Coyle TA et al (2012) Effects of climate change on global seaweed communities. J Phycol 48(5):1064–1078

Hernandez J-P, de Bashan LE, Rodriguez DJ, Rodriguez Y, Bashan Y (2009) Growth promotion of the freshwater microalga Chlorella vulgaris by the nitrogen-fixing, plant growth-promoting bacterium Bacillus pumilus from arid zone soils. Eur J Soil Biol 45(1):88–93

Herring P (2001) The biology of the deep ocean. Oxford University Press, New York

Howard DM, Howard PJA (1993) Relationships between CO~ 2 evolution, moisture content and temperature for a range of soil types. Soil Biol Biochem 25:1537

Ivanova J, Stoyancheva G, Pouneva I (2014) Lysis of Antarctic algal strains by bacterial pathogen. Antonie Van Leeuwenhoek 105(6):997–1005

Jansson JK, Hofmockel KS (2020) Soil microbiomes and climate change. Nat Rev Microbiol 18(1): 35–46

Jian S, Li J, Chen JI, Wang G, Mayes MA, Dzantor KE et al (2016) Soil extracellular enzyme activities, soil carbon and nitrogen storage under nitrogen fertilization: a meta-analysis. Soil Biol Biochem 101:32–43

Karigar CS, Rao SS (2011) Role of microbial enzymes in the bioremediation of pollutants: a review. Enzym Res 2011

Khan S, Malik A (2018) Toxicity evaluation of textile effluents and role of native soil bacterium in biodegradation of a textile dye. Environ Sci Pollut Res 25(5):4446–4458

Klekner V, Kosaric N (1992) Degradation of phenols by algae. Environ Technol 13(5):493–501

Kora AJ. Growth and metabolic characteristics of hydrocarbon degrading bacteria isolated from an oil refinery soil. 2018

Krivtsov V, Bezginova T, Salmond R, Liddell K, Garside A, Thompson J et al (2006) Ecological interactions between fungi, other biota and forest litter composition in a unique Scottish woodland. Forestry 79(2):201–216

Kumar A, Sharma S (2019) Microbes and enzymes in soil health and bioremediation. Springer, Singapore

Kumar A, Bisht BS, Joshi VD, Dhewa T (2011) Review on bioremediation of polluted environment: a management tool. Int J Environ Sci 1(6):1079–1093

Kuppusamy S, Maddela NR, Megharaj M, Venkateswarlu K (2020) An overview of total petroleum hydrocarbons. Total petroleum hydrocarbons, Springer, Cham. pp 1–27

Labbate M, Seymour JR, Lauro F, Brown MV (2016) Anthropogenic impacts on the microbial ecology and function of aquatic environments. Front Microbiol 7:1044

Landis WG, Durda JL, Brooks ML, Chapman PM, Menzie CA, Stahl RG Jr et al (2013) Ecological risk assessment in the context of global climate change. Environ Toxicol Chem 32(1):79–92

León-Santiesteban HH, Rodríguez-Vázquez R (2017) Fungal degradation of organochlorine pesticides. Microbe-induced degradation of pesticides. Springer, Cham. pp 131–149

Li T, Liu Y, Lin S, Liu Y, Xie Y (2019) Soil pollution management in China: a brief introduction. Sustainability 11(3):556

Lim S-L, Chu W-L, Phang S-M (2010) Use of Chlorella vulgaris for bioremediation of textile wastewater. Bioresour Technol 101(19):7314–7322

Łozowicka B, Kaczyński P, Wolejko E, Piekutin J, Sagitov A, Toleubayev K et al (2016) Evaluation of organochlorine pesticide residues in soil and plants from East Europe and Central Asia. Desalin Water Treat 57(3):1310–1321

Łozowicka B, Mojsak P, Kaczyński P, Konecki R, Borusiewicz A (2017) The fate of spirotetramat and dissipation metabolites in Apiaceae and Brassicaceae leaf-root and soil system under greenhouse conditions estimated by modified QuEChERS/LC–MS/MS. Sci Total Environ 603:178–184

Lukwambe B, Zhao L, Nicholaus R, Yang W, Zhu J, Zheng Z (2019) Bacterioplankton community in response to biological filters (clam, biofilm, and macrophytes) in an integrated aquaculture wastewater bioremediation system. Environ Pollut 254:113035

Madhav S, Ahamad A, Singh AK, Kushawaha J, Chauhan JS, Sharma S et al (2020) Water pollutants: sources and impact on the environment and human health. Sensors in water pollutants monitoring: role of material. Springer, Singapore. pp 43–62

Mahdi JE, Abbott LK, Pauli N, Solaiman ZM (2017) Biological indicators for soil health: potential for development and use of on-farm tests. Modern tools and techniques to understand microbes. Springer, Cham. pp 123–134

Malik Z, Ahmad M, Abassi GH, Dawood M, Hussain A, Jamil M (2017) Agrochemicals and soil microbes: interaction for soil health. Xenobiotics in the soil environment. Springer International, Cham, pp 139–152

Mallick I, Hossain ST, Sinha S, Mukherjee SK, Brevibacillus sp. (2014) KUMAs2, a bacterial isolate for possible bioremediation of arsenic in rhizosphere. Ecotoxicol Environ Saf 107:236–244

Melillo JM, Frey SD, DeAngelis KM, Werner WJ, Bernard MJ, Bowles FP et al (2017) Long-term pattern and magnitude of soil carbon feedback to the climate system in a warming world. Science 358(6359):101–105

Mesa V, Navazas A, González-Gil R, González A, Weyens N, Lauga B et al (2017) Use of endophytic and rhizosphere bacteria to improve phytoremediation of arsenic-contaminated industrial soils by autochthonous Betula celtiberica. Appl Environ Microbiol 83(8):e03411–e03416

Minich JJ, Morris MM, Brown M, Doane M, Edwards MS, Michael TP et al (2018) Elevated temperature drives kelp microbiome dysbiosis, while elevated carbon dioxide induces water microbiome disruption. PLoS One 13(2):e0192772

Mukherjee D (2019) Microbial interventions in soil and plant health for improving crop efficiency. Microbial interventions in agriculture and environment. Springer, Singapore. pp 17–47

Muñoz-Leoz B, Ruiz-Romera E, Antigüedad I, Garbisu C (2011) Tebuconazole application decreases soil microbial biomass and activity. Soil Biol Biochem 43(10):2176–2183

Nagelkerken I, Connell SD (2015) Global alteration of ocean ecosystem functioning due to increasing human CO2 emissions. Proc Natl Acad Sci 112(43):13272–13277

Naylor D, DeGraaf S, Purdom E, Coleman-Derr D (2017) Drought and host selection influence bacterial community dynamics in the grass root microbiome. ISME J 11(12):2691–2704

Ndiaye EL, Sandeno JM, McGrath D, Dick RP (2000) Integrative biological indicators for detecting change in soil quality. Am J Altern Agric:26–36

Osuolale O, Thompson L, Knight R, Gilbert J, Jansson J, Ackermann G, et al. (2017) A communal catalogue reveals Earth's multiscale microbial diversity

Pachauri RK, Allen MR, Barros VR, Broome J, Cramer W, Christ R, et al. (2014) Climate change 2014: synthesis report. Contribution of working groups I, II and III to the fifth assessment report of the intergovernmental panel on climate change: Ipcc

Pacheco D, Rocha AC, Pereira L, Verdelhos T (2020) Microalgae water bioremediation: trends and hot topics. Appl Sci 10(5):1886

Pal R, Das P, Chakrabarti K, Chakraborty A, Chowdhury A (2008) Side effects of pencycuron on nontarget soil microorganisms in waterlogged soil: field experiment. Appl Soil Ecol 38(2): 161–167

Parte SG, Mohekar AD, Kharat AS (2017) Microbial degradation of pesticide: a review. Afr J Microbiol Res 11(24):992–1012

Pignattelli S, Broccoli A, Piccardo M, Felline S, Terlizzi A, Renzi M (2021) Short-term physiological and biometrical responses of Lepidium sativum seedlings exposed to PET-made microplastics and acid rain. Ecotoxicol Environ Saf 208:111718

Raftery AE, Zimmer A, Frierson DMW, Startz R, Liu P (2017) Less than 2 C warming by 2100 unlikely. Nat Clim Chang 7(9):637–641

Ramakrishnan B, Megharaj M, Venkateswarlu K, Sethunathan N, Naidu R (2011) Mixtures of environmental pollutants: effects on microorganisms and their activities in soils. Rev Environ Contam Toxicol 211:63–120

Reid PC, Fischer AC, Lewis-Brown E, Meredith MP, Sparrow M, Andersson AJ et al (2009) Impacts of the oceans on climate change. Adv Mar Biol 56:1–150

Rohr JR, Johnson P, Hickey CW, Helm RC, Fritz A, Brasfield S (2013) Implications of global climate change for natural resource damage assessment, restoration, and rehabilitation. Environ Toxicol Chem 32(1):93–101

Romero-Olivares AL, Allison SD, Treseder KK (2017) Soil microbes and their response to experimental warming over time: a meta-analysis of field studies. Soil Biol Biochem 107:32–40

Schimel J, Balser TC, Wallenstein M (2007) Microbial stress-response physiology and its implications for ecosystem function. Ecology 88(6):1386–1394

Schindlbacher A, Rodler A, Kuffner M, Kitzler B, Sessitsch A, Zechmeister-Boltenstern S (2011) Experimental warming effects on the microbial community of a temperate mountain forest soil. Soil Biol Biochem 43(7):1417–1425

Schweitzer L, Noblet J (2018) Water contamination and pollution. Green chemistry. Elsevier, Boston, MA. pp 261–290

Sharma A, Kumar V, Kohli SK, Kaur R, Kaur T, Arora S et al (2020) Pesticide metabolism in plants, insects, soil microbes and fishes: an overview. Pesticides Crop Prod Physiol Biochem Action:35–53

Shukla PR, Skea J, Calvo Buendia E, Masson-Delmotte V, Pörtner HO, Roberts DC, et al. (2019) IPCC, 2019: Climate change and land: an IPCC special report on climate change, desertification, land degradation, sustainable land management, food security, and greenhouse gas fluxes in terrestrial ecosystems

Stancu MM (2020) Kerosene tolerance in Achromobacter and pseudomonas species. Ann Microbiol 70(1):1–13

Steinfeld B, Scott J, Vilander G, Marx L, Quirk M, Lindberg J et al (2015) The role of lean process improvement in implementation of evidence-based practices in behavioral health care. J Behav Health Serv Res 42(4):504–518

Stojić NŠS, Prokić D (2019) Soil pollution and remediation. In: Hussain CM (ed) Handbook of environmental materials management. Springer, Cham, pp 583–616

Sun H, Santalahti M, Pumpanen J, Köster K, Berninger F, Raffaello T et al (2015) Fungal community shifts in structure and function across a boreal forest fire chronosequence. Appl Environ Microbiol 81(22):7869–7880

Sviridov AV, Shushkova TV, Ermakova IT, Ivanova EV, Epiktetov DO, Leontievsky AA (2015) Microbial degradation of glyphosate herbicides. Appl Biochem Microbiol 51(2):188–195

Sylvestre S, Lessard P, De la Noüe J (1996) Performance D'un Photobioréacteur Utilisant La Cyanobactérie Phormidium Bohneri pour L'Enlèvement de L'Azote et du Phosphore removal performance of nitrogen and phosphorus compounds by a Photobioreactor using a biomass of cyanobacteria Phormidium Bohneri. Environ Technol 17(7):697–706

Tejada M, Gómez I, García-Martínez AM, Osta P, Parrado J (2011) Effects of Prochloraz fungicide on soil enzymatic activities and bacterial communities. Ecotoxicol Environ Saf 74(6): 1708–1714

Thurber AR, Sweetman AK, Narayanaswamy BE, Jones DOB, Ingels J, Hansman RL (2014) Ecosystem function and services provided by the deep sea. Biogeosciences 11(14):3941–3963

Treseder KK, Berlemont R, Allison SD, Martiny AC (2018) Drought increases the frequencies of fungal functional genes related to carbon and nitrogen acquisition. PLoS One 13(11):e0206441

Van Eerd LL, Hoagland RE, Zablotowicz RM, Hall JC (2003) Pesticide metabolism in plants and microorganisms. Weed Sci:472–495

Verma JP, Jaiswal DK, Sagar R (2014) Pesticide relevance and their microbial degradation: a-state-of-art. Rev Environ Sci Biotechnol 13(4):429–466

Vogel TM (2017) Natural bioremediation of chlorinated solvents. Handbook of bioremediation. CRC Press, Boca Raton, FL, pp 201–226

Wang H, Laughinghouse Iv HD, Anderson MA, Chen F, Willliams E, Place AR et al (2012) Novel bacterial isolate from Permian groundwater, capable of aggregating potential biofuel-producing microalga Nannochloropsis Oceanica IMET1. Appl Environ Microbiol 78(5):1445–1453

Wang S, Peng X, Zhong L, Tan J, Jing S, Cao X et al (2015) An ultralight, elastic, cost-effective, and highly recyclable superabsorbent from microfibrillated cellulose fibers for oil spillage cleanup. J Mater Chem A 3(16):8772–8781
Wang L, Xiao H, He N, Sun D, Duan S (2019) Biosorption and biodegradation of the environmental hormone nonylphenol by four marine microalgae. Sci Rep 9(1):1–11
Wang Z, Li T, Li Y, Zhao D, Han J, Liu Y et al (2020) Relationship between the microbial community and catabolic diversity in response to conservation tillage. Soil Tillage Res 196: 104431
Watanabe K, Takihana N, Aoyagi H, Hanada S, Watanabe Y, Ohmura N et al (2005) Symbiotic association in chlorella culture. FEMS Microbiol Ecol 51(2):187–196
Wołejko E, Jabłońska-Trypuć A, Wydro U, Butarewicz A, Łozowicka B (2020) Soil biological activity as an indicator of soil pollution with pesticides–a review. Appl Soil Ecol 147:103356
Xin H, Xuefeng MAYW (2001) Tolerance of Chlorella ellipsoidea and its removal of heavy metals. J Lake Sci 2
Yanxun S, Yani W, Hui Q, Yuan F (2011) Analysis of the groundwater and soil pollution by oil leakage. Procedia Environ Sci 11:939–944
Zak DR, Pregitzer KS, Burton AJ, Edwards IP, Kellner H (2011) Microbial responses to a changing environment: implications for the future functioning of terrestrial ecosystems. Fungal Ecol 4(6): 386–395
Zhan H, Feng Y, Fan X, Chen S (2018) Recent advances in glyphosate biodegradation. Appl Microbiol Biotechnol 102(12):5033–5043
Zozaya-Valdés E, Roth-Schulze AJ, Egan S, Thomas T (2017) Microbial community function in the bleaching disease of the marine macroalgae Delisea pulchra. Environ Microbiol 19(8): 3012–3024

Impact of Climate Change on Soil Microorganisms Regulating Nutrient Transformation

Pankaj Sharma, Seema Sangwan, Anita Kumari, Sushila Singh, and Harpreet Kaur

Abstract The increased human interventions have forced the global climatic conditions to change at a very faster pace and in a way that seems to be totally uncontrollable and highly impulsive. The elevated levels of carbon dioxide, continuously increasing temperature, altered precipitation patterns, altered moisture content of soils, and greater frequency of some extreme events are affecting every form of life in a significant way. These are affecting the plants systems by altering their geographical distribution, fitness, and productivity. Microbial systems, being the key drivers behind major ecological processes and nutrient cycles, are also being distressed by the continuously changing climatic conditions. The temperature sponsored alteration in global carbon cycling is expected to change the status of soil from "carbon sink" to "carbon source." The increased rate of microbial respiration and enzymatic activities also bring about quicker mineralization of soil organic matter that is leading toward a reduction in the organic carbon as well as nitrogen content of soil. The microbiological inhabitants are also declining in the uppermost layer of soil owing to the increased surface temperature. The climatic extremes are also known to negatively affect the plant-microbe symbiotic associations that further ensues in a reduced plant fitness. Reduction in the rhizospheric microbial count is leading to a decline in the microbial sequestration of carbon that further accounts for the reduced carbon inputs to the soil systems. As a whole, the altered environmental conditions are altering the microbial as well as plant habitats. The present chapter, therefore, highlights the impacts of changing climatic conditions in special context to soil microorganisms involved in different processes regulating nutrient transformation.

P. Sharma · S. Sangwan (✉) · H. Kaur
Department of Microbiology, CCS Haryana Agricultural University, Hisar, Haryana, India

A. Kumari
Department of Botany and Plant Physiology, CCS Haryana Agricultural University, Hisar, Haryana, India

S. Singh
Department of Chemistry, CCS Haryana Agricultural University, Hisar, Haryana, India

© The Author(s), under exclusive license to Springer Nature Singapore Pte Ltd. 2022
A. Vaishnav et al. (eds.), *Plant Stress Mitigators*,
https://doi.org/10.1007/978-981-16-7759-5_8

Keywords Carbon exchange · Nitrogen exchange · Microbial respiration · Plant-microbe interaction

Introduction

The global climatic conditions are changing at a very faster pace and in a totally unpredictable manner. Climate change is largely driven by anthropogenic activities which happens to be a prime global concern (Sharma et al. 2020; Singh et al. 2021). It lays a strong effect on the plant as well as microbial diversity of planet Earth either in a direct or an indirect way. It is also apparent that the estimated upsurges in carbon dioxide levels, temperature, and transformed precipitation patterns are foremost aftermaths of changing climatic conditions. These have augmented the intricacy and ambiguity to plant as well as agroecosystems that have further threatened their sustainable management (Singh et al. 2019). The changing climatic conditions lay a negative effect on every possible life form ranging from aquatic to terrestrial habitats. The anthropogenic emanation of greenhouse gases (GHGs) has largely contributed to these climatic changes. For instance, carbon dioxide (CO_2), a foremost greenhouse gas, has been augmented by almost 30% in the atmosphere only because of anthropogenic origins (Stocker 2014). The emission of greenhouse gases has become more prominent since the global industrialization. The sun rays directly falling on earth's crust are being trapped by GHGs like CO_2, nitrous oxide (N_2O), methane (CH_4), hydrofluorocarbons (HFCs), etc., that ultimately results in a hike in global temperature. More precisely, the concentration of atmospheric CO_2 is at peak and has touched a level of 410 ppm (http://scripssco2.ucsd.edu) at current that happens to be the major reason behind warming upshot. In addition, human activities are also leading to the increased deposition of atmospheric nitrogen that is also exerting strains on global agroecosystems. These changes affect plants, microorganisms, and other higher organisms from genes to ecosystem services (Singh et al. 2019). Climatic changes are causing a significant increase in the global temperature, resulting in water insufficiency, damage to vegetations, wildfire mutilation, soil destruction, melting of permafrost, and coastal deterioration along with reduction in yields of tropical crops (Rani et al. 2019; Sharma et al. 2021). These factors not only disturb the efficiency of plant systems but also alter the global distribution of plants. The nutrient status of plants is also deteriorating at a faster pace, for instance, the elevated levels of greenhouse gases force a decline in the nitrogen content of non-legume plants (Jablonski et al. 2002; Taub et al. 2008). The increasing global temperature is also affecting oceans as they act as sink for most of the supplementary heat engendered by human-induced climatic changes. The increment in surface temperatures and decline in the levels of dissolved oxygen has also put the aquatic life in a state of distress.

Apart from other forms of life, the life at microbiological scale also happens to be sensitive to changing climatic conditions. Since plants and other higher forms of life have evolved in a world of microbes, therefore, any alteration in the patterns of microbial life would affect the higher forms of life as well. The soil microbes are found to be sensitive toward increasing temperatures; therefore, the increasing temperature also targets their diversity and abundance as well. Plants are highly dependent on plant-microbe interactions for their growth, survival, as well as productivity. Therefore, the disturbance in assembly of microbial communities owing to the changing climatic conditions also disturbs the plant productivity and its resilience toward changing environmental conditions (Keeler et al. 2021; Maloy 2021; Zytynska 2021).

In addition, the microbial systems play central roles in various ecological processes like nutrient transformation and cycling, nutrient mobilization, enhancing the bioavailability of several nutrients, etc. The changing climatic conditions affect the microbes involved in such processes of extreme ecological significance, thereby disturbing the ecological functioning and roles of microbiota. Besides other climatic variations, increase in temperature happens to be the foremost factor affecting the functioning of microbial systems. The microbial systems experience a steeper decline in their biomass on increase in temperature. The elevated temperatures alter the fluidity and permeability of microbial membranes that in turn affects their viability (Kannojia et al. 2019). However, a slight increment in the soil temperature excites the actions of soil microbiota that further corroborates upsurge in overall rates of nitrification, phosphorus and nitrogen mineralization, and total soil respiration. Nevertheless, this reaction of microbial community to soil warming lasts for extremely short period of time. The impacts are found to be harsher when the increment in temperature is found to be higher. Adding more to the concern, the mounting concentration of atmospheric CO_2 has been projected to intensify the comprehensive surface temperatures amid 1.8 °C to 3.6 °C by the culmination of the twenty-first century (IPCC Climate Change 2007). Therefore, the situations will get more upsetting in the upcoming decades. Apart from this, such diverse ecosystem scale factors are potent enough to alter the patterns of microbial metabolism. However, the effects in case of microbial systems are hard to predict as in the case of plants and other organisms. The altered conditions also affect the rates of microbial respiration, plant-microbe interactions, and their ability to perform specific functions such as carbon sequestration along with a reduction in the degree of their ecological performances. The current chapter, therefore, highlights the impacts of altered climatic circumstances on the microorganisms and their activities. In addition, climatic changes driven effects on other life forms have also been touched, yet microbial functioning under altered environments remains the central idea.

Changing Climatic Conditions: A Global Perspective

The increasing human interventions in the environmental processes are greatly affecting the global climatic conditions. A perpetual upsurge in universal temperature, fluctuating precipitation forms, and larger rate of recurrence of a few extreme events are greatly affecting the climatic conditions. The agricultural emissions account for almost one quarter of greenhouse gases, yet the agricultural sector owes the potential to find a cure for comprehensive climate objectives. The extreme weather conditions are affecting every possible habitat including the oceans by their acidification and terrestrial habitats by inviting more intense droughts. The grievous effects range from the occurrence of extreme weather conditions, pest and ailment distribution, reduction in biodiversity, dilapidation of bionetworks, and water paucity that will go downhill as the globe warms. Oceans are at frontline to face the long-standing influences of climatic changes as they absorb the major proportion of heat generated by anthropogenic events. The altered climatic conditions will also lead to an increment in the sea surface temperatures and upsurge in sea levels. It is further speculated that the end of the twenty-first century would also mark a rise of 18–140 cm in the sea levels (IPCC 2007a, b, c). It will further strengthen tropical storms that would attain higher peak wind speeds. In addition, the climatic changes would lead to perilous drops in oceanic oxygen concentrations that would further lay a negative impact on the aquatic life (FAO 2019). The agronomic, forestry, and fishery divisions are designated to be dominant in this repute as they are not only distressed by climatic fluctuations along with their contribution to global GHG emanations but they also offer several prospects for lucrative mitigation opportunities. The changing climatic conditions also slow down the advancement toward the accomplishment of the Millennium Development Goals (MDGs), exclusively in special context to abatement in hunger as well as poverty along with the maintenance of environmental sustainability. Climatic change depressingly distresses the basic foundations of food production, for example, soil, water, and biodiversity. On a broader scale, it disturbs all the four confines of food security: food availability, food accessibility, the steadiness of food stock, and the capability of consumers to use food counting food safety along with its nutritive values (FAO 2008). The rapidly changing environmental conditions are reducing the agricultural productivity and also affecting the terrestrial distribution of different crops. Adding more to the concern, the agricultural sector itself is contributing to the changing climatic conditions. Agricultural sector makes 13.5% of universal GHG emanations or around 1.8 GtC eqv./yr. (6.6 Gt of CO2 eqv./yr), primarily as CH_4 and N_2O from rice production, enteric fermentation, fertilized soils, biomass burning, and fertilizer fabrication (IPCC 2007a, b). In addition, the change in land usage patterns epitomize 17.4% of overall GHG discharges (IPCC 2007b). The fishery sector contributes in CO_2 emissions from the harvesting as well as transportation that is assessed to be 0.05 Gt annually. The developing nations account for almost 75% of global agriculture and land use-based discharges (IPCC 2007c). The emissions from agriculture, forestry, and other land use (AFOLU) events added CO_2 (13%), CH_4 (44%), and N_2O equaling 81% from anthropological actions during 2007–2016 on a global basis, which was found to be equivalent to 23% of overall human-caused emanations of GHGs (IPCC 2019). The elevated levels of greenhouse gases mark significant increment in universal surface temperature as equated to the pre-industrial stages, which strongly affects the progressions making a count in desertification, land dilapidation, and food security. Alterations in such progressions invite hazards toward global food systems, living conditions, worth of land resources, infrastructures, and human as well as ecological health. The prevailing stages of comprehensive warming are inviting reasonable menaces from larger dryland water shortage, damage to vegetation, wildfire mutilation, soil erosion, melting of permafrost, dilapidation of coastal areas, and reduction in crop yields. However, the magnitude of risks is being speculated to become more severe with increasing temperatures marking them up to the levels of cascading risks (IPCC 2018). With an upsurge of 1.5 °C in the universal temperature, these dangers are anticipated to be higher, whereas at around 2 °C and 3 °C, these losses are projected to be much higher. The dryland inhabitants susceptible to water shortage, drought strength, and home deprivation are expected to touch the numbers of 178 million by 2050 at 1.5 °C, upsurging to 220 million at 2 °C, and 277 million at 3 °C warming, whereas Asia and Africa have been predicted to lead in numbers with people susceptible to enlarged desertification. The increased risk of wildfires is likely to affect North America, South America, Mediterranean, Southern Africa, and Central Asia. The tropics and subtropics are anticipated to be most susceptible to reduction in crop yields. Similarly, the land deterioration subsequent to rise in sea levels and strong cyclones has also been estimated to threaten survival and living conditions in cyclone-affected zones (IPCC 2019).

Effects of Climatic Change on Plants

The changing climatic conditions lay an impact on each and every living being on this planet (Kumar et al. 2020). Fluctuations in environmental conditions govern the global and terrestrial distribution of different crops and also affect their yields. For instance, the productivity of plants like maize and wheat grown in numerous lower-latitude areas has been reported to be reduced, whereas those grown in higher-latitude regions (for instance, maize, wheat, and sugar beets) have been found to be increased due to changing climatic conditions (FAO 2019). Extensively quoted approximations demonstrate that the period between 1980 and 2008 marked the global reductions of 5.5 and 3.8% in wheat and maize yields, respectively, as equated to their yield projections under stable climatic conditions (Lobell et al. 2011). There are many climatic factors that are known to shape the growth as well as yield of plant systems. These are usually comprised by the physical parameters like temperature, rainfall configurations, CO_2 levels, alterations in agricultural ecosystems, and the adaptive rejoinders of humanoid organizations. The temperature seems to be the major factor that is known to be altered by the altered environmental conditions owing to its easily noticeable approach. Its aftermaths on progression of plants systems are well comprehended only up to the optimal levels for the development of crops. The increase in temperature and levels of carbon dioxide may also prove to be beneficial to some crops but only up to a certain level. For instance, wheat as well as soybeans can experience enhancement in their yields with increasing CO_2 levels under optimum temperatures (Porter et al. 2014). Thus, the changing climatic conditions can also prove to be beneficial to the plant systems, but the drastic change in environmental conditions put the plant systems in a state of distress. However, it has also been proposed that until about 2030, the positive aspects of changing climatic conditions on the plant yields will counterbalance the negative ones on a global scale, and later the effects pronounced would be largely negative with any further acceleration in climatic modification. Thus, the last half of the twenty-first century will mark negative impressions on the yields of maize, wheat, as well as rice, and the effects will be more severe in the tropical regions as equated to the temperate regions (Porter et al. 2014; Challinor et al. 2014). The damaging effects of elevated tropospheric ozone on plant yields have also been reported, with approximations of losses ranging from 8.5 to 14% for soybean, 3.9 to 15% in wheat, and 2.2 to 5.5% in maize in 2000 (Porter et al. 2014). The mean crop yields are anticipated to be reduced by roughly 1% by the passage of every succeeding decade of climatic changes. This seems to be a smaller proportion, but it happens to be a highly significant fraction as a concomitant increment of around 14% is strongly required in the production every decade in the quest to fulfil the growing demands. Although the ecosystem services are affected in numerous ways by the changing climates, for instance, the equilibrium among plants and pests along with the upshots on pollinators, these happen to be tedious ones to measure and, thereby, are usually not taken into consideration for making future projections. The climatic change is also expected to alter the land use patterns and geographical distribution of crops. For instance, the aptness for potato plants is projected to surge in higher latitudes and higher tropical elevations by the end of the twenty-first century (Schafleitner et al. 2011). Similarly, the vital crops of Brazil like sugarcane and coffee are speculated to drift toward more favorable regions in the south (Pinto et al. 2007; Pinto et al. 2008). Alongside, the expansion in range of destructive pests like *Phytophthora cinnamomi* in Europe has also been projected (Bergot et al. 2004). Likewise, augmented generations under coffee nematodes and walnut pests have been predicted for Brazil and California, respectively, in changing climatic conditions (Ghini et al. 2008; Luedeling et al. 2011). Additionally, the larger extents of phenotypic as well as genotypic plasticity allied with the weeds comparative to the consistency inherent in large cropping systems would aid them in limiting crop yields to a larger degree with increasing concentrations of CO_2 (Ziska 2011). The chemical control of weeds will become less effective with increasing economic and environmental costs.

Effect of Climatic Change on Global Agricultural Systems

The last five decades have marked the increment in global population from 3 billion to 6.7 billion in a time period from 1959 to 2009. Surprisingly, it is anticipated to further upsurge by additional 50%, touching a mark of 9.1 billion by the mid of the twenty-first century. To meet the food demands of this much of populace, the agricultural, forestry, and fishery subdivisions would experience enormous pressure in the pursuit to offer food, feed, and fiber along with the provision of revenue, employment, and vital ecosystem amenities. On the other hand, the global agricultural systems are facing a challenge of stagnant crop yield due to changing climatic conditions. Therefore, the comprehensive retort toward changing climate must be articulated by keeping food and agricultural sectors at the center. The agricultural division represents almost 90% of the nations' Nationally Determined Contributions which evidently validates the sturdy demand for actions in response to climate changes (FAO 2019). The plants, animals, and ecosystems inhabiting a particular place have acclimatized toward the prevailing environments. Any alteration in climatic conditions will affect these in directions that seem hard to forecast exactly. However, different reports have also made attempts to address the biophysical influences of anticipated fluctuations explicitly on agroecosystems (FAO 2016). The effects on plant systems are ultimately reflected in the agricultural sector that are further interconnected across environmental, social, and economic dimensions. The climatic changes influence the above-mentioned four dimensions of food security, nutrition and human health, the agricultural sectors, natural resources, post-production stages, and human migration. These changes are further going to distress the agricultural production, and these influences are further anticipated to mount, with corresponding increment in worldwide temperatures. The global human residents of above 122 million may face extremes of poverty by 2030 due to changing climates and its upshots on the earnings of small-scale farming community (FAO 2016). The projected yield declines will generate food scarcity that will lead to increase in food prices. The process to attain the desired food production will increase the strain on natural resources and is very likely to alter the geography of production. The altered environments will amend the trends of plant as well as animal ailments that will further lead to increments in usage of chemicals meant for combatting these diseases. The increased temperatures will increase the risk of fungal growth that will further increase the susceptibility of cereals and pulses toward contamination with fungi and mycotoxins (FAO 2017). The projected climatic changes would also lead toward a reduction in the per capita food obtainability by above 3%, whereas the utilization of fruits as well as vegetables would reduce through almost 4% by 2050 (Springmann et al. 2016). The Intergovernmental Panel on Climate Change (IPCC) has also cautioned about the decline of 10–25% and more in crop yields by 2050 (IPCC 2014). The cereal prices are also expected to upsurge by 29% by 2050. Currently, the agricultural sector absorbs almost 26% of the financial repercussions of climatic catastrophes, which is projected to rise to 83% for drought specifically in developing countries. The climatic alterations also affect the availability of water resources. Currently, the shortage of water resources lays a strong effect on almost 40% of the global populace. The severity of situation can be taken into consideration by the fact that every 1 °C increment in temperature will push 500 million additional humans to confront a 20% dip in renewable resources of water. Another important consequence of global warming is the temperature-persuaded increment in food wastage. Presently, about 14% of food equivalent to monetary values of 400 billion USD is wasted after harvesting prior to its reach to the sellers. Adding more to the concern, the lost food makes a contribution of around 8% in the global emanations of GHGs. The changing climatic conditions and perpetually increasing global temperature will lead to more food losses that would account for more release of greenhouse gases (FAO 2019). In addition, the altered environmental conditions also affect the nutritional quality of agricultural produce.

Effect on Soil Microbiota Regulating Nutrient Transformation

The soil inhabiting microbiota oversees the bio-geochemical cycling of different macro- and micronutrients along with the other elements that are essential for the plant as well as animal growth. The understanding and prediction of the effect of changing climatic conditions on microbiological dwellers of soil and the ecosystem amenities provided by them put forward an imposing problem and a foremost opportunity as well. The climate would continue to change, and the environmental conditions would touch more extremes. Since microorganisms are involved in different nutrient transformation processes, they also have a crucial part to play in climate response, counting generation or utilization of GHGs like CO_2, CH_4, and N_2O. The microbial functioning is affected by all the types of changing climatic conditions, for instance, rise in global CO_2 concentrations, raised temperature, augmented drought, amplified rainfall, and higher fire occurrence (Jansson and Hofmockel 2020). Although the altered climatic conditions are known to affect the microbes engaged in cycling of all the nutrients, carbon and nitrogen cycles happen to be the major ones and are thereby taken into consideration.

Carbon Exchange

The microbiota inhabiting soil systems is found to be principally accountable for the cycling of soil organic carbon (SOC) pools. The enhanced mineralization of SOC by microbial systems under modified climatic conditions happens to be a major concern that further makes a significant contribution toward the release of GHGs (CO_2 and CH_4), thus aggravating the warming patterns. This speculation seems to be a grievous concern owing to the total quantity of carbon present in permafrost that is appraised to be approximately 3300 petagrams (Pg) which happens to be fivefold higher than the existing atmospheric pool of CO_2 (Cavallaro et al. 2018; Tarnocai et al. 2009). Soil-residing microbiota perform the dual roles of mineralization as well as immobilization of SOC. The net efflux of CO_2 and CH_4 from soil is governed by the balance between these two processes. The continuously changing climatic conditions have led to an upsurge in the heterotrophic respiration of soil organic carbon that further contributes to the CO_2 loads of the atmosphere (Bond-Lamberty et al. 2018). The exposure to increased temperatures marks changes in cellular membranes of bacteria that further trigger the expression of heat shock proteins, whereas in case of soil fungi warming can either stimulate or suppress the fungal biomass and activity depending on other factors like moisture and presence of other vegetation (Allison and Treseder 2008; Clemmensen et al. 2013). Climatic alteration-instigated intricacies reported in the microbial diversity of soil modify the physiochemical situations of soil ecosystems that eventually disturb productivity of plants (Bardgett et al. 2008).

The varying climatical situations affect the functioning of soil microflora directly and indirectly that feedback GHGs to the environment and aid in global warming. The direct impacts comprise the effect of elevated temperature, altered rainfall, and occurrence of dangerous climatical events on soil microbiota and GHG emanations, while the indirect impacts are generated as a consequence of climate-obsessed deviations in plant output and variety that modify the physicochemical environments of soil, its carbon supply, and the structural as well as functional components of microbial groups participating in disintegration progressions and carbon discharge from the soils. Climatic changes also affect the soil carbon sink as warming surge the release of CO_2 from soil to environment owing to increased microbial degradation of SOM. This kind of speeding up in carbon loss could meaningly impair the soil carbon cycle response if projected climatic variation circumstances are precise (Cox et al. 2000; Friedlingstein et al. 2006). In addition to it, the alterations recorded in climatical trends are further known to be amended by carbon apportionment to microbiological communities which inclusively touches their structures as well as dynamics, playing a decisive part in organic matter breakdown. Besides, the biotic mechanisms accountable for the regulation of this exchange along with the circulation of carbon amid inter-reliant community systems affect climatic variation via climate-ecosystem reaction and might increase the enduring possessions of local or universal environment (Heimann and Reichstein 2008).

In a similar way, drought and freezing also directly affect the microbial physiology along with the microbial diversity and richness, with significant penalties for bionetwork-level carbon and nutrient dynamics (Schimel and Mikan 2005). Increasing occurrences of drought limit the soil moisture content which in turn limit the rates of microbial respiration and further create a harmful reaction on microbiological breakdown and soil carbon loss. Contrary to this, the prevalence of droughts in marshlands and peatlands leads to the creation of conditions favorable for actions of microbial systems. The dropping of water table and introduction of oxygen into formerly anaerobic soils lead to upsurge in the actions of enzymes like phenol oxidases (Freeman et al. 2004; Zibilske and Bradford 2007).

The allocation and movement of carbon amid the earthly bionetworks serve as comprehensive carbon sink via its accretion in the alive vegetations, in microbiological biomass, and in soil. However, the release and absorption of greenhouse gases like CH_4, N_2O, and CO_2 regulate universal climate feedback leanings. The anthropogenetic activity-triggered climatic changes have largely affected the carbon-sink activity of global ecosystems. As carbon dioxide happens to be the principal substrate exploited as metabolic fuel by plant systems, therefore, the atmospheric CO_2 shapes the distribution of carbon present under the soil and correspondingly impacts the chemistry of root exudation. Such alterations potentially distress the plant-microbe beneficial interactions happening in the rhizosphere (Williams et al. 2018). However, the total carbon economy of any environment during changeable climatical situations is governed by the equilibrium amid respiration and photosynthesis (Aamir et al. 2019).

Nitrogen Exchange

The increment in the global temperatures is also going to affect the accessibility of nitrogen in soils by stimulating the decomposition and mineralization of SOM. Elevation in temperature leads to increment in the enzymatic activities involved in nitrogen cycle (Beier et al. 2008; Lukac et al. 2010). Soil warming can enhance the mineralization of nitrogen and may promote loss of nitrogen via leaching (Beier et al. 2008). Increment in the actions of enzymes participating in nitrogen cycling has been recorded in the occurrence of elevated temperature and adequate moisture levels (Sardans et al. 2008).

Nitrogen and carbon cycles are highly interconnected; therefore, any factor that affects the carbon cycle also touches the nitrogen cycle. In case of elevated concentrations of carbon dioxide, carbon assimilation in soil systems is expected to be increased due to upsurge in root exudation patterns that in turn brings out an enhancement in the microbial activities in the root zone. The increased carbon input as a consequence of elevated carbon dioxide levels has earlier been reported to alter the nitrogen dynamics in soil systems. It may either excite immobilization of microbial nitrogen or promote mineralization owing to the priming impact (Blagodatskaya and Kuzyakov 2008; Zak et al. 2000), relying on the constraint of carbon or nitrogen in microbiological groups and on the C/N proportion of substrate (Hodge et al. 2000).

The increment in the extent of nitrogen immobilization further marks a decrease in the capability of plant systems to retort to higher CO_2 levels that in turn results in a reduction in plant yield (Hu et al. 2001). The penalties of climatic changes are well reflected on the nitrogen turnover and on the microorganisms as well that bring out the nitrogen transformations in the ecosystems. Wang et al. (2016) also reported that soil nitrogen revenue was found to be responsive toward warming, condensed precipitation, and improved statistics of freeze-thaw cycles (Andrade-Linares et al. 2021).

In general, the assimilation of nitrogen becomes low with upsurge in CO_2 levels which further marks nitrogen losses and acidification. The increased levels of root exudates at higher CO_2 concentrations generally result in enhanced concentrations of organic acids. This leads to an increment in the microbial activities that further quicken organic matter mineralization. The intensity of losses rises with surge in the intensity of climatic changes (Raza et al. 2021). The climatic extremes also distress the microbiological content of soil systems. For instance, the acidity in soils limits the survival and persistence of nitrogen-fixing bacteria along with a drop in the action of nitrogenase enzyme (Ferguson et al. 2013). Comparatively higher root temperature is also known to affect the infectivity, nitrogen-fixing capability, and plant progression (Mohammadi et al. 2012). In addition, higher temperatures lead to loss of infectivity in rhizobial strains. Additionally, the high-temperature shock leads to plasmid curing in fast-growing strains which ultimately affects the establishment of symbiotic relationship (Zahran 2017). Higher temperatures are also known to affect the synthesis as well as secretion of inducers specific for *nod* genes (Hungria and Stacey 1997) which further alters the working of nodule predominantly synthesis of leghemoglobin, action of nitrogenase enzyme, and H_2 excretion along with the accelerated senescence of nodules (Hungria and Vargas 2000). Thus, the elevated temperature leads to a significant and total upsurge in the nitrogen mineralization and nitrification in terrestrial ecosystems. However, the results are found to contrary when the sole effect of carbon dioxide is evaluated on the nitrogen dynamics. Conversely, if the carbon dioxide-triggered global warming is also considered for studying the nitrogen dynamics under climatic changes, results may be different. Consequently, the outcomes of climatic changes on nitrogen dynamics may prove to be multifaceted, and the long-standing consequences on nitrogen maintenance and its utilization efficacy are still uncertain.

Agricultural systems happen to be the major producers of greenhouse gases. The agricultural emission of nitrous oxide is the major contributor toward global warming as N_2O holds the warming capacity of almost 300-fold higher than of CO_2. The annual release of nitrous oxide from agricultural sector is thus comparable to increasing the greenhouse effect as done by additional three billion tons of CO_2 (Reay 2015). Altering climatic conditions are further acknowledged to alter the patterns and dynamics of nitrous oxide release from the agroecosystems. The emissions from the soils are mainly due to the reduction of nitrate to nitrous oxide as a consequence of the actions of denitrifying bacteria in the absence of oxygen. In addition, the nitrogen cycling process is also found to be sensitive toward variation in other climatic changes like global warming-instigated alteration in moisture content of soil (Singh et al. 2011). The process of denitrification is sturdily connected to the soil moisture levels and to the obtainability of carbon source, to the nitrate supply, and to temperature. Elevated temperatures as well as the wetter soils due to higher precipitation will also endorse denitrification. As a whole, all the climatic variations are known to disturb the nitrogen cycle. The cycle is not only disturbed by a single variation but also by the combinations of variations. The elevated temperatures mark an increment in the speed of disintegration of soil organic matter that clearly indicates temperature-sponsored reduction in the content of organic nitrogen. Conversely, the reduction in soil moisture content is acknowledged for declining the rate of degradation process. Surprisingly, the process is known to be least affected by the blend of elevated temperatures and reduced moistures that could lead to no net alteration in disintegration (Bradbury and Powlson 1994).

Alteration in Microbial Respiration Rates and Enzymatic Activities

Soil respiration also happens to be a vital parameter for evaluating soil health as it is known to be completely interrelated with the SOM content. This can be assessed as either a function of carbon dioxide release or oxygen requirement (Arias et al. 2005; Dalal and Moloney 2000; Haynes 2008). Soil respiration, predominantly its temperature retort, is extensively recognized to be a critical connection amid climatic changes and the comprehensive carbon cycle (Wixon and Balser 2009). Soil respiration is known to be another major carbon flux after photosynthesis in earthly environments that may represent almost 70% of total yearly ecosystem respiration (Yuste et al. 2005).

Activities of some enzymes can be suggestively interrelated with a few soils' microbiological parameters like microbial respiration and biomass (McKenzie et al. 2015). The activity of dehydrogenase endoenzyme is often utilized as bioindicator for assessing the fertility of soil. A bigger intensity of respiration specifies the excess of substrate availability for microbial metabolism on incubation at elevated temperatures in in vivo conditions. Generally, increments in the activities of soil enzymes β-glucosidase, fluorescein diacetate hydrolase, and dehydrogenase are observed with a corresponding increment in the temperature of incubation. The activities of soil enzymes dehydrogenase and fluorescein diacetate hydrolase are generally used to assess the load of microbial activities in soil, whereas the enzyme β-glucosidase happens to be a carbon depolymerizing enzyme. The upsurge in the actions of these enzymes may be due to increased availability of substrate and microbial biomass at the higher temperature. Quite the reverse, the decline in the actions of enzymes aryl sulfatase and acid phosphomonoesterase as well might be attributable to their denaturation at elevated temperatures (Joergensen et al. 1990).

Soil respiration is found to be much responsive toward changing climatic conditions (Allen et al. 2011). Any processes, climatic changes, or management practices that reduce the inputs of organic carbon or augment outputs from the soils are going to adversely affect the soil health and changing climatic conditions (Bond-Lamberty and Thomson 2010). The increasing temperature is known to affect the soil respiration, and it is usually presumed that respiration rate becomes twofold with a rise of 10 °C in temperature. However, this parameter sounds good for seasonal variation of temperatures as under climatic extremes increased temperature is also known to distort the structure of biological molecules. In a similar way, the reduction in soil moisture can potentially bring about dormancy and/or death of microorganisms in addition to decline in their mobility (Orchard and Cook 1983), thus negatively affecting the microbial respiration. The global warming can possibly change the earthly environments from carbon basins to carbon sources. Soil warming upsurges the rates of soil respiration by almost 20%, where the forest bionetworks are known to be most responsive toward climatic warming (Yiqi and Zhou 2010; Rustad et al. 2001). In a similar way, during the precipitation extremes, the movement of CO_2 in soil pores subsequent to penetration of rainwater may also subsidize to quick growths in soil CO_2 outflow (Huxman et al. 2004).

Soil enzymes are recognized to be the most imperative indicators of soil health that are known to participate actively in the elemental transformations and biochemical functions regulating the dynamics of SOM (Bakshi and Varma 2010; Maddela et al. 2017). These enzymes are principally derived from soil faunae, dead as well as alive microorganisms, and roots and remains of plants and animals. The primary function of these enzymes is to catalyze different reactions counting the nutrient cycling and the disintegration of biological wastes along with the fabrication of

SOM (Burke et al. 2011; Mele 2011; Gul and Whalen 2016; Yang et al. 2012). These enzymes happen to be extracellular in nature with their prime function to speed up the metabolic reactions. Therefore, the activities of these enzymes define the proportion of energy derived by soil microorganisms by utilizing different organic as well as inorganic composites in the environment owing to their usage of extracellular enzymes for carrying out metabolic functions (Fuhrmann 2005). They also happen to be the mainspring behind the flow of matter and energy cycle in the agricultural systems. In general, the quantification of activities of soil enzymes happens to be a viable parameter for determining the rate of SOM decomposition that further reflects microorganisms' sponsored cycling of C, N, and P. The enzymes arylsulfatase, β-glucosidase, phosphatases, and ureases are emblematically utilized to track the corresponding transformations of S, C, P, and N, and in agrarian soils (Shan et al. 2008; Yang et al. 2012; Abd 2000). However, the action of these enzymes is also a subject to variation depending upon several factors like quality of soil, diversity as well as community configuration of plant systems, SOM, fabrication as well as exudation of soil enzymes, pH, temperature, and salt concentration of soil (Alshaal et al. 2017).

The changing climatic conditions affect the functioning of soil enzymes in a direct as well as indirect way. The direct effects correspond to the temperature- and moisture-sponsored inhibition of enzymatic activities, whereas the indirect effects occur via affecting microbial enzyme producers. In addition, the possible variations in plant and soil faunal communities also make a count among the different factors (Henry 2013). These enzymes are also found to be sensitive to seasonal climatic changes. Waldrop and Firestone (2006) reported reductions in the microbial biomass, enzyme activities, and microbial respiration on change of soil environments.

The N-degrading enzymes display lower compassions than C-deteriorating enzymes toward increased temperatures (Stone et al. 2012). At increased temperatures and nitrogen limitation in soil systems, the microbial systems shift toward the synthesis of enzymes carrying out N-degradation and lessen the fabrication of enzymes catalyzing C-degradation (Steinweg et al. 2013). In a similar way, the actions of enzymes like xylosidase, leucine aminopeptidase, β-glucosidase, acid phosphomonoesterase, cellobiohydrolase, and N-acetylglucosaminidase are also known to upsurge with rising temperature in the soil depths ranging from 0 to 5 cm. Surprisingly, the enzymes acid phosphomonoesterase as well as β-glucosidase are known to exhibit peak action in warmer environment (Steinweg et al. 2013). The elevated levels of carbon dioxide are also known to enhance the activity of soil enzymes like urease, acid phosphatase, xylanase, alkaline phosphatase, N-degrading enzymes, protease, invertase, and alkaline phosphomonoesterase. The other important factor, soil moisture content, is known to affect the circulation of substrates along with the dispersion of enzymes and their corresponding products that may consequently enforce diffusion restrictions on enzymes along with their corresponding substrates. The production of enzymes usually declines during the drought stress as a consequence of reduction in microbial biomass (Chatterjee and Saha 2018).

The effect of climatic changes on the enzymatic activities is more pronounced in the arid climatic regions. The different factors altered during the climatic changes lay a direct or indirect effect on the soil enzymatic activities. The enzymatic activities may show an increment with increase in moisture, temperature, or carbon dioxide level, but the increment is pronounced only up to the optimal levels. Thereafter, the enzymatic activities start declining when the climatic changes are known to touch the extremes.

Impact on Plant-Microbe Interactions

The changing climatic conditions lay a strong effect on the species and their communications that further result in progression of novel communities along with the modified ecosystem processes. The increase in level of greenhouse gases leads to an increment in the global atmospheric temperature that further brings a decline in soil water. The reduced soil water induces drought in different ranges of the globe that will eventually disturb the plant-microorganism interactions along with their groups and roles in the soils (Wahid et al. 2020). The changing climatic conditions affect the aboveground as well as belowground terrestrial ecosystem in a direct as well as indirect way. The aboveground effects are reflected on the abundance and diversity of plant populations, while the belowground effects amend the carbon pool along with the root framework that in turn will exert an indirect effect on the microbial load, its configuration, and microbial communities. All kinds of possible changes in the climatic conditions pose both straight and subsidiary effects on the physiology of plant systems and structure of soil microbial communities (Bardgett et al. 2013).

The climate change-prompted variations in the physiology of plant systems are also reflected in their root exudation patterns. Elevated levels of carbon dioxide often upsurge the carbon distribution of root zone and henceforth alter the configuration of root exudates. The alteration in root exudation patterns also distresses the working of advantageous microbiota and their inhabitation on the roots of plant systems that further affects the growth as well as health of plants. Alongside, the climatic changes are also known to force alterations in the signal molecules, C/N proportion, and accessibility of chemoattractants as well as nutrients (Kandeler et al. 2006; Haase et al. 2007).

Although higher levels of carbon dioxide stimulate the photosynthesis in plant systems, the enhanced levels would also augment the occurrence of thrilling climatic shocks like heatwaves and droughts that prove to be detrimental to plant development and produce (Wang et al. 2013; Gray and Brady 2016). In addition, increasing levels of carbon dioxide are also known to upsurge the occurrence of plant-pathogen communications with rhizosphere-inhabiting microbes (Eastburn et al. 2011; Gschwendtner et al. 2016) consequently upsetting the development, progression, and yield of plants.

The alteration in moisture content of soils as a consequence of rising temperature also affects the belowground microbes. However, inadequate rainfall is known to reduce the richness of different microorganisms like fungi, enchytraeids, and collembolan. Drought stress is also known to decrease the abundance and diversity of microorganisms. However, the reduction depends on the patterns of drought stress, for instance, a report claimed 40% decline in phylogenetic alpha-diversity on the exposure of plots to drought strain as equated to those receiving some prior drought exposures (Blankinship et al. 2011; Bouskill et al. 2013). The drought stress-instigated alteration in the soil microbial community may happen to be a modification in the comparative richness as opposed to complete elimination, and hence it gives appropriate specifics for any alterations in alpha diversity. It has been noticed that the exposure to drought stress marks an increase in the proportion of Gram-negative bacteria as paralleled to the Gram-positive (Fuchslueger et al. 2014, 2016; Chodak et al. 2015). For instance, on exposure to drought stress, a decline in the numbers of Gram-negative bacteria belonging to *Proteobacteria*, *Bacteroidetes*, and *Verrucomicrobia* has been recorded (Barnard et al. 2013; Acosta-Martinez et al. 2014; Yuste et al. 2014), whereas an augmentation in Gram-positive bacteria belonging to *Actinobacteria* and *Firmicutes* has been reported (Bouskill et al. 2013; Hartmann et al. 2017). The drought stress also marks the inhabitation of plant root surface by the microbes that owe the potential to produce ACC deaminase enzyme which further aid the plant systems to alter the usual progression of plant systems under drought stress by cleaving precursor of ethylene hormone (Glick et al. 2007). The enhancement of interactions between plants exposed to drought stress and microbial members capable of alleviating the similar stress also leads to enhancement in the plant forbearance of the stress. Such kind of plant-microbe interactions is also known to alter the root surface structure of plants (Cassán et al. 2001).

Agrarian production is principally governed by the persistence of pathogenic microbes and the position of plant diseases in any particular environ. The varying environmental conditions are boosting the disease incidence in crop plants owing to the changes in distribution formats, evolution of novel cultivars, as well as pathotypes that can further lead to the progression of epidemics (Yáñez-López et al. 2012). Altering climatical situations have also forced a significant increment in the virulence of different pests that further result in greater numeral of pest manifestations. The disease sternness of plant systems is suggestively affected by elevated temperature and extent of exposure confronted by plants (Evans et al. 2008). Alterations in temperatures, predominantly optimal temperatures, affect the progression of hosts and functional characteristics of pathogenic microbes and consequently shape the occurrence of disease progression (Suzuki et al. 2014; Ashoub et al. 2015). Furthermore, the variations assessed in the host biology and that of pathogens as well, on account of their temperature requirements, link to a hard and intricate temperament of disease consequences. Thus, the climate-induced changes amend the sensing as response mechanisms of soil microbiological communities along with the alterations in the behavioral responses of plant systems and, therefore, will absolutely moderate the plant-pathogen interactions.

The plant and microbial systems have adapted to particular environments, and their intercommunications advocating their interdependence have also evolved under a defined set of precise environmental conditions. Therefore, any particular change in the environmental conditions has the potential to disrupt any kind of plant-microbe interaction. The most studied and explored plant-microbe interaction is mutualism. The mutualistic communications intensely impact the assembly as well as steadiness of ecological groups to some extent as they can relate different classes to a mutual destiny, triggering annihilation cascades that can trail in a direct or indirect manner from the loss of inter-reliant species. Thus, the changing climatic conditions can directly distress the mutualistic grids by altering the abundance of mutualistic associates. Additionally, it can also distress the mutualistic associations in an indirect manner by changing the characters that construct communications (Encinas-Viso et al. 2012). The anthropogenic climatic changes alter the expression of characters, species dispersal, phenology, and eventually the evolution of species. Such upshots can sequentially modify frequency as well as strength of mutualistic interactions.

The augmented levels of CO_2 also affect the biomass accumulation in C_3 and C_4 plants. During the state of warming, a higher accrual of aboveground biomass has been reported in C_3 plants as equated to C_4 plants (Poorter and Navas 2003). These alterations of biomass accumulation further govern the association of plant systems with advantageous microorganisms predominantly arbuscular mycorrhizal (AM) fungi. In general, C_4 plants allot enhanced carbon to AM fungi for developing beneficial association; thus, evolutionary force favors the association of AM fungi with C_4 plants (Aamir et al. 2019). Therefore, it can be best assessed how the climatic changes disturb the biomass accumulation that in turn affects the symbiotic relationship of AM fungi with plant system. The symbiotic associations of host with *Glomus intraradices* and *Glomus mosseae* get disturbed under higher temperature, and this has also been confirmed experimentally (Baon et al. 1994; Monz et al. 1994). In addition to bipartite interactions, the climatic changes are also known to alter the tripartite mutualistic interconnections among the flowering florae, soil microorganisms, and insect pollinators like bees.

The alteration of interactions between rhizobia and leguminous plants is another example of the effect of fluctuating climatical regimes on mutualistic interconnections. Rhizobia owe the capability to enhance the seed germination by secreting plant hormones and are also known to aid the plant systems in flowering as well as fruiting. The prevalence of drought conditions makes the rhizobia prone to dehydration that further leads to cell death. However, some rhizobia attain the state of dormancy in the quest to avoid water loss. This dormancy leads to the inhibition of nitrogen fixation followed by demodulation of leguminous plants. Thus, it is quite clear that soil dryness weakens the rhizobium-legume mutualistic association (Keeler et al. 2021). In addition, the active microbial cells present in dry soils also find it difficult to establish interactions with plant roots since decline in moisture content of soils is also known to damagingly affect the signaling capabilities of soil bacteria as well as plants (Schimel 2018; Williams and de Vries 2020). Therefore, the rapidly changing climatic conditions are largely known to distress the plant-microbe interactions either in a direct or an indirect way that in turn affects the fitness, growth, and health of plant systems.

Decline in Carbon Sequestration

Soil acts as the principal pool of carbon as the carbon content of soil is found to be superior to that of vegetations and atmosphere collectively (Averill et al. 2014; Tan et al. 2014). The SOM is constituted by the dead and decayed animals and plants besides microbes, along with some inorganic configurations like lime and carbonates. The soil organic matter is also broadly classified into active that accounts for the 35% proportion and passive that makes a count of 65%. The active fraction is represented by living organisms and dead animals along with the residual plant parts that comprise effortlessly digestible proteins and carbohydrates. On the contrary, the passive part of SOM is represented by constituents that prove hard for the microbial decomposition. Primarily, the soil organic carbon finds its origin from the atmospheric carbon dioxide that is arrested by plants and autotrophs via photosynthesis into several organic composites like sugars and cellulose that further leads to the building up of their biomass. The organic matter present in soil contributes suggestively toward the quality as well as fertility of soils as it encompasses a proportion equaling to 58% of SOM (Chan 2008; Stockmann et al. 2013). Therefore, carbon loss from the soil systems would distress the environments as well as properties of soils which unswervingly impacts the agrarian production. This also leads to increment in the atmospheric levels of CO_2 that further leads to increments in the universal temperatures ultimately leading to a state of global warming.

Carbon sequestration is defined as a progression that adds to the soil organic carbon content by confiscating atmospheric carbon dioxide and introducing it into the soil systems. Apart from the plant resources, microorganisms also add up to the soil organic carbon content. The soil microbial anabolic processes contribute to the formation as well as stabilization of soil organic carbon pool (Ahmed et al. 2019). The soil microbiota not only release the carbon to environment via catabolic processes but also transform the exogenic carbon into a definite form of matter via anabolic pathways followed by its storage in the soil systems. Thus, the soil microbes have the potential to play the role of both contributors and decomposers of soil organic carbon (Liang and Zhu 2021). However, the sequestration of carbon by soil microorganisms is largely affected by changing climatic conditions. Several factors like soil organic matter content, moisture and pH, temperature, and different physicochemical possessions influence the activities of soil microbiota that further affect the formation, accretion, and maintenance of soil carbon finding its origin from microbial systems.

The high moisture content of soils and inadequate oxygen supply lead to a decline in the growth efficiency of microorganisms that does not prove to be conducive to the building up of microbial biomass along with the accretion of microbial necromass (Zheng et al. 2019). In addition, higher temperature and prevalence of drought in soil systems as a consequence of elevated levels of carbon dioxide are also known to reduce the growth of microorganisms. The elevation of soil temperature above the optimal temperatures required for the growth of microorganisms slows down the growth of microorganism, and a further increment stops the growth of microorganisms. All these factors make a count and can be held reasonable for the reduction in microbial sequestration of carbon. The altered climatic conditions also affect the microbial carbon sequestration in an indirect manner. The climate change-triggered reduction in the secretion of root exudates further leads to a reduction in the microbial population inhabiting the plant roots which indirectly signifies a reduction in carbon sequestration.

The climatic conditions are changing at a faster rate and the long-term maintenance of soil carbon pool as a function of soil microbial biomass doesn't seem to be conducive. The altered climate would also disturb the soil carbon of terrestrial ecosystems particularly originating from microbial resources. In addition, it would also lay a deep and long effect on the equilibrium amid carbon source as well as sink of soil systems (Liang et al. 2015). Therefore, the microbial sequestration of carbon in the soil systems seem to be negatively affected under the climate extremes.

Alteration in Microbial Distribution

It is a well-established fact that plant communities respond to the climatic changes and the response may lead to the alterations in the geographical distribution of plant systems. Several studies have also speculated about the possible shifts in the habitats of several plant species under the extremes of climatic conditions (Grabherr and Gottfried 1994; Parmesan and Yohe 2003; Walther et al. 2002). However, a large number of reports haven't addressed the capability of allied soil microbiota to change their host range or distribution for maintaining a positive or negative association with their host plants (van der Putten 2012). The microbes native to soil systems are known to be poor dispersers; therefore, their response toward altered climatic conditions is found to be in a different pace than plants. Adding more to the concern, the lack of availability of knowledge about the microbial dispersal at the level of local communities is very scarce. Moreover, the dispersal-sponsored amendments in the key microbial functions like decomposition are also very limited. Conversely, the shifts in dispersion capabilities among plants and microorganisms owe the potential to modify the plant establishment as well as its productivity along with the communications among plant systems in a community, for example, by modifications in the plant litter input eminence (Bever et al. 1997; Nunez et al. 2009; Bever et al. 2010).

Although it is largely acknowledged that microbiological groups also respond to the climatic variations, the speed as well as frequency at which isolated microbiological groups can acclimatize to climatic changes is generally unidentified. Consequently, the arising queries like the scale at which microbiological dispersal restraint begins to matter for ecosystem functioning and how rapidly will the microbial systems acclimatize to the varying climate still need to be addressed. The microbial communities inhabiting the soil might retort to the strain instigated by climatic changes via shifting their distribution in the soil systems. For instance, the higher soil surface temperatures may force the downward movement of soil microbiota deep in the soil profile in the quest of optimal thermal range. This kind of recategorization of microbiota in the soil systems can further modify the plant-microbe process intercommunications. However, the extent of change on microbe-microbe or plant-microbe interactions due to the direct and/or the indirect influences of climate changes that may still be considered pertinent for the ecosystem functioning still remains mysterious. Evans et al. (2011) evaluated the effect of climate changes on the distribution of microorganism by determining the copiousness of cyanobacteria, pico- and nanoeukaryotic algae, nanoflagellates, heterotrophic bacteria, and viruses using flow cytometry. In the waters of future sub-Antarctic zone, shallow mixed layers along with the elevated iron levels will raise the abundance of viral, bacterial, and cyanobacterial members. Conversely, waters of the Polar Frontal Zone displayed comparatively lower concentrations of autotrophic and heterotrophic microorganisms and viruses, revealing iron restriction in the region. In a similar way, Castro et al. (2010) also evaluated the consequence of several experimental climatic change drivers on soil microbiological groups. They noticed an increment in the bacterial count in warmed plots with elevated proportion of CO_2, whereas the bacterial abundance declined in warmed plots containing ambient CO_2 levels. Alterations in the rainfall amended the comparative richness of *Acidobacteria* and *Proteobacteria*, whereas *Acidobacteria* declined with an associated increment in the *Proteobacteria* in wet treatments as compared to the dry ones.

The major problem that the globe is facing today is global warming. The current climatic changes are marking an upsurge of high temperature actions (Stocker et al. 2013) along with the mismanagement of soil proficiencies, and unfortunate supervision of flora can further support desertification leading to the development of arid or semi-arid soils. The soils with declined plant cover are directly exposed to powerful solar radiation that leads to an increment in the temperatures of the upper layers of soil. These factors increase the soil temperature significantly higher than the optimum levels required by common mesophilic soil microbes. The temperature values above 40 °C are regularly observed with magnitude reaching 75 °C (Portillo et al. 2012; Gonzalez et al. 2015), and surprisingly, some of the researchers have also reported temperatures higher than 90 °C in deserts (McCalley and Sparks 2009). Consequently, the microbial activity in the upper soil layers, usually up to the top 5 cm, has been proposed to be extremely condensed during temperature extremes (Townsend et al. 1992; Conant et al. 2011). Subsequently, the microbes inhabiting the uppermost layer of soil might migrate to the lower layers of soil having optimal temperature for their sustenance.

The changing patterns of temperature are also known to amend the structure of microbiological community. For instance, Okubo et al. (2014) described noteworthy increments in the comparative loads of methanotrophs, namely, *Methylocystis* and *Methylosinus*, owing to increased temperature. The alteration in the soil microbial community in turn leads to an alteration in the plant-microbe interactions that further affect the fitness of plants (Singh et al. 2019).

Conclusion and Future Prospects

All the higher forms of life including the members of plant kingdom have originated in a microbial world. Plants as well as microorganisms have designed several mechanisms to improve their health. Plants as well as microbial forms of life have evolved in a definite environment, and they can tolerate the change in environmental conditions only up to a certain level. The change in environment besides the tolerance limit of microorganisms puts them in a state of stress that leads to a reduction in their working efficiency along with the ecological role assigned to them. The global climatic conditions are changing at a very faster pace and thus are continuously challenging the fitness as well as working efficiency of plants as well as microbial systems. Microbial systems are known to be the key drivers behind every possible ecological process. The climatic extremes are known to disturb these processes by affecting the functioning of microorganisms. The alteration in these processes is further known to disturb the productivity of plant systems that in turn may create a state of food insecurity in the near future by reducing the agricultural production. Therefore, restoring the climatic change-triggered damage to ecosystem and further stopping these continuously changing conditions can prove to be a useful tool in combatting this challenge. The restoration of cultivable and deteriorated lands owes the potential of removing around 51 gigatons of CO_2 from air that can further contribute in getting an enhanced production of food by a magnitude of 17.6 megatons on annual basis. In addition, agricultural inputs make 70% of freshwater extractions; therefore, the actions to reduce water usage in agricultural sector, too, without compromising the agricultural yields will also aid in achieving a milestone toward acclimatizing to fluctuating climatic conditions. Furthermore, the adoption of sustainable practices and afforestation along with the reduced human interventions can aid in restricting the changes in global climatic conditions.

References

Aamir M, Rai KK, Dubey MK, Zehra A, Tripathi YN, Divyanshu K, Samal S, Upadhyay RS (2019) Impact of climate change on soil carbon exchange, ecosystem dynamics, and plant–microbe interactions. In: Climate change and agricultural ecosystems. Woodhead Publishing, Cambridge, MA, pp 379–413

Abd SA (2000) Impacts of climate change on microbial activity in agricultural Egyptian soils. In: Climate change impacts on agriculture and food security in Egypt. Springer, Cham, pp 97–114

Acosta-Martinez V, Cotton J, Gardner T, Moore-Kucera J, Zak J, Wester D, Cox S (2014) Predominant bacterial and fungal assemblages in agricultural soils during a record drought/

heat wave and linkages to enzyme activities of biogeochemical cycling. Appl Soil Ecol 84:69–82

Ahmed AA, Odelade KA, Babalola OO (2019) Microbial inoculants for improving carbon sequestration in agroecosystems to mitigate climate change. Handbook of climate change resilience. Springer International Publishing, Cham, pp 1–21

Allen DE, Singh BP, Dalal RC (2011) Soil health indicators under climate change: a review of current knowledge. Soil health and climate change. Springer, Berlin, Heidelberg, pp 25–45

Allison SD, Treseder KK (2008) Warming and drying suppress microbial activity and carbon cycling in boreal forest soils. Glob Chang Biol 14(12):2898–2909

Alshaal T, El-Ramady H, Al-Saeedi AH, Shalaby T, Elsakhawy T, Omara AE, Gad A, Hamad E, El-Ghamry A, Mosa A, Amer M (2017) The rhizosphere and plant nutrition under climate change. Essential plant nutrients. Springer, Cham, pp 275–308

Andrade-Linares DR, Zistl-Schlingmann M, Foesel B, Dannenmann M, Schulz S, Schloter M (2021) Short term effects of climate change and intensification of management on the abundance of microbes driving nitrogen turnover in montane grassland soils. Sci Total Environ 780:146672

Arias ME, González-Pérez JA, González-Vila FJ, Ball AS (2005) Soil health: a new challenge for microbiologists and chemists. Int Microbiol 8(1):13–21

Ashoub A, Baeumlisberger M, Neupaertl M, Karas M, Brüggemann W (2015) Characterization of common and distinctive adjustments of wild barley leaf proteome under drought acclimation, heat stress and their combination. Plant Mol Biol 87(4–5):459–471

Averill C, Turner BL, Finzi AC (2014) Mycorrhiza-mediated competition between plants and decomposers drives soil carbon storage. Nature 505(7484):543–545

Bakshi M, Varma A (2010) Soil enzyme: the state-of-art. In: Soil enzymology. Springer, Berlin, Heidelberg, pp 1–23

Baon JB, Smith SE, Alston AM (1994) Phosphorus uptake and growth of barley as affected by soil temperature and mycorrhizal infection. J Plant Nutr 17(2–3):479–492

Bardgett RD, Freeman C, Ostle NJ (2008) Microbial contributions to climate change through carbon cycle feedbacks. ISME J 2(8):805–814

Bardgett RD, Manning P, Morriën E, De Vries FT (2013) Hierarchical responses of plant–soil interactions to climate change: consequences for the global carbon cycle. J Ecol 101(2):334–343

Barnard RL, Osborne CA, Firestone MK (2013) Responses of soil bacterial and fungal communities to extreme desiccation and rewetting. ISME J 7(11):2229–2241

Beier C, Emmett BA, Peñuelas J, Schmidt IK, Tietema A, Estiarte M, Gundersen P, Llorens L, Riis-Nielsen T, Sowerby A, Gorissen A (2008) Carbon and nitrogen cycles in European ecosystems respond differently to global warming. Sci Total Environ 407(1):692–697

Bergot M, Cloppet E, Pérarnaud V, Déqué M, Marçais B, Desprez-Loustau ML (2004) Simulation of potential range expansion of oak disease caused by Phytophthora cinnamomi under climate change. Glob Chang Biol 10(9):1539–1552

Bever JD, Dickie IA, Facelli E, Facelli JM, Klironomos J, Moora M, Rillig MC, Stock WD, Tibbett M, Zobel M (2010) Rooting theories of plant community ecology in microbial interactions. Trends Ecol Evol 25(8):468–478

Bever JD, Westover KM, Antonovics J (1997) Incorporating the soil community into plant population dynamics: the utility of the feedback approach. J Ecol 1:561–573

Blagodatskaya E, Kuzyakov Y (2008) Mechanisms of real and apparent priming effects and their dependence on soil microbial biomass and community structure: critical review. Biol Fertil Soils 45(2):115–131

Blankinship JC, Niklaus PA, Hungate BA (2011) A meta-analysis of responses of soil biota to global change. Oecologia 165(3):553–565

Bond-Lamberty B, Bailey VL, Chen M, Gough CM, Vargas R (2018) Globally rising soil heterotrophic respiration over recent decades. Nature 560(7716):80–83

Bond-Lamberty B, Thomson A (2010) Temperature-associated increases in the global soil respiration record. Nature 464(7288):579–582

Bouskill NJ, Lim HC, Borglin S, Salve R, Wood TE, Silver WL, Brodie EL (2013) Pre-exposure to drought increases the resistance of tropical forest soil bacterial communities to extended drought. ISME J 7(2):384–394

Bradbury NJ, Powlson DS (1994) The potential impact of global environmental change on nitrogen dynamics in arable systems. In: Soil responses to climate change. Springer, Berlin, Heidelberg, pp 137–153

Burke DJ, Weintraub MN, Hewins CR, Kalisz S (2011) Relationship between soil enzyme activities, nutrient cycling and soil fungal communities in a northern hardwood forest. Soil Biol Biochem 43(4):795–803

Cassán F, Bottini R, Schneider G, Piccoli P (2001) Azospirillum brasilense and Azospirillum lipoferum hydrolyze conjugates of GA20 and metabolize the resultant aglycones to GA1 in seedlings of rice dwarf mutants. Plant Physiol 125(4):2053–2058

Castro HF, Classen AT, Austin EE, Norby RJ, Schadt CW (2010) Soil microbial community responses to multiple experimental climate change drivers. Appl Environ Microbiol 76(4): 999–1007

Cavallaro N, Shrestha G, Birdsey R, Mayes MA, Najjar RG, Reed SC (2018) Second state of the carbon cycle report (SOCCR2): a sustained assessment report. Usgcrp 878:10–7930

Challinor AJ, Watson J, Lobell DB, Howden SM, Smith DR, Chhetri N (2014) A meta-analysis of crop yield under climate change and adaptation. Nat Clim Chang 4(4):287–291

Chan Y (2008) Increasing soil organic carbon of agricultural land. Primefact 735:1–5

Chatterjee D, Saha S (2018) Response of soil properties and soil microbial communities to the projected climate change. In: Advances in crop environment interaction. Springer, Singapore, pp 87–136

Chodak M, Gołębiewski M, Morawska-Płoskonka J, Kuduk K, Niklińska M (2015) Soil chemical properties affect the reaction of forest soil bacteria to drought and rewetting stress. Ann Microbiol 65(3):1627–1637

Clemmensen KE, Bahr A, Ovaskainen O, Dahlberg A, Ekblad A, Wallander H, Stenlid J, Finlay RD, Wardle DA, Lindahl BD (2013) Roots and associated fungi drive long-term carbon sequestration in boreal forest. Science 339(6127):1615–1618

Conant RT, Ryan MG, Ågren GI, Birge HE, Davidson EA, Eliasson PE, Evans SE, Frey SD, Giardina CP, Hopkins FM, Hyvönen R (2011) Temperature and soil organic matter decomposition rates–synthesis of current knowledge and a way forward. Glob Chang Biol 17(11): 3392–3404

Cox PM, Betts RA, Jones CD, Spall SA, Totterdell IJ (2000) Acceleration of global warming due to carbon-cycle feedbacks in a coupled climate model. Nature 408(6809):184–187

Dalal RC, Moloney D (2000) Sustainability indicators of soil health and biodiversity. Management for sustainable ecosystems. Centre for Conservation Biology, Brisbane, p 10108

Eastburn DM, McElrone AJ, Bilgin DD (2011) Influence of atmospheric and climatic change on plant–pathogen interactions. Plant Pathol 60(1):54–69

Encinas-Viso F, Revilla TA, Etienne RS (2012) Phenology drives mutualistic network structure and diversity. Ecol Lett 15(3):198–208

Evans C, Thomson PG, Davidson AT, Bowie AR, Van Den Enden R, Witte H, Brussaard CP (2011) Potential climate change impacts on microbial distribution and carbon cycling in the Australian Southern Ocean. Deep-Sea Res II Top Stud Oceanogr 58(21–22):2150–2161

Evans N, Baierl A, Semenov MA, Gladders P, Fitt BD (2008) Range and severity of a plant disease increased by global warming. J R Soc Interface 5(22):525–531

FAO (2008) Climate change and food security: a framework document. http://ftp.fao.org/docrep/fao/010/k2595e/k2595e00.pdf

FAO (2016) The state of food and agriculture: climate change, agriculture and food security. http://www.fao.org/3/a-i6030e.pdf

FAO (2017) FAO strategy on climate change. Rome (July 2017)

FAO (2019) FAO'S Work on Climate Change United Nations Climate Change Conference 2019

Ferguson B, Lin MH, Gresshoff PM (2013) Regulation of legume nodulation by acidic growth conditions. Plant Signal Behav 8(3):e23426

Freeman C, Ostle NJ, Fenner N, Kang H (2004) A regulatory role for phenol oxidase during decomposition in peatlands. Soil Biol Biochem 36(10):1663–1667

Friedlingstein P, Cox P, Betts R, Bopp L, von Bloh W, Brovkin V, Cadule P, Doney S, Eby M, Fung I, Bala G (2006) Climate–carbon cycle feedback analysis: results from the C4MIP model intercomparison. J Clim 19(14):3337–3353

Fuchslueger L, Bahn M, Fritz K, Hasibeder R, Richter A (2014) Experimental drought reduces the transfer of recently fixed plant carbon to soil microbes and alters the bacterial community composition in a mountain meadow. New Phytol 201(3):916–927

Fuchslueger L, Bahn M, Hasibeder R, Kienzl S, Fritz K, Schmitt M, Watzka M, Richter A (2016) Drought history affects grassland plant and microbial carbon turnover during and after a subsequent drought event. J Ecol 104(5):1453–1465

Fuhrmann JF (2005) Microbial metabolism. In: Sylvia DM, Hartel PG, Fuhrmann JJ, Zuberer DA (eds) Principles and applications of soil microbiology, 2nd edn. Pearson Prentice Hall, Upper Saddle River, NJ, pp 54–83

Ghini R, Hamada E, Pedro Júnior MJ, Marengo JA, Gonçalves RR (2008) Risk analysis of climate change on coffee nematodes and leaf miner in Brazil. Pesq Agrop Brasileira 43:187–194

Glick BR, Todorovic B, Czarny J, Cheng Z, Duan J, McConkey B (2007) Promotion of plant growth by bacterial ACC deaminase. Crit Rev Plant Sci 26(5–6):227–242

Gonzalez JM, Portillo MC, Piñeiro-Vidal M (2015) Latitude-dependent underestimation of microbial extracellular enzyme activity in soils. Int J Environ Sci Technol 12(7):2427–2434

Grabherr GM, Gottfried PH (1994) Climate effects on mountain plants. Nature 369:448–448

Gray SB, Brady SM (2016) Plant developmental responses to climate change. Dev Biol 419(1): 64–77

Gschwendtner S, Engel M, Lueders T, Buegger F, Schloter M (2016) Nitrogen fertilization affects bacteria utilizing plant-derived carbon in the rhizosphere of beech seedlings. Plant Soil 407(1): 203–215

Gul S, Whalen JK (2016) Biochemical cycling of nitrogen and phosphorus in biochar-amended soils. Soil Biol Biochem 103:1–5

Haase S, Neumann G, Kania A, Kuzyakov Y, Römheld V, Kandeler E (2007) Elevation of atmospheric CO2 and N-nutritional status modify nodulation, nodule-carbon supply, and root exudation of Phaseolus vulgaris L. Soil Biol Biochem 39(9):2208–2221

Hartmann M, Brunner I, Hagedorn F, Bardgett RD, Stierli B, Herzog C, Chen X, Zingg A, Graf-Pannatier E, Rigling A, Frey B (2017) A decade of irrigation transforms the soil microbiome of a semi-arid pine forest. Mol Ecol 26(4):1190–1206

Haynes RJ (2008) Soil organic matter quality and the size and activity of the microbial biomass: their significance to the quality of agricultural soils. In: Soil mineral microbe-organic interactions. Springer, Berlin, Heidelberg, pp 201–231

Heimann M, Reichstein M (2008) Terrestrial ecosystem carbon dynamics and climate feedbacks. Nature 451(7176):289–292

Henry HA (2013) Reprint of "soil extracellular enzyme dynamics in a changing climate". Soil Biol Biochem 56:53–59

Hodge A, Robinson D, Fitter A (2000) Are microorganisms more effective than plants at competing for nitrogen? Trends Plant Sci 5(7):304–308

Hu S, Chapin FS, Firestone MK, Field CB, Chiariello NR (2001) Nitrogen limitation of microbial decomposition in a grassland under elevated CO 2. Nature 409(6817):188–191

Hungria M, Stacey G (1997) Molecular signals exchanged between host plants and rhizobia: basic aspects and potential application in agriculture. Soil Biol Biochem 29(5–6):819–830

Hungria M, Vargas MA (2000) Environmental factors affecting N2 fixation in grain legumes in the tropics, with an emphasis on Brazil. Field Crop Res 65(2–3):151–164

Huxman TE, Snyder KA, Tissue D, Leffler AJ, Ogle K, Pockman WT, Sandquist DR, Potts DL, Schwinning S (2004) Precipitation pulses and carbon fluxes in semiarid and arid ecosystems. Oecologia 141(2):254–268

IPCC (2007a) Climate Change 2007: synthesis report. Contribution of working groups I, II and III to the fourth assessment report of the intergovernmental panel on climate change. Geneva, Switzerland

IPCC (2007b) Technical summary. In Climate Change 2007: Mitigation. Contribution of Working Group III to the Fourth Assessment Report of the Intergovernmental Panel on Climate Change. Cambridge, United Kingdom and New York, NY, USA, Cambridge University Press

IPCC (2007c) Agriculture. In Climate Change 2007: Mitigation. Contribution of Working Group III to the Fourth Assessment Report of the Intergovernmental Panel on Climate Change. Cambridge, United Kingdom and New York, NY, USA, Cambridge University Press

IPCC (2014). http://www.ipcc.ch/pdf/assessment-report/ar5/wg2/WGIIAR5-Chap7_FINAL.Pdf

IPCC (2018) In: Masson-Delmotte V, Zhai P, Pörtner H-O, Roberts D, Skea J, Shukla PR, Pirani A, Moufouma-Okia W, Péan C, Pidcock R, Connors S, Matthews JBR, Chen Y, Zhou X, Gomis MI, Lonnoy E, Maycock T, Tignor M, Waterfield T (eds). Global Warming of 1.5°C. An IPCC Special Report on the impacts of global warming of 1.5°C above pre-industrial levels and related global greenhouse gas emission pathways, in the context of strengthening the global response to the threat of climate change, sustainable development, and efforts to eradicate poverty. In Press

IPCC (2019) In: Shukla PR, Skea J, Buendia EC, Masson-Delmotte V, Pörtner H-O, Roberts DC, Zhai P, Slade R, Connors S, van Diemen R, Ferrat M, Haughey E, Luz S, Neogi S, Pathak M, Petzold J, Pereira JP, Vyas P, Huntley E, Kissick K, Belkacemi M, Malley J (eds) Climate Change and Land: an IPCC special report on climate change, desertification, land degradation, sustainable land management, food security, and greenhouse gas fluxes in terrestrial ecosystems. In press

IPCC Climate Change (2007) Synthesis report. Summary for policymakers. http:www.ipcc.ch

Jablonski LM, Wang X, Curtis PS (2002) Plant reproduction under elevated CO2 conditions: a meta-analysis of reports on 79 crop and wild species. New Phytol 156(1):9–26

Jansson JK, Hofmockel KS (2020) Soil microbiomes and climate change. Nat Rev Microbiol 18(1): 35–46

Joergensen RG, Brookes PC, Jenkinson DS (1990) Survival of the soil microbial biomass at elevated temperatures. Soil Biol Biochem 22(8):1129–1136

Kandeler E, Mosier AR, Morgan JA, Milchunas DG, King JY, Rudolph S, Tscherko D (2006) Response of soil microbial biomass and enzyme activities to the transient elevation of carbon dioxide in a semi-arid grassland. Soil Biol Biochem 38(8):2448–2460

Kannojia P, Sharma PK, Sharma K (2019) Climate change and soil dynamics: effects on soil microbes and fertility of soil. In: Climate change and agricultural ecosystems. Woodhead Publishing, Cambridge, MA, pp 43–64

Keeler AM, Rose-Person A, Rafferty NE (2021) From the ground up: building predictions for how climate change will affect belowground mutualisms, floral traits, and bee behavior. Climate Change Ecol 100013

Kumar R, Sharma P, Gupta RK, Kumar S, Sharma MM, Singh S, Pradhan G (2020) Earthworms for eco-friendly resource efficient agriculture. In: Resources use efficiency in agriculture. Springer, Singapore, pp 47–84

Liang C, Gutknecht J, Balser T (2015) Microbial lipid and amino sugar responses to long-term simulated global environmental changes in a California annual grassland. Front Microbiol 6:385

Liang C, Zhu X (2021) The soil microbial carbon pump as a new concept for terrestrial carbon sequestration. Sci China Earth Sci 1:1–4

Lobell DB, Schlenker W, Costa-Roberts J (2011) Climate trends and global crop production since 1980. Science 333(6042):616–620

Luedeling E, Steinmann KP, Zhang M, Brown PH, Grant J, Girvetz EH (2011) Climate change effects on walnut pests in California. Glob Chang Biol 17(1):228–238

Lukac M, Calfapietra C, Lagomarsino A, Loreto F (2010) Global climate change and tree nutrition: effects of elevated CO2 and temperature. Tree Physiol 30(9):1209–1220
Maddela NR, Golla N, Vengatampalli R (2017) Soil enzymes: influence of sugar industry effluents on soil enzyme activities. Springer International Publishing, Cham
Maloy S (2021) Climate change and microbes. In: The impacts of climate change. Elsevier, United Kingdom, pp 195–203
McCalley CK, Sparks JP (2009) Abiotic gas formation drives nitrogen loss from a desert ecosystem. Science 326(5954):837–840
McKenzie NJ, Baldock J, Balks MR, Camps Arbestain M, Condron LM, Elder-Ratutokarua M, Grundy MJ, Hewitt A, Kelliher F, Leys JF, McDowell RW (2015) Regional assessment of soil change in the Southwest Pacific. Status of the World's soil resources (SWSR) - Main report. FAO and ITPS, Rome, pp 476–519
Mele PM (2011) Soil biota, soil health and global change. In: Soil health and climate change. Springer, Berlin, Heidelberg, pp 155–177
Mohammadi K, Sohrabi Y, Heidari G, Khalesro S, Majidi M (2012) Effective factors on biological nitrogen fixation. Afr J Agric Res 7(12):1782–1788
Monz CA, Hunt HW, Reeves FB, Elliott ET (1994) The response of mycorrhizal colonization to elevated CO 2 and climate change in Pascopyrum smithii and Bouteloua gracilis. Plant Soil 165(1):75–80
Nunez MA, Horton TR, Simberloff D (2009) Lack of belowground mutualisms hinders Pinaceae invasions. Ecology 90(9):2352–2359
Okubo T, Tokida T, Ikeda S, Bao Z, Tago K, Hayatsu M, Nakamura H, Sakai H, Usui Y, Hayashi K, Hasegawa T (2014) Effects of elevated carbon dioxide, elevated temperature, and rice growth stage on the community structure of rice root—associated bacteria. Microbes Environ 29(2):184–190
Orchard VA, Cook FJ (1983) Relationship between soil respiration and soil moisture. Soil Biol Biochem 15(4):447–453
Parmesan C, Yohe G (2003) A globally coherent fingerprint of climate change impacts across natural systems. Nature 421(6918):37–42
Pinto HS, Zullo J Jr, Assad ED, Evangelista BA (2007) O aquecimento global e a cafeicultura brasileira. Embrapa Informática Agropecuária-Artigo em periódico indexado (ALICE) 65–72
Pinto HS, Assad ED, Junior JZ, Evangelista S, Otavian A, Ávila A, Evangelista B, Marin F, Junior C, Pellegrino GQ, Coltri P, Coral G (2008) Global warming and the new geography of agricultural production in Brazil. Centro de Pesquisas Meteorológicas e Climáticas Aplicadas à Agricultura (CEPAGRI) and Brazilian Agricultural Research Corporation (EMBRAPA), Sao Paulo, p 83
Poorter H, Navas ML (2003) Plant growth and competition at elevated CO2: on winners, losers and functional groups. New Phytol 157(2):175–198
Porter JR, Xie L, Challinor AJ, Cochrane K, Howden SM, Iqbal MM, Lobell DB, Travasso MI (2014) Food security and food production systems. In: Climate Change 2014: Impacts, Adaptation, and Vulnerability. Part A: Global and Sectoral Aspects. Contribution of Working Group II to the Fifth Assessment Report of the Intergovernmental Panel on Climate Change [Field CB, Barros VR, Dokken DJ, Mach KJ, Mastrandrea MD, Bilir TE, Chatterjee M, Ebi KL, Estrada YO, Genova RC, Girma B, Kissel ES, Levy AN, MacCracken S, Mastrandrea PR and White LL (eds.)]. Cambridge University Press, Cambridge/New York, NY, pp. 485–533
Portillo MC, Santana M, Gonzalez JM (2012) Presence and potential role of thermophilic bacteria in temperate terrestrial environments. Naturwissenschaften 99(1):43–53
Rani K, Sharma P, Kumar S, Wati L, Kumar R, Gurjar DS, Kumar D (2019) Legumes for sustainable soil and crop management. In: Sustainable management of soil and environment. Springer, Singapore, pp 193–215
Raza S, Zamanian K, Ullah S, Kuzyakov Y, Virto I, Zhou J (2021) Inorganic carbon losses by soil acidification jeopardize global efforts on carbon sequestration and climate change mitigation. J Clean Prod 128036

Reay D (2015) Terrestrial nitrogen and climate change. In: Nitrogen and climate change. Palgrave Macmillan, London, pp 85–102. https://doi.org/10.1057/9781137286963_7

Rustad LE, Campbell J, Marion G, Norby R, Mitchell M, Hartley A, Cornelissen J, Gurevitch J (2001) A meta-analysis of the response of soil respiration, net nitrogen mineralization, and aboveground plant growth to experimental ecosystem warming. Oecologia 126(4):543–562

Sardans J, Peñuelas J, Estiarte M (2008) Changes in soil enzymes related to C and N cycle and in soil C and N content under prolonged warming and drought in a Mediterranean shrubland. Appl Soil Ecol 39(2):223–235

Schafleitner R, Ramirez J, Jarvis A, Evers D, Gutierrez R, Scurrah M (2011) Adaptation of the potato crop to changing climates. In: Crop adaptation to climate change. John Wiley & Sons, Oxford, pp 287–297

Schimel JP (2018) Life in dry soils: effects of drought on soil microbial communities and processes. Annu Rev Ecol Evol Syst 49:409–432

Schimel JP, Mikan C (2005) Changing microbial substrate use in Arctic tundra soils through a freeze-thaw cycle. Soil Biol Biochem 37(8):1411–1418

Shan Q, Yu Y, Yu J, Zhang J (2008) Soil enzyme activities and their indication for fertility of urban forest soil. Front Environ Sci Eng China 2(2):218–223

Sharma MM, Sharma P, Kapoor D, Beniwal P, Mehta S (2021) Phytomicrobiome community: an agrarian perspective towards resilient agriculture. In: Plant performance under environmental stress. Springer, Cham, pp 493–534

Sharma P, Sharma MM, Kapoor D, Rani K, Singh D, Barkodia M (2020) Role of microbes for attaining enhanced food crop production. Microb Biotechnol Basic Res Appl:55–78

Singh BP, Cowie AL, Chan KY (2011) The nitrogen cycle: implications for management, soil health, and climate change. Soil health and climate change (soil biology). Springer, Berlin, Heidelberg, pp 107–129

Singh S, Sangwan S, Sharma P, Devi P, Moond M (2021) Nanotechnology for sustainable agriculture: an emerging perspective. J Nanosci Nanotechnol 21(6):3453–3465

Singh VK, Shukla AK, Singh AK (2019) Impact of climate change on plant–microbe interactions under agroecosystems. In: Climate change and agricultural ecosystems. Woodhead Publishing, Cambridge, MA, pp 153–179

Springmann M, Mason-D'Croz D, Robinson S, Garnett T, Godfray HC, Gollin D, Rayner M, Ballon P, Scarborough P (2016) Global and regional health effects of future food production under climate change: a modelling study. Lancet 387(10031):1937–1946

Steinweg JM, Dukes JS, Paul EA, Wallenstein MD (2013) Microbial responses to multi-factor climate change: effects on soil enzymes. Front Microbiol 4:146

Stocker T (2014) Climate change 2013: the physical science basis: working group I contribution to the fifth assessment report of the intergovernmental panel on climate change. Cambridge University Press, Cambridge

Stocker TF, Qin D, Plattner GK (2013) The physical science basis. Contribution of working group I to the fifth assessment report of the intergovernmental panel on climate change. Cambridge University Press, Cambridge

Stockmann U, Adams MA, Crawford JW, Field DJ, Henakaarchchi N, Jenkins M, Minasny B, McBratney AB, De Courcelles VD, Singh K, Wheeler I (2013) The knowns, known unknowns and unknowns of sequestration of soil organic carbon. Agric Ecosyst Environ 164:80–99

Stone MM, Weiss MS, Goodale CL, Adams MB, Fernandez IJ, German DP, Allison SD (2012) Temperature sensitivity of soil enzyme kinetics under N-fertilization in two temperate forests. Glob Chang Biol 18(3):1173–1184

Suzuki N, Rivero RM, Shulaev V, Blumwald E, Mittler R (2014) Abiotic and biotic stress combinations. New Phytol 203(1):32–43

Tan WF, Zhang R, Cao H, Huang CQ, Yang QK, Wang MK, Koopal LK (2014) Soil inorganic carbon stock under different soil types and land uses on the loess plateau region of China. Catena 121:22–30

Tarnocai C, Canadell JG, Schuur EA, Kuhry P, Mazhitova G, Zimov S (2009) Soil organic carbon pools in the northern circumpolar permafrost region. Glob Biogeochem Cycles 23(2). https://doi.org/10.1029/2008GB003327

Taub DR, Miller B, Allen H (2008) Effects of elevated CO2 on the protein concentration of food crops: a meta-analysis. Glob Chang Biol 14(3):565–575

Townsend AR, Vitousek PM, Holland EA (1992) Tropical soils could dominate the short-term carbon cycle feedbacks to increased global temperatures. Clim Chang 22(4):293–303

Van der Putten WH (2012) Climate change, aboveground-belowground interactions, and species' range shifts. Annu Rev Ecol Evol Syst 43:365–383

Wahid F, Sharif M, Ali A, Fahad S, Adnan M, Noor M, Mian IA, Khan IA, Alam M, Saeed M, Ilyas M (2020) Plant-microbes interactions and functions in changing climate. In: Environment, climate, plant and vegetation growth 2020. Springer, Cham, pp 397–419

Waldrop MP, Firestone MK (2006) Response of microbial community composition and function to soil climate change. Microb Ecol 52(4):716–724

Walther GR, Post E, Convey P, Menzel A, Parmesan C, Beebee TJ (2002) Ecological responses to recent climate change. Nature 416:389–395

Wang C, Chen Z, Unteregelsbacher S, Lu H, Gschwendtner S, Gasche R, Kolar A, Schloter M, Kiese R, Butterbach-Bahl K, Dannenmann M (2016) Climate change amplifies gross nitrogen turnover in montane grasslands of Central Europe in both summer and winter seasons. Glob Chang Biol 22(9):2963–2978

Wang L, Feng Z, Schjoerring JK (2013) Effects of elevated atmospheric CO2 on physiology and yield of wheat (Triticum aestivum L.): a meta-analytic test of current hypotheses. Agric Ecosyst Environ 178:57–63

Williams A, de Vries FT (2020) Plant root exudation under drought: implications for ecosystem functioning. New Phytol 225(5):1899–1905

Williams A, Pétriacq P, Beerling DJ, Cotton TE, Ton J (2018) Impacts of atmospheric CO2 and soil nutritional value on plant responses to rhizosphere colonization by soil bacteria. Front Plant Sci 9:1493

Wixon DL, Balser TC (2009) Complexity, climate change and soil carbon: a systems approach to microbial temperature response. Syst Res Behav Sci 26(5):601–620

Yáñez-López R, Torres-Pacheco I, Guevara-González RG, Hernández-Zul MI, Quijano-Carranza JA, Rico-García E (2012) The effect of climate change on plant diseases. Afr J Biotechnol 11(10):2417–2428

Yang L, Zhang Y, Li F (2012) Soil enzyme activities and soil fertility dynamics. In: Advances in citrus nutrition. Springer, Dordrecht, pp 143–156

Yiqi L, Zhou X (2010) Soil respiration and the environment. Elsevier, London

Yuste JC, Fernandez-Gonzalez AJ, Fernandez-Lopez M, Ogaya R, Penuelas J, Sardans J, Lloret F (2014) Strong functional stability of soil microbial communities under semiarid Mediterranean conditions and subjected to long-term shifts in baseline precipitation. Soil Biol Biochem 69: 223–233

Yuste JC, Nagy M, Janssens IA, Carrara A, Ceulemans R (2005) Soil respiration in a mixed temperate forest and its contribution to total ecosystem respiration. Tree Physiol 25(5):609–619

Zahran HH (2017) Legume-microbe interactions under stressed environments. In: Microbes for legume improvement. Springer, Cham, pp 301–339

Zak DR, Pregitzer KS, King JS, Holmes WE (2000) Elevated atmospheric CO2, fine roots and the response of soil microorganisms: a review and hypothesis. New Phytol 147(1):201–222

Zheng Q, Hu YT, Zhang SS, Noll L, Böckle T, Richter A, Wanek W (2019) Growth explains microbial carbon use efficiency across soils differing in land use and geology. Soil Biol Biochem 128:45–55

Zibilske LM, Bradford JM (2007) Oxygen effects on carbon, polyphenols, and nitrogen mineralization potential in soil. Soil Sci Soc Am J 71(1):133–139

Ziska LH (2011) Global climate change and carbon dioxide: assessing weed biology and management. Handbook on climate change and agroecosystems: impacts, adaptation, and mitigation. Imperial College Press, London, pp 191–208

Zytynska SE (2021) Embracing the complexity of plant-microbe-insect interactions under a changing climate for sustainable agriculture. Curr Opin Insect Sci 44:89–94

Climate Change: Current Scenario and Its Implications on Soil Health and Mitigation

Ranjana Yadav, Jitendra Kumar, and Kirpa Ram

Abstract Climate change affects natural and socioeconomic environments throughout the world although their vulnerability differs across regions and countries. The increased incidence of life-threatening events such as hurricanes, famines, warm waves, floods, recession of glaciers, and monsoon rain variability causes instability in food production, reduced water availability, loss of habitats, and displacement. Global change in climate will pressure global precipitation patterns, fluctuating precipitation quantity received and the distribution of rainfall in many locations. According to various estimations in precipitation, evaporation, and temperature, substantial changes occur in the turnover of soil organic matter. To tackle climate change, an interdisciplinary economic climate-resilient approach must be used and an agroforestry system for maximum land use must be outlined. This chapter reviewed the impact of climate change on soil and mitigation approaches.

Keywords Climate change · Crop yield · Mitigation · Soil well-being

Climate Change

Climate change impact has been observed worldwide. Climate change influences the natural and socioeconomic environments throughout the world although their vulnerability differs across regions and countries. Climate change is also an added risk to arid and semiarid and coastal regions. Changes in the environment have been discovered to create climate stress, as distinct to the usual pressures caused by activities within that environment, and not apparent in everyday life but do have subsidiary influences upon an individual's life. Extreme atmospheric heat raises soil

R. Yadav
Indian Plywood Industries Research and Training Institute, Centre, Mohali, Punjab, India

J. Kumar (✉)
Department of Biosciences-UIBT, Chandigarh University, Mohali, Punjab, India

K. Ram
Department of Botany, Baba Mastnath University, Rohtak, Haryana, India

© The Author(s), under exclusive license to Springer Nature Singapore Pte Ltd. 2022

A. Vaishnav et al. (eds.), *Plant Stress Mitigators*,
https://doi.org/10.1007/978-981-16-7759-5_9

heat, which enhances the chemical reactions along with dispersion processes in the mixture (Buol et al. 1990). Diaz et al. (1997) reported that no change would occur with regard to most physical variables. Extreme weather events such as droughts, floods, and storms are expected to become more common as warming trends and precipitation patterns accelerate (Desanker 2002). CO_2, CH_4, N_2O, H_2O, and O_3 constitute the Earth's atmosphere and include the chlorofluorocarbons and greenhouse gases. Many greenhouse gases are constituents of the natural atmosphere, and their atmospheric concentrations were estimated by the entirety of chemical, biological, and geophysical responses (National Research Council 2001). Global warming refers to any change in the environment over the time period, either caused by natural factors or human activity (Karmakar et al. 2016). As per the IPCC data, worldwide, average heat would climb between 1.1 and 6.4 °C by 2090–2099 (IPCC 2007; Brevik et al. 2017). The environmental change affects the soil quality, and as a result, agriculture production. Integrated soil fertility management technologies have the ability to halt soil degradation (Mutegi et al. 2018). Changes in the climate directly affect temperature, precipitation, and moisture change while indirectly affecting irrigation, crop rotation, and tillage practice. These all parameters ultimately affect soil functions (Hamidov et al. 2018).

The temperature in the soil also affects the speed of decomposition of organic matter, including the release and uptake of nutrients and plant metabolic processes. Different researchers predict how predicted variations in heat, rainfall, and absorption will induce a major change in soil nutrient rotation and green gas house variability, maintaining a reasonable carbon input to the soils from plants. Soil production is broadly regulated by three key variables: environment, growing conditions, and type of crop. Soil growth is primarily anticipated as a result of changes in soil humidity levels and rises in soil heat and CO_2 levels. Coping with and adapting to climate stress are therefore an issue of concern for experts worldwide; by implementing agricultural methods and techniques that mitigate negative impacts of rising or declining precipitation and extreme rising conditions, agriculture may respond to climate change.

Agriculture in mostly heavy rain areas is dependent on monsoon, demonstrating its reliance on the environment, and it is critical in respect to food security in two manners: generating the nutrition consumed by persons and employing 36% of the global workforce (Pareek 2017). Due to changing climate and varying precipitation, food security, health, education, and wealth become concerns. The following factors contribute to their susceptibility to climate change: adaptive capacity, sensitivity, and exposure (Mutegi et al. 2018). Higher temperatures hasten organic matter decomposition and soil degradation processes, as well as carbon, nitrogen, phosphorous, potassium, and sulfur cycling in the soil–plant–atmosphere system. The increase in nitrogen fixation is related to enhanced root growth (Anjali and Dhananjaya 2019). The mean temperature of every region witnesses a larger change with an increment of global warming (Collins et al. 2013; IPCC 2021). Climate change has many negative consequences, including water scarcity, soil fertility loss, and pest infestations in crops (Malhi et al. 2021).

Climate Change Effect on the Health of Soil

Wetter, denser soils hold heat and help stabilize the environment from temperature swings better than drier, looser soils. During the course of a day, the temperature in deserts can rise by more than 60 °F. These heat islands can have a temperature difference of up to 5 °C during the day and up to 20 °C at night. These microclimates can have diverse plant and animal communities around the surrounding area, and they are crucial for ecological niche creation (Brevik 2013; Sindelar 2015). The soils on earth contain more than double the carbon quantity in the atmosphere. According to researchers, global warming may disturb the conformation of the soil at the molecular level, which might have a considerable impact on carbon dioxide levels in the atmosphere (Herout and Shutterstock 2009). Therefore, for attaining potential yield or most yield, climate is a major role player (Karmakar et al. 2016). The potential effect on soil well-being coming about because of the environmental change is over natural substance stockpile, temperature systems, hydrology, and saltiness. Subsequent are the significant outcomes of worldwide environmental alteration on soil properties. Soil well-being and quality are a major concern in the green deal of the United Nations Sustainable Development Mission (Bonfante et al. 2020).

Physical Parameters of Soil

These enhancements normally consist of the expansion of natural substances as well as H_2O and O_2, but in some areas, modest persistent expansion of novel mineral ingredients occurs outside of dissolvable components that are conserved from groundwater. Materials broken down or suspended in water percolating through the profile or running off the surface alongside water and through permeable soil account for a majority of losses. The higher temperature, high and low limits of precipitation, expansion in CO_2 concentration, and their connections because of the environmental change are relied upon to impact a few soil physical processes, which put the soil at the critical point of salinization, diminished H_2O accessibility, and variations in C, N elements, supplement stockpiling in the soil, and decrease soil biodiversity. Soil health is influenced by the physical qualities and processes of the soil (Allen et al. 2011), which modify water transport, root penetration, and water congestion. Soil decomposition results in CO_2 release into the air and a fall in the C: N ratio, and these two consequences might be partially countered via increased root biochar and agriculture wastes as a result of plant response to elevated CO_2. The organic matter degradation rate is influenced by soil temperature. It came about into discharge, uptake of supplements, and plant metabolic cycles. Soil efficiency and supplement cycling are subsequently affected by the total and activity of soil microorganisms. Microorganisms present in the soil have two significant capacities; for example, they go about as specialists of supplement component change to stockpile C and supplements in living biomass going about responsible repository accessible supplement by quick revenue. Changes in C:N proportions in tree buildups that have found in the soil have an impact on soil microbiological periods and influence the creation of subsequent chemicals. Environmental change may expand the percentage of land degradation, limiting food creation even further. Rainfall expansions will speed up the pace of soil erosion, decreasing agricultural yield considerably much. Another way where erosion could speed up is through a reduction in rainfall, which could prompt drought and expanded danger of wind erosion (Parry et al. 1999). If the speed of erosion is not controlled, then constant depletion of soil will force farmers to abandon their lands. In this way, destruction is one of the threats to the production of foods in hotter climates. Other problems related to land degradation, for example, waterlogging, soil saltiness, and sodicity advancement, are arising because of quick land practice design and changes in land cover.

The soil surface is normally the extent of sand, sediment, and clay, which affect the environment directly. Significant soil measures along with the surface separation in the soil surface are considerably affected by the four probable environment situations (arid, semiarid, subhumid, and humid) (Scharpenseel et al. 1990; Brinkman and Brammer 1990).

Soil Edifice and Constancy

Soil structure is the course of action and association of fundamental and auxiliary elements in a soil mass that controls the percentage of H_2O and air in the soil. Cumulative constancy to outer energy like extreme rainfall and cultivation is dictated by soil edifice and constancy, scope of biochemical characteristics, and managing performance (Moebius et al. 2007). The natural substance, their amount and types, development tactics, and typical physical procedures conducted all have an impact on the structure's nature and class. A drop in soil natural matter levels causes a decrease in total strength, penetration rates, and a rise in shrinking and loss vulnerability (Bot and Benites 2005; Karmakar et al. 2016).

Void Fraction or Porosity

The capacity of the soil to accumulate root region water and air is determined by a void fraction, which is a proportion of the void areas in a material expressed as a percentage, and pore size dispersion (Reynolds et al. 2002). Soil superiority, mass thickness, miniature porosity, and particle pre-parameters are linked to porous structure qualities. Though soil porous structure and water discharge qualities straightforwardly impact a scope of soil files, including soil air circulation limit and plant accessible water limit. Furthermore, future environmental change situations (raised CO_2 and temperature, variable and outrageous precipitation occasions) may adjust root improvement and soil organic exercises. Soil void fraction and particle size appropriation and subsequent soil capacities are probably influenced in unforeseen ways. Forthcoming examinations on the well-being of soil relationship and environmental variation will need to be modified. Diminished microbial movement, decreased root development and exudates, lessening total solidness, and expanded precipitation forces where downpour beads sway cause surface fixing on sodic soils. It will prompt helpless yield development and build the odds of surface spillover.

Plant Available Water and Infiltration

The H_2O accessibility for development of crops and significant earth measures are administered as the scope of characteristics of the soil including porosity, field limit, lesser breaking point of plant accessible water (hence barring osmotic potential), and miniature pore stream and surface (Jarvis 2007; Reynolds et al. 2002). To evaluate agricultural consequences, plant accessible water volume was used as a component of soil (integrative) wellness testing. Furthermore, climate variability, particularly unpredictable and heavy rain or drought events, may cause soil accessible water and distribution to react quickly (Lal 1995). These accessible H_2O content in the earth may aid in alleviating the effect of serious precipitation and dry spell occasions or extreme destruction occasions (Salvador Sanchis et al. 2008).

Bulk Density

Bulk density is regularly evaluated to describe the condition of soil minimization in relation to land management practices (Hakansson and Lipiec 2000). Overall, mass density is contrarily connected with the organic matter of the soil or soil organic carbon (Weil and Magdoff 2004). The deficiency of organic form C content from expanded deterioration is due to raised temperature. After infrequent and intense rainfall and dry spell, land management practices and environmental changes pressures may cause bulk density expansion, making the soil more susceptible to compaction (Davidson and Janssens 2006; Birkas et al. 2009).

Root Deepness

The level of rooting is viewed as a significant marker of soil well-being since variations under this feature probably influence plant accessible water limit, subsoil saltiness, percentage of SOC, or different features that show physiochemical limitations in the soil profile (Arias et al. 2005; Birkas et al. 2009). During the delayed dry season, the effect of subsoil imperatives, e.g., saltiness and excessive Cl^{-1} concentrations, is probably more prominent on plant accessible water and efficiency (Dang et al. 2008; Rengasamy 2010).

Surface Cover of Soil

The surface area of the soil gives the scope of significant environmental capacities, including the security of soil's superficial water content. The surface area of soil supplements maintenance, carbon obsession, and, in certain occasions, nitrogen obsession and backing local seed propagation. Surface cover development, principally identified with sodicity, is utilized to portray soil well-being under environmental change (Box and Bruce 1996).

Soil Temperature

Profits and losses of solar energy at the surface, evaporation, heat transmission via the soil profile, and convection exchange by means of the development of gas and water make up the soil temperature systems (Karmakar et al. 2016). Soil temperature will speed up soil measures, fast deterioration of natural substance, expanded biological action, faster supplement discharge, and an increased nitrification rate. Nonetheless, the type of plant growing on the surface of the soil will influence its temperature. Climate transition or adaptation strategies may cause this to change.

Soil pH

The pH of the soil is influenced by the source material, lasting season, plants, and habitat. It is regarded as a significant indicator of soil well-being. Environmental changes like extreme heat, CO_2 enrichment, irregular rainfall, and climatic nitrogen accumulation would not expose the majority of soils to rapid pH shifts. Nonetheless, environmental change, on the other hand, will have an influence on natural organic substance quality, carbon, crop accessible water, and plant efficiency, all of which will alter soil pH (Reth et al. 2005).

Soil Electrical Conductivity

Soil dielectric strength is a proportion of salt fixation. It could also advise patterns in saltiness, agriculture execution, supplement cycling, and natural movement. It can act as a substitute for a portion of soil primary decline when combined with pH, especially in saline soils (Arnold et al. 2005). Dielectric properties have been utilized as a compound marker to illuminate soil natural superiority because of yield the board rehearses (Gil et al. 2009; Smith et al. 2002).

Sorption and Cation Exchange Capacity

Sorption and cation trade limits are viewed as significant characteristics, especially the maintenance of significant supplements Ca^{2+}, Mg^{2+}, and K^+ and immobilization of conceivably harmful cations Al^{3+} and Mn^{3+}. These properties would thus be able to be helpful pointers of soil well-being, exhibiting a soil's ability to ingest supplements just as pesticides and synthetic substances (Ross et al. 2008). Since CEC of coarse-finished soils and low-action earth soils is credited to that of soil organic matter (SOM), the expanding disintegration and loss of SOM because of raised temperature may prompt deficiency of the cation exchange capacity of the soil (Davidson and Janssens 2006). It might bring about expanded draining of base cations because of high and exceptional precipitation occasions, along these lines shipping alkalinity from soil to streams.

Plant Available Nutrients

Estimation of extractable supplements may give a sign of a soil's ability to help plant development; on the other hand, it might recognize basic or edge esteems for ecological danger evaluation (Dalal and Moloney 2000). Supplements, mainly nitrogen, are personally connected with soil natural carbon chain (Weil and Magdoff 2004) and thus variables of environmental change as extreme heat, varied rainfall, and barometrical nitrogen changes have an influence on nitrogen cycling and other accessible supplements like phosphorus and sulfur.

Biological Parameters of Soil

The soil microorganisms are versatile to changes under ecological conditions. Under states of environmental change, organic pointers structure an essential part in soil well-being appraisal. SOM and its components, surface carbon, breath, and soil organisms' nutrients were chosen as key organic pointers for the scope of the current investigation.

Soil Organic Matter

Organic soil nutrient is composed of both living and nonliving components; SOM is perhaps utmost unpredictable and mixed soil components, with varying characteristics, capacities, and high turnover (Weil and Magdoff 2004). It gives an additional provision with inputs to the charge soil qualities (charge), serves as a sink for the source of carbon and nitrogen, and controls P as well as S recycling to a degree. It has the capacity to shape composites with multivalent particles and natural mixtures.

It gives microbial and faunal territory and substrates, influencing total dependability, water maintenance, and pressure-driven properties (Haynes 2008). Land use practice that develops SOM will help in engrossing CO_2 from the climate, in this manner moderating a worldwide temperature alteration (Hoffland et al. 2020; Navarro-Pedreño et al. 2021). By expanding water stockpiling, SOM can assume a significant part in relief of flood consequences during heavy rainstorms by increasing water stockpile, while also storing water in the case of dry spells, therefore increasing soil versatility.

C:N Relation of Soil

Expanded heat and occasional rainstorm stimulate microbial action (disintegration). It might prompt diminished vegetative growth, exhaustion of soil carbon, and reduction in the C:N proportion (Anderson 1992; Rosenzweig and Hillel 2000; Lal 2004). Expanded biometrical CO_2 builds crop water consumption performance (WUE). It will build microbial activity per millimeter H_2O (Kimball 2003). A deterioration rate is more noteworthy than net essential production under expanded water shortfall. Dry spell instigated misfortunes of biomass; it lessens the yearly and perpetual vegetation. Management strategies incorporate conservation culturing rehearses.

Potentially Mineralizable C and N

The mineralizable natural matter goes about as an interface among autotrophic and heterotrophic living beings in the midst of the enhancement cycling advancement (Gregorich et al. 1994). In any case, the mineralizable natural matter might be valuable to evaluate soil well-being during environmental changes since it impacts supplement cycling in single seasons.

Soil Respiration

Soil decomposition is utilized as a natural marker of soil health since it is directly related to SOM substances. Soil respiration is an essential interface between environmental change and the worldwide carbon chain (Wixon and Balser 2009).

Soil Microbial Biomass

The live part of soil organic nutrient is microbiological metabolism. It is the most volatile C pool in soils and tricky pointer of changes in soil measures with linkages to soil supplement and energy elements as well as intervening in SOC component exchange (Saha and Mandal 2009). Notwithstanding, microbial soil biomass like volatile C is demonstrated to be responsive to momentary habitat fragmentation (Haynes 2008). Soil microbes controls the recycling of nutrients in soil which is necessity of soil and governs the plant growth and development (Jansson and Hofmockel 2020).

Enzyme Activity

Actions of soil enzymes demonstrate changes inside the biotic–edaphic framework because they are definitely attached with (1) cycling of supplements and (2) are handily estimated, (3) data on the microbial status and physicochemical soil conditions are combined, and (4) they have fast reaction in managing change (Garcia-Ruiz et al. 2009). Besides changing the amount and quality of ground C contribution by crops, raised CO_2 may invigorate enzyme activities. It has been observed that microbial enzyme exercise is engaged with natural C turnover, supplement cycling, and greenhouse gas emission.

Soil Temperature

By modifying radiant energy and protective activity, the surface deposits have a major impact on soil temperature. Reflection, soil and air warming, and soil water evaporation all help to modify the radiant energy. The brilliant deposit has a stronger reflection

Soil Structure and Soil Aggregation

Water stable totals help in keeping up great penetration rate, great design, shield in the problem of wind and water erosion. The natural matters are obtained from parasites, microorganisms, earthworms, and dissimilar structures through their nourishing and other vital activities. The very much accumulated soil has a more noteworthy water section at the surface, better air circulation, and water holding limit than inadequately totaled soil. The gluey substances that tight spot parts into totals are made to a great extent by the different living organic entities present in sound soil. Consequently, accumulation is expanded by rehearses that favor soil biota. Since limiting materials are defenseless to degradation through microbes' action, natural substances should be renewed to look after accumulation. Surprisingly, a lot of accumulated soil resists crusting because the water stable totals are less likely to crumble when a raindrop strikes them.

Density and Porosity

Culturing layer compactness is lesser in furrowed than unplowed (zone in the grass, low culturing territory, and so forth). Characteristic compaction happens in soils that are poor in organic substances and need to be released. However, rehearsing protection culturing to balance the compaction will be viable just when there is sufficient buildup, while escalated culturing may antagonistically impact the soil organisms, thus affecting the soil strength and permeability.

Soil Abrasion

Soil crusting affects germination, the rise of sprouts. It is triggered because of total scattering and soil's elements turning and revision through a rainstorm then evaporating. Protection culturing and interface residual help in ensuring scattering of the earth to expanding soaked pressure-driven conductivity. Expanded pressure-driven conductivity related to expanded invasion coming about because of protection culturing permits the soil profile to be all the more promptly loaded up with water. Furthermore, less dissipation is likewise upheld by protection culturing, and the profile can hold more water.

Mitigation Approaches

The protection culturing and deposit management help in the accompanying manner in affecting a portion of soil characteristics and minimizing the unfavorable impacts of environmental change on soil wellness (Sharma 2011). In 2008, the government launched a national climate change action plan (NAPCC) that recognizes various arrangements that immediately enhance the country's development and environmental change adaptability and prevention goals (Table 1).

Five models were employed to identify locations sensitive to different soil concerns in the Andalusia region as part of Mediterranean Land Evaluation Information System decision support system packages: Terraza, Cervatana, Sierra, Raizal, and Pantanal (Anaya-Romero et al. 2015). Plant-associated microbial communities promote plant development and resilience to a variety of abiotic and biotic stressors (Dubey et al. 2019). Crop rotation could be a management practices to reduce negative effect of Climate change (Saleem et al. 2020). After examining suggestions from different countries and observer groups, the IPCC Panel decided in 2016 to issue three distinct findings during the sixth assessment cycle. These findings are divided into four sections that look at greenhouse emission in farming, agriculture utilization, and long-term agroforestry (IPCC 2019) (Table 2).

Table 1 Objectives of National missions

S. No.	National mission	Purpose
1.	Jawaharlal Nehru National Solar Mission	Build up India as a global pioneer in solar energy
2.	National Mission for Enhanced Energy Efficiency	Development with environmental supportability by devising cost-effective and energy-proficient methodology
3.	National Mission on Sustainable Habitat	Enhance habitat sustainability through improvements in energy efficiency in construction, waste managing, and city advancement
4.	National Water Mission	Conservation of water, control wastage
5.	National Mission on Sustainable Agriculture	Agriculture change into environmentally reasonable climate-resilient creation framework and adventure to its greatest degree, guaranteeing food security
6.	National Mission for Sustaining the Himalayan Ecosystem	Sustaining and safeguarding the Himalayan glaciers and mountain ecosystem; addressing issues such as biodiversity, wildlife conservation, and livelihood
7.	National Mission for a Green India	Upgrade ecosystem facilities as carbon sequestration and capacity and so on
8.	National Mission on Strategic Knowledge for Climate Change	Innovative work for development to climate change

Table 2 Sections included in IPCC report 2019

Parts	Initiatives
1	Climate conditions and people in a warming world
2	Multiple choices for adaptability and preventions
3	Permitting different kinds of reactions
4	Quick response action

Conclusion

Agriculture can perform a crucial role in environmental change mitigation via agriculture systems to decrease GHG emissions and enhance soil C storage. The outflow of CO_2, CH_3, and N_2O can all be lowered via reduced biomass burning, enhanced agriculture system, and better organization of N, respectively. To mitigate climate change, economically viable climate-resilient technologies must be outlined using an interdisciplinary approach as agroforestry for maximum land use.

References

Allen DE, Singh BP, Dalal RC (2011) Soil health indicators under climate change: a review of current knowledge. In: Singh BP et al (eds) Soil health and climate change, Soil biology, vol 29. Springer, Berlin. https://doi.org/10.1007/978-3-642-20256-8-2

Anaya-Romero M, Abd-Elmabod SK, Munoz-Rojas M et al (2015) Evaluating soil threats under climate change scenarios in the Andalusia region, Southern Spain. Land Degrad Dev. https://doi.org/10.1002/ldr.2363

Anderson JM (1992) Responses of soils to climate change. Adv Ecol Res 22:163–210

Anjali MC, Dhananjaya BC (2019) Effect of climate change on soil chemical and biological properties - a review. Int J Curr Microbiol Appl Sci 8(2):1502–1512

Arias ME, Gonzalez-Perez JA, Gonzalez-Villa FJ, Ball AS (2005) Soil health-A new challenge for microbiologists and Chemists. Int Microbiol 8:13–21

Arnold SL, Doran JW, Schepers J, Wienhold B (2005) Portable probes to measure electrical conductivity and soil quality in the field. Commun Soil Sci Plant Anal 36:2271–2287

Birkas M, Dexter A, Szemok A (2009) Tillage induced soil compaction as a climate threat increasing stressor. Cereal Res Commun 37:379–382

Bonfante A, Basile A, Bouma J (2020) Targeting the soil quality and soil health concepts. Soil 6: 453–466

Bot A, Benites J (2005) The importance of soil organic matter: Key to drought-resistant soil and sustained food and production, FAO Soils Bull, vol 80. Food and Agriculture Organization of the United Nations, Rome, Italy, p 78

Box JE, Bruce RR (1996) The effect of surface cover on infiltration and soil erosion. In: Agassi M (ed) Soil erosion, conservation and rehabilitation. Marcel Dekker, Inc., New York, pp 107–118

Brevik EC (2013) The potential impact of climate change on soil properties and process and corresponding influence on food security. Agriculture 3:398–417. https://doi.org/10.3390/agriculture3030398

Brevik EC, Steffan JJ, Burgess LC, Cerdà A (2017) Links between soil security and the influence of soil on human health. In: Field D, Morgan C, McBratney A (eds) Global soil security, Progress in soil science series. Springer, Rotterdam, pp 261–274

Brinkman R, Brammer H (1990) The influence of a changing climate on soil properties. In: Proceedings of the transactions 14th international congress of soil science, Kyoto, Japan, pp 283–288

Buol SW, Sanchez PA, Kimble JM, Weed SB (1990) Predicted impact of climatic warming on soil properties and use. Am Soc Agron 53:71–82

Collins M, Knutti R, Arblaster J, Dufresne JL, Fichefet T, Friedlingstein P, Gao X, Gutowski WJ, Johns T, Krinner G, Shongwe M, Tebaldi C, Weaver AJ, Wehner M (2013) Long-term climate change: projections, commitments and irreversibility. In: Stocker TF, Qin D, Plattner G-K, Tignor M, Allen SK, Boschung J, Nauels A, Xia Y, Bex V, Midgley PM (eds) Climate change 2013: the physical science basis. Contribution of Working Group I to the fifth assessment report of the Intergovernmental Panel on Climate Change. Cambridge University Press, Cambridge, UK

Dalal RC, Moloney D (2000) Sustainability indicators of soil health and biodiversity. In: Hale P, Petrie A, Moloney D, Sattler P (eds) Management for sustainable ecosystems. Centre for Conservation Biology, Brisbane, pp 101–108

Dang YP, Dalal RC, Mayer D, McDonald M, Routley R, Schwenke GD et al (2008) High subsoil chloride concentrations reduce soil water extraction and crop yield on vertisol in northeastern Australia. Aust J Agric Res 59:321–330

Davidson EA, Janssens IA (2006) Temperature sensitivity of soil carbon decomposition and feedbacks to climate change. Nature 440:165–173

Desanker PV (2002) Impact of Climate Change on Africa, WWF Special Report (www.wwf.org)

Diaz RA, Margin GO, Travasso MI, Rodriguez RO (1997) Climate change and its impact on the properties of agricultural soils in Argentinean rolling pampas. Clim Res 9:25–30

Dubey A, Malla MA, Khan F, Chowdhary K et al (2019) Soil microbiome: a key player for conservation of soil health under changing climate. Biodivers Conserv 28:2405–2429

Garcia-Ruiz R, Ochoa V, Vinegla B, Hinojosa MB, Pena-Santiago R, Liebanas G et al (2009) Soil enzymes, nematode community and selected physico-chemical properties as soil quality indicators in organic and conventional olive-oil farming: influence of seasonality and site features. Appl Soil Ecol 41:305–314

Gil SV, Meriles J, Conforto C, Figoni G, Basanta M, Lovera E et al (2009) Field assessment of soil biological and chemical quality in response to crop management practices. World J Microbiol Biotechnol 25:439–448

Gregorich EG, Carter MR, Angers DA, Monreal CM, Ellert BH (1994) Towards a minimum data set to assess soil organic matter quality in agricultural soils. Can J Soil Sci 74:367–385

Hakansson I, Lipiec J (2000) A review of the usefulness of relative bulk density values in studies of soil structure and compaction. Soil Till Res 53:71–85

Hamidov A, Helming K, Bellocchi G, Bojar W, Dalgaard T et al (2018) Impact of climate change adaptation options on soil functions: a review of European case studies. Land Degrad Dev 29: 2378–2389

Haynes RJ (2008) Soil organic matter quality and the size and activity of the microbial biomass: their significance to the quality of agricultural soils. In: Huang Q, Huang PM, Violante A (eds) Soil mineral-microbe-organic interactions: theories and applications. Springer, Berlin, pp 201–230

Herout V, Shutterstock (2009) Will warmer soil be as fertile? Environ Health Perspect 117(2):59

Hoffland E, Kuyper TW, Comans RN, Creamer RE (2020) Eco-functionality of organic matter in soils. Plant Soil 455:1–22

IPCC (2007) In: Parry ML, Canziani OF, Palutikof JP, van der Linden PJ, Hanson CE (eds) Impacts, adaptation and vulnerability. Contribution of Working Group II to the fourth assessment report of the Intergovernmental Panel on Climate Change. Cambridge University Press, Cambridge, UK, p 976

IPCC (2019) Summary for policymakers. In: Shukla PR, Skea J, Calvo Buendia E, Masson-Delmotte V, Pörtner H-O, Roberts DC, Zhai P, Slade R, Connors S, van Diemen R, Ferrat M, Haughey E, Luz S, Neogi S, Pathak M, Petzold J, Portugal Pereira J, Vyas P, Huntley E, Kissick K, Belkacemi M, Malley J (eds) Climate change and land: an IPCC special report on climate change, desertification, land degradation, sustainable land management, food security, and greenhouse gas fluxes in terrestrial ecosystems. ISBN 978-92-9169-154-8

IPCC (2021) Summary for policymakers. In: Masson-Delmotte V, Zhai P, Pirani A, Connors SL, Péan C, Berger S, Caud N, Chen Y, Goldfarb L, Gomis MI, Huang M, Leitzell K, Lonnoy E, Matthews JBR, Maycock TK, Waterfield T, Yelekçi O, Yu R, Zhou B (eds) Climate change 2021: the physical science basis. Contribution of Working Group I to the sixth assessment report of the Intergovernmental Panel on Climate Change. Cambridge University Press, Cambridge, UK

Jansson JK, Hofmockel KS (2020) Soil microbiomes and climate change. Nat Rev Microbiol 18: 35–46. https://doi.org/10.1038/s41579-019-0265-7

Jarvis NJ (2007) A review of non-equilibrium water flow and solute transport in soil macropores: principles, controlling factors and consequences for water quality. Eur J Soil Sci 58:523–546

Karmakar R, Das I, Dutta D, Amitava R (2016) Potential effects of climate change on soil properties: a review. Sci Int 4(2):51–73. https://doi.org/10.17311/sciintl.2016.51.73

Kimball BA (2003) Response of plants to elevated atmospheric CO_2. Ind J Plant Physiol:18–24

Lal R (1995) Tillage systems in the tropics: management options and sustainability implications, FAO Soils Bull, vol 71. Food and Agriculture Organization of the United Nations, Rome, Italy, p 180

Lal R (2004) Soil carbon sequestration to mitigate climate change. Geoderma 123:1–22

Malhi GS, Kaur M, Kaushik P (2021) Impact of climate change on agriculture and its mitigation strategies: a review. Sustainability 13:1318. https://doi.org/10.3390/su13031318

Moebius BN, Van EHM, Schindelbeck RR, Idowu OJ, Clune DJ, Thies JE (2007) Evaluation of laboratory measured soil properties as indicators of soil physical quality. Soil Sci 172:895–912

Mutegi J, Ameru J, Harawa R, Kiwia A, Njue A (2018) Soil health and climate change: implications for food security in sub–Saharan Africa. Int J Dev Sustain 7(1):21–33

National Research Council (2001) Climate change science: an analysis of some key questions. The National Academies Press, Washington, DC, p 10.17226/10139
Navarro-Pedreño J, Almendro-Candel MB, Zorpas AA (2021) The increase of soil organic matter reduces global warming, myth or reality? Science 3:18. https://doi.org/10.3390/sci3010018
Pareek N (2017) Climate change impact on soils: adaptation and mitigation. MOJ Ecol Environ Sci 2(3):136–139. https://doi.org/10.15406/mojes2017.02.00026
Parry M, Rosenzweig C, Iglesias A, Fischer G, Livermore M (1999) Climate change and world food security: a new assessment. Glob Environ Chan 9:51–67
Rengasamy P (2010) Soil processes affecting crop production in salt-affected soils. Funct Plant Biol 37:613–620
Reth S, Reichstein M, Falge E (2005) The effect of soil water content, soil temperature, soil pH-value and root mass on soil CO_2 efflux-a modified model. Plant Soil 268:21–33
Reynolds WD, Bowman BT, Drury CF, Tan CS, Lu X (2002) Indicators of good soil physical quality: density and storage parameters. Geoderma 110:131–146
Rosenzweig C, Hillel D (2000) Soils and global climate change: challenges and opportunities. Soil Sci 165:47–56
Ross DS, Matschonat G, Skyllberg U (2008) Cation exchange in forest soils: the need for a new perspective. Eur J Soil Sci 59:1141–1159
Saha N, Mandal B (2009) Soil health-a precondition for crop production. In: Khan MS, Zaidi A, Musarrat J (eds) Microbial strategies for crop improvement. Springer, Heidelberg, pp 161–168
Saleem S, Levison J, Parker B, Martin R, Persaud E (2020) Impacts of climate change and different crop rotation scenarios on ground water nitrate concentrations in sandy aquifer. Sustainability 12:1153. https://doi.org/10.3390/su12031153
Salvador Sanchis MP, Torri D, Borselli L, Poesen J (2008) Climate effects on soil erodibility. Earth Surf Process Landf 33:1082–1097
Scharpenseel HW, Schomaker M, Ayoub A (1990) Soils on a warmer earth: effects of expected climate change on soil processes, with emphasis on the tropics and sub-tropics. Elsevier, Amsterdam, p 273
Sharma KL (2011) Adverse effects of climate change. Model training course on Impact of Climate Change in Rainfed Agriculture and Adaptation Strategies, 22–29 Nov 2011. CRIDA, Hyderabad, India, pp 68–84
Meghan Sindelar (2015). www.soils.org/IYS
Smith JL, Halvorson JJ, Bolton H (2002) Soil properties and microbial activity across a 500 m elevation gradient in a semi-arid environment. Soil Biol Biochem 34:1749–1757
Weil RR, Magdoff F (2004) Significance of soil organic matter to soil quality and health. In: Magdoff F, Weil RR (eds) Soil organic matter in sustainable agriculture. CRC Press, Boca Raton, pp 1–43. https://doi.org/10.1201/9780203496374.ch1
Wixon DL, Balser TC (2009) Complexity, climate change and soil carbon: a systems approach to microbial temperature response. Syst Res Behav Sci 26:601–620

Part II
Microbe Mediated Plant Stress Mitigation

Habitat-Imposed Stress Tolerance in Plants via Soil–Microbe Interactions

Arvinder Singh, Bhumika Arora, and Kirpa Ram

Abstract Plants, due to their static nature, are continuously put at risk to different kinds of habitat-imposed stresses, viz. drought, salinity, temperature, etc., which constrain their productivity besides growth and development. At the same time, they keep on communicating with the soil microbiota in diverse ways under natural conditions to overcome these stresses by modifying their physiological and molecular pathways. The stress-induced changes generally alter plants' proteomics, transcriptomics, phenomics, and metabolomics, which, in turn, affect the rhizospheric conditions due to changes in the nutrient, mineral, and metabolite composition of root and shoot exudates secreted in the soil. Flavonoids, coumarins, and other organic compounds serve as plant signals to shape the structure and composition of microbiomes that interact with the host plant. The role of several rhizospheric occupants like symbiotic fungi (Arbuscular Mycorrhiza) and bacteria (nitrogen-fixing and plant growth-promoting rhizobacteria) in plant stress tolerance by direct and indirect mechanisms has been well documented. These mechanisms in mitigating the effects of multiple stresses involve reinforcing the plant defense system (through the production of allopathic compounds, HCN, etc.), enhancing the heat shock proteins and phytohormone production along with inducing genes related to plant stress. The identification, isolation, and use of stress-tolerant rhizospheric microbial strains under habitat-imposed stress have the potential to solve the universal problem of food security and also to nourish soil health. However, the questions regarding the formulation of the effective consortia of microbes (SynComs), their synchronization, and delivery into the field to overcome the harmful effects of changing environment need to be addressed. As microbe–plant interactions are very complex, system biology may play a crucial role in enhancing our knowledge to understand these complex relationships.

Keywords Abiotic stresses · Plant–microbe interactions · PGPR · AMF

A. Singh (✉) · B. Arora
Department of Botany, Akal University, Bathinda, Punjab, India
e-mail: arvinder_bot@auts.ac.in

K. Ram
Department of Botany, Baba Mastnath University, Rohtak, Haryana, India

© The Author(s), under exclusive license to Springer Nature Singapore Pte Ltd. 2022
A. Vaishnav et al. (eds.), *Plant Stress Mitigators*,
https://doi.org/10.1007/978-981-16-7759-5_10

Introduction

As far as the growth and decline in the yield of agricultural crops are concerned, the stresses created because of man-made activities and the natural phenomena are the major culprits (Grayson 2013). Plants are incessantly put at risk to an unforeseeable combination of various stresses, which may occur as a disturbance or in a chronic manner (Slama et al. 2015). The abiotic stresses comprise physical or chemical stresses such as drought or water-logging, salinity/alkalinity, extreme temperatures (cold, frost, or heat), anaerobiosis, heavy metals, UV radiations, and nutrient imbalance (Wang et al. 2003; Hirel et al. 2007; Bailey-Serres and Voesenek 2008; Wani et al. 2016). These stresses generally impact plants from morphological levels to molecular levels at each of their developmental stage. It is the intensity of the stresses to which plants are exposed determines whether the plants will suffer severe toxicity—the outcome of which may be impairment of the tissues or even death, or a less-obvious chronic damage resulting in decreased productivity (Singh 2016). Imbalance in hormonal and nutritional status, physiological disorders, disease susceptibility, etc. retard the growth of the plant under unfavorable environmental conditions. To face these stressful conditions, the plants have acquired a number of pathways/mechanisms during their evolution period (Yolcu et al. 2016). These improvisations facilitate the normal physicochemical processes under unpropitious external situations by altering the cell metabolism (Shao et al. 2008; Massad et al. 2012).

The close alliance of the plants with varieties of the microorganisms, inhabiting rhizosphere, endosphere, and other plant parts such as leaf surfaces is collectively known as microbiome (Liu et al. 2017). For long time, researchers have shown great interest in studying plant–microbe interactions. The microbes provide plants the fundamental support in acquiring nutrients and tolerating a number of abiotic stresses (Turner et al. 2013). However, unlike plants, the microbiome is dynamic in nature, and its structure and composition get changed in response to stresses and external stimuli (Timm et al. 2018). The interactions between plants and microbes elicit different localized and systemic responses resulting in the improvement of plants' metabolism to endure the abiotic stresses (Nguyen et al. 2016). Under stressed conditions, a large number of bacteria belonging to different genera such as *Arthrobacter*, *Rhizobium*, *Pseudomonas*, *Enterobacter*, *Serratia*, and *Azospirillum* along with actinobacteria *Streptomyces* sp. have been identified as plant growth-promoting rhizobacteria (PGPR) to provide endurance to the host plant (Choudhary et al. 2011; Etesami and Beattie 2017). Vivid responses of plants toward various inanimate stresses as a consequence of plant–microbe interactions have been studied at numerous levels, viz. morphological, physicochemical, and molecular levels (Farrar et al. 2014). Furthermore, the biological data have also been generated using multiomics approaches to provide a deeper insight into these interaction mechanisms, chalking out the relationship between the changes at the level of genes and proteins, and the tolerance responses toward various abiotic stresses (Kissoudis et al. 2014). In the present chapter, we will summarize the consequences of abiotic stresses and the responses generated by the plant with regard to molecular, biochemical, and physiological mechanisms. Moreover, a detailed discussion on the microbe-mediated stress alleviation methods will also be carried out.

Implications of Abiotic Stresses

In general, any plant for its normal growth and development needs some fundamental requirements, viz. water, light, carbon dioxide, and nutrient. If the concentrations of these fall below or above the optimum level in nature, it hinders plant growth, reproduction, and development. An adverse environment comprises a myriad of abiotic stress conditions such as elevated carbon dioxide (eCO_2), salinity, water stress, temperature extremities, and ozone layer depletion, which may affect the crop production at varying degrees; however, their prompt effects on the field include growth reductions and over- and down-expression of some stress-responsive biochemical, physiological, and molecular processes (Kamanga and Mndala 2019). Plants have the ability to perceive these stresses, which, in turn, provokes varied responses that support their survival (Jiang et al. 2016). The different stresses affect the plants in different ways, as described below (Fig. 1).

Water Stress

Under the conditions of the limiting supply of water to the roots, or the intense transpiration rate, plants experience water stress; the condition generally arises due to drought or under high soil salinity. The impact of this stressful condition at various levels, viz. biochemical, physiological, and morphological levels in plants, is well documented (Zwicke et al. 2015). The most conspicuous effect of all the unfavorable conditions (including water stress) initially observed at the cellular level and thereafter morphological and physiological symptoms are seen. Water stress in plants results in growth inhibition and reproductive failure due to a decrease in cell enlargement, which may be attributed to the lower water potential and cell turgidity (Wallace et al. 2016). Furthermore, water limitations for longer time led to reduction in photosynthetic efficiency, lipid peroxidation, overproduction of reactive chemical species like superoxide, singlet oxygen, peroxides, etc., and enhanced apoptosis (Deeba et al. 2012; Gill and Tuteja 2010). Reduced leaf size, reduced seed number, size and viability, suppressed root growth, and delayed flowering and fruiting are vivid in the plants grown under a water-stressed environment (Xu et al. 2016). Thus, the plants have developed ways to restrict water consumption under its scarcity until dissenting conditions exist (Osakabe et al. 2013).

Fig. 1 Implications of different abiotic stresses in plants

Cold Stress

Depending on the thermal niche and their adaptations, the plants can be tolerant or sensitive to cold. The membrane lesions resulting from the cellular dehydration are considered as a primary cause of the freezing injury besides other secondary factors contributing toward cold-induced cellular mutilation (Bhandari and Nayyar 2014). Whereas ROS production causes membrane damage, the intracellular ice crystal formed at freezing temperature results in cell rupturing due to its adhesions with cell membranes and wall (Olien and Smith 1977). Extreme crystal formation results in increased electrolyte leakage and changes in the membrane lipid phase and eventually causes the plants to die (Ritonga and Chen 2020). Moreover, protein denaturation occurring in plants at low temperatures (Guy et al. 1998) may further exacerbate cellular damage. Under severe conditions of cold stress due to chilling (0–15 °C) and freezing (<0 °C) temperatures, sensitive crops like chickpea, soybean, cucurbits, etc. may suffer huge yield losses (Thakur et al. 2010). This loss can be attributed to the delay heading, which leads to pollen sterility under the chilling temperature at the reproductive phase of the plant (Yadav 2010). Furthermore, cold stress also leads to slow germination, dwarfed seedlings, chlorosis, withering, and reduced tillering.

Heat Stress

A persistent rise in the temperature far away from the limit that a plant cannot tolerate and causes damage to its growth and development indicates the condition of heat stress. Due to a constant rise in the earth's temperature, the plants are continuously facing heat stress besides other abiotic stresses. As different stages of the life cycle are dependent on temperature in plants, the effect of heat stress can be sensed early from the seed germination stage, which can be reflected in the form of molecular, morpho-anatomical, and physicochemical alteration in the plants; a temperature difference of 1 °C than its ambient temperature can affect plant height, tiller number, and seeds per tiller (Shafiei Masouleh et al. 2019). High temperature influences the leaf expansion and elongation of the internode, promotes abortion of flower buds, and alters fertility processes. It also affects various physiological processes directly or indirectly, like scorching of leaves and stems, modulating the levels of hormones, leaf abscission and senescence, and fruit damage, resulting in an enormous decrease in the crop yield (Hemantaranjan et al. 2014). At the cellular level, high temperature results in denaturation and aggregation of proteins and also increases the membrane fluidity. Furthermore, at the molecular level, enzyme inactivation, translational inhibition, protein degradation, and membrane disintegration are the adverse effects of slow heat damage. It also affects conjointly the microtubule organization and spindle elongation (Hemantaranjan et al. 2014).

Metal Stress

A number of anthropogenic activities such as industrialization, fertilizer applications, and mining and smelting operations, besides natural sources such as volcanic eruptions, aerosols, etc., release heavy metals (density greater than 5 g/cm^3), viz. Hg, Fe, As, Mn, Zn, Ni, Pb, and Cd in aquifers and soil (Wuana and Okieimen 2011; Basheer 2018). The ambulant heavy metals added in the soils from anthropogenic sources are easily available to the plants. Once these get entered into the plant vegetative tissues, these can hamper their multiple physicochemical processes, which eventually affect the health of human beings (Singh et al. 2015). Heavy metals such as Zn, Cu, and Mn are needed as micronutrients by the plants at low concentrations, but the enhanced level of such metals than the tolerance limits in soil negatively affects plant metabolism, becoming toxic to the plants. On the other side, the metals like Cd, Pb, As, and Ni always remain nonnutritional and toxic components for most of the organisms (He et al. 2013). The nutrient imbalance encountered by the plants growing under metal stress is considered due to interference in the taking up of essential micro- and macroelements by the plants. Lipid peroxidation resulting from the plants' exposure to heavy metals leads to cell membrane deterioration—one of the most deleterious effects in plants (Yadav 2010). The other effects of heavy metal stress include protein oxidation, enzyme inactivation, and damage to genetic material, which are the results of over-accumulation of ROS and methylglyoxal (MG) (Gill 2014). The enzyme inactivation or protein denaturation due to metal stress disrupts the substitution reaction of essential metal ions in biomolecules, which further disintegrates the cell membrane and leads to changes in metabolic reactions, viz. photosynthesis, respiration, homeostasis, and eventually plant cell death (Hossain et al. 2012; Adrees et al. 2015).

Salinity Stress

Salt stress is one of the abiotic stresses of paramount importance that reduces crop productivity and is a cause of the dereliction of land for agricultural purposes (Dodd and Pérez-Alfocea 2012). High salinity has been shown to affect about 20% of the world's cultivated land area, 8% of the global land, and nearly 50% of the irrigated land (Zhu 2001). As a consequence of toxic effects of Na^+ and Cl^- ions and osmotic stress, salinity reduces the growth and development in plants. The hypertonic behavior of the soil solution due to excess salts retards the water absorption potential of plants, resulting in poor growth of the plants. This effect of salinity is known as the osmotic effect or water deficit effect. If a considerable amount of salt enters the plant via a transpiration stream, this damages the cells of transpiring leaves and further slows down the growth. The salinity effects can be noticed at any stage of the plants' life cycle, such as seedling, vegetative, or maturity. However, at a given salinity level, the responses in terms of molecular, biochemical, and physiological alterations can vary among different plants. These alterations include dwarfism, increased succulence, earlier lignification, reduction in number and size of stomata, etc. (Nawaz et al. 2010). Several growth aspects related to reproduction, which include blooming of flowers, pollination and fertilization, and fruit growth and maturation, are also affected by salinity (Singh 2016). The reduced growth and yield of agricultural crops might be ascribed to the formation of ROS, nutrient ion and osmotic imbalance, and ion toxicity due to the piling up of sodium and chloride ions under uninterrupted exposure of plants to the saline environment (Rai et al. 2019).

Light Intensity Stress

Light, being also an important environmental component, affects the plants' growth and morphogenesis besides other physicochemical processes (Zoratti et al. 2014). The plants keep on facing the change in the light intensity of several orders of magnitude during the daytime. Light intensity that is too high or too low can affect several plant processes such as seedling development, chlorophyll development, phototropism movement, photoperiodism, photomorphogenesis, circadian clock and metabolic processes, flowering, growth, and development (Singhal et al. 2017). Under high light conditions, some plants such as strawberries fall to thrive because the light radiation slows down the rate of photosynthesis (Guo et al. 2006); however, elongated leaves with more leaf area and plant height can be observed at low light intensity (Setiawati et al. 2018). In excess light, plants generate ROS, which may further destruct cell membrane structural disorganization and lipid peroxidation.

Responses of Plants Toward Abiotic Stresses

The unpredictable variations in the physical surroundings of the plants may hamper their growth and development. To sense, respond, and adapt to unfavorable growing conditions, plants have developed a number of sophisticated and efficient mechanisms during their evolutionary period. Responses of plants toward these adverse conditions can be seen at multiple levels of organization, i.e., from molecular, cellular, biochemical, morphological, anatomical, to physiological levels. In agronomic research, understanding the various response mechanisms and knowing the tolerance limit of the abiotic stress by plants are gaining wide importance nowadays (Atkinson and Urwin 2012). In the following sections, we will deal in detail the different responses and the mechanisms endorsed by the plants to face the extreme stress conditions (Table 1).

Table 1 Varied plant responses under abiotic stress conditions

S. No.	Stress	Physiological and biochemical responses	Molecular responses	References
1	Drought stress	Accumulation of LEA proteins; increased activities of antioxidative enzyme SOD, CAT, APX, and POX; increased ascorbate content; stomatal closure; increased accumulation of polyamines and proline; production of ROS; decreased efficiency of rubisco	Enhanced expression of ABA biosynthetic gene; stress-responsive gene expression, i.e., DREB, WRKY, and NAC; upregulation of mitogen MAPKs and SnRKs	Golldack et al. (2011), Pan et al. (2003), Manzi et al. (2016)
2	Heat stress	Increased activity of GSH and GR; increased synthesis of AsA and glycinebetaine (GB); increased rate of sugar metabolism; decreased photosynthetic efficiency	Activation of heat stress-responsive genes to produce heat shock proteins (HSPs)	Li et al. (2011), Islam et al. (2018), Wu et al. (2008)
3	Salinity stress	Accumulation of LEA proteins; increased antioxidative enzyme activities, viz. SOD, CAT, and POX; accumulation of proline, sugars (glucose, fructose), and sugar alcohols (mannitol); photosynthesis impairment; more synthesis of plant growth hormones such as ABA, SA, IAA, and CK	Induction of SOS stress signaling pathway; overexpression of transcriptional factors—NAC, DREB/CBF; stress-responsive gene expression, i.e., bZIP	Shi et al. (2002), Chaves et al. (2009), Wani and Kumar (2015)
4	Cold stress	Accumulation of dehydrins; enhanced activity of antioxidative enzyme—SOD, CAT, APX, POX, and GR; accumulation of sugars (D-glucose, D-glucose-6-P and maltose); increased ABA production	Activation of transcription factors—DREB/CBFs; expression of cold stress-related genes, viz. *STA1*, *MdHY5*, *RDM4*	Chan et al. (2016), Wang et al. (2019), Lee et al. (2006)
5	Heavy metal stress	Increase in antioxidative enzyme activities, i.e., SOD, CAT, APX, POX, and GR; production of phytochelatins (PCs), glutathione, and α-tocopherol	Activation of transcription factors such as *bHLH*, *bZIP*, *AP2/ERF*; MAPK signaling cascade	Singh et al. (2016), Tiwari et al. (2017), Saba et al. (2013)

Physiological and Biochemical Responses

All types of environmental stresses are responsible for producing ROS and RCS (reactive carbonyl species) by uncoupling the enzymes and metabolic pathways. The common ROS responsible for oxidative stress are superoxide radical (O^{-}_2) singlet oxygen, hydrogen peroxide (H_2O_2), hydroxyl radical, malondialdehyde ($CH_2(CHO)_2$), and methylglyoxal (CH_3COCHO) (Asada 2006). However, these are not only the toxins that need to be expelled but also the signaling molecules necessary for a number of physiological activities including stress resistance. Peroxisomes, plastids, and mitochondria are considered as the primary source of ROS production, which cause destruction of macromolecules like proteins, lipids, carbohydrates, etc. and create disastrous conditions for the cell. Fortunately, the plants possess a sophisticated ROS scavenging system in the form of antioxidants, both enzymatic and nonenzymatic, which help in tolerating stressful conditions. A number of nonenzymatic metabolites include betalains, carotenoids, flavonoids, and tocopherols with antioxidant properties (Zhao et al. 2011). On the other hand, superoxide dismutase (SOD), catalase (CAT), various peroxidases (PODs), and glutathione reductase (GR) are the enzymatic antioxidants that play a paramount role in cellular defense. Glutathione *S*-transferase (GST) along with ferritins, the other important enzymes, has also been reported to participate in cell detoxification (Sharma et al. 2012). Being the primary defense system against ROS, SODs convert O^{-}_2 radicals into H_2O_2 for further reduction into water by catalases and peroxidases. There has also been shown a positive correlation in terms of increased activity between SOD and H_2O_2 degrading enzymes like CAT and PODs under different stress conditions like metal, water, and salinity (Davis and Swanson 2001; Koca et al. 2007. Catalase and APXs, the main enzymes ensuring H_2O_2 removal, assist the plants in opposing drought, salt, and high light conditions (Cao et al. 2017). Furthermore, GR and GSH were also reported to play an important role in chilling and metal and heat stress tolerance in different crops like rice and mulberry (Tewari et al. 2006; Kumar and Trivedi 2018). Although the activities of the aforementioned specialized enzymes with antioxidant properties get increased under different stress conditions, the increment can be seen significantly higher in tolerant genotype over the sensitive one (Rani et al. 2013).

In order to lessen the effect of oxidative stress, a number of aforementioned nonenzymatic antioxidants have also been evolved in the plants. Enhanced synthesis of ascorbic acid (AsA) under heat stress has been correlated with lower ROS production (Xu et al. 2006). Furthermore, the plants' exposure to heavy metal stress has led to the alleviated level of AsA, which plays a major role in detoxifying the ROS (Parmar et al. 2013). Similar elevation in the AsA content, in drought stress conditions, in *Picea* has also been revealed by Yang et al. (2008). Under abiotic stresses, ascorbate peroxidase (APX) utilizes AsA to reduce H_2O_2 to water and generate MDA (monodehydroascorbic acid) in ascorbic acid–glutathione cycle (Pan et al. 2003). The lipophilic antioxidant vitamin-E or α-tocopherol is also synthesized in the plants to scavenge free radicals in combination with other antioxidants

(Massacci et al. 2008). Tocopherol acts as a free radical trap, reacting and then reducing RO•, ROO•, and RO* lipid radicals at the membrane water interface to ROH• to ROH*. α-Tocopherol has been reported to protect the structure and functions of PSII. Phytohormones like ABA, jasmonic, and salicylic acid also act as mediators of drought, salt, and osmotic stress. ABA, among these, acts as a central regulator to impart abiotic stress resistance in plants (Wani and Kumar 2015). The stomatal closure to regulate the loss of water under drought conditions is controlled mainly by ABA (Manzi et al. 2016). Similarly, ABA has been proved in enhancing the defense-related gene expression against pathogen attack through regulation of stomatal aperture (Lim et al. 2015). It has also been shown to increase the rate of sugar metabolism and its transportation under heat stress to spikelets, which provides plants an advantage to survive under stressed conditions (Islam et al. 2018). ABA may crosstalk with other hormones such as SA, IAA, and BRs to provide additive heat tolerance in plants. BRs have been reported to make ABA content double under heat stress (Kurepin et al. 2008). A few reports have described the synergistic interplay of ABA and IAA in regulating plant growth and survival under drought conditions (Du et al. 2013).

In order to acclimatize to abiotic stresses, plants accumulate several other biomolecules such as protective osmoprotectants like sugars, trehalose, and proline (Hayat et al. 2012; Ilhan et al. 2015); proteins like HSPs; dehydrins; LEA (late embryogenesis abundant) proteins (Lipiec et al. 2013); and glycine and betaine (Wang et al. 2010; Chen and Murata 2011) that are safe and do not intervene the plant processes. Proline, an imino acid, gets accumulated under various abiotic stresses, viz. salt stress, drought stress, etc., which indicate its role in stress tolerance (Ahmad and Sharma 2008). The increased proline level in plants has been correlated with the decrease in mitochondrial electron transport activity under environmental stress (Saradhi et al. 1995). Proline improves the plant performance under stressful conditions by its antioxidant, osmoprotective, and metal chelating properties (Farago and Mullen 1979). Proline, besides playing a role in osmotic adjustment, also participates in maintaining the photosynthetic properties (Carpena et al. 2003), regulating the cytosolic acidity (Gajewska and Skłodowska 2008), stabilizing organelles and macromolecules (John et al. 2008), and protecting enzyme denaturation (Gajewska and Skłodowska 2008). Similarly, in response to cold, drought, or salinity stress, a large group of hydrophilic proteins known as late embryogenesis abundant (LEA) proteins start accumulating in various tissues of the plants at high concentrations (Campos et al. 2013). LEA proteins have multiple roles in drought tolerance. These work in combination with trehalose to inhibit protein aggregation during water starvation (Goyal et al. 2005). Likewise, the expression of many LEA proteins has been shown to be regulated by ABA, a key hormone in dehydration (Zamora-Briseño and de Jiménez 2016). LEA proteins, belonging to group 2, get accumulated in plants in response to water deficit and cold conditions (Close 1997). Several studies have also reported the accumulation of various mono- and disaccharides such as glucose, levulose, sucrose, trehalose, and sugar derivatives like mannitol, pinitol, etc. under abiotic stresses (Parida and Das 2005; Geissler et al. 2009). The main roles of the modified sugar alcohols have been described in osmotic adaptation, carbon storage, and radical scavenging (Messedi et al. 2006; Lee et al. 2008; Ahmad and Sharma 2008). The disaccharide sugar trehalose, being a good osmoprotectant and osmolyte, is known to protect biomembranes and intracellular proteins and reduce the aggregation of denatured proteins in many organisms under numerous inanimate stresses (Penna 2003). Polyamines (PAs), the amino groups containing polyvalent compounds, are also ascribed a decisive role to carry out plant processes such as acquisition of the embryogenic potential by the differentiated somatic cell (Silveira et al. 2013), apoptosis (Kim et al. 2013), fruit maturation and ripening (Gil-Amado and Gomez-Jimenez 2012), and vascular differentiation (Tisi et al. 2011). Different responses can be seen in different plant species to polyamine levels under stressed conditions (Shao et al. 2015). Whereas some reveal piling up of polyamines as a stress response, the others remain unaffected with respect to endogenous polyamine levels when exposed to severe conditions. The frequently occurring polyamines in plants belonging to higher groups are putrescine, spermidine, and spermine; diamines, diaminopropane, and cadaverine are less occurring in nature (Mustafavi et al. 2018). Stress-induced polyamines regulate antioxidant systems or repress ROS production by modulating the mechanism of ROS homeostasis via direct or indirect ways (Zhang et al. 2015). Polyamines have also been presented to lessen the effects of heavy metal stress on lipid peroxidation, thus protecting the membrane stability (Janda et al. 2018). Under severe temperature in the surrounding of the plants, the positive effects of PAs can be observed in terms of enhancing photosynthesis and elevated antioxidant capacity and osmotic adjustment capability in these (Tian 2012). LMW phytochelatins (PCs) produced in response to HM stress with the help of phytochelatin synthase (PCS) possess a great binding ability to heavy metals when they occur at concentrations that are toxic for plants. Thus, PC production can be taken as biosignatures for primary observation of HM stress in crop species (Saba et al. 2013).

Molecular Responses

For alleviating the stress in plants, first it becomes imperative to know how the molecular machinery and its networks operate under these conditions. By their upregulation under different abiotic stress conditions, several genes have been reported to combat the abiotic stress conditions, which lead to plant acclimatization (Tuteja 2009). Stress-induced genes not only defend the cells from stress by synthesizing chaperones or LEA proteins but also regulate cell signaling in stressed conditions. The activation of the stress-responsive genes provokes different independent or interlinked signal transduction pathways, which contribute to the detoxification of ROS, protein and enzyme reactivation, and reinstatement of cellular homeostasis, which often regulate different responses for the plant stress development (Kaur and Gupta 2005; Ciarmiello et al. 2011). The gene regulation can be observed at transcriptional, post-transcriptional, and post-translational levels; however, the first one is still the key regulatory node. Once the stress is sensed by the plants, gene regulation factors, viz. transcription factors (TFs), histone acetyltransferases (HATs), alternative splicing factors, and miRNAs, become active in fine-tuning the defense system (Stockinger et al. 2001; Zhang et al. 2013; Laloum et al. 2018). Several reports have described the function of a number of transcription factors (TFs), viz. ABA-responsive element (ABRE)-binding factors, WRKYs, zinc-finger proteins, HSFs, and dehydration-responsive element-binding (DREB) proteins toward stress responses (Mizoi et al. 2012; Sah et al. 2016; Bai et al. 2018). The heat shock proteins (HSPs) synthesized by HSFs impart resistance against heat stress and play an important role in achieving thermotolerance in plants, animals, and microorganisms (Kotak et al. 2007). Similarly, manifestation of a number of WRKY proteins involving in plant drought and salt stress responses has been reported by Golldack et al. (2011). Recently, Wu et al. (2019) have revealed the increased drought tolerance in *Oryza sativa* in which the overexpression of WRKY11 protein was observed by these workers. Furthermore, the role of alternative splicing in regulating the gene expression has also been shown in *A. thaliana* grown under frigid conditions (Lee et al. 2006); in this case, the importance of *STA1* gene encoding a nuclear pre-mRNA splicing factor has been revealed by the workers. Besides the above-mentioned factors/mechanisms, various post-translational modifications (PTMs) like sumoylation, phosphorylation, and ubiquitination have shown to play an important role in plant responses toward abiotic stresses. Among these PTMs, phosphorylation and de-phosphorylation are more crucial as various enzyme cascades, like MAP kinases and SNF-1-related protein kinases, get activated through phosphorylation of their specific amino acid residues under water deprivation and osmotic stress conditions (Zhu 2002). Likewise, the activation of ABREs by transcription factors like bZIP-proteins and DREB proteins results in expressing the dehydrins—a drought stress tolerance effector.

Plant–Microbe Interactions to Mitigate Abiotic Stresses

Being an inseparable part of the living ecosystem, interactions of microbes with plants in nature modulate the plant's localized and systemic mechanisms to come up with better protection under unpropitious external environmental conditions. It is very crucial to study and interpret these interactions to understand their role in providing defense against various habitat-imposed abiotic stresses, which are due to continuous changes in the climatic and edaphic factors. Fairly, a good number of microorganisms have been shown to have innate genetic and metabolic capabilities to relieve inanimate stresses in plants (Gopalakrishnan et al. 2015). In their natural environment, microorganisms show different kinds of interactions with one another, such as commensalism, parasitism, amensalism, saprophytism, or symbiosis, which influence plant growth and development for their sustenance (Berendsen et al. 2012). These interactions are dynamic, not static. The "cry for help" hypothesis has suggested the recruitment of specific microbes by plants under given stress conditions to alleviate its effect (Neal et al. 2012; Nishida and Suzaki 2018). The microbes associated with plants can be categorized into rhizosphere microorganisms, rhizoplane microorganisms, and endophytic microorganisms depending upon their niche. Whereas rhizosphere and rhizoplane microbes reside in the vicinity and surface of the roots respectively, the endophytes inhabit the interior of tissues (Sturz et al. 2000). The microorganisms that are of great benefit for plants come from the rhizospheric zone and include mainly the rhizobacteria and MF (mycorrhizal fungi). Both groups help the plants not only in nutrient acquisition by mutualistic interactions but also in producing signaling compounds that assist in improved plant growth and abiotic stress tolerant capability (Fig. 2). Generally, the abiotic stress tolerance capability in plants via plant–microbe interactions involves various mechanisms such as (a) reinforcing the antioxidant defense system, enhanced secondary metabolite as well as HSP production, and adjusting the levels of phytohormones, and (b) elevating the expression of stress-responsive genes like *HKT1*, sequestering as well as decreasing the heavy metal motility in the soil through siderophores, and solubilizing the micro-and macronutrients for easy access to plants (Yang et al. 2009; Etesami and Beattie 2017). A number of findings have made it clear that plants can modify their rhizosphere and endosphere microbiome under biological and nonbiological stress conditions to acquire more stress-tolerant microbes for their growth and instatement (Hardoim et al. 2008; Noori et al. 2018). In the past few decades, studies on the importance of microbes in regulating the plants' responses toward abiotic stresses have been the focal area of research presuming it as an economical, eco-friendly, and successful way to tackle these adverse conditions (Nadeem et al. 2014; Souza et al. 2015). A lot of rhizopheric inhabitants pertaining to different genera like *Pseudomonas*, *Azotobacter*, *Azospirillum*, *Bacillus*, and *Trichoderma* have been well characterized for their mitigatory role toward the multiple kinds of inanimate stresses (Sorty et al. 2016; Sahoo et al. 2014; Ahmad et al. 2015). How the microbe interaction with the plants supplements their natural ability to combat stressful environmental conditions will be discussed in the following sections.

Microbe-Mediated Alleviation of Drought Stress

Drought, a major abiotic stress in stepped climatic condition, poses devastating consequences on the growth and productivity of cultivated agricultural crops (Etesami and Maheshwari 2018). By 2050, it is anticipated to affect 50% of the arable agricultural land (Etesami and Beattie 2017). Under water stress conditions, a plethora of microbes like *Actinobacteria* or some monoderms in the rhizospheric zone and/or in the endosphere of plant roots have suggested a coadaptive strategy between plant and microbes in particular stress (TerHorst et al. 2014; Santos-Medellín et al. 2017; Fitzpatrick et al. 2018). Different mechanisms are employed by these microorganisms inhabiting the rhizosphere to avoid undesirable consequences of drought on existing flora; however, the same holds true in the case of other abiotic stresses as well. For such soil microbes, which induce a positive

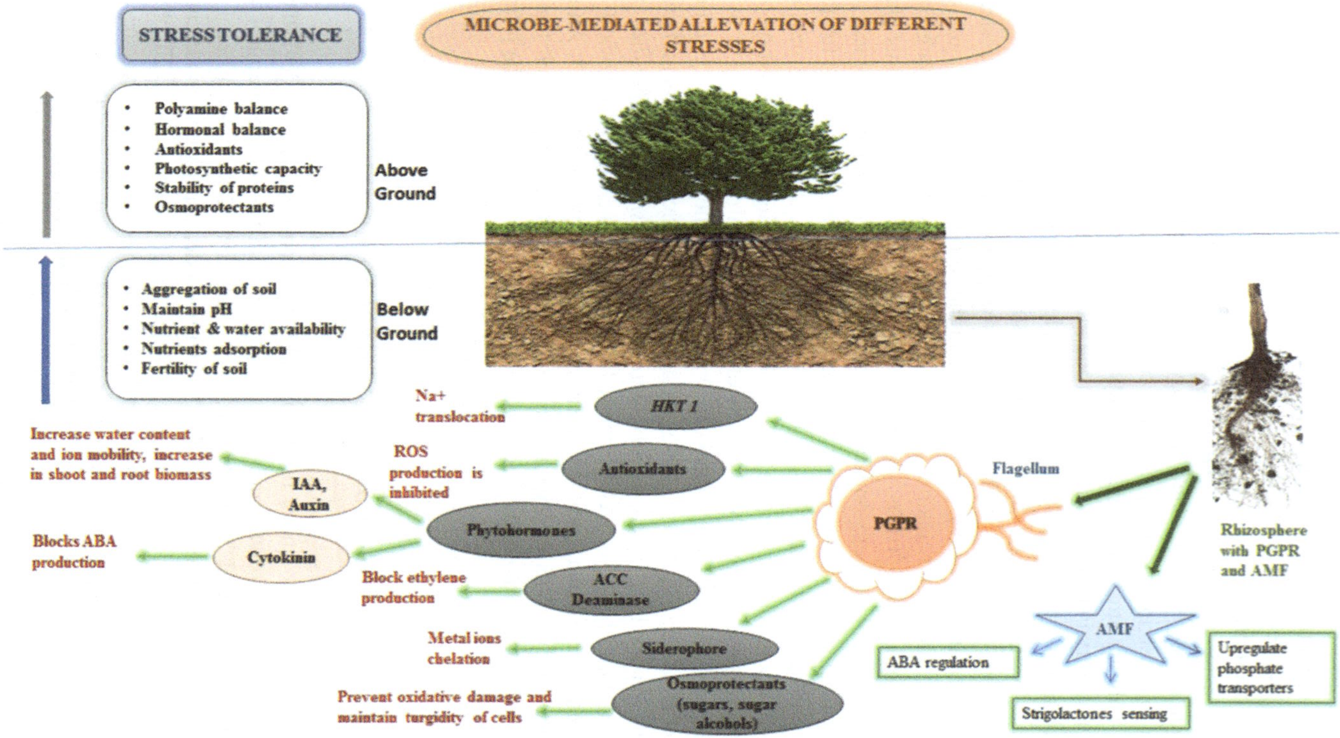

Fig. 2 Natural interaction of PGPR (plant growth-promoting rhizobacteria) and AMF (arbuscular mycorrhizal fungi) with plants and their roles in mitigating abiotic stresses via different mechanisms

response in the plants under harsh environmental conditions, the term plant growth promoter (PGP) is nicely quoted. It is well known that the auxins such as IAA, 2,4-D, or NAA in a proper concentration have a stimulatory effect on cell elongation, which further results in the root initiation and its growth; however, the higher concentration may have a negative impact on its growth (Sorty et al. 2016). The same has been proved true for the increased synthesis of ethylene during water stress (Etesami and Maheshwari 2018). Drought-resistance bacteria generally augment the plant tolerance by modifying the plant growth regulators, viz. GA, IAA, ethylene, and abscisic acid. For example, the increased number of root tips and surface area of the root was exhibited in plants through IAA production by PGPRs (*Azospirillum*, *Bacillus*, *Pseudomonas*), thus augmenting water and nutrient procurement by the plants, which helped them to overcome water scarcity (Naveed et al. 2014). PGPRs possessing the ACC-deaminase enzymes have also been proven to degrade the ethylene producing precursor 1-aminocyclopropane-1-carboxylic acid (ACC) to NH_3 and α-ketobutyric acid, thus waning stress ethylene levels (Glick 2014). Moreover, under severe drought conditions, the lettuce co-inoculated with *Pseudomonas* sp. and *Glomus mossae* has shown an augmentation in an antioxidant enzyme—catalase—to alleviate the damage caused by oxidative stress (Kohler et al. 2009). At the molecular level, the increased transcriptional level of drought-responsive genes in *Arabidopsis* inoculated with bacteria *Paenibacillus polymyxa* improved the drought tolerance (Timmusk and Wagner 1999). Whereas the water deficit had negative impacts on the anatomical features of the plants under regulated conditions, the wheat co-inoculated with AM fungi and *Pseudomonas* species revealed the modifications in mesophyll and phloem tissues and the thickness of the epidermis, thereby lessening the effects of drought stress (El-Afry et al. 2012). The synthesis of different osmoprotectants, viz. proline, glycine betaine (GB), and trehalose, is a prime response in plants under water stress conditions (Etesami and Maheshwari 2018). Drought-resistant bacteria such as *Bacillus thuringiensis* under water-deficit conditions also exudate these kinds of osmolytes, which in harmony with plant-generated osmolytes stimulate the growth of plants (Ortiz et al. 2015). AM fungi associated with plants also play a critical role in drought stress resistance by altering plant hormonal physiology, improving the efficiency of photosystem II and the Calvin cycle (Ruiz-Lozano et al. 2015). Based on these studies, it could be suggested to isolate and introduce drought-resistant microbial consortium in the field to boost the metabolic plasticity of plants and to support the plant growth under water-deficit conditions.

Microbe-Mediated Alleviation of Salinity Stress

Studies on salinity stress have demonstrated poor plant growth under these conditions because of decreased mobilization of the nutrients, imbalance hormonal content, genesis of free radicals like ROS, and ionic toxicity (Rai et al. 2019). The plants harbor the potential microbes to deal with the salinity stress and assist plants in withstanding the saline conditions (Kumar et al. 2020). A number of ways, including microbe-assisted technologies, have been explained to improve salinity tolerance and productivity in plants under saline environments. Various direct and indirect mechanisms can be employed by PGP microbes under salt stress to trigger plant growth. These may include biofilm formation, phytohormone production, N_2-fixation, nutrient mobilization, ACC-deaminase activity, and production of siderophore (Egamberdieva and Kucharova 2009; Ansari et al. 2019). An alternative approach to chemical fertilizers may be the siderophore producing bacteria due to simultaneously mitigating the salt-stress effects and increasing the availability of Fe in saline soils (Ferreira et al. 2019). *Pseudomonas* sp. and *Acinetobacter* sp. in oats and barley have been shown to increase the synthesis of IAA and enzyme ACC deaminase in soil afflicted with salt (Chang et al. 2014). These auxins result in root growth initiation due to their growth-promoting effect with regards to cell elongation. The other important causes for microbe-assisted plant growth promotion under salt stress have been the reduction in the ethylene level due to the production of ACC-deaminase enzyme in them (Bhise et al. 2017). In the case of maize, the stress tolerance against salinity was attributed to a reduction in electrolyte leakage, osmotic potential, elevated proline level, and selective K^+ ions uptake in the presence of *Rhizobium* and *Pseudomonas* (Bano and Fatima 2009). Similarly, besides its ability to secrete exopolysaccharides (EPS), the excellent colonizing ability of *Pseudomonas* sp. PMDzncd2003 with monocot roots has led to increased salt tolerance in rice plants (Sen and Chandrasekhar 2014). EPS is secreted as slimy material, which binds with the soil and eventually creates a protective shell around soil aggregates, and because of this property of EPS-secreting microbes, the plants display salinity resistance (Ansari et al. 2019). Furthermore, increased EPS generation in response to salt stress also favors biofilm formation, which adds on plant adaptation to salinity by reinforcing soil structure and physicochemical characteristics (Banerjee et al. 2019). Jha and Subramanian (2014), in their study on salt-sensitive rice lines, have reported the reduction in lipid peroxidation and the activities of superoxide dismutase involving the two rhizobacteria—*Pseudomonas pseudoalcaligenes* and *Bacillus pumilus*—during salinity stress. Besides rhizobacteria, mycorrhizal fungi also serve as key plant growth promoters. An extensive network of very fine hyphae in these fungi increases the overall efficiency of the plants in nutrient uptake by the roots under stress conditions. Moreover, these endophytes may induce salinity tolerance in plants by elevating the antioxidant level (Ruppel et al. 2013). Endophytic fungi also maintain the ionic homeostasis by modulating ion accumulation, thus restricting the Na^+ transport to leaves and ensuring a lower ratio of Na^+:K^+ in the cytosol of the plants (Gupta et al. 2020). So, for mollifying salt stress in glycophytes, co-inoculation of PGPR and plant-associated fungi could serve as potential tools.

Microbe-Mediated Alleviation of Heavy Metal Stress

Heavy metal (HM)-contaminated soil is generally remediated using chemical and physical methods, which are costly, time consuming, and nonsustainable in the long run; phytoremediation has been suggested as an alternative to these processes (Singh 2016). Furthermore, plant–microbe associations may add another dimension to the latter strategy targeting the management of HM stress. PGPR help in solubilizing the heavy metals and make them available for the hyper-accumulators to uptake and compartmentalize these in the vacuoles, thus directly assisting the plant-based remediating technology (Vymazal and Brezinova 2015). Varied methodologies can be employed by microbes allied with plants for detoxifying the HMs from contaminated sites. Microbes inhabiting the rhizosphere possess the potential to produce antifungal compounds, compatible solutes, and soil enzymes along with cytokinin, auxin, GA, and ABA production (Mustafa et al. 2019), which are prominent participants in reducing the toxic effects of HMs and also enhance the productivity of crops grown in heavy metal-affected soil (Naveed et al. 2020). The inoculation experiment in *B. napus* using the PGPR strains such as *Pseudomonas fluorescens* (Pf), *Variovorax* sp. (Va), and *Bacteroidetes bacterium* (Ba) in combination and separately has revealed the improved plant growth in Cd- and Zn-affected soil (Dąbrowska et al. 2017). Similarly, Weyens et al. (2013) reported the promotion in the root growth of rapeseed plants when the Cd-contaminated soil was augmented with metal-tolerant *Bacillus* sp. RJ16 and *Pseudomonas* sp. Siderophore (a low molecular weight organic compound) producing microbes are also useful in reducing the damaging effects of the HMs by chelating these and increasing the uptake of nutrients by the plants (Saha et al. 2016). HM-tolerant bacterial strains, viz. *Microbacterium* sp., *Achromobacter* sp., and *Rhodococcus erythropolis*, have been reported to enhance the fresh weight of *Trifolium* grown in the soil contaminated with zinc and cadmium (Hassan et al. 2017). This improvement in plant biomass was attributed to their properties to synthesize IAA, siderophores, and ACC-deaminase under metal-stressed conditions. In the same way, arsenic resistance bacteria from *Pteris vittata* have been reported to be an energetic siderophore producer, which increases nutrition acquisition by solubilizing P and results in enhanced plant growth (Ghosh et al. 2015). Metal-resistant rhizobacteria with the ability to produce organic acids like gluconic acid, citric acid, and oxalic acid can reduce the deleterious effects of heavy metals where the acids were shown to chelate these HMs (Etesami 2018). PGPRs also possess the potential to fix atmospheric N_2 and transfer it to plants (via symbiotic association) growing under metal stress conditions (Nonnoi et al. 2012). Another way of promoting plant growth can include induced systemic resistance (ISR) mediated by endophytic bacteria under HM stress (Ma et al. 2016). Besides bacteria, AM fungi in symbiotic association with plants have also been shown to help in metal stress tolerance. It was reported that AM in a mutualistic relationship with ryegrass resulted in immobilization of HMs in soil, which further protected the plants from the adverse effects of these metals as their translocation to different parts of the plant get considerably retarded (Takács and Vörös 2003). Similarly, Huang et al. (2005) have reported the role of *Glomus mosseae* in binding the metals to organic matter, which decreased the bioavailability of excessive HMs for the plant, thereby limiting the possibility of metal uptake by the host plant. Overall, the microbial diversity not only promotes the sweeping up of HMs from the polluted environment by assisting the plant's phytoremediation ability but also stimulates the growth and development in plants.

Microbe-Mediated Alleviation of Temperature Stress

For a given plant–microbe interaction, the effect of an environmental condition would be expected on both the plant and the microbe. A few degrees of elevation in temperature, compared with the optimal range for growth, has long been known to suppress the plant morphogenesis and has been of a major concern across the globe (Balasubramanian et al. 2006). However, some rhizobacteria and endophytes could alleviate the harmful impacts of heat stress on plants and render host plants an ability to grow at different temperatures. A fascinating example can be presented by the mutualism between the tropical panic grass *Dichanthelium lanuginosum* and the fungus *Curvularia protuberate*, which permits both to survive at elevated soil temperatures; however, neither the plant nor the fungus could survive at this condition (Márquez et al. 2007). All organisms, plants, or microbes generally respond to unexpected heat stress by synthesizing a specific group of proteins referred to as HSPs. Inoculation of sorghum seedling, after its exposure to elevated temperature, with a heat-stable *Pseudomonas aeruginosa* strain AMK-P6 resulted in enhanced fresh weight and increased sugar, amino acid, and chlorophyll content in the crop (Ali et al. 2009). In the same way, McLellan et al. (2007) have reported the enhanced heat stress tolerance in *Arabidopsis thaliana* when inoculated with the rhizosphere fungus *Paraphaeosphaeria quadriseptata*. It was shown as a result of the induction of HSP101 and HSP70—the conserved components of the stress response. A recent study by Mukhtar et al. (2020) has also depicted the importance of heat stress-tolerant bacteria *Bacillus cereus* on tomato plant growth. Bacterial augmented enhanced root and shoot growth, leaf surface area, and fresh and dry weight were attributed to the production of ACC deaminase as well as extracellular polymeric substances (EPS) that significantly decreased the unfavorable effects of heat stress on tomatoes. Similarly, an inoculated soybean plant with thermotolerant *B. cereus* strain SA1 showed improved biomass and chlorophyll content under heat stress conditions. Antioxidant analysis of the inoculated soybean plant revealed an increase in APX, SOD, and glutathione content vis-à-vis control. The overexpression of stress-responsive *LAX3* and *AKT2* proteins in SA1-inoculated soybean plants also resulted in decreased ROS production and altered auxin and ABA stimuli, which are critical in plants under heat stress (Khan et al. 2020). A few bacterial species such as *P. cedrina*, *Brevundimonas terrae*, and *Arthrobacter nicotianae* habituated for low temperature with multifunction plant growth-promoting ability were also reported by Yadav et al. (2014). The PGPR isolated from root nodules of low temperature growing pea plants has been shown to possess efficient biofertilizer ability (Meena et al. 2015). Enhanced biomass and nutrient uptake of wheat seedlings grown in cold temperature was reported when seed bacterization was carried out with cold-tolerant *Serratia marscescens* strain SRM and *Pantoea dispersa* strain 1A (Selvakumar et al. 2007). These species were later on reported to exhibit several characteristics such as IAA and siderophores production, phosphate solubilization, etc. at 15 and 4 °C by these workers.

Conclusion

In most of the past studies, plant-associated microbes have been evaluated to play a major role in bestowing resistance to plants toward abiotic stresses. The microorganisms might involve rhizoplane, rhizosphere, and endophytes (bacteria, AM fungi, etc.), which work via provoking osmotic response, phytohormone production, and induction of novel genes in plants. However, using rDNA technology, stress-tolerant crop varieties can be evolved, but it is a time-consuming process, whereas inoculating plants with a consortium of microbes or a particular strain to ameliorate stresses might prove as a cost-cutting and eco-friendly strategy with its availability over a shorter time span. Hence, it could be suggested to carry out collaborative research in the future in this direction to develop the most effective PGPRs. It may include an amalgam of scientists from various scientific fields like microbiologists, molecular biologist, agronomists, etc.

References

Adrees M, Ali S, Rizwan M et al (2015) Mechanisms of silicon-mediated alleviation of heavy metal toxicity in plants: a review. Ecotoxicol Environ Saf 119:186–197

Ahmad P, Sharma S (2008) Salt stress and phytobiochemical responses of plants. Plant Soil Environ 54:89–99

Ahmad P, Hashem A, Abd-Allah EF, Alqarawi AA, John R, Egamberdieva D et al (2015) Role of *Trichoderma harzianum* in mitigating NaCl stress in Indian mustard (*Brassica juncea* L) through antioxidative defense system. Front Plant Sci 6:868

Ali SZ, Sandhya V et al (2009) *Pseudomonas* sp. strain AKM-P6 enhances tolerance of sorghum seedlings to elevated temperatures. Biol Fertil Soils 46(1):45–55

Ansari FA, Ahmad I, Pichtel J (2019) Growth stimulation and alleviation of salinity stress to wheat by the biofilm forming *Bacillus pumilus* strain FAB10. Appl Soil Ecol 143:45–54

Asada K (2006) Production and scavenging of reactive oxygen species in chloroplasts and their functions. Plant Physiol 141(2):391–396

Atkinson NJ, Urwin PE (2012) The interaction of plant biotic and abiotic stresses: from genes to the field. J Exp Bot 63(10):3523–3543

Bai Y, Sunarti S et al (2018) The role of tomato *WRKY* genes in plant responses to combined abiotic and biotic stresses. Front Plant Sci 9:801

Bailey-Serres J, Voesenek LA (2008) Flooding stress: acclimations and genetic diversity. Annu Rev Plant Biol 59:313–339

Balasubramanian S et al (2006) Potent induction of *Arabidopsis thaliana* flowering by elevated growth temperature. PLoS Genet 2:e106
Banerjee A, Sarkar S, Orellana SC, Bandopadhyay R (2019) Exopolysaccharides and biofilms in mitigating salinity stress: the biotechnological potential of halophilic and soil-inhabiting PGPR microorganisms. In: Microorganisms in saline environments: strategies and functions, pp 133–146
Bano A, Fatima M (2009) Salt tolerance in *Zea mays* (L) following inoculation with *Rhizobium* and *Pseudomonas*. Biol Fertil Soils 45:405–413
Basheer AA (2018) Chemical chiral pollution: impact on the society and science and need of the regulations in the 21st century. Chirality 30(4):402–406
Berendsen RL, Pieterse CM, Bakker PA (2012) The rhizosphere microbiome and plant health. Trends Plant Sci 17(8):478–486
Bhandari K, Nayyar H (2014) Low temperature stress in plants: an overview of roles of cryoprotectants in defense in physiological mechanisms and adaptation strategies in plants under changing environment, vol 1, 3rd edn. Springer, New York, pp 193–265
Bhise KK, Bhagwat PK, Dandge PB (2017) Synergistic effect of *Chryseobacterium gleum* sp. SUK with ACC deaminase activity in alleviation of salt stress and plant growth promotion in *Triticum aestivum* L. 3. Biotech 7:105
Campos F, Cuevas-Velazquez C et al (2013) Group 1 LEA proteins, an ancestral plant protein group, are also present in other eukaryotes, and in the archeae and bacteria domains. Mol Gen Genomics 288:503–517
Cao S, Du XH, Li LH, Liu YD, Zhang L, Pan X et al (2017) Overexpression of populus tomentosa cytosolic ascorbate peroxidase enhances abiotic stress tolerance in tobacco plants. Russ J Plant Physiol 64:224–234
Carpena RO, Vazquez S, Esteban E, Fernandez-Pascual M, Rosario de Felipe M, Zornoza P (2003) Cadmium–stress in white lupin: effects on nodule structure and functioning. Plant Physiol Biochem 161:911
Chan Z et al (2016) RDM4 modulates cold stress resistance in Arabidopsis partially through the CBF-mediated pathway. New Phytol 209:1527–1539
Chang P et al (2014) Plant growth-promoting bacteria facilitate the growth of barley and oats in salt-impacted soil: implications for phytoremediation of saline soils. Int J Phytoremediation 16(7–12):1133–1147
Chaves MM, Flexas J, Pinheiro C (2009) Photosynthesis under drought and salt stress: regulation mechanisms from whole plant to cell. Ann Bot 103(4):551–560
Chen TH, Murata N (2011) Glycine betaine protects plants against abiotic stress: mechanisms and biotechnological applications. Plant Cell Environ 34:1–20
Choudhary DK, Sharma KP, Gaur RK (2011) Biotechnological perspectives of microbes in agro-ecosystems. Biotechnol Lett 33:1905–1910
Ciarmiello LF, Woodrow P, Fuggi A, Pontecorvo G, Carillo P (2011) Plant genes for abiotic stress. In: Shanker AK, Venkateswarlu B (eds) Abiotic stress in plants—mechanisms and adaptations. InTech, Rijeka, Croatia, pp 283–308
Close TJ (1997) Dehydrins: a commonality in response of plants to dehydration and low temperature. Physiol Plant 100:291–296
Dąbrowska G, Hrynkiewicz K, Trejgell A, Baum C (2017) The effect of plant growth-promoting rhizobacteria on the phytoextraction of Cd and Zn by *Brassica napus* L. Int J Phytoremediation 19:597–604
Davis DG, Swanson HR (2001) Activity of stress-related enzymes in the perennial weed leafy spurge (Euphorbia esula L.). Environ Exp Bot 46:95–108
Deeba F, Pandey AK, Ranjan S et al (2012) Physiological and proteomic responses of cotton (*Gossypium herbaceum* L.) to drought stress. Plant Physiol Biochem 53:6–18
Dodd IC, Pérez-Alfocea F (2012) Microbial amelioration of crop salinity stress. J Exp Bot 63(9): 3415–3428

Du H, Wu N et al (2013) Carotenoid deficiency impairs ABA and IAA biosynthesis and differentially affects drought and cold tolerance in rice. Plant Mol Biol 83(4–5):475–488

Egamberdieva D, Kucharova Z (2009) Selection for root colonising bacteria stimulating wheat growth in saline soils. Biol Fertil Soil 45:563–571

El-Afry MM, El-Nady MF, Abdelmonteleb EB, Metwaly MMS (2012) Anatomical studies on drought-stressed wheat plants (*Triticum aestivum* L.) treated with some bacterial strains. Acta Biol Szeged 56(2):165–174

Etesami H (2018) Bacterial mediated alleviation of heavy metal stress and decreased accumulation of metals in plant tissues: mechanisms and future prospects. Ecotoxicol Environ Saf 147:175–191

Etesami H, Beattie GA (2017) Plant-microbe interactions in adaptation of agricultural crops to abiotic stress condition. Probiot Plant Health:163–200

Etesami H, Maheshwari DK (2018) Use of plant growth promoting rhizobacteria (PGPRs) with multiple plant growth promoting traits in stress agriculture: action mechanisms and future prospects. Ecotoxicol Environ Saf 156:225–246

Farago ME, Mullen WA (1979) A possible copper–proline complex from the roots of *Armeria maritima*. Inorg Chim Acta 32:L93–L94

Farrar K, Bryant D, Cope-Selby N (2014) Understanding and engineering beneficial plant–microbe interactions: plant growth promotion in energy crops. Plant Biotechnol J 12:1193–1206

Ferreira MJ, Silva H, Cunha A (2019) Siderophore-producing rhizobacteria as a promising tool for empowering plants to cope with iron limitation in saline soils: a review. Pedosphere 29:409–420

Fitzpatrick CR, Copeland J, Wang PW, Guttman DS et al (2018) Assembly and ecological function of the root microbiome across angiosperm plant species. Proc Natl Acad Sci U S A 115(6): E1157–E1165

Gajewska E, Skłodowska M (2008) Differential biochemical responses of wheat shoots and roots to nickel stress: antioxidative reactions and proline accumulation. Plant Growth Regul 54:179

Geissler N, Hussin S, Koyro HW (2009) Interactive effects of NaCl salinity and elevated atmospheric CO_2 concentration on growth, photosynthesis, water relations and chemical composition of the potential cash crop halophyte *Aster tripolium* L. Environ Exp Bot 65:220–231

Ghosh P, Rathinasabapathi B, Ma LQ (2015) Phosphorus solubilization and plant growth enhancement by arsenic-resistant bacteria. Chemosphere 134:1–6

Gil-Amado JA, Gomez-Jimenez MC (2012) Regulation of polyamine metabolism and biosynthetic gene expression during olive mature-fruit abscission. Planta 235:1221–1237

Gill SS, Tuteja N (2010) Reactive oxygen species and antioxidant machinery in abiotic stress tolerance in crop plants. Plant Physiol Biochem 48:909–930

Glick BR (2014) Bacteria with ACC deaminase can promote plant growth and help to feed the world. Microbiol Res 169(1):30–39

Golldack D, Lüking I, Yang O (2011) Plant tolerance to drought and salinity: stress regulating transcription factors and their functional significance in the cellular transcriptional network. Plant Cell Rep 30:1383–1391

Gopalakrishnan S, Sathya A, Vijayabharathi R, Varshney RK, Gowda CL, Krishnamurthy L (2015) Plant growth promoting rhizobia: challenges and opportunities. 3Biotech 5:355–377

Goyal K, Walton LJ, Tunnacliffe A (2005) LEA proteins prevent protein aggregation due to water stress. Biochem J 388:151–157

Grayson M (2013) Agriculture and drought. Nature 501:S1. https://doi.org/10.1038/501S1a

Guo YP, Guo DP, Zhou HF, Hu MJ, Shen YG (2006) Photoinhibition and xanthophyll cycle activity in bayberry (*Myrica rubra*) leaves induced by high irradiance. Photosynthetica 44:439–446

Gupta S et al (2020) Alleviation of salinity stress in plants by endophytic plant-fungal symbiosis: current knowledge, perspectives and future directions. Plant Soil. https://doi.org/10.1007/s11104-020-04618-w

Guy C, Haskell D et al (1998) Association of proteins with the stress 70 molecular chaperones at low temperature: evidence for the existence of cold labile proteins in spinach. Cryobiology 36(4):301–314

Hardoim PR, van Overbeek LS, vanElsas JD (2008) Properties of bacterial endophytes and their proposed role in plant growth. Trend Microbiol 16(10):463–471

Hassan TU, Bano A, Naz I (2017) Alleviation of heavy metals toxicity by the application of plant growth promoting rhizobacteria and effects on wheat grown in saline sodic field. Int J Phytoremediation 19:522–529

Hayat S, Hayat Q, Alyemeni MN, Wani AS, Pichtel J, Ahmad A (2012) Role of proline under changing environments: a review. Plant Signal Behav 7(11):1456–1466

He H, Ye Z, Yang D, Yan J, Xiao L, Zhong T et al (2013) Characterization of endophytic *Rahnella* sp. JN6 from *Polygonum pubescens* and its potential in promoting growth and Cd, Pb, Zn uptake by *Brassica napus*. Chemosphere 90:1960–1965

Hemantaranjan A, Bhanu AN et al (2014) Heat stress responses and thermotolerance. Adv Plant Agric Res 1(3):62–70

Hirel B, Le Gouis J, Ney B, Gallais A (2007) The challenge of improving nitrogenuse efficiency in crop plants: towards a more central role for genetic variability and quantitative genetics within integrated approaches. J Exp Bot 58:2369–2387

Hossain MA, Piyatida P, Teixeira da Silva JA, Fijuta M (2012) Molecular mechanism of heavy metal toxicity and tolerance in plants: central role of glutathione in detoxification of reactive oxygen species and methylglyoxal and in heavy metal chelation. J Bot 2012:1–37

Huang Y, Tao S, Chen YJ (2005) The role of arbuscular mycorrhiza on change of heavy metal speciation in rhizosphere of maize in wastewater irrigated agriculture soil. J Environ Sci 17(2): 276–280

Ilhan S, Ozdemir F, Bor M (2015) Contribution of trehalose biosynthetic pathway to drought stress tolerance of *Capparis ovata* Desf. Plant Biol 17:402–407

Islam RM, Baohua F et al (2018) Role of abscisic acid in thermal acclimation of plants. J Plant Biol 61(5):255–264

Jha Y, Subramanian RB (2014) PGPR regulate caspase-like activity, programmed cell death, and antioxidant enzyme activity in paddy under salinity. Physiol Mol Biol Plant 20:201–207

Jiang QY, Zhuo F, Long SH et al (2016) Can arbuscular mycorrhizal fungi reduce Cd uptake and alleviate Cd toxicity of *Lonicera japonica* grown in Cd-added soils? Sci Rep 6:21805

John R, Ahmad P, Gadgil K, Sharma S (2008) Effect of cadmium and lead on growth, biochemical parameters and uptake in *Lemna polyrrhiza* L. Plant Soil Environ 54:262

Kamanga R, Mndala L (2019) Crop abiotic stresses and nutrition of harvested food crops: a review of impacts, interventions and their effectiveness. Afr J Agric Res 14(3):118–135

Kaur N, Gupta AK (2005) Signal transduction pathways under abiotic stresses in plants. Curr Sci 88:1771–1780

Khan MA, Asaf S, Khan AL et al (2020) Thermotolerance effect of plant growth-promoting Bacillus cereus SA1 on soybean during heat stress. BMC Microbiol 20:175

Kim NH, Kim BS, Hwang BK (2013) Pepper arginine decarboxylase is required for polyamine and gamma-aminobutyric acid signaling in cell death and defense response. Plant Physiol 162: 2067–2083

Kissoudis C, van de Wiel C, Visser RGF, van der Linden G (2014) Enhancing crop resilience to combined abiotic and biotic stress through the dissection of physiological and molecular crosstalk. Front Plant Sci 5:207

Koca M, Bor M, Ozdemir F, Turkan I (2007) The effect of salt stress on lipid peroxidation, antioxidative enzymes and proline content of sesame cultivars. Environ Exp Bot 60:344–351

Kohler J, Hernandez JA, Caravaca S, Roldan A (2009) Induction of antioxidant enzymes is involved in the greater effectiveness of a PGPR versus AM fungi with respect to increasing the tolerance of lettuce to severe salt stress. Environ Exp Bot 65:245–252

Kotak S, Larkindale J et al (2007) Complexity of the heat stress response in plants. Curr Opin Plant Biol 10:310–316

Kumar S, Trivedi PK (2018) Glutathione S-transferases: role in combating abiotic stresses including arsenic detoxification in plants. Front Plant Sci 9:751

Kumar A, Singh S, Gaurav AK, Srivastava S, Verma JP (2020) Plant growth-promoting bacteria: biological tools for the mitigation of salinity stress in plants. Front Microbiol 11:1216

Kurepin LV, Qaderi MM et al (2008) A rapid effect of applied brassinolide on abscisic acid concentrations in *Brassica napus* leaf tissue subjected to short-term heat stress. Plant Growth Regul 55:165–167

Laloum T, Martín G, Duque P (2018) Alternative splicing control of abiotic stress responses. Trends Plant Sci 23:140–150

Lee BH, Kapoor A, Zhu J, Zhu JK (2006) STABILIZED1, a stress-upregulated nuclear protein, is required for pre-mRNA splicing, mRNA turnover, and stress tolerance in *Arabidopsis*. Plant Cell 18:1736–1749

Lee G, Carrow RN, Duncan RR, Eiteman MA, Rieger MW (2008) Synthesis of organic osmolytes and salt tolerance mechanisms in *Paspalum vaginatum*. Environ Exp Bot 63:19–27

Li S, Li F, Wang J, Zhang WEN, Meng Q, Chen THH, Yang X (2011) Glycinebetaine enhances the tolerance of tomato plants to high temperature during germination of seeds and growth of seedlings. Plant Cell Environ 34(11):1931–1943

Lim CW, Baek W, Jung J, Kim JH, Lee SC (2015) Function of ABA in stomatal defense against biotic and drought stresses. Int J Mol Sci 16:15251–15270

Lipiec J, Doussan C, Nosalewicz A, Kondracka K (2013) Effect of drought and heat stresses on plant growth and yield: a review. Int Agrophys 27:463–477

Liu H et al (2017) Effects of Jasmonic acid signalling on the wheat microbiome differ between body sites. Sci Rep 7:41766

Ma Y et al (2016) Inoculation of *Brassica oxyrrhina* with plant growth promoting bacteria for the improvement of heavy metal phytoremediation under drought conditions. J Hazard Mater 320: 36–44

Manzi M, Lado J, Rodrigo MJ, Arbona V, Gómez-Cadenas A (2016) ABA accumulation in water-stressed citrus roots does not rely on carotenoid content in this organ. Plant Sci 252:151–161

Márquez LM, Redman RS, Rodriguez RJ, Roossinck MJ (2007) A virus in a fungus in a plant: three-way symbiosis required for thermal tolerance. Science 315:513–515

Massad TJ, Dyer LA, Vega CG (2012) Cost of defense and a test of the carbon-nutrient balance and growth-differentation balance hypotheses for two co-occurring classes of plant defense. PLoS One 7:7554

Massacci A, Nabiev SM, Pietrosanti L, Nematovn SK, Chernikova TN, Thor K, Leipner J (2008) Response of the photosynthetic apparatus of cotton (Gossypium hirsutum) to the onset of drought stress under field conditions studied by gas-exchange analysis and chlorophyll fluorescence imaging. Plant Physiol Biochem 46(2):189–195

McLellan CA et al (2007) A rhizosphere fungus enhances *Arabidopsis* thermotolerance through production of an HSP90 inhibitor. Plant Physiol 145(1):174–182

Meena RK et al (2015) Isolation of low tempertutre surviving plant growth promoting rhizobacteria (PGPR) from pea (*Pisum sativum* L.) and documentation of their plant growth promoting traits. Biocatal Agric Biotechnol 4:806–811

Messedi D, Slama I, Labidi N, Ghnaya T, Savouré A, Soltani A, Abdelly C (2006) Proline metabolism in *Sesuvium portulacastrum* under salinity and drought. In: Öztürk M, Waisel Y, Khan MA, Görk G (eds) Biosaline agriculture and salinity tolerance in plants. Birkhäuser, Basel, pp 65–72

Mizoi J, Shinozaki K, Yamaguchi-Shinozaki K (2012) AP2/ERF family transcription factors in plant abiotic stress responses. Biochim Biophys Acta 1819:86–96

Mukhtar T, Rehman S, Smith D et al (2020) Mitigation of heat stress in Solanum lycopersicum L. by ACC-deaminase and exopolysaccharide producing Bacillus cereus: effects on biochemical profiling. Sustainability 12

Mustafa S, Kabir S, Shabbir U, Batool R (2019) Plant growth promoting rhizobacteria in sustainable agriculture: from theoretical to pragmatic approach. Symbiosis 78:115–123

Mustafavi SH, Badi HN, Sekara A, Al E (2018) Polyamines and their possible mechanisms involved in plant physiological processes and elicitation of secondary metabolites. Acta Physiol Plant 40:102

Nadeem SM, Ahmad M, Zahir ZA, Javaid A, Ashraf M (2014) The role of mycorrhizae and plant growth promoting rhizobacteria (PGPR) in improving crop productivity under stressful environments. Biotechnol Adv 32:429–448

Naveed M, Hussain MB, Zahir ZA, Mitter B, Sessitsch A (2014) Drought stress amelioration in wheat through inoculation with *Burkholderia phytofirmans* strain PsJN. Plant Growth Regul 73(2):121–131

Naveed M, Bukhari SS, Mustafa A, Ditta A, Alamri S, El-Esawi MA, Rafique M, Ashraf S, Siddiqui MH (2020) Mitigation of nickel toxicity and growth promotion in sesame through the application of a bacterial endophyte and zeolite in nickel contaminated soil. Int J Environ Res Public Health 17(23):8859

Nawaz K et al (2010) Fatality of salt stress to plants: morphological, physiological and biochemical aspects. Afr J Biotechnol 9(34):5475–5480

Neal AL, Ahmad S, Gordon-Weeks R, Ton J (2012) Benzoxazinoids in root exudates of maize attract *Pseudomonas putida* to the rhizosphere. PLoS One 7(4):e35498

Nguyen D, Rieu I, Mariani C, van Dam NM (2016) How plants handle multiple stresses: hormonal interactions underlying responses to abiotic stress and insect herbivory. Plant Mol Biol 91:727–740

Nishida H, Suzaki T (2018) Nitrate-mediated control of root nodule symbiosis. Curr Opin Plant Biol 44:129–136

Nonnoi F, Chinnaswamy A, de la Torre VSG, de la P�na TC, Lucas MM, Pueyo JJ (2012) Metal tolerance of rhizobial strains isolated from nodules of herbaceous legumes (*Medicago* spp. and *Trifolium* spp.) growing in mercury-contaminated soils. Appl Soil Ecol 61:49–59

Noori F, Etesami H, Zarini HN et al (2018) Mining alfalfa (*Medicago sativa* L.) nodules for salinity tolerant non-rhizobial bacteria to improve growth of alfalfa under salinity stress. Ecotoxicol Environ Saf 162:129–138

Olien CR, Smith MN (1977) Ice adhesions in relation to freeze stress. Plant Physiol 60:499–503

Ortiz N, Armadaa E, Duque E, Roldánc A, Azcóna R (2015) Contribution of arbuscular mycorrhizal fungi and/or bacteria to enhancing plant drought tolerance under natural soil conditions: effectiveness of autochthonous or allochthonous strains. J Plant Physiol 174:87–96

Osakabe Y et al (2013) Osmotic stress responses and plant growth controlled by potassium transporters in *Arabidopsis*. Plant Cell 25(2):609–624

Pan XY, Wang GX, Yang HM, Wei XP (2003) Effect of water defi cits on within-plot variability in growth and grain yield of spring wheat in north west china. Field Crop Res 80:195–205

Parida AK, Das AB (2005) Salt tolerance and salinity effects on plants: a review. Ecotoxicol Environ Saf 60:324–349

Penna S (2003) Building stress tolerance through overproducing trehalose in transgenic plants. Trends Plant Sci 8:355–357

Rai A, Borpatragohain B, Sahoo S (2019) Role of plant-microbe interactions on abiotic stress tolerance in plants: a review. Int J Agric Plant Sci 1(4):25–31

Ritonga FN, Chen S (2020) Physiological and molecular mechanism involved in cold stress tolerance in plants. Plants 9:560

Ruiz-Lozano JM et al (2015) Arbuscular mycorrhizal symbiosis induces strigolactone biosynthesis under drought and improves drought tolerance in lettuce and tomato. Plant Cell Environ 39(2): 441–452

Ruppel S, Franken P, Witzel K (2013) Properties of the halophyte microbiome and their implications for plant salt tolerance. Funct Plant Biol 40:940–951

Saba H, Jyoti P, Neha S (2013) Mycorrhizae and phytochelators as remedy in heavy metal contaminated land remediation. Int Res J Environ Sci 2(1):74–78

Sah SK, Reddy KR, Li J (2016) Abscisic acid and abiotic stress tolerance in crop plants. Front Plant Sci 7:571

Saha M, Sarkar S, Sarkar B, Sharma BK, Bhattacharjee S, Tribedi P (2016) Microbial siderophores and their potential applications: a review. Environ Sci Pollut Res 23:3984–3999

Sahoo RK, Ansari MW, Pradhan M et al (2014) A novel Azotobacter vinellandii (SRIAz3) functions in salinity stress tolerance in rice. Plant Sig Behav 9:e29377

Santos-Medellín C, Edwards J, Liechty Z, Nguyen B, Sundaresan V (2017) Drought stress results in a compartment-specific restructuring of the rice root-associated microbiomes. MBio 8(4): e00764–e00717

Saradhi PP, AliaArora S, Prasad KVSK (1995) Proline accumulates in plants exposed to UV radiation and protects them against UV induced peroxidation. Biochem Biophys Res Commun 209:1

Selvakumar G, Kundu S, Joshi P et al (2007) Characterization of a cold-tolerant plant growth-promoting bacterium Pantoea dispersa 1A isolated from a sub-alpine soil in the North Indian Himalayas. World J Microbiol Biotechnol 24:955–960

Sen S, Chandrasekhar CN (2014) Effect of PGPR on growth promotion of rice (*Oryza sativa* L.) under salt stress. Asian J Plant Sci Res 4:62–67

Setiawati T, Ayalla A, Nurzaman MA, Mutaqin Z (2018) Influence of light intensity on leaf photosynthetic traits and alkaloid content of Kiasahan (*Tetracera scandens* L.). IOP Conf Ser Earth Environ Sci 166:12–25

Shafiei Masouleh SS, Sassine Y, Aldine NJ (2019) The role of organic solutes in the osmotic adjustment of chilling stressed plants (vegetable, ornamental and crop plants). Ornam Hortic 25(4):434–442

Shao HB, Chu LY, Jaleel CA, Zhao CX (2008) Water-deficit stress— induced anatomical changes in higher plants. Crit Rev Biol 331:215–225

Shao CG, Wang H, Yu-Fen BI (2015) Relationship between endogenous polyamines and tolerance in *Medicago sativa* L. under heat stress. Acta Agrestia Sin 23:1214–1219

Sharma P, Jha AB, Dubey RS, Pessarakli M (2012) Reactive oxygen species, oxidative damage, and antioxidative defense mechanism in plants under stressful conditions. J Bot:217037

Shi H, Quintero FJ, Pardo JM, Zhu JK (2002) The putative plasma membrane Na^+/H^+ antiporter SOS1 controls long-distance Na^+ transport in plants. Plant Cell 14(2):465–477

Silveira V, De Vita AM et al (2013) Morphological and polyamine content changes in embryogenic and non-embryogenic callus of sugarcane. Plant Cell Tissue Organ Cult 114:351–364

Singh VP, Singh S, Kumar J, Prasad SM (2015) Investigating the roles of ascorbate-glutathione cycle and thiol metabolism in arsenate tolerance in ridged Luffa seedlings. Protoplasma 252: 1217–1229

Singh A (2016) Varied responses and tolerant mechanisms towards salinity stress in plants. Int J Plant Soil Sci 11(5):1–13

Singh S, Parihar P, Singh R, Singh VP, Prasad SM (2016) Heavy metal tolerance in plants: role of transcriptomics, proteomics, metabolomics, and ionomics. Front Plant Sci 6:1143

Singhal RK, Kumar V, Kumar S, Choudhary BL (2017) High light stress response and tolerance mechanism in plant. Interdisc J Contemp Res 4(1)

Slama I, Abdelly C et al (2015) Diversity, distribution and roles of osmoprotective compounds accumulated in halophytes under abiotic stress. Ann Bot 115:433–447

Sorty AM, Meena KK, Choudhary K, Bitla UM, Minhas PS, Krishnani KK (2016) Effect of plant growth promoting bacteria associated with halophytic weed (*Psoralea corylifolia* L.) on germination and seedling growth of wheat under saline conditions. Appl Biochem Biotechnol 180: 872–882

Souza RD, Ambrosini A, Passaglia LMP (2015) Plant growth-promoting bacteria as inoculants in agricultural soils. Genet Mol Biol 38:401–419

Stockinger EJ, Mao Y et al (2001) Transcriptional adaptor and histone acetyltransferase proteins in *Arabidopsis* and their interactions with CBF1, a transcriptional activator involved in cold-regulated gene expression. Nucleic Acid Res 29:1524–1533

Sturz AV, Christie BR, Nowak J (2000) Bacterial endophytes: potential role in developing sustainable systems of crop production. Crit Rev Plant Sci 19:1–30

Tajti J, Janda T, Majláth I, Szalai G, Pál M (2018) Comparative study on the effects of putrescine and spermidine pre-treatment on cadmium stress in wheat. Ecotoxicol Environ Safety 148:546–554

Takács T, Vörös I (2003) Effect of metal non-adapted arbuscular mycorrhizal fungi on Cd, Ni and Zn uptake by ryegrass. Acta Agron Hung 51:347–354

TerHorst CP, Lennon JT, Lau JA (2014) The relative importance of rapid evolution for plant-microbe interactions depends on ecological context. Proc R Soc B Biol Sci 281(1785): 20140028

Tewari RK, Kumar P, Sharma PN (2006) Antioxidant responses to enhanced generation of superoxide anion radical and hydrogen peroxide in the copper-stressed mulberry plants. Planta 223:1145–1153

Thakur P, Kumar S, Malik JA, Berger JD, Nayyar H (2010) Cold stress effects on reproductive development in grain crops. Environment 67:429–443

Tian J (2012) Physiological regulation function and proteomics research of exogenous spermidine on alleviating high temperature stress the of cucumber seedlings. Nanjing Agric Univ:7–21

Timm CM, Carter KR et al (2018) Abiotic stresses shift belowground *Populus*-associated bacteria toward a core stress microbiome. mSystems 3:e00070–e00017

Timmusk S, Wagner EGH (1999) The plant growth-promoting rhizobacterium *Paenibacillus polymyxa* induces changes in *Arabidopsis thaliana* gene expression, a possible connection between biotic and abiotic stress responses. Mol Plant-Microbe Interact 12:951–959

Tisi A, Federico R, Moreno S et al (2011) Perturbation of polyamine catabolism can strongly affect root development and xylem differentiation. Plant Physiol 157:200–215

Tiwari S, Lata C, Singh Chauhan P, Prasad V, Prasad M (2017) A functional genomic perspective on drought signalling and its crosstalk with phytohormone-mediated signalling pathways in plants. Curr Genomics 18:469–482

Turner TR, James EK, Poole PS (2013) The plant microbiome. Genome Biol 14:209

Tuteja N (2009) Integrated calcium signaling in plants. In: Baluska F, Mancuso S (eds) Signaling in plants. Springer, Heidelberg, Germany, pp 29–49

Vymazal J, Brezinova T (2015) Accumulation of heavy metals in above-ground biomass of *Phragmites australis* in horizontal flow constructed wetlands for wastewater treatment: a review. Chem Eng J 290:232–242

Wallace JG, Zhang XC, Beyene Y et al (2016) Genome-wide association for plant height and flowering time across 15 tropical maize populations under managed drought stress and well-watered conditions in sub-Saharan Africa. Crop Sci 56:2365–2378

Wang W, Vinocur B, Altman A (2003) Plant responses to drought, salinity and extreme temperatures: towards genetic engineering for stress tolerance. Planta 218:1–14

Wang GP, Li F, Zhang J et al (2010) Overaccumulation of glycine betaine enhances tolerance of the photosynthetic apparatus to drought and heat stress in wheat. Photosynthetica 48:30–41

Wang Y, Mao Z, Jiang H, Zhang Z, Chen XA (2019) A feedback loop involving MdMYB108L and MdHY5 controls apple cold tolerance. Biochem Biophys Res Commun 512:381–386

Wani SH, Kumar V (2015) Plant stress tolerance: engineering ABA: a potent phytohormone. Transcriptomics 3:113

Wani SH, Kumar V, Shriram V, Sah SK (2016) Phytohormones and their metabolic engineering for abiotic stress tolerance in crop plants. Crop J 4:162–176

Weyens N, Beckers B, Schellingen K, Ceulemans R, Croes S, Janssen J, Haenen S, Witters N, Vangronsveld J (2013) Plant associated bacteria and their role in the success or failure of metal phytoextraction projects: first observations of a field-related experiment. Microb Biotechnol 6: 288–299

Wu X, Shiroto Y, Kishitani S, Ito Y, Toriyama K (2008) Enhanced heat and drought tolerance in transgenic rice seedlings overexpressing OsWRKY11 under the control of HSP101 promoter. Plant Cell Rep 28:21–30

Wu T, Zhang M, Zhang H, Huang K, Chen M, Chen C, Yang X, Li Z, Chen H, Ma Z et al (2019) Identification and characterization of EDT1 conferring drought tolerance in rice. J Plant Biol 62: 39–47

Wuana RA, Okieimen FE (2011) Heavy metals in contaminated soils: a review of sources, chemistry, risks and best available strategies for remediation. Int Scholar Res Net. https://doi.org/10.5402/2011/402647

Xu S, Li J, Zhang X, Wei H, Cui L (2006) Effects of heat acclimation pretreatment on changes of membrane lipid peroxidation, antioxidant metabolites, and ultrastructure of chloroplasts in two cool-season turfgrass species under heat stress. Environ Exp Bot 56:274–285

Xu Z, Jiang Y, Jia B, Zhou G (2016) Elevated-CO_2 response of stomata and its dependence on environmental factors. Front Plant Sci 7:657

Yadav SK (2010) Cold stress tolerance mechanisms in plants agronomy for sustainable development. Agron Sustain Dev 30:515–527

Yadav J et al (2014) Evaluation of PGPR and different concentration of phosphorus level on plant growth, yield and nutrient content of rice (*Oryza sativa* L.). Ecol Eng 62:123–128

Yang Y, Han C, Liu Q, Bo L, Wang J (2008) Effect of drought and low light on growth and enzymatic antioxidant system of Picea asperata seedlings. Acta Physiol Plant 30:433–440

Yang J, Kloepper JW, MinRyu C (2009) Rhizosphere bacteria help plants tolerate abiotic stress. Trend Plant Sci 14(1):1–4

Yolcu S, Ozdemir F, Güler A, Bor M (2016) Histone acetylation influences the transcriptional activation of POX in *Beta vulgaris* L. and *Beta maritima* L. under salt stress. Plant Physiol Biochem 100:37–46

Zamora-Briseño JA, de Jiménez ES (2016) A LEA 4 protein up-regulated by ABA is involved in drought response in maize roots. Mol Biol Rep 43:221–228

Zhang X, Cai J, Wollenweber B, Liu F, Dai T, Cao W, Jiang D (2013) Multiple heat and drought events affect grain yield and accumulations of high molecular weight glutenin subunits and glutenin macropolymers in wheat. J Cereal Sci 57:134–140

Zhang QH, Wang M et al (2015) PtrABF of *Poncirus trifoliata* functions in dehydration tolerance by reducing stomatal density and maintaining reactive oxygen species homeostasis. J Exp Bot 66:5911–5927

Zhao SZ et al (2011) Growth regulator-induced betacyanin accumulation and dopa-4,5-dioxygenase (DODA) gene expression in euhalophyte Suaeda salsa calli *in vitro*. Cell Dev Biol Plant 47:391–398

Zhu JK (2001) Plant salt tolerance. Trends Plant Sci 6(2):66–71

Zhu JK (2002) Salt and drought stress signal transduction in plants. Annu Rev Plant Biol 53:247–273

Zoratti L, Karppinen K, Escobar AL, Häggman HM (2014) Light-controlled flavonoid biosynthesis in fruits. Front Plant Sci 5(534):534

Zwicke M, Picon-Cochard C, Morvan-Bertrand A et al (2015) What functional strategies drive drought survival and recovery of perennial species from upland grassland? Ann Bot 116:1001–1015

Arbuscular Mycorrhizal Fungi: A Potential Candidate for Nitrogen Fixation

Monika, N. Yadav, Mamta, N. Kumar, A. Kumar, S. Devi, V. Kaur, S. Kumar, and S.S. Arya

Abstract Arbuscular mycorrhizal fungi (AMF) and rhizobia are the two most important plant symbionts. The function of rhizobia in nitrogen fixation is well known. Apart from bacteria, AMF also assist plants and provide benefits in many ways. AMF are one of the types of mycorrhiza having specialized branched hyphae structure called arbuscules. They show association with about 80% of land plant species, play an important role in nutrient uptake, and improve soil texture and quality. AMF also provide tolerance to abiotic stress and protection against pathogens. They also enhance nutritional value in fruits and accumulate secondary metabolites in plants. The main role of AMF is the uptake of immobile nutrients like phosphate and zinc. However, nitrogen uptake by AMF is still controversial. Various studies have demonstrated a pathway of nitrogen transport by AMF to the host plant, during which ERM (Extraradical mycelium) of AMF takes up nitrogen in the form of NH_4^+ and NO_3^-. These forms are converted to arginine and transferred to IRM (Intraradical mycelium), and finally, they are transported to the host plant. Genes like AAP and RIPTR2 are also discovered, which facilitate the uptake of different organic nitrogen forms. AMF and rhizobia also show similarities in their signaling pathways known as "common symbiotic pathways," which are also helpful in a better understanding of AMF function. In a few studies, it is indicated that AMF enhance Ca^{2+} uptake, which boosts up the process of nodulation and nitrogen fixation indirectly in leguminous plants. On the other hand, some studies suggest that nodulation suppresses AMF colonization due to competition between rhizobia

Monika · N. Yadav · Mamta · N. Kumar · A. Kumar · S.S. Arya (✉)
Department of Botany, Maharshi Dayanand University, Rohtak, Haryana, India

S. Devi
Department of Botany & Plant Physiology, CCS Haryana Agricultural University, Hisar, Haryana, India

V. Kaur
Division of Germplasm Evaluation, ICAR-National Bureau of Plant Genetic Resources, New Delhi, India

S. Kumar
Department of Environmental Science, Maharishi Dayanand University, Rohtak, Haryana, India

© The Author(s), under exclusive license to Springer Nature Singapore Pte Ltd. 2022
A. Vaishnav et al. (eds.), *Plant Stress Mitigators*,
https://doi.org/10.1007/978-981-16-7759-5_11

and AMF for carbohydrates. Further research is required to understand the complete nitrogen-fixing pathway by AMF. In this chapter, various roles of AMF as a symbiont, their similarities with bacterial symbionts, and their current knowledge with nitrogen uptake are summarized.

Keywords Symbiont · Organic nitrogen · Arginine · ERM · IRM · Controversial

Introduction

Have you ever decorated a pine on Christmas eve or admired a beautiful orchid flower or ever enjoyed the shade of an oak tree? If so, then you must know about the hidden world of the mycorrhizal fungi that made possible the existence of these beautiful plants. We can say that mycorrhiza is a **hidden partner** or a **down to earth fertilizer** to plants. Mycorrhiza is a mutual symbiotic association between plant roots and fungi. In this association, plants provide carbohydrates to fungi, and in return, fungi provide nutrients like phosphorus, nitrogen, and other nutrients to plants.

The mycorrhizal association was established between the Ordovician and Devonian eras (Stubblefield et al. 1987). It played an important role in the initial colonization of plants on land for the first time; in fact, the survival of land plants was possible because of mycorrhiza, and even today 90% of plant families show mycorrhizal associations (Hibbett et al. 2000). Based on the position of fungal hyphae in the tissue of plant roots, mycorrhiza is traditionally divided into three major groups: ectomycorrhiza, ectendomycorrhiza, and endomycorrhiza (Fig. 1).

Ectomycorrhizas: They mainly belong to fungal classes Basidiomycetes and Ascomycetes. These are characterized by an intercellular network of hyphae called Hartig net, which envelops epidermal and cortical cells of plant roots. Hartig net helps in the exchange of substances between fungus and plants.

Ectendomycorrhizas: Characterized by the presence of both intracellular hyphae penetrating the cortex cells and intercellular Hartig net.

Endomycorrhizas: As the name indicates, "endo" means "within." These fungi are able to penetrate within the root cells. They form specialized feeding structures like arbuscules or large, food-storing structures called vesicles. Endomycorrhiza is further classified into five types: arbutoid, ericoid, orchidoid, monotropoid, and vesicular arbuscular mycorrhiza (VAM).

Arbutoid Mycorrhizas: They belong to order Ericales; like ectomycorrhizas, they form Hartig net or fungal sheath around root cells, but they also have the ability to penetrate the root cells. In the literature, they are represented as a type of ectendomycorrhiza.

Ericoid Mycorrhizas: They are able to penetrate the cortical cells of plant roots, but they form hyphal coils instead of arbuscules.

Monotropoid Mycorrhizas: They belong to the family Monotropoideae, which is a subfamily within the family Ericaceae. They also form Hartig net like arbutoid mycorrhiza. They have a unique feature called fungal pegs, which are an invasion of hyphae originating from Hartig net into the epidermal cell wall.

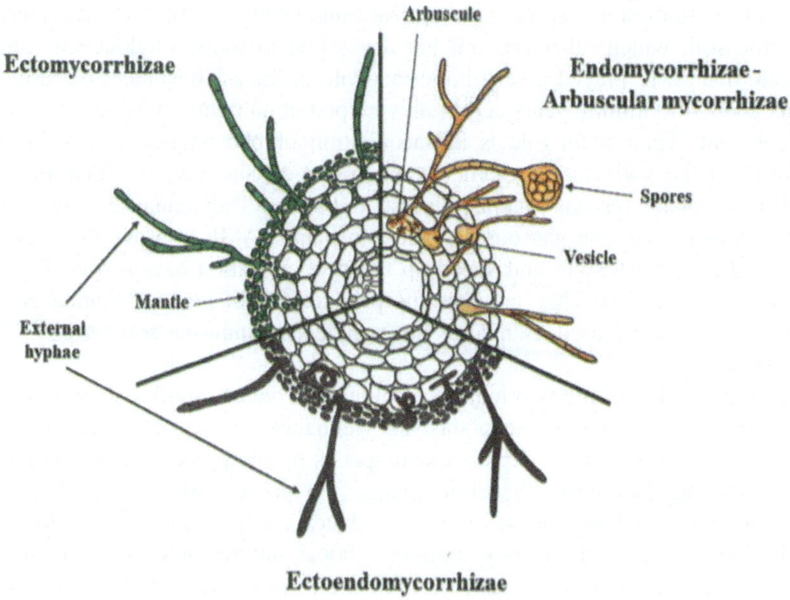

Fig. 1 Different types of mycorrhiza (Ganugi et al. 2019)

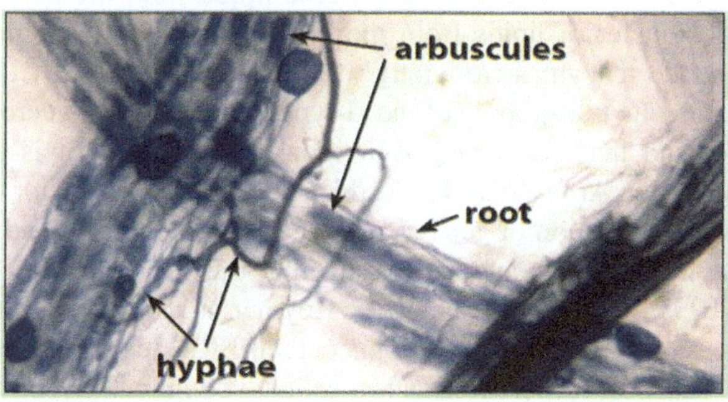

Fig. 2 Stained AMF. (https://www.soilquality.org.au/factsheets/arbuscular-mycorrhizas-s-a)

Orchidoid Mycorrhizas: Some orchids are not able to photosynthesize during the seedling stage, or they are completely nonphotosynthetic. Hence, they depend on fungal partners for nutrients to grow. The orchid parasitizes the fungus that invades its roots.

Arbuscular Mycorrhizas: These mycorrhizas penetrate their hyphae into plant root cells and form arbuscules. Arbuscules are tree-shaped, branched hyphae structures. They are functional sites for nutrient exchange between roots and hyphae (Fig. 2). AM are obligate symbionts that show mutualistic symbiosis with about 80% of land plant species. They belong to the phylum Glomeromycota.

The mycelium of arbuscular mycorrhizal fungi (AMF) is able to obtain nutrients from the soil, which otherwise are not accessible to roots. Phylogenetic studies indicate that AMF played a very important role in the establishment of early land plants about 400 million years ago as they helped in nutrient uptake in a very harsh environment. Their main role is the acquisition of phosphorus, which has poor mobility in the soil, and, to some extent, nitrogen and various other minerals. AMF make plants resistant against disease and stress. They enhance water uptake, photosynthetic rate, and gaseous exchange in plants. AMF improve the quality of soil by altering its texture and speed up the decomposition of soil organic matter (Thirkell et al. 2017). They enhance the productivity of plants (Gianinazzi et al. 2010). They can be used as natural biofertilizers to improve soil fertility (Ortas 2012).

AMF and rhizobia are two major symbionts for plants. Studies show that AMF boost nodulation in leguminous plants. They enhance Ca^{2+} uptake, which increases nod gene induction and its expression. It speeds up the process of nodulation. As AMF increase the nutrient level in plants, the process of photosynthesis also increases and a larger proportion of photosynthates is available to nodules (Mortimer et al. 2008). That is how they indirectly boost up the process of nodulation. According to Sakamoto et al. (2013), nodulation in the soybean and alfalfa resists AMF, suggesting that nodulation suppresses AMF colonization. It may happen because of competition between AMF and rhizobia for carbohydrates. Recently, new evidence is found that AMF also receive lipids from the host plant besides carbohydrates. Transfer of lipids from the host plant to AMF requires some host genes that encode enzymes for fatty acid biosynthesis, ABCG transporters (ATP-binding cassette transporters of the G-type), and GPAT (glycerol-3-phosphate acyltransferase) and that generate MAG (monoacylglycerol). It is estimated that RAM1 (Required for Arbuscular Mycorrhiza1) transcription factor in the host is responsible for the induction of genes required for AMF function and lipid transfer. However, complete pathway and aspects of lipid transfer remain to be discovered (Bravo et al. 2017; Luginbuehl et al. 2017; Keymer et al. 2017; Jiang et al. 2017).

AMF have a special feature to interconnect their ERM (extraradical mycelium) to plants of the same or different species called "common mycorrhizal network (CMN)." CMN provides a better way for the movement of nutrients and photosynthates within the fungal hyphae network and interconnected plants. However, complete knowledge of nutrient exchange and redistribution is still unknown. Recently, a new role of "common mycorrhizal network" was discovered. CMN helps plants to transfer the defense signal to neighboring plants. This is demonstrated in *Solanum lycopersicon* infected by the fungus *Alternaria solani*. A few defense-related genes are seen upregulated in neighboring uninfected plants. It is assumed that defense signals were transferred through CMN (Lanfranco et al. 2017).

AMF symbiosis has an impact on other parts of the plant and physiology beyond root apparatus. Gerlach et al. (2015) observed the change in leaf elemental composition and secondary metabolite accumulation in AMF-inoculated plants. It is found that AM symbiosis has the potential to increase nutritional values in fruits. An increase in lycopene, carotenoid, and other volatile compounds is observed in

AMF-inoculated plants (Giovannetti et al. 2012). AMF influence vegetative as well as reproductive traits; therefore they have major agronomical implications.

Evolution and Taxonomy of AMF

Primarily, plants evolved in a freshwater environment; later, they adapted themselves to grow on terrestrial lands. Early plants were nonphotosynthetic; later on, they developed chloroplast by endosymbiosis (by engulfment of a photosynthetic cyanobacterium). To adapt themselves to this new environment, they need protection from solar radiation and a well-developed vascular system for water uptake. They lack special absorptive organs like roots to acquire water and nutrients from the soil. Then, fungal symbiosis helped them to colonize on land mass. AMF are found in intimate association with the roots of higher plants since the evolution of land plants. In fact, it was AMF that provided nutrition to early land plants via their hyphae. AMF belong to the class Glomeromycetes of the phylum Glomeromycota and have around 250 species, 13 families, and 19 genera. The phylum is found worldwide in almost all major terrestrial biomes, i.e., bryophytes, pteridophytes, gymnosperms, and angiosperms.

Spores of AMF are comparatively larger than other fungi, and the size ranges between 50 and 500 μm. Hyphae have no septum. Because of the absence of septum, they were included in the phylum Zygomycota for a very long time, but recently, they were separated from the phylum Zygomycota. The sexual growth phase is absent in AMF. They lack classical meiosis and recombination process, but they exchange their genetic material through hyphal fusion (anastomoses), and that is how they change their genome and produce new genetic diversity. Favorable conditions are required for spore germination. AMF penetrate in the host cell and form specialized structures called "vesicles" and "arbuscules."

AMF and Abiotic Stresses

Abiotic stress hinders plant growth. It has been increasing day by day due to climate change and excessive use of chemical fertilizers. It is predicted that freshwater scarcity, climate change, and excessive use of chemical fertilizers will increase the impact of abiotic stress on the plant. Hence, there is an urgency to develop new ways that are eco-friendly and resilient to abiotic stress. AMF are a way out to this problem, are natural symbionts, and provide essential nutrients to the plant and increase growth. They are commonly known as bio-fertilizers. Various studies show that AMF help in mitigating the abiotic stress. The major abiotic stress occurring in the plant include drought, temperature, salinity, and heavy metals.

The problem of soil salinization is increasing day by day worldwide. It suppresses the assimilation rate and vegetative development and reduces the plant yield

(Hasanuzzaman et al. 2013). It enhances the generation of ROS (reactive oxygen species) in plants (Ahanger et al. 2018). ROS are very toxic and injurious to plants. El-Nashar (2017) observed an increased in the growth rate, WUE (water use efficiency), and leaf water potential in plants due to AMF. Borde et al. (2010) reported improvement in fresh and dry weights and leaf area index in the *Allium sativa* plant under saline conditions. AMF protect plants from both cold and heat stress. It is reported that plants with AMF grow better at a low temperature than non-AMF-inoculated plants (Chen et al. 2013). AMF can maintain moisture in the host plant and strengthen the plant immune system. They increase plant secondary metabolite production and protein content that supports plants to combat cold stress conditions. Heat stress decreases plant growth as it inhibits seed germination and causes wilting and burning of leaves. It damages plant fruits and reduces plant yield (Hasanuzzaman et al. 2013). AMF-inoculated plants resist heat stress better than non-AMF-inoculated plants.

Drought stress reduces the transpiration rate and induces oxidative stress in plants. It imparts detrimental effects on plants as it affects enzyme activity, ion uptake, and nutrient assimilation in plants. AMF increase osmotic adjustments and enhance proline accumulation and glutathione levels in plants (Rani 2016). AMF have the potentiality of bioremediation, meaning that they have the ability to accumulate or withdraw toxic metal ions and hence they protect plants from metal toxicity (Chen et al. 2018). They may chelate metal ions with other substances or may store them in vacuoles (Punamiya et al. 2010). AMF hyphae can bind toxic metals and reduce their chances of damaging the plants.

Common Symbiotic Pathway

AMF and rhizobia are the two most important symbionts in plants. Phylogenetic studies suggest that AMF symbiosis (AM) has originated much earlier than rhizobial symbiosis/root nodule symbiosis (RNS). Rhizobia fulfill the demand for nitrogen in legumes and AMF help in the uptake of nutrients like phosphorus, zinc, and nitrogen. Symbiosis is a molecular mechanism that involves a number of genes, receptors, phytohormones, messengers, etc. Both these symbiosis (AM and RNS) mechanisms show similarities in signaling pathways as they have some common set of genes; that is why the SYM pathway is defined as a common SYM pathway (CSM).

During phosphate deficiency plants release some carotenoid-based phytohormones—strigolactone—in the rhizosphere. These are chemically labile molecules and serve as a signal for AMF. After signal detection, AMF metabolism is activated and it stimulates hyphal growth and branching; however, it inhibits the growth of fungal pathogens. Recently, a new transporter NOPE1 is discovered in *Zea mays* and rice. It indicates that strigolactones not only are signal molecules in AMF but also trigger the release of Myc factors from AMF. Similarly, in legumes, flavonoids and isoflavonoids are released from plants, and they induce secretion of Nod factors from bacteria, which act as signaling molecules. Both Myc and Nod factors are lipochitooligosaccharides in nature. CSP is mainly described in three steps: perception, transmission, and transcription. During perception, Myc and Nod factor signals are perceived by host cells with the help of LysM receptors-like kinases such as OsCERK1, LYK3 (*Medicago truncatula*; Zhang et al. 2015), and SlLYK10 (*Solanum lycopersicum*; Buendia et al. 2016). DMI2/SYMRK proteins are essential for signal perception during endosymbiosis. They act as coreceptors to unidentified AMF Myc receptors.

In the next step, i.e., transmission, the enzyme HMG-CoA interacts with DMI2/SYMRK and produces mevalonate. It is assumed that mevalonate is a secondary messenger, and it transmits perception signals from the plasma membrane to the nucleus, and it increases calcium spiking after interaction with CASTOR and POLLUX/DMI1. CASTOR is a K^+ channel that enables K^+ efflux to balance Ca^{2+} influx by CNGC15. Activation of both CASTOR/POLLUX/DMI1 and CNGC15 is required for calcium oscillations (Fig. 3).

The last step of CSP is transcription in which nuclear calcium spiking induces the association of Ca^{2+} calmodulin with CCaMK (calcium and calmodulin-dependent kinase) and it leads to conformational change in kinase and phosphorylates CYCLOPS and IPD3 protein. Now, CCaMK and CYCLOPS complexes interact with GRAS (GIBBERELLIC-ACID INSENSITIVE, REPRESSOR of GAI, and SCARECROW) domain regulatory protein such as DELLA protein, which further induces the expression of the RAM1 (Reduced Arbuscular Mycorrhiza 1) gene (Fig. 3). Expression of the RAM1 gene is essential for AM symbiosis (MacLean et al. 2017).

Though we have acknowledged the pathway of common symbiosis in the last few years, a detailed understanding remains to be yet achieved. We lack information about many receptors and signaling molecules. There is a necessity to achieve appropriate knowledge about both symbiotic partners.

Although most of the plant families perform symbiotic association with AMF, there are certain families like Brassicaceae (*Arabidopsis thaliana*, cabbage, cauliflower) and Chenopodiaceae that lack some important gene responsible for symbiosis. These are not able to establish AM association. They are deprived of benefits from both rhizobia and AMF. They obtain the necessary nutrients by themselves.

Emphasis is going on inserting nitrogen-fixing genes in cereals and nonlegume plants (Fig. 4). It is a big challenge to transfer symbiotic machinery to cereals as it requires a detailed understanding of the complete symbiotic pathway, but it would definitely decrease the cost, time, and expenses of nitrogen fertilizers.

All three symbiotic pathways, i.e., RNS (root nodule symbiosis), AM (arbuscular mycorrhizal), and ARS (actinorhizal symbiosis), show resemblance in the invasion of bacterial and fungal partners in host cells. *PPA (prepenetration apparatus) in AMF symbiosis and *PIT (pre-infection thread) in RNS and ARS show similarities in their existence. A detailed study of these mechanisms will help in introducing nitrogen-fixing symbionts in nonlegume plants (Venkateshwaran et al. 2013).

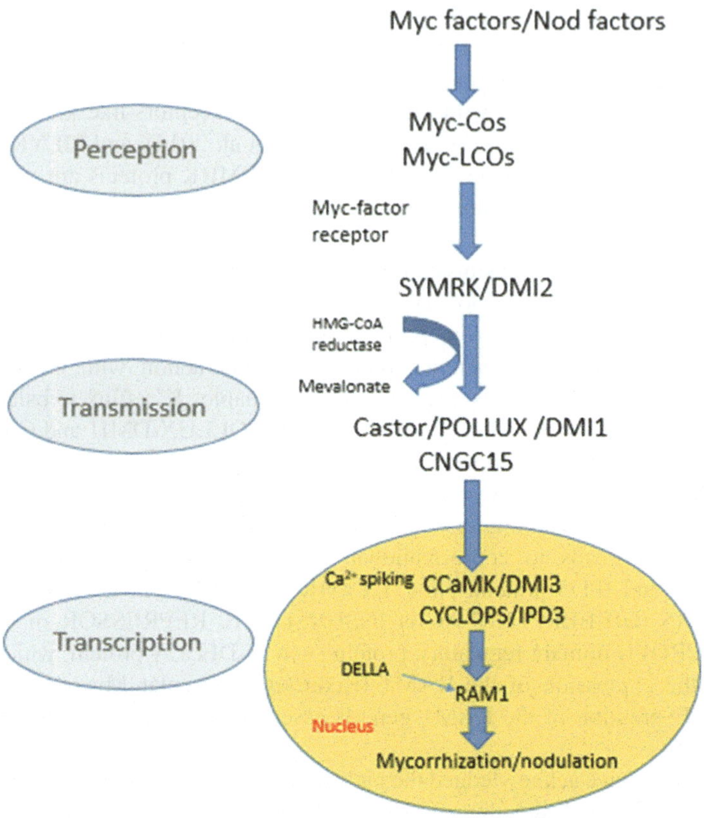

Fig. 3 Common symbiotic pathway (CSP)

*PPA: During symbiosis, AMF associate with the host with the help of PPA. It is a cytoplasmic column consisting of ER, microtubule, and microfilaments. This helps fungus to enter and grow inside host cell.

*PIT: Its formation takes place in legumes. In legumes, bacteria are able to invade root cells through infection thread (IT). IT is a tubular compartment formed by host cells. Bacteria divide within IT and then they reach to the inner cortex of cortical cells through PIT. Like PPA, PIT also forms by cytoplasmic bridges.

- In our environment, nitrogen is present in both inorganic as well organic forms.
- **Inorganic forms**: Nitrogen, ammonia, and nitrate.
- **Organic forms**: Amino acids and nucleic acids.

(continued)

- These forms undergo transformation from one form to another during physical and biological processes.
- During biological nitrogen fixation, molecular nitrogen is reduced into ammonia. This happens with the help of nitrogen-fixing organisms.
- **Ammonification**: When organic nitrogen present in soil is converted into ammonium ions by bacteria and fungi, it is known as ammonification.
- **Nitrification**: This process occurs within two steps. In the first step, oxidation of ammonium ion occurs and it is converted into nitrite. Some nitrifying bacteria carried out this step like *Nitrosomonas*, *Nitrospira*, and *Nitrosococcus*.

$$2NH_4^+ + 3O_2 \rightarrow 2NO_2^- + 4H^+ + 2H_2O \quad (1)$$

- In the next step, further oxidation of nitrite into nitrate occurs. It happens with the help of nitrifying bacteria of genera *Nitrospira*, *Nitrobacter*, and *Nitrococcus*.

$$2NO_2^- + O_2 \rightarrow 2NO_3^- \quad (2)$$

- **Denitrification**: In this process, with the help of anaerobic denitrifying bacteria, NO_3^- is converted into N_2.

$$2NO_3^- \rightarrow 2NO_2^- \rightarrow 2NO \rightarrow N_2O \rightarrow N_2 \quad (3)$$

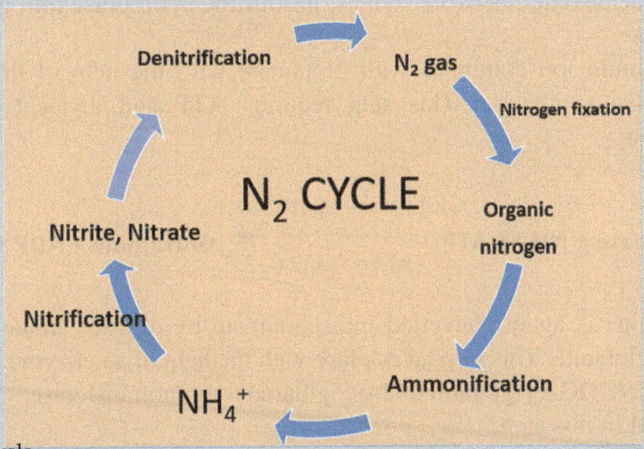

Nitrogen cycle

- **Assimilation of nitrate or ammonia**: Plants obtain nitrogen from the soil either in the form of nitrate or ammonium ions. Nitrate is transported to roots by an active process through H^+-ATPase pump, against a

(continued)

concentration gradient, and then nitrate is reduced to ammonium ions. It is accomplished in two steps.

- In the first step, nitrate is reduced to nitrite in the presence of enzyme nitrate reductase. Nitrite reductase enzyme is composed of two identical subunits, each one containing three prosthetic group: FAD, Heme, and Mo containing organic molecule (pterin). It requires NAD(P)H as a reducing agent (Fig. 5).
- In the second step, in the presence of nitrite reductase, nitrite is reduced to ammonia. Nitrite reductase is a monomeric protein and composed of two prosthetic groups: siroheme and Fe_4S_4. It requires reduced ferredoxin as a reducing agent in photosynthetic cells and NAD(P)H in nonphotosynthetic cells (Figs. 6 and 7).

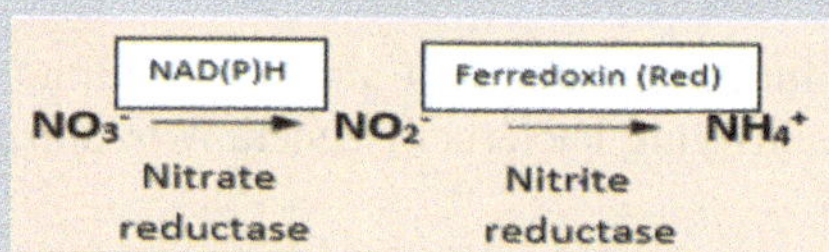

(4)

- Nitrate is the most available form of nitrogen to plants; plants also take up nitrogen in the form of ammonium ion directly from the soil. But ammonia does not accumulate anywhere in plants because of its toxic nature as it inhibits ATP formation by acting as an uncoupling agent. It also disturbs the pH gradient in both chloroplast and mitochondria. So it is rapidly converted into amino acid by plants. This conversion of ammonia into amino acids requires two enzymes: **glutamine synthetase and glutamate synthase.**
- Ammonium ion combines with glutamate with the help of the enzyme glutamine synthetase. This step requires ATP and divalent cation as cofactors.

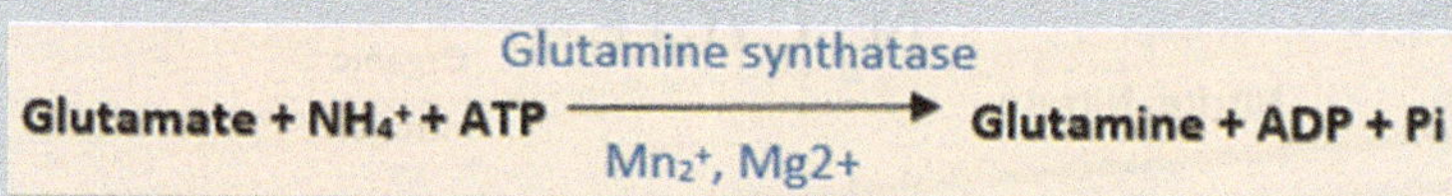

(5)

- Glutamine is again converted into glutamate by transfer amide group to α-ketoglutarate. This step takes place with the help of an enzyme glutamate synthase/GOGAT/glutamine-2-oxoglutarate aminotransferase. NADH is required in this step.

$$\text{Glutamine} + \alpha\text{-Ketoglutarate} \overset{\text{NADH, Gogat}}{\rightarrow} 2\ \text{Glutamate} \quad (6)$$

- Out of two glutamates, one is required for maintaining first reaction and second is converted into protein or other molecule.

(continued)

- **Nitrogenase enzyme complex** plays a very important role in nitrogen fixation in plants. It has two components: MoFe protein (molybdenum–iron protein complex)/dinitrogenase and dinitrogenase reductase (Fig. 8).
- MoFe protein component: It consists of four subunits, meaning that it is tetramer and each subunit has two Fe-Mo clusters.
- Dinitrogenase reductase: It is a dimer of identical 30 kDa subunits attached by the 4Fe-4S cluster.
- Nitrogenase enzyme complex is very sensitive to nitrogen, and it gets inactivated by oxygen. Hence, anaerobic conditions are required for the enzyme to work. In anaerobic organisms like BGA (blue green algae) or cyanobacteria, anaerobic conditions are generated with specialized structure or cells called heterocysts. These are specialized cells in which oxygen producing photosystem or Photosystem II is absent.
- **Nitrogen Fixing Prokaryotes**:
  - Various types of bacteria are present in soil for nitrogen fixation. These work in different conditions. Some are aerobic, and others are anaerobic (Fig. 9). They may be phototrophs, chemotrophs, or photolithotrophs. They may form nodules in plant roots or not.
- **Nodule Formation in Leguminous Plants**:
  - This process occurs with the help of Gram-negative bacteria: Rhizobium plant roots secrete some chemicals like flavonoids and homoserine, which attract these bacteria. A special adhesion protein called rhicadhesin is present on the surface of bacteria, which helps in plant–bacterium attachment. Interaction with flavonoids activates NodD protein, inducing transcription of nodABC genes. These nod genes encode enzymes that help in the synthesis of nodulation factors. The Nod factor triggers plant cell division and root hair curling. Cell division induces the formation of root nodules. Bacteria digest the root hair cell wall and invade in roots. Now, formation of infection thread occurs (Fig. 10).

    Now, infection thread extends in cortical cells of root and infects other cells. These infected cells are called nodule primordium. Bacteria continue to divide and stimulate nodule formation. Due to rapid multiplication, some group of bacteria become nonmotile and are called bacteriods. Bacteriods are surrounded by a membrane called peribacteriod membrane. Nitrogen is reduced to ammonia in bacteriods with the help of nitrogenase enzyme. As the nitrogenase enzyme is very sensitive to oxygen, the oxygen level is controlled by leghemoglobin. It gives nodule a pink color.

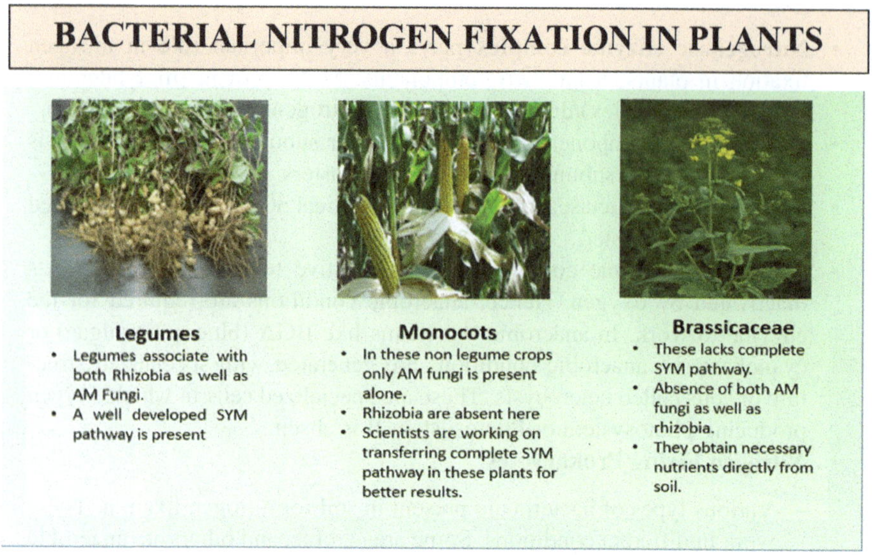

Fig. 4 A comparative analysis between legumes, monocots, and brassicaceae

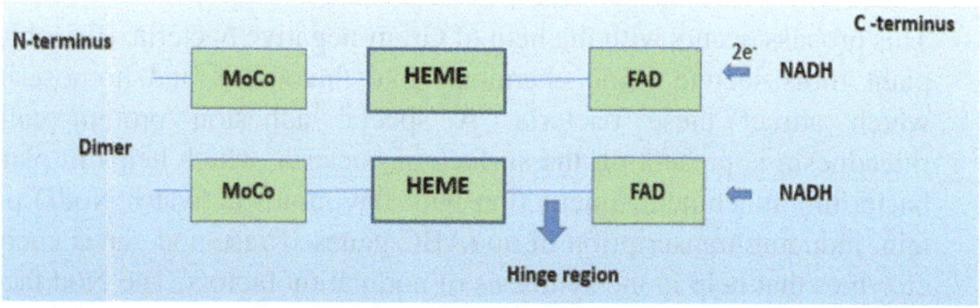

Fig. 5 Nitrate reductase

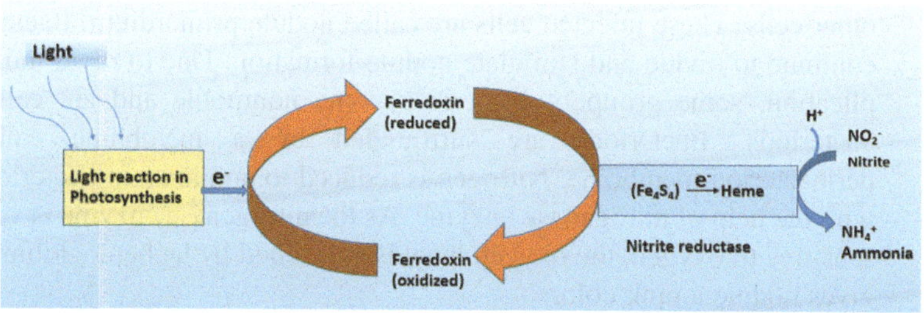

Fig. 6 Nitrite reductase

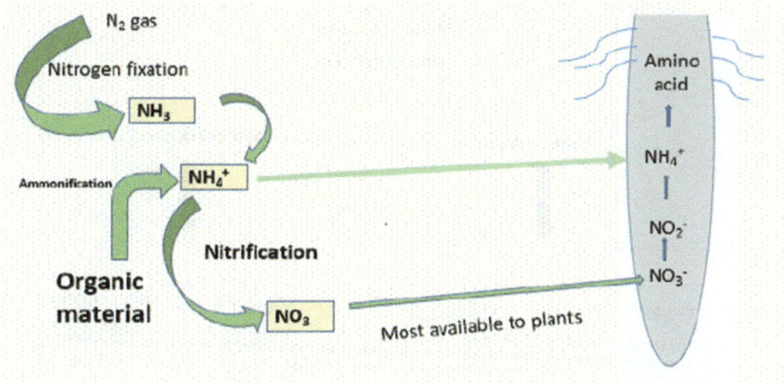

Fig. 7 Various steps during nitrogen fixation

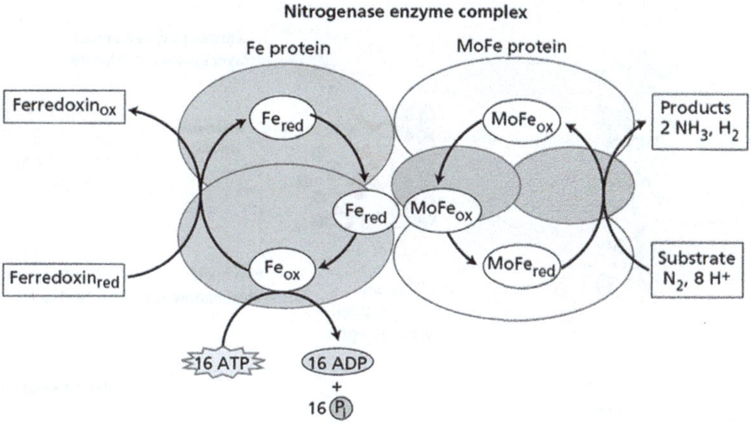

Fig. 8 Nitrogenase enzyme complex. (Source: Taiz and Zeiger 2003)

Uptake of Nitrogen by AMF

AMF explore nutrients in the soil and then transport them to the roots of the host plant. They play the main role in the uptake of immobile nutrients such as phosphorus and zinc. However, the role of AMF in nitrogen uptake is still unclear. For plants, nitrogen uptake is easy through mass flow and diffusion. However, there may be situations like drought or acidic conditions. When nitrogen content is very low in the soil, AMF may play a significant role in nitrogen uptake. The role of AMF in the absorption of mobile nitrogen is of less significance than the uptake of immobile phosphorus.

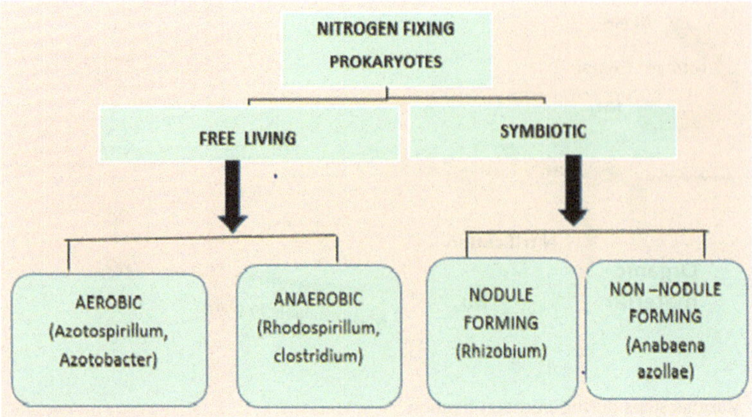

Fig. 9 Nitrogen fixing Prokaryotes

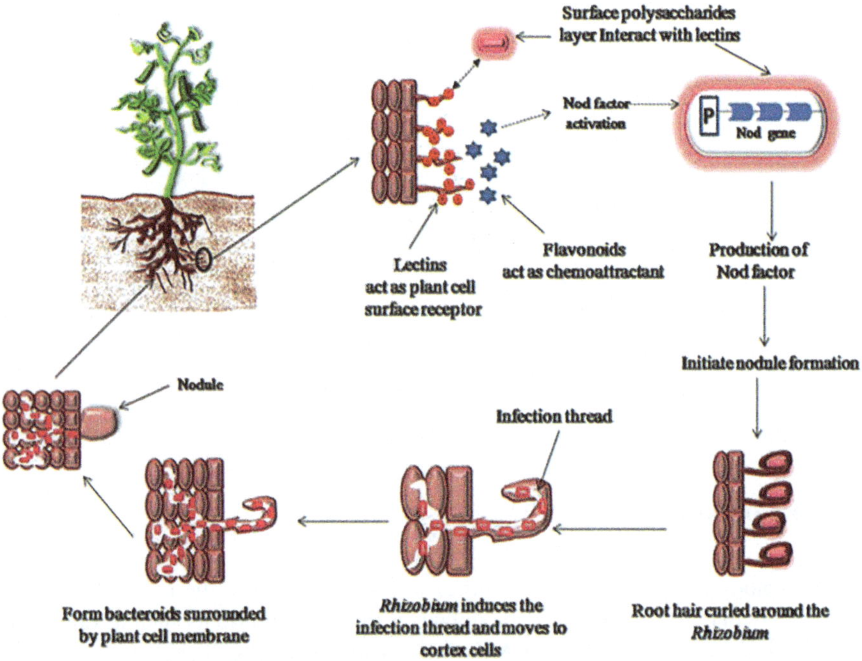

Fig. 10 Nodule formation in legumes. (Source: Singh et al. 2019)

In soil, both inorganic and organic forms of nitrogen are found. Inorganic nitrogen is found in the form of nitrate and ammonium ions, and the organic form includes urea, amino acids, amines, and peptides.

Uptake of Inorganic Nitrogen by AMF

AMF have two types of mycelium based on their interaction with different environments: IRM (intraradical mycelium) and ERM (extraradical mycelium). ERM comes across or experiences environmental conditions like soil moisture, pH, nutrient availability, etc. At the same time, IRM grows in an environment controlled by plant homeostasis. The ERM obtains nitrate (NO_3^-), ammonium (NH_4^+), and amino acids (AAs) from the external medium (Fig. 11). The complete uptake and transport of different forms of nitrogen are still unknown. GintAMT1 (a NH_4^+ transporter) is discovered from *G. intraradices*. In ERM, NH_4^+ is assimilated via the GS (glutamine synthase)/GOGAT (glutamine oxoglutarate aminotransferase) pathway, and uptake of NO_3^- takes place via nitrate and nitrite reductases. Now, these assimilated nitrogen forms are further incorporated into amino acids, especially arginine. Arginine is first accumulated into ERM, and then it is transported into IRM. In IRM, arginine is broken down into urea and ornithine, which are further converted into NH_4^+ and amino acids in the presence of enzymes urease and ornithine aminotransferase (OAT). NH_4^+ is transported to the host via ammonium transporters (AMT) or it can be incorporated into other free amino acids.

Uptake of Organic Nitrogen

The uptake of organic nitrogen by AMF is still a matter of debate. AMF lack saprophytic capabilities, so it requires microbial communities for decomposition of

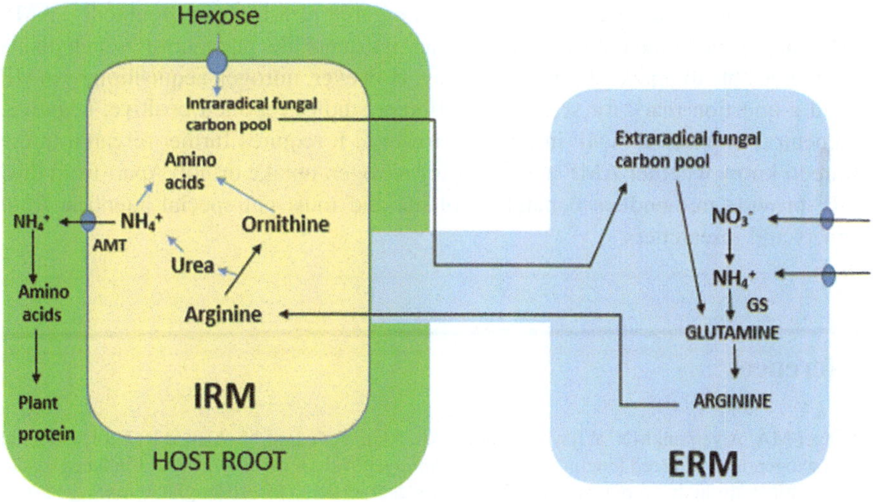

Fig. 11 Nitrogen uptake by AMF

organic matter and then AMF act for nitrogen mineralization (Bücking and Kafle 2015). AMF possess some special transporters like AAP (amino acid permease) for the uptake of organic nitrogen like amino acids. Similarly, RIPTR2 is a dipeptide transporter, which helps in the acquisition of peptides (a form of organic nitrogen). These are found in ERM. Enzymes like nitrite reductase, nitrate reductase, glutamate dehydrogenase, and glutamate synthase act in a similar way during nitrogen assimilation as in bacterial nitrogen fixation described previously in the chapter, except for some differences like the position of activity of enzymes in plants. For example, nitrate reductase activity is higher in roots and shoots of AM plants than nonmycorrhizal plants (Bücking and Kafle 2015). The activity of NO_3^- reductase mainly occurs in leaves in non-AMF plants, whereas it is primarily found in roots of AMF plants.

Conclusion

The world population is increasing at a very rapid pace, so does the consumption of food. To enhance the production of crops, inorganic fertilizers are being used. They possess a great threat to the environment and the health of humans. It is high time to find an alternative for inorganic fertilizers. AMF are one such solution. AMF help plants in nutrient uptake and provide resistance against diseases and stress. The AMF hyphal network improves soil texture. AMF help in the aggregation of soil particles and prevent soil erosion. They inhibit the leaching of nutrients and decrease the risk of groundwater contamination. They are harmless and effective technique to enhance soil fertility and productivity. In spite of their vast potential, unfortunately AMF are not well known among farmers and are deprived of their benefits. We should promote AMF as an alternative to inorganic fertilizers. Undoubtedly, AMF play a major role in the transport of necessary nutrients like phosphorus which plants may not be able to uptake during deficiency. However, nitrogen acquisition by AMF is still a question mark for scientists. Different studies reported positive, negative, and neutral impacts of AMF in nitrogen fixation. It requires further research in the future to know whether AMF are useful for nitrogen uptake or not. Apart from this, AMF provide tremendous benefits to plants and must get special attention from farmers and researchers.

References

Ahanger MA, Alyemeni MN, Wijaya L, Alamri SA, Alam P, Ashraf M, Ahmad P (2018) Potential of exogenously sourced kinetin in protecting *Solanum lycopersicum* from NaCl-induced oxidative stress through up-regulation of the antioxidant system, ascorbate-glutathione cycle and glyoxalase system. PLoS One 13(9):e0202175

Borde M, Dudhane M, Jite PK (2010) AM fungi influences the photosynthetic activity, growth and antioxidant enzymes in Allium sativum L. under salinity condition. Notulae Scientia Biologicae 2(4):64–71

Bravo A, Brands M, Wewer V, Dörmann P, Harrison MJ (2017) Arbuscular mycorrhiza-specific enzymes FatM and RAM 2 fine-tune lipid biosynthesis to promote development of arbuscular mycorrhiza. New Phytol 214(4):1631–1645

Bücking H, Kafle A (2015) Role of arbuscular mycorrhizal fungi in the nitrogen uptake of plants: current knowledge and research gaps. Agronomy 5(4):587–612

Buendia L, Wang T, Girardin A, Lefebvre B (2016) The LysM receptor-like kinase Sl LYK 10 regulates the arbuscular mycorrhizal symbiosis in tomato. New Phytol 210(1):184–195

Chen S, Jin W, Liu A, Zhang S, Liu D, Wang F et al (2013) Arbuscular mycorrhizal fungi (AMF) increase growth and secondary metabolism in cucumber subjected to low temperature stress. Sci Hortic 160:222–229

Chen M, Arato M, Borghi L, Nouri E, Reinhardt D (2018) Beneficial services of arbuscular mycorrhizal fungi–from ecology to application. Front Plant Sci 9:1270

El-Nashar YI (2017) Response of snapdragon (Antirrhinum majus L.) to blended water irrigation and arbuscular mycorrhizal fungi inoculation: uptake of minerals and leaf water relations. Photosynthetica 55(2):201–209

Ganugi P, Masoni A, Pietramellara G, Benedettelli S (2019) A review of studies from the last twenty years on plant–arbuscular mycorrhizal fungi associations and their uses for wheat crops. Agronomy 9(12):840

Gerlach N, Schmitz J, Polatajko A, Schlueter U, Fahnenstich H, Witt S et al (2015) An integrated functional approach to dissect systemic responses in maize to arbuscular mycorrhizal symbiosis. Plant Cell Environ 38(8):1591–1612

Gianinazzi S, Gollotte A, Binet MN, van Tuinen D, Redecker D, Wipf D (2010) Agroecology: the key role of arbuscular mycorrhizas in ecosystem services. Mycorrhiza 20(8):519–530

Giovannetti M, Balestrini R, Volpe V, Guether M, Straub D, Costa A et al (2012) Two putative-aquaporin genes are differentially expressed during arbuscular mycorrhizal symbiosis in Lotus japonicus. BMC Plant Biol 12(1):186

Hasanuzzaman M, Nahar K, Alam M, Roychowdhury R, Fujita M (2013) Physiological, biochemical, and molecular mechanisms of heat stress tolerance in plants. Int J Mol Sci 14(5):9643–9684

Hibbett DS, Gilbert LB, Donoghue MJ (2000) Evolutionary instability of ectomycorrhizal symbioses in basidiomycetes. Nature 407(6803):506–508

Jiang Y, Wang W, Xie Q, Liu N, Liu L, Wang D et al (2017) Plants transfer lipids to sustain colonization by mutualistic mycorrhizal and parasitic fungi. Science 356(6343):1172–1175

Keymer A, Pimprikar P, Wewer V, Huber C, Brands M, Bucerius SL et al (2017) Lipid transfer from plants to arbuscular mycorrhiza fungi. elife 6:e29107

Lanfranco L, Bonfante P, Genre A (2017) The mutualistic interaction between plants and arbuscular mycorrhizal fungi. Microbiol Spectr 4(6):727–747

Luginbuehl LH, Menard GN, Kurup S, Van Erp H, Radhakrishnan GV, Breakspear A et al (2017) Fatty acids in arbuscular mycorrhizal fungi are synthesized by the host plant. Science 356(6343):1175–1178

MacLean AM, Bravo A, Harrison MJ (2017) Plant signalling and metabolic pathways enabling arbuscular mycorrhizal symbiosis. Plant Cell 29(10):2319–2335

Mortimer PE, Pérez-Fernández MA, Valentine AJ (2008) The role of arbuscular mycorrhizal colonization in the carbon and nutrient economy of the tripartite symbiosis with nodulated *Phaseolus vulgaris*. Soil Biol Biochem 40(5):1019–1027

Ortas I (2012) The effect of mycorrhizal fungal inoculation on plant yield, nutrient uptake and inoculation effectiveness under long-term field conditions. Field Crop Res 125:35–48

Punamiya P, Datta R, Sarkar D, Barber S, Patel M, Das P (2010) Symbiotic role of Glomus mosseae in phytoextraction of lead in vetiver grass [*Chrysopogon zizanioides* (L.)]. J Hazard Mater 177(1–3):465–474

Rani V (2016) Endophytes isolated from the plants of Himachal Pradesh and their importance. Int J Curr Microbiol App Sci 5(9):105–110

Sakamoto K, Ogiwara N, Kaji T (2013) Involvement of autoregulation in the interaction between rhizobial nodulation and AM fungal colonization in soybean roots. Biol Fertil Soils 49(8): 1141–1152

Singh, P., Rajput, R. S., Ram, R. M., Singh, H. B. (2019). A Deeper Insight into the Symbiotic Mechanism of Rhizobium spp. from the Perspective of Secondary Metabolism. In Secondary Metabolites of Plant Growth Promoting Rhizomicroorganisms (pp. 265-291). Springer, Singapore.https://plantlet.org/nitrogen-fixation-root-nodule-formation/

Stubblefield SP, Taylor TN, Seymour RL (1987) A possible endogonaceous fungus from the Triassic of Antarctica. Mycologia 79(6):905

Taiz L, Zeiger E (2003) Assimilation of mineral nutrients. In: Plant physiology, first Indian reprint. Panima Publishing Corp., New Delhi, pp 259–282

Thirkell TJ, Charters MD, Elliott AJ, Sait SM, Field KJ (2017) Are mycorrhizal fungi our sustainable saviours? Considerations for achieving food security. J Ecol 105(4):921–929

Venkateshwaran M, Volkening JD, Sussman MR, Ané JM (2013) Symbiosis and the social network of higher plants. Curr Opin Plant Biol 16(1):118–127

Zhang X, Ren BH, Wu SL, Sun YQ, Lin G, Chen BD (2015) Arbuscular mycorrhizal symbiosis influences arsenic accumulation and speciation in Medicago truncatula L. in arsenic-contaminated soil. Chemosphere 119:224–230

IAA Biosynthesis in Bacteria and Its Role in Plant-Microbe Interaction for Drought Stress Management

Ees Ahmad, Pawan K. Sharma, and Mohd Saghir Khan

Abstract Globally, the availability of water for irrigation is decreasing for agricultural practices, including different growing crops. The productivity of different crops is declining as the time duration of drought conditions is increasing. The plants have been adapting different strategies including morphological changes, modulation on physiological process and maintaining the osmotic potential of cell to counter the drought stress condition. Diverse microbes associated with plants reported as plant growth-promoting rhizobacteria (PGPR), having PGP activity including P-solubilization, indole acetic acid (IAA) synthesis, siderophore production and antifungal activity. While, some PGPR having additional characteristics i.e. nitrogen fixing activity and PGPR without N-fixing ability, can postulated the improvement of different parameter for crop (root biomass, shootbiomass, crop yield, root architecture) leading to enhancement in capability to drought stress conditions by different mechanisms. Additionally, the modification of root and shoot of plants by physio-biochemical activity of inole-3-acetic acid (IAA) secreted by diverse microbes in rhizosphere during plant–microbe interaction is a key process to help for mitigation of drought stress in deferent crops. The IAA-secreting bacteria have operated five different pathways for biosynthesis of IAA by utilizing tryptophan as only known precursor. While some IAA-producing bacteria also synthesize IAA without tryptophan by operating tryptophan-independent pathway. The plant–microbe interaction is one of the main physiological and biochemical processes in rhizosphere where IAA has key role and involves in crosstalk between plants and microbes. Moreover, root colonization process is also a part of plant–microbe interaction in which the rhizobacteria aggressively colonize the root by biochemical signalling between rhizobacteria and plants. The root colonizing bacteria secrete various enzymes and wide range of metabolites that can help plants in improving the

E. Ahmad (✉) · P. K. Sharma
ICAR-National Bureau of Agriculturally Important Microorganisms, Maunath Bhanjan, Uttar Pradesh, India

M. S. Khan
Department of Agricultural Microbiology, Faculty of Agricultural Sciences, Aligarh Muslim University, Aligarh, Uttar Pradesh, India

© The Author(s), under exclusive license to Springer Nature Singapore Pte Ltd. 2022
A. Vaishnav et al. (eds.), *Plant Stress Mitigators*,
https://doi.org/10.1007/978-981-16-7759-5_12

tolerance capacity under drought conditions. In conclusion, IAA biosynthesis from various bacteria helps in root colonization during plant–microbe interaction procedure, which leads to mitigate the drought stress in different plants by modulation in physiological and biochemical characteristics of plants.

Keywords Plant–microbe interaction · Drought tolerance · Tryptophan-independent pathway · IAA biosynthesis pathway · Drought tolerance mechanism

Introduction

Auxin is main plant growth hormone which participates in different stages of plant development and involves in responses to abiotic and biotic stress (Pieterse et al. 2009; Zhang et al. 2021). Additionally, indole-3-acetic acid (IAA) is considered biochemical molecules which are considered as biologically active form of auxin, and its content is strictly regulated in plants through the interaction of many pathways connecting IAA biosynthesis, hydrolysis, oxidation, and conjugation to carbohydrate and amino acids with IAA molecules. The capability of synthesizing IAA of plant-related microorganisms influenced the endogenous auxin pool present in host (Yang et al. 2007; Kong et al. 2015; Di et al. 2016; Morffy and Strader 2020). Auxins are often believed to play a key function in crosstalk between plant and bacterial signals during establishment of plant–microbe interaction (Spaepen 2015; O'Banion et al. 2020). The scientific study reveals that plant growth-promoting bacteria (PGPR) including IAA-producing bacteria use multiple IAA biosynthetic pathways in many circumstances to synthesize IAA for multiple functions including the plant–microbe interaction. However, the bacterial strains having multiple pathways for IAA biosynthesis reflect the importance of IAA in active participation and functioning of plant growth-promoting activity (Duca et al. 2014; Shao et al. 2015). Genetically all bacteria can be divided in two types of bacteria: (1) IAA-deficient and (2) IAA-overproducing strains. While, all the IAA biosynthetic pathways can be summarized as tryptophan-dependent and tryptophan-independent on the basis of utilized precursor for IAA biosynthesis (Duca et al. 2014). To date, no genes or enzymes related to the tryptophan-independent pathway have been identified (Ahmad et al. 2020). However, many enzymes, genes and cofactors are studied involved in the tryptophan-dependent pathway operated in diverse microbes (Patten and Glick 1996; Ahmad et al. 2020). In continuation, if tryptophan is utilizing by microbes as precursor molecule for IAA biosynthesis, the microbes synthesized IAA by operating five major pathways, namely (1) indole-3-acetonitrile pathway (IAN), (2) indole-3-pyruvate pathway (IPy), (3) indole-3-acetamide pathway, (4) indole-3-acetaldoxime pathway (IAOx), and (5) tryptamine (TAM) pathway, while if without utilizing tryptophan the microbes synthesize IAA by operating by tryptophan-independent pathway (Ostin et al. 1999; Sitbon et al. 2000; Cohen et al. 2003; Wang et al. 2015; Ahmad et al. 2020). Biosynthesis of IAA via IAM pathway is a key characteristic of various pathogenic bacteria, while IPA pathway is operated for biosynthesis of IAA in plant growth-promoting bacteria (Patten and Glick 1996;

Yang et al. 2007). These two main routes use different intermediates and many enzymes are involved to produce IAA in different microbes. Therefore, the IPA and IAM pathways are thought to be involved in different processes of plant developmental stages, i.e. formation of tumour, nodule formation or colonization of plant roots (Brandl et al. 2001; Ahmad et al. 2013; Mashiguchi et al. 2019).

While IAA is most important plant hormone involved in development and regulation of different stages of plant growth, but in contrast IAA synthesized from *P. indica* after early root colonization of barley plant does not improve or stimulate the biomass production of barley after early root colonization (Hilbert et al. 2012). However, in this study, it was not clear that IAA biosynthesis in barley plant triggered by *P. indica*.

Biosynthesis Pathways of IAA Operated in Bacteria

After qualitative and quantitative analysis of IAA from diverse bacterial species, different IAA biosynthetic pathways have been characterized. There is very high similarity of different IAA biosynthetic pathways operated in bacteria and plants. Till date, tryptophan is the only known precursor for IAA production in bacteria and plants. On the basis of different intermediate molecules five different IAA biosynthesis pathways are identified where tryptophan is converted to IAA via different intermediates. Here, the current status of different IAA biosynthetic pathways, including different intermediate molecules, proteins and genes involved in these pathways, has been discussed.

Indole-3-Acetamide (IAM) Pathway

In two phases, the indole-3-acetamide pathway transforms tryptophan to IAA. The first enzyme of IAM pathway, L-tryptophan 2-monooxygenase act on tryptophan and convert to indole-3-acetaamide by oxidative decarboxylation in the first step. The second step is oxidative deamination in which the enzyme indole-3-acetamide hydrolase covert the indole-3-acetamide to ammonia and IAA (Fig. 1). Interestingly, this ammonia can be used by plant as nitrogen source by assimilation process. During tryptophan decarboxylation, molecular oxygen is used as oxidant by cofactor flavin adenine dinucleotide associated with tryptophan 2-monooxygenase (Trp) enzyme (Emanuele and Fitzpatrick 1995; Sobrado and Fitzpatrick 2003; Gaweska et al. 2013; Asano and Yasukawa 2019). In the previous study, the enzyme Trp-monooxygenase showed low affinity ($Km = 50$ μM) to L-tryptophan after extraction and purification from *P. syringae* pv. *savastanoi* (Hutcheson and Kosuge 1985; McClerklin et al. 2018). Additionally, 25 μM indole-3-acetaamide and 230 μM IAA inhibit the 50% activity of Trp-monooxygenase (Emanuele et al. 1995). It is very clear that IAA is synthesized by this pathway when tryptophan

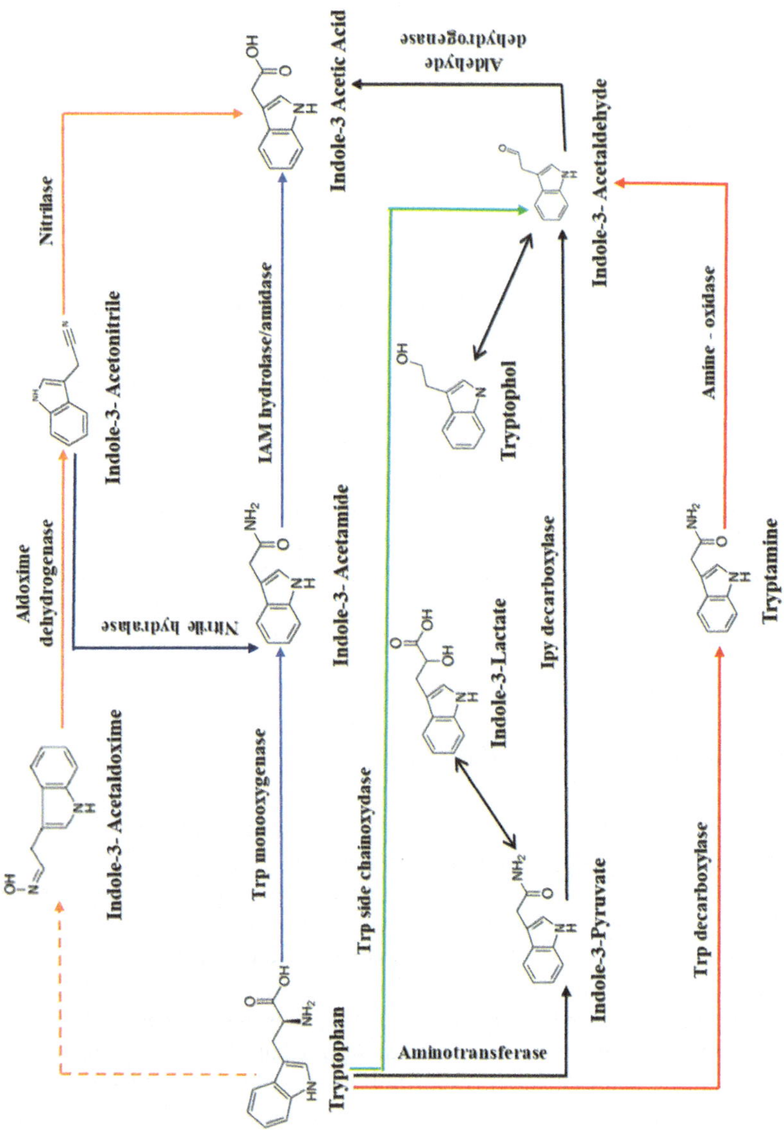

Fig. 1 IAA biosynthetic pathway operating in bacteria

concentration is higher than cellular concertation (~20 μM) and also sufficient amount of tryptophan is available for protein synthesis (Emanuele et al. 1995). Indole-3-acetamide hydrolase hydrolyse the carbon-nitrogen (C=N) bonds and convert to nitriles (R-C≡N) and amides (R-C=O (NH_2)). In vitro the indole-3-acetamide is like a natural source for the enzyme indole-3-acetamide hydrolase extracted and purified from *A. tumefaciens* as it has low Km value (1.2 μM) for indole-3-acetamide. Furthermore, the other substrates including indole-3-acetonitrile and phenyl acetamide are converted to IAA and phenylacetic acid, respectively, by indole-3-acetamide hydrolase (Kemper et al. 1985; Estenson et al. 2018; Liu et al. 2019).

IAA induces gall formation and development of tumours in plants, when microbes or phytopathogens associated with plant roots, synthesized the IAA by IAM pathway operated in phytopathogens. For example, *P. agglomerans* pv. *gypsophilae* secreted the IAA by using IAM pathway and infected and formed gall in gypsophila through virulence mechanism (Manulis et al. 2018; Mohabi et al. 2017; Mashiguchi et al. 2019). While in olive tree the *P. syringae* pv. *savastanoi* infected and formed gall by synthesizing IAA from operating IAM pathway (Comai and Kosuge 1980). In other example, the *Agrobacterium tumefaciens* simulate the proliferation of gall of its host plant by transfer of IAM pathway genes to host plant (Thomashow et al. 1984).

Indole-3-Pyruvate (IPy) Pathway

For IAA biosynthesis in many bacteria, indole-3-pyruvate is one of the major routes. Aminotransferase enzyme catalyses the first step of this pathway in which L-tryptophan is converted to α-ketoglutarate by transferring the amino group of L-tryptophan (Fig. 1). In this biochemical reaction the α-carbon of amino group from tryptophan transfer to α-ketoglutarate (Imada et al. 2017). However, the biochemical reaction catalysed by aminotransferase in IPy pathway may be specific to tryptophan (Helinck et al. 2004; Ostrowski et al. 2020). Additionally, different intermediates of α-keto acid from aromatic amino acids converted by these enzymes from lactic acid bacteria contributes the specific flavours in different chesses (Ardö 2006; Zuljan et al. 2016; Li et al. 2020). Many aminotransferases are found in cell of lactic acid bacteria that can initiate the catabolism of different aromatic amino acids and these enzymes characterized with wide range of substrate specificity can utilize leucine, methionine and all three essential aromatic amino acids (Kittell et al. 1989; Yvon et al. 1997; Liu et al. 2008; Galili et al. 2016). Specifically, in-vitro study reveals, that the aminotransferase have been showed low affinity and high Km value (3.3 mM) for L-tryptophan, when this aminotransferase enzyme was extracted from *E. cloacae*. In contrast, the concentration of tryptophan is much lower in bacterial cell so without exogenous supply the IAA is not synthesized by operating the IPy pathway (Koga et al. 1994; Ryu and Patten 2008; Li et al. 2018).

In the second step of this pathway, indole-3-pyruvate is converted to IAA which is catalysed by the enzyme indole-3-pyruvate decarboxylase through non-oxidative decarboxylation process. The enzyme has catalytic properties including biosynthesis of amino acids (acetolactate synthase) or breakdown process (indolepyruvate decarboxylases) or alcohol synthesis (pyruvate decarboxylases). So, the family of this enzyme is decarboxylases which is dependent on thiamine diphosphate. Due to large size (homotetramer) of this enzyme, it adjusted the indole ring of substrate and converted to indole-3-pyruvate (Schütz et al. 2003b; Kneen et al. 2011; Harris et al. 2018). For catalytic activity and substrate binding of this enzyme the tetramerization is required which is formed by binding of thiamine diphosphate and Mg^{2+} with each subunit (Koga et al. 1992; Schütz et al. 2003b; Andrews et al. 2014). In decarboxylation process the keto-carbonyl group binds with thiamine diphosphate and after protonation the indoleacetaldehyde is formed (Pohl et al. 2004). The *ipdC* gene of *E. cloacae* involved in IAA biosynthesis and encoded for indolepyruvate decarboxylase (Koga et al. 1992). While other bacteria are also reported for involvement of *ipdC* gene in IAA biosynthesis via IPy pathway (Costacurta et al. 1994; Brandl and Lindow 1996; Somers et al. 2005; Fedorov et al. 2010).

Generally, decarboxylase enzymes for α-keto acid have wide range of substrate specificity. In vitro study reveals that indole-3-pyruvate decarboxylase purified from *E. cloacae* can also reduce the benzoylformate with low catalytic property as compared to indole-3-pyruvate (Schütz et al. 2003a; Versees et al. 2007). Additionally, decarboxylation reaction of pyruvate and indolepyruvate carried by pyruvate decarboxylase which was extracted and purified from *Saccharomyces cerevisiae*. Overall, in this IAA biosynthesis pathway the indole-3-pyruvate decarboxylase can convert only few amino acid to indole-3-pyruvate via substitution mechanism in active side of this enzyme (Patten et al. 2013). While another study on *P. polymyxa* E681confirms that the enzyme's primary function is decarboxylation of indole-3-pyruvate but not in IAA synthesis pathway (Phi et al. 2008).

Conversion of IAA from indole-3-acetaldehyde by oxidation reaction is the last step of this pathway. While it is not clear how this oxidation reaction is catalysed by bacterial oxidases. Theoretically, the oxygen associated with indole-3-acetaldehyde oxidase acts as electron acceptor and coverts indole-3-acetaldehyde to hydrogen peroxide and IAA (Koshiba et al. 1996; Seo et al. 1998; Suzuki et al. 2003; McClerklin et al. 2018). However, in yeast the Ehrlich pathway operated for IAA biosynthesis in which the $NADP^+$ or NAD^+ with aldehyde dehydrogenase acts as electron acceptor and catalyses the reaction (Hazelwood et al. 2006; Li et al. 2011). Interestingly, a bifunctional enzyme characterized in thermoacidophilic archaeon catalyses the oxidation of indole-3-acetaldehyde and decarboxylation of indole-3-pyruvate (Wakagi et al. 2002).

Indole-3-Acetotonitrile/Indole-3-Aldoxime (IAN/IAOx) Pathway

However, the biosynthesis of IAA via operating indole-3-acetonitrile pathway in bacteria has been less studied but some of the enzymes involved in this pathway has been characterized. In this pathway, IAA is converted from tryptophan through two intermediates: (1) indole-3-acetaldoxime and (2) indole-3-acetonitrile (Fig. 1). In bacteria the enzyme is not characterized which have involved in biosynthesis of indole-3-acetaldoxime, but in *Arabidopsis* plant, the cytochrome P450 monooxygenase enzyme had been involved in synthesis of indole-3-acetaldoxime (Mikkelsen et al. 2000). Indole-3-acetonitrile, the second intermediate compound of this pathway, has nitrile (-C≡N) functional group converted from dehydration reaction of aldoxime dehydratase and aldoxime (-CH=NOH) functional group. The enzyme involved in this step is aldoxime dehydratase extracted and purified from *Bacillus* sp. OxB-1. The enzyme is characterized as phenylacetaldoxime dehydratase containing heme group, and for activation flavin mononucleotide is required (Kato et al. 2000). This enzyme has highest affinity for phenylacetaldoxime but it also catabolizes the indoleacetaldoxime and many other arylacetaldoxime with low affinity (Kato et al. 2000). Additionally, aldoxime dehydratase extracted from *Rhodococcus globerulus* catalyses both indole- and phenyl-acetaldoxime (Xie et al. 2003). Furthermore, the functional group, nitriles (-C≡N) are transformed to carboxylic acids and ammonia by nitrilase enzyme, while in other reaction mechanisms this conversion occurs in two steps. In the first step, the nitrile is hydrated to amide and finally carboxylic acid is formed by nitrile hydratase and amidase, respectively (Patten et al. 2013). The activity of indole-3-acetontrile hydratase is also reported in *Rhodococcus* (Xie et al. 2003), *Agrobacterium* and *Rhizobium* (Xie et al. 2003).

Generally, nitrilases have large number of substrate specificity but it showed highest affinity to aromatic nitriles (O'Reilly and Turner 2003). For example, nitrilase from *P. fluorescens* DSM 7155 has showed more affinity for phenylacetonitrile among tested arylacetonitriles (Layh et al. 1998). The bacteria were capable to grow without nitrogen source and the phenylacetonitrile used by bacteria, as only nitrogen source in grown culture medium, while nitrilase activity was suppressed by ammonia present in culture medium (Layh et al. 1998). In another example, *P. syringae* B728a synthesized the IAA in supplemented indole-3-acetonitrile grown media without nitrogen source (Howden et al. 2009). In a connecting pathway the indole-3-acetamide is converted to indole-3-acetontrile and finally synthesizes the IAA (Fig. 1).

Tryptamine (TAM) Pathway

The tryptamine pathway is first identified and characterized in *Bacillus cereus* by activity of tryptophan decarboxylase (Perley and Stowe 1966). The L-tryptophan is converted to tryptamine by decarboxylation process catalysed by tryptophan decarboxylase. Later the conversion of tryptamine to IAA was confirmed in *Azospirillum* by exogenous supply of tryptamine in growth media (Hartmann et al. 1983). In contrast, the tryptamine was isolated as endogenous compound in plants. The role of tryptamine is controversial in plants as it has similar growth regulating property to IAA (Winter 1966). The encoding genes of enzyme, tryptophan decarboxylases, also been characterized and cloned from different plants, demonstrating that IAA is synthesized via tryptamine pathway in plants (Spaepen et al. 2007). Further, the tryptamine is first converted to *N*-hydroxyl-tryptamine, which is catalysed by flavin monooxygenase (YUCCA), and finally converted to indole-3-acetaldoxime and indole-3-acetaldehyde in plants (Zhao et al. 2001). While in bacteria indole-3-acetaldehyde is directly converted from tryptamine catalysed by amine oxidase without forming *N*-hydroxyl-tryptamine (Hartmann et al. 1983). The tryptamine pathway occurs in some organs of plants but it is possible that in all organs of plants or all plant species does not exist tryptamine pathway (Tivendale et al. 2014).

Tryptophan-Independent Pathway

This pathway is first proposed after analysis of *Arabidopsis thaliana* mutant which cannot synthesize the tryptophan but this mutant increased the concentration of IAA conjugates (Normanly et al. 1993) as tryptophan is only known precursor of IAA biosynthesis. Till date, no enzyme has been characterized for this pathway. In bacteria this pathway was reported in *A. brasilense* by using radioactive labelled precursor of IAA. The study reveals that major IAA was synthesized via tryptophan-independent pathway while only less than 2% IAA was synthesized via tryptophan-dependent (IAM) pathway (Prinsen et al. 1993). Recently, Ahmad et al. (2020) developed a method for screening the IAA biosynthesis via tryptophan-independent pathway. By using this screening method IAA produced by *M. aloeveare* DCB-20 in Tris-minimal broth supplemented with casein acid hydrolysate via tryptophan-independent pathway.

So we need improvement in analytical techniques for qualitative and quantitative assay for detection of IAA and intermediate compounds of different IAA biosynthetic pathway as well as techniques involved in functional genomics for rapid analysis of genome and proteome analysis. This will provide more exhaustive and comprehensive knowledge on different aspects of biosynthesis pathways of IAA present in bacteria.

Role of IAA in Plant–Microbe Interaction: An Overview

The role of IAA in plant–microbe interaction have been now diversified, as increasing the study of IAA biosynthesis of diverse group of bacteria and further its effect on root and shoot growth of plant, root colonization, root architecture, inhibiting the growth of phytopathogenic fungi in soil. Earlier, the IAA synthesis was associated with tumour, nodule and gall formation in plants. However, now it is well established that not only phytopathogenic (gall inducing) bacteria but also plant growth-promoting rhizobacteria have the capability to produce the IAA. While plants have the ability to synthesize IAA via different biosynthetic pathways, the question arises why bacteria have different metabolic pathways and diverse genes for biosynthesis of IAA. The production of IAA by bacteria and its role on associated plants remain ambiguous with few exceptions. In this part, we address the production of IAA by bacteria and different strategies for development of associated plants.

The indole-3-acetamide pathway genes of *A. tumefaciens* are incorporated in T-DNA region (tumour-inducing) of plasmid (Ti plasmid) and then transferred to root cells of plant and incorporated with chromosome. After incorporation, the genes of IAA biosynthesis pathway expressed and produced the IAA for proliferation of plant cell and formation of crown gall tumours (Thomashow et al. 1984; Mashiguchi 2019). While genes of IAA biosynthetic pathway are not transferred to host cells which previously received the genes of IAA biosynthetic pathway (Comai et al. 1982; Manulis et al. 1991; Lemcke et al. 2000; Verma and Mathur 2011). Additionally, the gall formation on different plants such as oleander, olive, gypsophila and privet is also induced by IAA-producing *P. syringae* and *P. agglomerans* phytopathogens. If gall formation capability was reduced or abolished in *P. agglomerans* and *P. syringae* infected plants, the mutant strains of *P. agglomerans* and *P. syringae* from this plant, would not have IAA-secreting capability due to inactivation of functional genes of IAA biosynthetic pathways (Smidt and Kosuge 1978; Surico et al. 1984; Yamada 1993; Iacobellis et al. 1994). Furthermore, the other virulence factors for nontumorigenic infection in plants is functional gene of indole-3-acetamide (IAM) pathway. A study reveals that IAA synthesis by *D. dadantii iaaMH* mutant minimizes the rot disease of *Saintpaulia ionantha* plant (African violet) by reducing the synthesis of pectinase which lysed the cell wall of host plant (Yang et al. 2007). The type III secretion system along with activation by GacSA-mediated are involved in degradation of the cell walls of host plant by exudation of extracellular enzymes (Yang et al. 2007).

Surprisingly, various plant growth-promoting rhizobacteria (PGPR) which aggressively colonize the plant roots in the rhizosphere having capability the IAA biosynthesis. When genes of IAA biosynthetic pathways are inactivated or disrupted in plant mutant, then IAA synthesized from PGPR attributed the root-promoting activity to plant (Barbieri et al. 1986; Dobbelaere et al. 1999; Patten and Glick 2002; Ahmad et al. 2013). Mutants of *Sinorhizobium meliloti*, *A. brasilense*, and *P. putida* secreted high amount of IAA and were reported as IAA-overproducing strains. After colonization of *Medicago*, wheat and canola roots, these strains are improving the architecture of lateral roots by synthesizing IAA from *Sinorhizobium meliloti, A. brasilense, P. putida* respectively (Barbieri and Galli 1993; Mayak et al. 1997; Bianco and Defez 2010; Ahmad et al. 2013; Liu et al. 2016; Mehmood et al. 2020). The plants having improved root system (increased surface area) are more capable to obtain the water and minerals from rhizospheric soil while it also helps to anchor the plants in rhizosphere. In contrast, some high IAA-producing bacteria inhibit the root elongation of primary roots. For example, *P. fluorescens* CHA0 mutant inhibits the growth of primary roots of cucumber and wheat plants by overproduction of IAA (Beyeler et al. 1997). Similarly, the IAA-producing wild-type strain of *P. putida* GR12-2 has positive effect on root elongation of canola seedlings while mutant of this strain constrains the primary growth of canola seedlings (Xie et al. 1996). A linear relation was established between retardation in root growth of seedling of sugar beet and the amount of accumulated IAA in bacterial isolates from rhizosphere (Loper and Schroth 1986).

Earlier, the role of IAA was established when low amount of IAA was applied exogenously and root elongation was improved significantly; this experiment indicates that IAA is directly involved in root growth of plants (Thimann and Lane 1938). The concentration of IAA varies between 10 and 12 mM for development of different stages of plant roots (Evans et al. 1994; Patten et al. 2013). Higher concentration of IAA indirectly retards the root growth by stimulating the ethylene level in plants and high level of ethylene inhibits the elongation of plant root (Peck and Kende 1995). So, the amount of IAA secretion in vitro by rhizospheric bacteria can influence or interfere the growth of plants root. In addition, the phenylacetic acid is also classified as weak auxin than IAA which can help in root elongation of plants (Wightman and Lighty 1982). However, IAA and phenylacetic acid synthesized by some bacteria via IPy pathway and stimulated the expression of indole-3-pyruvate decarboxylase by utilizing the tryptophan as precursor molecule (Ryu and Patten 2008). The amount of synthesized phenylacetic acid is not measured but the concentration of IAA was analysed in growth medium after synthesis by bacterial strain (Ryu and Patten 2008). The root elongation and intermodal segments of *Phaseolus vulgaris* plant were observed to be higher when combination of phenylacetic acid and IAA were applied while only application of phenylacetic acid gives weaker response for root elongation and intermodal segments (Small and Morris 1990). In another study more growth was recorded in lateral root of pea seedlings after application of phenylacetic acid in comparison to IAA application (Wightman et al. 1980). The phenylacetate can be transported through transport proteins by mechanism in which phenylacetate dissociated in anionic form and binds to transporter protein of auxins. While, after synthesis of phenylacetate in organs of plant cell, it remains dissociated, but after dissociation it can be diffuses into the cytoplasm of plant cell then bind to transporter protein for further transport (Löbler and Klämbt 1985). Additionally, the polar transportation of IAA is prevented by inhibiting activity of carrier for auxin efflux which has no role in phenylacetate transportation (Morris and Johnson 1987; Ding et al. 2012).

The concentration of IAA produced by PGPR can determine the level of attributed PGP activity of these PGPR. The concentration gradients of IAA can determine the root development of plants and other biological functions associated with IAA. While, the gradients of IAA in plant organs are influenced by IAA transport (Dhonukshe et al. 2008; Simon and Petrášek 2011). Formation of new primordia in root resulted after division of founder cells. While founder cells of lateral roots formed in response to IAA accumulation in root cells of pericycle, the IAA can help in formation of enhances the lateral root formation (Dubrovsky et al. 2008). So, now it is established that the stimulation of lateral root by phenylacetate mediated inhibition of polar IAA transport in plants.

IAA-secreting bacteria promote the root colonization process in rhizosphere through suppressing the defence mechanism of plants. The IAA-plant signalling involved in this process resulted in the susceptibility of bacterial infection to plant (Kazan and Manners 2009). The disease symptom by infection of *P. syringae* DC3000 was decreased when IAA-responsive genes and IAA signalling F-box were downregulated in *Arabidopsis thaliana* (Navarro et al. 2006). Simultaneously, increase in IAA biosynthesis and addition of exogenous IAA application increase 20-fold colonizing capability of *P. syringae* and also disease symptom rises in plants (Navarro et al. 2006; Wang et al. 2007). Similarly, the infection of *Xanthomonas oryzae* was increased after exogenous application of IAA in rice plants (Ding et al. 2008). The expansins may be activated on root surfaces by secretion of IAA by bacteria which resulted in the loosening of cell wall protein, and through weakening of extracellular matrix the bacterial colonization had improved (Ding et al. 2008). In other mechanisms the root colonization enhanced after plant-auxin signalling by decreasing the defence system of plants. While IAA secreted from PGPR can inhibit the growth of phytopathogen and can directly protect the plants from different infectious disease.

IAA Biosynthesis for Plant–Microbe Interaction in Response to Drought Stress

Drought is one of the major environmental conditions which creates the severe stress in plants. The stress condition impact negatively on plant development and plant growth and finally crop yield is decreased. The climate change is major contributor for changing the weather patterns, resulted low rain and created drought stress for plants (Han et al. 2017; Smith et al. 2020). In recent decades, the researchers focus on investigating the role of IAA in drought stress. Globally, among farmland total 28% dry land (water stress) is available for production of different crops (Blumwald 2000). To overcome the drought conditions, plants have adapted many strategies including the alteration in root structure for extraction of more water from soil, synthesizing different biomolecules for maintaining the osmotic potential and crosstalk between plant–microbe by signalling molecule. The regulation of IAA biosynthesis is also one of the strategies critically required for adaptation from drought stress. In an example, drought tolerance capability of alfalfa plant has been increased after increasing the concentration of IAA (Defez et al. 2017). Similarly, expression of YUC7 gene increased, which is involved in IAA biosynthesis in *Arabidopsis* and resulted in the drought tolerance improvement (Lee et al. 2012). Moreover, expression of IAA synthesizing gene YUC6 in tomato and potato is also linked to improvement of drought tolerance (Kim et al. 2013). Interestingly, the IAA synthesis process is not directly linked to drought tolerance through by expression of YUC6 gene but thiol reductase found in YUC6, reduces the reactive oxygen species (ROS) and IAA can involve indirectly in drought stress (Cha et al. 2015). However, increase in concentration of IAA in *Arabidopsis* can directly affect drought tolerance through different mechanisms, including improvement in architecture of plant root, modification in different metabolic pathway, decrease in ethylene concentration and regulation and expression of genes involved in ABA biosynthesis (Shi et al. 2014).

In a study, revealed that the drought tolerance capability have been increased in tomato when concentration of IAA was increased in rhizosphere, which resulted in increase in the surface area and biomass of plant root (Moles et al. 2018). Additionally, the surface area and biomass of plant root increased by improvement in lateral roots of plant. Synthesis of specific molecules and alteration of growth of shoot and root are the strategies for adaptation mechanism by plant in drought conditions. Mechanistically, extraction of water from dried rhizospheric soil requires increase in root surface area (Davies et al. 2002). In opposite some study reveals that the drought tolerance of rice increased after decreased IAA concentration in root and also the retardation in gene expression of *YUC* under drought conditions (Du et al. 2013; Naser and Shani 2016). To check the water storage in soil the root acts as sensors and also root system is essentially required for uptake of water (Davies et al. 2002). The transition zone of root-shoot is now focusing for expression and regulation of YUC gene in drought condition. The transition zone (TZ) acts as platform for receiving the signal of hormonal crosstalk and implicating the drought tolerance capacity of plant (Kong et al. 2018). At molecular level, transcription of *YUC* genes increases twofold in roots during drought condition but no increase was recorded of *YUC* genes in leaves of the same plant in drought conditions (Blakeslee et al. 2019).

IAA-Producing Bacteria for Mitigation of Drought Conditions

Generally, bacteria synthesize phytohormone IAA by utilizing the tryptophan as precursor. Mechanistically, IAA is involved in different biological processes of plants including cell division, root initiation, geotropism, cell enlargement, apical dominance, transpiration, phototropism and root growth inhibition (Ahmad et al. 2013). Eighty percent of colourable microbes from rhizosphere have the capacity to synthesize IAA (Patten and Glick 2002). The IAA-producing bacteria from different genera including *Pseudomonas*, *Rhizobium*, *Serratia*, *Azotobacter*, *Bradyrhizobium*,

Mesorhizobium, *Xanthomonas*, *Azospirillum*, *Bacillus*, *Enterobacter* etc. have been reported which help in PGP activity for different plants (Spaepen et al. 2007; Ahmad et al. 2013; Patten et al. 2013; Ahmad et al. 2020). The plants can survive in water stress condition due to co-inoculation of IAA-producing *Pseudomonas putida* with plants (Marulanda et al. 2009). *Bacillus subtilis* secreted volatile organic compounds which can help in upregulation of auxin homeostasis genes; the resulted growth improvement was recorded in *Arabidopsis* plant (Zhang et al. 2007). The xylem architecture of coleoptile in wheat plant is modified and improved morphologically to survive in severe drought conditions when wheat plants were co-inoculated with *Azospirillum* (Pereyra et al. 2012). The IAA biosynthesis was improved in *B. subtilis* and *Azospirillum* in drought conditions by upregulation of genes involved in indole-3-pyruvate decarboxylase pathway through modulation of IAA genes (Barnawal et al. 2012). Similarly, the IAA signalling response changed in chickpea plants inoculated with *P. putida* through modulation in F-BOX 2(AFB2) IAA signalling (Jatan et al. 2018).

Involvement of IAA to Change the Different Physiological Processes in Drought Stress

Modification in Root Architecture for Water Uptake from Rhizospheric Soil

To mitigate the drought stress in plants the alteration and improvement of root structure is the most important biological process (Bacon et al. 2002; Huang et al. 2014). Root architecture includes the length and diameter of root, distribution of lateral root, structure of primary root and topology system of plant roots (de Dorlodot et al. 2007; Vacheron et al. 2013). The morphological plasticity is the main character of roots which is exhibited in different physical conditions of soil (Bengough et al. 2006; Tuberosa 2012). The improved root architecture of different crops is developed in response to drought condition in soybeans, chickpea, maize and wheat (Tuberosa et al. 2003; Landi et al. 2010; Sadok and Sinclair 2011; Varshney et al. 2011; Wasson et al. 2012).

The inoculation of PGPR has been reported for root growth promotion and modification in root structure of different plants (Kloepper 1992; Ngumbi 2011; Ahmad et al. 2013). Furthermore, the improved root architecture by bacteria may increase the total surface area of root and subsequently lead to enhancement of nutrient and water uptake by plant which shows positive growth on plants (Ahmad et al. 2013; Timmusk et al. 2014). In other example, root biomass was increased by 10% after co-inoculation of *Alcaligenes faecalis* on maize plants as compared to non-inoculated maize plants in drought stress condition created in plant growth chamber (Naseem and Bano 2014). The root biomass of Kaleo and Mazurka cultivars was increased by 40% and 47%, respectively, after inoculated with

Enterobacter sp. as compared to non-inoculated maize plant under drought condition (Naveed et al. 2014). Moreover, identification of ideal root systems of plant in drought conditions will be required for further study linked to root structure–drought stress relation.

Shoot Growth Under Drought Stress

Shoot growth inhibition is one of the primary responses in drought stress conditions. The inhibition of shoot growth resulted in minimizing the leaf surface area which can decrease the water loss during drought conditions (Sinclair and Muchow 2001; Neumann 2008; Skirycz and Inzé 2010). Additionally, the essential solutes can be diverted for different physiological functions including maintaining the osmotic potential after inhibition of shoot growth during drought stress (Neumann 1995; Aachard et al. 2006). So, shoot growth inhibition mechanism is considered as adaptation mechanism that helps in mitigation of plants during drought stress conditions (Neumann 2008). While shoot growth of many plants has improved after inoculation with PGPR in drought conditions. The shoot biomass was increased by 50% in pepper plants inoculated with *Bacillus licheniformis* K11 as compared to non-inoculated plants in drought stress conditions. The improvement of shoot growth after treatment of PGPR in different crops including sorghum, sunflower, wheat, green gram, mung bean, and maize are also reported (Arzanesh et al. 2011; Saravanakumar et al. 2011; Castillo et al. 2013; Sarma and Saikia 2014; Sandhya et al. 2010).

Maintaining the Osmotic Potential in Drought Stress

Accumulation of inorganic and organic solutes (compatible solutes) in cells by active process in response to water stress condition is referred as osmotic adjustment (Nilsen and Orcutt 1996; Kiani et al. 2007). The solutes maintain the osmotic potential of plant cell without decreasing water present in these cells (Serraj and Sinclair 2002). These compounds are (1) Sugars—sucrose, (2) non-essential amino acids—proline (3) organic acids—malate (4) inorganic ions—Ca^{2+}, Na^{+} (5) polyols-mannitol and (6) glycine betaine that maintain the osmotic potential during water stress condition. The adjustment of osmotic potential in plant cell during drought stress is the main adaptation mechanism to overcome drought-induced stress in plants (Farooq et al. 2009). This mechanism protects proteins, enzymes, cellular organelles and cellular membrane against damage by oxidation process (Hoekstra and Buitink 2001; Huang et al. 2014).

The concentration of proline has increased in different organs of plants during water stress condition after inoculation with drought-tolerant bacteria (Farooq et al. 2008). In other many examples, the study reveals that proline content was increased after inoculation with drought-tolerant PGPR with plants including maize, sorghum, potato, mung bean and *Arabidopsis* (Sandhya et al. 2010; Gururani et al. 2013; Grover et al. 2014; Sarma and Saikia 2014; Cohen et al. 2015). The proline was accumulated three- to –fourfold in cucumber leaf to protect the dehydration during water stress condition after inoculation with *Bacillus subtilis* SM21 (Wang et al. 2012). Simultaneously, the accumulation of soluble sugars and free amino acids in plants is also one of the strategies to overcome water stress condition.

Conclusion

The drought tolerance capacity of plants influenced by plant–microbe interaction mechanism established that IAA has key role in drought tolerance of plants. Moreover, the function and role of IAA-producing bacteria must be assessed, for their colonizing capacity, PGP activity and stress (drought) tolerance ability under water stress condition and their effect on different crops under pot and field conditions. Finally, the integrated testing of different PGP bacteria including IAA-producing bacteria and different crops for water stress tolerance can help in modern agriculture practices to overcome the effect of climate change on agriculture.

References

Aachard P, Cheng H, De Grauwe L, Decat J, Schoutteten H, Moritz T, Van Der Straeten D, Peng J, Harberd NP (2006) Integration of plant responses to environmentally activated phytohormonal signals. Science 311:91–94

Ahmad E, Khan MS, Zaidi A (2013) ACC deaminase producing Pseudomonas putida strain PSE3 and Rhizobium leguminosarum strain RP2 in synergism improves growth, nodulation and yield of pea grown in alluvial soils. Symbiosis 61:93–104

Ahmad E, Sharma SK, Sharma PK (2020) Deciphering operation of tryptophan-independent pathway in high indole-3-acetic acid (IAA) producing *Micrococcus aloeverae* DCB-20. FEMS Microbiol Lett 67:fnaa190

Andrews FH, Rogers MP, Paul LN, McLeish MJ (2014) Perturbation of the monomer–monomer interfaces of the benzoylformate decarboxylase tetramer. Biochemistry 53:4358–4367

Ardö Y (2006) Flavour formation by amino acid catabolism. Biotechnol Adv 24:238–242

Arzanesh MH, Alikhani HA, Khavazi K, Rahimian HA, Miransari M (2011) Wheat (*Triticum aestivum* L.) growth enhancement by *Azospirillum* sp. under drought stress. World J Microbiol Biotechnol 27:197–205

Asano Y, Yasukawa K (2019) Identification and development of amino acid oxidases. Curr Opin Chem Biol 49:76–83

Bacon MA, Davies WJ, Mingo D, Wilkinson S (2002) Root signals. In: Waisel Y, Eshel A, Kafkafi U (eds) Plant roots: the hidden half. Monticello, New York, pp 461–470

Barbieri P, Galli E (1993) Effect on wheat root development of inoculation with an *Azospirillum brasilense* mutant with altered indole-3-acetic acid production. Res Microbiol 144:69–75

Barbieri P, Zanelli T, Galli E, Zanetti G (1986) Wheat inoculation with *Azospirillum brasilense* Sp6 and some mutants altered in nitrogen fixation and indole-3-acetic acid production. FEMS Microbiol Lett 36:87–90

Barnawal D, Bharti N, Maji D, Chanotiya CS, Kalra A (2012) 1-Aminocyclopropane-1-carboxylic acid (ACC) deaminase-containing rhizobacteria protect *Ocimum sanctum* plants during waterlogging stress via reduced ethylene generation. Plant Physiol Biochem 58:227–235

Bengough AG, Bransby MF, Hans J, McKenna SJ, Roberts TJ, Valentine TA (2006) Root responses to soil physical conditions; growth dynamics from field to cell. J Exp Bot 57:437–447

Beyeler M, Michaux P, Keel C, Haas D (1997) Effect of enhanced production of indole-3-acetic acid by the biological controlagent Pseudomonas fluorescens CHA0 on plant growth. pp 310–311. In A. Ogoshi, K. Kobayashi, Y. Homma, F. Kodama, N.Kondo, and S. Akino (Eds.). Plant Growth-Promoting Rhizobacteria: Present Status and Future Prospects. Organization for Economic Cooperation and Development, Paris

Bianco C, Defez R (2010) Improvement of phosphate solubilisation and *Medicago* plant yield by an indole-3-acetic acid-overproducing strain of *Sinorhizobium meliloti*. Appl Environ Microbiol 76:4626–4632

Blakeslee JJ, Spatola Rossi T, Kriechbaumer V (2019) Auxin biosynthesis: spatial regulation and adaptation to stress. J Exp Bot 70:5041–5049

Blumwald E (2000) Sodium transport and salt tolerance in plants. Curr Opin Cell Biol 12:431–434

Brandl MT, Lindow SE (1996) Cloning and characterization of a locus encoding an indolepyruvate decarboxylase involved in indole-3-acetic acid synthesis in *Erwinia herbicola*. Appl Environ Microbiol 62:4121–4128

Brandl MT, Quinones B, Lindow SE (2001) Heterogeneous transcription of an indole acetic acid biosynthetic gene in *Erwinia herbicola* on plant surfaces. Proc Natl Acad Sci U S A 98:3454–3459

Castillo P, Escalante M, Gallardo M, Alemano S, Abdala G (2013) Effects of bacterial single inoculation and co-inoculation on growth and phytohormone production of sunflower seedlings under water stress. Acta Physiol Plant 35:2299–2309

Cha JY, Kim WY, Kang SB et al (2015) A novel thiol-reductase activity of Arabidopsis YUC6 confers drought tolerance independently of auxin biosynthesis. Nat Commun 6:8041

Cohen JD, Slovin JP, Hendrickson AM (2003) Two genetically discrete pathways convert tryptophan to auxin: more redundancy in auxin biosynthesis. Trends Plant Sci 8:197–199

Cohen A, Bottini R, Pontin M, Berli F, Moreno D, Boccanlandro H, Travaglia C, Picocoli P (2015) *Azospirillum brasilense* ameliorates the response of *Arabidopsis thaliana* to drought mainly via enhancement of ABA levels. Physiol Plant 153:79–90

Comai LU, Kosuge TS (1980) Involvement of plasmid deoxyribonucleic acid in indoleacetic acid synthesis in Pseudomonas savastanoi. J Bacteriol 143:950–957

Comai L, Surico G, Kosuge T (1982) Relation of plasmid DNA to indoleacetic acid production in different strains of *Pseudomonas syringae* pv. *savastanoi*. J Gen Microbiol 128:2157–2163

Costacurta A, Keijers V, Vanderleyden J (1994) Molecular cloning and sequence analysis of an *Azospirillum brasilense* indole-3-pyruvate decarboxylase gene. Mol Gen Genet 243:463–472

Davies WJ, Wilkinson S, Loveys B (2002) Stomatal control by chemical signaling and the exploitation of this mechanism to increase water use efficiency in agriculture. New Phytol 153:449–460

de Dorlodot S, Forster B, Pages L, Price A, Tuberosa R, Draye X (2007) Root system architecture: opportunities and constraints for genetic improvement of crops. Trend Plant Sci 12:474–481

Defez R, Andreozzi A, Dickinson M, Charlton A, Tadini L, Pesaresi P, Bianco C (2017) Improved drought stress response in alfalfa plants nodulated by an IAA over-producing rhizobium strain. Front Microbiol 8:2466

Dhonukshe P, Tanaka H, Goh T, Ebine K, Mähönen AP, Prasad K, Blilou I, Geldner N, Xu J, Uemura T, Chory J, Ueda T, Nakano A, Scheres B, Friml J (2008) Generation of cell polarity in plants links endocytosis, auxin distribution and cell fate decisions. Nature 456:962–966

Di DW, Zhang C, Luo P, An CW, Guo GQ (2016) The biosynthesis of auxin: how many paths truly lead to IAA? Plant Growth Regul 78:275–285

Ding X, Cao Y, Huang L, Zhao J, Xu C, Li X, Wang S (2008) Activation of the indole-3-acetic acid-amido synthetase GH3-8 suppresses expansin expression and promotes salicylate- and jasmonate independent basal immunity in rice. Plant Cell 20:228–240

Ding Z, Wang B, Moreno I et al (2012) ER-localized auxin transporter PIN8 regulates auxin homeostasis and male gametophyte development in *Arabidopsis*. Nat Commun 3:941

Dobbelaere S, Croonenborghs A, Thys A, Vande Broek A (1999) Phytostimulatory effect of *Azospirillum brasilense* wild type and mutant strains altered in IAA production on wheat. Plant Soil 212:155–164

Du H, Wu N, Chang Y, Li X, Xiao J, Xiong L (2013) Carotenoid deficiency impairs ABA and IAA biosynthesis and differentially affects drought and cold tolerance in rice. Plant Mol Biol 83:475–488

Dubrovsky JG, Sauer M, Napsucialy-Mendivil S, Ivanchenko MG, Friml J, Shishkova S, Celenza J, Benková E (2008) Auxin acts as a local morphogenetic trigger to specify lateral root founder cells. Proc Natl Acad Sci U S A 105:8790–8794

Duca D, Lorv J, Patten CL, Rose D, Glick BR (2014) Indole-3-acetic acid in plant–microbe interactions. Antonie VanLeeuwenhoek. 106:85–125

Emanuele JJ, Fitzpatrick PF (1995) Mechanistic studies of the flavoprotein tryptophan 2-monooxygenase. 2. pH and kinetic isotope effects. Biochemistry 34:3716–3723

Emanuele JJ, Heasley CJ, Fitzpatrick PF (1995) Purification and characterization of the flavoprotein tryptophan 2-monooxygenase expressed at high levels in Escherichia coli. Arch Biochem Biophys 316:241–248

Estenson K, Hurst GB, Standaert RF, Bible AN, Garcia D, Chourey K, Doktycz MJ, Morrell-Falvey JL (2018) Characterization of Indole-3-acetic acid biosynthesis and the effects of this phytohormone on the proteome of the plant-associated microbe Pantoea sp. YR343. J Proteome Res 17:1361–1374

Evans ML, Ishikawa H, Estelle MA (1994) Responses of *Arabidopsis* roots to auxin studied with high temporal resolution: comparison of wild type and auxin-response mutants. Planta 194:215–222

Farooq M, Basra SM, Wahid A, Cheema ZA, Cheema MA, Khaliq A (2008) Physiological role of exogenously applied glycinebetaine to improve drought tolerance in fine grain aromatic rice (Oryza sativa L.). J Agro Crop Sci 194:325–333

Farooq M, Wahid A, Kobayashi N, Fujita D, Basra SMA (2009) Plant drought stress: effects, mechanisms and management. Agron Sustain Dev 29:185–212

Fedorov DN, Doronina NV, Trotsenko YA (2010) Cloning and characterization of indolepyruvate decarboxylase from *Methylobacterium extorquens* AM1. Biochem Mosc 75:1435–1443

Galili G, Amir R, Fernie AR (2016) The regulation of essential amino acid synthesis and accumulation in plants. Annu Rev Plant Biol 67:153–178

Gaweska HM, Taylor AB, Hart PJ, Fitzpatrick PF (2013) Structure of the flavoprotein tryptophan 2-monooxygenase, a key enzyme in the formation of galls in plants. Biochemistry 52:2620–2626

Grover M, Madhubala R, Ali SZ, Yadav SK, Venkateswarlu B (2014) Influence of Bacillus spp. strains on seedling growth and physiological parameters of sorghum under moisture stress conditions. J Basic Microbiol 54:951–961

Gururani MA, Upadhyaya CP, Baskar V, Venkatesh J, Nookaraju A, Park SW (2013) Plant growth-promoting rhizobacteria enhance abiotic stress tolerance in *Solanum tuberosum* through inducing changes in the expression of ROS-scavenging enzymes and improved photosynthetic performance. J Plant Growth Regul 32:245–258

Han EH, Petrella DP, Blakeslee JJ (2017) 'Bending' models of halotropism: incorporating protein phosphatase 2A, ABCB transporters, and auxin metabolism. J Exp Bot 68:3071–3089

Harris DM, Berrue F, Kerr R, Patten CL (2018) Metabolomic analysis of indolepyruvate decarboxylase pathway derivatives in the rhizobacterium *Enterobacter cloacae*. Rhizosphere 6:98–111

Hartmann A, Singh M, Klingmüller W (1983) Isolation and characterization of *Azospirillum* mutants excreting high amounts of indoleacetic acid. Can J Microbiol 29:916–923
Hazelwood LA, Tai SL, Boer VM, de Winde JH, Pronk JT, Daran JM (2006) A new physiological role for Pdr12p in *Saccharomyces cerevisiae*: export of aromatic and branched-chain organic acids produced in amino acid catabolism. FEMS Yeast Res 6:937–945
Helinck S, Le Bars D, Moreau D, Yvon M (2004) Ability of thermophilic lactic acid bacteria to produce aroma compounds from amino acids. Appl Environ Microbiol 70:3855–3861
Hilbert M, Voll LM, Ding Y, Hofmann J, Sharma M, Zuccaro A (2012) Indole derivative production by the root endophyte Piriformospora indica is not required for growth promotion but for biotrophic colonization of barley roots. New Phytologist196:520–534
Hoekstra FA, Buitink J (2001) Mechanisms of plant desiccation tolerance. Trends Plant Sci 8:431–438
Howden AJ, Rico A, Mentlak T, Miguet L, Preston GM (2009) *Pseudomonas syringae* pv. syringae B728a hydrolyses indole-3-acetonitrile to the plant hormone indole-3-acetic acid. Mol Plant Pathol 10:857–865
Huang B, DaCosta M, Jiang Y (2014) Research advances in mechanisms of turfgrass tolerance to abiotic stresses: from physiology to molecular biology. Crit Rev Plant Sci 33:141–189
Hutcheson SW, Kosuge T (1985) Regulation of 3-indoleacetic acid production in Pseudomonas syringae pv. savastanoi. Purification and properties of tryptophan 2-monooxygenase. J Biol Chem 260:6281–6287
Iacobellis NS, Sisto A, Surico G, Evidente A, DiMaio E (1994) Pathogenicity of *Pseudmonas syringae* subsp. *savastanoi* mutants defective in phytohormone production. J Phytopathol 140: 238–248
Imada EL, de Oliveira AL, Hungria M, Rodrigues EP (2017) Indole-3-acetic acid production via the indole-3-pyruvate pathway by plant growth promoter *Rhizobium tropici* CIAT 899 is strongly inhibited by ammonium. Res Microbiol 168:283–292
Jatan R, Chauhan PS, Lata C (2018) *Pseudomonas putida* modulates the expression of miRNAs and their target genes in response to drought and salt stresses in chickpea (*Cicer arietinum* L.). Genomics. https://doi.org/10.1016/j.ygeno.2018.01.007
Kato Y, Nakamura K, Sakiyama H, Mayhew SG, Asano Y (2000) Novel heme-containing lyase, phenylacetaldoxime dehydratase from *Bacillus* sp. strain OxB-1: purification, characterization, and molecular cloning of the gene. Biochemistry 39:800–809
Kazan K, Manners JM (2009) Linking development to defense: auxin in plant-pathogen interactions. Trends Plant Sci 14:373–382
Kemper E, Wafenschmidt S, Weiler EW, Rausch T, Schröder J (1985) T-DNA-encoded auxin formation in crown-gall cells. Planta 163:257–262
Kiani SP, Talia P, Maury P, Grieu P, Heinz R, Perrault A, Nishinakamasu V, Hopp E, Gentzbittel L, Paniego N, Sarrafi A (2007) Genetic analysis of plant water status and osmotic adjustment in recombinant inbred lines of sunflower under two water treatments. Plant Sci 172:773–787
Kim I, Baek D, Park HC et al (2013) Overexpression of Arabidopsis YUCCA6 in potato results in high-auxin developmental phenotypes and enhanced resistance to water deficit. Mol Plant 6: 337–349
Kittell BL, Helinski DR, Ditta GS (1989) Aromatic aminotransferase activity and indole acetic acid production in *Rhizobium meliloti*. J Bacteriol 171:5458–5466
Kloepper JW (1992) Plant growth-promoting rhizobacteria as biological control agents. In: Metting FB (ed) Soil microbial ecology: applications in agricultural and environmental management. Marcel Dekker Inc., New York, pp 255–274
Kneen MM, Stan R, Yep A, Tyler RP, Saehuan C, McLeish MJ (2011) Characterization of a thiamin diphosphate-dependent phenylpyruvate decarboxylase from *Saccharomyces cerevisiae*. FEBS J 278:1842–1853
Koga J, Adachi T, Hidaka H (1992) Purification and characterization of indolepyruvate decarboxylase. A novel enzyme for indole-3-acetic acid biosynthesis in *Enterobacter cloacae*. J Biol Chem 267:15823–15828

Koga J, Syōono K, Ichikawa T, Adachi T (1994) Involvement of L-tryptophan aminotransferase in indole-3-acetic acid biosynthesis in *Enterobacter cloacae*. Biochim Biophys Acta (BBA) Protein Struct Mol Enzymol 1209:241–247

Kong W, Li Y, Zhang M, Jin F, Li J (2015) A novel Arabidopsis microRNA promotes IAA biosynthesis via the indole-3-acetaldoxime pathway by suppressing SUPERROOT1. Plant Cell Physiol 56:715–726

Kong X, Liu G, Liu J, Ding Z (2018) The root transition zone: a hot spot for signal crosstalk. Trends Plant Sci 23:403–409

Koshiba T, Saito E, Ono N, Yamamoto N, Sato M (1996) Purification and properties of flavin- and molybdenum-containing aldehyde oxidase from coleoptiles of maize. Plant Physiol 110:781–789

Landi P, Giuliani S, Salvi S, Ferri M, Tuberosa R, Sanguineti MC (2010) Characterization of root-yield-1.06: a major constitutive QTL for root and agronomic traits in maize across water regimes. J Exp Bot 61:3553–3562

Layh N, Parratt J, Willetts A (1998) Characterization and partial purification of an enantioselective arylacetonitrilase from *Pseudomonas fluorescens* DSM 7155. J Mol Catal 5:467–474

Lee M, Jung JH, Han DY, Seo PJ, Park WJ, Park CM (2012) Activation of a flavin monooxygenase gene YUCCA7 enhances drought resistance in Arabidopsis. Planta 235:923–938

Lemcke K, Prinsen E, van Onckelen H, Schmülling T (2000) The ORF8 gene product of *Agrobacterium rhizogenes* TL-DNA has tryptophan 2-monooxygenase activity. Mol Plant-Microbe Interact 13:787–790

Li S, Wen J, Jia X (2011) Engineering Bacillus subtilis for isobutanol production by heterologous Ehrlich pathway construction and the biosynthetic 2-ketoisovalerate precursor pathway overexpression. Appl Microbiol Biotechnol 91(3):577–589

Li M, Guo R, Yu F, Chen X, Zhao H, Li H, Wu J (2018) Indole-3-acetic acid biosynthesis pathways in the plant-beneficial bacterium *Arthrobacter pascens* ZZ21. Int J Mol Sci 19:443

Li J, Huang Q, Zheng X, Ge Z, Lin K, Zhang D, Chen Y, Wang B, Shi X (2020) Investigation of the lactic acid bacteria in kazak cheese and their contributions to cheese fermentation. Front Microbiol 11:228

Liu L, Zhong S, Yang R, Hu H, Yu D, Zhu D, Hua Z, Shuldiner AR, Goldstein R, Reagan WJ, Gong DW (2008) Expression, purification, and initial characterization of human alanine aminotransferase (ALT) isoenzyme 1 and 2 in High-five insect cells. Protein Expr Purif 60:225–231

Liu Y, Chen L, Zhang N, Li Z, Zhang G, Xu Y, Shen Q, Zhang R (2016) Plant-microbe communication enhances auxin biosynthesis by a root-associated bacterium, *Bacillus amyloliquefaciens* SQR9. Mol Plant-Microbe Interact 29:324–330

Liu WH, Chen FF, Wang CE, Fu HH, Fang XQ, Ye JR, Shi JY (2019) Indole-3-acetic acid in Burkholderia pyrrocinia JK-SH007: enzymatic identification of the indole-3-acetamide synthesis pathway. Front Microbiol 10:2559

Löbler M, Klämbt D (1985) Auxin-binding protein from coleoptile membranes of corn (*Zea mays* L.) purification by immunological methods and characterization. J Biol Chem 260:9848–9853

Loper JE, Schroth MN (1986) Influence of bacterial sources of indole-3-acetic acid on root elongation of sugar beet. Phytopathology 76:386–389

Manulis S, Valinski L, Gafni Y, Hershenhorn J (1991) Indole-3-acetic acid biosynthetic pathways in *Erwinia herbicola* in relation to pathogenicity on *Gypsophila paniculata*. Physiol Mol Plant Pathol 39:161–171

Manulis S, Haviv-Chesner A, Brandl MT, Lindow SE, Barash I (2018) Differential involvement of indole-3-acetic acid biosynthetic pathways in pathogenicity and epiphytic fitness of Erwinia herbicola pv. gypsophilae. Mol Plant-Microbe Interact 11:634–642

Marulanda A, Barea JM, Azcón R (2009) Stimulation of plant growth and drought tolerance by native microorganisms (AM fungi and bacteria) from dry environments: mechanisms related to bacterial effectiveness. J Plant Growth Regul 28:115–124

Mashiguchi K, Hisano H, Takeda-Kamiya N, Takebayashi Y, Ariizumi T, Gao Y, Ezura H, Sato K, Zhao Y, Hayashi KI, Kasahara H (2019) *Agrobacterium tumefaciens* enhances biosynthesis of two distinct auxins in the formation of crown galls. Plant Cell Physiol 60:29–37

Mayak S, Tirosh T, Glick BR (1997) The influence of plant growth promoting rhizobacterium *Pseudomonas putida* GR12-2 on the rooting of mung bean cuttings. In: Ogoshi A, Kobayashi K, Homma Y, Kodama F, Kondo N, Akino S (eds) Plant growth-promoting rhizobacteria: present status and future prospects. OECD, Paris, pp 313–315

McClerklin SA, Lee SG, Harper CP, Nwumeh R, Jez JM, Kunkel BN (2018) Indole-3-acetaldehyde dehydrogenase-dependent auxin synthesis contributes to virulence of *Pseudomonas syringae* strain DC3000. PLoS Pathog 14:e1006811

Mehmood A, Hussain A, Irshad M, Hamayun M, Iqbal A, Tawab A, Khan N (2020) Yucasin and cinnamic acid inhibit IAA and flavonoids biosynthesis minimizing interaction between maize and endophyte *Aspergillus nomius*. Symbiosis 81:149–160

Mikkelsen MD, Hansen CH, Wittstock U, Halkier BA (2000) Cytochrome P450 CYP79B2 from *Arabidopsis* catalyzes the conversion of tryptophan to indole-3-acetaldoxime, a precursor of indole glucosinolates and indole-3-acetic acid. J Biol Chem 275:33712–33717

Mohabi S, Kalantar-Neyestanaki D, Mansouri S (2017) Inhibition of quorum sensing-controlled virulence factor production in *Pseudomonas aeruginosa* by Quercus infectoria gall extracts. Iran J Microbiol 9:26

Moles TM, Mariotti L, De Pedro LF, Guglielminetti L, Picciarelli P, Scartazza A (2018) Drought induced changes of leaf-to-root relationships in two tomato genotypes. Plant Physiol Biochem 128:24–31

Morffy N, Strader LC (2020) Old Town Roads: routes of auxin biosynthesis across kingdoms. Curr Opin Plant Biol 55:21–27

Morris DA, Johnson CF (1987) Regulation of auxin transport in pea (*Pisum sativum* L.) by phenylacetic acid: inhibition of polar auxin transport in intact plants and in stem segments. Planta 172:408–416

Naseem H, Bano A (2014) Role of plant growth-promoting rhizobacteria and their exopolysaccharide in drought tolerance in maize. J Plant Interact 9:689–701

Naser V, Shani E (2016) Auxin response under osmotic stress. Plant Mol Biol 91:661–672

Navarro L, Dunoyer P, Jay F, Arnold B, Dharmasiri N, Estelle M, Voinnet O, Jones JD (2006) A plant miRNA contributes to antibacterial resistance by repressing auxin signaling. Science 312: 436–439

Naveed M, Mitter B, Reichenauer TG, Wieczorek K, Sessitsch A (2014) Increased drought stress resilience of maize through endophytic colonization by *Burkholderia phytofirmans* PsJN and *Enterobacter* sp. FD 17. Environ Exp Bot 97:30–39

Neumann PM (1995) The role of cell wall adjustment in plant resistance to water deficits. Crop Sci 35:1258–1266

Neumann PM (2008) Coping mechanisms for crop plants in drought-prone environments. Ann Bot 101:901–907

Ngumbi EN (2011) Mechanisms of olfaction in parasitic wasps: analytical and behavioral studies of response of a specialist (*Microplitis croceipes*) and a generalist (*Cotesia marginiventris*) parasitoid to host-related odor. Ph.D. Dissertation, Auburn University, Auburn

Nilsen ET, Orcutt DM (1996) The physiology of plants under stress. Wiley, New York

Normanly J, Cohen JD, Fink GR (1993) Arabidopsis thaliana auxotrophs reveal a tryptophan-independent biosynthetic pathway for indole-3-acetic acid. Proceedings of the National Academy of Sciences, 90:10355–10359

O'Banion BS, O'Neal L, Alexandre G, Lebeis SL (2020) Bridging the gap between single-strain and community-level plant-microbe chemical interactions. Mol Plant-Microbe Interact 33:124–134

O'Reilly C, Turner PD (2003) The nitrilase family of CN hydrolysing enzymes - a comparative study. J Appl Microbiol 95:1161–1174

Ostin A, Ilic N, Cohen JD (1999) An *in vitro* system from maize seedlings for tryptophan-independent indole-3-acetic acid biosynthesis. Plant Physiol 119:173–178

Ostrowski M, Ciarkowska A, Dalka A, Wilmowicz E, Jakubowska A (2020) Biosynthesis pathway of indole-3-acetyl-myo-inositol during development of maize (*Zea mays* L.) seeds. J Plant Physiol 245:153082

Patten CL, Glick BR (1996) Bacterial biosynthesis of indole-3-acetic acid. Can J Microbiol 42:207–220

Patten CL, Glick BR (2002) Role of *Pseudomonas putida* indole acetic acid in development of the host plant root system. Appl Environ Microbiol 68:3795–3801

Patten CL, Blakney AJ, Coulson TJ (2013) Activity, distribution and function of indole-3-acetic acid biosynthetic pathways in bacteria. Crit Rev Microbiol 39:395–415

Peck SC, Kende H (1995) Sequential induction of the ethylene biosynthetic enzymes by indole-3-acetic acid in etiolated peas. Plant Mol Biol 28:293–301

Pereyra MA, Garcia P, Colabelli MN, Barassi CA, Creus CM (2012) A better water status in wheat seedlings induced by *Azospirillum* under osmotic stress is related to morphological changes in xylem vessels of the coleoptile. Appl Soil Ecol 53:94–97

Perley JE, Stowe BB (1966) The production of tryptamine from tryptophan by *Bacillus cereus* (KVT). Biochem J 100:169–174

Phi QT, Park YM, Ryu CM, Park SH, Ghim SY (2008) Functional identification and expression of indole-3-pyruvate decarboxylase from *Paenibacillus polymyxa* E681. J Microbiol Biotechnol 18:1235–1244

Pieterse CM, Leon-Reyes A, Van der Ent S, VanWees SC (2009) Networking by small-molecule hormones in plant immunity. Nat Chem Biol 5:308–316

Pohl M, Sprenger GA, Müller M (2004) A new perspective on thiamine catalysis. Curr Opin Biotechnol 15:335–342

Prinsen E, Costacurta A, Michiels K, Vanderleyden J, Van Onckelen H (1993) *Azospirillum brasilense* indole-3-acetic acid biosynthesis – evidence for a non tryptophan-dependent pathway. Mol Plant-Microbe Interact 6:609–6615

Ryu RJ, Patten CL (2008) Aromatic amino acid-dependent expression of indole-3-pyruvate decarboxylase is regulated by TyrR in *Enterobacter cloacae* UW5. J Bacteriol 190:7200–7208

Sadok W, Sinclair TR (2011) Crops yield increase under water-limited conditions: review of recent physiological advances for soybean genetic improvement. Adv Agron 113:313–337

Sandhya V, Ali SZ, Grover M, Reddy G, Venkateswarlu B (2010) Effect of plant growth promoting *Pseudomonas* spp. on compatible solutes, antioxidant status and plant growth of maize under drought stress. Plant Growth Regul 62:21–30

Saravanakumar D, Kavino M, Raguchander T, Subbian P, Samiyappan R (2011) Plant growth promoting bacteria enhance water stress resistance in green gram plants. Acta Physiol Plant 33: 203–209

Sarma R, Saikia R (2014) Alleviation of drought stress in mung bean by strain *Pseudomonas aeruginosa* GGRJ21. Plant Soil 377:111–126

Schütz A, Golbik R, Tittmann K, Svergun DI, Koch MH, Hübner G, König S (2003a) Studies on structure-function relationships of indolepyruvate decarboxylase from *Enterobacter cloacae*, a key enzyme of the indole acetic acid pathway. Eur J Biochem 270:2322–2331

Schütz A, Sandalova T, Ricagno S, Hübner G, König S, Schneider G (2003b) Crystal structure of thiamindiphosphate-dependent indolepyruvate decarboxylase from *Enterobacter cloacae*, an enzyme involved in the biosynthesis of the plant hormone indole3-acetic acid. Eur J Biochem 270:2312–2321

Seo M, Akaba S, Oritani T, Delarue M, Bellini C, Caboche M, Koshiba T (1998) Higher activity of an aldehyde oxidase in the auxin overproducing superroot1 mutant of *Arabidopsis thaliana*. Plant Physiol 116:687–693

Serraj R, Sinclair TR (2002) Osmolyte accumulation: can it really help increase crop yield under drought condition. Plant Cell Environ 25:331–341

Shao J, Li S, Zhang N, Cui X, Zhou X, Zhang G, Shen Q, Zhang R (2015) Analysis and cloning of the synthetic pathway of the phytohormone indole-3-acetic acid in the plant-beneficial *Bacillus amyloliquefaciens* SQR9. Microb Cell Factories 14:1–3

Shi H, Chen L, Ye T, Liu X, Ding K, Chan Z (2014) Modulation of auxin content in Arabidopsis confers improved drought stress resistance. Plant Physiol Biochem 82:209–217

Simon S, Petrášek J (2011) Why plants need more than one type of auxin. Plant Sci 180:454–460

Sinclair TR, Muchow RC (2001) System analysis of plant traits to increase grain yield on limited water supplies. Agron J 93:263–270

Sitbon F, Astot C, Edlund A, Crozier A, Sandberg G (2000) The relative importance of tryptophan-dependent and tryptophan-independent biosynthesis of indole-3-acetic acid in tobacco during vegetative growth. Planta 211:715–721

Skirycz A, Inzé D (2010) More from less: plant growth under limited water. Curr Opin Biotechnol 21:197–203

Small DK, Morris DA (1990) Promotion of elongation and acid invertase activity in *Phaseolus vulgaris* L. internode segments by phenylacetic acid. Plant Growth Regul 9:329–340

Smidt M, Kosuge T (1978) The role of indole-3-acetic acid accumulation by alpha methyl tryptophan-resistant mutants of *Pseudomonas savastanoi* in gall formation on oleanders. Physiol Plant Pathol 13:203–214

Smith S, Zhu S, Joos L, Roberts I, Nikonorova N, Dai Vu L, Stes E, Cho H, Larrieu A, Xuan W, Goodall B (2020) The CEP5 peptide promotes abiotic stress tolerance, as revealed by quantitative proteomics, and attenuates the AUX/IAA equilibrium in Arabidopsis. Mol Cell Proteomics 19:1248–1262

Sobrado P, Fitzpatrick PF (2003) Analysis of the role of the active site residue Arg98 in the flavoprotein tryptophan 2-monooxygenase, a member of the L-amino oxidase family. Biochemistry 42:13826–13832

Somers E, Ptacek D, Gysegom P, Srinivasan M, Vanderleyden J (2005) Azospirillum brasilense produces the auxin-like phenylacetic acid by using the key enzyme for indole-3-acetic acid biosynthesis. Appl Environ Microbiol 71:1803–1810

Spaepen S (2015) Plant hormones produced by microbes. In: Principles of plant-microbe interactions. Springer, Cham, pp 247–256

Spaepen S, Versées W, Gocke D, Pohl M, Steyaert J, Vanderleyden J (2007) Characterization of phenylpyruvate decarboxylase, involved in auxin production of *Azospirillum brasilense*. J Bacteriol 189(21):7626–7633

Surico G, Comai L, Kosuge T (1984) Pathogenicity of strains of *Pseudomonas syringae* pv. *savastanoi* and theri indoleacetic acid deficient mutants on olive and oleander. Phytopathology 74:490–493

Suzuki S, He Y, Oyaizu H (2003) Indole-3-acetic acid production in Pseudomonas fluorescens HP72 and its association with suppression of creeping bentgrass brown patch. Curr Microbiol 47:0138–0143

Thimann KV, Lane RH (1938) After-effects of treatment of seed with auxin. Am J Bot 25:535–543

Thomashow LS, Reeves S, Thomashow MF (1984) Crown gall oncogenesis: evidence that a T-DNA gene from the *Agrobacterium* Ti plasmid pTiA6 encodes an enzyme that catalyzes synthesis of indoleacetic acid. Proc Natl Acad Sci U S A 81:5071–5075

Timmusk S, Abd El-Daim IA, Copolovici L, Tanilas T, Kannaste A, Behers L, Niinemets U (2014) Drought-tolerance of wheat improved by rhizosphere bacteria from harsh environments: enhanced biomass production and reduced emissions of stress volatiles. PLoS One 9:e96086

Tivendale ND, Ross JJ, Cohen JD (2014) The shifting paradigms of auxin biosynthesis. Trends Plant Sci 19:44–51

Tuberosa R (2012) Phenotyping for drought tolerance of crops in the genomic era. Front Physiol 3: 10–20

Tuberosa R, Salvi S, Sanguineti MC, Maccaferri M, Giuliani S, Landi P (2003) Searching for quantitative trait loci controlling root traits in maize: a critical appraisal. Plant Soil 255:35–54

Vacheron J, Desbrosses G, Bouffaud M, Touraine B, Moenne-Loccoz Y, Muller D, Legendre L, Wisniewski-Dye F, Prigent-Combaret C (2013) Plant growth-promoting rhizobacteria and root system functioning. Front Plant Sci 4:356

Varshney RK, Pazhamala L, Kashiwagi J, Gaur PM, Krishnamurthy L, Hoisington D (2011) Genomics and physiological approaches for root trait breeding to improve drought tolerance in chickpea (*Cicer arietinum* L.). In: Costa de Oliveira A, Varshney RK (eds) Root genomics. Springer, The Netherlands, pp 233–250

Verma P, Mathur AK (2011) *Agrobacterium tumefaciens*-mediated transgenic plant production via direct shoot bud organogenesis from pre-plasmolyzed leaf explants of *Catharanthus roseus*. Biotechnol Lett 33:1053–1060

Versees W, Spaepen S, Vanderleyden J, Steyaert J (2007) The crystal structure of phenylpyruvate decarboxylase from *Azospirillum brasilense* at 1.5 Å resolution: implications for its catalytic and regulatory mechanism. FEBS J 274:2363–2375

Wakagi T, Fukuda E, Ogawa Y, Kino H, Matsuzawa H (2002) A novel bifunctional molybdo-enzyme catalyzing both decarboxylation of indolepyruvate and oxidation of indole acetaldehyde from a thermoacidophilic archaeon, *Sulfolobus* sp. strain 7. FEBS Lett 510:196–200

Wang D, Pajerowska-Mukhtar K, Culler AH, Dong X (2007) Salicylic acid inhibits pathogen growth in plants through repression of the auxin signaling pathway. Curr Biol 17:1784–1790

Wang C, Yang W, Wang C, Gu C, Niu D, Liu H, Wang Y, Guo J (2012) Induction of drought tolerance in cucumber plants by a consortium of three plant growth-promoting rhizobacterium strains. PLoS One 7:e52565

Wang B, Chu J, Yu T, Xu Q, Sun X, Yuan J, Xiong G, Wang G, Wang Y, Li J (2015) Tryptophan-independent auxin biosynthesis contributes to early embryogenesis in Arabidopsis. Proc Natl Acad Sci U S A 112:4821–4826

Wasson AP, Richards RA, Chatrath R, Misra SC, Prasad SVS, Rebetzke GJ, Kirkegaard JA, Christopher J, Watt M (2012) Traits and selection strategies to improve root systems and water uptake in water-limited wheat crops. J Exp Bot 63:3485–3498

Wightman F, Lighty DL (1982) Identification of phenylacetic acid as a natural auxin in the shoots of higher plants. Physiol Plant 55:17–24

Wightman F, Schneider EA, Thimann KV (1980) Hormonal factors controlling the initiation and development of lateral roots. II. Effects of exogenous growth factors on lateral root formation in pea roots. Physiol Plant 49:304–314

Winter A (1966) A hypothetical route for the biogenisis of IAA. Planta 71:229–239

Xie SX, Kato Y, Komeda H, Yoshida S, Asano Y (2003) A gene cluster responsible for alkylaldoxime metabolism coexisting with nitrile hydratase and amidase in *Rhodococcus globerulus* A-4. Biochemistry 42:12056–12066

Xie H, Pasternak JJ, Glick BR (1996) Isolation and characterization of mutants of the plant growth-promoting rhizobacterium Pseudomonas putida GR12-2 that overproduce indoleacetic acid. Current Microbiol 32:67–71

Yamada T (1993) The role of auxin in plant-disease development. Annu Rev Phytopathol 31:253–273

Yang S, Zhang Q, Guo J, Charkowski AO, Glick BR, Ibekwe AM, Cooksey DA, Yang CH (2007) Global effect of indole-3-acetic acid biosynthesis on multiple virulence factors of *Erwinia chrysanthemi* 3937. Appl Environ Microbiol 73:1079–1088

Yvon M, Thirouin S, Rijnen L, Fromentier D, Gripon JC (1997) An aminotransferase from *Lactococcus lactis* initiates conversion of amino acids to cheese flavor compounds. Appl Environ Microbiol 63:414–419

Zhang H, Kim MS, Krishnamachari V, Payton P, Sun Y, Grimson M, Paré PW (2007) Rhizobacterial volatile emissions regulate auxin homeostasis and cell expansion in *Arabidopsis*. Planta 226:839

Zhang A, Yang X, Lu J, Song F, Sun J, Wang C, Lian J, Zhao L, Zhao B (2021) OsIAA20, an Aux/IAA protein, mediates abiotic stress tolerance in rice through an ABA pathway. Plant Sci 308:110903

Zhao Y, Christensen SK, Fankhauser C, Cashman JR, Cohen JD, Weigel D, Chory J (2001) A role for flavin monooxygenase-like enzymes in auxin biosynthesis. Science 291:306–309

Zuljan FA, Mortera P, Alarcón SH, Blancato VS, Espariz M, Magni C (2016) Lactic acid bacteria decarboxylation reactions in cheese. Int Dairy J 62:53–62

Role of Plant–Microbe Interactions in Combating Salinity Stress

Rachna Bhateria, Rimmy, Yogita, and Sunil Kumar

Abstract In modern agriculture, a plant suffers from many abiotic factors, which include, but are not limited to, stress caused due to heavy metals, UV radiations, extreme temperature conditions, high salinity, etc. Microbial activity is drastically affected due to soil salinity. Microbial conditions, toxic effects created due to ions, and also the osmotic stress lead to diminished growth of plants. Microorganisms are bestowed with an array of metabolic capabilities to lessen abiotic stress. The interaction of plants with the microbe is an intrinsic part of an ecosystem. They are innate partners that regulate indigenous and corporal mechanisms in plants to provide defense under severe external conditions. Plant–microbe interaction involves multiplex operations, i.e., biochemical, molecular, and physiological processes within the plant cellular system. Under sustained pressure of expanding climatic adaptations, it is vital to define and interpret the relationship between plants and microbes to save plants from abiotic stresses. Plant growth-promoting rhizobacteria (PGPR) living in consociation with roots of the plant affect their yield and susceptibility. PGPR colonize the plant's rhizosphere and also aid in enhancing the growth of the plant and help reduce diseases caused due to pathogenic fungi, viruses, nematodes, and bacteria. In this chapter, the physiological and molecular responses of a plant against salinity, such as mycorrhizal fungi and plant growth-promoting bacteria (PGPB), which contribute to the growth and development of the plant under salinity stress, are discussed.

Keywords *Azospirillum* · Plant growth-promoting bacteria · Phytohormones · Salinity

---

R. Bhateria (✉) · Rimmy · Yogita · S. Kumar
Department of Environmental Sciences, Maharshi Dayanand University, Rohtak, Haryana, India

© The Author(s), under exclusive license to Springer Nature Singapore Pte Ltd. 2022
A. Vaishnav et al. (eds.), *Plant Stress Mitigators*,
https://doi.org/10.1007/978-981-16-7759-5_13

Introduction

The term salinity may be defined as "the weight in grams of the dissolved inorganic matter in 1 kg of water." This is the main abiotic stress, which has a negative outcome on the quality and yield of the crop. Globally, salinity affects about 800 MH area of land, which is more than 6% of the total global region (Munns and Tester 2008). Not a large area of land is being cultivated, but out of cultivated land, a significant area is under stress due to salinity problems. Currently, 230 MH of land is irrigated; out of that, 45 MH is tolerating the salt stress (19.5%). An area of 1500 MH comes under dryland agriculture, and an area of 32 MH bears the salinity problem (2.1%). Salinity stress is the exertion of osmotic stress on plants when grown in a saline environment. The presence of soluble salts in soil such as magnesium, sodium, potassium, bicarbonate, chloride, sulfate, calcium, and carbonate is termed salinization. The sodium chloride is present in the soils in higher percentage in comparison with other salts; moreover, the chloride ions present in combination are much toxic for the plant as they retard the growth of the plant.

A poor drainage system is the main reason behind the salinity problem. Excess salt content outside as well as inside the plant hinders plant growth. Various regions have their specific soil type, which consists of varied salt concentrations. Salinity can be defined by the "measurement of the electrical conductivity of the soil of that specific area." Soils with electrical conductivity (EC) 20 mM or more are saline or are affected with salt. Table 1 shows the units for measuring salinity.

Improper drainage system and less leaching of salts cause salt accumulation in soil. Different salts like bicarbonates, sulfates, magnesium chlorides, carbonates, calcium, sodium, and potassium are present in irrigated water. The soils with the salinity problem produce a low yield. The salinity is of various kinds, such as groundwater salinity, irrigation salinity, and transient salinity. At the global level, saline stress is one of the major environmental stress and is the most severe one that affects soil, plant yield and growth, and living beings. Salt stressed soils have a high concentration of soluble salts that are present in between the macro and micro pore spaces. For sufficient food production, there is a need to increase the area under irrigation. Overuse of groundwater has been seen to solve the problem of irrigation.

Table 1 Units for measuring salinity and conversion factors

Measurement and units	Application	1 dS/m is equal to	Equivalent units
Conductivity (dS/m)	Soils	1	1 dS/m = 1 mS/cm = 1 mmho/cm
Conductivity (μS/cm)	Irrigation and river water	1000 μS/cm	1 μS/cm = 1 μmho/cm
Total dissolved salts (mg/L)	Irrigation and river water	640 mg/L (approx.)	1 mg/L = 1 mg/kg = 1 ppm
Molarity of NaCl (mM)	Laboratory	10 mM	1 mM = 1 mmol/L

It has resulted in rising water table, which had been at several meters deep before irrigation started. It has finally resulted in increased salinity. The major problem related to salinity is that it raises the amount of soluble salts in the rhizospheric zone that influence the population nearby and also the plant productivity. Salinity reduces plant growth and causes osmotic stress and less crop productivity in arid and semiarid areas. Accumulation of excess ions affects metabolic disturbances in processes where these are required in fewer amounts. For example, the building up of chloride ions inhibits nitrate activity during the process of photosynthesis (Xu et al. 2000). Excess salt accumulation in cells causes cell death and dehydration. High salinity reduces the rate of leaf production, which results in plant death. Salinity in soils is a big problem in the field of agriculture. Arid and semiarid areas have saline soil, which becomes less capable for agriculture purposes. Thus, these areas are dependent on irrigation for their crop production, which further causes the problem of secondary salinization, which includes about 20% of the irrigated land. In soils, salts are present in the form of ions. Weathering of soils is the major source of these ions. The other sources may include fertilizers and irrigation water and many times, these ions also migrate toward the upper surface (soil) from the groundwater. Under less precipitation conditions, the ions do not leach and thus accumulate in the soils, resulting in soil salinity. Every soil type has some soluble salts, which are absorbed by the plants as plant nutrients, but too much accumulation strongly suppresses the growth of the plant. Every year, the area under affected soils is increasing day by day with the increase or introduction of irrigation. "Salinization is known to be the major threats for human beings as well as natural resources which affects approximately 1 billion ha worldwide" (Yensen 2008). Approximately, 7 MH of land in India is under salinity stress (Patel et al. 2011), especially in the region of the Indo-Gangetic Plain, which almost covers the states like Punjab, Uttar Pradesh, Haryana, and also a few parts of Rajasthan. There are area-wise differences in the soils affected with salinity problem in a continent. In regions with more precipitation, the tropical nations like South America, Australia, and Africa, salt ions like Na^+ and Cl^- ions dominate the water bodies with less salinity. Likewise in arid or dry areas/subtropics, NaCl dominates with high salinity surface waters, with higher evaporation. The imbalance of salts in inland as well as groundwater determines the exchange of nutrients between the water column and sediments (del Giorgio and Bouvier 2002). As the salinity of freshwater increases, it eliminates the biodiversity of the freshwater, which even includes the death of wetland plants and trees, and it also eliminates the fishes and stimulates deadly cyanobacterial blooms. A dead or a dying tree at the boundary line of a wetland is a good indication of a vanishing wetland.

Impact of Salinity on Plants

High salinity in soil declines crop production and ecological balance of the concerned area. Salinity results in the problem of soil erosion, lesser agricultural output, and less income generation. Salinity arises as a result of complex interactions among morphological, physiological, and biochemical processes. Approximately, every plant process like reproductive development, vegetative growth, and germination is influenced by salinity. Plants grown in saline soils suffer osmotic stress, oxidative stress, ion toxicity, and nutrient deficiency, which further results in limiting water uptake from soil. Salt accumulation in soils appreciably decreases the phosphorus (P) uptake by the plant as P ions get precipitated with Ca^+ ions. The soils that are contaminated with toxic elements may be harmful to the plants that are too much sensitive at a much low concentration of these elements. Varieties of salts act as the phytonutrients, and the higher concentration of these salts may imbalance the nutrient level of plants or the uptake of nutrients. Saline soils affect the process of photosynthesis by reducing the leaf area, stomata conductance, and chlorophyll content and also lessen photosystem II efficiency. The saline atmosphere or increase in salinity results in declining the plant growth for the reason that low soil osmotic stress, particularly ions, results in nutrition imbalances or their amalgamation. This adversely affects the development and growth of plant physiologically, biochemically, and at the molecular level (Tester and Davenport 2003). For evaluating the tolerance of plants toward salt stress, their survival and development is measured as it integrates the up- or downregulation of different functions inside plants. Osmotic equilibrium is crucial for the plant to nurture in salinity because its failure results in cell dehydration, turgidity loss, and cell death. Obstruction in delivering the photosynthetic hormones in growing tissues badly affects plant growth. As potassium ions are replaced by sodium ions and also Na^+ and Cl^- induced conformational changes in proteins, this results in ion toxicity in biochemical mechanisms. Potassium ions cannot be replaced by another metal ion, and they also act as the cofactor for different enzymes. Also, high concentrations of potassium ions are required for protein synthesis (Zhu 2002). The imbalance in the metabolic activity created due to ion toxicity leads to oxidative stress. The problem of salinity becomes more intense and affects plant growth in the reproduction phase. Recent research suggested that the salt stress in plants unfavorably affects the development and growth in plants, seedling growth, seed germination, DNA, enzyme activity, RNA, mitosis, and protein synthesis (Javid et al. 2011).

Plant Growth-Promoting Bacteria (PGPB)

Kloepper and Schroth (1978) were first to explain the PGPB (Fig. 1) as soil bacteria colonizing the plant root, followed by seed inoculation and hence boosting plant growth. *Pantoea* and *Azospirillum* were defined more recently as PGPB, which have

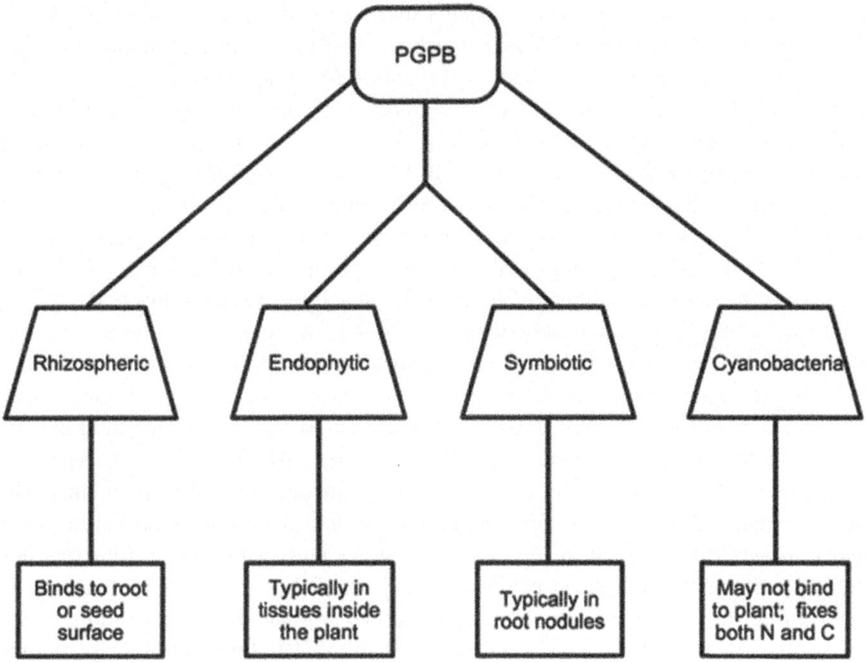

Fig. 1 Different types of plant growth-promoting bacteria (Glick 2015)

the potential for better yield and growth of plant species with ecological and agronomic importance. Microorganisms have a major role to play in plant growth promotion, nutrient balance, and controlling diseases. Microorganisms inhabit the rhizosphere of plants and support plant development via different mechanisms. Studies suggested that using PGPB is a good substitute for ease in plant stress due to salinity (Yao et al. 2010), and using microorganisms for managing abiotic and biotic stresses is gaining importance. *Azospirillum* is PGPB that results in the better yield and development of plants with ecological as well as agronomic importance. There are different mechanisms involved that stimulate the effect of PGPB on plant growth along with the potential of producing many phytohormones and vital biomolecules, nitrogen fixation, and enhancing water and mineral uptake by plants. Microbial population living in the soil is in symbiotic relationships with the plant, which helps boost up the growth and defense mechanism in plants in exchange of food resources. PGPB facilitate nutrient availability like phosphorus, nitrogen, and iron. These nutrients are crucial for plant development. Some nutrients are available in abundance, but they are in the form that plant cannot uptake. For example, nutrients like iron, nitrogen, and phosphorus are present in soils in high amounts but in the form that the plant cannot use. Therefore, these available nutrients are converted into usable form by these PGPB. In soils with higher amounts of phosphates, mostly these are available in insoluble forms or forms that cannot be utilized by the plants, thus not supporting the growth of the plant. Availability of phosphorus in limited concentration in soils and also its essentiality often limit plant growth. Plant growth-promoting fungi like mycorrhizae and PGPB have an important characteristic of soil mineralization and solubilization of phosphorus.

For stimulation of growth of plant roots and shoots, these plant growth-promoting bacteria produce plant hormones, such as gibberellins, cytokinins, and auxins in exchange to get food from the plant. PGPB have the potential to protect the plant from pathogens through microbial antagonism. PGPB are used in the production of antibiotics as well as antifungal metabolites and also out-compete pathogens for nutrients. PGPR trigger plant growth hormones, production of siderophores, antioxidant system, and enhancement of nutritional capacity in plants (Glick 2012).

Bacteria like PGPB have the potential to support the plant via induced systemic resistance (ISR) so that the plant can defend itself from the pathogens. Actually, this ISR initiates a signal itself in the plant, which activates the plant's defense system. The mechanism for plant growth by PGPB is shown in Fig. 2. In different biotic and abiotic conditions, the efficiency of PGPB varies. Abiotic factors hamper the advantageous properties and effectiveness of the introduced PGPB inoculants. The plants inoculated with PGPB generate much root hair and attain essential nutrients and minerals efficiently from the soil. Salt stress causes trouble in minerals and nutrient uptake through roots at the cell level, and this has an important role in determining the extent of salt stress tolerance by a plant. Bacteria residing in the rhizosphere boost up nitrogen concentration. Using PGPB, the roots of the plants can partly overcome the antagonism and also the membrane damage or both.

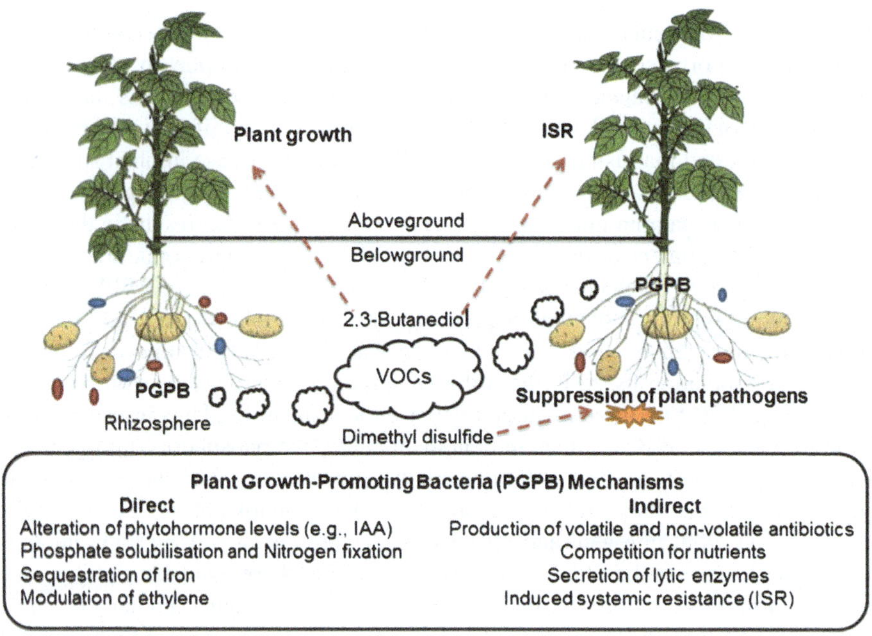

Fig. 2 Mechanism for plant growth by PGPB (Vespermann et al. 2007)

Antagonistic surroundings are harmful to the root microbiome and to the better functioning of the introduced PGPR inoculants. Research done in this area revealed that the plants affected by salinity are highly susceptible to pathogens (Triky-Dotan et al. 2005). It has been studied that salinity badly affects the colony formation of bacteria in the rhizosphere (Sato and Jiang 1996). Some salt-tolerant bacteria, because of their proliferation and their persistence specifically in semiarid regions, can cope up in the plant's rhizosphere (Egamberdieva et al. 2013). Using the technology of PGPB as an essential part of agricultural practice is the need of the hour. PGPB have been used successfully worldwide, and it is estimated that this practice will grow more. With the developing technology and the cheaper agricultural chemicals, using PGPB has a little but increasing niche in the advancement of organic agriculture. Additionally, it is rational to look forward to the increased use of PGPB for different phytoremediation practices. In different biotic and abiotic conditions, the PGPB behave inconsistently. The main characteristics of the introduced PGPR inoculants may be that they get badly affected under abiotic factors. Rigorous research is in progress to get better plant growth, accept diverse abiotic stresses, and defend plants from soil-borne pathogens by means of plant growth-promoting rhizobacteria (PGPR), which have immense capability to boost agriculture productivity. Root-related microbes, as well as endophytes, directly associate together and aid in vital physiological processes, particularly nutrient acquirement, and promote plant health under abiotic stresses (Abd Allah et al. 2015). An induced systemic resistance (ISR) in plants alongside foliar pathogens by PGPR has been reported (Choudhary et al. 2007). A number of studies recommended that a few PGPR induce systemic tolerance in plants via high antioxidant responses at the enzymatic level and metabolite accretion (Hashem et al. 2015). The antioxidant defense system allows scavenging the reactive oxygen species (ROS) for dynamic balance (Ahanger et al. 2014).

Azospirillum

Since the last 30 years, the bacterial genus *Azospirillum* has been the topic of research as it enhances crop production and also promotes growth in the plant. Much research has been done since the 1990s on the potential of this genus to increase in plant development and agricultural productivity through certain mechanisms, jointly with fixing nitrogen, to develop the overall response of the plant to inoculation. This genus is deeply studied for plant growth promotion since it possesses the capacity for plant growth and a huge arrangement of possible mechanisms. Not a single mechanism but a combination of different mechanisms participates in plant growth promotion by *Azospirillum*. The mechanism involved depends upon specific species of plants, the strain of *Azospirillum* chosen, and, most importantly, surrounding conditions during the interaction. After its discovery in the 1970s by Johana Döbereiner and others, the genus *Azospirillum*, as plant-associated bacteria (Döbereiner and Day 1976), is defined by two major properties: fixing atmospheric nitrogen and producing phytohormones. *Azospirillum* is mainly studied among PGPB, exclusive of rhizobia, and reached commercialization in many countries (Hartmann and Bashan 2009).

Three facts without any dispute are the following:

1. Different strains of *Azospirillum* help in fixing the atmospheric nitrogen, but a little amount of it reaches the plant.
2. Phytohormones are produced by several strains in vitro, but transferring hormones is limited.
3. In almost every case, a positive response has been collected for inoculating plants with the PGPB, but the result is not noticed in the case of beneficial benefit only (Díaz-Zorita and Fernández-Canigia 2009).

Inoculation of PGPB promotes root elongation. Inoculated plants absorb more minerals and water and are more vigorous and greener.

Azospirillum: Role in Plant Growth and Production of Phytohormones Under Saline Conditions

With the over-increasing demand for agricultural land, saline soils are found to be unsuitable for crop production, but they are generally cultivated by the marginal farmers who grow salt-tolerant crops. They do not use chemical fertilizers; instead, they opt for bio-fertilizers as they are effortlessly affordable and an alternative to chemical fertilizers for the improvement of crop yields. To uphold plant development in saline environments, the application of PGPR is a highly renowned advancement (Bacilio et al. 2004). The association of microbes in agriculture has opened a path to substitute or lessen the use of chemicals that were earlier used for the development of plant growth (Burdman et al. 2000; Dobbelaere et al. 2003). In agricultural biotechnology, bacteria and fungi were found to be the suppressors of pathogens and promoters of plant growth (Berg 2009). The widely considered and utilized PGPB is *Azospirillum*, with surrounding bacteria having a tremendous potential to uphold the growth of plant species (Bashan and De-Bashan 2010a, b; Fukami et al. 2016). Plant improvement is keeping pace and synchronized by the action of a number of phytohormones like brassinosteroids (BRs), gibberellins (GAs), ethylene, cytokinins (CKs), auxins (indole-3-acetic acid (IAA)), and abscisic acid (ABA) that organize numerous processes in the sessile plant. The above hormones either remain or move from respective sites of production in order to control genetic development alterations and respond to environmental stimuli (Davies 2004). Hormones therefore have a fundamental function in showing the reaction of plants toward abiotic stress so that plants can survive under harsh stressed conditions and result in retarded development in such a manner that plants can focus their resources in mitigating the stress (Skirycz and Inzé 2010). Consequently, abiotic stresses can result in changes in signal transduction related to plant growth and stress hormones and changes at the production and distribution level, which may enhance particular defensive mechanisms.

Salinity is also responsible for the lack of essential nutrients and amplification in sodium ion concentration (Zahedi et al. 2012). Phytohormones have vital functions in the augmentation, enlargement, and responses during stress (Shaterian et al. 2005a, b). Under unfavorable harsh conditions, plants have various cellular defense mechanisms. Production of phytohormones like zeatin, auxins, and gibberellins during salinity stress can be suppressed in the leaves and roots of various plants. In *Azospirillum brasilense*, nitrogenase activity and synthesis can be inhibited under saline conditions (Tripathi et al. 2002). Tripathi et al. (1998) investigated that under salinity, attuned solutes such as trehalose, proline, betaine, glycine, and glutamate get accumulated in *Azospirillum* sp. In the case of *Azospirillum brasilense*, proline has its role in osmotic stress. A rise in osmotic stress shifts dominant osmolyte from glutamate to proline. The genus *Azospirillum* helps in colonizing hundreds of plant species in order to achieve considerable productivity as well as growth and development of plants underneath agronomic environments (Bashan and De-Bashan 2010a, b). On the basis of available data, it can be shown that there is a relation between the production of hormones like cytokinins (CKs) (Tien et al. 1979), abscisic acid (ABA) (Cohen et al. 2008), ethylene (Et) (Strzelczyk et al. 1994), gibberellins (GAs) (Bottini et al. 1989), auxins (Prinsen et al. 1993), and further plant growth regulators. Generation of phytohormones is one among various plant growth-promoting activities of *Azospirillum*. Perrig et al. (2007) revealed that for 20 years in Argentina, two strains of *Azospirillum brasilense* (Cd and Az39) having different parameters and biosynthetic phytohormones were used for inoculant formulation. The concentration of abscisic acid, indole-3-acetic acid, gibberellic acid, zeatin (Z), and ethylene can be measured using GC-MS and HPLC. Therefore, indole-3-acetic acid (IAA) is the ultimate hormone produced (Hartmann et al. 1983), while some others were indole-3-butyric acid (IBA) (Falik et al. 1989), indole-3-methanol (Crozier et al. 1988), unrevealed indole compounds, gibberellins (Bottini et al. 1989), abscisic acid (ABA) (Kolb and Martin 1985), and cytokinins (Horemans et al. 1986). Studies also support that there are elevated levels of IAA and IBA production in roots of inoculated maize in comparison to uninoculated control (Falik et al. 1989). Beneficial uses in the plant due to the release of phytohormones are the following:

- Root length alterations
- Root hair enhancement (Morgenstern and Okon 1987)
- Branched root hairs (Jain and Patriquin 1984)
- Production of lateral roots
- Increased differentiation and division of meristematic tissues in cells

Production of Phytohormones

The peculiar characteristic of many microorganisms or PGPB is to form phytohormones, particularly species of *Azospirillum* that excite and smoothen the progress of plant growth (Tsavkelova et al. 2006). Besides the production of plant hormones, *Azospirillum* is capable to form the amino acids along with polyamines in culture media (Thuler et al. 2003). Among the different phytohormones, indoles, primarily indole-3-acetic acid, and gibberellins (GAs) of several kinds may play a larger role. The different strains of these bacteria alter the structure and the rate of metabolism in plants and consequently the more retention of water and of minerals as well. In *Chlorella vulgaris*, a unicellular microalga, these phytohormones produce populations with larger cells (de Bashan et al. 2008). IAA is the majorly studied plant hormone with carboxymethyl group, which is heterocyclic in nature and is associated with the family of auxin phytohormones. This plays a significant role in different mechanisms involved in the physiology of the plant. The phytohormones like auxins are the reason for the division processes in plants and also the differentiation of plant cells. Phytohormones related to auxins cause the amplification in the formation of xylem and root, vegetative growth, fluorescence, and tropism processes along with fructification of plants. It is also known to cause deleterious effects on the photosynthesis process and also on the different biological processes of metabolite production, reaction forming the pigments and resistance to biotic stress factors. Primarily to increase the metabolism and growth of microalgae of the genus *Chlorella* for wastewater treatment, in marine surroundings also *Azospirillum* is used as an inoculant chiefly (Hernandez et al. 2006). For mass production, dissolved nutrients, pH, and toxic molecules have impacts on microalgae. High pH causes the disturbance in the cell cycle of microalgae, resulting in lessening and their decreased population. Culturing the microalgae in the company of *A. brasilense* removes the negative impact (de Bashan et al. 2005). In some crops, sometimes inoculating plants in light intensity creates stress and causes an inhibitory effect. It has been observed that the amount of chlorophyll a and b was augmented after inoculation of *A. brasilense* in wheat seedlings. This resulted in more green plants without stress (Bashan et al. 2006).

Auxins

Auxins represent a cluster of chemical compounds that have the capability to enhance cell enlargement in the subapical region of the stem. They are abundantly found in nature, occurring as auxin molecules, indole-3-acetic acid (IAA). Plant hormones like amino acids and polyamines were synthesized by *Azospirillum* spp. in culture (Thuler et al. 2003). Among all hormones, gibberellins and indole acetic acid (IAA) play a vital role.

IAA

Various experiments on IAA production by different cultures of *Azospirillum* reflected that their synthesis depends on the nature of culture media and accessibility of tryptophan as a pioneer. *A. brasilense* Cd synthesized the maximum amount of IAA among the experimented strains (approx. 380 μmol/L) (El-Khawas and Adachi 1999). Moreover, it was found that pH has a considerable effect on the quantity of IAA produced (Ona et al. 2003).

Gibberellins

Azospirillum spp. were found to be beneficial for plants as they can be advantageous for the production of gibberellins. Enhancement in the density of root hairs can be a result of the application of gibberellins, which shows related outputs like *Azospirillum* inoculation. *A. lipoferum* USA 5b is a gibberellin-producing strain; when cultured in the presence of glucosyl ester or glucoside of gibberellin A20, both the conjugates were hydrolyzed. These in vitro outcomes support the assumption that promotes development in plants by *Azospirillum* inoculation, which results in an amalgamation of both gibberellin production and gibberellin-glucoside/glucosyl ester deconjugation by the bacterium (Piccoli et al. 1997). Involvement of GA A3 produced by *Azospirillum* spp. in promoting the growth of maize was also recommended (Lucangeli and Bottini 1997).

Ethylene

Ethylene is a different plant hormone that is recognized to control many processes like fruit ripening, leave abscission, etc. In addition, at an elevated concentration, ethylene exerts defoliation and causes premature senescence that ultimately leads to retardation of root and stem growth (Li et al. 2006). During the maximum time span of plant development, minimal ethylene is produced. Ethylene functions a foremost part in seed germination by breaking the seed dormancy; yet, an elevated intensity of ethylene concentration inhibits succeeding root elongation. Sometimes, an excessive amount of ethylene is produced in response to stress caused due to biological and environmental factors (Fig. 3). Overcoming ethylene levels, mainly by reducing them, can cope up with a considerable monetary loss in agriculture.

Cytokinins

Cytokinins are synthesized in distinct culture environments by numerous rhizospheric bacteria (Barea et al. 1976), including *Azospirillum* (Cacciari et al. 1989; Horemans et al. 1986). Cytokinins released through bacteria may positively or

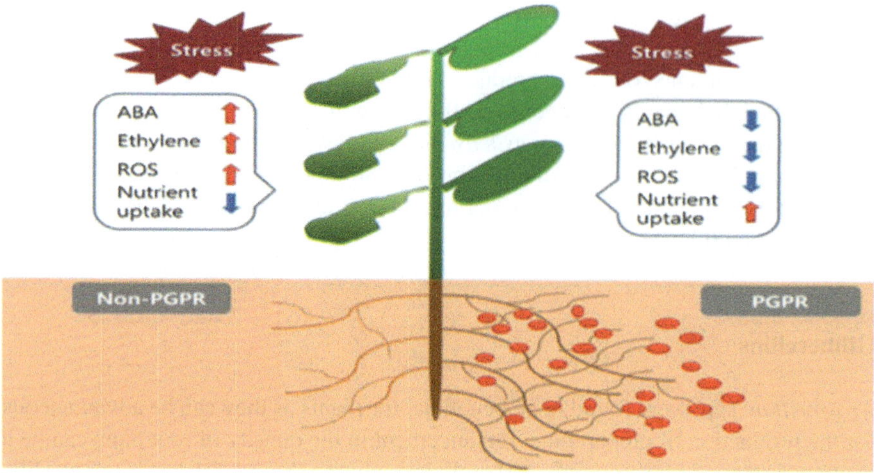

Fig. 3 PGPR mechanisms against abiotic stress (Kang et al. 2014)

negatively influence plant augmentation. Out of primary results of *Azospirillum*-inoculated plants, it is questionable if cytokinins have the ability to change the root morphology, which was pragmatic in numerous *Azospirillum* inoculation models, or auxin and gas together combine to produce the observed effect.

Improvement in Mineral and Water Uptake

Water and nutrients are necessary for the appropriate plant performance, hampering plant growth deficiency and yield production. Salinity affects vital nutrient uptake adversely; therefore, overcoming the deprived availability of soil nutrients, *Azospirillum* assists their host plant in adsorption of essential minerals, thus promoting plant growth. Adverse environmental stress conditions can affect the status of mineral nutrition and their ability to adapt to abiotic stress factors and fail to overcome stress. Plant growth, development, and yield can be altered by nutritional imbalance. Imbalances can be the consequence of salinity on nutrient accessibility, uptake, or transport inside the plant due to physiological retardation of a particular nutrient that fulfills the need of the plant for a particular nutrient (Grattan and Grieve 1994). Nutritional disorders arising from salinity can affect crop performance. From total mineral nutrients absorbed by plants, nitrogen in varied form accounts for about 80%. Nitrogen deficiency can be a growth-limiting factor (Marschner 1995). It is shown that salinity reduces phosphate accumulation and uptake in plants and crops by reducing its availability, which consequently leads to phosphate deficiency symptoms (Sharpley et al. 1992). Moreover, upholding of ample amount of K^+ is crucial for plant continued existence in saline environments (Botella et al. 1997). Similarly, in maize, inoculation of *Azospirillum* can increase the level of proline

(Kandowangko et al. 2009) and free amino acids during environmental stress (Sandhya et al. 2010). In plants, the accretion of proline is known to be the best modification induced by salinity and water stress (Verbruggen and Hermans 2008). Kandowangko et al. (2009) found that leaf proline content can be increased by inoculation of *Azospirillum* in corn.

In maximum species, root hairs (as enhanced root systems) are common phenotypic expressions, as shown by *Azospirillum* inoculation. Therefore, enhanced root expansion and function lead to better mineral and water uptake. Improved mineral uptake results in increment of root system volume (Morgenstern and Okon 1987; Murty and Ladha 1988). Inoculation may encourage accessibility of ions in the soil by serving the plant hunt to limiting nutrients (Lin et al. 1983). Due to repetitive inoculation of *Azospirillum* in plants, the plant root morphology alters and has been accredited to the synthesis of plant growth-promoting substances, mainly auxins (Spaepen et al. 2008).

In greenhouse hydroponic mechanisms, *A. brasilense* inoculation in *Sorghum bicolor* can augment the count and extent of adventitious roots by 33–40% comparatively to non-inoculated controls like high rate of growth, the former appearance of roots, and an elevated pace of individual roots (Sarig et al. 1992). However, rising (Kapulnik et al. 1981, 1985) or declining (Kucey 1988) inoculation of various root parameters can affect many aspects of foliage. The above alterations in plants can be accredited to positive bacterial outcomes on mineral adsorption; improvement in uptake of NO_3, NH_4^+, PO_4^{2-}, K^+, Rb^+, and Fe^{2+} and various other micronutrients by *Azospirillum* (Jain and Patriquin 1984; Lin et al. 1983) was anticipated by enhancement in foliar dry matter and accretion of minerals in stems and leaves. Higher mineral content and yield can be reached by transferring these minerals to the spikes and panicles during the reproductive period (Ogut and Er 2006). To support the concept of enhanced mineral uptake by inoculated roots is justified by the *Azospirillum*-inoculated wheat roots showing increased proton efflux activity (Bashan et al. 1989). It was observed that ions in plant roots should be in balance with proton efflux activity as they are directly dependent on each other (described below). Some of the researchers showed that accumulation of minerals and N is due to inoculation, while some reported that improved augmentation of wheat and soybeans was not inevitably due to a common improvement of mineral uptake (Bashan and Levanony 1990). Moreover, inoculation not only improves mineral uptake but also enhances water content in sorghum plants under abiotic stress. When inoculated plants were compared with non-inoculated ones, it was observed that inoculated plants had inferior canopy temperature, had elevated leaf water potential, had extra water in their foliage, and were under little stress. *Azospirillum*-inoculated plants can penetrate deeper soil layers to absorb water, and their extraction of soil moisture was also high. Hence, an increase in the yield of sorghum was principally attributed to the enhanced exploitation of soil moisture.

Mitigation of Salinity Stress in Plants by *Azospirillum*

Currently, agricultural scientists and farmers confront a major challenge of soil salinity. Na^+ and Cl^- ions in saline soil can be toxic and affect the microbial activities and growth of plants. The reason behind the natural or primary salinity is the increase in the level of salts in surface water or in soil. Anthropogenic salinity is due to the increased amount of dissolved salts in the soil and high irrigation processes responsible for retarded plant growth and abandoned agricultural land. Various cultivated crops like horticultural or cereal crops known for human/animal nutrition are prone to salinity stress (>4 dS/m), and it leads to an imbalance nutrition in the plant (Chinnusamy et al. 2005; Mantri et al. 2012). Studies conducted in the past reveal that seed germination is inhibited in a range of crops like wheat (Egamberdieva 2009), rice (Xu et al. 2011), maize (Khodarahmpour et al. 2012), and soybean (Essa 2002) due to salinity. Jamil et al. (2006) reported from their research that there is a significant change in the percentage and rate of seed germination, shoot length, and seedling root of paniculate amaranthus, sugar beet, pak-choi, and cabbage. In their preceding work observation, they have shown that salt concentration increase results in a reduction of shoot length (50%) and root length (7%) in bean seedling grown in xenobiotic sand system (Egamberdieva 2011). A review of literature has shown the hormonal imbalance (Prakash and Parthapasenan 1990), protein metabolism alterations (Dantas et al. 2005), nucleic acid metabolic enzyme inhibition (Arbona et al. 2005), and nutrient uptake failure due to salinity. The major reason behind these changes is the salt ion toxicity (Munns 2002; Tavakkoli et al. 2011) and the osmotic effect of soil (Shirokova et al. 2000). Therefore, structured stability of soil constituents is one of the most important factors to be considered as it controls the growth of plants.

With an increase in understanding of rhizosphere biology, a particular genus of microorganisms called a PGPR has been discovered that colonize the plant's roots and has the potential to enhance plant augmentation and expansion (Bianciotto et al. 2001). Different bacterial strains like *Rhizobium*, *Azatobactor*, *Bacillus*, *Bradyrhizobium*, *Azospirillum*, etc. have been identified from stressed environments. Inspite of developing mechanisms for their own stress tolerance, these PGPR (plant growth-promoting rhizobacteria) also provide tolerance to plants at certain levels of tolerance toward salinity stress. PGPR interaction with various crops under saline environments has reduced the poor growth and hence improved the plant survival and performance in harsh conditions. The use of root-associated bacteria interacting with plants for stress mitigation has opened a path to a newfangled and highly developed technique for mitigating salinity. Various studies conducted in the past revealed that plants can resist harsh environmental conditions like that of salinity or drought when they were inoculated with significant microbes. Therefore, these strains can lead to sustainable agricultural technologies in a stressful environment (Glick et al. 2007; Dodd and Perez-Alfocea 2012). PGP strains show direct and indirect mechanisms to combat salinity and promote plant growth. Moreover, under saline conditions, biofilm synthesized is protective against increased ruinous effects

(Kasim et al. 2016). The direct mechanisms to overcome salinity include nitrogen fixation, formation of phytohormones (e.g., cytokinin, gibberellins, ethylene, and auxin), siderophore production, and nutrient mobilization (Hayat et al. 2010). By using a range of metabolic actions and mechanisms of *Azospirillum* inoculation leads to the increased surface area, root number and root length, and hence uptake of nutrients. Alternatively, oblique pathways consist of lowering the rate of disease-causing plant pathogens. Hence, to summarize the above explanations, the overall mechanism can be categorized as induction of systemic resistance, production of lytic enzymes, phytohormones, and antifungal metabolites, rise in plant nutrient availability, and decrease in stress generated by ethylene production (Penrose et al. 2001).

With the increased application of inconsequential water for irrigation, excessive fertilizers and desert formation processes have enhanced the rate of salinity in the soil. Therefore, it has become inevitable to mention that salinity ultimately hampers the growth and development of crop plants. Extensive research has been carried out by various scientists to find alternative biological pathways for mitigating abiotic salinity stress. Inoculation of *Azospirillum* strains can give a solution to the problem. Prior findings (Bashan and Holguin 1997) stated that the widespread strains of *Azospirillum* used in agriculture can tolerate a high salinity up to 2%. The research reveals that *A. mazonense* has the lowest while *A. halopraeferans* has the highest salt resistance. *A. halopraeferans* can tolerate beyond 3% NaCl, i.e., seawater salinity. There is a significant increase in K, chlorophyll, Ca, protein contents, and soluble saccharides of maize when inoculated *Azospirillum* at NaCl concentration up to 1.2 MPa as compared to without NaCl, i.e., control maize (Hamdia and El-Komy 1997). Improvement in salinity stress in maize results from various operating mechanisms (Fig. 4); i.e., amino acid concentration increases while proline concentration decreases on exposure to NaCl or to *Azospirillum*. Thereafter, inoculation stimulates the nitrate reductase and nitrogenase activity in roots and shoots (Hamdia et al. 2004). Lettuce seed inoculated with *Azospirillum* over control has shown comparatively good germination and vegetative development in saline environments (Barassi et al. 2006). Nia et al. (2012) have observed that salinity-tolerant strains of *Azospirillum* can uphold plant development and boost grain and dry weight of wheat under saline conditions. ABA, a plant hormone, plays an important function in salinity stress in the course of acidification of apoplast in maize plant. Rhizosphere is more vulnerable to colonized microbes in comparison with the remaining areas of the soil. Therefore, these bacteria in the rhizosphere can perk up root and shoot augmentation, stress tolerance in plants, seed germination, and nutrient uptake and control various diseases. *Azospirillum brasilense* in seedlings of bean is capable of promoting root branching and enhancing the secretion of flavonoids and lipochitooligosaccharides (Dardanelli et al. 2008).

The next common cellular mechanism for adaptation or mitigation of salinity stress is an intracellular accumulation of organic solutes, i.e., osmolytes. For instance, inoculation of *A. brasilense* in sorghum can lower the adverse effects of osmotic stress (Bashan and Holguin 1997). Moreover, it was demonstrated that *A. brasilense* can tolerate up to 200 mmol/L NaCl in medium, lacking decline in

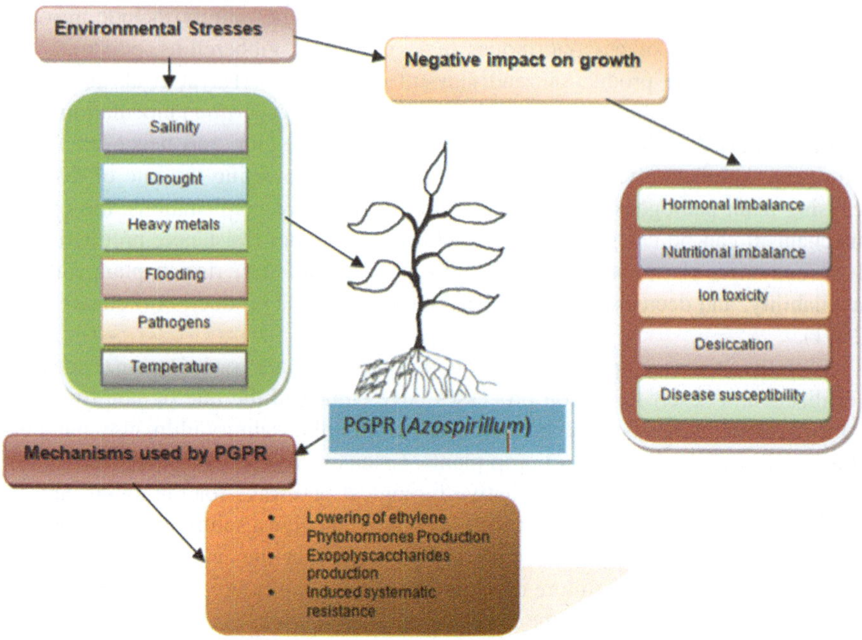

Fig. 4 Mechanisms used by *Azospirillum* (a plant growth-promoting rhizobacterium) for enhancing plant growth under stress

growth. *A. brasilense* can elevate its glutamate concentration in response to salinity stress at 24 h and K^+ at 48 h (Rivarola et al. 1998). However, it is known that *Azospirillum* sp. can mount up solutes like trehalose, glycine, glutamate, proline, and betaine to facilitate soil salinity changes (Choi and Gal 1998).

Inoculating the seeds of wheat with *A. brasilense* can help it to grow under severe salt (NaCl) and can lead to shoot elongation. In shoots of elongated plants, relative water content and fresh weight/dry weight ratio were found to be high in comparison to control under stress (Creus et al. 1997). Correspondingly, in elevated NaCl concentration, *A. lipoferum*-inoculated wheat has the ability to overcome detrimental results of NaCl (Bacilio et al. 2004). Moreover, it was observed that lettuce seeds inoculated with *Azospirillum* show better germination and vegetative development (than non-inoculated controls) on exposure to sodium chloride (NaCl) (Barassi et al. 2006). Alteration in plant root structure is the possible outcome of *Azospirillum* inoculation. Inoculation with *Azospirillum* can promote root elongation, development of lateral and adventitious root hairs, and branching of root hairs; some strains may possibly be available in many plant species, thus growing and getting a better root system. It is usually acceptable that such developmental changes in the morphology of roots are accelerated by phytohormones, perhaps supported by their allied molecules.

Conclusion

Technology expansion leads to some fruitful conclusions as *Azospirillum* inoculation paved a path for improvement in agronomic activities. The significance of *Azospirillum* under saline conditions has been discussed. Various crop plants have benefited from different strains of *Azospirillum*, which enhances the growth and development of plants. As *Azospirillum* spp. are categorized as PGPR, these species occupy the rhizosphere region of roots. Moreover, *Azospirillum* spp. synthesize various phytohormones to combat saline conditions, which enhance the mechanism and mode of actions of plant augmentation. Phytohormones like ABA, gibberellins, cytokinins, zeatin, etc. are responsible for the water and mineral uptake enhancement. *Azospirillum* spp. provide better growth of roots by branching of adventitious roots and increase the rate of meristematic cell division. Therefore, *Azospirillum* is one of the most important applications that works under saline conditions and helps plants survive under stress. A number of research studies have been put forward to support the *Azospirillum* inoculation in crops such as wheat, maize, sorghum, etc. under high-grade saline conditions resulting in growth promotion.

References

Abd Allah EF, Hashem A, Alqarawi A, Bahkali AH, Alwhibi MS (2015) Enhancing growth performance and systemic acquired resistance of medicinal plant *Sesbania sesban* (L.) Merr using arbuscular mycorrhizal fungi under salt stress. Saudi J Biol Sci 22:274–283

Ahanger MA, Hashem A, Abd Allah EF, Ahmad P (2014) *Arbuscular mycorrhiza* in crop improvement under environmental stress. In: Ahmad P, Rasool S (eds) Emerging technologies and management of crop stress tolerance, vol 2. Academic, New York, pp 69–95

Arbona V, Marco AJ, Iglesias DJ, Lopez-Climent MF, Talon M, Gomez-Cadenas A (2005) Carbohydrate depletion in roots and leaves of salt-stressed potted *Citrus clementina* L. Plant Growth Regul 46:153–160

Bacilio M, Rodriguez HM, Moreno M, Hernandez PJ, Bashan Y (2004) Mitigation of salt stress in wheat seedlings by a gfp-tagged *Azospirillum lipoferum*. Biol Fertil Soil 40:188–193

Barassi CA, Ayrault G, Creus CM, Sueldo RJ, Sobrero MT (2006) Seed inoculation with *Azospirillum* mitigates NaCl effects on lettuce. Sci Hortic 109:8–14

Barea JM, Navarro M, Montoya E (1976) Production of plant growth regulators by rhizosphere phosphate-solubilizing bacteria. J Appl Bacteriol 40:129–134

Bashan Y, De-Bashan LE (2010a) How the plant growth-promoting bacterium *Azospirillum* promotes plant growth—a critical assessment. In: Advances in agronomy, vol 108. Academic, Burlington, pp 77–136

Bashan Y, De-Bashan LE (2010b) How the plant growth-promoting bacterium promotes plant growth—a critical assessment. Adv Agron 108:77–136

Bashan Y, Holguin G (1997) *Azospirillum*-plant relationships: environmental and physiological advances Can J Microbiol 43:103–121

Bashan Y, Levanony H (1990) Current status of *Azospirillum* inoculation technology: Azospirillum as a challenge for agriculture. Can J Microbiol 36:591–608

Bashan Y, Levanony H, Mitiku G (1989) Changes in proton efflux of intact wheat roots induced by *Azospirillum brasilense* Cd. Can J Microbiol 35:691–697

Bashan Y, Bustillos JJ, Leyva LA, Hernandez JP, Bacilio M (2006) Increase in auxiliary photoprotective photosynthetic pigments in wheat seedlings induced by *Azospirillum brasilense*. Biol Fertil Soils 42:279–285

Berg G (2009) Plant–microbe interactions promoting plant growth and health: perspectives for controlled use of microorganisms in agriculture. Appl Microbiol Biotechnol 84:11–18

Bianciotto V, Andreotti S, Balestrini R, Bonfante P, Perotto S (2001) Extracellular polysaccharides are involved in the attachment of *Azospirillum brasilense* and *Rhizobium leguminosarum* to arbuscular mycorrhizal structures. Eur J Histochem 45:39–49

Botella MA, Martinez V, Pardines J, Cerdá A (1997) Salinity induced potassium deficiency in maize plants. J Plant Physiol 150(1–2):200–205. https://doi.org/10.1016/S0176-1617(97)80203-9

Bottini R, Fulchieri M, Pearce D, Pharis R (1989) Identification of gibberellins A1, A3, and Iso-A3 in cultures of *A. lipoferum*. Plant Physiol 90:45–47

Burdman S, Jurkevitch E, Okon Y (2000) Recent advances in the use of plant growth promoting bacteria in agriculture in microbe interaction in agricultural forestry. Plant Sci 2:250–299

Cacciari I, Lippi D, Pietrosanti T, Pietrosanti W (1989) Phytohormone-like substances produced by single and mixed diazotrophic cultures of *Azospirillum spp.* and *Arthrobacter*. Plant Soil 115: 151–153

Cassán F, Vanderleyden J, Spaepen S (2014) Physiological and agronomical aspects of phytohormone production by model plant-growth-promoting rhizobacteria (PGPR) belonging to the genus *Azospirillum*. J Plant Growth Regul 33:440–459

Chinnusamy V, Jagendorf A, Zhu JK (2005) Understanding and improving salt tolerance in plants. Crop Sci 45:437–448

Choi YJ, Gal SW (1998) Effects of osmoprotectants on the growth and nitrogenase activity of *Rhizobium* and *Azospirillum* under osmotic stress. Agric Chem Biotechnol 41:53–59

Choudhary DK, Prakash A, Johri BN (2007) Induced systemic resistance (ISR) in plants: mechanism of action. Indian J Microbiol 47:289–297

Cohen A, Bottini R, Piccoli P (2008) *Azospirillum brasilense* Sp 245 produces ABA in chemically-defined culture medium and increases ABA content in Arabidopsis plants. Plant Growth Regul 54:97–103

Creus CM, Sueldo RJ, Barassi CA (1997) Shoot growth and water status in *Azospirillum* -inoculated wheat seedlings grown under osmotic and salt stresses. Plant Physiol Biochem (Paris) 35: 939–944

Crozier A, Arruda P, Jasmim JM, Monteiro AM, Sandberg G (1988) Analysis of indole-3-acetic acid and related indoles in culture medium from *Azospirillum lipoferum* and *Azospirillum brasilense*. Appl Environ Microbiol 54:2833–2837

Dantas BF, Sa Ribeiro L, Aragao CA (2005) Physiological response of cowpea seeds to salinity stress. Rev Brasil Semen 27(1):144–148

Dardanelli MS, De Cordoba FJF, Espuny MR, Carvajal MAR, Diaz MES, Serrano AMG, Okon Y, Megias M (2008) Effect of *Azospirillum brasilense* coinoculated with *Rhizobium* on *Phaseolus vulgaris* flavonoids and Nod factor production under salt stress. Soil Biol Biochem 40:2713–2721

Davies PJ (2004) Plant hormones: biosynthesis, signal transduction, action. Kluwer Academic Publishers, Dordrecht

de Bashan LE, Antoun H, Bashan Y (2005) Cultivation factors and population size control uptake of nitrogen by the microalgae *Chlorella vulgaris* when interacting with the microalgae growth-promoting bacterium *Azospirillum brasilense*. FEMS Microbiol Ecol 54:197–203

de Bashan LE, Antoun H, Bashan Y (2008) Involvement of indole-3-acetic-acid produced by the growth-promoting bacterium *Azospirillum* spp. in promoting growth of Chlorella vulgaris. J Phycol 44:938–947

del Giorgio PA, Bouvier TC (2002) Linking the physiologic and phylogenetic successions in free-living bacterial communities along an estuarine salinity gradient. Limnol Oceanogr 47(2): 471–486

Díaz-Zorita M, Fernández-Canigia MV (2009) Field performance of a liquid formulation of *Azospirillum brasilense* on dryland wheat productivity. Eur J Soil Biol 45:3–11

Dobbelaere S, Vanderleyden J, Okon Y (2003) Plant growth promoting effects of diazotrophs in rhizosphere. CritRev Plant Sci 22:107–149

Döbereiner J, Day L (1976) Associative symbiosis in tropical grasses: characterization of microorganisms and dinitrogen fixing sites. In: Newton WE, Nyman CJ (eds) Proceedings first international symposium on nitrogen fixation. Washington State University Press, Pullman, pp 518–538

Dodd IC, Perez-Alfocea F (2012) Microbial alleviation of crop salinity. J Exp Bot 63:3415–3428

Egamberdieva D (2009) Alleviation of salt stress by plant growth regulators and IAA producing bacteria in wheat. Acta Phys Plant 31:861–864

Egamberdieva D (2011) Survival of Pseudomonas extremorientalis TSAU20 and P. chlororaphis TSAU13 in the rhizosphere of common bean (Phaseolus vulgaris) under saline conditions. Plant Soil Environ 57(3):122–127

Egamberdieva D, Berg G, Lindström K, Räsänen LA (2013) Alleviation of salt stress of symbiotic *Galega officinalis* L. (goat's rue) by co-inoculation of *rhizobium* with root colonising *Pseudomonas*. Plant Soil 369(1):453–465

El-Khawas H, Adachi K (1999) Identification and quantification of auxins in culture media of *Azospirillum* and *Klebsiella* and their effect on rice roots. Biol Fertil Soils 28:377–381

Essa TA (2002) Effect of salinity stress on growth and nutrient composition of three soybean (Glycine max (L.) Merrill) cultivars. J Agron Crop Sci 188(2):86–93

Falik E, Okon Y, Epstein E, Goldman A, Fischer M (1989) Identification and quantification of IAA and IBA in *Azospirillum brasilense*-inoculated maize roots. Soil Biol Biochem 21:147–153

Fukami J, Nogueira MA, Araujo RS, Hungria M (2016) Accessing inoculation methods of maize and wheat with *Azospirillum brasilense*. AMB Express 6:1

Glick BR (2012) Plant growth-promoting bacteria: mechanisms and applications. Scientifica 2012: 963401

Glick BR (2015) Beneficial plant-bacterial interactions, vol 243. Springer, Heidelberg

Glick BR, Todorovic B, Czarny J, Cheng ZY, Duan J, McConkey B (2007) Promotion of plant growth by bacterial ACC deaminase. Crit Rev Plant Sci 26:227–242

Grattan SR, Grieve CM (1994) Mineral nutrient acquisition and response by plants grown in saline environments. In: Pessarakli M (ed) Handbook of plant and crop stress, 2nd edn. Marcel Dekker, New York, pp 203–226. https://doi.org/10.1201/9780824746728.ch9

Hamdia MA, El-Komy HM (1997) Effect of salinity, gibberellic acid and *Azospirillum* inoculation on growth and nitrogen uptake of Zea mays. Biol Plant 40:109–120

Hamdia ABE, Shaddad MAK, Doaa MM (2004) Mechanisms of salt tolerance and interactive effects of *Azospirillum brasilense* inoculation on maize cultivars grown under salt stress conditions. Plant Growth Regul 44:165–174

Hartmann A, Bashan Y (2009) Ecology and application of *Azospirillum* and other plant growth-promoting bacteria (PGPB)—special issue. Eur J Soil Biol 45:1–2

Hartmann A, Singh M, Klingmüller W (1983) Isolation and characterization of *Azospirillum* mutants excreting high amounts of indoleacetic acid. Can J Microbiol 29:916–923

Hashem A, Abd-Allah EF, Alqarawi AA, Aldebasi A, Egamberdieva D (2015) Arbuscular mycorrhizal fungi enhance salinity tolerance of *Panicum turgidum* Forssk by altering photosynthetic and antioxidant pathways. J Plant Interact 10:230–242

Hayat R, Ali S, Amara U, Khalid R, Ahmed I (2010) Soil beneficial bacteria and their role in plant growth promotion: a review. Ann Microbiol 60(4):579–598. https://doi.org/10.1007/s13213-010-0117-1

Hernandez JP, de Bashan LE, Bashan Y (2006) Starvation enhances phosphorus removal from wastewater by the microalga Chlorella spp. co-immobilized with *Azospirillum brasilense*. Enzym Microb Technol 38:190–198

Horemans S, Koninck K, Neuray J, Hermans R, Vlassak K (1986) Production of plant growth substances by *Azospirillum* sp. and other rhizophere bacteria. Symbiosis 2:341–346

Jain DK, Patriquin DG (1984) Root hair deformation, bacterial attachment, and plant growth in wheat–*Azospirillum* associations. Appl Environ Microbiol 48:1208–1213

Jamil M, Lee DB, Jung KY, Ashraf M, Lee SC, Rhal ES (2006) Effect of salt (NaCl) stress on germination and early seedling growth of four vegetables species. J Cent Eur Agric 7:273–282

Javid MG, Sorooshzadeh A, Moradi F, Sanavy Seyed AMM, Allahdadi I (2011) The role of phytohormones in alleviating salt stress in crop plants. Aust J Crop Sci 5(6):726–734

Kandowangko NY, Suryatmana G, Nurlaeny N, Simanungkalit RDM (2009) Proline and abscisic acid content in droughted corn plant inoculated with *Azospirillum* sp. and arbuscular mycorrhizae fungi. Hayati J Biosci 16(1):15–20

Kang SM, Waqas M, Khan AL, Lee IJ (2014) Plant-growth-promoting rhizobacteria: potential candidates for gibberellins production and crop growth promotion. Agron Sustain Dev 34(4): 737–752

Kapulnik Y, Kigel J, Okon Y, Nur I, Henis Y (1981) Effect of *Azospirillum* inoculation on some growth parameters and N content of wheat, sorghum and panicum. Plant Soil 61:65–70

Kapulnik Y, Gafni R, Okon Y (1985) Effect of *Azospirillum* spp. inoculation on root development and NO_3 uptake in wheat (Triticum aestivum cv. Miriam) in hydroponic systems. Can J Bot 63: 627–631

Kasim WA, Gaafar RM, Abou-Ali RM, Omar MN, Hewait HM (2016) Effect of biofilm forming plant growth promoting rhizobacteria on salinity tolerance in barley. Ann Agric Sci 61(2): 217–227

Khodarahmpour Z, Ifar M, Motamedi M (2012) Effects of NaCl salinity on maize (Zea mays L.) at germination and early seedling stage. Afr J Biotechnol 11:298–304

Kloepper JW, Schroth MN (1978) Plant growth-promoting rhizobacteria on radishes. In: Station de pathologie vegetale et phyto-bacteriologie (ed) Proceedings of the 4th international conference on plant pathogenic bacteria, vol II. Gilbert-Clarey, Tours, France, pp 879–882

Kolb W, Martin P (1985) Kolb Response of plant roots to inoculation with Azospirillum brasilense and to application of indoleacetic acid. In: Klingmuller W (ed) Azospirillum III, genetics, physiology, ecology. Springer, Berlin Heidelberg New York, pp 215–221

Kucey RMN (1988) Alteration of size of wheat root systems and nitrogen fixation by associative nitrogen-fixing bacteria measured under field conditions. Can J Microbiol 34:735–739

Li YQ, Zhao HL, Yi XY, Zuo XA, Chen YP (2006) Dynamics of carbon and nitrogen storages in plant–soil system during desertification process in horqin sandy land. Huan Jing Ke Xue 27(4): 635–640

Lin W, Okon Y, Hardy RWF (1983) Enhanced mineral uptake by Zea mays and Sorghum bicolor roots inoculated with *Azospirillum brasilense*. Appl Environ Microbiol 45:1775–1779

Lucangeli C, Bottini R (1997) Effects of *Azospirillum* spp. on endogenous gibberellin content and growth of maize (Zea mays L.) treated with uniconazole. Symbiosis 23:63–72

Mantri N, Patade V, Penna S, Ford R, Pang E (2012) Abiotic stress responses in plants: present and future. In: Ahmad P, Prasad MNV (eds) Abiotic stress responses in plants: metabolism, productivity and sustainability. Springer, New York, pp 1–19

Marschner H (1995) Mineral nutrition of higher plant, 2nd edn. Academic, New York, p 889

Morgenstern E, Okon Y (1987) Promotion of plant growth and NO_3 - and Rbþ uptake in Sorghum bicolor X Sorghum sudanense inoculated with *Azospirillum brasilense*—Cd. Arid Soil Res Rehabil 1:211–217

Munns R (2002) Comparative physiology of salt and water stress. Plant Cell Environ 25:239–250

Munns R, Tester M (2008) Mechanisms of salinity tolerance. Annu Rev Plant Biol 59:651–681

Murty MG, Ladha JK (1988) Influence of *Azospirillum* inoculation on the mineral uptake and growth of rice under hydroponic conditions. Plant Soil 108:281–285

Nia SH, Zarea MJ, Rejali F, Varma A (2012) Yield and yield components of wheat as affected by salinity and inoculation with *Azospirillum* strains from saline or non-saline soil. J Saudi Soc Agric Sci 11(2):113–121

Ogut M, Er F (2006) Micronutrient composition of field-grown dry bean and wheat inoculated with *Azospirillum* and *Trichoderma*. J Plant Nutr Soil Sci 169:699–703

Ona O, Smets I, Gysegom P, Bernaerts K, Impe JV, Prinsen E, Vanderleyden J (2003) The effect of pH on indole-3-acetic acid (IAA) biosynthesis of *Azospirillum brasilense* sp7. Symbiosis 35: 199–208

Patel BB, Patel BB, Dave RS (2011) Studies on infiltration of saline–alkali soils of several parts of Mehsana and Patan districts of North Gujarat. J Appl Technol Environ Sanit 1(1):87–92

Penrose DM, Moffatt BA, Glick BR (2001) Determination of 1-aminocyclopropane-1-carboxylic acid (ACC) to assess the effects of ACC deaminase-containing bacteria on roots of canola seedlings. Can J Microbiol 47:77–80

Perrig D, Boiero ML, Masciarelli OA, Penna C, Ruiz OA, Cassán FD, Luna MV (2007) Plant-growthpromoting compounds produced by two agronomically important strains of Azospirillum brasilense, and implications for inoculant formulation. Appl Microbiol Biotechnol 75(5): 1143–1150. https://doi.org/10.1007/s00253-007-0909-9. Epub 2007 Mar 8. PMID: 17345081

Piccoli P, Lucangeli CD, Schneider G, Bottini R (1997) Hydrolysis of (17, 17–2 H2) gibberellin A20-glucoside and (17, 17–2 H2) gibberellin A20-glucosyl ester by *Azospirillum lipoferum* cultured in a nitrogen-free biotin-based chemically-defined medium. Plant Growth Regul 23: 179–182

Prakash L, Parthapasenan G (1990) Interactive effect of NaCl salinity and gibberellic acid on shoot growth, content of abscisic acid and gibberellin like substances and yield of rice (Oryza sativa). Plant Sci 100:173–181

Prinsen E, Costacurta A, Michiels K, Vanderleyden J, Van Onckelen H (1993) *Azospirillum brasilense* indole-3-acetic acid biosynthesis: evidence for a non-tryptophan dependent pathway. Mol Plant Microbe Interact 6:609–615

Rivarola V, Castro S, Mori G, Jofre E, Fabra A, Garnica R, Balegno H (1998) Response of *Azospirillum brasilense* Cd to sodium chloride stress. Antonie Leeuwenhoek 73:255–261

Sandhya V, Ali SZ, Grover M, Reddy G, Venkateswarlu B (2010) Effect of plant growth promoting *Pseudomonas* spp. on compatible solutes, antioxidant status and plant growth of maize under drought stress. Plant Growth Regul 62(1):21–30. https://doi.org/10.1007/s10725-010-9479-4

Sarig S, Okon Y, Blum A (1992) Effect of *Azospirillum brasilense* inoculation on growth dynamics and hydraulic conductivity of Sorghum bicolor roots. J Plant Nutr 15:805–819

Sato K, Jiang JY (1996) Gram-negative bacterial flora on the root surface of wheat (*Triticum aestivum*) grown under different soil conditions. Biol Fertil Soils 23:273–281. https://doi.org/10.1007/BF00335955

Sharpley AN, Meisinger JJ, Power JF, Suarez DL (1992) Root extraction of nutrients associated with long-term soil management. In: Hatfiedl JL, Stewart BA (eds) Limitations to plant growth, Adv Soil Sci, vol 19. Springer, New York, pp 151–217. https://doi.org/10.1007/978-1-4612-2894-3_6

Shaterian J, Georges F, Hussain A, Waterer D, De Jong H, Tanino KK (2005a) Root to shoot communication and abscisic acid in calreticulin (CR) gene expression and salt-stress tolerance in grafted diploid potato clones. Environ Exp Bot 53:323–332

Shaterian J, Waterer D, De-Jong H, Tanino KK (2005b) Differential stress response to NaCl salt application in early and late maturing diploid potato (*Solanum* sp.) clones. Environ Exp Bot 54: 202–212

Shirokova Y, Forkutsa I, Sharafutdinova N (2000) Use of electrical conductivity instead of soluble salts for soil salinity monitoring in Central Asia. Irrig Drain Syst 14:199–205

Skirycz A, Inzé D (2010) More from less: plant growth under limited water. Curr Opin Biotechnol 21:197–203

Spaepen S, Dobbelaere S, Croonenborghs A, Vanderleyden J (2008) Effects of *Azospirillum brasilense* indole-3-acetic acid production on inoculated wheat plants. Plant Soil 312:15–23

Strzelczyk E, Kamper M, Li C (1994) Cytokinin-like substances and ethylene production by *Azospirillum* in media with different carbon sources. Microbiol Res 149:55–60

Tavakkoli E, Fatehi F, Coventry S, Rengasamy P, McDonald GK (2011) Additive effects of Na^+ and Cl^- ions on barley growth under salinity stress. J Exp Bot 62:2189–2203

Tester M, Davenport R (2003) Na^+ tolerance and Na^+ transport in higher plants. Ann Bot 91:503–507

Thuler DS, Floh EIS, Handro W, Barbosa HR (2003) Plant growth regulators and amino acids released by *Azospirillum* sp in chemically defined media. Lett Appl Microbiol 37:174–178

Tien TM, Gaskins MH, Hubell DH (1979) Plant growth substances produced by *Azospirillum brasilense* and their effect on the growth of pearl millet (*Pennisetum americanum* L). Appl Environ Microbiol 37:1016–1024

Triky-Dotan S, Yermiyahu U, Katan J, Gamliel A (2005) Development of crown and root rot disease of tomato under irrigation with saline water. Phytopathology 95:1438–1444

Tripathi AK, Mishra MB, Tripathi P (1998) Salinity stress responses in the plant growth promoting rhizobacteria, *Azospirillum lipoferum* spp. J Biol Sci 23:463–471

Tripathi AK, Nagarajan T, Verma SC, Le Rudulier D (2002) Inhibition of biosynthesis and activity of nitrogenases in *Azospirillum brasilense* Sp7 under salinity stress. Curr Microbiol 44(5): 363–367. https://doi.org/10.1007/s00284-001-0022-8

Tsavkelova EA, Klimova SY, Cherdyntseva TA, Netrusov AI (2006) Microbial producers of plant growth stimulators and their practical use: a review. Appl Biochem Microbiol 42:117–126

Verbruggen N, Hermans C (2008) Proline accumulation in plants: a review. Amino Acids 35:753–759

Vespermann A, Kai M, Piechulla B (2007) Rhizobacterial volatiles affect the growth of fungi and *Arabidopsis thaliana*. Appl Environ Microbiol 73:5639

Xu G, Magen H, Tarchirky J, Kafkafy U (2000) Advances in chloride nutrition in plants. Adv Agron 68:97–150

Xu GY, Rocha PS, Wang ML, Xu ML, Cui YC, Li LY, Zhu YX, Xia X (2011) A novel rice calmodulin-like gene, OsMSR2, enhances drought and salt tolerance and increases ABA sensitivity in *Arabidopsis*. Planta 234:47–59

Yao L, Wu Z, Zheng Y, Kaleem I, Li C (2010) Growth promotion and protection against salt stress by *Pseudomonas putida* Rs-198 on cotton. Eur J Soil Biol 46:49–54

Yensen NP (2008) Halophyte uses for the twenty-first century. In: Khan MA, Weber DJ (eds) Ecophysiology of high salinity tolerant plants. Springer, Dordrecht, pp 367–396

Zahedi AM, Fazeli I, Zavareh M, Dorry H, Gerayeli N (2012) Evaluation of the sensitive components in seedling growth of common bean (*Phaseolus vulgaris* L.) affected by salinity. Asian J Crop Sci 4:159–164. https://doi.org/10.3923/ajcs.2012.159.164

Zhu JK (2002) Salt and drought stress signal transduction in plants. Annu Rev Plant Biol 53:247–273

Beneficial Rhizobacteria Unveiling Plant Fitness Under Climate Change

Sharon Nagpal, Kamalpreet Singh Mandahal, Kailash Chand Kumawat, and Poonam Sharma

Abstract The growing world population along with rapid industrialization has resulted in climate fluctuation leading to varied biotic and abiotic stresses affecting crop productivity. Biotic stresses, for example, pathogens and abiotic fluctuations, viz., drought, salinity, temperature, heavy metals, etc., have resulted in depletion in the crop yield and productivity at farmer's field. The effect of combined stress is known to neutralize the effect of each other proving ineffective in decreasing the crop yield. The exploration of green alternatives, i.e., plant growth-promoting rhizobacteria (PGPR) to combat the harmful effects of climate change on crops, is the recent area of research. The potential role of PGPR can assist in their employment in sustainable agriculture providing efficient methods to investigate the adaptation of plants to varied stresses and the appropriate tolerance mechanisms, which can be further harnessed to develop genetically engineered plants able to rest any climatic fluctuations. PGPR, apart from directly promoting plant growth via improved nutrient uptake, are known to confer indirect benefits to crops such as production of antioxidants, siderophores, and osmoprotectants, along with enzymes that degrade the cell wall and thus acting as potential antagonists against the diverse forms of stresses through many overlapping approaches. The modern "omics" technologies are widely utilized by researchers to understand the stress tolerance mechanisms in crops. The chapter elucidates the comprehensive role of PGPR and the allied mechanisms, which improve crop production during unfavorable climatic conditions.

S. Nagpal (✉)
Department of Microbiology, Punjab Agricultural University, Ludhiana, Punjab, India

K. S. Mandahal · P. Sharma
Department of Plant Breeding and Genetics, Punjab Agricultural University, Ludhiana, Punjab, India

K. C. Kumawat
Department of Microbiology, Punjab Agricultural University, Ludhiana, Punjab, India

Department of Industrial Microbiology, Jacob Institute of Biotechnology and Bioengineering, Sam Higginbottom Institute of Agriculture, Technology and Sciences (SHUATS), Prayagraj, Uttar Pradesh, India

© The Author(s), under exclusive license to Springer Nature Singapore Pte Ltd. 2022
A. Vaishnav et al. (eds.), *Plant Stress Mitigators*,
https://doi.org/10.1007/978-981-16-7759-5_14

Keywords Abiotic · Biotic · Climate · Omics · PGPR

Introduction

Climate can be defined as weather conditions prevailing in a particular region, recorded as air pressure, temperature, precipitation, humidity, wind speed, etc., throughout the year, averaged over a period of years. Reports published by NASA and IPCC (Intergovernmental Panel on Climate Change) state a surge of 1.14 °C in average temperatures of the globe. This seemingly small change has had dire effects on global climate and weather conditions. The worldwide occurrence of droughts and heatwaves is predicted to increase in numbers as well as severity. The Arctic Ocean could actually be ice free in the near future. Oceans comprise two-thirds of the earth, and while the level of ocean is rising, this is concomitant with an increase in temperatures of the waters. Shrinking ice sheets, retreating glaciers, and decreased snow covers are also among the major visible impacts of climate change. Weather extremes are also recorded to be more frequent in occurrence. Increased anthropological emission of greenhouse gases is the main reason for climate change phenomenon, carbon dioxide being the primary culprit. As of September 2020, CO_2 levels have spiked to 415 ppm, the highest recorded in human history.

Climatic conditions are the deciding factors behind the choice and productivity of crops in a geographical region. Crops are grown in niches best suited for their cultivation. Crop growth, development, and economic yield are highly influenced by the micro- as well macro-climate of the agroecosystems. Agricultural ecosystems are quite complex in nature owing to the interaction of various species; different ecological processes take place at distinct spatial scales along with interactions between management practices and ecological processes (Loeuille et al. 2013). On account of these complex interactions, seeming trivial changes in the climate can have drastic impacts on total biomass production, economic yields, and the quality of produce. Reduction in yields as an impact of global warming has been reported by several researchers, viz., Asseng et al. (2013), Lesk et al. (2016), Lobell and Gourdji (2012), Zhao et al. (2016), and Zhao et al. (2017).

Climatic and weather anomalies resulting from global warming have led to a spike in the occurrence of stress conditions both biotic and abiotic, individually as well as in an interactive manner; the latter have more severe implications on plant metabolism and general health (Mahalingam 2015; Mittler 2006; Prasad et al. 2011; Prasch and Sonnewald 2013; Ramegowda and Senthil-Kumar 2015). Plants are predisposed to biotic stresses, viz., diseases and pests. The situation is aggravated when certain abiotic stresses occur in concurrence with biotic ones (Coakley et al. 1999; McDonald et al. 2009; Peters et al. 2014). For instance, minor pests/pathogen of today may emerge as a major threat of tomorrow (Duveiller et al. 2007). Weed–crop competition under drought conditions is another case of abiotic–biotic stress interaction. Higher water use efficiency of weeds worsens the conditions for plants (Patterson 1995; Valerio et al. 2013; Ziska et al. 2010). Plant metabolism and physiology along with defense responses are altered under stress conditions, further exposing the plants (Scherm and Coakley 2003). However, the interactions between various stresses may not always be additive as consequences of these depend upon the nature of stresses interacting (Atkinson et al. 2013; Pandey et al. 2015; Ramu et al. 2016).

Legumes belong to the family Fabaceae, comprising 200 genera with a total of approximately 20,000 species (Lewis 2005). Out of these, pulses or grain legumes (e.g., chickpea, field pea, faba bean, soybean, etc.) are cultivated as field crops along with some fodder legumes, such as alfalfa (lucerne), clovers, etc. These play a crucial role in agroecosystem stability and serve a wide range of services for humans. Grain legumes are rich in proteins and satisfy the protein needs of vegetarian communities. Green manuring, atmospheric nitrogen fixation, improved soil structure (McCallum et al. 2004; Rochester 2011), and alteration of microbial compositions of the topsoil in favor of plants are some salient features associated with legumes. Deep-rooted legumes prevent nitrate contamination of groundwater and possess the potential to avert dryland salinity (Angus et al. 2001; Entz et al. 2001; Lefroy et al. 2001). Legumes aid in reclamation and remediation along with revegetation of degraded lands (De Faria et al. 2011; Thrall et al. 2005). Legumes form the core of sustainable farming systems owing to their capability of atmospheric nitrogen fixation. A certain amount of fixed nitrogen is also available for succeeding crops; hence, N-fertilizer demand of two crop seasons is reduced. A lower input demand of the legumes translates to a smaller carbon footprint, and thus, these crops could be at the forefront in battling climate change.

The rhizosphere is the zone of the highest degree of microbial activity surrounding the plant root system, bustling with interactions between microbes and roots. Among microbes, bacteria are the primary dwellers of the rhizosphere. Plant growth-promoting rhizobacteria (PGPR) are a subgroup of the soil bacteria and inhabit the rhizosphere, acting as plant growth promoters both directly and indirectly (Ahemad and Kibret 2014). These are root colonizers and have been proven to act in an antagonistic manner with plant pathogens in the soil. These bacteria exhibit a synergistic effect on plant metabolism and hence growth under both optimal and suboptimal conditions. These interactions are pivotal for sustainable agriculture owing to the biological nature as opposed to the use of agrochemicals. Hormonal imbalance, ion toxicity, increased pathogen/pest susceptibility, nutritional imbalance, etc. are some of the factors jeopardizing normal plant growth under suboptimal conditions. PGPR present a potential solution to alleviate these inimical influences. These bacteria aid in the regulation of nutritional as well as hormonal balance, improve nutrient availability, and have been shown to induce resistance against various pathogens.

Current climate change scenarios will exert heavy environmental stresses on the agricultural systems worldwide (Pachauri et al. 2014). With the passage of this century, complications will increase as arable lands are going to be claimed by rising oceans, desertification, salinization, erosion, etc. This implies maintenance of the same level of productions under suboptimal conditions. A system of utilizing PGPR for mitigating the adverse impacts of climate change on grain as well as other legumes and combating resulting biotic and abiotic stresses is a sustainable solution that could be pivotal in maintaining production levels, thus ensuring global food and fiber security. The chapter primarily deals with the potential role of rhizobacteria and their interaction with legumes, which could help in plant survival and achieving optimal yields under suboptimal conditions.

Different Stresses Encountered by Plants

Stress is any factor that delimits the productivity of crop plants. Under stress conditions, plants are unable to complete normal metabolic activities; as a result, the production is capped at levels lower than the genetic potential of the plant. During the period of its growth, starting from germination to maturity, plants face a variety of stresses, which are categorized into biotic and abiotic stresses. These are briefly discussed in the current section. Normal plant physiology is affected under stress conditions, which takes a toll on productivity as well as yield. Plant–pathogen/pest and plant–PGPR interactions and communications are also affected along with crop physiology under the stress conditions (Scherm and Coakley 2003).

Abiotic Stress

Nonliving factors such as water deficit, lodging, high salt concentrations, heavy metal contaminations, etc., which have an antagonistic impact on plants, are categorized under abiotic stresses. Abiotic stresses comprise any nonliving factors that have an adverse impact on the life of a specific environment. Abiotic stresses are generally more deleterious if and when the occurrence is in combinations (Mittler 2006). Water stress, temperature stress, salinity, and heavy metal contamination are major abiotic stresses. These stresses often occur in combinations with each other.

Salinity

Various factors impact plant growth among which soil and its characteristics play a crucial role. Increasing soil salinity is a primary factor leading to the decline of arable lands worldwide (Ilangumaran and Smith 2017). Any soil whose saturation extract has electrical conductivity levels above 4 dS mL^{-1} at 25 °C in the root zone along with more than 15% exchangeable Na^+ shall be classified as saline soil (Shrivastava and Kumar 2015). Approximately 15% of the arid and semiarid soils fall under the saline category, while a total of 40% of the irrigated lands are saline (Orhan 2016). An annual salinity increase of 10% has been documented, and it is projected that nearly half of the world's arable area will be impacted by salinity by the mid of twenty-first century (Jamil et al. 2012). Saline soils occur naturally as a result of gradual deposition of natural salts in soil and groundwater. Anthropogenic activities are the secondary source of causing salinity (Shrivastava and Kumar 2015). Around

800 mha of agricultural land worldwide has been affected by salinity (Yasin et al. 2018).

Excess quantity of salts in soil solution has a far-reaching and cumulative effect on plant health (Artiola et al. 2019). Metabolic irregularities are a common occurrence under saline conditions owing to ion imbalance and the associated toxic effects. High ion concentrations hinder normal osmosis leading to water stress and therefore cellular toxicity. Under hypertonic conditions, plant cells accrue ions, particularly Na^+ and Cl^-, inside the plasma membrane leading to water imbalance (Ilangumaran and Smith 2017). Salinity affects the water and mineral absorption by the roots (Jung and McCouch 2013). In leguminous plants, the plant growth, as well as nodulation and biomass, is affected adversely (Han and Lee 2005). Higher salt concentrations lead to osmotic, ionic, and oxidative stress, which hamper the attainment of the full genetic potential of plants (Rahneshan et al. 2018). Nutrient deficiency, disturbances in nitrogen and carbon assimilation pathways, decelerating rates of photosynthesis, increased generation of ROS, etc. are triggered under salt stress conditions (Hashem et al. 2016; Pan et al. 2019). Implications of salt stress on certain soil pH, rate of decomposition, availability of nutrients, soil microbial biodiversity, and availability of water are known to induce drought-like conditions (Mishra and Arora 2018; Bulgari et al. 2019). Concomitant occurrence of drought and salinity, particularly in arid and semiarid zones, has been reported by Attia et al. (2020).

Drought

Drought is among the most severe, destructive, yield-limiting, and widespread abiotic stresses around the globe. Legumes particularly are more affected by drought. Over the last four decades, drought has been responsible for around 10% yield reduction in cereals (Lesk et al. 2016). Vinocur and Altman (2005) have forecasted that as much as half of the global arable acreage is going to face water scarcity by 2050. Simply put, drought is an atmospheric or weather phenomenon categorized by lack of water availability (Shao et al. 2008; Dai 2013). A reduction in the amount of free water for plant roots resulting from low soil moisture can be defined as drought (Keyvan 2010; Dai 2013). Under dry summer conditions, the rate of transpiration from plant leaves becomes higher than the water uptake by the plant, leading to a negative water balance. Normal soil moisture allows the replenishment of water lost through transpiration. However, when water deficit occurs in soil, plants are unable to recover the lost water. Water deficit upsets the osmotic potentials and ion balances of plant cells. Also, nutrient uptake is inversely impacted as most of the plant nutrients are acquired by roots in the form of a solution aided by water uptake. The worldwide occurrence of drought has escalated in intensity in the previous decades, severely affecting the world's food production. The impact of drought on yields depends on the intensity, duration, and growth stage of the crop. Drought occurrence is common in semiarid and arid regions. Definitions of drought change depend upon the descriptive variable and fall under four categories (Wilhite and Glantz 1985):

1. Meteorological drought: No precipitation for a given time period.
2. Hydrological drought: Deficit of water in surface and subsurface resources for meeting the requirement of the water management system.
3. Socioeconomic drought: Failure of water resources to meet water demands of society immediately dependent.
4. Agricultural drought: Delimiting soil moisture adversely impacting crop production.

Temperature

Temperature range is a deciding factor behind crop growth, and yields are particularly affected by temperatures at the time of fruit set. Variations in temperature are of spatial as well as temporal nature. Temperature stresses, either high or low, are equally injurious to plant health, and the degree of inflicted damage depends upon exposure duration, rate of changes, and growth stage of the plant exposed to stress. Temperature stresses can be classified into three categories:

1. High-temperature stress (long-term exposure to >27 °C at reproductive or vegetative stage, short-term exposure to >37 °C at vegetative stage).
2. Chilling stress (0–15 °C).
3. Freezing stress (<0 °C).

Exposure of seedlings to temperatures above a threshold leads to retarded growth, necrosis, lower photosynthate assimilation, and even death. A significant reduction in O_2 evolution with an increase in temperature was recorded in *Arabidopsis* and tobacco (Murakami et al. 2000). Severe necrosis was observed in tobacco seedlings exposed to high-temperature treatments. Reproductive stages are the most vulnerable to temperature stress, and fertility is impaired at high temperatures. In rice, spikelet fertility dropped to 18–71% from 95% when exposed to 38 °C temperature (Jagadish et al. 2010). Exposure to high temperatures at the time of grain maturation has a detrimental impact on grain filling as well as quality in cereals. Grain starch in wheat was reduced upon exposure to high temperature after anthesis. This reduction was associated with a drop in amylopectin production and changes in shape, size, and structure of the starch granule (Liu et al. 2011). Increased temperatures are associated with higher evapotranspiration losses and therefore lower net photosynthate assimilation.

Chilling and freezing stresses are associated with growth retardation and pollen sterility. Degradation of starch granules in mature pollen, inhibition of pollen germination, meager elongation of pollen tubes, and pollen sterility have been documented to be induced by chilling stress (Shinada et al. 2013). Exposure to temperatures of 12 °C leads to a spike in levels of ABA in chilling-sensitive rice. The increased ABA concentrations have been correlated with the induction of pollen sterility (Ji et al. 2011). Impacts of freezing stress resemble those of water stress, i.e., dehydration and therefore osmotic stress as the available water decreases. Physical injuries in the cells are also caused by prolonged exposure to freezing temperatures due to the formation of ice crystals. Low temperatures under freezing stress lead to the denaturation of protein molecules. The integrity of lipid membranes is also affected under freezing conditions because of the altered physical state of lipid, lipid organization, and fluidity (Wolkers et al. 2007).

Contaminants

The industrial revolution and related activities such as mining have certain deleterious effects on the environment and biosphere. Over a period of time, these activities have led to a buildup of heavy metals in various terrestrial ecosystems worldwide. Heavy metals are nonbiodegradable and persistent in nature, having atomic mass higher than 20 and a density above 5 g cm^{-3}. Heavy metals are known to exert mutagenic, genotoxic, and cytotoxic effects on plants and animals alike, via tainted food chains, and soil and water sources. Heavy metal contamination is responsible for severe environmental and health problems in all living forms. Plant growth and development are also adversely affected by heavy metal contamination of the soil (Tiwari and Lata 2018). These contaminants destroy soil texture, alter pH values, and interfere with nutrient uptake by plants. Various molecular and physiological activities are adversely affected by the presence of heavy metals, invariably leading to yield penalties (Panuccio et al. 2009; Hassan et al. 2017). Certain heavy metals, namely Co, Cu, Mo, Zn, Mn, and Ni, are treated as essential elements for completion of biological processes and developmental pathways (Salla et al. 2011; Shahid et al. 2015). However, other heavy metals such as arsenic, cadmium, mercury, and lead can highly reduce crop productivity when present in supraoptimal levels (Xiong et al. 2014; Pereira et al. 2015). These elements are known to cause morphological abnormalities and metabolic disorders, as a result of which reactive oxygen species are produced. These ROS disrupt the redox homeostasis of plants, and this imbalance of redox status is the major reason behind heavy metal toxicity in plants (Gill and Tuteja 2010; Pourrut et al. 2011; Ibrahim et al. 2015; Shahid et al. 2015).

Biotic Stress

Unfavorable conditions for survival and normal growth and development of plants, which are the result of interference of any living entity, can be termed biotic stresses. The living organisms could be pathogens such as fungi, bacteria, viruses, etc., or they could be insect pests, arachnids, or weeds. Biotic stresses differ from abiotic stresses in the sense that biotic stress agents directly deprive hosts of nutrients leading to sub-normal growth and, under extreme infestations, death of the hosts. Biotic stresses are a major cause of pre- and postharvest losses in the case of agricultural commodities. Globally, 10–28%, 25–41%, 20–41%, 8–21%, and

11–32% losses have been reported in wheat, rice, maize, potato, and soyabean, respectively, due to various pests as well as pathogens (Savary et al. 2019).

Insects are one of the major biotic stress agents and are accountable for a great deal of crop losses globally. Losses due to pests are classified as either direct losses or indirect losses; the direct ones equate to the total quantity of dry matter consumed, while indirect losses are those caused by contaminating stored and processed foods. Although an expenditure of more than $30 billion annually is done on insecticides, still insects cause 30–40% damage to food crops. These are highly versatile in adaptability to man-made agroecosystems. The problem is further aggravated by monoculture and input-intensive farming, which attracts more insects and provides favorable conditions for the proliferation of insect populations. The rate of evolution of biotypes, capable of evolution against various bottlenecks such as insecticides, resistant plants, is quite fast in insects.

Plants are always exposed to a variety of potential microbial pathogens such as fungi, bacteria, oomycetes, nematodes, etc. Legumes, like all the cultivated species, are adversely affected by diseases. Viral diseases such as mosaics (yellow mosaic disease) and necrotic diseases (e.g., faba bean necrotic yellows virus (FBNYV)) could cause complete crop failures under certain conditions (Haq et al. 2010). Fungal diseases such as wilt, anthracnose, blights, rots, rusts, etc. reduce yields significantly.

PGPR and Stress Tolerance

PGPR role in facilitating plant growth and development has been clearly established. Nitrogen fixation, sequestered iron (siderophores), phosphate in a solubilized form, and phytohormone are some instances of direct stimulation of plant growth by PGPR. On the other hand, indirect stimulation can be seen as competing with and thwarting pathogenic microbes from the rhizosphere, antibiosis, and inducing systemic acquired resistance (SAR). The aforementioned functions are attained via specific enzymes that participate in certain pathways and thereby are responsible for alterations at both molecular and physiological levels. One such enzyme, 1-amino cyclopropane-1-carboxylate (ACC) deaminase, has been studied in detail, and its role in the regulation of phytohormone ethylene has been clearly elucidated. Bacterial strains having ACC deaminase can alleviate the negative impacts of ethylene to some extent (Podile and Kishore 2007).

Signal Exchange Between Plant Roots and PGPR

PGPR have been reported to produce phytohormones that control plant growth. These phytohormones also act in the capacity of molecular signals in the plant body

(Fahad et al. 2015). PGPR-produced phytohormones have been shown to improve plant response under stress conditions, thereby boosting plant health. Auxins are produced by many of the rhizobacteria (Omer et al. 2004; Gupta et al. 2015), and their role in root growth and architecture has been reported by Jha and Saraf (2015) and Vacheron et al. (2013), respectively. Auxins produced by PGPR are known to provoke changes in the transcription of the cell wall and defense-related and hormone-related genes (Spaepen et al. 2014). Synthesis of gibberellins and cytokinins by the rhizobacteria has been reported by Gupta et al. (2015) and Kumar et al. (2015). Ethylene production under stress conditions is a crucial player in deciding the stress response of plants. ACC deaminase is secreted by some PGPR, which keeps the ethylene levels below the threshold of plant growth reduction (Ahemad and Kibret 2014; Heydarian et al. 2016; Pérez-Montaño et al. 2014; Ruzzi and Aroca 2015).

The bacteria produce some other volatile organic compounds (VOCs) and secondary metabolites, which improve stress tolerance in plants. For example, the polyamine spermidine was reported to improve root characters in *Arabidopsis* (Zhou et al. 2016), and HCN (Kumar et al. 2015) produced by PGPR is documented to keep the level of harmful microbes under check. Lumichrome and riboflavin (Dakora et al. 2015), lipo-chitooligosaccharides, and thuricin 17 (Subramanian et al. 2016; Subramanian and Smith 2015; Zipfel and Oldroyd 2017) are some of the signals involved in plant–microbe communications. Root exudates such as amino acids, organic acids, and sugars along with secondary metabolites such as flavonoids, phenolic compounds, terpenes, etc. also act as signal molecules.

PGPR Boost Plant Growth Under Stress Environments

PGPR and their role in climate-resilient agriculture have gained momentum in recent years. Since PGPR are of natural origin, no restrictions or biosafety hurdles are against their use. PGPR are a sustainable and organic solution to mitigate the ill effects of various stresses on plants (Table 1).

Abiotic Stress Tolerance Associated with PGPR

Drought tolerance in chickpea was imparted by *Pseudomonas putida* MTCC5279 (Tiwari et al. 2016). The said tolerance was imparted by accumulating osmolytes, namely glycine and proline, modulation of membrane integrity, and also due to scavenging ability of reactive oxygen species produced. Similarly, when bacterial metabolite thuricin 17 (*Bacillus thuringiensis* NEB17) was used to treat soybean plants under water stress conditions, it improved root growth, nodule biomass, and therefore total nitrogen content of the plant (Prudent et al. 2015). Maize plants treated with *Alcaligenes faecalis* (AF3), when subjected to drought conditions, showed an increase of 10% in terms of root length over the control plants. The

Table 1 Plant growth-promoting rhizobacteria conferring resistance to abiotic stress in different crops

Stress	Crop	PGPR	References
Salinity	*Cicer arietinum*	*P. pseudoalcaligenes* *P. putida*	Patel et al. (2012)
		A. brasilense	Hamaoui et al. (2001)
	Glycine max	*Glomus etunicatum*	Sharifi et al. (2007)
	Oryza sativa	*Alcaligenes faecalis* *Bacillus pumilus* *Ochrobactrum* sp.	Bal et al. (2013)
		A. brasilense	Dardanelli et al. (2008)
	Triticum aestivum	*Bacillus safensis* *Ochrobactrum pseudogregnonense*	Chakraborty et al. (2013)
		Pseudomonas putida *Enterobacter cloacae* *Serratia ficaria* *P. fluorescens*	Nadeem et al. (2013)
		P. pseudoalcaligenes *B. pumilus*	Jha et al. (2013)
	Vicia faba	*A. brasilense*	Hamaoui et al. (2001)
	Zea mays	*Azotobacter chroococcum*	Rojas-Tapias et al. (2012)
Drought	*Cicer arietinum*	Consortia of *Bacillus* isolate 23-B *Pseudomonas* 6-P *Mesorhizobium ciceri*	Sharma et al. (2013)
	Glycine max	*P. putida* H-2–3	Sang-Mo et al. (2014)
	Lycopersicon Esculentum	*Bacillus polymyxa*	Shintu and Jayaram (2015)
	Oryza sativa	Consortia containing *P. jessenii*, R62, *P. synxantha*, R81, *A. nitroguajacolicus* strain YB3, strain YB5	Gusain et al. (2015)
	P. vulgaris	*A. brasilense*	German et al. (2000)
	Pisum sativum	*Variovorax paradoxus* 5C-2	Belimov et al. (2009)
		Pseudomonas fluorescens Biotype G (ACC-5)	Zahir et al. (2008)
	Triticum aestivum	ACC deaminase-producing rhizobacteria	Shakir et al. (2012)
		Bacillus thuringiensis AZP2	Timmusk et al. (2014a, b)

(continued)

Table 1 (continued)

Stress	Crop	PGPR	References
Heavy metals	*Cicer arietinum*	*Acinetobacter* sp. nbri05	Srivastava and Singh (2014)
	Medicago lupulina	*Sinorhizobium meliloti* CCNWSX0020	Kong et al. (2015)
	Triticum aestivum	*Klebsiella* sp. CIK-502	Ahmad et al. (2014)
	Zea mays	*Klebsiella* sp. CIK-502	Ahmad et al. (2014)
		Ralstonia eutropha *Chryseobacterium humi*	Moreira et al. (2014)
		Acinetobacter sp. RG30 *Pseudomonas putida* GN04	Rojas-Tapias et al. (2014)

increased root lengths associate with improved water uptake and therefore stress tolerance (Naseem and Bano 2014). An improvement of the root mass to the tune of 70% and 47% in Mazurka cultivar and 58% and 40% in Kaleo cultivar was reported by Naveed et al. (2014) when inoculated with *Burkholderia phytofirmans* (PsJN) and *Enterobacter* sp. (FD), respectively. PGPR strain 9K was reported to enhance root growth by 43.3% in maize exposed to drought conditions (Yasmin et al. 2017). Likewise, *Bacillus thuringiensis* strain AZP2 was used by Timmusk et al. (2014a, b) for inoculation of wheat plants. They reported a better root network in the treated plants under drought conditions. *Bacillus licheniformis* strain K11 was used for inoculation of peeper plants, and biomass advantage of 50% was observed over noninoculated plants when subjected to water deficit (Lim and Kim 2013). Drought tolerance associated with PGPR has been documented in quite many crops. Some examples are green gram (Saravanakumar et al. 2011), maize (Naveed et al. 2014; Naseem and Bano 2014; Sandhya et al. 2010), mung bean (Sarma and Saikia 2014), sorghum (Grover et al. 2014), sunflower (Castillo et al. 2013), and wheat (Arzanesh et al. 2011; Kasim et al. 2013).

ACC deaminase is the primary enzyme for diminishing salt stress effects. Wang et al. (2016) reported increased biomass production in pea plants upon inoculation with *Variovorax paradoxus* 5C-2 under high salt conditions. The pea plants showed higher photosynthesis, better electron transport, and lower xylem pressure and stomatal resistance. The same results were observed by Belimov et al. in 2009. Salt stress mitigating properties have also been demonstrated by *Pseudomonas putida* UW4 (Grichko and Glick 2001) and *Achromobacter piechaudii* (Mayak et al. 2004) in tomato. Halo-tolerant PGPR have gained importance as organic tools to alleviate the noxious impacts of high salt fixations and improve the development of plants, at the same time remediating the corrupted saline soils through elicitation of certain molecular and physiological mechanisms. Several halotolerant soil microbes, for example, *Alcaligenes*, *Arthrobacter*, *Azospirillum*, *Bacillus*, *Burkholderia*, *Enterobacter*, *Flavobacterium*, *Pseudomonas*, and *Rhizobium*, are known to enhance salt pressure tolerance in crops. Nautiyal et al. (2013)

inoculated rice plants exposed to saline conditions with *Bacillus amyloliquefaciens* SN13. The inoculated plants demonstrated better tolerance and growth under salt pressure. Application of *Bacillus aquimaris* to wheat seeds under salt stress resulted in improved imbibition. Higher concentrations of reducing and soluble sugars and lower Na^+ concentrations were observed for the inoculated plants, along with increased biomass production (Upadhyay and Singh 2015). Similarly, *Capsicum annuum* inoculated with *Pantoea dispersa* and *Azospirillum brasilense* exhibited improved root dry weight under saline conditions (del Amor and Cuadra-Crespo 2012). Trehalose 6-phosphate gene overexpression has been associated with *Paenibacillus polymyxa* and *Rhizobium tropici* inoculations in *Phaseolus vulgaris* (Figueiredo et al. 2008). This overexpression led to an increase in nodulation and therefore improved N content correlated with yield benefits.

Improved low-temperature stress tolerance was reported in potatoes by inoculation of *Burkholderia phytofirmans* (Bensalim et al. 1998). Likewise, Barka et al. (2006) also used the same bacterium for the treatment of the Chardonnay cultivar of grapes. They observed improved physiological activity and plant growth under low (4 °C) as well as ambient (26 °C) temperatures.

Maize inoculated with *Bacillus polymyxa* strain BcP26, *Mycobacterium* strain *phlei* MbP18, and *Pseudomonas alcaligenes* strain PsA15 exhibited healthier plants and improved nutrient uptake under heavy metal contaminations (Egamberdiyeva 2007). Potters et al. (2007) reported the implications of *Klebsiella mobilis* strain CIAM 880 in bioremediation of heavy metal lands. The microbe was able to reduce cadmium content to half along with a yield increase of 120% in barley when cultivated under cadmium contamination. Burd et al. (1998) reported the beneficial effect of *Kluyvera ascorbata* strain SUD165 on *Brassica napus* cultivated under heavy metal contaminations. Inoculation of Indian mustard by *E. coli* triggered GSH synthetase overexpression in the cytosol. Increased tolerance and accumulation of Cd were achieved due to overexpression of the enzyme (Mosa et al. 2016). Transformed and encapsulated ACC deaminase bacterium was effective in promoting the growth of canola in creosote-contaminated soils (Reed and Glick 2005). *Pseudomonas* sp. A3R3, which is a Ni-resistant bacterium, reportedly improved *Brassica juncea* growth when cultivated under soils high in Ni (Ma et al. 2011).

Biotic Stress Tolerance Associated with PGPR

The fungus *Rhizoctonia solani* is controlled by the bacterium *Bacillus amyloliquefaciens* (SN13) by means of heightened defense response (Srivastava et al. 2016). *Bacillus* spp. were used to inoculate cotton (*Gossypium hirsutum*) plants, and the inoculation resulted in increased jasmonic acid production and gossypol with a corresponding reduction in feeding by *Spodoptera exigua* larvae (Zebelo et al. 2016). Viral resistance induction has been reported by inoculation with *Enterobacter asburiae* BQ9 against tomato yellow leaf curl virus (Li et al. 2016). Kumar et al. (2016) have reported that *Peanibacillus lentimorbus* B-30488 inoculations in tobacco cv White barley resulted in decreased RNA accumulation of cucumber mosaic virus by 91% as compared to control. A concomitant surge in pathogenesis-related and stress tolerance gene expression was also reported.

The microbial ACC deaminase is reported to incite resilience against *Scelerotium rolfsii* responsible for southern curse sickness of tomato. Inoculated plants exhibited balanced ethylene pathway along with systemic resistance induction, as reported by Dixit et al. (2016). *P. putida* IsoF and *Serratia liquefaciens* MG1, which produce acyl-homoserine lactones (AHL), were proven to induce systemic resistance against *Alternaria alternata* in tomato. This induction is a result of acyl-homoserine lactone production by the said microbe, and a lower degree of ISR is observed in AHL-null strains (mutant) (Schuhegger et al. 2006). Teplitski et al. (2000) reported that AHL mimicking exudates from plant roots also promote beneficial microbes. The bacterium *Bacillus thuringiensis* has been used for insecticidal properties. Recently, use of *Bacillus sphaericus* and *Bacillus pumilus* for controlling white grubs was demonstrated by Coy et al. (2019). HCN production by *Pseudomonads* that showed antagonism against *Helicoverpa* was reported by Patel et al. (2010).

Synergistic effects of *Glomus*, fluorescent *Pseudomonas*, and *Trichoderma* in controlling *Fusarium* wilt of potato were reported by Srivastava et al. (2010). Similarly, defense-related gene expression in *Cucumis sativus* was observed by Alizadeh et al. (2013) under the influence of the bacterium *Pseudomonas* sp. Ps14 and fungal microbe *Trichoderma harzianum* strain Tr6. Kloepper et al. (1980) reported that *Pseudomonas putida* B10 produces a siderophore "Pseudobactin," which showed inhibitory action against *Fusarium oxysporum*. The potential of ACC deaminase against pathogens has been explored by many researchers. Yuquan et al. (1999) isolated ACC deaminase bacterium, which exhibited a high degree of antagonism against *Fusarium oxysporum*. Bacterial biocontrol agents were genetically transformed with *P. putida* strain UW4 for the addition of the ACC deaminase gene. The transformed bacteria were more effective at curbing *Pythium ultimum* growth and development (Wang et al. 2000). Similar reports of the enzyme in countering pathogen pressure have been documented by Donate-Correa et al. (2005): *P. fluorescens* acting antagonistically against *F. oxysporum* and *F. proliferatum* in vivo. Likewise, Pandey et al. (2005) observed antagonism against *R. solani* and *S. sclerotium* by an endophytic bacterium *Burkholderia* sp. An enhanced degree of resistance against both pathogen and drought was imparted to the *Arabidopsis* by symbiotic relations with the bacterium *Paenibacillus polymyxa* (Timmusk and Wagner 1999). Also, in addition to salt resistance, *Fusarium* and *Blumeria* infections were also countered in barley by *Piriformospora indica* (Waller et al. 2005).

Strategies to Alleviate Biotic and Abiotic Stresses

Interaction with beneficial plant microbes boosts the strength, development, availability, and assimilation of nutrients and enhances the competence of plants against several phytopathogens (Patel et al. 2017). Bioagents produce several hormones such as auxins, cytokinin, gibberellins, and ACC deaminase, which can act to confer benefits to the crops by either growth stimulation or by resisting the stress conditions. Recent studies have led to the development of resistance to stress in plants under harsh environments by beneficial interactions with rhizospheric bioagents due to their positive effect on plant growth. Many research groups have highlighted the strategies for alleviating stress-induced detrimental effects on plants (Evelin et al. 2009; Saharan and Nehra 2011; Nadeem et al. 2014; Nagpal et al. 2020). For instance, the application of ethylene inhibitors like amino silver ion (Ag^{+}), cobalt ion (Co^{2+}), and ethoxy vinyl glycine (AVG) reduced high levels of ethylene produced under stress and in turn enhanced the plant growth (Nadeem et al. 2014). However, these expensive agrochemicals are toxic to humans, and also the persistence of these chemicals affects soil fertility and health. Despite increased yield, the utilization of such chemicals resulted in diminishing net income of the farmers (Ahmadi et al. 2009). Other adverse effects such as root desiccation under drought or specific ion toxicity due to salinity may not be controlled by the usage of these chemicals.

PGPR influence crop yield by interacting with plant roots by varied direct and indirect mechanisms. Unique properties of microbes such as tolerance to extreme environmental stresses, their genetic complexity, and synergistic effect with crop plants can be explored. Agricultural production can be enhanced by understanding the adaptation and response mechanisms for stress tolerance, which could then also be engineered into the plants.

Plant Adaptations

Resistance is defined as the ability to survive and sustain growth during periods of stress. Prolonged stress conditions in the vicinity of the plants trigger the accumulation of reactive oxygen species (ROS), which act as markers for the harmful effect encountered by plants. ROS can damage cells through lipid and protein oxidation of lipids and proteins, damaging nucleic acids, chlorophyll bleaching, and eventually death of the cell (Apel and Hirt 2004). Enzymes like ascorbate peroxidase, catalase, glutathione reductase, and superoxide dismutase are produced in plants as machinery for self-defense (Abdel Latef 2011; Ashraf and Ali 2008). Tolerant species reported higher concentrations of antioxidative enzymes, which play a significant part in providing resistance to plant against environmental stress (Gill and Tuteja 2010). Antioxidants such as cellular redox buffers, glutathione, carotenoids, tocopherols, flavonoids, ascorbate, etc. are nonenzymatic in nature (Apel and Hirt 2004). Alterations in soil water capacity due to high salt concentrations in salinity stress leads to nutritional and hormonal imbalance in plants. In such situations, the plant body accumulates solutes of both organic and inorganic nature like proline, trehalose, polyols, and glycine betaine to protect it from the stress-induced lethal effects. The plant undergoes osmotic adjustments, limits water loss, and dilutes the toxic ion concentrations to combat harsh environments (Ashraf et al. 2013). Furthermore,

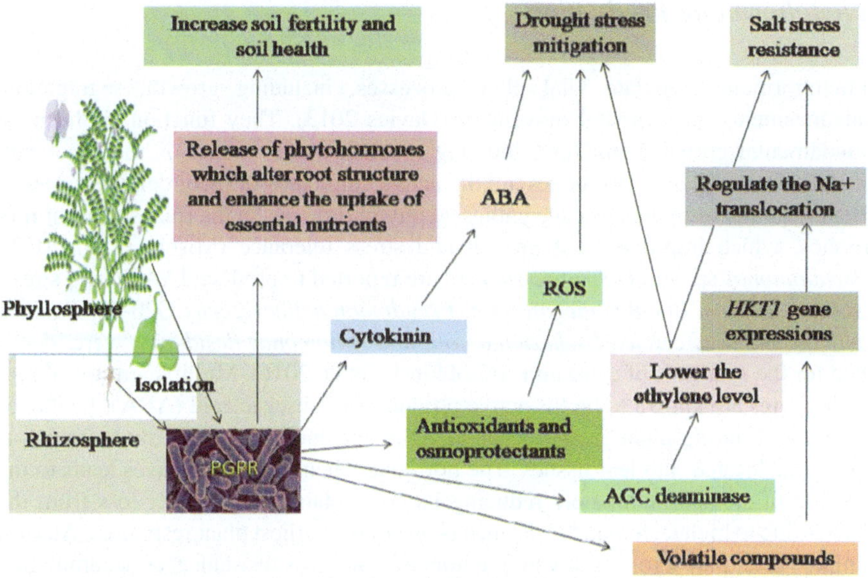

Fig. 1 Mechanism of plant growth promoting rhizobacteria to counter salinity and drought stress with improved soil health. *ABA* abscisic acid, *ROS* reactive oxygen species

construction of guarding barriers such as viscous materials like exopolysaccharides, lignin and suberin, release of antimicrobial phytoalexins, induction of hydrolytic enzymes, hypersensitive reactions, etc. also contribute to plant protection (Gonzalez-Teuber et al. 2010).

Plants have adapted many different mechanisms to cope with drought stress, including adaptations leading to altered morphology, water capacity optimization, osmotic modification, and induction of antioxidants, which have the potential to combat the effect of ROS and to unregulate the expression profiles of stress-response genes (Chaves et al. 2003; Farooq et al. 2009; Huang et al. 2014). These adaptations of plants against drought stress are divided into three groups: (1) drought escape, where plant undergoes dormancy before the onset of any drought condition (Farooq et al. 2009); (2) drought prevention and maintaining flexible phenotype by sustaining the water requirements throughout the stress period by minimizing water loss by transpiration or increasing water uptake from soil (Blum 2005); and (3) tolerance to drought, which signifies optimal plant activity even under harsh environments, which takes place by either the processes such as osmotic modifications, inducing high root viability, stabilization of membranes, and the accumulation of stress response proteins and osmolytes, which in turn enhance the structural stability of plant cells (Huang et al. 2014). Many plant metabolites are the fundamental arsenals of defense to varied stresses (Fig. 1). Systemic acquired resistance (SAR) induction in plant systems is brought about by pathogenic microorganisms. Pathogen-related (PR) proteins and salicylic acid are thought to be correlated.

Phytohormone Production

Phytohormones regulate vital plant processes, including growth, segregation, advancement, and stomatal movement (Davies 2013). They function as the most fundamental growth regulators, showing a considerable impact on the secondary metabolism of plants. Hormones are the active stimulants of plant defense systems. Beneficial microorganisms can synthesize phytohormones in the rhizosphere or root tissues, which improve plant growth and stress tolerance (Mishra et al. 2017). *Pseudomonas* sp. and *Ochrobactrum* sp. are reported to produce IAA during stress. In other studies, *Bacillus megaterium*, *Pseudomonas fluorescens* G20-18, *Bacillus subtilis*, *Bacillus cereus*, *Escherichia coli*, and *Halomonas desiderata* were identified for the synthesis of cytokinins (Großkinsky et al. 2016). Multiple groups of root endophytes are known to be the active producers of abscisic acid (ABA). Incidence of drought during plant growth leads to ABA accumulation in the shoot, which is later translocated into leaf tissues. The hormone when present in leaves leads to the closure of stomata, ultimately reducing its conductance. Less water loss from the leaves due to higher stomatal resistance is one of the earliest plant responses. Auxins, unlike ABA, play a role as downregulators of water loss resistance, as accumulation of hormones such as IAA and ABA results in late embryogenesis. Cytokinin (zeatin) and gibberellin (GA3) levels fall rapidly when water deficit hits (Vandana et al. 2020). Lwin et al. (2016) found that application of IAA-producing bacteria to maize seedlings considerably enhanced the fresh root weight and length, dry root weight, and the number of adventitious roots under water deficit conditions. The functions mediated by PGPR are beneficial for the plants to survive under diverse environmental stresses (Paul and Lade 2014; Oosten et al. 2017). Thus, the IAA-producing rhizobacteria can be used to considerably enhance the yield of various crops under stressed conditions.

ACC Deaminase Production

1-Aminocyclopropane 1-carboxylate (ACC) in the plant roots is metabolized by ACC deaminase into α-ketobutyrate and ammonia, ultimately monitoring the ethylene levels, which if increased in cells may inhibit the growth of plants. Ethylene is known to function in fruit ripening and abscission of leaves in plants (Kumar et al. 2019). Plants are known to synthesize ACC when exposed to varied biotic and abiotic stresses (Nagpal et al. 2019). PGPR having ACC deaminase activity that colonize plant roots are reported to show high tolerance to varied stresses (Naveed et al. 2008). Any kind of biotic stress encountered by plants leads to ethylene induction and activates certain stress response pathways so that no change takes place in the plant productivity and optimum growth is attained (Saleem et al. 2007). An abrupt increase in the ethylene concentrations is sensed by the appropriate cellular receptors, which late switch on the cellular machinery against the stress

(Jouyban 2012). ACC deaminase and nitric oxide-producing bacteria like *Azospirillum* are involved in changing the root morphology in plants by reducing ethylene level and functioning as the stress markers (Molina-Favero et al. 2008). Important plant physiological pathways such as signal transduction are under ethylene regulation. Plant synthesizes proline under biotic stresses, which facilitates the stabilization of subcellular structures by scavenging of free radicals and osmotic adjustment (Kumar et al. 2019). However, in the pepper plant, treatment with *Arthrobacter* and *Bacillus* sp. caused proline accumulation even under no stress conditions (Sziderics et al. 2007).

Bacterial species growing on nitrogen as the sole source of nutrients are known to express ACC deaminase gene (Glick 1995). Extensive research on the phylogeny of all the discovered *acdS* genes (Hontzeas et al. 2005) revealed that various genes encoding ACC deaminase evolved by means of horizontal gene transfer (HGT). Blaha et al. (2006) reported that ACC deaminase was horizontally transferred to bacteria from Proteobacteria. A phylogenetic analysis on the *acdS* gene in proteobacteria revealed the evolution of *acdR* from *acdS* by HGT (Prigent-Combaret et al. 2008). Singh et al. (2015) documented that wheat plants exhibited salinity tolerance and ACC deaminase activity after inoculation with *Klebsiella* sp. Similarly, canola plants when inoculated with *Pseudomonas putida* showed increased growth and also ACC deaminase activity, as stated by Li and Glick (2001). The plants inoculated with *Enterobacter cloacae* HSNJ4 showed prominent tolerance to stress under saline conditions. The bacteria-mediated digestion of ACC into α-ketobutyrate and ammonia ultimately lowered the production of ethylene in the plant roots (Saleem et al. 2015). Bacterial isolates from the tomato rhizosphere showed plant growth-promoting activities even under saline conditions. These isolates were namely *Pseudomonas fluorescens*, *Pseudomonas stutzeri*, and *Pseudomonas aeruginosa*. These strains produced various plant hormones including ACC deaminase, which helped the plant to thrive the stress conditions (Bal et al. 2013). Tittabutr et al. (2015) found that a PGPR *Sinorhizobium* sp. BL3 is capable of producing ACC deaminase and positively effecting the symbiosis and nodulation in the mungbean. However, the absence of this enzyme activity accelerates nodule senescence, although its presence had a neutral effect on the lifespan of nodules. Saleem et al. (2015) reported that rhizobacteria are capable of producing ACC deaminase, viz., species of *Rhizobium*, *Pseudomonas*, *Rhizopus*, and *Aspergillus* helped velvet bean to combat water stress. Saikia et al. (2018) studied the effect of ACC deaminase producing PGPR and *Rhizobium* and observed augmented biomass and germination count, along with root and shoot length under induced drought.

Altered Rooting and Shoot Structures

Rooting architecture is an important trait conferring crop endurance to varied stresses (Huang et al. 2014). Since initial contact with any fluctuation occurring in the soil is of the roots, the root topology has a major role to play in the resistance strategies. Characteristics such as root arrangement, number, and length are detrimental to plant survival under stress. It integrates with root system architecture (Vacheron et al. 2013). Plasticity exhibited by plant roots aids them in combating unfavorable conditions through alterations in both physical and chemical attributes (Tuberosa 2012; Yu et al. 2007). Increased roots with less diameter and deep branching are some of the attributes upholding plant productivity under stress conditions (Comas et al. 2013). In many crops like soybean (*G. max*) (Sadok and Sinclair 2011), chickpea (*C. arietinum* L.) (Varshney et al. 2011), maize (*Z. mays*) (Hund et al. 2011), and wheat (*T. aestivum* L.) (Wasson et al. 2012), a profound and productive root system is correlated with tolerance to drought. Plants with increased root number with a small diameter under drought stress show increased conductance of water by improving the surface area in soil vicinity and by expanding their reach into the large volume of soil (Comas et al. 2013). Plants with deeper and prolific root systems show greater tolerance to drought stress compared to the plants with fewer roots (Gowda et al. 2011). PGPR-treated plants have been reported to alter the root architecture and promote root growth (Ngumbi 2011). Furthermore, the researchers have reported that the root structural alterations induced by beneficial bacteria lead to an increased total root surface area along with improved nutrient and water uptake, with a positive effect on plant growth (Timmusk et al. 2014a, b). Naseem and Bano (2014) documented that maize seeds treated with the *Alcaligenes faecalis* (AF3) showed 10% increase in root length under drought stress after 3 weeks of planting under pot conditions. PGPR aid in tolerance against drought stress via the alteration and enhancement of root parameters. Extensive studies focusing on the interrelation between improvements in root structures mediated by PGPR under drought leading to the resilience in plants need to be undertaken. Pertinent studies related to root structure modeling in soil needs to be undertaken in the case of drought stress (Dunbabin et al. 2013).

Inhibition of shoot growth is one of the key responses to drought stress, which benefits plants by limiting the leaf area available for water loss through evaporation (Inzé 2010). Also, retardation in the growth of shoot causes redirection of vital solutes to the region encountering stress rather than the growth regions. Inhibition of shoot growth is an efficient strategy that aids in providing resistance to stresses, especially drought; hence, it is considered an adaptive strategy (Aachard et al. 2006). But major limitations to this resistance strategy are the depletion in plant size and ultimately yield of the crop (Claeys and Inzé 2013). In the case of crop plants that are under moderate drought conditions, inhibition of shoot growth is a counter-productive response. There may not be any ill effect on the survival of plant, but the yield would be compromised if using this strategy. Thus, it is advantageous for the development of crop varieties that are able to maintain almost normal shoot growth rates during moderate drought (Neumann 2008). PGPR-treated plants typically increase shoot growth. Consequently, under drought stress, increased crop productivity is observed in plants treated with beneficial strains of PGPR. For example, Vardharajula et al. (2011) reported that corn plants inoculated with plant growth-promoting rhizobacteria like *Bacillus* spp. improved shoot growth and plant biomass under drought stress. In drought stress conditions, PGPR-treated wheat showed 78% increased plant dry weight than control plants (Timmusk et al. 2014a, b).

Osmotic Adjustments and Relative Water Content

In plant leaves, relative water content (RWC) is considered one of the best measurement metrics of the plant water status since it participates in the crucial metabolic activities of plants. Reduction in RWC leads to turgor loss and diminished cell growth and reduced plant development (Castillo et al. 2013). Species best suited to dry conditions have elevated RWCs (Jarvis and Jarvis 1963). The rise in relative water content must therefore be regarded as a significant strategy for increasing drought tolerance. Relative water content can thus be a crucial parameter when screening the PGPR for their drought tolerance. In fact, multiple studies examining PGPR and their ability to make plants resilient to drought have screened the RWC in inoculated and un-inoculated plants and have reported that PGPR inoculation is known to increase water content in plants over control plants. For instance, Grover et al. (2014) observed 24% improved relative water content with the inoculation of *Bacillus* sp. KB 129 in drought conditions in sorghum. Similar findings have been reported by Bano et al. (2013) and Naveed et al. (2014) in maize.

Drought resilience is brought about by osmotic adjustment in the cellular structures protecting the enzymes, proteins, and organelles from harsh conditions (Huang et al. 2014). Vigorous increase in solutes of organic and inorganic origin (compatible solutes), in response to drought stress, is an osmotic adjustment process occurring in cells (Kiani et al. 2007). These osmolytes (betaine, sucrose, polyols, organic acids, and nonprotein amino acids) help in maintaining the cell turgor and reduce the water potential in the plants without affecting the actual water content in the internal tissues. Proline is one of the major osmolyte acting as an osmotic modifier in plants during drought stress (Huang et al. 2014). It helps stabilize vital cellular structures such as proteins and membranes and free radicals and aid in cell redox buffering (Hayat et al. 2012). Many research groups have documented an upsurge in proline concentration during drought stress in pea (*Pisum sativum* L.) (Alexieva et al. 2001), chickpeas (*Cicer arietinum* L.) (Mafakheri et al. 2010), rice (*Oryza sativa* L.) (Lum et al. 2014), and soybean (Silvente et al. 2012).

Osmotic Balance and Ion Homeostasis

The water capacity has a direct impact on hydraulic conductivity and plant transpiration during stress conditions (PGPR), which ultimately affects the stomatal opening. The *Bacillus megaterium*-infected maize plant showed increased root water conductivity as opposed to the noninoculated plants by the increased expression of aquaporin proteins of plasma membrane origin (Marulanda et al. 2009). Beneficial microbes assist the host plant in the production of osmoprotective and signaling molecules for phytohormones against salinity stress (Cheng et al. 2007). The rice crop, inoculated with *Bacillus amyloliquefaciens* strain SN13, demonstrated better salt tolerance, according to Nautiyal et al. (2013), in hydroponic as well as soil conditions by altering the expression of 14 genes against the unfavorable conditions. Four genes, viz., SOS1, ethylene receptive element-binding protein (EREBP), somatic embryonic receptor-like kinase SERK1, and NADP-malic enzyme (NADP-Me2), were upregulated conferring resistance. High salt stress in soil triggers decreased water potential and affects seed germination.

Bacillus aquimaris treatment in wheat showed increased total and reducing sugars aiding in seed imbibitions under high salt concentrations. Inoculated seeds hence showed a boost in plant biomass with high macronutrients such as nitrogen, phosphorus, and potassium and lower content of sodium in leaf (Upadhyay and Singh 2015). The concentration of salt soluble in the soil during salinity raises and reduces the water potential. The movement of water takes place from high to low potential; hence, it is difficult for the plants to absorb water from a water deficit soil environment (Dodd et al. 2010). The high level of salt also competes with nutrients in the soil and decreases their absorption. Furthermore, the plant accumulates into the root cell a few toxic ions (Na^+ and Cl_2), which get transported in leaves, altering the stomatal conductance and carbon availability in the leaves and decreasing the process of photosynthesis (Ilangumaran and Smith 2017). The buildup of toxic ions in the soil is declined by using beneficial PGPR, which function by modifying the physiology of the host plant such as ion transportation and production of exopolysaccharides as structural barriers. In addition, microbes play an immediate role in enhancing nutrients (micro- and macronutrients) for plant acquisition by producing organic acids lowering the pH of the soil (Dodd and Pérez-Alfocea 2012). Zuccarini and Okurowska (2008) reported an increase in the K^+/Na^+ ratio as a result of positive microbes and plant interactions under salinity stress. Near the roots, the bacterial synthesized exopolysaccharides interact with Na^+ cations and reduce their apoplastic flow to leaf tissue (Ashraf et al. 2004). Beneficial rhizobacteria-secreted 1-aminocyclopropane-1-carboxylic acid deaminase (ACCD) improves plant nutrient status with enhanced K/Na^+ in salt-affected maize (Nadeem et al. 2009) and tomato (Mayak et al. 2004).

Molecular Mechanisms of Stress Tolerance

The analysis of gene expression makes it possible to recognize and compare an organism's "holistic" responses in terms of its ever-changing environment. The best way to do such research is to test the transcriptome, which comprises the entire transcript series at a certain stage of development. Different technologies such as microarrays based on hybridization and extending to the sequencing of RNA are noteworthy (Trewavas 2006; Wang et al. 2009). The stress tolerance of *Arabidopsis thaliana* in the condition of low water availability was enhanced by inoculation of

Paenibacillus polymyxa B2 as observed at the transcriptional level. In one of the studies conducted by Timmusk and Wagner (1999), the upregulation of early response genes (ERD15) was documented when compared to un-inoculated control plants under drought stress. The identification of six stress proteins expressed with varied potential under the condition of sewage stress with the application of *Bacillus licheniformis* K11 in the pepper was carried out using techniques such as 2D polyacrylamide gel electrophoresis (2D-PAGE) and differential display polymerase chain reaction (DD-PCR). Amplification in different stress proteins (Cadhn, VA, sHSP and CaPR-10) by 1.5-fold was noticed by Lim and Kim (2013) in PGPR-inoculated plants as compared to the control plants. Resistance against drought stress as conferred by a combination of *B. amyloliquefaciens* 5113 and *Azospirillum brasilense* strain N040 was confirmed by a positive real-time PCR (RT-PCR) reaction of the positively expressed genes such as APX1, SAMS1, and HSP17.8, which are active stress response genes in wheat. The results were further confirmed by the high ascorbate enzyme activity observed in inoculated wheat plants.

Downregulation of drought response genes was noticed in *Arabidopsis thaliana* plants inoculated with *P. chlororaphis* O6 over noninoculated plants when observed by microarray analysis. The high degree of organic or inorganic salts in the soil prevents plants from achieving their complete potential both physiologically and genetically, ultimately reducing plant growth, yield, and grain quality (Atkinson and Urwin 2012). Based on plant response to high salt conditions, they are divided into halophytes, which can thrive at high salt concentrations, and glycophytes, which are susceptible to high levels of salt. Most of the crops are glycophytes, hence leading to severe losses in crop productivity at extensively high salt concentrations (Gupta and Huang 2014). High concentrations of salt in the soil cause loss in plant productivity through accelerated osmotic, ionic, and oxidative stresses in plants (Rahneshan et al. 2018).

The abrupt change in the temperature results in the synthesis of special type of polypeptides, collectively called heat shock proteins (HSPs), which aid in the degradation of the proteins that have lost their function and result in maintaining the structure of proteins. HSPs constitute a group of chaperons, viz., GroEL, DnaK, DnaJ, GroES, ClpB, ClpA, ClpX, small heat shock proteins (sHSPs), and proteases. Ali et al. (2009) reported heat shock protein induction with the inoculation of *P. aeruginosa* AMK-P6 (thermotolerant strain). All the stress-responsive genes upregulated by the PGPR inoculation in different crops are summarized in Table 2.

The plant cell is known to build up Na^+ and Cl_2 ions during their encounter with the hypertonic state of the soil. These ions are highly toxic to plants and lead to water imbalance and homeostasis of ions in the plant cell matrix (Ilangumaran and Smith 2017). Excess of salt interferes with major plant physiological and metabolic activities that result in reduced crop productivity, leaf area, internode length, and concomitant increase in thickness of leaves and root necrosis (Rahneshan et al. 2018). There have been recent attempts to incorporate new genes into plants in order to overcome various environmental stresses using the approach of genetic engineering. In the same direction, in the past few decades, significant focus has been devoted to the introduction to crop plants of unique genes accountable for the

Table 2 Stress-responsive genes and their relative functions mediated by PGPR in crops

Stress type	Plant species	PGPR inoculants	Stress gene involved	Function	References
	Arabidopsis thaliana	*Bacillus subtilis*, *Paenibacillus younginensis*, *Enterobacter*, *Pseudomonas putida*	*LOX2*, *DREB2b*, *RD29A*, *RD29B*, *RAB18*, *AtERD15*, *AtRSA1*, *AtVQ9*, *AtWRKY8*	Jasmonic acid, proline synthesis, defense response pathways, ROS detoxification, ion homeostasis	Chu et al. (2019), Sukweenadhi et al. (2015), Kim et al. (2014)
	Rice (*Oryza sativa*)	*Bacillus amyloliquifaceins* and root-associated rhizobacteria	*NADP-Me2*, *EREBP*, *SERK*, *SOS1*, *BADH*, *RAB18*	Na^+/H^+ antiporter, oxidative carboxylation of L-malate, ion homeostasis, intracellular protein transport	Jha et al. (2014), Nautiyal et al. (2013)
Salinity	Soybean (*Glycine max*)	*Pseudomonas simiae* strain AU, *Bacillus firmis* SW5	*POX*, *CAT*, vegetative storage protein (*VSP*), nitrate reductase (*NR*), *GmVSP*, *GmPHD2*, *GmbZIP62*, *GmWRKY54*, *GmOLPb*, *CHS*	Production of antioxidant enzymes such as peroxidase, catalase, etc.	Vaishnav et al. (2016), El-Esawi et al. (2018)
	Solanum lycopersicum	*Bacillus megaterium*, *Pseudomonas putida*	*MT2*, *GR1*, *Toc* GTPase	Synthesis of metallothionein glutathione reductase enzyme and regulation of chloroplast import machinery	Zameer et al. (2016), Yan et al. (2014)
	Maize (*Zea mays*)	*Bacillus amyloliquefaciens SQR9*	*RBCS*, *RBCL*, *HKT1*, *NHX1*, *NHX2*, *NHX3*	Photosynthesis, Na^+ sequestration	Chen et al. (2016)
	Wheat (*Triticum aestivum*)	*Bacillus safensis*, *Dietzia natronolimnaea*, *Arthrobacter nitroguajacolicus*	Expansions, endotransglucosylase, sulfur-rich thionin, adenosylmethionine decarboxylase, *TaABARE*, *TaOPR1*, *AA0618700*, *APX*, *GPX*	Stress response regulators, cell wall precursors, phenylpropanoid pathway	Safdarian et al. (2019), Chakraborty et al. (2018), Bharti et al. (2016)

Drought	*Arabidopsis thaliana*	*Pseudomonas chlororaphis* and *Paenibacillus polymyxa*	Jasmonic acid genes, *VSP1*, *PR*-1, *HEL*, ethylene response gene, salicylic acid regulator, early response to dehydration protein (*ERD15*)	Storage protein in vegetative organs, activation of antioxidants, amino acids, soluble sugars, stomatal opening and closing	Cho et al. (2013), Timmusk and Wagner (1999)
	Pepper	*Bacillus licheniformis*	*Cadhn*, *VA*, *Shsp*, *CaPr-10*	Rapid cell division and differentiation, decrease cell water potential, maintain hydrostatic pressure in cells	Lim and Kim (2013)
	Sugarcane	*Gluconoacetobacter diazotrophicus*	ABA-dependent signaling genes	Proteins and compatible osmolyte synthesis	Vandana et al. (2020)
	Wheat	*Azospirillum brasilense* and *Bacillus amyloliquefaciens 5113*	*APX1*, *SAMS1*, *HSP 17.8*	ROS signaling, chloroplast protection, synthesis of scavenging enzymes	Kasim et al. (2013)
Temperature	Winter wheat	Plant growth-promoting rhizobacteria	*StSy* and *PAL*	Synthesis of resveratrol and SA	Gaudet et al. (2003)
	Arabidopsis sp.	*Bacillus* sp.	CBF transcription factors	Accumulation of defense-related gene transcripts, and stress-related metabolites	Gilmour et al. (2000)
	Red table grape (*Vitis vinifera*)	*Burkholderia phytofirmans* PsJN	*LOX* (LOX-encoding), *VcSTS* and *VcPAL*	Synthesis of oxylipins and reactive oxygen species (ROS) regulation	Theocharis et al. (2012)

upregulation of genes encoding enzymes such as ACC deaminase similar to that produced by beneficial microorganisms.

Genetic Approaches for Engineering PGPR for Stress Resilience

Plants express different stress regulation mechanisms depending on the presence and absence of any environmental fluctuations such as the proline production during high salts and water loss conditions, heat shock protein expression, photosynthesis, antioxidative boost, hormonal fluctuations, osmolytic reactions, etc. Plants are distinctive in their response to multiple stresses than the individual stress encountered (Atkinson and Urwin 2012). The plant resistance mechanisms against various stresses remain unclear, which needs further studies. "Omics" methods with the combination of bioinformatics and computational approaches, which include genomics, transcriptomics, proteomics, metabolomics, and phenomics, have helped in understanding the mechanism of plant tolerance and aid in the development of plants able to survive these stresses. The combination of stresses, both biotic and abiotic, is known to worsen the situation such that the exposure of plants to abiotic stresses weakens the plant, which makes it more susceptible to the diverse phytopathogens around it. Sometimes, the presence of both abiotic and biotic stresses simultaneously in plants would neutralize the impact of each other, leading to a complete positive or neutral effect. The growth of alfalfa (*Medicago truncatula*) is normally impaired in drought and high ozone stress, but it was observed that the combination of both the stresses in the alfalfa plant made it resistant to such stresses without affecting its growth (Pandey et al. 2017). Vandana et al. (2020) documented that plants engineered with the imt1 gene, encoding myo-inositol-*o*-methyltransferase (inositol biosynthesis), were more resilient to drought conditions than the wild plants.

Genomics deals with gene sequence, intragenic sequence, genetic structure, and annotation (Duque et al. 2013). Gene research includes numerous methods starting from DNA extraction, purification, PCR amplification, sequencing, and the assembly of the obtained sequences (Ramegowda et al. 2014). Functional genomics and its methods help classify functional genes that participate in abiotic and biotic stress responses in plants and also reveal the interactions between networking genes, which are involved in regulated stress tolerance (Ramegowda et al. 2014). In various plant varieties, researchers have identified varied genes involved in stress resistance using the technique of functional genomics (Arpat et al. 2004; Micheletto et al. 2007). The NAC-domain, OSNAC10 gene, was identified by Jeong et al. (2010) using functional genomics, and it was documented that this gene was constitutively regulated by promoters such as GOS2, showing positive results under drought stress making the plant tolerant to the stress.

Transcriptomics is an investigation into the complete collection of RNA transcripts produced by the cell under certain conditions. Transcriptomic comparison enables the identification of various stress-related genes that are expressed differently in a particular part of the plant. The technology for transcriptomics has provided a clear insight into a variety of stress reactions and their tolerance mechanisms. It offers a deeper understanding of genes and plant–pathogen relation for abiotic stress amelioration. For transcriptomic studies, a range of technologies are used based on hybridization and sequencing methodologies (Tan et al. 2009). The transcriptomic study of plants helped classify significant transcripts and connections between different physiological processes on exposure to salt stress (Maathius 2006). Temperature and salt stress are known to confer a negative impact on the expression of photosynthetic genes with concomitant upregulation of the stress gene transcription factors (Evers et al. 2012). The transcriptome analysis of *Artemisia annua* genes encoding the survival of the plant in diverse stress (salinity, temperature, drought, etc.) was documented by Vashisth et al. (2018).

The proteomics covers identification, biochemical properties, and functionality of proteins along with their altered expression, quantity, and structure in response to the external stimulus. Different proteins act in varying responses to different stresses at both the pre-transcriptional and translation stages. These proteins expressed during stress are completely useful in the production of plant stress tolerance. The proteomics research therefore provides an incentive to discover new proteins and pathways linked to crop response under stress. There have been a few proteomic studies on different abiotic stresses in plants, including salt stress (Nam et al. 2012), drought (Castillejo et al. 2008), and high temperature (Rollins et al. 2013). Plant biologists have used proteomic approaches to investigate the response of saline stress in *Arabidopsis thaliana*, *Nicotiana tobaccum*, and *Agrostis stolonifera* (Xu et al. 2011). The proteomics strategy was used by Passamani et al. (2017) to find salt stress in two cultivars of sugarcane, and proteins actively participating in calcium-based protein kinase, photosystem I, phospholipase D, and glyceraldehyde-3-phosphate dehydrogenase were found to be abundant in salt-tolerant types of micropropagated sugarcane.

Metabolomics is widely used for the determination and quantification of metabolites in various life processes (Deshmukh et al. 2014). As a wide variety of metabolites can be synthesized by plants in order to adjust the stress conditions, recognizing and quantifying such metabolites can give a better understanding of stress biology (Badjakov et al. 2012). Different techniques such as nuclear magnetic resonance (NMR) or mass spectrometry, Fourier-transform ion cyclotron resonance mass spectrometry, or Fourier-transform infrared (FT-IR) spectroscopy, Fourier transform infrared (FT-IR), gas chromatography-MS, liquid chromatography-mass spectrometry (LC-MS), capillary electrophoresis-mass spectrometry (CE-MS), and FT-IR spectroscopy can be used for detection and quantification of the metabolites (Bagati et al. 2018). Researchers have used a metabolomics approach to understand the metabolic effects of salinity in a number of crops and plant-related species of sea lavender (*Limonium latifolium*), rice (Hirai et al. 2004), tomato (Lenz et al. 2011), and grapevine (Cramer et al. 2011).

Phenomics deals with the acquiring of multidimensional phenotypic data in an organism (Houle et al. 2010). It expands our understanding of mechanisms of stress tolerance in crop plants. The linkage between phenomics and various omics technologies, such as genomics, transcriptomics, or metabolomics, helps evaluate the performance of plants under different stress conditions and fills the gap between plant genetics and vital characteristics required for the efficient productivity of crops (Singh et al. 2018). Plants are responsive to fluctuations in temperatures through alterations in their physiology and biochemical status, which can be observed by studying the phenotypic variations in them (Singh et al. 2018). To study the effect of two-phased salinity stress (osmotic and tissue-tolerant phase) on plants, a phenomics approach can be practicable. The plants sensitive to the osmotic fluctuations are more prone to high salts in comparison to the ones tolerant to an excess of NaCl. The negative impact of the salt stress on plants can then be monitored regularly using infrared thermography (Sirault et al. 2009), similar to that described by James and Sirault (2012) in wheat.

Apart from positive regulation, negative regulation mediated by microRNAs (miRNAs), which control the expression of mRNA, is vital to study. These RNAs are 22-bp short conserved sequences that target cleavage or translational repression of mRNA transcripts playing a role in plant growth. In addition, miRNAs play an important role in response to varied stresses (Kruszka et al. 2012). These miRNAs modify their expression profiles during any stress encountered by plants and hence control the plant development (Mendoza-Soto et al. 2012). Various types of miRNA participate in different types of abiotic and biotic stress response. In *Arabidopsis*, miR169 was updated to respond to both cold and drought conditions (Li et al. 2008; Zhou et al. 2008). Likewise, the expressions of different salt-responsive miRNAs are modified in response to the salt stress, which function by affecting the working of the transcription factors (Ding et al. 2009).

Lipidomics has emerged as the distinctive technology that enables the extensive study of cellular lipidome together with cellular signals, membrane structures, modulating transcription and translation, cellular and intercellular interactions, and the composition of various cellular structures. Cellular components are rich in lipids and their derivatives that aid in carbon storage, compartmentalization, protection, and maintaining the overall physiological state of the plant. Alteration in the lipid membrane is one of the strategies employed by plants to resist abiotic stresses.

Various researchers have employed lipidomics approaches to understand the response of plants to various biotic and abiotic stresses. Chen et al. (2013) noted that, for adjusting to the cold stress, *Arabidopsis* plants could change their lipid membrane structures. The lipidomics studies of the thyme plants (*Thymus serpyllum* L. and *Thymus vulgaris* L.) (drought tolerant and sensitive) demonstrated that the levels of galactolipids and phospholipids were reduced in the sensitive ones while the resistant plants showed a boost in these molecules, suggesting their role in some sort of stress signaling (Moradi et al. 2017).

The "omics" and its major technologies are closely dependent on the stream of bioinformatics. Various "omics" methods result in large-scale and high-throughput data, which need clear analysis and visualization, the work undertaken by the field of bioinformatics. Therefore, all the computations generated using "omics" tools are closely connected to powerful bioinformatic methods for analyzing and generating an output (El-Metwally et al. 2014). The reliable, quick, and error-free analysis of biological data is the key role of bioinformatics tools and computational biology (Orozco et al. 2013).

Conclusion and Future Perspective

Today, modern agriculture faces multiple stresses caused by biological and abiotic influences. The different stresses discussed in this chapter accentuate ethylene biosynthesis, which mostly inhibits plant growth by means of several molecular mechanisms. In the current context, it is important to control plant ethylene by using PGPR resilient to varied stresses. However, many beneficial aspects of PGPR have not been thoroughly exploited, such as their role in salinity, drought, waterlogging, biocontrol, temperature, nutrient stresses, and nodulation in legumes. Commercially, their use in agriculture could prove useful and could be a sound step toward sustainable cultivation and conservation. It allows scientific researchers to generate reliable results that can be reproduced efficiently.

In biotic and abiotic stress responses in plants, the part of "omics" technology is inevitable. The quality control of crops under stress is an important aspect of efficient agriculture. Thus, using different "omic" methods, the establishment of networks of interactions between genes, proteins, and metabolites involved in the stress reaction mechanism has provided a forum for further study. A potential, successful application may include improving drought and/or salt tolerance with the involvement of transgenic plants through the identification of autotolerance mediated by osmoprotecting enzymes and genes in the host plant. The knowledge of molecular signals and pathways that drive beneficial plant–microbial interactions is still limited, and the relationship between phytohormones in PGPR-inoculated plants and their cumulative impact on stress throughout the plant is less understood. Microbiological techniques have become the most effective resource for remediating and increasing the sustainability of stressed soils as a result of inefficient conventional methods. In order to increase crop production and monitor the stress pandemic caused by world climate change, this new technology needs to be explored further.

References

Aachard P, Cheng H, De Grauwe L, Decat J, Schoutteten H, Moritz T, Van Der Straeten D, Peng J, Harberd NP (2006) Integration of plant responses to environmentally activated phytohormonal signals. Science 311:91–94

Ahemad M, Kibret M (2014) Mechanisms and applications of plant growth promoting rhizobacteria: current perspective. J King Saud Univ Sci 26(1):1–20

Ahmad I, Akhtar MJ, Zahir ZA, Naveed M, Mitter B, Sessitsch A (2014) Cadmium-tolerant bacteria induce metal stress tolerance in cereals. Environ Sci Pollut Res 21(18):11054–11065

Ahmadi N, Mibus H, Serek M (2009) Characterization of ethylene-induced organ abscission in F1 breeding lines of miniature roses (*Rosa hybrida* L.). Postharvest Biol Technol 52:260–266

Alexieva V, Sergiev I, Mapelli S, Karanov E (2001) The effect of drought and ultraviolet radiation on growth and stress markers in pea and wheat. Plant Cell Environ 24:1337–1344

Ali SZ, Sandhya V, Grover M, Kishore N, Rao LV, Venkateswarlu B (2009) *Pseudomonas* sp. strain AKM-P6 enhances tolerance of sorghum seedlings to elevated temperatures. Biol Fertil Soils 46:45–55

Alizadeh H, Behboudi K, Ahmadzadeh M, Javan-Nikkhah M, Zamioudis C, Pieterse CM, Bakker PA (2013) Induced systemic resistance in cucumber and *Arabidopsis thaliana* by the combination of *Trichoderma harzianum* Tr6 and *Pseudomonas* sp. Ps14. Biol Control 65(1):14–23

Angus JF, Gaul RR, Peoples MB, Stapper M, van Hawarden AF (2001) Soil water extraction by dryland crops, annual pastures and lucerne in south-eastern Australia. Aust J Agric Res 52:183–192

Apel K, Hirt H (2004) Reactive oxygen species: metabolism, oxidative stress, and signal transduction. Annu Rev Plant Biol 55:373–399

Arpat A, Waugh M, Sullivan JP et al (2004) Functional genomics of cell elongation in developing cotton fibers. Plant Mol Biol 54:911–929

Artiola JF, Walworth JL, Musil SA, Crimmins MA (2019) Soil and land pollution. In: Environmental and pollution science. Academic, San Diego, pp 219–235

Arzanesh MH, Alikhani HA, Khavazi K, Rahimian HA, Miransari M (2011) Wheat (*Triticum aestivum* L.) growth enhancement by *Azospirillum* sp. under drought stress. World J Microbiol Biotechnol 27:197–205

Ashraf M, Ali Q (2008) Relative membrane permeability and activities of some antioxidant enzymes as the key determinants of salt tolerance in canola (*Brassica napus* L.). Environ Exp Bot 63:266–273

Ashraf M, Berge SH, Mahmood OT (2004) Inoculating wheat seedling with exopolysaccharide producing bacteria restricts sodium uptake and stimulates plant growth under salt stress. Biol Fertil Soils 40:157–162

Ashraf MA, Asif M, Zaheer A, Malik A, Ali Q, Rasool M (2013) Plant growth promoting rhizobacteria and sustainable agriculture: a review. Afr J Microbiol Res 7:704–709

Asseng S, Ewert F, Rosenzweig C, Jones JW, Hatfield JL, Ruane AC, Boote KJ, Thorburn PJ, Rötter RP, Cammarano D, Brisson N (2013) Uncertainty in simulating wheat yields under climate change. Nat Clim Chang 3(9):827–832

Atkinson NJ, Urwin PE (2012) The interaction of plant biotic and abiotic stresses: from genes to the field. J Exp Bot 63(10):3523–3543. https://doi.org/10.1093/jxb/ers100

Atkinson NJ, Lilley CJ, Urwin PE (2013) Identification of genes involved in the response to simultaneous biotic and abiotic stress. Plant Physiol 162:2028–2041

Attia H, Alamer KH, Ouhibi C, Oueslati S, Lachaal M (2020) Interaction between salt stress and drought stress on some physiological parameters in two pea cultivars. Int J Bot 16:1–8

Badjakov I, Kondakova V, Atanassov A (2012) Metabolomics: current view on fruit quality in relation to human health. In: Benkeblia N (ed) Sustainable agriculture and new biotechnologies. CRC Press, Boca Raton, pp 303–319

Bagati S, Mahajan R, Nazir M, Dar AA, Zargar SM (2018) "Omics": a gateway towards abiotic stress tolerance. In: Zargar SM, Zargar MY (eds) Abiotic stress-mediated sensing and signaling in plants: an omics perspective. Springer, Singapore, pp 1–46

Bal HB, Nayak L, Das S, Adhya TK (2013) Isolation of ACC deaminase producing PGPR from rice rhizosphere and evaluating their plant growth promoting activity under salt stress. Plant Soil 366:93–105

Bano Q, Ilyas N, Bano A, Zafar N, Akram A, Hassan F (2013) Effect of *Azospirillum* inoculation on maize (*Zea mays* L.) under drought stress. Pak J Bot 45:13–20

Barka EA, Nowak J, Clement C (2006) Enhancement of chilling resistance of inoculated grapevine plantlets with a plant growth-promoting rhizobacterium, *Burkholderia phytofirmans* strain PsJN. Appl Environ Microbiol 72:7246–7252

Belimov AA, Dodd IC, Hontzeas N, Theobald JC, Safronova VI, Davies WJ (2009) Rhizosphere bacteria containing 1-aminocyclopropane-1-carboxylate deaminase increase yield of plants grown in drying soil via both local and systemic hormone signalling. New Phytol 181:413–423

Bensalim S, Nowak J, Asiedu SK (1998) A plant growth promoting rhizobacterium and temperature effects on performance of 18 clones of potato. Am J Potato Res 75:145–152

Bharti N, Pandey SS, Barnawal D, Patel VK, Kalra A (2016) Plant growth promoting rhizobacteria *Dietzi anatronolimnaea* modulates the expression of stress responsive genes providing protection of wheat from salinity stress. Sci Rep 6:34768

Blaha D, Prigent-Combaret C, Mirza MS, Moënne-Loccoz Y (2006) Phylogeny of the 1-aminocyclopropane-1-carboxylic acid deaminase-encoding gene *acdS* in phytobeneficial and pathogenic Proteobacteria and relation with strain biogeography. FEMS Microbiol Ecol 56:455–470

Blum A (2005) Drought resistance, water-use efficiency, and yield potential—are they compatible, dissonant, or mutually exclusive? Aust J Agric Res 56:1159–1168

Bulgari R, Franzoni G, Ferrante A (2019) Biostimulants application in horticultural crops under abiotic stress conditions. Agronomy 9(6):306

Burd GI, Dixon DG, Glick BR (1998) A plant growth-promoting bacterium that decreases nickel toxicity in seedlings. Appl Environ Microbiol 64:3663–3668

Castillejo M, Maldonado AM, Ogueta S, Jorrín JV (2008) Proteomic analysis of responses to drought stress in sunflower (Helianthus annuus) leaves by 2DE gel electrophoresis and mass spectrometry. Proteomics J 1:59–71

Castillo P, Escalante M, Gallardo M, Alemano S, Abdala G (2013) Effects of bacterial single inoculation and co-inoculation on growth and phytohormone production of sunflower seedlings under water stress. Acta Physiol Plant 35:2299–2309

Chakraborty N, Ghosh R, Ghosh S, Narula K, Tayal R, Datta A, Chakraborty S (2013) Reduction of oxalate levels in tomato fruit and consequent metabolic remodeling following overexpression of a fungal oxalate decarboxylase. Plant Physiol 162(1):364–378

Chakraborty U, Chakraborty BN, Dey PL, Chakraborty AP (2018) *Bacillus safensis* from wheat rhizosphere promotes growth and ameliorates salinity stress in wheat. Indian J Biotechnol Pharm 17:466–79

Chaves MM, Maroco JP, Pereira JS (2003) Understanding plant responses to drought—from genes to the whole plant. Funct Plant Biol 30:239–264

Chen D, Yan X, Xu J, Su X, Li L (2013) Lipidomic profiling and discovery of lipid biomarkers in *Stephanodiscus* sp. under cold stress. Metabolomics 9:949–959

Chen L, Liu Y, Wu G, VeronicanNjeri K, Shen Q, Zhang N, Zhang R (2016) Induced maize salt tolerance by rhizosphere inoculation of *Bacillus amyloliquefaciens* SQR9. Physiol Plant 158: 34–44

Cheng Z, Park E, Glick BR (2007) 1-Aminocyclopropane-1-carboxylate (ACC) deaminase from *Pseudomonas putida* UW4 facilitates the growth of canola in the presence of salt. Can J Microbiol 53:912–918

Cho SM, Kang BR, Kim YC (2013) Transcriptome analysis of induced systemic drought tolerance elicited by *Pseudomonas chlororaphis* O6 in *Arabidopsis thaliana*. Plant Pathol J 29:209

Chu TN, Tran B, Van Bui L, Hoang M (2019) Plant growth-promoting rhizobacterium *Pseudomonas* PS01 induces salt tolerance in *Arabidopsis thaliana*. BMC Res Notes 12:11

Claeys H, Inzé D (2013) The agony of choice: how plants balance growth and survival under water-limiting conditions. Plant Physiol 162:1768–1779

Coakley SM, Scherm H, Chakraborty S (1999) Climate change and plant disease management. Annu Rev Phytopathol 37:399–426

Comas LH, Becker SR, Cruz VV, Byrne PF, Dierig DA (2013) Root traits contributing to plant productivity under drought. Front Plant Sci 4:442

Coy RM, Held DW, Kloepper JW (2019) Rhizobacterial treatments of tall fescue and bermuda grass increases tolerance to damage from white grubs. Pest Manag Sci 75(12):3210–3217

Cramer G, Urano K, Delrot S et al (2011) Effect of abiotic stress on plant: a systems biology perspective. BMC Plant Biol 11:163

Dai A (2013) Increasing drought under global warming in observations and models. Nat Clim Chang 3(1):52

Dakora F, Matiru V, Kanu A (2015) Rhizosphere ecology of lumichrome and riboflavin, two bacterial signal molecules eliciting developmental changes in plants. Front Plant Sci 6:700

Dardanelli MS, de Cordoba FJF, Espuny MR, Carvajal MAR, Diaz MES, Serrano AMG (2008) Effect of *Azospirillum brasilense* coinoculated with rhizobium on *Phaseolus vulgaris* flavonoids and Nod factor production under salt stress. Soil Biol Biochem 40(11):2713–2721

Davies P (2013) Plant hormones: physiology, biochemistry and molecular biology. Springer Science & Business Media, New York, NY

De Faria SM, Resende AS, Saggin Júnior OJ, Boddey RM (2011) Exploiting mycorrhizae and *rhizobium* symbioses to recover seriously degraded soils. In: Polacco JC, Todd CD (eds) Ecological aspects of plant nitrogen metabolism. Wiley-Blackwell, New York, pp 195–214

del Amor FM, Cuadra-Crespo P (2012) Plant growth-promoting bacteria as a tool to improve salinity tolerance in sweet pepper. Funct Plant Biol 39(1):8290

Deshmukh R, Sonah H, Patil G, Chen W, Prince S, Mutava R, Vuong T, Valliyodan B, Nguyen HT (2014) Integrating omic approaches for abiotic stress tolerance in soybean. Front Plant Sci 5:1–12

Ding D, Zhang L, Wang H, Liu Z, Zhang Z, Zheng Y (2009) Differential expression of miRNAs in response to salt stress in maize roots. Ann Bot 103:29–38

Dixit R, Agrawal L, Gupta S, Kumar M, Yadav S, Chauhan PS (2016) Southern blight disease of tomato control by 1-aminocyclopropane-1-carboxylate (ACC) deaminase producing *Paenibacillus lentimorbus* B-30488. Plant Signal Behav 11:e1113363

Dodd IC, Pérez-Alfocea F (2012) Microbial amelioration of crop salinity stress. J Exp Bot 63(9): 3415–3428

Dodd IC, Zinovkina NY, Safronova VI, Belimov AA (2010) Rhizobacterial mediation of plant hormone status. Ann Appl Biol 157:361–379

Donate-Correa J, Leon-Barrios M, Perez-Galdona R (2005) Screening for plant growth-promoting rhizobacteria in *Chamaecytisus proliferus* (tagasaste), a forage tree-shrub legume endemic to the Canary Islands. Plant Soil 266:261–272

Dunbabin VM, Postma JA, Schnepf A, Pagès L, Javaux M, Wu L, Leitner D, Chen YL, Rengel Z, Diggle AJ (2013) Modelling root–soil interactions using three-dimensional models of root growth, architecture and function. Plant Soil 372:93–124

Duque AS, de Almeida AM, da Silva AB et al (2013) Abiotic stress responses in plants: unraveling the complexity of genes and networks to survive. In: Vahdati K, Leslie C (eds) Abiotic stress plant responses and applications in agriculture. IntechOpen, Rijeka, Croatia, pp 49–101

Duveiller E, Singh RP, Nicol JM (2007) The challenges of maintaining wheat productivity: pests, diseases and potential epidemics. Euphytica 157:417–430

Egamberdiyeva D (2007) The effect of plant growth promoting bacteria on growth and nutrient uptake of maize in two different soils. Appl Soil Ecol 36:184–189

El-Esawi MA, Alaraidh IA, Alsahli AA, Alamri SA, Ali HM, Alayafi AA (2018) *Bacillus firmus* (SW5) augments salt tolerance in soybean (*Glycine max* L.) by modulating root system architecture, antioxidant defense systems and stress-responsive genes expression. Plant Physiol Biochem 132:375–84

El-Metwally S, Ouda OM, Helmy M (2014) Next generation sequencing technologies and challenges in sequence assembly. Springer, New York

Entz MH, Bullied WJ, Forster DA, Gulden R, Vessey JK (2001) Extraction of subsoil nitrogen by alfalfa, alfalfa-wheat and perennial grass systems. Agron J 93:495–503

Evelin H, Kapoor R, Giri B (2009) Arbuscular mycorrhizal fungi in alleviation of salt stress: a review. Ann Bot 104:1263–1280

Evers D, Legay S, Lamoureux D, Hausman JF, Hoffmann L, Renaut J (2012) Towards a synthetic view of potato cold and salt stress response by transcriptomic and proteomic analyses. Plant Mol Biol 78:503–514

Fahad S, Hussain S, Bano A, Saud S, Hassan S, Shan D (2015) Potential role of phytohormones and plant growth-promoting rhizobacteria in abiotic stresses: consequences for changing environment. Environ Sci Pollut Res 22:4907–4921

Farooq M, Wahid A, Kobayashi N, Fujita DBSMA, Basra SMA (2009) Plant drought stress: effects, mechanisms and management. In: Sustainable agriculture. Springer, Dordrecht, pp 153–188

Figueiredo MVB, Burity HA, Martınez CR, Chanway CP (2008) Alleviation of drought stress in the common bean (Phaseolus vulgaris L.) by co-inoculation with *Paenibacillus polymyxa* and *Rhizobium tropici*. Appl Soil Ecol 40:182–188

Gaudet DA, Laroche A, Frick M, Huel R, Puchalski B (2003) Cold induced expression of plant defensin and lipid transfer protein transcripts in winter wheat. Physiol Plant 117(2):195–205

German MA, Burdman S, Okon Y, Kigel J (2000) Effects of *Azospirillum brasilense* on root morphology of common bean (*Phaseolus vulgaris* L.) under different water regimes. Biol Fertil Soil 32:259–264

Gill SS, Tuteja N (2010) Reactive oxygen species and antioxidant machinery in abiotic stress tolerance in crop plants. Plant Physiol Biochem 48:909–930

Gilmour SJ, Sebolt AM, Salazar MP, Everard JD, Thomashow MF (2000) Overexpression of the Arabidopsis CBF3 transcriptional activator mimics multiple biochemical changes associated with cold acclimation. Plant Physiol 124(4):1854–1865

Glick BR (1995) The enhancement of plant growth by free living bacteria. Can J Microbiol 41:109–117

Gonzalez-Teuber M, Pozo MJ, Muck A, Svatos A, Adame-Alvarez RM, Heil M (2010) Glucanases and chitinases as causal agents in the protection of Acacia extrafloral nectar from infestation by phytopathogens. Plant Physiol 152:1705–1715

Gowda VRP, Henry A, Yamauchi A, Shashidhar HE, Serraj R (2011) Root biology and genetic improvement for drought avoidance in rice. Field Crops Res 122:1–13

Grichko VP, Glick BR (2001) Amelioration of flooding stress by ACC deaminase-containing plant growth-promoting bacteria. Plant Physiol Biochem 39:11–17

Großkinsky DK, Tafner R, Moreno MV, Stenglein SA, De Salamone IEG, Nelson LM, Roitsch T (2016) Cytokinin production by *Pseudomonas fluorescens* G20-18 determines biocontrol activity against *Pseudomonas syringae* in Arabidopsis. Sci Rep 6:23310

Grover M, Madhubala R, Ali SZ, Yadav SK, Venkateswarlu B (2014) Influence of *Bacillus* spp. strains on seedling growth and physiological parameters of sorghum under moisture stress conditions. J Basic Microbiol 54:951–961

Gupta B, Huang B (2014) Mechanism of salinity tolerance in plants: physiological, biochemical, and molecular characterization. Int J Genom. https://doi.org/10.1155/2014/701596

Gupta G, Parihar SS, Ahirwar NK, Snehi SK, Singh V (2015) Plant growth promoting rhizobacteria (PGPR): current and future prospects for development of sustainable agriculture. J Microb Biochem Technol 7:96–102

Gusain YS, Singh US, Sharma AK (2015) Bacterial mediated amelioration of drought stress in drought tolerant and susceptible cultivars of rice (*Oryza sativa* L.). Afr J Biotechnol 14:764–773

Hamaoui B, Abbadi JM, Burdman S, Rashid A, Sarig S, Okon Y (2001) Effects of inoculation with *Azospirillum brasilense* on chickpeas (*Cicer arietinum*) and faba beans (*Vicia faba*) under different growth conditions. Agronomie 21(6–7):553–560

Han SH, Lee JH (2005) An overview of peak-to-average power ratio reduction techniques for multicarrier transmission. IEEE Wirel Commun 12(2):56–65

Haq QM, Arif A, Malathi VG (2010) Engineering resistance against Mungbean Yellow Mosaic India Virus using antisense RNA. Indian J Virol 21:82–85

Hashem A, AbdAllah E, Alqarawi A, Al-Huqail A, Shah M (2016) Induction of osmoregulation and modulation of salt stress in *Acacia gerrardii* Benth. by arbuscular mycorrhizal fungi and *Bacillus subtilis* (BERA 71). Biomed Res Int 2016:e6294098

Hassan TU, Bano A, Naz I (2017) Alleviation of heavy metals toxicity by the application of plant growth promoting rhizobacteria and effects on wheat grown in saline sodic field. Int J Phytoremediation 19:522–529

Hayat S, Hayat Q, Alyemeni MN, Wani AS, Pichtel J, Ahmad A (2012) Role of proline under changing environments. Plant Signal Behav 7:1456–1466

Heydarian Z, Yu M, Gruber M, Glick BR, Zhou R, Hegedus DD (2016) Inoculation of soil with plant growth promoting bacteria producing 1-aminocyclopropane-1-carboxylate deaminase or expression of the corresponding *acds* gene in transgenic plants increases salinity tolerance in *Camelina sativa*. Front Microbiol 7:1966

Hirai MY, Yano M, Goodenowe DB, Kanaya S, Kimura T, Awazuhara M, Arita M, Fujiwara T, Saito K (2004) Integration of transcriptomics and metabolomics for understanding of global responses to nutritional stresses in Arabidopsis thaliana. Proc Natl Acad Sci U S A 101:10205–10210

Hontzeas N, Richardson AO, Belimov AA, Safranova VI, Abu-Omar MM, Glick BR (2005) Evidence for horizontal gene transfer (HGT) of ACC deaminase genes. Appl Environ Microbiol 71:7556–7568

Houle D, Govindaraju DR, Omholt S (2010) Phenomics: the next challenge. Nat Rev Genet 11: 855–866

Huang XF, Chaparro JM, Reardon KF, Zhang R, Shen Q, Vivanco JM (2014) Rhizosphere interactions: root exudates, microbes, and microbial communities. Botany 92:267–275

Hund A, Reimer R, Messmer R (2011) A consensus map of QTLs controlling the root length of maize. Plant Soil 344:143–158

Ibrahim MA, Khan PR, Hegazy SS, Hashim EA, Azamal H, Ansari MKA (2015) Improving the phytoextraction capacity of plants to scavenge heavy-metal infested sites. Environ Rev 23:1–22

Ilangumaran G, Smith DL (2017) Plant growth promoting rhizobacteria in amelioration of salinity stress: a systems biology perspective. Front Plant Sci 8:1768

Jagadish SVK, Muthurajan R, Oane R, Wheeler TR, Heuer S, Bennett J, Craufurd PQ (2010) Physiological and proteomic approaches to address heat tolerance during anthesis in rice (Oryza sativa L.). J Exp Bot 61(1):143–156

James RA, Sirault XR (2012) Infrared thermography in plant phenotyping for salinity tolerance. Methods Mol Biol 913:173–189

Jamil M, Bashir S, Anwar S, Bibi S, Bangash A, Ullah F, Rha ES (2012) Effect of salinity on physiological and biochemical characteristics of different varieties of rice. Pak J Bot 44:7–13

Jarvis PG, Jarvis MS (1963) The water relations of tree seedlings. IV. Some aspects of the tissue water relations and drought resistance. Physiol Plant 16:501–516

Jeong JS, Kim YS, Baek KH, Jung H, Ha SH, Do Choi Y, Kim JK (2010) Root-specific expression of OsNAC10 improves drought tolerance and grain yield in rice under field drought conditions. Plant Physiol 153(1):185–197

Jha CK, Saraf M (2015) Plant growth promoting rhizobacteria (PGPR): a review. E3 J Agric Res Dev 5:108–119

Jha M, Chourasia S, Sinha S (2013) Microbial consortium for sustainable rice production. Agroecol Sustain Food Syst 37(3):340–362

Jha Y, Sablok G, Subbarao N, Sudhakar R, Fazil MT, Subramanian RB, Squartini A, Kumar S (2014) Bacterial-induced expression of RAB18 protein in Orzya sativa salinity stress and insights into molecular interaction with GTP ligand. J Mol Recognit 27:521–527

Ji X, Dong B, Shiran B (2011) Control of abscisic acid catabolism and abscisic acid homeostasis is important for reproductive stage stress tolerance in cereals. Plant Physiol 156:647–662

Jouyban Z (2012) Ethylene biosynthesis. Tech J Eng Appl Sci 1:100–107

Jung JK, McCouch S (2013) Getting to the roots of it: genetic and hormonal control of root architecture. Front Plant Sci 4:186

Kasim W, Osman M, Omar M, Abd El-Daim I, Bejai S, Meijer J (2013) Control of drought stress in wheat using plant-growth promoting rhizobacteria. J Plant Growth Regul 32:122–130
Keyvan S (2010) The effects of drought stress on yield, relative water content, proline, soluble carbohydrates and chlorophyll of bread wheat cultivars. J Anim Plant Sci 8(3):1051–1060
Kiani SP, Talia P, Maury P, Grieu P, Heinz R, Perrault A, Nishinakamasu V, Hopp E, Gentzbittel L, Paniego N, Sarrafi A (2007) Genetic analysis of plant water status and osmotic adjustment in recombinant inbred lines of sunflower under two water treatments. Plant Sci 172:773–787
Kim K, Jang YJ, Lee SM, Oh BT, Chae JC, Lee KJ (2014) Alleviation of salt stress by Enterobacter sp. EJ01 in tomato and Arabidopsis is accompanied by up-regulation of conserved salinity responsive factors in plants. Mol Cell 37:109–117
Kloepper JW, Leong J, Teintze M, Schroth MN (1980) Enhanced plant growth by siderophores produced by plant growth-promoting rhizobacteria. Nature 286:885–886
Kong ZY, Mohamad OA, Deng ZS, Liu XD, Glick BR, Wei GH (2015) Rhizobial symbiosis effect on the growth, metal uptake, and antioxidant responses of *Medicago lupulina* under copper stress. Environ Sci Pollut Res 22:12479–12489
Kruszka K, Pieczynski M, Windels D (2012) Role of microRNAs and other sRNAs of plants in their changing environments. J Plant Physiol 169:1664–1672
Kumar A, Bahadur I, Maurya B, Raghuwanshi R, Meena V, Singh D (2015) Does a plant growth promoting rhizobacteria enhance agricultural sustainability. J Pure Appl Microbiol 9:715–724
Kumar S, Chauhan PS, Agrawal L, Raj R, Srivastava A, Gupta S (2016) *Paenibacillus lentimorbus* inoculation enhances tobacco growth and extenuates the virulence of cucumber mosaic virus. PLoS One 11:e0149980
Kumar A, Patel JS, Meena VS, Srivastava R (2019) Recent advances of PGPR based approaches for stress tolerance in plants for sustainable agriculture. Biocatal Agric Biotechnol 20:101271
Latef AAA (2011) Influence of arbuscular mycorrhizal fungi and copper on growth, accumulation of osmolyte, mineral nutrition and antioxidant enzyme activity of pepper (*Capsicum annuum* L.). Mycorrhiza 21:495–503
Lefroy EC, Stirzaker RJ, Pate JS (2001) The influence of tagasaste (*Chamaecytisus profliferus* Link.) trees on the water balance of an alley cropping system on deep sand in south-western Australia. Aust J Agric Res 52:235–246
Lenz T, Fischer JJ, Dreger M (2011) Probing small molecule-protein interactions: a new perspective for functional proteomics. J Proteome 75:100–115
Lesk C, Rowhani P, Ramankutty N (2016) Influence of extreme weather disasters on global crop production. Nature 529(7584):84–87
Lewis GP (2005) Legumes of the world. Royal Botanic Gardens, Kew
Li J, Glick BR (2001) Transcriptional regulation of the Enterobacter cloacae UW4 1-aminocyclopropane-1-carboxylate (ACC) deaminase gene (acdS). Can J Microbiol 47:359–367
Li WX, Oono Y, Zhu J, He XJ, Wu JM, Iida K, Lu XY, Cui X, Jin H, Zhu JK (2008) The Arabidopsis NFYA5 transcription factor is regulated transcriptionally and post-transcriptionally to promote drought resistance. Plant Cell 20:2238–2251
Li H, Ding X, Wang C, Ke H, Wu Z, Wang Y (2016) Control of tomato yellow leaf curl virus disease by *Enterobacter asburiae* BQ9 as a result of priming plant resistance in tomatoes. Turk J Biol 40:150–159
Lim JH, Kim SD (2013) Induction of drought stress resistance by multi-functional PGPR *Bacillus licheniformis* K11 in pepper. Plant Pathol J 29:201
Liu HC, Liao HT, Charng YY (2011) The role of class A1 heat shock factors (HSFA1s) in response to heat and other stresses in *Arabidopsis*. Plant Cell Environ 34:738–751
Lobell DB, Gourdji SM (2012) The influence of climate change on global crop productivity. Plant Physiol 160(4):1686–1697
Loeuille N, Barot S, Georgelin E, Kylafis G, Lavigne C (2013) Eco-evolutionary dynamics of agricultural networks: implications for sustainable management. In: Advances in ecological research, vol 49. Academic, San Diego, pp 339–435

Lum MS, Hanafi MM, Rafii YM, Akmar ASN (2014) Effect of drought stress on growth, proline and antioxidant enzyme activities of upland rice. J Anim Plant Sci 24:1487–1493
Lwin KM, Myint MM, Tar T, Aung WZM (2016) Isolation of plant hormone (indole-3-acetic acid - IAA) producing rhizobacteria and study on their effects on maize seedling. Int J Chem Sci 50: 158–177
Ma Y, Rajkumar M, Luo Y, Freitas H (2011) Inoculation of endophytic bacteria on host and non-host plants—effects on plant growth and Ni uptake. J Hazard Mater 195:230–237
Maathius FJM (2006) The role of monovalent cation transporters in plant responses to salinity. J Exp Bot 57:1137–1147
Mafakheri A, Siosemardeh A, Bahramnejad B, Struik PC, Sohrabi Y (2010) Effect of drought stress on yield, proline and chlorophyll contents in three chickpea cultivars. Aust J Crop Sci 4:580–585
Mahalingam R (2015) Consideration of combined stress: a crucial paradigm for improving multiple stress tolerance in plants. In: Combined stresses in plants. Springer International Publishing, Cham, pp 1–25
Marulanda A, Barea JM, Azcón R (2009) Stimulation of plant growth and drought tolerance by native microorganisms (AM Fungi and Bacteria) from dry environments: mechanisms related to bacterial effectiveness. J Plant Growth Regul 28:115–124
Mayak S, Tirosh T, Glick BR (2004) Plant growth-promoting bacteria that confer resistance to water stress in tomatoes and peppers. Plant Sci 166:525–530
McCallum MH, Kirkegaard JA, Green T, Cresswell HP, Davies SL, Angus JF, Peoples MB (2004) Improved subsoil macroporosity following perennial pastures. Aust J Exp Agric 44:299–307
McDonald A, Riha S, DiTommasob A, DeGaetanoa A (2009) Climate change and the geography of weed damage: analysis of U.S. maize systems suggests the potential for significant range transformations. Agric Ecosyst Environ 130:131–140
Mendoza-Soto AB, Sánchez F, Hernández G (2012) MicroRNAs as regulators in plant metal toxicity response. Front Plant Sci 3:105
Micheletto S, Rodriguez-Uribe L, Hernandez R et al (2007) Comparative transcript profiling in roots of *Phaseolus acutifolius* and *P. vulgaris* under water deficit stress. Plant Sci 73:510–520
Mishra J, Arora NK (2018) Secondary metabolites of fluorescent pseudomonads in biocontrol of phytopathogens for sustainable agriculture. Appl Soil Ecol 125:35–45
Mishra J, Singh R, Arora NK (2017) Alleviation of heavy metal stress in plants and remediation of soil by rhizosphere microorganisms. Front Microbiol 8:1706
Mittler R (2006) Abiotic stress, the field environment and stress combination. Trends Plant Sci 11: 15–19
Molina-Favero C, Creus CM, Simontacchi M, Puntarulo S, Lamattina L (2008) Aerobic nitric oxide production by *Azospirillum brasilense* Sp245 and its influence on root architecture in tomato. Mol Plant-Microbe Interact 21:1001–1009
Moradi P, Mahdavi A, Khoshkam M, Iriti M (2017) Lipidomics unravels the role of leaf lipids in thyme plant response to drought stress. Int J Mol Sci 18:2067
Moreira H, Marques AP, Franco AR, Rangel AO, Castro PM (2014) Phytomanagement of Cd-contaminated soils using maize (*Zea mays* L.) assisted by plant growth-promoting rhizobacteria. Environ Sci Pollut Res 21(16):9742–9753
Mosa KA, Saadoun I, Kumar K, Helmy M, Dhankher OP (2016) Potential biotechnological strategies for the cleanup of heavy metals and metalloids. Front Plant Sci 7:303
Murakami Y, Tsuyama M, Kobayashi Y, Kodama H, Iba K (2000) Trienoic fatty acids and plant tolerance of high temperature. Science 287:476–479
Nadeem SM, Zahir ZA, Naveed M, Arshad M (2009) Rhizobacteria containing ACC-deaminase confer salt tolerance in maize grown on salt affected fields. Can J Microbiol 55:1302–1309
Nadeem SM, Zahir ZA, Naveed MNS (2013) Mitigation of salinity induced negative impact on the growth and yield of wheat by plant growth-promoting rhizobacteria in naturally saline conditions. Ann Microbiol 63(1):225–232

Nadeem SM, Ahmad M, Zahir ZA, Javaid Ashraf M (2014) The role of mycorrhizae and plant growth promoting rhizobacteria (PGPR) in improving crop productivity under stressful environments. Biotechnol Adv 32:429–448

Nagpal S, Sharma P, Kumawat KC (2019) Assessment of native single and dual inoculants of *Mesorhizobium* sp. and endophytic rhizobacteria for plant growth promotion in chickpea. Agric Res J 56:746–751

Nagpal S, Sharma P, Sirari A, Gupta RK (2020) Coordination of *Mesorhizobium* sp. and endophytic bacteria as elicitor of biocontrol against *Fusarium* wilt in chickpea. Eur J Plant Pathol. https://doi.org/10.1007/s10658-020-02062-1

Nam MH, Huh SM, Kim KM, Park WW, Seo JB, Cho K, Kim DY, Kim BG, Yoon II (2012) Comparative proteomic analysis of early salt stress responsive proteins in roots of SnRK2 transgenic rice. Proteome Sci 10:25

Naseem H, Bano A (2014) Role of plant growth-promoting rhizobacteria and their exopolysaccharide in drought tolerance of maize. J Plant Interact 9:689–701

Nautiyal CS, Srivastava S, Chauhan PS, Seem K, Mishra A, Sopory SK (2013) Plant growth-promoting bacteria *Bacillus amyloliquefaciens* NBRISN13 modulates gene expression profile of leaf and rhizosphere community in rice during saltstress. Plant Physiol Biochem 66:1–9

Naveed M, Khalid M, Jones DL, Ahmad R, Zahir ZA (2008) Relative efficacy of *Pseudomonas* spp., containing ACC-deaminase for improving growth and yield of maize (*Zea mays* L.) in the presence of organic fertilizer. Pak J Bot 40:1243–1251

Naveed M, Mitter B, Reichenauer TG, Wieczorek K, Sessitsch A (2014) Increased drought stress resilience of maize through endophytic colonization by *Burkholderia phytofirmans* PsJN and *Enterobacter* sp FD17. Environ Exp Bot 97:30–39

Neumann PM (2008) Coping mechanisms for crop plants in drought-prone environments. Ann Bot 101:901–907

Ngumbi EN (2011) Mechanisms of olfaction in parasitic wasps: analytical and behavioral studies of response of a specialist (*Microplitis croceipes*) and a generalist (Cotesia marginiventris) parasitoid to host-related odor. Ph.D. Dissertation, Auburn University, Auburn

Omer ZS, Tombolini R, Broberg A, Gerhardson B (2004) Indole-3-acetic acid production by pink-pigmented facultative methylotrophic bacteria. Plant Growth Regul 43:93–96

Oosten MJV, Pepe O, Pascale SD, Silletti S, Maggio A (2017) The role of biostimulants and bioeffectors as alleviators of abiotic stress in crop plants. Chem Biol Technol Agric 4. https://doi.org/10.1186/s40538-017-0089-5

Orhan F (2016) Alleviation of salt stress by halotolerant and halophilic plant growth-promoting bacteria in wheat (*Triticum aestivum*). Braz J Microbiol 47(3):621–627

Orozco A, Morera J, Jiménez S, Boza R (2013) A review of bioinformatics training applied to research in molecular medicine, agriculture and biodiversity in Costa Rica and Central America. Brief Bioinform 14:661–670

Pachauri RK, Allen MR, Barros VR, Broome J, Cramer W, Christ R, Church JA, Clarke L, Dahe Q, Dasgupta P, Dubash NK (2014) Climate change 2014: synthesis report. Contribution of Working Groups I, II and III to the fifth assessment report of the Intergovernmental Panel on Climate Change

Pan J, Peng F, Xue X, You Q, Zhang W, Wang T (2019) The growth promotion of two salt-tolerant plant groups with PGPR inoculation: a meta-analysis. Sustainability 11(2):378

Pandey P, Kang SC, Maheshwari DK (2005) Isolation of endophytic plant growth promoting *Burkholderia* sp. MSSP from root nodules of *Mimosa pudica*. Curr Sci 89:170–180

Pandey P, Ramegowda V, Senthil-Kumar M (2015) Shared and unique responses of plants to multiple individual stresses and stress combinations: physiological and molecular mechanisms. Front Plant Sci 6:723

Pandey P, Irulappan V, Bagavathiannan MV et al (2017) Impact of combined abiotic and biotic stresses on plant growth and avenues for crop improvement by exploiting physio-morphological traits. Front Plant Sci 8:537

Panuccio MR, Sorgona A, Rizzo M, Cacco G (2009) Cadmium adsorption on vermiculite, zeolite and pumice: batch experiment studies. J Environ Manag 90:364–374

Passamani LZ, Barbosa RR, Reis RS, Heringer AS, Rangel PL, Santa-Catarina C, Grativol C, Veiga CF, Souza-Filho GA, Silveira V (2017) Salt stress induces changes in the proteomic profile of micro-propagated sugarcane shoots. PLoS One 12:e0176076

Patel JS, Raju SVS, Sarma BK, Singh R (2010) Screening and biochemical characterization of certain *Pseudomonas fluorescens* isolates against *Helicoverpa armigera* (Hubner). DMMB 1: 45–54

Patel D, Jha CK, Tank N, Saraf M (2012) Growth enhancement of chickpea in saline soils using plant growth-promoting rhizobacteria. J Plant Growth Regul 31(1):53–62

Patel JS, Kharwar RN, Singh HB, Upadhyay RS, Sarma BK (2017) *Trichoderma anasperellum* (T42) and *Pseudomonas fluorescens* (OKC)-enhances resistance of pea against *Erysiphe pisi* through enhanced ROS generation and lignifications. Front Microbiol 8:306

Patterson DT (1995) Effects of environmental stress on weed/crop interaction. Weed Sci 43:483–490

Paul D, Lade H (2014) Plant-growth-promoting rhizobacteria to improve crop growth in saline soils: a review. Agron Sustain Dev 34:737–752

Pereira SIA, Barbosa L, Castro PML (2015) Rhizobacteria isolated from a metal polluted area enhance plant growth in zinc and cadmium-contaminated soil. Int J Environ Sci Technol 12: 2127–2142

Pérez-Montaño F, Alías-Villegas C, Bellogín R, Del Cerro P, Espuny M, Jiménez-Guerrero I (2014) Plant growth promotion in cereal and leguminous agricultural important plants: from microorganism capacities to crop production. Microbiol Res 169:325–336

Peters K, Breitsameter L, Gerowitt B (2014) Impact of climate change on weeds in agriculture: a review. Agric Sustain Dev 34:707–721

Podile AR, Kishore GK (2007) Plant growth-promoting rhizobacteria. In: Plant-associated bacteria. Springer, Dordrecht, pp 195–230

Potters G, Pasternak TP, Guisez Y, Palme KJ, Jansen MAK (2007) Stress induced morphogenic responses: growing out of trouble? Trends Plant Sci 12:98–105

Pourrut B, Jean S, Silvestre J, Pinelli E (2011) Lead-induced DNA damage in *Vicia faba* root cells: potential involvement of oxidative stress. Mutat Res 726:123–128

Prasad PVV, Pisipati SR, Momcilovic I, Ristic Z (2011) Independent and combined effects of high temperature and drought stress during grain filling on plant yield and chloroplast EF-Tu expression in spring wheat. J Agron Crop Sci 197:430–441

Prasch CM, Sonnewald U (2013) Simultaneous application of heat, drought, and virus to Arabidopsis plants reveals significant shifts in signaling networks. Plant Physiol 162:1849–1866

Prigent-Combaret C, Blaha D, Pothier JF, Vial L, Poirier MA, Wisniewski-Dye F (2008) Physical organization and phylogenetic analysis of *acdR* as a leucine-responsive regulator of the 1-aminocyclopropane-1-carboxylate deaminase gene acdS in phytobeneficial *Azospirillum lipoferum* 4B and other Proteobacteria. FEMS Microbiol Ecol 65:202–219

Prudent M, Salon C, Souleimanov A, Emery RN, Smith DL (2015) Soybean is less impacted by water stress using *Bradyrhizobium japonicum* and thuricin-17 from *Bacillus thuringiensis*. Agric Sustain Dev 35(2):749–757

Rahneshan Z, Nasibi F, Moghadam AA (2018) Effects of salinity stress on some growth, physiological, biochemical parameters and nutrients in two pistachio (*Pistacia vera* L.) rootstocks. J Plant Interact 13(1):73–82. https://doi.org/10.1080/17429145.2018.1424355

Ramegowda V, Senthil-Kumar M (2015) The interactive effects of simultaneous biotic and abiotic stresses on plants: mechanistic understanding from drought and pathogen combination. J Plant Physiol 176:47–54

Ramegowda V, Mysore KS, Senthil-Kumar M (2014) Virus-induced gene silencing is a versatile tool for unraveling the functional relevance of multiple abiotic-stress-responsive genes in crop plants. Front Plant Sci 5:323

Ramu VS, Paramanantham A, Ramegowda V, Mohan-Raju B, Udayakumar M, Senthil-Kumar M (2016) Transcriptome analysis of sunflower genotypes with contrasting oxidative stress tolerance reveals individual- and combined-biotic and abiotic stress tolerance mechanisms. PLoS One 11:e0157522

Reed MLE, Glick BR (2005) Growth of canola (*Brassica napus*) in the presence of plant growth-promoting bacteria and either copper or polycyclic aromatic hydrocarbons. Can J Microbiol 51: 1061–1069

Rochester IJ (2011) Sequestering carbon in minimum-tilled clay soils used for irrigated cotton and grain production. Soil Tillage Res 112:1–7

Rojas-Tapias D, Moreno-Galvan A, Pardo-Diaz S, Obando M, Rivera D, Bonilla R (2012) Effect of inoculation with plant growth-promoting bacteria (PGPB) on amelioration of saline stress in maize (*Zea mays*). Appl Soil Ecol 61:264–272

Rollins JA, Habte E, Templer SE, Colby T, Schmidt J, Von Korff M (2013) Leaf proteome alterations in the context of physiological and morphological responses to drought and heat stress in barley (*Hordeum vulgare* L.). J Exp Bot 64:3201–3212

Ruzzi M, Aroca R (2015) Plant growth-promoting rhizobacteria act as biostimulants in horticulture. Sci Hortic 196:124–134

Sadok W, Sinclair TR (2011) Crops yield increase under water-limited conditions: review of recent physiological advances for soybean genetic improvement. Adv Agron 113:313–337

Safdarian M, Askari H, Shariati JV, Nematzadeh G (2019) Transcriptional responses of wheat roots inoculated with *Arthrobacter nitroguajacolicusto* salt stress. Sci Rep 9:1792

Saharan BS, Nehra V (2011) Plant growth promoting rhizobacteria: a critical review. Life Sci Med Res 2011:21

Saikia J, Sarma RK, Dhandia R, Yadav A, Bharali R, Gupta VK, Saikia R (2018) Alleviation of drought stress in pulse crops with ACC deaminase producing rhizobacteria isolated from acidic soil of Northeast India. Sci Rep 8:1–16

Saleem M, Arshad M, Hussain S, Bhatti AS (2007) Perspective of plant growth promoting rhizobacteria (PGPR) containing ACC deaminase in stress agriculture. J Ind Microbiol Biotechnol 34:635–648

Saleem AR, Bangash N, Mahmood T, Khalid A, Centritto M, Siddique MT (2015) Rhizobacteria capable of producing ACC deaminase promote growth of Velvet Bean (*Mucuna pruriens*) under water stress condition. Int J Agric Biol 17:663–667

Salla V, Hardaway CJ, Sneddon J (2011) Preliminary investigation of *Spartina alterniflora* for phytoextraction of selected heavy metals in soils from Southwest Louisiana. Microchem J 97: 207–212

Sandhya V, Ali SZ, Grover M, Reddy G, Venkateswarlu B (2010) Effect of plant growth promoting *Pseudomonas* spp. on compatible solutes, antioxidant status and plant growth of maize under drought stress. Plant Growth Regul 62:21–30

Sang-Mo K, Radhakrishnan R, Khan AL, Min-Ji K, Jae-Man P, Bo-Ra K, Dong-Hyun S, In-Jung L (2014) Gibberellin secreting rhizobacterium, *Pseudomonas putida* H-2-3 modulates the hormonal and stress physiology of soybean to improve the plant growth under saline and drought conditions. Plant Physiol Biochem 84:115–124

Saravanakumar D, Kavino M, Raguchander T, Subbian P, Samiyappan R (2011) Plant growth promoting bacteria enhance water stress resistance in green gram plants. Acta Physiol Plant 33: 203–209

Sarma R, Saikia R (2014) Alleviation of drought stress in mung bean by strain *Pseudomonas aeruginosa* GGRJ21. Plant Soil 377:111–126

Savary S, Willocquet L, Pethybridge SJ, Esker P, McRoberts N, Nelson A (2019) The global burden of pathogens and pests on major food crops. Nat Ecol Evol 3(3):430–439

Scherm H, Coakley SM (2003) Plant pathogens in a changing world. Australas Plant Pathol 32: 157–165

Schuhegger R, Ihring A, Gantner S, Bahnweg G, Knappe C, Vogg G (2006) Induction of systemic resistance in tomato by N-acyl-L-homoserine lactone-producing rhizosphere bacteria. Plant Cell Environ 29:909–918

Shahid M, Khalid S, Abbas G, Shahid N, Nadeem M, Sabir M (2015) Heavy metal stress and crop productivity. In: Hakeem KR (ed) Crop production and global environmental issues. Springer International Publishing, Cham, pp 1–25

Shakir MA, Asghari B, Arshad M (2012) Rhizosphere bacteria containing ACC deaminase conferred drought tolerance in wheat grown under semi-aridclimate. Soil Environ 31:108–112

Shao HB, Chu LY, Jaleel CA, Zhao CX (2008) Water-deficit stress-induced anatomical changes in higher plants. C R Biol 331(3):215–225

Sharifi M, Ghorbanli M, Ebrahimzadeh H (2007) Improved growth of salinity-stressed soybean after inoculation with salt pre-treated mycorrhizal fungi. J Plant Physiol 164:1144–1151

Sharma P, Khanna V, Kumar PI (2013) Efficacy ofaminocyclopropane-1-carboxylic acid (ACC)-deaminase-producing rhizobacteria in ameliorating water stress in chickpea under axenic conditions. Afr J Microbiol Res 7:5749–5757

Shinada H, Iwata N, Sato T, Fujino K (2013) Genetical and morphological characterization of cold tolerance at fertilization stage in rice. Breed Sci 63:197–204

Shintu PV, Jayaram KM (2015) Phosphate solubilising bacteria (*Bacillus polymyxa*) an effective approach to mitigate drought in tomato (*Lycopersicon esculentum* mill). Trop Plant Res 2:17–22

Shrivastava P, Kumar R (2015) Soil salinity: a serious environmental issue and plant growth promoting bacteria as one of the tools for its alleviation. Saudi J Biol Sci 22(2):123–131

Silvente S, Sobolev AP, Lara M (2012) Metabolite adjustment in drought tolerant and sensitive genotypes in response to water stress. PLoS One 7:e38554

Singh RP, Jha P, Jha PN (2015) The plant-growth-promoting bacterium *Klebsiella* sp. SBP-8 confers induced systemic tolerance in wheat (*Triticum aestivum*) under salt stress. J Plant Physiol 184:57–67

Singh B, Mishra S, Bohra A, Joshi R, Siddique KH (2018) Crop phenomics for abiotic stress tolerance in crop plants. In: Wani SH (ed) Biochemical, physiological and molecular avenues for combating abiotic stress tolerance in plants. Elsevier Inc, Amsterdam, pp 277–296

Sirault XRR, James RA, Furbank RT (2009) A new screening method for osmotic component of salinity tolerance in cereals using infrared thermography. Funct Plant Biol 36:970–977

Skirycz A, Inzé D (2010) More from less: plant growth under limited water. Curr Opin Biotechnol 21:197–203

Spaepen S, Bossuyt S, Engelen K, Marchal K, Vanderleyden J (2014) Phenotypical and molecular responses of *Arabidopsis thaliana* roots as a result of inoculation with the auxin-producing bacterium *Azospirillum brasilense*. New Phytol 201:850–861

Srivastava S, Singh N (2014) Mitigation approach of arsenic toxicity in chickpea grown in arsenic amended soil with arsenic tolerant plant growth promoting *Acinetobacter* sp. Ecol Eng 70(70): 146–153

Srivastava R, Khalid A, Singh US, Sharma AK (2010) Evaluation of arbuscular mycorrhizal fungus, fluorescent *Pseudomonas* and *Trichoderma harzianum* formulation against *Fusarium oxysporum* f. sp. *lycopersici* for the management of tomato wilt. Biol Control 53:24–31

Srivastava S, Bist V, Srivastava S, Singh PC, Trivedi PK, Asif MH (2016) Unraveling aspects of *Bacillus amyloliquefaciens* mediated enhanced production of rice under biotic stress of *Rhizoctonia solani*. Front Plant Sci 7:587

Subramanian S, Smith DL (2015) Bacteriocins from the rhizosphere microbiome - from an agriculture perspective. Front Plant Sci 6:909

Subramanian S, Souleimanov A, Smith DL (2016) Proteomic studies on the effects of lipo-chitooligosaccharide and thuricin 17 under unstressed and salt stressed conditions in *Arabidopsis thaliana*. Front Plant Sci 7:1314

Sukweenadhi J, Kim YJ, Choi ES, Koh SC, Lee SW, Kim YJ, Yang DC (2015) *Paenibacillus yonginensis* DCY84T induces changes in *Arabidopsis thaliana* gene expression against aluminum, drought, and salt stress. Microbiol Res 172:7–15

Sziderics AH, Rasche F, Trognitz F, Sessitsch A, Wilhelm E (2007) Bacterial endophytes contribute to abiotic stress adaptation in pepper plants (*Capsicum annuum* L.). Can J Microbiol 53: 1195–1202

Tan KC, Ipcho SV, Trengove RD, Oliver RP, Solomon PS (2009) Assessing the impact of transcriptomics, proteomics and metabolomics on fungal phytopathology. Mol Plant Pathol 10:703–715

Teplitski M, Robinson JB, Bauer WD (2000) Plants secrete substances that mimic bacterial N-acyl homoserine lactone signal activities and affect population density-dependent behaviors in associated bacteria. Mol Plant-Microbe Interact 13:637–648

Theocharis A, Bordiec S, Fernandez O, Paquis S, Dhondt-Cordelier S, Baillieul F, Barka EA (2012) Burkholderia phytofirmans PsJN primes Vitis vinifera L. and confers a better tolerance to low nonfreezing temperatures. Mol Plant-Microbe Interact 25(2):241–249

Thrall PH, Millsom DA, Jeavons AC, Waayers M, Harvey GR, Bagnall DJ, Brockwell J (2005) Seed inoculation with effective root-nodule bacteria enhances revegetation success. J Appl Ecol 42:740–751

Timmusk S, Wagner EGH (1999) The plant-growth-promoting rhizobacterium *Paenibacillus polymyxa* induces changes in *Arabidopsis thaliana* gene expression: a possible connection between biotic and abiotic stress responses. Mol Plant-Microbe Interact 12(11):951959

Timmusk S, Abd El-Daim IA, Copolovici L, Tanilas T, Kannaste A, Behers L, Niinemets U (2014a) Drought-tolerance of wheat improved by rhizosphere bacteria from harsh environments: enhanced biomass production and reduced emissions of stress volatiles. PLoS One 9: e96086

Timmusk S, Islam A, Abd El D, Lucian C, Tanilas T, Kannaste A (2014b) Drought-tolerance of wheat improved by rhizosphere bacteria from harsh environments: enhanced biomass production and reduced emissions of stress volatiles. PLoS One 9:1–13

Tittabutr P, Sripakdi S, Boonkerd N, Tanthanuch W, Minamisawa K, Teaumroong N (2015) Possible role of 1-aminocyclopropane-1-carboxylate (ACC) deaminase activity of *Sinorhizobium* sp. BL3 on symbiosis with mung bean and determinate nodule senescence. Microbes Environ 30:310–320

Tiwari S, Lata C (2018) Heavy metal stress, signaling, and tolerance due to plant-associated microbes: an overview. Front Plant Sci 9:452

Tiwari S, Lata C, Chauhan PS, Nautiyal CS (2016) *Pseudomonas putida* attunes morphophysiological, biochemical and molecular responses in *Cicer arietinum* L. during drought stress and recovery. Plant Physiol Biochem 99:108–117

Trewavas A (2006) A brief history of systems biology: "every object that biology studies is a system of systems". Plant Cell 18(10):24202430

Tuberosa R (2012) Phenotyping for drought tolerance of crops in the genomic era. Front Physiol 3: 10–20

Upadhyay S, Singh D (2015) Effect of salt-tolerant plant growth-promoting rhizobacteria on wheat plants and soil health in a saline environment. Plant Biol 17(1):288–293

Vacheron J, Desbrosses G, Bouffaud M, Touraine B, Moenne-Loccoz Y, Muller D, Legendre L, Wisniewski-Dye F, Prigent-Combaret C (2013) Plant growth-promoting rhizobacteria and root system functioning. Front Plant Sci 4:356

Vaishnav A, Kumari S, Jain S, Varma A, Tuteja N, Choudhary DK (2016) PGPR-mediated expression of salt tolerance gene in soybean through volatiles under sodium nitroprusside. J Basic Microbiol 56:1274–1288

Valerio M, Lovelli S, Perniola M, Di Tommaso T, Ziska L (2013) The role of water availability on weed-crop interactions in processing tomato for southern Italy. Acta Agric Scand Sect B 63:62–68

Vandana UK, Singha B, Gulzar ABM, Mazumdar PB (2020) Molecular mechanisms in plant growth promoting bacteria to resist environmental stress in plants. In: Molecular aspects of plant beneficial microbes in agriculture. Elsevier, Amsterdam, pp 221–233. https://doi.org/10.1016/B978-0-12-818469-1.00019-5

Vardharajula S, Ali SZ, Grover M, Reddy G, Bandi V (2011) Drought-tolerant plant growth promoting *Bacillus* spp.: effect on growth osmolytes, and antioxidant status of maize under drought stress. J Plant Interact 6:1–14

Varshney RK, Pazhamala L, Kashiwagi J, Gaur PM, Krishnamurthy L, Hoisington D (2011) Genomics and physiological approaches for root trait breeding to improve drought tolerance in chickpea (*Cicer arietinum* L.). In: Costa de Oliveira A, Varshney RK (eds) Root genomics. Springer, The Netherlands, pp 233–250

Vashisth D, Kumar R, Rastogi S, Patel VK, Kalra A, Gupta MM, Gupta AK, Shasany AK (2018) Transcriptome changes induced by abiotic stresses in *Artemisia annua*. Sci Rep 8:3423

Vinocur B, Altman A (2005) Recent advances in engineering plant tolerance to abiotic stress: achievements and limitations. Curr Opin Biotechnol 16:123–132

Waller F, Achatz B, Baltruschat H, Fodor J, Becker K, Fischer M, Heier T, Hückelhoven R, Neumann C, Von Wettstein D, Franken P (2005) The endophytic fungus *Piriformospora indica* reprograms barley to salt-stress tolerance, disease resistance, and higher yield. Proc Natl Acad Sci U S A 102(38):13386–13391

Wang CKE, Glick BR, Defago G (2000) Effect of transferring 1-aminocyclopropane-1-carboxylic acid (ACC) deaminase genes into *Pseudomonas fluorescens* strain CHA0 and its *gacA* derivative CHA96 on their growth-promoting and disease-suppressive capacities. Can J Microbiol 46:898–907

Wang Z, Gerstein M, Snyder M (2009) RNA-Seq: a revolutionary tool for transcriptomics. Nat Rev Genet 10(1):57

Wang A, Dodd IC, Belimov AA, Jiang F (2016) Rhizosphere bacteria containing 1-aminocyclopropane-1-carboxylate deaminase increase growth and photosynthesis of pea plants under salt stress by limiting Na^+ accumulation. Funct Plant Biol 43:161–172

Wasson AP, Richards RA, Chatrath R, Misra SC, Prasad SVS, Rebetzke GJ, Kirkegaard JA, Christopher J, Watt M (2012) Traits and selection strategies to improve root systems and water uptake in water-limited wheat crops. J Exp Bot 63:3485–3498

Wilhite DA, Glantz MH (1985) Understanding the drought phenomenon: the role of definitions. Water Int 10:111–120

Wolkers WF, Balasubramanian SK, Ongstad EL, Zec HC, Bischof JC (2007) Effects of freezing on membranes and proteins in LNCaP prostate tumor cells. Biochim Biophys Acta 1768(3): 728–736

Xiong T, Leveque T, Shahid M, Foucault Y, Mombo S, Dumat C (2014) Lead and cadmium phytoavailability and human bioaccessibility for vegetables exposed to soil or atmospheric pollution by process ultrafine particles. J Environ Qual 43:1593–1600

Xu C, Jiang Z, Huang B (2011) Nitrogen deficiency-induced protein changes in immature and mature leaves of creeping bentgrass. J Am Soc Hortic Sci 136:399–407

Yan J, Smith MD, Glick BR, Liang Y (2014) Effects of ACC deaminase containing rhizobacteria on plant growth and expression of Toc GTPases in tomato (*Solanum lycopersicum*) under salt stress. Botany 92:775–781

Yasmin H, Nosheen A, Naz R, Bano A, Keyani R (2017) L-tryptophan-assisted PGPR-mediated induction of drought tolerance in maize (Zea mays L.). J Plant Interact 12(1):567–578

Yasin NA, Akram W, Khan WU, Ahmad SR, Ahmad A, Ali A (2018) Halotolerant plant growth promoting rhizobacteria modulate gene expression and osmolyte production to improve salinity tolerance and growth in *Capsicum annum* L. Environ Sci Pollut Res 25(23):23236–23250

Yu GR, Zhuang J, Nakayama K, Jin Y (2007) Root water uptake and profile soil water as affected by vertical root distribution. Plant Ecol 189:15–30

Yuquan X, Rong S, Zhixing L (1999) Quickly screening a strain of *Pseudomonas* B8 with both ACC deaminase activity and antagonism against *Fusarium oxysporum* 33(2)

Zahir ZA, Munir A, Asghar HN, Shahroona B, Arshad M (2008) Effectiveness of rhizobacteria containing ACC-deaminase for growth promotion of peas (*P. sativum*) under drought conditions. J Microbiol Biotechnol 18:958–963

Zameer M, Zahid H, Tabassum B, Ali Q, Nasir IA, Saleem M, Butt SJ (2016) PGPR potentially improve growth of tomato plants in salt-stressed environment. Turk J Agric Food Sci Technol 4: 455–463

Zebelo S, Song Y, Kloepper JW, Fadamiro H (2016) Rhizobacteria activates (C)-delta-cadinene synthase genes and induces systemic resistance in cotton against beet armyworm (*Spodoptera exigua*). Plant Cell Environ 39:935–943

Zhao C, Piao S, Wang X, Huang Y, Ciais P, Elliott J, Huang M, Janssens IA, Li T, Lian X, Liu Y (2016) Plausible rice yield losses under future climate warming. Nat Plants 3(1):1–5

Zhao C, Liu B, Piao S, Wang X, Lobell DB, Huang Y, Huang M, Yao Y, Bassu S, Ciais P, Durand JL (2017) Temperature increase reduces global yields of major crops in four independent estimates. Proc Natl Acad Sci U S A 114(35):9326–9331

Zhou X, Wang G, Sutoh K, Zhu JK, Zhang W (2008) Identification of cold-inducible micro RNAs in plants by transcriptome analysis. Biochim Biophys Acta 1779:780–788

Zhou C, Ma ZY, Zhu L, Xiao X, Xie Y, Zhu J (2016) Rhizobacterial strain *Bacillus megaterium* bofc15 induces cellular polyamine changes that improve plant growth and drought resistance. Int J Mol Sci 17:976

Zipfel C, Oldroyd GE (2017) Plant signalling in symbiosis and immunity. Nature 543:328–336

Ziska LH, Tomecek MB, Gealy DR (2010) Evaluation of competitive ability between cultivated and red weedy rice as a function of recent and projected increases in atmospheric CO_2. Agron J 102:118–123

Zuccarini P, Okurowska P (2008) Effects of mycorrhizal colonization and fertilization on growth and photosynthesis of sweet basil under salt stress. J Plant Nutr 31(3):497–513

Microorganisms: The Viable Approach for Mitigation of Abiotic Stress

Kirpa Ram, S. Devi, A. Singh, V. Kaur, J. Kumar, and S.S. Arya

Abstract Abiotic strains such as salinity, drought, and extreme heat are the major rate-limiting factor for growth, yield, and crop productivity. In the direction to raise crop growth and yield, it is very important to develop cost-effective approaches for the management of abiotic stress. The number of strategies is proposed in such a direction by which mitigation of stresses by microorganism application has grown fast and more resilient. Microorganisms are the natural soil microflora having great metabolic aptitudes to the promotion of growth and diminished abiotic stresses. Recent investigation has reported that soil microbial association may have mechanisms for mitigation of abiotic strains viz. salinity, drought and temperature stress, and heavy metals. Plant microorganism communications comprise very complex mechanisms under stress conditions by connecting the molecular, biochemical, and physiological processes. Microbes are effectively mitigating the assessment of abiotic stresses through the formation of biofilm and production of exopolysaccharides. Species of *Azospirillum*, *Pseudomonas*, and *Bacillus* induce tolerance under drought, salinity, and nutrient-deficit conditions. Strains of *Trichoderma* enhance the tolerance to salinity, and drought has also been reported. *Glomus* sp. enhance nitrogen, phosphorous uptake, accumulation of compatible osmolytes, and structural changes under abiotic stress. This book chapter pointed

K. Ram
Department of Botany, Baba Mastnath University, Rohtak, Haryana, India

S. Devi
Department of Botany & Plant Physiology, CCS Haryana Agricultural University, Hisar, Haryana, India

A. Singh
Department of Botany, Akal University, Bathinda, Punjab, India

V. Kaur
National Bureau of Plant Genetic Resources, New Delhi, India

J. Kumar
Department of Biosciences-UIBT, Chandigarh University, Mohali, Punjab, India

S.S. Arya (✉)
Department of Botany, Maharshi Dayanand University, Rohtak, Haryana, India

© The Author(s), under exclusive license to Springer Nature Singapore Pte Ltd. 2022

A. Vaishnav et al. (eds.), *Plant Stress Mitigators*,
https://doi.org/10.1007/978-981-16-7759-5_15

the plant–microorganism relationship under abiotic stresses and an overview of current knowledge on how the plant–microorganism relationship helps in improving abiotic strain condition.

Keywords Microorganism · Abiotic stress · Plant–microorganism relationship · Stress tolerance

Introduction

Fluctuating environmental conditions generating abiotic stresses are among the major rate restrictive influences for deterioration in agronomic productivity (Grayson 2013). According to the FAO, only 3.5% of the total cultivable land area is not affected by any environmental stresses. Water scarcity has pretentious 64% of the worldwide land area, whereas 13% (flooding), 6% (salinity), 9% (mineral deficiency), 15% (acidic soils), and 57% (cold) were influenced by these stress (Cramer et al. 2011). Water, air, and nutrients are the basic requirements for optimum growth, development, and reproduction of plants. Unfavorable conditions (above and below the optimum levels) of water, salts, and temperature had limits on plant growth and development (Jiang et al. 2016). The most understandable effect of hostile events manifests itself first and foremost at the cellular level, after which physiological symptoms become apparent. Various environmental stresses such as water stress (Osakabe et al. 2014), low or high light intensities (Xu et al. 2016), freezing (cold) injury (Pareek et al. 2010), and edaphic factors such as acidity, salinity, and alkalinity of soils (Bromham et al. 2013), pollutant contamination, and anthropogenic trepidations (Emamverdian et al. 2015) adversely affect the growth and development. In any stressful environment, the interaction of microbes and plants is critical for the adaptation and survival of the partnership relationship between microorganism and plants (Nadeem et al. 2014).

The role of microbes to attenuate abiotic stresses in plants has been a great concern in the last few years (de Souza et al. 2015). Microbe-mediated induced systemic tolerance (IST) physical and chemical changes in plants stimulate microbe-mediated tolerance to abiotic stress. Microbes with their impending inherent metabolic and genetic abilities contribute to diminishing the abiotic stresses in the crops (Gopalakrishnan et al. 2015). The rhizospheric incumbent belonging from the genera *Pseudomonas*, *Azotobacter*, *Azospirillum*, *Rhizobium*, *Pantoea*, *Bacillus*, *Enterobacter*, *Bradyrhizobium*, *Methylobacterium*, *Burkholderia*, *Trichoderma*, and cyanobacteria is acted as a plant growth stimulator and multiple varieties of abiotic stress extenuator (Sorty et al. 2016; Ahmad et al. 2015; Pandey et al. 2016).

In recent years, there has been a significant deal of interest in the role of microorganisms in mitigating abiotic stressors in plants (de Souza et al. 2015). Physical and chemical changes in plants enhance microbe-mediated abiotic stress tolerance. Microbes, with their innate metabolic and genetic abilities, help to the reduction in abiotic stressors in crops (Gopalakrishnan et al. 2015). *Pseudomonas*, *Azospirillum*, *Pantoea*, *Enterobacter*, *Bacillus*, *Bradyrhizobium*, *Burkholderia*,

Trichoderma, Azotobacter, Methylobacterium, Rhizobium, and *cyanobacteria* are rhizospheric incumbents that act as plant growth stimulators and abiotic stress extenuators (Sorty et al. 2016; Ahmad et al. 2015; Pandey et al. 2016).

Under the effect of various abiotic stimuli, microbes were able to increase the synthesis of antioxidant enzymes, the accumulation of appropriate osmolytes, the improvement of germination, and the expression of stress-responsive genes (Sorty et al. 2018). A variety of fungi, bacteria, and actinomycetes directly interact with the host plant. The microorganism also improves plant nutrition under abiotic stress conditions by changing soil properties, such as plant-beneficial fungi and mycorrhiza in the root rhizosphere, which stimulates soil accretion and colonization (Hayat et al. 2010; Wang et al. 2012). This chapter discusses the constructive relationships between plants and the accompanying soil microflora in abiotically challenged situations, as well as the role of plant microbes in mitigating various abiotic pressures.

Effects of Stress on Plants

Stress never encounters plants alone, every time it emanates, and it also brings with a combination of another abiotic stress. The effect of any digit of stressors at a particular point varies greatly with determinant factors such as exposer duration, growth stage, and susceptibility (Sorty et al. 2018). Furthermore, many times one stress is antagonistic to adverse effects of each other (Iyer et al. 2013). Water and salt stress during wheat seedlings altered the impact of high temperature by enhancing the antioxidative defense system in wheat seedlings (Keleş and Öncel 2002). Plants showing under drought, temperature stress, or a combination of drought and temperature stress have been shown to accumulate sucrose and other sugars such as maltose and glucose (Rizhsky et al. 2004; Koussevitzky et al. 2008).

Effects of Drought Stress on Plants

Drought is a multifaceted stress that can negatively affect cell division, elongation, and differentiation through turgor loss, altered enzyme actions, decreased photosynthetic activity, and ultimately quantity and quality of crop development and production (Nezhadahmadi et al. 2013). Increased buildup of reactive oxygen species is caused by a decrease in the occupancy of the photosynthetic electron transport chain (Ghosh and Xu 2014). Photoinhibition enhanced the amount of ABA in drought-stressed plants, which primes stomatal closure for dehydration prevention and the generation of defense bodies (e.g., compatible osmolytes) and aided membrane structure maintenance (Verslues et al. 2006). Drought stress is one of the most damaging impacts that wreaks havoc on plant growth and development, and it can result in considerable losses in plant productivity (Ciais et al. 2005). Drought effects vary according on the strength and duration of the drought, the growth phase and plant variety, and the inscription of earlier stress experiences in plants (Fleta-Soriano and Munné-Bosch 2016). Plant turgor and water potential decrease in dehydration circumstances; as a result, cells are unable to complete their normal metabolic tasks, resulting in decreased cell expansion and growth (Rahdari and Hoseini 2012).

Effects of Salt Stress on Plants

The process of increased saline content viz. NaCl, Na_2CO_3, $MgSO_4$, Na_2SO_4, $CaSO_4$, $MgCl_2$, and KCl within the soil and electrical conductivity 4 dS m^{-1} (around 40 mM NaCl) at 25 °C designates the term salinity or salinization (Tavakkoli et al. 2010; Shrivastava and Kumar 2015). Excess salt concentration causes a number of adverse effects in plants at diverse stages of growth from germination, vegetative, flowering, and among them the significantly influenced plant growth stages and development (Liang et al. 2018). Saline environment altered the physiological and metabolic processes in plants and severity of stress depending on the environmental factors, growth conditions, and plant growth stage irrigation management strategies (Gupta and Huang 2014; Kaleem et al. 2018). Roots established a connecting relationship between the soil and plants (Acosta-Motos et al. 2017). Accumulation of lethal ions (e.g., Na^+) at cellular level instigating necrosis and chlorosis reduced the cellular metabolic activity (Panuccio et al. 2014), causing secondary stresses viz. osmotic stress and oxidative stress and damaging the membrane lipids content, nucleic acid, and protein (Perez-Lopez et al. 2009). Salt stress impairs root number, root diameter, root length, and root hair number growth by dropping elongation and cell division (Duan et al. 2015). Accumulation of Na^+ inhibits a numbers of enzyme activity (Das et al. 2015), antagonistic with uptake of other ions (Paul and Lade 2014), seed germination (Xue et al. 2004), carbohydrate partition (Carillo et al. 2011), augmented chlorophyll-degrading enzyme (chlorophyllase) activity (Saravanavel et al. 2011), conduction of stomata (Hanin et al. 2016), disruption of thylakoid membrane, increase in the reactive oxygen species activity (Parida and Das 2005), reduction in the plant lipid content (Keshtehgar et al. 2013), and increase in desirable osmolytes concentration (Singh et al. 2000).

Effects of Heat Stress on Plants

Intergovernmental Panel on Climate Change report from 1880 to 2012 showed increase in global average surface temperature stretched 0.85 °C, and report also predicted that annual daily temperature would be raised around 1–3 °C by the mid of twenty-first century (Meehl and Tebaldi 2004). Seasonal high temperature is more pronounced effect on vegetative growth and development which in turns decline the reproductive growth and crop yield (Zinn et al. 2010; Varshney et al. 2011). High temperature adversely affects seed vigor and germination (Teixeira et al. 2019), stem and leave abscission and senescence, and inhibition of shoot, root, and fruit growth (Wahid et al. 2007; Tian et al. 2009). Impose of heat stress causes various physiological alterations in plants viz. membrane disfunctioning and increased membrane lipid fluidity (Chen et al. 2010), protein aggregation and denaturation (Mittler et al. 2012), slower activity of chloroplast and mitochondria enzymes and imbalance in the metabolic homeostasis, and production of antioxidants and abscisic acid. Increased air temperature causes increased transpiration, which reduces photosystem II (PSII) activity leading to decreased photosynthesis and, eventually, oxidative loss of the photosynthetic electron transport chain (Feng et al. 2014). Photosystem II suppression reduces variable chlorophyll fluorescence (Greer and Weedon 2012), increases thylakoid damage, loss of grana stacking, and grana swelling (Downs et al. 2013), decreases carbon metabolism (Song et al. 2014), and nitrogen-fixing ability in pulse crops (Bansal et al. 2014). Heat causes stomatal closure, which influences intercellular carbon dioxide concentration, leaf gas exchange, and CO_2 assimilation rates (McAdam et al. 2016; Bindumadhava et al. 2015).

Effects of Cold/Chilling Stress on Plants

Low temperature is a major ecological impact that causes epidermal damage and limits the distribution, plant survivability, and severe damages in crop productivity (Rudell et al. 2011; Theocharis et al. 2012; Wu et al. 2015). Over the past few years, regularity, duration, and hardness of low-temperature stresses have steadily increased in many areas of the world (Sanghera et al. 2011). Freezing and cold suffering plants routinely show severe morphological and physiological responses, comprising cellular structure damage and decrease in processes like rate of evaporation, chlorophyll content, stomatal conduction, and photosynthetic rate (Ogweno et al. 2009; Schreiber et al. 2013). Exposer of cold/freezing causes reduction in superoxide dismutase (SOD) activity, and PSII quantum yield, however, an increase in dark respiration, and abscisic acid content (Yoshikawa et al. 2007; Ogweno et al. 2009).

Effects of Heavy Metal Stress on Plants

Industrialization has also become a serious concern as heavy metals accumulation has increased in the soil over a few years which affects the soil and crop productivity (Shahid et al. 2015). Accumulation of heavy metal damages soil properties by affecting soil pH, electrical conductivity, soil texture, and physiological and biochemical properties of soil (Hassan et al. 2017). Cu, Mn, Co, Zn, Ni, and Mo are vital element for vigorous living activities and developmental pathways (Shahid et al. 2015), whereas synergetic relationship of these elements that accumulate Pb, As, Hg,

Al, Cd, Be, and Cr had determinate effects on crop yield toward high magnitude when absorption increases beyond supra-optimal levels (Pierart et al. 2015; Xiong et al. 2014). Metal toxicity causes enzymatic breakdown, metabolic disorder, morphological irregularities, and crop yield reduction (Amari et al. 2017) and ultimately leads to increase in the production of reactive oxygen molecules like as O_2^-, H_2O_2, and OH^-, consequently interruption of the cell redox homeostasis (Pourrut et al. 2011; Ibrahim et al. 2015; Verma et al. 2017).

Role of Microbes to Stress Management

A diverse group of bacteria created the plant root system, forming a complex rhizosphere that controls plant development through metabolism and plant interconnections (Schmidt et al. 2014). Plant-related microorganism communities around the root region not only improve drought stress, but also promote growth and yield of plants (Cherif et al. 2015). A slight manipulation in soil rhizosphere in *Brassica rapa* showed that smaller microbial communities, reduced chlorophyll content, less flowers, and fewer are productive as compared with crop growing on soil with diverse microbial populations (Lau and Lennon 2011). *S. subrutilus* Wbn2-11, *P. polymyxa* Mc5Re-14, and *B. subtilis* Co1-6 (Egypt Gram-positive strains) and *S. plymuthica* 3Re4-18, *S. rhizophila* P69, and *P. fluorescens* L13-6-12 (European Gram-negative strains) are applied during seedling stage that showed enhancement of the biological active secondary metabolite apigenin-7-*O*-glucoside in chamomile under stress condition. The beneficial microbiome associated with roots and plant tissues alleviates plant stress through a variety of metabolic processes (Berg et al. 2013). Growth-promoting microorganism directly enhances micro-nutrient absorption and affects phyto-hormones homeostasis, or secondarily encourages immune system of plant against phyto-pathogens (Balloi et al. 2010).

Microorganism-Based Management of Drought Stress

Microorganisms used in drought-stressed plants increase growth and nutrient management by colonizing the plant's endorhizosphere and promoting plant development through a variety of direct and indirect processes (Grover et al. 2011). Plant drought tolerance caused by microorganisms may involve the production of gibberellic acid, abscisic acid (ABA), indole acetic acid, and cytokinins, as well as persuaded systemic tolerance, reduced root ethylene levels, and the formation of bacterial exopolysaccharides (Kim et al. 2013; Timmusk et al. 2014). IAA generated by *Azospirillum* sp. improved plant drought resistance (Dimkpa et al. 2009). *A. brasilense* produces nitric oxide gas in tomato plants, which acts as a signaling molecule in the IAA-generating pathway and aids in adventitious root formation (Molina-Favero et al. 2008). When compared to noninoculated controls of

Phaseolus vulgaris L., *A. brasilense* Cd inoculation increased root surface and specific area, specific root length (German et al. 2000), auxin (Dimkpa et al. 2009), absolute water content, and proline content (Casanovas et al. 2002). The *Azospirillum* and wheat interaction increased leaf water content and decreased leaf water potential, which increased IAA production, root growth, and lateral root formation, and so increased water and nutrient use efficiency in a drought situation (Arzanesh et al. 2011). *B. thuringiensis* improves IAA nutrition, metabolic activity, and plant physiology in *Lavandula* dentate plants (Armada et al. 2014).

Inoculation of gibberellins excreting rhizobacterium *P. putida* H2-3 in soybean plants (Sang-Mo et al. 2014), *Azospirillum lipoferum* in maize plants (Cohen et al. 2009), and *Phyllobacterium brassicacearum* strain STM196 in *Brassica napus* (Bresson et al. 2013) improved water stress tolerance finally plant growth. Inoculation of cytokinin-producing PGPR (*Bacillus subtilis*) in *Platycladus orientalis* seedlings with ABA increased the stomatal conductance and convening drought stress resistance (Liu et al. 2013). *R. phaseoli* (strain MR-2), *R. leguminosarum* (strain LR-30), and *M. ciceri* (strains CR-30 and CR-39) in wheat crop increase auxin synthesis and improved tolerance to water stress and the growth and biomass (Hussain et al. 2014).

A volatile metabolite 2*R*,3*R*-butanediol produced by *Pseudomonas chlororaphis* O6 colonized with Arabidopsis roots facilitate stomatal closer, prevent water loss, and provide resistance to drought stress (Cho et al. 2008). Pea plants inoculated with *Variovorax paradoxus* 5C-2, containing ACC deaminase under soil-dehydrating situation enhanced the seed number, seed yield, nodulation and restoration and ripening of seed under drought condition (Dodd et al. 2005; Arshad et al. 2008).

Pisum sativum with *P. fluorescens* biotype G (ACC-5) induced longer roots (Zahir et al. 2008), *Achromobacter piechaudii* ARV8 in pepper and tomato plant reduced production of ethylene (Mayak et al. 2004), *Bacillus* isolate (23-B) and *Pseudomonas* (6-P) with *M. ciceris* in chickpea improved proline accumulation, germination, and biomass of seedling (Sharma et al. 2013), *Bacillus thuringiensis* AZP2 in wheat plant improved photosynthetic efficiency (Timmusk et al. 2014), and strain of *B. licheniformis* (K11) increased the countenance of stress-related genes such as VA, sHSP, Cadhn, and CaPR-10 in pepper plants (Hui and Kim 2013).

Microorganism-Based Management of Salt Stress

The tolerance to salinity and the availability of certain nutrients are two selection pressures, both of which are required for bacteria to colonize the rhizosphere (Matilla et al. 2007). These can live in the cellular space of diverse tissues and plant organs without causing an outward infection indication or having a harmful effect on the host (Weyens et al. 2009; Kandel et al. 2017). Several studies comparing rhizosphere and microbial endophytic communities show a significant difference in species assemblages, but also that endophytic bacteria and their societies have several distinctive characteristics (Kushwaha et al. 2020). These microbial communities are entwined in osmoregulating halophytes, either directly or indirectly, which aid their survival in salinity stress circumstances. The diversity, abundance, and evenness of rhizosphere bacterial communities appear to be greater than that of endophytic populations (Huang 2018). When the tissue and rhizospheric soil of the plant are taken into account, bacterial phyla differ in the former and *acidobacteria*, *bacteroidets*, and *planctomycetes*, with *proteobacteria* and *chloroflexi predominating* (Kandel et al. 2017). Because of their rhizosphere and rhizoplane selection of soil bacteria, Liu et al. (2017) argued that rooted plants operate as "gatekeepers." In addition to bacteria, mycorrhizal symbioses have been found to have an important role in improving plant nutrition, particularly when environmental challenges are present (Qin et al. 2017).

Salt-resistant bacteria include mainly the strains from *Pseudomonas*, *Bacillus*, *Enterobacter*, *Streptomyces*, *Agrobactium*, *Ochrobactrum*, and *Klebsiella* (Sharma et al. 2016).

Microorganism-Based Management of Heat Stress

High temperatures in tropical and subtropical areas not only affect directly but also increase soil temperature that leads to demises crop production and microbial colonization (Grover et al. 2011). Denaturation and degradation of proteins are the basic reasons of cellular damage caused by elevated temperatures. Heat shock proteins are a form of polypeptide that is produced in response to a rapid increase in temperature in all organisms. Chaperones (such as DnaK, GroEL, GroES, DnaJ, ClpB, ClpA, sHSP, and ClpX) make up heat shock proteins and played a role in the proper folding of proteins denatured by stress, and proteases are required to break down irrevocably denatured proteins (Munchbach et al. 1999). High-temperature exposer shows induction of HSPs in the thermotolerant *P. putida* strain AKMP7 and *P. aeruginosa* strain AMK-P6 isolated from a semi-arid area (Ali et al. 2009). Under high-temperature situation in pigeon pea *Pseudomonas*, AKM-P6 strain diminishes membrane injury and chlorophyll structure and improves cellular metabolic proteins amino acid, proline, and sugars (Ali et al. 2009) and *Glomus etunicatum* improve transpiration rate, stomatal conductance, and photosynthetic rate under high-temperature condition in maize (Zhu et al. 2011). According to Grover et al. (2010), certain bacterial strains influence plant resistance to high temperatures. *Pseudomonas* sp. thermotolerance strain NBRI0987 in sorghum seedlings increased plant biomass and synthesized high molecular weight proteins in the leaves (Grover et al. 2010).

Microorganism-Based Management of Cold/Chilling

Cold stress is caused by physical and chemical changes that result in low temperatures among biological molecules, with freezing (0 °C) and cooling (10–15 °C) circumstances enforcing plant growth processes (Körner 2006). A number of bacteria, including *Erwinia herbicola* and *Pseudomonas syringae*, protect plants from frost damage by causing the production of ice crystals at subzero temperatures (2–10 °C) when water would otherwise stay supercooled and liquid (Skirvin et al. 2000). Competitive exclusion of microbial isolates (*P. syringae* and *P. fluorescens*) with naturally occurring or genetically produced "ice-minus" mutations is thought to be a viable method of freezing management (Lindow and Leveau 2002).

There are many microorganism strains that protect the plant in cold temperatures by altering physical and metabolic changes viz. *Glomus mosseae* enhance concentration of reducing sugars in wheat (Paradis et al. 1995) and *G. etunicatum* inoculation in *Z. mays* improve antioxidant potential and decrease lipid peroxidation in AMF inoculated plants at various temperatures, with the majority of benefits occurring at 5 °C (Zhu et al. 2010). Mishra et al. (2011) found that *Pseudomonas* sp. strains under cold stress maintain membrane integrity, relative amount of water and biomass of root and shoot in *T. aestivum,* and *B. phytofirmans* strain *V. vinifera* crop enhance trehalose (T6P) when plant exposed to 4 °C (Fernandez et al. 2012). *Piriformospora indica*, *Epicoccum nigrum*, *Chaetomium globosum*, *Glomus versiforme*, and *R. irregularis* showed positive effects on photosynthesis, biomass, uptake phosphorus, and turgor regulation under cold stress (5–25 °C) in barley species (Murphy et al. 2014; Hajiboland et al. 2019). Plants' ability to tolerate cold can be increased by exposing them to low, nonfreezing conditions. *Cuvularia* sp., an endophytic fungus isolated from *Dichanthelium lanuginosum*, is thermotolerant to temperatures, when grown on geothermal soil (Redman et al. 2011).

Microorganism-Based Management of Heavy Metal Stress

Plant productivity is heavily impacted by heavy metals, and their pollution of farm land has had an adverse effect on human health. Metal-chelating bacteria affect the absorption of various metals, i.e., Zn, Cu, and Fe (Dimkpa et al. 2009). Inoculation of *Klebsiella mobilis*, Cd-contaminated soil-grown barley plants developed more grain and had lower Cd levels (Idris et al. 2004). Surfactant rhamnolipid was secreted by *P. aeruginosa* that showed specificity for Cd and Pb, and *Leptospirillum ferrooxidans* and *Thiobacillus ferrooxidans* oxidized Fe and S (Sand et al. 1992). *Pseudomonas stutzeri* AG259 produce silver-based single crystals, which minimize metal toxicity (Klaus-Joerger et al. 2001). Jiang and coworkers in 2017 found a class of rhizosphere bacteria from the genera *Cupriavidus*, *Pseudomonas*, *Acinetobacter*, *and Bacillus* defended *Boehmeria nivea* from excess Pb, Cu, and Cd concentrations.

Bacterial community in heavy metal-polluted environments is dominated by *Firmicutes, Proteobacteria, and Actinobacteria*, with *Bacillus*, *Pseudomonas*, and *Arthrobacter* being the most often discovered taxa (Pires et al. 2017). The legume–rhizobia symbiosis is well regarded for its ability to detoxify heavy metals and increase the condition of polluted soils (Checcucci et al. 2017). Heavy metal infected soils, and the most widely identified fungi are *Ascomycota* and *Basidiomycota* (Narendrula-Kotha and Nkongolo 2017). The IAA-producing strain *B. subtilis* SJ-101 boosted the growth of *Brassica juncea* in Ni-contaminated soil (Zaidi et al. 2006). Indole acetic acid-generating strains of *B. subtilis* were reported to stimulate fast root elongation in *B. juncea* in Cd-contaminated soil (Belimov et al. 2005). In addition to IAA and ACC deaminase, phosphate solubilizers, siderophore developers, and nitrogen-fixing HMT-PGP bacteria help plant development and root production by enhancing nutrient delivery and modifying heavy metal bioavailability (Gupta et al. 2014).

Conclusion

When a plant is stressed, it lacks access to essential nutrients; in this situation, microorganisms make it possible for them to communicate with the plant. Microorganisms produce bioactive chemicals that improve the supply of nutrients and water to the plant. Microorganisms have a strong root colonization capability, which helps them to cope with environmental pressures by developing adaptive defense mechanisms, thereby increasing crop yield and productivity. Designers can produce stress-tolerant crop varieties using genetic modification and plant breeding, but this is a time-consuming process, while microbial vaccines are a relatively cost-effective and environmentally friendly way to reduce stress in soon-to-be-available plants. Some microbial species and strains can be useful in assessing plant stress tolerance, adaptation, and processes that survive in stressed plants, among other things. Biotechnology in agriculture could benefit from the use of microorganisms from stressed conditions.

References

Acosta-Motos JR, Ortuno MF, Bernal-Vicente A, Diaz-Vivancos P, Sanchez-Blanco MJ, Hernandez JA (2017) Plant responses to salt stress: adaptive mechanisms. Agronomy 7:18

Ahmad P, Hashem A, Abd-Allah EF, Alqarawi AA, John R, Egamberdieva D et al (2015) Role of *Trichoderma harzianum* in mitigating NaCl stress in Indian mustard (*Brassica juncea* L) through antioxidative defense system. Front Plant Sci 6:868

Ali SkZ, Sandhya V, Grover M, Kishore N, Rao LV, Venkateswarlu B (2009) Pseudomonas sp. strain AKM-P6 enhances tolerance of sorghum seedlings to elevated temperatures. Biol Fertil Soils 46:45–55

Amari T, Ghnaya T, Abdelly C (2017) Nickel, cadmium and lead phytotoxicity and potential of halophytic plants in heavy metal extraction. S Afr J Bot 111:99–110

Armada E, Roldan A, Azcon R (2014) Differential activity of autochthonous bacteria in controlling drought stress in native Lavandula and Salvia plants species under drought conditions in natural arid soil. Microb Ecol 67:410–420

Arshad M, Sharoona B, Mahmood T (2008) Inoculation with Pseudomonas spp. containing ACC deaminase partially eliminate the effects of drought stress on growth, yield and ripening of pea (*P. sativum* L.). Pedosphere 18:611–620

Arzanesh MH, Alikhani HA, Khavazi K, Rahimian HA, Miransari M (2011) Wheat (*Triticum aestivum* L.) growth enhancement by *Azospirillum* sp. under drought stress. World J Microbiol Biotechnol 27:197–205

Balloi A, Rolli E, Marasco R, Mapelli F, Tamagnini I, Cappitelli F et al (2010) The role of microorganisms in bioremediation and phytoremediation of polluted and stressed soils. Agrochimica 54:353–369

Bansal M, Kukreja K, Sunita S, Dudeja SS (2014) Symbiotic effectivity of high temperature tolerant mungbean (*Vigna radiata*) rhizobia under different temperature conditions. Int J Curr Microbiol Appl Sci 3:807–821

Belimov AA, Hontzeas N, Safronova VI, Demchinskaya SV, Piluzza G, Bullitta S et al (2005) Cadmium-tolerant plant growth promoting rhizobacteria associated with the roots of Indian mustard (*Brassica juncea* L. Czern). Soil Biol Biochem 37:241–250

Berg G, Zachow C, Müller H, Philipps J, Tilcher R (2013) Next-generation bio-products sowing the seeds of success for sustainable agriculture. Agronomy 3:648–656

Bindumadhava H, Nair RM, Easdown W (2015) Physiology of mungbean accessions grown under saline and high temperature conditions. AVRDC-annual report, No. 74199, Tainan, Weihai Shanhua Carpet Group Co., Ltd., pp 1–46

Bresson J, Varoquaux F, Bontpart T, Touraine B, Vile D (2013) The PGPR strain *Phyllobacterium brassicacearum* STM196 induces a reproductive delay and physiological changes that result in improved drought tolerance in Arabidopsis. New Phytol 200:558–569

Bromham L, Saslis-Lagoudakis CH, Bennett TH, Flowers TJ (2013) Soil alkalinity and salt tolerance: adapting to multiple stresses. Biol Lett 9:20130642

Carillo P, Annunziata MG, Pontecorvo G, Fuggi A, Woodrow P (2011) Salinity stress and salt tolerance. In: Shanker A, Venkateswarlu B (eds) Abiotic stress in plants, mechanisms and adaptations. IntechOpen, London, pp 21–38

Casanovas EM, Barassi CA, Sueldo RJ (2002) *Azospirillum* inoculation mitigates water stress effects in maize seedlings. Cereal Res Commun 30:343–350

Checcucci A, Bazzicalupo M, Mengoni A (2017) Exploiting nitrogen-fixing rhizobial symbionts genetic resources for improving phytoremediation of contaminated soils. In: Naser A, Anjum A, Gill SS, Tuteja N (eds) Enhancing cleanup of environmental pollutants: biological approaches, vol 1. Springer International Publishing, Cham, pp 275–288

Chen J, Wang P, Mi H-L, Chen G-Y, Xu D-Q (2010) Reversible association of ribulose-1,5-bisphosphate carboxylase/oxygenase activase with the thylakoid membrane depends upon the ATP level and pH in rice without heat stress. J Exp Bot 61:2939–2950

Cherif H, Marasco R, Rolli E, Ferjani R, Fusi M, Souss A et al (2015) Oasis desert farming selects environment-specific date palm root endophytic communities and cultivable bacteria that promote resistance to drought. Environ Microbiol Rep 7:668–678

Cho SM, Kang BR, Han SH, Anderson AJ, Park JY, Lee YH et al (2008) 2R,3R-Butanediol, a bacterial volatile produced by *Pseudomonas chlororaphis* O6, is involved in induction of systemic tolerance to drought in *Arabidopsis thaliana*. Mol Plant Microb Interact 21:1067–1075

Cohen AC, Travaglia CN, Bottini R, Piccoli PN (2009) Participation of abscisic acid and gibberellins produced by endophytic *Azospirillum* in the alleviation of drought effects in maize. Botanique 87:455–462

Ciais P, Reichstein M, Viovy N et al (2005) Europe-wide reduction in primary productivity caused by the heat and drought in 2003. Nature 437:529–533. https://doi.org/10.1038/nature03972

Cramer GR, Urano K, Delrot S, Pezzotti M, Shinozaki K (2011) Effects of abiotic stress on plants: a systems biology perspective. BMC Plant Biol 11:163

Das P, Nutan KK, Singla-Pareek SL, Pareek A (2015) Understanding salinity responses and adopting 'omics-based' approaches to generate salinity tolerant cultivars of rice. Front Plant Sci 6:712

de Souza PC, Morey AT, Castanheira GM, Bocate KP, Panagio LA, Ito FA, Furlaneto MC, Yamada-Ogatta SF, Costa IN, Mora-Montes HM, Almeida RS (2015) Tenebrio molitor (Coleoptera: Tenebrionidae) as an alternative host to study fungal infections. J Microbiol Methods 118:182–186

Dimkpa C, Weinand T, Asch F (2009) Plant-rhizobacteria interactions alleviate abiotic stress conditions. Plant Cell Environ 32:1682–1694

Dodd IC, Belimov AA, Sobeih WY, Safronova VI, Grierson D, Davies WJ (2005) Will modifying plant ethylene status improve plant productivity in water-limited environments? 4th International Crop Science Congress

Downs CA, McDougall KE, Woodley CM, Fauth JE, Richmond RH, Kushmaro A, Gibb SW, Loya Y, Ostrande KR, Kramarsky-Winter E (2013) Heat-stress and light-stress induce different cellular pathologies in the symbiotic dinoflagellate during coral bleaching. PLoS One 8:771–773

Duan L, Sebastian J, Dinneny JR (2015) Salt-stress regulation of root system growth and architecture in Arabidopsis seedlings. Methods Mol Biol 1242:105–122

Emamverdian A, Ding Y, Mokhberdoran F, Xie Y (2015) Heavy metal stress and some mechanisms of plant defense response. Sci World J 2015:756120

Feng B, Liu P, Li G, Dong ST, Wang FH, Kong LA, Zhang JW (2014) Effect of heat stress on the photosynthetic characteristics in flag leaves at the grain-filling stage of different heat-resistant winter wheat varieties. J Agron Crop Sci 200:1931–2250

Fernandez O, Vandesteene L, Feil R, Baillieul F, Lunn JE, Clément C (2012) Trehalose metabolism is activated upon chilling in grapevine and might participate in *Burkholderia phytofirmans* induced chilling tolerance. Planta 236:355–369

Fleta-Soriano E, Munné-Bosch S (2016) Stress memory and the inevitable effects of drought: a physiological perspective. Front Plant Sci 7

German MA, Burdman S, Okon Y, Kigel J (2000) Effects of Azospirillum brasilense on root morphology of common bean (*Phaseolus vulgaris* L.) under different water regimes. Biol Fertil Soil 32:259–264

Ghosh D, Xu J (2014) Abiotic stress responses in plant roots: a proteomics perspective. Front Plant Sci 1:1–13

Gopalakrishnan S, Sathya A, Vijayabharathi R, Varshney RK, Gowda CL, Krishnamurthy L (2015) Plant growth promoting rhizobia: challenges and opportunities. Biotech 5:355–377

Grayson M (2013) Agriculture and drought. Nature 501:S1

Greer DH, Weedon MM (2012) Modelling photosynthetic responses to temperature of grapevine (*Vitis vinifera* cv. Semillon) leaves on vines grown in a hot climate. Plant Cell Environ 35:1050–1064

Grover P, Rekhadevi P, Danadevi K, Vuyyuri S, Mahboob M, Rahman M (2010) Genotoxicity evaluation in workers occupationally exposed to lead. Int J Hyg Environ Health 213(2):99–106

Grover M, Ali SkZ, Sandhya V, Venkateswarlu B (2011) Role of microorganisms in adaptation of agricultural crops to abiotic stresses. World J Microbiol Biotechnol 27:1231–1240

Gupta B, Huang B (2014) Mechanism of salinity tolerance in plants: physiological, biochemical, and molecular characterization. Int J Genomics 701596:1

Gupta DK, Chatterjee S, Datta S, Veer V, Walther C (2014) Role of phosphate fertilizers in heavy metal uptake and detoxification of toxic metals. Chemosphere 108:134–144

Hajiboland R, Joudmand A, Aliasgharzad N, Tolrá R, Poschenrieder C (2019) Arbuscular mycorrhizal fungi alleviate low-temperature stress and increase freezing resistance as a substitute for acclimation treatment in barley. Crop Pasture Sci 70:218–233

Hanin M, Ebel C, Ngom M, Laplaze L, Masmoudi K (2016) New insights on plant salt tolerance mechanisms and their potential use for breeding. Front Plant Sci 7:1787

Hassan TU, Bano A, Naz I (2017) Alleviation of heavy metals toxicity by the application of plant growth promoting rhizobacteria and effects on wheat grown in saline sodic field. Int J Phytoremediation 19:522–529

Hayat R, Ali S, Amara U, Khalid R, Ahmed I (2010) Soil beneficial bacteria and their role in plant growth promotion: a review. Ann Microbiol 60:579–598

Huang Y (2018) Comparison of rhizosphere and endophytic microbial communities of Chinese leek through high-throughput 16S rRNA gene Illumina sequencing. J Integr Agric 17:359–367

Hui LJ, Kim SD (2013) Induction of drought stress resistance by multifunctional PGPR *Bacillus licheniformis* K11 in pepper. Plant Pathol J 29:201–208

Hussain MB, Zahir ZA, Asghar HN, Asghar M (2014) Exopolysaccharides producing *Rhizobia ameliorate* drought stress in wheat. Int J Agric Biol 16:3–13

Ibrahim MA, Khan PR, Hegazy SS, Hashim EA, Azamal H, Ansari MKA et al (2015) Improving the phytoextraction capacity of plants to scavenge heavy-metal infested sites. Environ Rev 23: 1–22

Idris R, Trifonova R, Puschenreiter M, Wenzel WW, Sessitsch A (2004) Bacterial communities associated with flowering plants of the Ni hyperaccumulator *Thlaspi goesingense*. Appl Environ Microbiol 70(5):2667–2677

Iyer NJ, Tang Y, Mahalingam R (2013) Physiological, biochemical and molecular responses to combination of drought and ozone in Medicago truncatula. Plant Cell Environ 36:706–720

Jiang QY, Zhuo F, Long SH, Zhao HD, Yang DJ, Ye ZH et al (2016) Can arbuscular mycorrhizal fungi reduce Cd uptake and alleviate Cd toxicity of *Lonicera japonica* grown in Cd-added soils? Sci Rep 6:21805

Jiang J, Pan C, Xiao A, Yang X, Zhang G (2017) Isolation, identification, and environmental adaptability of heavy-metal-resistant bacteria from ramie rhizosphere soil around mine refinery. Biotech 7(1):5

Kaleem F, Shabir G, Aslam K, Rasu S, Manzoor H, Shah SM, Khan AR (2018) An overview of the genetics of plant response to salt stress: present status and the way forward. Appl Biochem Biotechnol 186:306–334

Kandel SL, Joubert PM, Doty SL (2017) Bacterial endophyte colonization and distribution within plants. Microorganisms 5(4):E77

Keleş Y, Öncel I (2002) Response of antioxidative defence system to temperature and water stress combinations in wheat seedlings. Plant Sci 163:783–790

Keshtehgar A, Rigi K, Vazirimehr MR (2013) Effects of salt stress in crop plants. Int J Agric Crop Sci 5:2863–2867

Kim YC, Glick B, Bashan Y, Ryu CM (2013) Enhancement of plant drought tolerance by microbes. In: Aroca R (ed) Plant responses to drought stress. Springer, Berlin

Klaus-Joerger TK, Joerger R, Olsson E, Granqvist CG (2001) Bacteria as workers in the living factory: metal accumulating bacteria and their potential for material science. Trends Biotechnol 19(1):15–20

Körner C (2006) Significance of temperature in plant life. In: Morison JIL, Morecroft MD (eds) Plant growth and climate change. Blackwell Publishing, Oxford, UK, pp 48–69

Koussevitzky S, Suzuki N, Huntington S, Armijo L, Sha W, Cortes D et al (2008) Ascorbate peroxidase 1 plays a key role in the response of Arabidopsis thaliana to stress combination. J Biol Chem 283:34197–34203

Kushwaha P, Kashyap PL, Bhardwaj AK, Kuppusamy P, Srivastava AK, Tiwari RK (2020) Bacterial endophyte mediated plant tolerance to salinity: growth responses and mechanisms of action. World J Microbiol Biotechnol 36(2):26

Lau JA, Lennon JT (2011) Evolutionary ecology of plant–microbe interactions: soil microbial structure alters selection on plant traits. New Phytol 192:215–224

Liang W, Ma X, Wan P, Liu L (2018) Plant salt-tolerance mechanism: a review. Biochem Biophys Res Commun 495:286–291

Lindow SE, Leveau JH (2002) Phyllosphere microbiology. Curr Opin Biotechnol 13:238–243
Liu F, Xing S, Ma H, Du Z, Ma B (2013) Cytokinin producing, plant growth promoting rhizobacteria that confer resistance to drought stress in *Platycladus orientalis* container seedlings. Appl Microbiol Biotechnol 97:9155–9164
Liu H, Carvalhais LC, Crawford M, Singh E, Dennis PG, Pieterse CMJ et al (2017) Inner plant values: diversity, colonization and benefits from endophytic bacteria. Front Microbiol 8:2552
Matilla MA, Espinosa-Urgel M, Rodrıguez-Herva JJ, Ramos JL, Ramos González M (2007) Genomic analysis reveals the major driving forces of bacterial life in the rhizosphere. Genome Biol 8:R179
Mayak S, Tirosh T, Glick BR (2004) Plant growth-promoting bacteria confer resistance in tomato plants to salt stress. Plant Physiol Biochem 42:565–572
McAdam SAM, Brodribb TJ, Ross JJ (2016) Shoot-derived abscisic acid promotes root growth. Plant Cell Environ 39:652–659
Meehl G, Tebaldi C (2004) More intense, more frequent, and longer lasting heat waves in the 21st century. Science 305:994–997. https://doi.org/10.1126/science.1098704
Mittler R, Finka A, Goloubinoff P (2012) How do plants feel the heat? Trend Biochem Sci 37 (3):118–125
Mishra PK, Bisht SC, Ruwari P, Selvakumar G, Joshi GK, Bisht JK et al (2011) Alleviation of cold stress in inoculated wheat (*Triticum aestivum* L.) seedlings with psychrotolerant *Pseudomonas* from NW Himalayas. Arch Microbiol 193:497–513
Molina-Favero C, Creus CM, Simontacchi M, Puntarulo S, Lamattina L (2008) Aerobic nitric oxide production by Azospirillum brasilense Sp245 and its influence on root architecture in tomato. Mol Plant Microb Interact 2:1001–1009
Munchbach M, Nocker A, Narberhaus F (1999) Multiple small heat shock proteins in rhizobia. J Bacteriol 181:83–90
Murphy BR, Doohan FM, Hodkinson TR (2014) Yield increase induced by the fungal root endophyte *Piriformospora indica* in barley grown at low temperature is nutrient limited. Symbiosis 62:29–39
Nadeem SM, Ahmad M, Zahir ZA, Javaid A, Ashraf M (2014) The role of mycorrhizae and plant growth promoting rhizobacteria (PGPR) in improving crop productivity under stressful environments. *Biotechnol Adv* 32:429–448
Narendrula-Kotha R, Nkongolo KK (2017) Microbial response to soil liming of damaged ecosystems revealed by pyrosequencing and phospholipid fatty acid analyses. PLoS One 12:e0168497
Nezhadahmadi A, Prodhan ZH, Faruq G (2013) Drought tolerance in wheat. Sci World J 1:1–12
Ogweno JO, Song XS, Hu WH, Shi K, Zhou YH, Yu JQ (2009) Detached leaves of tomato differ in their photosynthetic physiological response to moderate high and low temperature stress. Sci Hortic 123(1):17–22
Osakabe Y, Osakabe K, Shinozaki K, Tran LSP (2014) Response of plants to water stress. Front Plant Sci 5:86
Pandey V, Ansari MW, Tula S, Yadav S, Sahoo RK, Shukla N et al (2016) Dose-dependent response of Trichoderma harzianum in improving drought tolerance in rice genotypes. Planta 243:1251–1264
Panuccio MR, Jacobsen SE, Akhtar SS, Muscolo A (2014) Effect of saline water on seed germination and early seedling growth of the halophyte quinoa. AoB Plants 6:plu047
Paradis R, Dalpé Y, Charest C (1995) The combined effect of arbuscular mycorrhizas and short-term cold exposure on wheat. New Phytol 129:637–642
Pareek A, Sopory SK, Bohnert HK, Govindjee (2010) Abiotic stress adaptation in plants: physiolgical, molecular and genomic foundation. Springer, Dordrecht, p 526
Parida AK, Das AB (2005) Salt tolerance and salinity effects on plants: a review. Ecotoxicol Environ Saf 60:324–349
Paul D, Lade H (2014) Plant-growth-promoting rhizobacteria to improve crop growth in saline soils: a review. Agron Sustain Dev 34:737–752

Perez-Lopez U, Robredo A, Lacuesta M, Sgherri C, Munoz-Rueda A, Navari-Izzo F, Mena-Petite A (2009) The oxidative stress caused by salinity in two barley cultivars is mitigated by elevated CO_2. Physiol Plant 135:29–42

Pierart A, Shahid M, Séjalon-Delmas N, Dumat C (2015) Antimony bioavailability: knowledge and research perspectives for sustainable agricultures. J Hazard Mater 289:219–234

Pires C, Franco AR, Pereira SIA, Henriques I, Correia A, Magan N et al (2017) Metal(loid)-contaminated soils as a source of culturable heterotrophic aerobic bacteria for remediation applications. Geomicrobiol J 34:1–9

Pourrut B, Jean S, Silvestre J, Pinelli E (2011) Lead-induced DNA damage in *Vicia faba root* cells: potential involvement of oxidative stress. Mutat Res 726:123–128

Qin Y, Pan X, Kubicek C, Druzhinina I, Chenthamara K, Labbe J et al (2017) Diverse plant-associated Pleosporalean fungi from saline areas: ecological tolerance and nitrogen-status dependent effects on plant growth. Front Microbiol 8:158

Rahdari P, Hoseini SM (2012) Drought stress: a review. Int J Agron Plant Prod 3:443–446

Redman RS, Kim YO, Woodward CJDA, Greer C, Espino L, Doty SL et al (2011) Increased fitness of rice plants to abiotic stress via habitat adapted symbiosis: a strategy for mitigating impacts of climate change. PLoS One 6:e14823

Rizhsky L, Liang HJ, Shuman J, Shulaev V, Davletova S, Mittler R (2004) When defense pathways collide. The response of Arabidopsis to a combination of drought and heat stress. Plant Physiol 134:1683–1696

Rudell DR, Buchanan DA, Leisso RS, Whitakeer BD, Mattheis JP, Zhu YM et al (2011) Ripening, storage temperature, ethylene action, and oxidative stress alter apple peel phytosterol metabolism. Phytochemistry 72(11–12):1328–1340

Sand W, Rohde K, Sabotke B, Zenneck C (1992) Evaluation of *Leptospirillum ferrooxidans* for leaching. Appl Environ Microbiol 58(1):85–92

Sanghera GS, Wani SH, Hussain W, Singh NB (2011) Engineering cold stress tolerance in crop plants. Curr Genomics 12(1):30–43

Sang-Mo K, Radhakrishnan R, Khan AL, Min-Ji K, Jae-Man P, Bo-Ra K, Dong-Hyun S, In-Jung L (2014) Gibberellin secreting *rhizobacterium, Pseudomonas putida* H-2-3 modulates the hormonal and stress physiology of soybean to improve the plant growth under saline and drought conditions. Plant Physiol Biochem 84:115–124

Saravanavel R, Ranganathan R, Anantharaman P (2011) Effect of sodium chloride on photosynthetic pigments and photosynthetic characteristics of Avicennia officinalis seedlings. Recent Res Sci Technol 3:177–180

Schmidt R, Köberl M, Mostafa A, Ramadan EM, Monschein M, Jensen KB, Bauer R, Berg G (2014) Effects of bacterial inoculants on the indigenous microbiome and secondary metabolites of chamomile plants. Front Microbiol 5:64

Schreiber SG, Hamann A, Hacke UG, Thomas BR (2013) Sixteen years of winter stress: an assessment of cold hardiness, growth performance and survival of hybrid poplar clones at a boreal planting site. Plant Cell Environ 36:419–428

Shahid M, Khalid S, Abbas G, Shahid N, Nadeem M, Sabir M et al (2015) Heavy metal stress and crop productivity. In: Hakeem KR (ed) Crop production and global environmental issues. Springer International Publishing, Cham, pp 1–25

Sharma P, Khanna V, Kumar PI (2013) Efficacy of aminocyclopropane-1-carboxylic acid (ACC)-deaminase-producing rhizobacteria in ameliorating water stress in chickpea under axenic conditions. Afr J Microbiol Res 7:5749–5757

Sharma S, Kulkarni J, Jha B (2016) Halotolerant Rhizobacteria promote growth and enhance salinity tolerance in peanut. Front Microbiol 13(7):2791

Shrivastava P, Kumar R (2015) Soil salinity: a serious environmental issue and plant growth promoting bacteria as one of the tools for its alleviation. Saudi J Biol Sci 22:123–131

Singh SK, Sharma HC, Goswami AM, Datta SP, Singh SP (2000) In vitro growth and leaf composition of grapevine cultivars as affected by sodium chloride. Biol Plant 43:283–286

Skirvin RM, Kohler E, Steiner H, Ayers D, Laughnan A, Norton MA, Warmund M (2000) The use of genetically engineered bacteria to control frost on strawberries and potatoes. Whatever happened to all of that research? Sci Hortic 84:179–189

Song Y, Chen Q, Ci D, Shao X, Zhang D (2014) Effects of high temperature on photosynthesis and related gene expression in poplar. BMC Plant Biol 14:111

Sorty AM, Meena KK, Choudhary K, Bitla UM, Minhas PS, Krishnani KK (2016) Effect of plant growth promoting bacteria associated with halophytic weed (Psoralea corylifolia L.) on germination and seedling growth of wheat under saline conditions. Appl Biochem Biotechnol 180: 872–882

Sorty AM, Bitla UM, Meena KK, Singh NP (2018) Role of microorganisms in alleviating abiotic stresses. In: Panpatte DG, Jhala YK, Shelat HN, Rajababu YV (eds) Microorganisms for green revolution, microorganisms for sustainability, vol VII. Singapore, Springer Nature, pp 115–128

Tavakkoli E, Rengasamy P, McDonald GK (2010) High concentrations of Na^+ and Cl^- ions in soil solution have simultaneous detrimental effects on growth of faba bean under salinity stress. J Exp Bot 61:4449–4459

Teixeira PJPL et al (2019) Beyond pathogens: microbiota interactions with the plant immune system. Curr Opin Microbiol 49:7–17

Theocharis A, Bordiec S, Fernandez O, Paquis S, Dhondt-Cordelier S, Baillieul F, Clément C, Barka EA (2012) *Burkholderia phytofirmans* PsJN primes *Vitis vinifera* L. and confers a better tolerance to low non freezing temperatures. Mol Plant-Microbe Interact 25(2):241–249

Tian J, Belanger FC, Huang B (2009) Identification of heat stress responsive genes in heat-adapted thermal Agrostis scabra by suppression subtractive hybridization. J Plant Physiol 166:588–601

Timmusk S, Islam A, Abd El D, Lucian C, Tanilas T, Kannaste A et al (2014) Drought-tolerance of wheat improved by rhizosphere bacteria from harsh environments: enhanced biomass production and reduced emissions of stress volatiles. PLoS One 9:1–13

Verslues PE, Agarwal M, Katiyar-Agarwal S, Zhu J, Zhu J-K (2006) Methods and concepts in quantifying resistance to drought, salt and freezing, abiotic stresses that affect plant water status. Plant J 45:523–539

Varshney RK, Bansal KC, Aggarwal PK, Datta SK, Craufurd PQ (2011) Agricultural biotechnology for crop improvement in a variable climate: hope or hype? Trends Plant Sci 16:363–371

Verma S, Verma PK, Meher AK, Bansiwal AK, Tripathi RD, Chakrabarty D (2017) A novel fungal arsenic methyltransferase, *WaarsM reduces* grain arsenic accumulation in the transgenic rice plant. J Hazard Mater 344:626–634

Wahid A, Galani S, Ashraf M, Foolad MR (2007) Heat tolerance in plants: an overview. Environ Exp Bot 61:199–223

Wang C, Yang W, Wang C, Gu C, Niu D, Liu H et al (2012) Induction of drought tolerance in cucumber plants by a consortium of three plant growth promoting rhizobacterium strains. PLoS One 7:e52565

Weyens N, van der Lelie D, Taghavi S (2009) Exploiting plant-microbe partnerships to improve biomass production and remediation. Trend Biotechnol 27:591–598

Wu ZJ, Li XH, Liu ZW, Li H, Wang YX, Zhuang J (2015) Transcriptome-based discovery of AP2/ERF transcription factors related to temperature stress in tea plant (*Camellia sinensis*). Funct Integr Genomics 15(6):741–752

Xiong T, Leveque T, Shahid M, Foucault Y, Mombo S, Dumat C (2014) Lead and cadmium phytoavailability and human bioaccessibility for vegetables exposed to soil or atmospheric pollution by process ultrafine particles. J Environ Qual 43:1593–1600

Xu Z, Jiang Y, Jia B, Zhou G (2016) Elevated-CO_2 response of stomata and its dependence on environmental factors. Front Plant Sci 7:657

Xue ZY, Zhi DY, Xue GP, Zhang H, Zhao YX, Xia GM (2004) Enhanced salt tolerance of transgenic wheat (*Triticum aestivum* L.) expressing a vacuolar Na^+/H^+ antiporter gene with improved grain yields in saline soils in the field and a reduced level of leaf Na^+. Plant Sci 167: 849–859

Yoshikawa H, Honda C, Kondo S (2007) Effect of low-temperature stress on abscisic acid, jasmonates, and polyamines in apples. Plant Growth Regul 52(3):199–206

Zahir ZA, Munir A, Asghar HN, Shahroona B, Arshad M (2008) Effectiveness of rhizobacteria containing ACC-deaminase for growth promotion of peas (*P. sativum*) under drought conditions. J Microbiol Biotechnol 18:958–963

Zaidi S, Usmani S, Singh BR, Musarrat J (2006) Significance of *Bacillus* subtilis strain SJ 101 as a bioinoculant for concurrent plant growth promotion and nickel accumulation in *Brassica juncea*. Chemosphere 64:991–997

Zhu X-C, Song F-B, Liu TD, Liu SQ (2010) Arbuscular mycorrhizae reducing water loss in maize plants under low temperature stress. Plant Signal Behav 5:591–593

Zhu XC, Song FB, Liu SQ, Liu TD (2011) Effects of arbuscular mycorrhizal fungus on photosynthesis and water status of maize under high temperature stress. Plant Soil 53(1):247–273

Zinn KE, Tunc-Ozdemir M, Harper JF (2010) Temperature stress and plant sexual reproduction: uncovering the weakest links. J Exp Bot 61(7):1959–1968

Emerging Microbe-Mediated Advanced Technology to Mitigate Climatic Stresses in Plants and Soil Health: Current Perspectives and Future Challenges

Kailash Chand Kumawat, Sharon Nagpal, Anamika Chattopadhyay, and Poonam Sharma

Abstract Soil health conservation is fundamental to agricultural sustainable development and is an important key factor in agroecosystem productivity. However, soil microbial resources are currently under tremendous pressure from a number of anthropogenic activities, including climate change. Changing climatic conditions adds to the challenges and dynamics of agriculture, rhizospheric microbiomes, and sustainability. Rhizospheric microbial communities help plant growth and enhance tolerance to various environmental stresses. The relationship between the distribution of rhizospheric microbial biodiversity and the functioning of ecosystems is critical for understanding the response of ecosystems to changes in the environment. Microbial taxa play a vital and unquestionable role in biogeochemical cycling, plant growth, and carbon sequestration in the context of global climate change. Modern genomic methods show an enormous potential for uncultivated diversity and changing bacteria population associated with sensitive and disease resistance plants and understand how climate change affects microbiomes. This chapter explores the current state of knowledge on how climate change affects microbial soil ecosystems and interactions between plants and microbes, as well as the potential ways in which rhizospheric microbiomes can be harnessed to mitigate adverse impacts of climate change. We also highlighted in this chapter the role of metagenomics to open the black box of the soil.

K. C. Kumawat (✉)
Department of Microbiology, Punjab Agricultural University, Ludhiana, Punjab, India

Department of Industrial Microbiology, Jacob Institute of Biotechnology and Bio-engineering, Sam Higginbottom University of Agriculture, Technology and Sciences, Prayagraj, Uttar Pradesh, India
e-mail: kailash-cobsmb@pau.edu; kailash.kumawat@shiats.edu.in

S. Nagpal · A. Chattopadhyay
Department of Microbiology, Punjab Agricultural University, Ludhiana, Punjab, India

P. Sharma
Department of Plant Breeding and Genetics, Punjab Agricultural University, Ludhiana, Punjab, India

© The Author(s), under exclusive license to Springer Nature Singapore Pte Ltd. 2022
A. Vaishnav et al. (eds.), *Plant Stress Mitigators*,
https://doi.org/10.1007/978-981-16-7759-5_16

Keywords Agriculture sustainability · Carbon sequestration · Climate change · Plant–microbe interaction · Plant growth-promoting rhizobacteria

Introduction

The present-day research primarily prioritizes the environmental conservation, global warming reduction, and increased food production to meet the demand and need of food worldwide. According to the Food and Agriculture Organization of the United Nations, in order to fulfill the needs of probable world population of about ten billion by 2050, it is essential to upsurge the global production of food by approximately 70%. To meet the demands, the global supply of food should necessarily increase sustainably provided the conditions of growing competition for natural resources (Thornton and Herrero 2010; Sergaki et al. 2018). The major effects of amplified anthropogenic undertakings have been observed on the environment specifically as a broad and sustainable farming (Fang et al. 2018; Malla et al. 2018a, b). To depict the challenges of increasing agricultural land productivity to yield more food supplies in an eco-friendly way amid climate change, the concept of "sustainable intensification" has been identified (Chowdhary et al. 2018). Industrialization, rapidly growing population, and their civilization are changing man's relationship with the environment and ecology drastically and are promoting climate changes worldwide (IPCC 2007). Parameters of climate change not only affect the microorganisms but also drastically affect the macroorganisms as well and stand a serious issue disturbing the life on the planet (Compant et al. 2010; Sergaki et al. 2018). Climatic changes induce alterations in the configuration, composition, abundance, and pursuit of plant microbiome.

Effects of changing climate on interactions between plants, soil, and microorganisms can be direct and indirect (Abhilash et al. 2013; Bojko and Kabala 2017). It alters the organization of its community, its function, and relative abundance as the soil microbiome depicts diverse physiological diversity, growth rate, and temperature sensitivity in them (Bagri et al. 2018). Certain direct and indirect consequences of changing climatic condition change are depicted in Fig. 1. The indirect consequences of climate change on the soil–microbial populations intervened by plants are far sturdier than direct consequences on belowground structure of microbial populations. Variations in microbiome assembly and configuration may change the functioning of ecosystem and richness of organisms, which regulates the crucial and specific procedures (Schimel and Schaeffer 2012). Although extensive studies have been performed on plant species migration in response to climate change, most of them failed to elaborate the potential of plant and soil allied microbes to alter their arrays as to sustain their association with plant and soil communities (Hameed et al. 2014).

Soil is a highly assorted and multifaceted habitation with the colonization of numerous species of microbes (Bardgett and Van Der Putten 2014). Microbes in rhizospheric have many functions like biogeochemical cycling and protecting plants from detrimental effects of environmental stresses (Hashem et al. 2017; Nagpal et al.

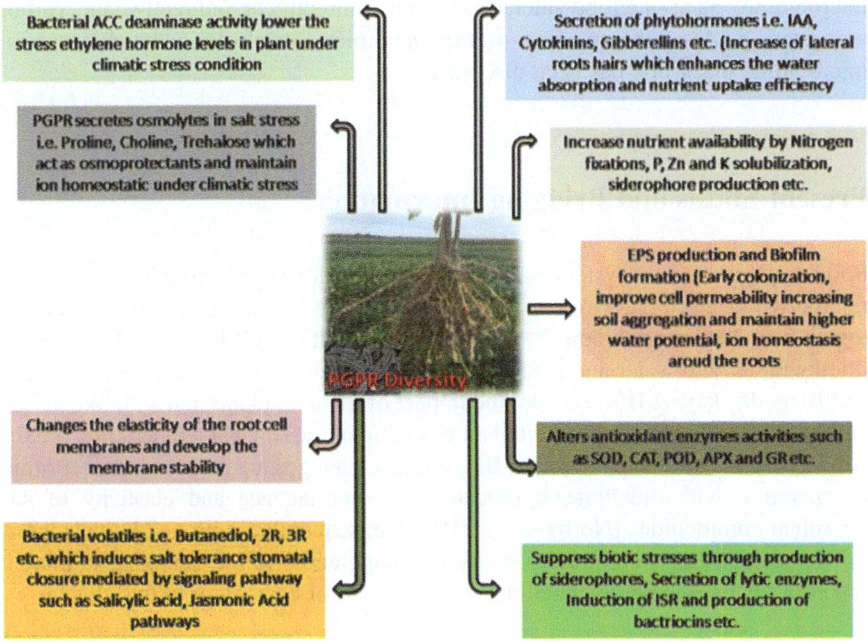

Fig. 1 Systematic mechanisms of PGPR under climatic stress conditions

2019, 2020). Although rigorous agriculture performances intensify the production of crops, it has harmful up shots on physical and biological properties of soils. Those performances affect the range, composition, and structure of the soil microbiome, which in turn affects the agricultural ecosystems functioning. Soil microbial communities have various influences on agroecosystems such as impacting the crop productivity, procurement of nutrients, and operations of ecosystem on usually tilled and nontilled soils (Kumar et al. 2015). The agricultural lands are generally supplied with macronutrients by chemical fertilizers, which have deleterious effects on both microbial communities in agroecosystems. Compared to chemical and synthetic fertilizers, the application of biofertilizers and biopesticides has proved to boost the development of plants better, thus serving in environmentally sustainable crop production. Plant growth-promoting rhizobacteria, an assemblage of varied rhizospheric microorganisms, synthesize various bioactive substances that promote plant growth and provide protection from pathogens as well (Dubey et al. 2018). Several researches demonstrated that rhizospheric interactions with microbiomes are mediated directly or indirectly through secretion of root exudates. Conversely, the recent developments in the advanced approaches have helped the scientific community to have a better understanding of such type of interactions at community level. Those researches majorly focus on the diversity inhabiting in different environments.

These kinds of studies have chiefly focused on recognizing the diversity inhabiting different environments. However, to identify the signals involved in the interspecies interactions, analysis of plant–microbe interactions at a functional level is necessary. In this chapter, we have discussed how the changing climate affects soil microbiome, plant–microbe interactions either directly or indirectly, and carbon sequestering. Also, the function of metagenomics in unlocking the rhizospheric microbiome black box has been discussed.

Present Status and Bridging Information

Soil is a diversified ecosystem on the planet along with a correlating body of various microbes like bacteria, viruses, protozoa, fungi, and archaea, which are denoted together as "soil microbiome." Abiotic environment of soil is assorted with pores filled with air and water and patchy resources that function as hot spots (Kuzyakov and Blagodatskaya 2015). Besides the impact of fauna in plants and soil, like insects and earthworms, and changes in moisture content of the soil and temperature, the soil atmosphere is extremely vigorous. But, climatic changes are inducing greater limits of alteration with indefinite significances on permanence and elasticity of soil microbial communities (Norby et al. 2016). Consequently, improved knowledge of microbial behaviors that discuss ecosystem suppleness to the changing climate is required for foreseeing and handling the responses of ecosystem to this change.

Role of Rhizospheric Microbiome in Improving the Soil Health Amid Changing Climatic Conditions

Microorganisms are one of the most varied and prolific type of organisms (Bar-On et al. 2018). Microorganisms play a vital role in nourishing and stabilizing the biosphere as sustenance of life depends on continuous and microbial adjudicated conversion of matter, in both aquatic and terrestrial environments (Vibha and Neelam 2012). Role of microbes is uncountable in biogeochemical cycling and maintaining sustainability (Curtis and Sloan 2005). They also play a role in medicine and drug manufacture, sewage discharge treatment, and bioremediation. Soil is a major stimulating and vital medium of all the natural environments, and this multifaceted environment of soil is responsible for the diversified community of species present in it. Soil microorganisms include bacteria, fungi, and archaea, and they have a varied and significant role in ecosystem functioning like cycling elements like nitrogen and carbon. Apart from influencing the assembly and functioning of ecosystems, cycling of nutrients also augments the soil with the capabilities that deliver diverse amenities to the humankind. Soil microbes and their associated utilities regulate the efficiency of agroecosystems (Van Der Heijden et al. 2008).

As stated previously, sustainable agriculture depends upon the vigor of soil and its microbial range. Consequently, current researches should give attention to organizing soil microbiome. Application of beneficial microbes to improve plant growth features and soil health while minimizing environmental impact is a critical function for long-term food and energy production. It is probable that major part of the microbial range is unexplored. That diversity is like treasure troves for enhanced and advanced biotechnological expansions and its usage in agriculture, mining, food materials, medicines, and environmental fortification. Identifying the major factors that managing the ecosystem and recognizing their beneficial aspects is a challenging task. Currently, researchers worldwide are finding it challenging to maximize the utilities of microbiomes under limited natural and man-made activities. This includes novel species of pests and pathogens, use of chemicals, and climate change, as the sevenths constantly cause menace in the stability of productions in agriculture (Callaway 2016). To achieve the goals in terms of increasing the productivity to fulfill the world food demands by 2050, prompt and instant solutions are necessary (Castro et al. 2012; Tilman et al. 2011). This can be achieved by organizing the soil microbiome for growing resistance against several biotic and abiotic stresses (Vorholt et al. 2017) and to augment assimilation of essential nutrients. Hiruma et al. (2016) presented an unutilized group of prospects to deal with issues related to sustainability in agroecosystems under climate change (Busby et al. 2017). An improved outlook of complex interfaces among plant genotypic variations; microbiome configurations and various environmental influences offer crucial evidence in sustainable agriculture.

Mechanism of Interactions Between Plant Roots and Microbes

As plants have a static lifestyle, they need to improve their health in their biotic environments. Pathogens and beneficial microbes are found to be the chief reasons that influence the plant health as they proliferate in the host plants. Constant with this, the microorganisms and their host plants have coevolved and developed a succession of techniques that control the consequences of the interactions between them (Oldroyd 2013). Plants depend upon tendency of roots to interact with a diversity of microorganisms. The interactions of bacteria and fungi rely on their association with plants, and their regulation by discharge of root exudes is often true. Roots exude like flavonoids, quinones, *p*-hydroxy acids, and cytokinins are enormously potential low molecular weight compounds found in the rhizosphere.

The principal stage in microbial colonization of roots is production of chemotactic reaction to various exudes released by the roots of plants. The various kinds of root exudes comprise organic compounds and amino acids (Zheng and Sinclair 1996). Terrestrial plants experience several complicated interactions. Some of them are chemical, physical, and biological. These interactions occur between the plant roots and rhizosphere. The different plant- and root-associated rhizospheric interactions comprise root–microbe, root–insect, and root–root interactions. Rhizospheric microbiome is a key factor for plant growth promotion and seed vigor of crop plant because it assists the host plant in protection against phytopathogens and nutrient uptake and provides tolerance to abiotic stress (Sasse et al. 2018). Harmonized communications among microorganisms with their hosts are crucial and significant in enhancing the plant development and in sustaining suitable soil conditions.

Significance of root exudes in facilitating biological communications in the arena of plant biology has been acknowledged recently (Baetz and Martinoia 2014). An instance of varied connotations for a chemical indicator is the exudation of isoflavones by roots of soybean, which allure a mutualistic organism (*Bradyrhizobium japonicum*) and a pathogenic organism (*Phytophthora sojae*) (Sasse et al. 2018). Chemotaxis influenced by root exudes drove the establishment of *Pseudomonas fluorescens* in tomato crop (De Weert et al. 2002). Few endophytes exhibit the growth in chemotaxis upon other endophytes in the existence of root exudes of rice. The consortia are highly affected by both soluble and in-soluble root exudes, polysaccharides layer surrounding the roots, etc. Among all different rhizospheric microbial diversity, the PGPRs achieve exclusive and extraordinary consideration because of their varied functions like hormones and beneficial enzyme production, efficacious root colonization, and nutrient solubilization, which help in maintaining sustainability in agroecosystems. Understanding and knowing the ecology, growth-enhancing factors, their mechanisms of action, and application of the naturally occurring microorganisms carry immense importance in terms of plant growth.

Over the past few years, enormous amount of steps were taken to comprehend various types of plant–microbe interactions. There are now a vast number of evidences, which support the fact that plants have the ability to alter and form their microbiome by plant–microbe interactions beneath the ground (Chaparro et al. 2014). The most indigenous lines of crop plants also display their capacity to change the range of microbial assemblages in rhizospheric soil (Valverde et al. 2016). Symbiotic relationship between plants and microorganisms may be observed as an assimilated ecological entity called holobiont (Vandenkoornhuyse et al. 2015). Those complementary microbial communities are recognized to differentiate the metabolic activities of root exudates (Rasmann and Turlings 2016) and in the rate of uptake of essential plant nutrients (Bell et al. 2015). Variations in genotypes and phenotypes traits of plants support microbiomes, which help in increasing the plant nutrient availability, combat pathogens, and improve plant health, growth, and performance as well.

Plants, Rhizospheric Microbes, and Climatic Conditions

Consistent changes in the climatic conditions can alter the spread and responses of species. Significances of estimated change in climate may lead to increase in temperature and consequent pattern of rainfall, which may bring uncertainty and intricacy to the agricultural systems and may have a negative impact on managing the sustainable agriculture. Climate change immensely affects the crop's eminence and dynamics of implications present among crops, diseases, and pests as well. Instabilities in climatic influences such as precipitation, temperature, and solar radiations have abundant potentials to have an impact on production of crops (Classen et al. 2015). These naturally occurring populations are intricate and habitually comprise of microorganisms with diverse tolerances to temperature and dispersion capacity. Moreover, interactions between diverse communities can be favorable, pathogenic, or can have less or no purposeful effect and these interactions tend to alter with change in category of environmental stress. Several researches performed by various scientists (Zhou et al. 2015; Langley and Hungate 2014; Gottfried et al. 2012) portrayed that to change the species scattering and the functioning of terrestrial ecosystems, shifts in interaction of species interactions with regard to climate change have been observed. However, fewer studies focused on soil microorganisms (De Vries et al. 2012). Rhizospheric microorganisms link with host plants in numerous ways, which contribute in sustaining and determining different mechanisms of ecosystem (Hassani et al. 2018). In the present chapter, we conferred the effects of climate change on interactions between soil, plants, and microorganisms through direct and indirect mechanisms. The direct impacts of change in climatic conditions on composition and function of the microbial composition have been broadly studied by various scientists (Henry 2013). However, the indirect impacts by instability of interactions between soil, plants, and microorganisms have been considered in a lesser scale. These interactions ensure abundant prospective to facilitate certain vital practices like plant community composition, mineralization, and important functions of shifts in ecological communication (Steinauer et al. 2015).

Direct Effects of the Changing Climate on Plants and Rhizospheric Microbial Communities

Composition and functioning of rhizospheric microbiome get affected with change in climatic conditions and show extreme variations in growth rates, their metabolic activities, and temperature sensitivity (Zhang et al. 2018b). A study conducted by DeAngelis et al. (2015) concluded that prolonged increase in temperate forest soil can make alterations in rhizospheric microbial diversity. Thus, with rise in 5 °C in temperate forests, range of microorganisms present in soil, for instance, bacteria, will change and that will in turn lead to increase in the bacterial-to-fungal ratio (Bintanja 2018). Consequences of climate change such as global warming bluntly amend the rate of respiration in microbes present in soil because of their temperature sensitivity. Therefore, the effect of raised temperature in the metabolism of microorganisms gained a substantial amount of consideration in current years (Gao et al. 2018). Yet, there are certain questions like "what are the effects of changing microbial groups on various roles like decomposition of organic matter?", "how temperature, moisture and their interface, disturb only certain microbial populations such as methanogens, in a community?", and "what are the mechanisms that determinations the net ecosystem feedback of microbial metabolism to the changing climate?" that needs to be answered. Thus, approaches such as manipulating microbial populations and factorial heating up besides temperature ascents (like elevation) are recommended for finding answers to these questions. Another valuable method for answering the stated questions is to practice mutual soils and plant transplants in addition to environment ascents.

Indirect Effects of the Changing Climate on Plants and Rhizospheric Microbial Communities

Change in climate not only changes the phenology of plants but also alters the distribution of plant microbiome species (Classen et al. 2015). As the microbes residing in soil are considered to be poor in dispersals, they react to the changing climate at varied degrees when likened to their host plants (Chen et al. 2015). Conversely, as per our knowledge, different dispersal potentials of microbes and plants can have an effect on productivity of plants, and inception of the new species of plants and their interfaces. The plants that exhibit successful establishment in new arrays are known to produce complex level compounds related to plant protection such as polyphenols (Agrawal and Weber 2015).

Global rise in temperature affects crops, weeds, insect pest, and diseases. Thirty-four percent of crop loss occurs because of invasion of weeds, 18% because of insects, and 16% due to diseases. Studies show that climate change possesses the ability to upsurge deleterious impact of weeds, insects, and disease-causing pathogens already present in the agroecosystems. Some estimated effects also include the following:

- Some species of weed grow better than the crop in elevated temperature and CO_2 levels.
- Higher temperature intensifies the insect pest occurrence because warm temperature accelerates the life cycles that lessen the time expended by them in weak life phases.
- Increase in temperature also supports the weeds, pathogens, and insects to survive in winters and promote their northward expansion.
- Prolonged life cycle of crops allows increase in pest populations because more number of generations of insect pests can be produced in one growing season.
- Temperature and moisture stress related to heating climate lead to more vulnerability to various disease infestation in crops, and its range also affects the production of livestock.

Climate Change Mitigation and Food Security by Microbiome Engineering and Synthetic Biotechnology

Agricultural practices endorse assimilated purposeful plant, and soil microbiomes also promote plant and soil eminence that helps sustaining or even increasing crop yields while alleviating the effects of climate change. Organic practices of farming are one of the best methods for reducing chemical usage and maintaining soil and plant microbiome population without disrupting them. But such practices do not necessarily prevent the microbes from mechanical disruption or develop the microbial community to efficiently manage the soil moisture and nutrient cycles. Practices for enhanced crop yield in conjugation with management methods and microbial community are needed, which, along with yield increment, are expected to reduce input cost and improve the services of ecosystem. Natural variation adapted by soil microbial populations and the requirement of discrepancy in organization practices may provide valuable socioeconomic responses theoretically by the growing the native agriculture specialist demand and increase opportunities for diversified small-scale farms. Ecological researches provide sufficient information about the detrimental effects of changing climatic conditions like rise in levels of global warming and depletion of ozone layer and diversity of rhizospheric microbiome. Evolutionary researches have predicted the capability of artificial microbes and microbial associations over time (Ji and Bever 2016). Nevertheless, multi-location field trials and meticulous experiments may be utilized to evaluate the effects of different external dynamics such as elevated level of carbon dioxide, drought and plant's genotype, and soil properties on microbial populations.

Also, the current emerging technologies such as synthetic biotechnology and its application to both microorganisms and plants can structure the microorganisms artificially along with their potential to improve environmental superiority and plant vigor (Peng et al. 2016). It is specific how modification and genetic engineering of microorganisms are possible for modification of the conventional features of communications between plants and microorganisms and the microbial community composition, which improves plant vigor. The plant engineering potential by alterations in their microbiome structure has been studied in the previous decade (Sharma et al. 2017; Kamthan et al. 2016; Orozco-Mosqueda et al. 2018). Certain significant microorganisms such as arbuscular mycorrhizal fungi have multinucleated genetic systems that are not suitable for engineering purpose.

Another approach for the production of microbial communities for the enhancement of crop productivity is by combing un-engineered species of microbes isolated from their natural habitat. This method can be assisted by combining the fields of system biology, evolutionary biology, and ecology. Harmful effects of climate change as per the ecological studies (e.g., raised CO_2 and ozone levels) can have detrimental effects on arrangement and assortment of soil microbiome (Kato et al. 2014). Conversely, controlled experiments and large-scale field trials can be performed to evaluate the influence of several external factors such as raised CO_2 levels, drought, soil properties, and plant genotype on microbiome composition.

Microorganisms for Plant Productivity, Soil Health, and Plant Disease Management

Soil vigor can be defined as the ability of soil to provide an array of agronomic and ecosystems utilities and amenities to sustain quality and health of the environment, enhance biological productivity, and stimulate the health of plants and animals (Lal 2016). Vigorous soils are known to be the establishment ground for sustainable and dynamic agroecosystems and it can be maintained by nominal disruption in soil profile, shielding the topsoil by cultivating additional plants, growing the range of different species of plants by poly-cultures, improving soil microbiome, and growing cover crops (Brussaard et al. 2007). Effective implementation of microorganisms aids to maintain the soil condition and enhance water-retaining capacity, root progression, accessibility, and cycling of crucial elements, carbon storage, purifying contaminants, and enhancement of crop productivity as well (Nannipieri et al. 2017). Enormous studies have been performed on beneficial soil microbes and their application in most of the agriculturally significant plants, but incorporation of these microbes in the field of agriculture restricted the progress of efficient disease management practices to a certain level. Plant-associated microbiome that is considered to be beneficial provides new opportunities in obtaining long-term benefits in the field of disease control and management, which result in increased crop productivity (Bonanomi et al. 2018). Along with plant protection against pest and diseases, soil microorganisms being a vital component of soil organic content provide nourishment to the plants. A microbe present in soil can be exhausted by various agricultural performances; however, this can be reduced by taking many steps to enhance the quality of soil. Certain soil microorganisms are vital for absorption of nutrients in plants as they provide nitrogen and aid in phosphorus solubilization making the nutrient available for uptake. The microbiome also aids in deprivation of recalcitrant organic matter and mineral weathering that supplies carbon in root exudes and several other rhizodeposits (Van Der Heijden et al. 2008).

Different species of microbes co-residing in the roots of plants benefit the plant in various ways. Few studies on arbuscular mycorrhizal fungus and PGPRs in a model grassland concluded that both the organisms were in a mutualistic relationship with each other and resulted in increased diversity of the plants. They also enriched seedling establishment and improved nutrient uptake. So, efficient variations in root microbiome and their symbiosis can help in the uptake of several restraining elements (Vyas et al. 2018). Few scientists examined the significance of crop rotations through aiming the function of soil microbiome diversity in upholding the health of soil and enhancing crop yield (Dias and Antunes 2014; Venter et al. 2016; Hou et al. 2018). Next-generation sequencing for the advancement of GenBank database of rhizospheric microbiome has the potential to confirm a better foundation of advanced indicators of soil health for improved sustainable agriculture cropping system (Dias and Antunes 2014).

The untenable exploitation of chemical inputs is causing interference in biogeochemical cycles for essential nutrients. Their excessive usage is altering the mechanism of biogeochemical cycles leading to soil degradation, emission of greenhouse gases, and eutrophication (Steffen et al. 2015). Nitrogen fertilizers are produced using Haber Bosch process, which is an energy-intensive method. This production procedure depends upon fossil fuel and hence adds to global warming and depletion of natural resources, which in turn adds to climatic changes (Erisman et al. 2013). As the usage of mineral fertilizers has extreme significances, there is a need for alternate approaches for nourishing plant nutrition and soil health with least involvement of mineral fertilizers (Foley et al. 2011). One of the ways to minimize the usage of those fertilizers is via application of organic amendments, and by augmenting plants with beneficial and explicit root-linked microorganisms that mineralize/solubilize the organic elements unavailable to them. Rhizospheric microbiome closely relates to the physical properties of soil and directly affects the processing of the ecosystem; their incidence, profusion, and assortment have often been recommended as bioindicators for determining the health of soil (Lu et al. 2014). Several practices in cycling of nitrogen are not clear yet, but certain microbes have been identified that functions in mediating the capacity of soil to intake greenhouse gases like N_2O (Jones et al. 2014). The bacteria that cause anaerobic oxidation of ammonium are capable of making a vital contribution to agroecosystems by checking the loss of nitrogen (Nie et al. 2015).

Current researches explain that plant symbiotic fungi and arbuscular mycorrhiza can not only decrease the extent of essential macro- and micro-nutrients percolated from rhizosphere but also likewise possess the capability to effectively search for phosphorus sources in the soil and enhance plant nourishment (Cavagnaro et al. 2015). Therefore, these fungi possess the prospective to enhance efficiency of nutrient usage in agroecosystems (Kumar et al. 2015). Soil is a prevalent source of organic carbon on the planet, and the soil carbon sequestration potential is determined by the biological processes taking place in the soil. Thus, the factors regulating the storage of carbon and its discharge from soils have major significance (Amundson et al. 2015). There is an instantaneous requirement for further sustainable agricultural performances proficient of producing greater yields, which is difficult to achieve without multifaceted synchronization between ecology, soil science, agronomy, genetics, and economics and without the complete commitment of farmers as well (Kumar et al. 2016). So, apart from engineering of microbes, other technologies are evolving that will help in advancement of our understanding the responses of microbial interaction with plants under changing climatic conditions.

Efforts to Manipulate Soil Microbiome

Our growing cognizance about the effects on soil microbiome with regard to climate change is leading to an urgent requirement of harnessing soil microbial competencies to diminish the adverse significances of the climatic changes. These interests may differ from direct administration of soil microbiome to indirect administration of their utilities through alterations in land organization methods and use of bioinoculants acting as probiotics for the environment.

Carbon Sequestration

Carbon shares present in the atmosphere may reduce by their sequestration into steady and nongaseous forms by varied biotic and abiotic practices. Through acclimatization of atmospheric CO_2, carbon enters into the soil. This is carried out majorly by plants and autotrophic microbes in the soil. The amount of photosynthate liberated by plants in the rhizosphere through mycorrhizal fungi, scraped root cap cells, or root exudations is extensive (up to 20%) (Averill et al. 2014).

The presence of carbon in the soil fuels free-living and symbiotic organisms that disperse the carbon via soil medium. Biochemical transformations of carbon through microbial contribution and succeeding exchange between communities result in cycling of bioavailable types of carbon to persevere in biologically unavailable forms. The ability of soils to carbon sequestering is directly proportional to the biodiversity present in it. This sequestration involves intensive activities and contribution of soil fungi and bacteria in the production of carbon polymers that aids in facilitating the development of soil aggregations and obstructs soil carbon in that course. Carbon storage approaches including excavation of unexplored biochemical ability of soil microbial community are being explored for different reactions that result in increasing the soil carbon deposition. Particular microbe or microbes coexisting in a consortium catalyze these reactions and help the pathways of carbon decomposition to yield much stable and recalcitrant end product (Hicks et al. 2017). Otherwise, manipulation of soil microbiome can be performed in situ by adding amendments to improve their process of obtaining and storing carbon in the soil. Like, insistent carbon formed by microbial deposits can be stored in more intense layers of soil (Schmidt et al. 2011). Microbial deposits are macromolecules formed by extracellular polymeric compounds or necrotic biomass, which persists in the soil (Kapilan et al. 2018). Another opportunity to explore is using pyrolyzed carbon, which is also known as biochar, as a source of enhancement for sequestration of soil carbon in its steady form (Jansson et al. 2010). Stability is dependent on the respiration of constituents present in biochar aided by soil microbes. Plants and soil rhizospheric microbiome interactions can be influenced and facilitated by carbon storage in soil (Jansson et al. 2018; Wallenstein 2017). For instance, deposition of root exudates could be improved via enhancing the plant descent of carbon to rhizosphere, where its transformation takes place and gets converted into steady metabolites deposited in microbial biomass (Jansson et al. 2018). In this situation, genetic modification of plants can be performed for the selection of beneficial rhizospheric microbes that traps explicit root exudates released by plants. Forthcoming strategies could lead toward developing a method for genetically controlling the distribution of photosynthate to enhance the plant–microorganism–soil system to achieve soil carbon deposition and desired plant yield (Jansson et al. 2018). To evade the application of genetically modified plants to the field, association between soil microbiologists and plant breeders is required to develop the superior combination of definite beneficial microbes with particular genotype of plants (Lakshmanan et al. 2017).

Microorganisms as Beneficial Plant Inoculants

Rhizospheric plant growth-promoting (PGP) bacteria and fungi can be very helpful in various ways. They aid in mitigating the adverse significances of drought by augmenting the plant growth in stressful situations (Naylor and Coleman-Derr 2018). PGP microbes could be used as biofertilizers by seed applications, liquid formulations, or granular supplements in the field. The best instance of a PGP strain is inoculants of *Rhizobium* spp. They help in biological fixation of nitrogen in legumes. Presently, the research interests are inclining toward traditional application of microbes as biopesticides and biofertilizers to exploit their beneficial properties and mitigate the deleterious effects of chemicals and of climate change (Compant et al. 2010). Several studies have been performed to explore the beneficial properties of PGP microbes to deal with drought-related stress in crops (Table 1) (Lakshmanan et al. 2017). For instance, certain beneficial bacteria release extracellular polymeric compounds, which result in the formation of hydrophobic biofilms that help in protecting the plants from desiccation (Naylor and Coleman-Derr 2018). They also produce phytohormones that enhance the growth of plants and accumulate osmolytes and other shielding components (Lakshmanan et al. 2017; Vurukonda et al. 2016a, b). For instance, certain bacteria produce auxins as plant growth regulator in the rhizosphere, which results in enhancement in root formation (Armada et al. 2015) and that helps enhancing water uptake mitigating water stress (Lakshmanan et al. 2017). Rhizospheric microbes also secrete metabolites that have the ability to accrue in plant cells to lessen osmotic strain (Casanovas et al. 2002; Pereyra et al. 2012). Tolerance to drought stress and procurement of nutrients can be improved by associating with favorable AM fungi, for instance, by amending the production of specific molecules known as aquaporins by plants, which decrease the water stress (Kapilan et al. 2018; Quiroga et al. 2017). Harmonizing the demands of plants with supply of microbial nitrogen also helps in reducing microbial N_2O assembly. For instance, AM fungi may be utilized to obtain ammonium and lessen N_2O production. Additional biological approach to lessen N_2O emission is to inoculate N_2O consuming microbial communities (Itakura et al. 2012) or hindering the process of nitrification by using biological inhibitors to inhibit the oxidation pathway of ammonia (Subbarao et al. 2009).

Collected, these instances elucidate how valuable characteristics of soil microbes can be exploited to uphold a sustainable ecosystem amid climate change. It is notable that the scientific breakthrough required for food and agriculture by 2030 has been published by the US National Academy of Science, which includes strategies for altering soil microbial community to enhance crop productivity amid climate

Table 1 PGPR interaction effects on different crop plants under climatic changes conditions

Microbial inoculants	Plant species	Mechanisms and actions	References
Raoultella planticola Rs-2; *Streptomyces* sp. strain PGPA39	Cotton	ACC deaminase activity and production of IAA	Wu et al. (2012)
Haererohalobacter (JG-11), *Brachybacterium saurashtrense* (JG-06), and *Brevibacterium casei* (JG-08)	Groundnut	Increase K^+/Na^+ and Ca^{2+}, ratio, and the accumulation of P and N	Shukla et al. (2012)
Bacillus amyloliquefaciens SQR9; *Azospirillum* spp.; *Pseudomonas putida* FBKV2	Maize	Production of phytohormone and exopolysaccharides; enhancing water retention; regulating the diffusion of carbon sources and induction of plant stress tolerance and defense genes	Chen et al. (2016), Fukami et al. (2017), Vurukonda et al. (2016a, b)
Bacillus aryabhattai strains MDSR7, MDSR11, and MDSR14, *Bradyrhizobium* sp. LSBR-3, *Leclercia* sp. LSE-1, *Pseudomonas* sp. LSE-2	Soybeanand wheat	Increased shoot dry weight, plant height, root dry weight, and zinc assimilation in seeds	Ramesh et al. (2014); Khande et al. (2017), Kumawat et al. (2019a, 2019b)
Enterobacter cloacae and *Bacillus drentensis*	Mung bean	Modified stomatal conductance, transpiration rate, water relations, and synthesis of photosynthetic pigments	Mahmood et al. (2016)
Arthrobacter protophormiae (SA3) and Dietzia natronolimnaea (STR1)	Wheat	Enhance photosynthetic efficiency; increase indole-3-acetic acid; modulating expression of a regulatory component (CTR1) of the ethylene signaling pathway, stress-related antioxidant genes, and DREB2 transcription factor	Barnawal et al. (2017), Bharti et al. (2016)
P. agglomerans RSO6 and RSO7 *B. aryabhattai* RSO25	Rice	Enzyme activities related to oxidative stress induced such as ascorbate peroxidase, guaiacol peroxidase, glutathione reductase, superoxide dismutase	García-Cristobal et al. (2015), Paul et al. (2006)
Enterobacter sp. UPMR18	Okra	ROS-scavenging enzymes	Sheikh et al. (2016)
Burkholderia cepacia SE4, *Promicromonospora* sp. SE188, and *Acinetobacter calcoaceticus* SE370	*Cucumis sativus*	Reduced activities of catalase, peroxidase, polyphenol oxidase	Kang et al. (2014)

change. Other research priorities include understanding of biological pathways sustaining decomposition of soil organic carbon and greenhouse gas production to develop better practices to avoid loss of carbon from soil. The present frontier describes how soil microbiome responds physiologically. This information will enable estimations of the effects of climatic changes on soil functioning and favorable attributes of the soil microbiome to aid in mitigating the undesirable significances of climate change.

Evolving Recent Technologies for Understanding of Plant–Microbial Interaction Responses Under Changing Climatic Conditions

Techniques that are commonly used to understand microbial community dynamics are terminal restriction fragment length polymorphism (TRFLP), phospholipid fatty acid analysis (PLFA), denaturing gel gradient electrophoresis (DGGE), and amplified ribosomal DNA restriction analysis (ARISA). While these molecular techniques have assisted scientists to comprehend configurations of composition of rhizospheric microbiome at stiff level (Gray et al. 2011; Bell-Dereske et al. 2017), appraising the reactions of each taxon and offer restricted understanding in functional shifts as well. Nevertheless, with progression in sequencing tools and molecular omics, scientists have started exploring the interactions between host and microbes in much deeper level. By application of different omic tools like metagenomics, proteomics, transcriptomics, and metabolomics, understanding of the microbial–community dynamics has become advanced, causing better understanding in prospect of taxonomy, genetics, and functions of the microbial populations (Muller et al. 2013; Dubey et al. 2018; Malla et al. 2018a, b). Much advanced methods such as stable probing of isotopes are available and have helped scientists to evaluate the dynamic assortment in assembly of utilities (Zhang et al. 2018a). Currently, amplicon sequencing is developed to be a common technique for the taxonomical characterization of bacterial configurations in the ecosystem (Sanschagrin and Yergeau 2014).

The 16S and 18S rRNA gene sequencing technique produces huge amount of data sets, which provide us with evidence regarding the occurrence and permeation of a species, but yield very less information about the functions and potential of those species (Fierer et al. 2012). Due to this, shotgun metagenomic usage is being preferred to discover and comprehend the structural and functional configuration of the microbial populations in a particular habitat. Though shotgun sequencing provides functional prospective, it lacks the complexity related to amplicon sequencing, and because of that, less profuse taxa remain unnoticed (Zhou et al. 2015). Among stall, the newly developed technologies for examining the soil microbiome dynamics, the most crucial thing, is the sampling of microbial communities for determining taxonomy and function of these microbes. Microorganisms interact at soil cumulative scale, with considerable differences observed through soil aggregates (Lombard et al. 2011). To understand the pattern of interaction between microbial populations with each other and with their respective host plants, upcoming studies should focus more on the diversity and the function of microbes. Along with these issues, developing technologies for understanding the microbial communities, and analyzing and exploring them in large numbers is essential (Zhou et al. 2015). There are some bioinformatic techniques and software, which assist in processing and analyzing the data. Few of them are MOTHUR (Schloss et al. 2009), MG-RAST (Glass and Meyer 2011), MEGAN (Huson and Weber 2013), Galaxy portal (Goecks et al. 2010), QIIME (Caporaso et al. 2010), and Functionalize R (Kristiansson et al. 2009). Another promising area for research and technology application is understanding the relationships between plants and rhizospheric microbial diversity. Microbial diversity is abundant, and it is closely connected through the roots of crop plants (Hol et al. 2013). Divisions of those groups infiltrate plant roots and inhabit the internal spaces present in their host plants; but to determine the method of colonization, to explore the molecular signaling pathways that allow microorganisms to escape the plants from phytopathogens through defense mechanisms and penetrate their roots is challenging. Implementation of modern sequencing techniques has various benefits such as they are economic, precise, and prompt in sequencing whole genomes. Understanding and exploring the interactions between plants and microbes in molecular level allow alterations in microbiome for enhancing the function of plants and the ecosystem. It will allow the researchers to categorize and identify the microbial assortment present in the root endosphere as well and will aid in formulating the capabilities of microbial communities for the enhancement of carbon storing and distribution, improving plant vigor, and reducing greenhouse gases.

Metagenomic Tools for Unlocking the Black Box

Rhizospheric microbiome plays a significant contribution in enhancing crop yield, nutrient cycling, carbon sequestration, and restoring sustainability in the environment (Dubey et al. 2018). Soil microbiome determines the health of ecosystem, and it is attaining a speedy recognition and implication in the field of ecology, and still, there is a lot of unmapped information about the function of these microorganisms in operating the environmental stress conditions. Constructing definite connotation between diverse microbial associations can exploit its remunerations, and binding the influence of microbial taxa can prove to be beneficial in dealing with global climate changes. Current studies conducted by various researches (Castro et al. 2010; Blaser et al. 2016; Lau et al. 2017) depicted the reaction of rhizospheric microbes to the stress due to climatic conditions. In aquatic (Deemer et al. 2016) and terrestrial ecosystems (Barant et al. 2016), the microorganisms have a great influence and contribution to the ecosystem by their function as plant mutualists, pathogens, or detritivorous organisms inducing the turnover of greenhouse gases, thus alleviating the climate change. Soils in forests are regarded as hot spots for rhizospheric microbial communities, and microbiota has a great effect on vigor and efficiency of trees. They play an important role in facilitating biogeochemical cycles in the forest ecosystem (Baldrian et al. 2012). These microbial groups are allocated into various types such as symbionts, decomposers, and pathogens (Baldrian 2017). But, the taxonomic and efficient classification and the synergistic effects between these microbial communities are ambiguous. These methods can, however, be demonstrated by metagenomic studies. The results obtained from metagenomic studies are an influential and potential approach to reveal ecological and evolutionary forces, prompting the microbial characters within a fluctuating environment.

Lately, metagenomic researches have considerably enhanced our knowledge rhizospheric microbiome. Recently, research carried out by various scientists recommended that microorganisms have a natural capability to sequester CO_2 through various mechanisms (Yuan et al. 2012; Gougoulias et al. 2014; Hicks et al. 2017). Likewise, some researchers discussed the importance of microorganisms in alleviating the detrimental effects of greenhouse gases (Nazaries et al. 2013; Xu et al. 2013; Gupta and Prakash 2014). They suggested that soil microbial respiration can provide a link to know how soil microbiome checks and inhibits climate change. The naturally arising varied microbial communities are capable of utilizing different global warming gases via different metabolic pathways. To understand the detailed mechanisms involved in those pathways requires modern high-throughput approaches. Metagenomic approaches for the study of rhizospheric microbiome can be of great help in understanding and restoring the functioning of the ecosystem. These approaches provide valuable information regarding taxonomic, functional, and genetic characteristics of soil microbiota (Singh et al. 2010). It is well known that next-generation sequencing approaches (amplicon or metagenomic sequencing) are far more accurate and greater than the traditional ones and are capable to be of substantial use in checking climatic changes.

Metagenomics is a unique and promising ground of science that is based on genomic analysis of samples of DNA obtained from the environment. This method helps in characterizing the microbiome taxonomically, metabolically, and functionally and helps in understanding their relationship with their habitat without the requirement of culturing. The advancement in sequencing, bioinformatics, and metagenomic techniques is growing rapidly and is transforming our understanding of the arrangement and utility of the "black box" (microbial communities), revealing genetic diversity and discovering novel bioremediation and biogeochemical pathways.

It is aforementioned that microorganisms colonize most peculiar parts of environment and play crucial roles in the functioning of ecosystem, industries, food and agriculture, etc. Therefore, recognizing the structure, function, diversity, and stability of the microbial communities is essential to enhance our knowledge on sustainability, community establishment, and evolution of life on earth. Conversely, receiving such type of evidence is complex, due to the reason that most of the microorganisms ($\geq$90%) are un-cultivable (Lewis et al. 2010). Previously, a standard modification has been witnessed in the field of metagenomics to the application of cross-sectional examinations and longitudinal assisted by advancements in high-performance in silico tools and DNA sequencing. These technologies have helped in evaluating the microbial diversity and their functions in a broader level, thus letting a systematic investigation of the largely unexplored microbial taxa. The relative metagenomic researches of diverse microbial populations have generated substantial perceptions to the dispersal of gene families through diverse ecosystems besides the function of detailed efficient qualities in adapting to different environmental circumstances (Delmont et al. 2011).

Current Perspectives and Future Challenges

Experts in the field of molecular biology, chemistry, and biochemistry can help in clarifying the understanding of detailed interfaces in nutrient cycling and in structure and functioning of microbial communities. Metabolic and genetic variations among microbial populations can be shown with the help of nucleic acid sequencing (Lucero et al. 2011). Genome sequencing, metabolomics, and proteomics of plants allied microbes develop our knowledge on the functions of microbiome such as influencing plant productivity, inducing tolerance against salinity and drought, promoting growth, providing resistance against pathogens, and nutrient cycling. Microbial genomes useful in the development of germplasm can be sequenced for evaluating genes that regulate host specificity, pathogenesis, and biotic and abiotic stress tolerance (Kuldau and Bacon 2008). Metagenomes related to host plant in a specific environment can disclose the whole metabolic capacity of microbial communities for the determination of plant production and varied types of carbon sequestration such as C fixation and biomineralization regulated by soil microorganisms. Databases like National Center for Biotechnology Information and Cyberinfrastructure for Advanced Microbial Ecology Research and Analysis (CAMERA) make assemblages of ecological and metagenomic data publically available, which in permutation with associated research papers may offer introductory material for fine-scale simulation models. By connecting them with simulations performed at higher scales, these models may be utilized for forecasting the responses of microbial populations with regard to management practices like tillage and irrigation, and also to recognize the influence of these reactions at higher scales.

References

Abhilash PC, Dubey RK, Tripathi V, Srivastava P, Verma JP, Singh HB (2013) Remediation and management of POPs-contaminated soils in a warming climate: challenges and perspectives. Environ Sci Pollut Res 20:5879–5885

Agrawal AA, Weber MG (2015) On the study of plant defense and herbivore using comparative approaches: how important are secondary plant compounds. Ecol Lett 18:985–991

Amundson R, Berhe AA, Hopmans JW, Olson C, Sztein AE, Sparks DL (2015) Soil and human security in the 21st century. Science. https://doi.org/10.1126/science.1261071

Armada E, Azcon R, Lopez-Castillo OM, Calvo-Polanco M, Ruiz-Lozano JM (2015) Autochthonous arbuscular mycorrhizal fungi and *Bacillus thuringiensis* from a degraded Mediterranean area can be used to improve physiological traits and performance of a plant of agronomic interest under drought conditions. Plant Physiol Biochem 90:64–74

Averill C, Turner BL, Finzi AC (2014) Mycorrhiza mediated competition between plants and decomposers drives soil carbon storage. Nature 505:543–545

Baetz U, Martinoia E (2014) Root exudates: the hidden part of plant defense. Trends Plant Sci 19: 90–98

Bagri DS, Upadhyaya DC, Kumar A, Upadhyaya CP (2018) Overexpression of PDX-II gene in potato (*Solanum tuberosum* L.) leads to the enhanced accumulation of vitamin B6 in tuber tissues and tolerance to abiotic stresses. Plant Sci 272:267–275

Baldrian P (2017) Forest microbiome: diversity, complexity and dynamics. FEMS Microbiol Rev. https://doi.org/10.1093/femsre/fuw040

Baldrian P, Kolairik M, Stursova M, Kopecky J, Valaskova V, Vetrovsky T, Zifcakova L, Snajdr J, Ridl J, Vlcek C, Voriskova J (2012) Active and total microbial communities in forest soil are largely different and highly stratified during decomposition. ISME J. https://doi.org/10.1038/ismej.2011.95

Barant S, Radbata D, Oberweis D et al (2016) Abdominal necrotizing fasciitis after caesarean delivery. Rev Med Brux 37:178–182. https://doi.org/10.1126/science.1097396

Bardgett RD, Van Der Putten WH (2014) Belowground biodiversity and ecosystem functioning. Nature 515:505–511

Barnawal D, Bharti N, Pandey SS, Pandey A, Chanotiya CS, Kalra A (2017) Plant growth promoting rhizobacteria enhances wheat salt and drought stress tolerance by altering endogenous phytohormone levels and TaCTR1/TaDREB2 expression. Physiol Plant 161:502–514

Bar-On YM, Phillips R, Milo R (2018) The biomass distribution on earth. Proc Natl Acad Sci U S A. https://doi.org/10.1073/pnas.1711842115

Bell CW, Asao S, Calderon F, Wolk B, Wallenstein MD (2015) Plant nitrogen uptake drives rhizosphere bacterial community assembly during plant growth. Soil Biol Biochem 85:170–182

Bell-Dereske L, Takacs-Vesbach C, Kivlin SN (2017) Leaf entophytic fungus interacts with precipitation to alter belowground microbial communities in primary successional dunes. FEMS Microbiol Ecol 93:1–16

Bharti N, Pandey SS, Deepti B, Patel VK, Kalra A (2016) Plant growth promoting rhizobacteria Dietzia natronolimnaea modulates the expression of stress responsive genes providing protection of wheat from salinity stress. Sci Rep 6:34768. https://doi.org/10.1038/srep34768

Bintanja R (2018) The impact of Arctic warming on increased rainfall. Sci Rep 8:16001. https://doi.org/10.1038/s41598-018-34450-3

Blaser MJ, Cardon ZG, Cho MK, Dangl JL, Donohue TJ, Green JL, Knight R, Maxon ME, Northen TR, Pollard KS, Brodie EL (2016) Toward a predictive understanding of earth's microbiomes to address 21st century challenges. MBio. https://doi.org/10.1128/mBio.00714-16

Bojko O, Kabala C (2017) Organic carbon pools in mountain soils-sources of variability and predicted changes in relation to climate and land use changes. CATENA. https://doi.org/10.1016/j.catena.2016.09.022

Bonanomi G, Lorito M, Vinale F, Woo SL (2018) Organic amendments, beneficial microbes, and soil microbiota: toward a unified framework for disease suppression. Annu Rev Phytopathol 56: 1–20

Brussaard L, Ruiter PC, Brown GG (2007) Soil biodiversity for agricultural sustainability. Agric Ecosyst Environ 121:233–244

Busby PE, Soman C, Wagner MR, Friesen ML, Kremer J, Bennett A, Morsy M, Eisen JA, Leach JE, Dangl JL (2017) Research priorities for harnessing plant microbiomes in sustainable agriculture. PLoS Biol 15:1–14

Callaway E (2016) Devastating wheat fungus appears in Asia for first time. Nature 532:421–422

Caporaso JG, Kuczynski J, Stombaugh J, Bittinger K, Bushman FD, Costello EK, Fierer N, Peña AG, Goodrich JK, Gordon JI, Huttley GA, Kelley ST, Knights D, Koenig JE, Ley R, Lozupone

CA, McDonald D, Muegge BD, Pirrung M, Reeder J, Sevinsky JR, Turnbaugh PJ, Walters WA, Widmann J, Yatsunenko T, Zaneveld J, Knight R (2010) QIIME allows high throughput community sequencing data. Nat Methods 7:335–336. https://doi.org/10.1038/nmeth0510-335

Casanovas EM, Barassi C, Sueldo RJ (2002) *Azospirillum* inoculation mitigates water stress effects in maize seedlings. JSTOR 30:343–350

Castro HF, Classen AT, Austin EE, Norby RJ, Schadt CW (2010) Soil microbial community responses to multiple experimental climate change drivers. Appl Environ Microbiol 76:999–1007

Castro HF, Classen AT, Austin EE, Crawford KM, Schadt CW (2012) Development and validation of a citrate synthase directed quantitative PCR marker for soil bacterial communities. Appl Soil Ecol 61:69–75. https://doi.org/10.1016/j.apsoil.2012.05.007

Cavagnaro TR, Bender SF, Asghari HR, Van Der Heijden MGA (2015) The role of arbuscular mycorrhizas in reducing soil nutrient loss. Trends Plant Sci 20:283–290

Chaparro JM, Badri DV, Vivanco JM (2014) Rhizosphere microbiome assemblage is affected by plant development. ISME J 8:790–803. https://doi.org/10.1038/ismej.2013.196

Chen D, Cheng J, Chu P, Hu S, Xie Y, Tuvshintogtokh I, Bai Y (2015) Regional-scale patterns of soil microbes and nematodes across grasslands on the Mongolian plateau: relationships with climate, soil, and plants. Ecography 38:622–631

Chen Y, Evans J, Feldlaufer M (2016) Horizontal and vertical transmission of viruses in the honey bee, *Apis mellifera*. J Invertebr Pathol 92:152–159

Chowdhary K, Kumar A, Sharma S, Pathak R, Jangir M (2018) *Ocimum* sp.: source of biorational pesticides. Ind Crop Prod 122:686–701

Classen AT, Sundqvist MK, Henning JA, Newman GS, Moore JAM, Cregger MA, Moorhead LC, Patterson CM (2015) Direct and indirect effects of climate change on soil microbial and soil microbial–plant interactions: what lies ahead? Ecosphere 6:1–21. https://doi.org/10.1890/ES15-00217.1

Compant S, Van Der Heijden MGA, Sessitsch A (2010) Climate change effects on beneficial plant–microorganism interactions. FEMS Microbiol Ecol 73:197–214. https://doi.org/10.1111/j.1574-6941.2010.00900.x

Curtis TP, Sloan WT (2005) Exploring microbial diversity-a vast below. Science 309:1331–1333

De Vries FT, Liiri ME, Bjørnlund L, Bowker MA, Christensen S, Setala HM, Bardgett RD (2012) Land use alters the resistance and resilience of soil food webs to drought. Nat Clim Chang 2: 276–280

De Weert S, Vermeiren H, Mulders IHM, Kuiper I, Hendrickx N, Bloemberg GV, Vanderleyden J, Mot RD, Lugtenberg BJJ (2002) Flagella-driven chemotaxis towards exudate components is an important trait for tomato root colonization by *Pseudomonas fluorescens*. Mol Plant 15(11): 1173–1180. https://doi.org/10.1094/MPMI.2002.15.11.1173

DeAngelis KM, Pold G, Topçuoglu BD, Diepen LTA, Varney RM, Blanchard JL, Melillo J, Frey SD (2015) Long-term forest soil warming alters microbial communities in temperate forest soils. Front Microbiol. https://doi.org/10.3389/fmicb.2015.00104

Deemer BR, Harrison JA, Li S, Beaulieu JJ, Del ST, Barros N, Bezerra-Neto JF, Powers SM, Santos MA, Vonk JA (2016) Greenhouse gas emissions from reservoir water surfaces: a new global synthesis. Bioscience 66:949–964

Delmont TO, Malandain C, Prestat E, Larose C, Monier JM, Simonet P, Vogel TM (2011) Metagenomic mining for microbiologists. ISME J 5:1837–1843. https://doi.org/10.1038/ismej.2011.61

Dias T, Antunes PM (2014) Accounting for soil biotic effects on soil health and crop productivity in the design of crop rotations. J Sci Food Agric 95(3):447–454. https://doi.org/10.1002/jsfa.6565

Dubey A, Kumar A, Abd-Allah EF, Hashem A, Khan ML (2018) Growing more with less: breeding and developing drought resilient soybean to improve food security. Ecol Indic 105:425–437. https://doi.org/10.1016/j.ecolind.2018.03.003

Erisman JW, Galloway JN, Seitzinger S, Bleeker A, Dise NB, Petrescu AMR, Leach AM, Vries W (2013) Consequences of human modification of the global nitrogen cycle. Philos Trans R Soc B Biol Sci. https://doi.org/10.1098/rstb.2013.0116

Fang J, Yu G, Liu L, Hu S, Chapin FS (2018) Climate change, human impacts, and carbon sequestration in China. Proc Natl Acad Sci U S A 115(16):4015–4020. https://doi.org/10.1073/pnas.1700304115

Fierer N, Lef JW, Adams BJ, Nielsen UN, Bates ST, Lauber CL, Owens S, Gilbert JA, Wall DH, Caporaso JG (2012) Cross-biome metagenomic analyses of soil microbial communities and their functional attributes. Proc Natl Acad Sci U S A 109:21390–21395. https://doi.org/10.1073/pnas.1215210110

Foley JA, Ramankutty N, Brauman KA, Cassidy ES, Gerber JS, Johnston M, Mueller ND, O'Connell C, Ray DK, West PC, Balzer C, Bennett EM, Carpenter SR, Hill J, Monfreda C, Polasky S, Rockstrom J, Sheehan J, Siebert S, Tilman D, Zaks DPM (2011) Solutions for a cultivated planet. Nature 478:337–342

Fukami J, Ollero FJ, Megías M, Hungria M (2017) Phytohormones and induction of plant stress tolerance and defense genes by seed and foliar inoculation with *Azospirillum brasilense* cells and metabolites promote maize growth. AMB Exp 7:153. https://doi.org/10.1186/s13568-017-0453-7

Gao D, Peng B, Edith B, Frank H, Qu G, Sun J, Liu J, Zheng J, Zhang L, Jiang P, Fan Z (2018) Small and transient response of winter soil respiration and microbial communities to altered snow depth in a mid-temperate forest. Appl Soil Ecol 130:40–49

García-Cristobal J, García-Villaraco A, Ramos B, Gutierrez-Mañero J, Lucas JA (2015) Priming of pathogenesis related-proteins and enzymes related to oxidative stress by plant growth promoting rhizobacteria on rice plants upon abiotic and biotic stress challenge. J Plant Physiol 188:72–79

Glass EM, Meyer F (2011) The metagenomics RAST server: a public resource for the automatic phylogenetic and functional analysis of metagenomes. In: Handbook of molecular microbial ecology. I: Metagenomics and complementary approaches. Wiley, Hoboken, pp 325–331. https://doi.org/10.1002/9781118010518.ch37

Goecks J, Nekrutenko A, Taylor J, Galaxy J (2010) Galaxy: a comprehensive approach for supporting accessible, reproducible, and transparent computational research in the life sciences. Genome Biol. https://doi.org/10.1186/gb-2010-11-8-r86

Gottfried M, Pauli H, Futschik A et al (2012) Continent-wide response of mountain vegetation to climate change. Nat Clim Chang 2:111–115

Gougoulias C, Clark JM, Shaw LJ (2014) The role of soil microbes in the global carbon cycle: tracking the below-ground microbial processing of plant-derived carbon for manipulating carbon dynamics in agricultural systems. J Sci Food Agric 94:2362–2371

Gray SB, Classen AT, Kardol P, Miller ZYRM (2011) Multiple climate change factors interact to alter soil microbial community structure in an old-field ecosystem. Soil Sci Soc Am J 75:2217. https://doi.org/10.2136/sssaj2011.0135

Gupta C, Prakash DG (2014) Role of microbes in combating global warming. Int J Pharm Sci Lett 4:359–363

Hameed A, Wu QS, Abd-Allah EF, Hashem A, Kumar A, Lone HA, Ahmad P (2014) Role of AM fungi in alleviating drought stress in plants. In: Miransari M (ed) Use of microbes for the alleviation of soil stresses. Springer, New York

Hashem A, Abd-Allah EF, Alqarawi AA, Radhakrishnan R, Kumar A (2017) Plant defense approach of *Bacillus subtilis* (Bera 71) against *Macrophomina phaseolina* (tassi) Goid in mung bean. J Plant Interact 12:390–401

Hassani MA, Duran P, Hacquard S (2018) Microbial interactions within the plant holobiont. Microbiome 6:58

Henry HAL (2013) Reprint of "Soil extracellular enzyme dynamics in a changing climate". Soil Biol Biochem 56:53–59

Hicks N, Vik U, Taylor P, Ladoukakis E, Park J, Kolisis F, Jakobsen KS (2017) Using prokaryotes for carbon capture storage. Trends Biotechnol 35:22–32. https://doi.org/10.1016/j.tibtech.2016.06.011

Hiruma K, Gerlach N, Sacristán S, Nakano RT, Hacquard S, Kracher B, Neumann U, Ramírez D, Bucher M, O'Connell RJ, Schulze-Lefert P (2016) Root endophyte *Colletotrichum tofieldiae* confers plant fitness benefits that are phosphate status dependent. Cell 165:464–474. https://doi.org/10.1016/j.cell.2016.02.028

Hol WHG, Bezemer TM, Biere A (2013) Getting the ecology into interactions between plants and the plant growth promoting bacterium *Pseudomonas fluorescens*. Front Plant Sci. https://doi.org/10.3389/fpls.2013.00081

Hou PF, Chien CH, Chiang-Hsieh YF, Tseng KC, Chow CN, Huang HJ, Chang WC (2018) Paddy-upland rotation for sustainable agriculture with regards to diverse soil microbial community. Sci Rep. https://doi.org/10.1038/s41598-018-26181-2

Huson DH, Weber N (2013) Microbial community analysis using MEGAN. In: DeLong EF (ed) Methods in enzymology. Elsevier, San Diego, pp 465–485

IPCC (2007) Climate change 2007 synthesis report. IPCC, Geneva

Itakura M, Uchida Y, Akiyama H, Hoshino YT, Shimomura Y, Morimoto S, Tago K, Wang Y, Hayakawa C, Uetake Y, Sánchez C, Eda S, Hayatsu M, Minamisawa K (2012) Mitigation of nitrous oxide emissions from soils by *Bradyrhizobium japonicum* inoculation. Nat Clim Chang 3:208–212

Jansson C, Tuskan GA, Wullschleger SD, Kalluri UC (2010) Phyto-sequestration: carbon bio-sequestration by plants and the prospects of genetic engineering. Bioscience 60:685–696

Jansson C, Vogel J, Hazen S, Brutnell T, Mockler T (2018) Climate-smart crops with enhanced photosynthesis. J Exp Bot 69:3801–3809

Ji B, Bever JD (2016) Plant preferential allocation and fungal reward decline with soil phosphorus: implications for mycorrhizal mutualism. Ecosphere. https://doi.org/10.1002/ecs2.1256

Jones CM, Spor A, Brennan FP, Spor A, Brennan FP, Breuil MC, Bru D, Lemanceau P, Griffiths B, Hallin S, Philippot L (2014) Recently identified microbial guild mediates soil N_2O sink capacity. Nat Clim Chang 4:801–805

Kamthan A, Chaudhuri A, Kamthan M, Datta A (2016) Genetically modified (GM) crops: milestones and new advances in crop improvement. Theor Appl Genet 129:1639–1655

Kang S, Khan AL, Waqas M, You Y, Kim J, Hamayun M, Lee I (2014) Plant growth-promoting rhizobacteria reduce adverse effects of salinity and osmotic stress by regulating phytohormones and antioxidants in Cucumis sativus. J Plant Interact 9(1):673–682

Kapilan R, Vaziri M, Zwiazek JJ (2018) Regulation of aquaporins in plants under stress. Biol Res 51:4. https://doi.org/10.1186/s40659-018-0152-0

Kato S, Yoshida R, Yamaguchi T, Sato T, Yumoto I, Kamagata Y (2014) The effects of elevated CO_2 concentration on competitive interaction between aceticlastic and syntrophic methanogenesis in a model microbial consortium. Front Microbiol. https://doi.org/10.3389/fmicb.2014.00575

Khande R, Sushil KS, Ramesh A, Mahaveer PS (2017) Zinc solubilizing *Bacillus* strains that modulate growth, yield and zinc biofortification of soybean and wheat. Rhizosphere 4:126–138

Kristiansson E, Hugenholtz P, Dalevi D (2009) Shotgun functionalize R: an R-package for functional comparison of metagenomes. Bioinformatics 25:2737–2738

Kuldau G, Bacon C (2008) Clavicipitaceous endophytes: their ability to enhance resistance of grasses to multiple stresses. Biol Control 46:57–71

Kumar A, Sharma S, Mishra S (2016) Evaluating effect of arbuscular mycorrhizal fungal consortia and *Azotobacter chroococcum* in improving biomass yield of *Jatropha curcas*. Plant Biosyst 150:1056–1064

Kumar A, Sharma S, Mishra S, Dames JF (2015) Arbuscular mycorrhizal inoculation improves growth and antioxidative response of *Jatropha curcas* (L.) under Na_2SO_4 salt stress. Plant Biosyst. https://doi.org/10.1080/11263504.2013.845268

Kumawat KC, Sharma P, Sirari A, Singh I, Gill BS, Singh U, Saharan K (2019a) Synergism of *Pseudomonas aeruginosa* (LSE-2) nodule endophyte with *Bradyrhizobium* sp. (LSBR-3) for improving plant growth, nutrient acquisition and soil health in soybean. World J Microbiol Biotechnol 35(3):1–17

Kumawat KC, Sharma P, Singh I, Sirari A, Gill BS (2019b) Co-existence of *Leclercia adecarboxylata* (LSE-1) and *Bradyrhizobium* sp. (LSBR-3) in nodule niche for multifaceted effects and profitability in soybean production. World J Microbiol Biotechnol 35(11):1–17

Kuzyakov Y, Blagodatskaya E (2015) Microbial hotspots and hot moments in soil: concept & review. Soil Biol Biochem 83:184–199

Lakshmanan V, Ray P, Craven KD (2017) Toward a resilient, functional microbiome: drought tolerance-alleviating microbes for sustainable agriculture. Methods Mol Biol 1631:69–84

Lal R (2016) Soil health and carbon management. Food Energy Secur 5:212–222

Langley JA, Hungate BA (2014) Plant community feedbacks and long-term ecosystem responses to multifactored global change. AoB Plants. https://doi.org/10.1093/aobpla/plu035

Lau JA, Lennon JT, Heath KD (2017) Trees harness the power of microbes to survive climate change. Proc Natl Acad Sci U S A. https://doi.org/10.1073/pnas.1715417114

Lewis K, Epstein S, D'Onofrio A, Ling LL (2010) Uncultured microorganisms as a source of secondary metabolites. J Antibiot (Tokyo) 63:468–476. https://doi.org/10.1038/ja.2010.87

Lombard N, Prestat E, van Elsas JD, Simonet P (2011) Soil-specific limitations for access and analysis of soil microbial communities by metagenomics. FEMS Microbiol Ecol 78:31–49. https://doi.org/10.1111/j.1574-6941.2011.01140.x

Lu JX, Burton SD, Xu YS, Buchko GW, Shaw WJ (2014) The flexible structure of the K24S28 region of Leucine-Rich Amelogenin Protein (LRAP) bound to apatites as a function of surface type, calcium, mutation and ionic strength. Front Physiol 100:101

Lucero ME, Unc A, Cooke P, Dowd SS (2011) Endophyte microbiome diversity in micro-propagated *Atriplex canescens* and *Atriplex torreyi var griffithsii*. Public Library of Science One 6:e17693

Mahmood S, Daur I, Al-Solaimani SG, Ahmad S, Madkour MH, Yasir M, Hirt H, Ali S, Ali Z (2016) Plant growth promoting rhizobacteria and silicon synergistically enhance salinity tolerance of mung bean. Front Plant Sci 7:876. https://doi.org/10.3389/fpls.2016.00876

Malla MA, Dubey A, Yadav S, Hashem A, Abd Allah EF (2018a) Exploring the human microbiome: the potential future role of next-generation sequencing in disease diagnosis and treatment. Front Immunol 9:1–23

Malla MA, Dubey A, Yadav S, Kumar A, Hashem A, Abd Allah EF (2018b) Understanding and designing the strategies for the microbe mediated remediation of environmental contaminants using omics approaches. Front Microbiol 9. https://doi.org/10.3389/fmicb.2018.01132

Muller EEL, Glaab E, May P, Vlassis N, Wilmes P (2013) Condensing the omics fog of microbial communities. Trends Microbiol 21:325–333

Nagpal S, Sharma P, Kumawat KC (2019) Assessment of native single and dual inoculants of *Mesorhizobium* sp. and endophytic rhizobacteria for plant growth promotion in chickpea. Agric Res J 56:746–751

Nagpal S, Sharma P, Sirari A, Gupta RK (2020) Coordination of *Mesorhizobium* sp. and endophytic bacteria as elicitor of biocontrol against *Fusarium* wilt in chickpea. Eur J Plant Pathol. https://doi.org/10.1007/s10658-020-02062-1

Nannipieri P, Ascher J, Ceccherini MT, Landi L, Pietramellara G, Renella G (2017) Microbial diversity and soil functions. Eur J Soil Sci. https://doi.org/10.1111/ejss.4_12398

Naylor D, Coleman-Derr D (2018) Drought stress and root-associated bacterial communities. Front Plant Sci 8:2223

Nazaries L, Murrell JC, Millard P, Baggs L, Singh BK (2013) Methane, microbes and models: fundamental understanding of the soil methane cycle for future predictions. Environ Microbiol 15:2395–2417. https://doi.org/10.1111/1462-2920.12149

Nie S, Li H, Yang X, Zhang Z, Weng B, Huang F, Zhu GB, Zhu YG (2015) Nitrogen loss by anaerobic oxidation of ammonium in rice rhizosphere. ISME J 9:2059–2067

Norby RJ, Kauwe MGD, Domingues TF, Duursma RA, Ellsworth DS, Goll DS, Lapola DM, Luus KA, MacKenzie AR, Medlyn BE, Pavlick R, Rammig A, Smith B, Thomas R, Thonicke K, Walker AP, Yang X, Zaehle S (2016) Model data synthesis for the next generation of forest free-air CO_2 enrichment (FACE) experiments. New Phytol 209:17–28

Oldroyd GED (2013) Speak, friend, and enter: signalling systems that promote beneficial symbiotic associations in plants. Nat Rev Microbiol 11:252–263

Orozco-Mosqueda MDC, Rocha-Granados MDC, Glick BR, Santoyo G (2018) Microbiome engineering to improve biocontrol and plant growth-promoting mechanisms. Microbiol Res 208:25–31

Paul D, Dineshkumar N, Nair S (2006) Proteomics of a plant growth-promoting rhizobacterium, *Pseudomonas fluorescens* MSP-393, subjected to salt shock. World J Microbiol Biotechnol 22(4):369–374

Peng X, Gilmore SP, O'Malley MA (2016) Microbial communities for bio-processing: lessons learned from nature. Curr Opin Chem Eng 14:103–109

Pereyra MA, García P, Colabelli MN, Barassi CA, Creus CM (2012) A better water status in wheat seedlings induced by *Azospirillum* under osmotic stress is related to morphological changes in xylem vessels of the coleoptile. Appl Soil Ecol 53:94–97

Quiroga G, Erice G, Aroca R, Chaumont F, Ruiz-Lozano JM (2017) Enhanced drought stress tolerance by the arbuscular mycorrhizal symbiosis in a drought sensitive maize cultivar is related to a broader and differential regulation of host plant aquaporins than in a drought-tolerant cultivar. Front Plant Sci 8:1056

Ramesh A, Sharma SK, Sharma MP, Yadav N, Joshi OP (2014) Inoculation of zinc solubilizing *Bacillus aryabhattai* strains for improved growth, mobilization, and biofortification of zinc in soybean and wheat cultivated in Vertisols of Central India. Appl Soil Ecol 73:87–96

Rasmann S, Turlings TCJ (2016) Root signals that mediate mutualistic interactions in the rhizosphere. Curr Opin Plant Biol 32:62–68

Sanschagrin S, Yergeau E (2014) Next-generation sequencing of 16S ribosomal RNA gene amplicons. J Vis Exp. https://doi.org/10.3791/51709

Sasse J, Martinoia E, Northen T (2018) Feed your friends: do plant exudates shape the root microbiome? Trends Plant Sci. https://doi.org/10.1016/j.tplants.2017.09.003

Schimel JP, Schaeffer SM (2012) Microbial control over carbon cycling in soil. Front Microbiol. https://doi.org/10.3389/fmicb.2012.00348

Schloss PD, Westcott SL, Ryabin T, Hall JR, Hartmann M, Hollister EB, Lesniewski RA, Oakley BB, Parks DH, Robinson CJ, Sah JW, Stres B, Thallinger GG, Horn DJV, Weber CF (2009) Introducing mothur: open-source, platform-independent, community-supported software for describing and comparing microbial communities. Appl Environ Microbiol 75:7537–7541. https://doi.org/10.1128/AEM.01541-09

Schmidt MWI, Torn MS, Abiven S, Dittmar T, Guggenberger G, Janssens IA, Kleber M, Kogel-Knabner I, Lehmann J, Manning DAC, Nannipieri P, Rasse DP, Weiner S, Trumbore SE (2011) Persistence of soil organic matter as an ecosystem property. Nature 478:49–56

Sergaki C, Lagunas B, Lidbury I, Gifford ML, Schafer P (2018) Challenges and approaches in microbiome research: from fundamental to applied. Front Plant Sci. https://doi.org/10.3389/fpls.2018.01205

Sharma S, Kaur R, Singh A (2017) Recent advances in CRISPR/Cas mediated genome editing for crop improvement. Plant Biotechnol Rep 11:193–207

Sheikh HH, Hossain K, Halimi MS (2016) Plant growth promoting rhizobacteria enhances salinity stress tolerance in okra through ROS-scavenging enzymes. Biomed Res Int 2016:6284547. https://doi.org/10.1155/2016/6284547

Shukla PS, Agarwal PK, Jha B (2012) Improved salinity tolerance of *Arachis hypogaea* (L.) by the interaction of halotolerant plant growth promoting rhizobacteria. J Plant Growth Regul 31(2): 195–206

Singh BK, Bardgett RD, Smith P, Reay DS (2010) Microorganisms and climate change: terrestrial feedbacks and mitigation options. Nat Rev Microbiol 8:779–790

Steffen W, Richardson K, Rockstrom J, Cornell SE, Fetzer I, Bennett EM, Biggs R, Carpenter SR, Vries W, de Wit CA, Folke C, Gerten D, Heinke J, Mace GM, Persson LM, Ramanathan V, Reyers B, Sörlin S (2015) Supplementary materials, planetary boundaries: guiding human development on a changing planet. Science. https://doi.org/10.1126/science.1259855

Steinauer K, Tilman D, Wragg PD, Cesarz S, Cowles JM, Pritsch K, Reich PB, Weisser WW, Eisenhauer N (2015) Plant diversity effects on soil microbial functions and enzymes are stronger than warming in a grassland experiment. Ecology 96:99–112

Subbarao GV, Nakahara K, Hurtado MP, Ono H, Moreta DE, Salcedo AF, Yoshihashi AT, Ishikawa T, Ishitani M, Ohnishi-Kameyama M, Yoshida M, Rondon M, Rao IM, Lascano CE, Berry WL, Ito O (2009) Evidence for biological nitrification inhibition in Brachiaria pastures. Proc Natl Acad Sci U S A 106:17302–17307

Thornton PK, Herrero M (2010) Potential for reduced methane and carbon dioxide emissions from livestock and pasture management in the tropics. Proc Natl Acad Sci U S A. https://doi.org/10.1073/pnas.0912890107

Tilman D, Balzer C, Hill J, Befort BL (2011) Global food demand and the sustainable intensification of agriculture. Proc Natl Acad Sci U S A. https://doi.org/10.1073/pnas.1116437108

Valverde A, De Maayer P, Oberholster T, Henschel J, Louw MK, Cowan D (2016) Specific microbial communities associate with the rhizosphere of *Welwitschia mirabilis*, a living fossil. PLoS One. https://doi.org/10.1371/journal.pone.0153353

Van Der Heijden MGA, Bardgett RD, Van Straalen NM (2008) The unseen majority: soil microbes as drivers of plant diversity and productivity in terrestrial ecosystems. Ecol Lett 11:296–310

Vandenkoornhuyse P, Quaiser A, Duhamel M, Van AL, Dufresne A (2015) The importance of the microbiome of the plant holobiont. New Phytol 206:1196–1206

Venter ZS, Jacobs K, Hawkins HJ (2016) The impact of crop rotation on soil microbial diversity: a metaanalysis. Pedobiologia (Jena) 59:215–223

Vibha B, Neelam G (2012) Importance of exploration of microbial biodiversity. ISCA J Biol Sci 1: 78–83

Vorholt JA, Vogel C, Carlstrom CI, Muller DB (2017) Establishing causality: opportunities of synthetic communities for plant microbiome research. Cell Host Microbe 22:142–155

Vurukonda SSKP, Vardharajula S, Shrivastava M, Ali SKZ (2016a) Multifunctional *Pseudomonas putida* strain FBKV2 from arid rhizosphere soil and its growth promotional effects on maize under drought stress. Rhizosphere 1:4–13

Vurukonda SSKP, Vardharajula S, Shrivastava M, Ali SKZ (2016b) Enhancement of drought stress tolerance in crops by plant growth promoting rhizobacteria. Microbiol Res 184:13–24

Vyas P, Kumar D, Dubey A, Kumar A (2018) Screening and characterization of *Achromobacter xylosoxidans* isolated from rhizosphere of *Jatropha curcas* L. (energy crop) for plant growth promoting traits. J Adv Res Biotechnol. https://doi.org/10.15226/2475-4714/3/1/00134

Wallenstein MD (2017) Managing and manipulating the rhizosphere microbiome for plant health: a systems approach. Rhizosphere 3:230–232

Wu F, Wan JHC, Wu S, Wong M (2012) Effects of earthworms and plant growth promoting rhizobacteria (PGPR) on availability of nitrogen, phosphorus and potassium in soil. J Plant Nutr Soil Sci 175(3):423–433

Xu S, Reuter T, Gilroyed BH, Tymensen L, Hao Y, Hao X, Belosevic M, Leonard JJ, McAllister TA (2013) Microbial communities and greenhouse gas emissions associated with the biodegradation of specified risk material in compost. Waste Manag. https://doi.org/10.1016/j.wasman.2013.01.036

Yuan H, Ge T, Chen C, Donnell AGO, Wu J (2012) Significant role for microbial autotrophy in the sequestration of soil carbon. Appl Environ Microbiol 78:2328–2336

Zhang D, Wang C, Li X, Yang X, Zhao L, Liu L, Zhu C, Ruihong L (2018a) Linking plant ecological stoichiometry with soil nutrient and bacterial communities in apple orchards. Appl Soil Ecol 126:1–10. https://doi.org/10.1016/j.apsoil.2017.12.017

Zhang L, Xie Z, Zhao R, Zhang Y (2018b) Plant, microbial community and soil property responses to an experimental precipitation gradient in a desert grassland. Appl Soil Ecol 127:87–95

Zheng XY, Sinclair JB (1996) Chemotactic response of *Bacillus megaterium* strain B153-2-2 to soybean root and seed exudates. Physiol Mol Plant Pathol 48:21–35

Zhou J, He Z, Yang Y, Deng Y, Tringe SG, Alvarez-Cohen L (2015) High throughput metagenomic technologies for complex microbial community analysis: open and closed formats. MBio. https://doi.org/10.1128/mBio.02288-14

Part III
Advances in Plant Stress Mitigation

Biotechnological Approaches for Mitigation and Adaptation of Climate Change

Shikha Kumari, Sushila Saini, and Geeta Dhania

Abstract Changing climate is a biggest threat to mankind in the present times. Rise in average temperature, disturbed rainfall patterns, and occurrence of pest have affected the agricultural production. The change in climate has given rise to such conditions, which affect the environment and human health. Reduction in greenhouse gas emission is the key to mitigate climate change. Agricultural activities also contribute to GHG emission; thus, the use of energy-efficient farming is a good solution. The use of white biotechnology and green biotechnological tools and methods such as GM crops, biofertilizers, biofuels, mycobiotechnology, and biochar can respond positively for mitigating climate variability.

Keywords Biotechnology · Climate change · Transgenic · Biochar · Biofuels

Introduction

Climate change is the change or variation in the climate of any area over a long period of time. Solar cycles, volcanic eruptions, continental drift, atmospheric gases, natural internal processes, and many other human activities have resulted in climate change. Greenhouse gas emission is another major cause leading to climate change and global warming. Greenhouse gases also called as GHGs cause warming in the earth's atmosphere by preventing the radiation from being reflected into the outer space. The major GHGs are carbon dioxide (CO_2), methane (CH_4), nitrous oxide (N_2O), hydrofluorocarbons (HFCs), perfluorocarbons (PFCs), and sulfur hexafluoride (SF_6). The concentration of GHGs has been increasing drastically in the environment; reasons behind this are industrialization, urbanization, and other

S. Kumari · G. Dhania (✉)
Department of Environmental Sciences, Maharshi Dayanand University, Rohtak, Haryana, India

S. Saini
Department of Botany, Janta Vidya Mandir Ganpat Rai Rasiwasia (JVMGRR) College, Charkhi Dadri, Haryana, India

© The Author(s), under exclusive license to Springer Nature Singapore Pte Ltd. 2022
A. Vaishnav et al. (eds.), *Plant Stress Mitigators*,
https://doi.org/10.1007/978-981-16-7759-5_17

natural processes. About 25% greenhouse gas emission is from agricultural activities. Rice fields are observed to be the major source of CH_4 and N_2O (Mtui 2011; Sims 2014; IPCC 2014).

Extreme climate has a negative impact on our major food crops (wheat, rice, maize, and soybean). The impact of climate change on major crop yields was analyzed, and it was seen that the yield in wheat, rice, soybean, and maize was reduced by about 6%, 3.2%, 3.1%, and 7.4%, respectively, with increase in temperature (Zhao et al. 2017; Ito et al. 2018). Maleki et al. (2013) analyzed that drought stress greatly influenced the yield of soybean and a 42% weight reduction occurred at the grain filling stage. Rising temperature, shortage of water, declining soil fertility, and increased salinity are major concerns for agronomists, and the magnitude of their impact varies according to region and cropping pattern.

Climate change mitigation and adaptation to the changing climate are indispensable to diminish the abovementioned severe impacts of climate change (Bakshi 2003; Stringer et al. 2009). Mitigation of climate change can be achieved by reforestation, reducing GHGs emission, and reducing use of fossil fuels. Biotechnology approaches play an important role in adaptation and mitigation to climate change through various approaches. Tissue culture, recombinant DNA techniques, molecular markers, and genomic science are helpful in genetic modification using transgenes for developing GM crops, which are better adapted to various biotic and abiotic stresses without compromising for yield (Qaim et al. 2003; Lybbert and Sumner 2010).

Change in Earth's Climate and Green Biotechnology

Human deeds are responsible for a drastic change in the climate, which is affecting the civilization either directly (extreme weather events) or indirectly (change in agriculture, ecosystem, and human settlements) (Fig. 1). Agricultural production in any country depends heavily on the climatic conditions, and in present scenario, rising temperature, shortage of water, declining soil fertility, and increased salinity are putting extreme challenges before farmers. Various abiotic and biotic stresses like water logging, high temperature, extreme cold, aridity, salinity, insects, and pests are greatly influencing growth of the plant resulting in reduced yield (Gornall et al. 2010; Ackerly et al. 2010; Ashraf et al. 2018; Benevenuto et al. 2017).

For the management of these disasters, there are requirements of new varieties, which are disease- and pest-resistant, highly nutritious, and able to bear droughts, floods, and saline soils. Along with the conventional breeding technologies, plant biotechnology involving genetic modification of present varieties can be employed to increase resilience of existing crops against changing environments (Fig. 2). Green biotechnology can play a paramount role by recovery of many useful genes and their reincorporation in our food crops so that these can adapt to rapid climate change. Presently, the use of excessive fertilizer is causing a great rise of GHG emissions so new farming methods, use of biofertilizer, and newer varieties are

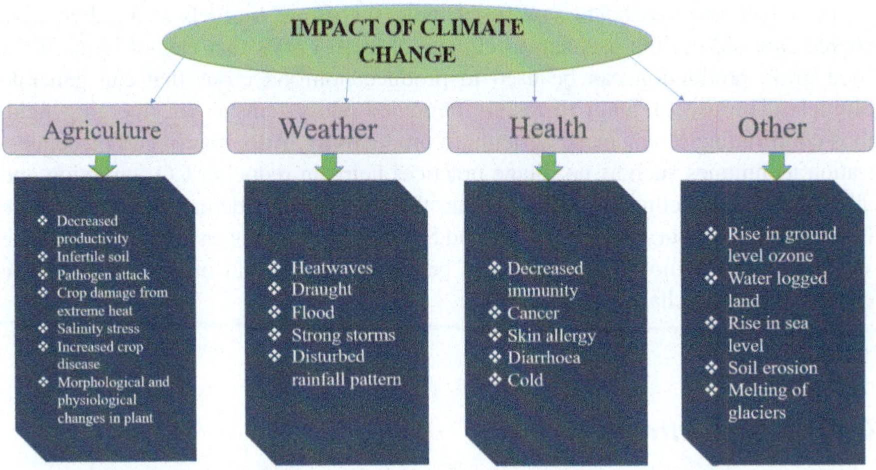

Fig. 1 Influence of climate variation on weather, agriculture, and human well-being

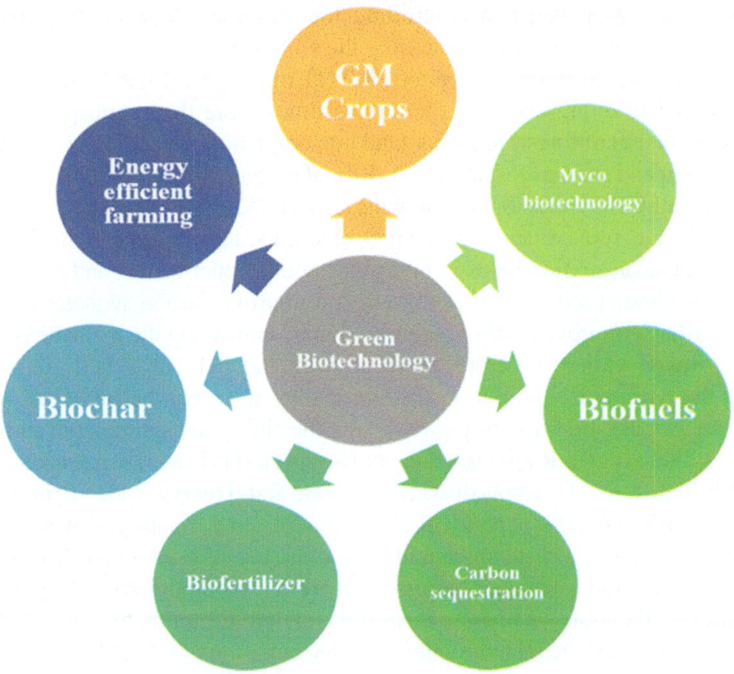

Fig. 2 Biotechnology in addressing climate variability

required, which give more yields with minimum input of chemical fertilizer and water to avoid their overuse (Fawcett and Towery 2003; Kumar et al. 2015; Tesfahun 2018).

To reduce our dependence on fossil fuels, the use of biofuels as an alternative source can address the problem to climate change. Lands that cannot be used for food grain production can be used to produce nonfood crops that can generate biofuels. The use of fungi to restore degraded ecosystems is an emerging field, thus contributing to enhanced agriculture productivity. Adoption of carbon sequestration techniques such as no tillage practices helps in reducing CO_2 emission and allows better recycling of soil nutrients, thus saving fuel demand for agriculture (Qaim 2009; Treasury 2009; Lybbert and Sumner 2010). This chapter deals with the various novel technologies, which can be used to adapt and mitigate the adverse effects of climate change on agriculture.

Carbon Sequestration

Carbon sequestration involves the removal of CO_2 (produced from combustion or other industrial activities) from the atmosphere and storing it in land and water. Soil carbon sequestration helps in controlling the increased concentration of carbon dioxide in the environment by increasing the soil organic carbon content. Soil as a major sink of carbon stores more than 2500 GT of carbon, which is four times than carbon stored in living organisms and three times more than atmospheric carbon. Due to conversion of forests and grassland into croplands, we are losing much of this stored carbon and soil has lost about 30–50% of carbon present initially contributing about ¼ of all GHG emissions by humans (Lal 2004; Lal et al. 2007).

Plant residue, roots, litter, and manures determine the concentration of carbon in soil. Carbon sequestration helps in improving soil productiveness and enhancement in yield. Methods used in control soil erosion improve carbon sequestration. Conservation tillage improves carbon sequestration by enhancing the methane consumption. Improved variety of crops prepared with the help of genetic engineering reduces the requirement of tillage. Carbon sequestration can be improved with the help of genetically modified crops possessing the ability to capture more carbon and convert it into oxygen. Roundup-ready TM soybean (GM crop) can sequester about 63.859 billion tons of carbon dioxide (Fawcett and Towery 2003; Brimner et al. 2004; Kleter et al. 2008). Nutrient management, crop rotation, cover cropping, mulching, biochar, reforestation and afforestation, use of organic fertilizers, reversing land degradation, and water management are few methods, which help in carbon sequestration. By adoption of these measures, soil has the ability to sequester C more than 1.35 GT/year, more than the anthropogenic emission of GHGs by transportation and it takes about 25–100 of years for the soil to reach the saturation level where the carbon removal through sequestration is stopped (West and Post 2002).

Biofuels

Human-induced climate change by release of GHGs has recently become a matter of global concern. GHGs possess the tendency of vibrating and rotating in a certain frequency range, due to which they are capable of trapping the energy from a similar frequency range of light waves. For mitigation of climate crisis, it is necessary to reduce the emission of GHGs. The major sources of GHG emission are burning of fossil fuels to generate electricity and to power the vehicles. The use of biofuels instead of traditional fossil fuels (petrol) for transport can help control the GHG emissions. Biofuels reduce the reliance on other countries for fossil fuels (Hiroki 2005; Bradshaw et al. 2005; Soccol et al. 2005; Kumar 2008; Food and Agriculture Organization (FAO) 2013).

Biofuel is the hydrocarbon-rich fuel obtained from the biomass. The use of food crops such as cereals, sugars, and oilseeds for the generation of ethanol (first-generation biofuels) has generated a debate of food vs. fuel as its negative impact on food supplies, and these food crops require high fertilizer and water inputs. So, the focus has been shifted to second-generation lignocellulosic biofuels obtained from forestry and agriculture residue or from perennial grasses such as miscanthus and switchgrass, which require less input and can be grown on marginal and degraded lands, thus reducing pressure on main cropland. Further biotechnology aids in the development of synthetic organisms for better transformation of ligno-cellulosic material into biofuels. Nowadays, microalgae and seaweeds have shown better prospective to generate huge volumes of biomass, which has the potential to be utilized for the production of third-generation biofuels. The use of innovative techniques for using algae as an alternate fuel can be helpful in the future. Algae and cyanobacteria show promising results as they can fix a large amount of GHGs, grow rapidly, and are nontoxic, and like first- and second-generation, biofuels do not require arable land for their production (Fig. 3). These cyanobacteria can be manipulated genetically for more production of lipids (triacylglycerol), which by pyrolysis can be transformed into bio-oil (Michael et al. 2011; Azapagic and Stichnothe 2011; Gajraj et al. 2018; Saini et al. 2018).

Biochar

Biochar is obtained from agrowaste and forestry residue, through gasification, pyrolysis, and hydrothermal carbonization (Matovic 2011; Osman 2020a, b). CO_2 is uptaken by plants during their growth; this CO_2 is processed into biochar, which may be applied to soil to improve soil fertility. Biochar can store carbon for thousands of years as carbon is resistant to decay. Thus, it can play a potential role in reducing GHGs by sequestering carbon in soil and making the system carbon-negative.

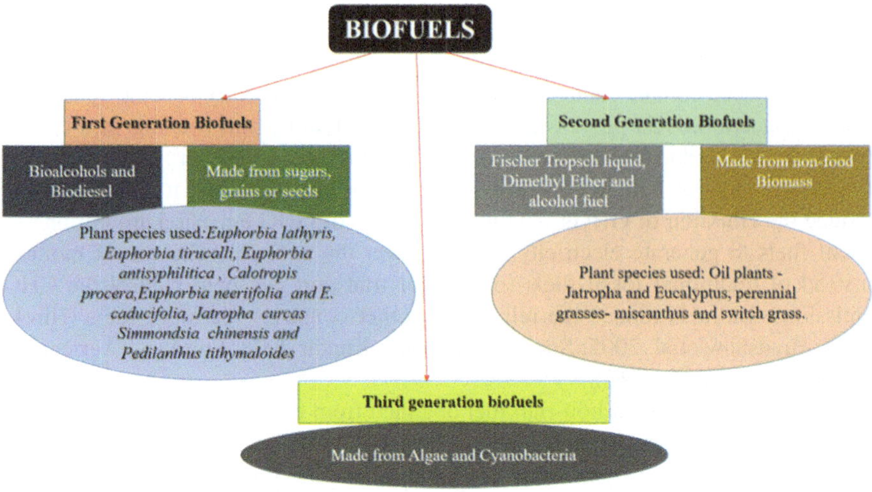

Fig. 3 An overview of different kinds of biofuels

The potential of biochar to remove carbon varies in different studies, ranging from 0.3 to 2 Gt CO_2 /year with investment of about \$90 to \$120/t CO_2. Reduction in the emission of CH_4 and N_2O from the atmosphere with the help of biochar is observed by many researchers (Chen 2019; Schmidt 2019; Semida 2019; Purakayastha 2019.

The benefits of application of biochar to the soil are improved nutrient cycling, reduction in nutrient leaching, enhanced water and nutrient retention, control of pathogens, and stimulation of activities of soil microbes. Physical and chemical properties of feedstock used and various processing methods adopted affect the efficiency of biochar (Semida 2019; Xiao 2019).

Energy-Efficient Farming

Agriculture is dependent on the use of energy from traditional resources of fossil fuels. Agriculture requires water for irrigation; for irrigation and water supply, electricity is used. The amount of electricity required to pump out water increases as the water table goes down. Water use efficiency helps to reduce the stress on water resources caused due to climate change. This in turn promotes energy-efficient farming.

Synthetic fertilizers are widely used to improve the fertility of soil, and pesticides are used for the eradication of pests. These fertilizers and pesticides are produced synthetically in the industries, which require a lot of chemicals and electricity. In the USA, 30% of the entire energy required for agricultural production is utilized in the production of nitrogen-based synthetic fertilizer (Pimentel 1980). The use of fossil fuels can be reduced by avoiding the application and reducing the demand of synthetic fertilizers and pesticides (Pimentel et al. 1973; Berardi 1978; Pimentel 1980; Lockeretz et al. 1981). One of the best solutions for saving energy used in agriculture is organic farming, which mainly relies on the use of organic fertilizers and products. Organic farming avoids the use of synthetic fertilizers and pesticides.

Green biotechnology promotes energy-efficient farming by developing genetically modified crops that lead to better production with lower inputs of water and fertilizers (Woods et al. 2010; Pelletier et al. 2011). Nonchemical methods are used for weed control, and biofertilizers are used as a source of essential nutrients required by the plant. Biofertilizers ameliorate plant–soil relation and boost soil health. Soil carbon sequestration is another effective method, which manages the carbon dioxide concentration (Cardi 2016; West and Post 2002; Ahmad et al. 2011). Further, the use of agricultural waste for the production of biofuels such as bioethanol and biodiesel reduces the pollutant emission from transport sector. Sugarcane, rapeseed, and Jatropha have been successfully used for producing biofuel (Michael et al. 2011). Utilization of agroresidue for these purposes reduces the emission caused due to burning of such waste. Energy-efficient farming thus helps in mitigation of climate change and also reduces the emission of greenhouse gases.

Mycobiotechnology

Mycobiotechnology is the field of biotechnology in which fungus is used for restoration of the degraded ecosystem and solving different environmental problems. The use of fungus for mitigation of climate change is a new approach. Regeneration of degraded forest and land is successfully achieved using both ectomycorrhizal (ECM) and endomycorrhizal (also known as arbuscular mycorrhizal fungi) symbiotic fungi; these fungi were used as inoculants along with actinomycetes. The use of fungi aids in the restoration of the ecosystem damaged due to natural or anthropogenic reasons (Sunagawa 2015).

Arbuscular mycorrhiza (AM) is found to affect soil and plant relationship. AM develops external hyphae and also produces enzymes, thereby increasing the nutrient uptake. It is capable of increasing the phosphorus content and helps the plants to acquire nitrogen from soil and thereby improves the plant health (Terrer et al. 2016; Mei et al. 2019). Nitrogen is highly needed by the plants in order to respond to high levels of CO_2 in the atmosphere (Weber et al. 2011). AM colonization helped in improving the salt tolerance in maize, clover, and mung bean. Inoculation with AM helped to endure salt stress in zucchini squash. Negative effect of saline stress on onion (*Allium cepa* L.) and basil (*Ocimum basilicum* L.) was controlled by AM inoculation (Banat et al. 2010; Belimov et al. 2009).

The fungi found in association with the plants are affected by the nitrogen-containing fertilizers. Burning of fossil fuels and excess usage of nitrogen-based fertilizers cause nitrogen pollution, which in turn affects the population of fungi. Excess of nitrogen makes ectomycorrhizal fungi less abundant in forest (Averill et al. 2018). Loss of these fungal communities increases the amount of carbon in soil leading to climate change processes. Abundance of AMF is found in warm and tropical forests in the areas where decomposition of CO_2 is very high; on the other hand, the areas where deposition of CO_2 is less are occupied by ECM. The mycorrhizal communities are a very reliable indicator of climate change (Yefetto et al. 2009).

Industrial Biotechnology

Industrial biotechnology (White Biotechnology) employs microbes and enzymes for making bio-based products such as detergents, paper, textile, chemicals, and pulp. The raw material used is renewable and aids in reducing the greenhouse gas emission. Industrial biotechnology saves energy during the production process and significantly lowers the emission of GHGs, which in turn help in mitigation of climate change. Industrial biotechnology by reducing the energy consumption and achieving energy efficiency has the potential to curtail 2.5 billion tons of CO_2 emission/year. Bio-based products being independent to petroleum-based fuels are carbon-negative as they have the capability to sequester carbon in themselves. Their application in daily life ranges from their use in pharmaceutical industry, bioplastics in packaging material, bio-based car seats, carpets, and also personal care products such as cosmetics made from algae. Biochemicals, biomaterials, and biopolymers may occupy a major market share in future. Industrial biotechnology is still in its infancy stage, and as it matures, it will provide further sustainable solutions to the environment.

OECD's and World Wide Fund for Nature (WWF) report has confirmed the impact of industrial biotechnology in cutting CO_2 emissions to build a greener economy. Significant role is played by biotechnology in any agreement related to climate change. Industrial biotechnology has been recognized as a Key Enabling Technology (KET) by the European Commission. Linking the KETs and work done to mitigate climate change would help the European Commission for fulfilling its financing obligations made under international climate change agreements (Ogunseitan 2003; Mtui 2007). Presently, industrial biotechnology suffers from paucity of funds, so an international collaboration and financial investment are required to meet the expectations of this sector.

Biofertilizer

Exponential growth of human population, urbanization, and industrialization has a very bad impact on the environment. The biggest challenge is to provide sufficient food to the rising population, and it has resulted in enormous use of chemical fertilizer and pesticides resulting in enhanced emission of greenhouse gases (Glick

2012; Savci 2012). Drought and salinity resulted by change in climate affect productivity of soil, thus leading to more use of fertilizers (Perrott et al. 1992; Steinshamn et al. 2004). Sustainable crop production is a major global challenge. Biofertilizers are substances or microorganism, which improves the nutrient uptake and stress tolerance in plants. Biofertilizers improve drought tolerance, improve plant health, increase salt tolerance, and they are capable of accessing nutrients from soil deposits (Vessey 2003; Arora 2013). Microorganisms perform specific interaction with soil and plants, and these interactions enhance plant's growth and productivity (Antoun and Prevost 2005; Adesemoye and Kloepper 2009; Ahmad et al. 2011; Lau and Lennon 2011).

The microorganisms used as biofertilizer are bacteria, fungi, and arbuscular mycorrhiza (Jones and Hinsinger 2008). AM (arbuscular mycorrhiza), phosphate solubilizers, nitrogen fixers and amalgamation of phosphate solubilizers, and nitrogen fixers microbes are good biofertilizers (Vessey 2003; Tawaraya et al. 2006). A few examples of AMF are *Entrophospora colombiana*, *Glomus caledonium*, and *G. mosseae*, phosphate solubilizer microbes are *Arthrobacter chlorophenolicus*, *Bacillus firmus*, *Burkholderia caryophylli*, *Enterobacter asburiae*, *Microbacterium arborescens*, *and Staphylococcus saprophyticus*, and nitrogen fixers are *Anabaena azollae*, *Aulosira fertilissima*, *Azolla caroliniana*, *Azospirillum brasilense*, *Azotobacter brasilense*, *Bacillus polymyxa*, *Brevundimonas diminuta*, *and Nostoc muscorum* (Tawaraya et al. 2006; Abbaspour et al. 2012; Abdel Latef et al. 2016).

Plant growth-promoting rhizobacteria (PGPR) affect plant growth in a positive way by producing plant hormones, improving absorption of nutrient, mediating nitrogen fixation, phosphate solubilization, and iron sequestering (Rabalais et al. 1998; Tawaraya et al. 2006). PGPR act on soil and aid in nitrogen fixation and improve the overall growth of the plant. PGPR synthesize and excrete indole-3-acetic acid (IAA), which gets adsorbed to the surface of roots and in turn stimulates the cell proliferation and elongation of plants (Mantelin and Touraine 2004; Peix et al. 2001). Extensive root growth in plants is possible if bacteria having ACC (1-aminocyclopropane-1-carboxylate) deaminase that breaks down ACC to ammonia are used to treat the plants; this helps the plant in resisting stress sources. Decreased ACC reduces ethylene levels, thus helps in stimulating physiological changes at molecular level and benefiting PGPR in performing specific functions (Glick et al. 1998, 2007). Application of *Achromobacter piechaudii* ARV8 having ACC deaminase activity to tomato and pepper seedlings increased their water use efficiency (WUE) in saline environment, helped them to tolerate water stress, and increased phosphorus and potassium intake resulting in improvement in the yield (Mayak et al. 2004). Saline resistance of the groundnut plants was increased with the help of *Pseudomonas fluorescens* strain TDK1 (Saravanakumar and Samiyappan 2007). A similar result was observed on maize plant under saline conditions after using ACC deaminase bacteria *Variovorax paradoxus* and ACC deaminase bacteria in peas (*Pisum sativum* L.) showed remarkable positive impact on leaf area, transpiration, and shoot biomass.

Genetically Modified Crops

Biotechnology utilizes living organisms, biological systems to produce or modify products, or processes for specific use. GM crops have an important role in agricultural processes. The advancement in biotechnology allows the manipulation of the genes of any organisms or plant species and inserting in plant species to obtain the desired trait (Conway 2012; Godfray and Garnett 2014). Tissue culture and molecular techniques such as DNA fingerprinting marker-assisted selection (MAS), molecular diagnostics, and genetic engineering/genetic modification are widely used as biotechnology tools for the improvement of crops (Cardi 2016). Genetically modified organisms (GMOs) created through biotechnology have shown promising results in many food crops (Fig. 4).

Pathogen-resistant variety of wheat was developed by introducing TaPIE1 in wheat; it also made the plant tolerant to chilling stress (Zhu et al. 2014). With the help of marker-assisted selection, rice flash flood tolerance gene Sub1A was introduced into commercial indica rice, which improved the gain yield in India and the Philippines (Dar et al. 2014; International Rice Research Institute 2015). Transgenic rice crop having the CaMsrB2 gene was developed, which performs well in stressed conditions (Dhungana et al. 2015). For making rice tolerant to dehydration, salinity, and chilling, a new variety of rice was created in which the OsMYB2 gene was overexpressed (Yang et al. 2012). Transgenic wheat and tobacco varieties developed with an expression of TaPIMP1 gene showed better tolerance against drought and salinity, and also, the resistance to pathogen attack was improved (Zhang et al. 2012; Liu et al. 2011).

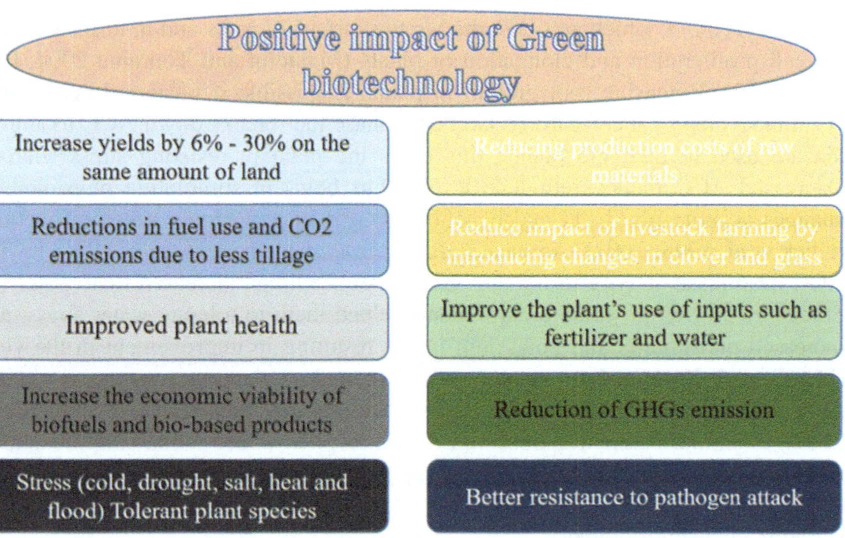

Fig. 4 Advantages of using green biotechnology. (Source: Wang et al. 2003; Brink et al. 1998; Brookes and Barfoot 2009; Ogunseitan 2003; Treasury 2009)

Table 1 Various transgenic crops and their tolerance against biotic and abiotic stress

Transgenic crop	**Biotic stress**	**References**
Bt maize, Bt cotton	Resistant to insect attack; lepidoptera and diptera	Zhe and Mithcell (2011)
Transgenic canola and soybean	Herbicide-resistant	May et al. (2005), Bonny (2008)
GM cassava, potatoes, bananas	Resistant to viruses, bacteria, and fungi	Mneney et al. (2001), Van Camp (2005)
Transgenic crops	**Abiotic stress**	**References**
GM cassava, millet, and sunflower	Draught	Manavalan et al. (2012)
Soybean	Salinity and freeze	Cao et al. (2013)
Transgenic rice	Salinity, chilling, and dehydration	Yang et al. (2012)

Isolation of GmMYB76 gene in soybean using *Agrobacterium* model enhanced its tolerance against salinity and freeze (Cao et al. 2013). Similarly, isolation of gene ZmMYB30 from maize improves stress tolerance against salinity (Chen et al. 2017). ZmWRKY33 gene in maize has been induced for ABA stress, freeze, salinity, and drought (Li et al. 2013). Overexpression of CmWRKY1 in *Chrysanthemum* increased dehydration tolerance (Fan et al. 2016). The MdSIMYB1 gene isolated from apple was used to develop the transgenic varieties resistant to drought, cold, and salinity stress (Wang et al. 2014). Resistance against dehydration and TMV was increased in tobacco crops by developing transgenic tobacco with an overexpressed GmERF3 gene. It also improved the tolerance of plants toward salinity stress (Zhang et al. 2009) (Table 1).

The two major genetically modified crops in India include Bt cotton and Bt brinjal. Bt Brinjal was developed to minimize the use of pesticides and chemical fertilizers, as these hybrid crops are resistant to the attack of pest and insects. Genetic modification in cotton is carried out to improve insect resistance and herbicide tolerance of the plant. The cotton hybrids MECH-12 Bt, MECH-162 Bt, and MECH-184 Bt were first approved for commercial cultivation in India (Barwale et al. 2004). Studies have been conducted to analyze the performance of the Bt cotton. It was found that the yield of Bt cotton was 40% higher than that of non-Bt cotton crops. Also, the incidence of bollworm attack was significantly less in Bt cotton plants (James 2008). Thus, agricultural biotechnology has the capability to combat the negative impact of climate change by producing new varieties resistant to biotic and abiotic stresses.

Conclusion

The various biotechnological tools discussed in the paper can positively contribute to the mitigation of climate change. Green biotechnology improves the crop adaptability and improves tolerance against both abiotic and biotic stress. The use of biofertilizers and biochar improves the soil fertility, increases crop yield, and also balances the nutrient cycling. Carbon sequestration maintains the soil carbon ratio and also controls the emission of carbon dioxide. The use of biofuels, biofertilizers, and industrial biotechnology reduces the burden on natural resources and in turn the GHG emission is reduced. The utilization of "genetically modified organisms" (GMOs) possesses the potential to improve agriculture, food quality, food nutrition, and health. Proper application of modern biotechnology will help in protecting our environment from the negative impacts of climate change.

References

Abbaspour H, Saeidi-Sar S, Afshari H, Abdel-Wahhab MA (2012) Tolerance of Mycorrhiza infected Pistachio (*Pistacia vera* L.) seedling to drought stress under glasshouse conditions. J Plant Physiol 169:704–709. https://doi.org/10.1016/j.jplph.2012.01.014

Abdel Latef AAH, Hashem A, Rasool S, AbdAllah EF, Alqarawi AA, Egamberdieva D (2016) Arbuscular mycorrhizal symbiosis and abiotic stress in plants: a review. J Plant Biol 59:407–426. https://doi.org/10.1007/s12374-0160237-7

Ackerly DD, Loarie SR, Cornwell WK, Weiss SB, Hamilton H, Branciforte R, Kraft NJB (2010) The geography of climate change: implications for conservation biogeography. Divers Distrib 16:476–487

Adesemoye AO, Kloepper JW (2009) Plant–microbes interactions in enhanced fertilizer-use efficiency. Appl Microbiol Biotechnol 85(1):1–12

Ahmad MN, Mokhtar MN, Baharuddin AS, Hock LS, Ali SRA, Abd-Aziz S, Rahman NAA, Hassan MA (2011) Changes in physicochemical and microbial community during co-composting of oil palm frond with palm oil mill effluent anaerobic sludge. Bioresources 6(4):4762–4780

Antoun H, Prevost D (2005) Ecology of plant growth promoting rhizobacteria. In: PGPR: Biocontrol and biofertilization. Springer, Dordrecht, pp 1–38

Arora NK (2013) Plant microbe symbiosis: fundamentals and advances. Plant microbe symbiosis: fundamentals and advances. Springer Science+Business Media, Dordrecht

Ashraf MA, Akbar A, Askari SH, Iqbal M, Rasheed R, Hussain I (2018) Recent advances in abiotic stress tolerance of plants through chemical priming: an overview. Advances in seed priming. Springer, Berlin/Heidelberg, Germany, pp 51–79

Averill C, Dietze MC, Bhatnagar JM (2018) Continental-scale nitrogen population in shifting forest mycorrhizal association and soil carbon stocks. Glob Chang Biol 24(10):4544–4533

Azapagic A, Stichnothe H (2011) Assessing sustainability of biofuels. In: Azapagic A, Perdan S (eds) Sustainable development in practice: case studies for engineers and scientists, 2nd edn. John Wiley & Sons, Chichester, UK

Bakshi A (2003) Potential adverse health effects of genetically modified crops. J Toxicol Environ Health 6(B):211–226

Banat IB, Franzetti A, Gandolfi I, Martinotti MG, Fracchia L, Smyth TJ, Marchant R (2010) Microbial biosurfactants production, applications and future potential. Appl Microbiol Biotechnol 87:427–444

Barwale RB, Gadwal VR, Zehr U, Zehr B (2004) Prospects for Bt cotton technology in India. AgBioforum 7(1&2):23–26

Belimov AA, Dodd IC, Hontzeas N, Theobald JC, Safronova VI, Davies WJ (2009) Rhizosphere bacteria containing 1-aminocyclopropane-1-carboxylate deaminase increase yield of plants grown in drying soil via both local and systemic hormone signaling. New Phytol 181:413–423

Benevenuto RF, Agapito-Tenfen SZ, Vilperte V, Wikmark OG, VanRensburg PJ, Nodari RO (2017) Molecular responses of genetically modified maize to abiotic stresses as determined through proteomic and metabolomic analyses. PLoS One 12:e0173069
Berardi GM (1978) Organic and conventional wheat production: examination of energy and economics. Agro-Ecosystems 4:367–376
Bonny S (2008) Genetically modified glyphosate-tolerant soybean in USA: Adoption factors, impacts and prospects. Rev Agron Sustain Dev 28:21–32
Bradshaw J, Chen Z, Garg A, Gomez D, Rogner H-H (2005) Sources of CO_2. IPCC special report on carbon dioxide capture and storage
Brimner TA, Gallivan GJ, Stephenson GR (2004) Influence of herbicide-resistant canola on the environmental impact of weed management. Pest Manag Sci 61(1):47–52
Brink JA, Woodward BR, Da Silva E (1998) Plant Biotechnology: a tool for development in Africa. Electron J Biotechnol 1(3):14–15
Brookes G, Barfoot P (2009) Global impact of biotech crops: income and production effects, 1996–2007. J Agrobiotechnol Manage Econ 12(2):184–208
Cao ZH, Zhang SZ, Wang RK, Zhang RF, Hao YJ (2013) Genome wide analysis of the apple MYB transcription factor family allows the identification of MdoMYB121 gene conferring abiotic stress tolerance in plants. PLoS One 8:e69955
Cardi T (2016) Cisgenesis and genome editing: combining concepts and efforts for a smarter use of genetic resources in crop breeding. Plant Breed 135:139–147
Chen H (2019) Upcycling food waste digestate for energy and heavy metal remediation applications. Resource Conserv Recycl X(3):100015. https://doi.org/10.1016/j.rcrx.2019.100015
Chen Y, Cao Y, Wang L, Li L, Yang J, Zou M (2017) Identification of MYB transcription factor genes and their expression during abiotic stresses in maize. Biol Plant 62:1–9
Conway G (2012) One billion hungry, can we feed the world. Cornell University Press, Ithaca, NY
Dar MH, Singh S, Singh US, Zaidi NW, Ismail AM (2014) Stress tolerant rice varieties—making headway in India. SATSA Mukhapatra Annu Tech 18:1–14
Dhungana SK, Kim BR, Son JH, Shin DH (2015) Comparative study of CaMsrB2 gene containing drought-tolerant transgenic rice (*Oryza sativa* L.) and non-transgenic counterpart. J Agron Crop Sci 201:10–16
Fan Q, Song A, Jiang J, Zhang T, Sun H, Wang Y, Chen S, Chen F (2016) CmWRKY1 enhances the dehydration tolerance of chrysanthemum through the regulation of ABA-associated genes. PLoS One 11:e0150572
Fawcett R, Towery D (2003) Conservation tillage and plant biotechnology: how new technologies can improve the environment by reducing the need to plow. CT Information Center, USA. (http://www.ctic.purdue.edu/CTIC/Biotech.html)
Food and Agriculture Organization (FAO) (2013) Biofuels and the sustainability challenge: a global assessment of sustainability issues, trends and policies for biofuels and related feedstocks. Food and Agriculture Organization of the United Nations, Rome, Italy
Gajraj RS, Singh GP, Kumar A (2018) Third-generation biofuel: algal biofuels as a sustainable energy source. In: Kumar A, Ogita S, Yuan-YeuYau (eds) Biofuels: greenhouse gas mitigation and global warming. Next generation biofuels and role of biotechnology. Springer, India, pp 292–307
Glick BR (2012) Plant growth-promoting bacteria: mechanisms and applications. Scientifica (Cairo)
Glick BR, Penrose DM, Li J (1998) A model for the lowering of plant ethylene concentrations by plant growth-promoting bacteria. J Theor Biol 190(1):63–68
Glick BR, Todorovic B, Czarny J, Cheng Z, Duan J, McConkey B (2007) Promotion of plant growth by bacterial ACC deaminase. Crit Rev Plant Sci 26(5–6):227–242
Godfray HCJ, Garnett T (2014) Food security and sustainable intensification. Philos Trans Royal Soc B: Biol Sci 369:20120273

Gornall J, Betts R, Burke E, Clark R, Camp J, Willett K, Wiltshire A (2010) Implications of climate change for agricultural productivity in the early twenty-first century. Philos Trans Royal Soc B-Biol Sci 365:2973–2989

Hiroki H (2005) Life cycle GHG emission analysis of power generation systems: Japanese case. Elsevier Energy 30:2042–2056

International Rice Research Institute (2015) Climate change-ready rice. IRRI Press, Manila, Philippines

IPCC (2014) Climate Change 2014: impacts, adaptation, and vulnerability. In: Houghton J et al (eds) Intergovernmental panel on climate change. Cambridge University Press, Cambridge, UK

Ito R, Vasconcelos HL, Feeley KJ (2018) Global climate change increases risk of crop yield losses and food insecurity in the tropical Andes. Glob Chang Biol 24:e592–e602

James C (2008) Global status of commercialized biotech/GM crops. ISAAA Briefs No. 39. International Service for the Acquisition of Agri-Biotech Applications, Ithaca, NY

Jones DL, Hinsinger P (2008) The rhizosphere: complex by design. Plant Soil 3(12):1–6

Kleter GA, Harris C, Stephenson G, Unsworth J (2008) Comparison of herbicide regimes and the associated potential environmental effects of glyphosate-resistant crops versus what they replace in Europe. Pest Manag Sci 64:479–488

Kumar A (2008) Bioengineering of crops for biofuels and bioenergy. In: Kumar A, Sopory S (eds) Recent advances in plant biotechnology. I.K. International, New Delhi, pp 346–360

Kumar S, Bansode R, Malav MK, Malav L (2015) Role of agricultural biotechnology in climate change mitigation. Int J Appl Pure Sci Agric 1(4):139–145

Lal R (2004) Soil carbon sequestration to mitigate climate change. Geoderma 123:1–22

Lal R, Follett RF, Stewart BA, Kimble JM (2007) Soil carbon sequestration to mitigate climate change and advance food security. Soil Sci 172(12):943–956

Lau JA, Lennon JT (2011) Evolutionary ecology of plant–microbe interactions: soil microbial structure alters selection on plant traits. New Phytol 192(1):215–224

Li H, Gao Y, Xu H, Dai Y, Deng D, Chen J (2013) ZmWRKY33, a WRKY maize transcription factor conferring enhanced salt stress tolerances in Arabidopsis. Plant Growth Regul 70:207–216

Liu H, Zhou X, Dong N, Liu X, Zhang H, Zhang Z (2011) Expression of a wheat MYB gene in transgenic tobacco enhances resistance to Ralstonia solanacearum and to drought and salt stresses. Funct Integr Genomics 11:431–443

Lockeretz W, Shearer G, Kohl DH (1981) Organic farming in the corn belt. Science 211:540–547

Lybbert T, Sumner D (2010) Agricultural technologies for climate change mitigation and adaptation in developing countries: policy options for innovation and technology diffusion. ICTSD-IPC Platform on Climate Change, Agriculture and Trade, Issue Brief No. 6, International Centre for Trade and Sustainable Development, Geneva, Switzerland and International Food & Agricultural Trade Policy Council, Washington DC, USA

Maleki A, Naderi A, Naseri R, Fathi A, Bahamin S, Maleki R (2013) Physiological performance of soybean cultivars under drought stress. Bullet Environ Pharmacol Life Sci 2:38–44

Manavalan LP, Chen X, Clarker J, Salmeron J, Nguyen HT (2012) RNAi-mediated disruption of squalene synthase improves drought tolerance and yield in rice. J Exp Bot 63:163–175

Mantelin S, Touraine B (2004) Plant growth-promoting bacteria and nitrate availability: impacts on root development and nitrate uptake. J Exp Bot 55:27–34

Matovic D (2011) Biochar as a viable carbon sequestration option: global and Canadian perspective. Energy 36:2011–2016. https://doi.org/10.1016/j.energy.2010.09.031

May MJ, Gillian Champion GT, Dewar AM, Qi A, Pidgeon JD (2005) Management of genetically modified herbicide-tolerant sugar beets for spring and autumn environmental benefit. Proc Royal Soc Biol Sci 272(1559):111–119

Mayak S, Tirosh T, Glick BR (2004) Plant growth-promoting bacteria that confer resistance to water stress in tomatoes and peppers. Plant Sci 166(2):525–530

Mei L, Yang X, CaoH ZT, Guo J (2019) Arbuscular mycorrhizal fungi alter plant and soil C:N:P Stoichiometries under warming and nitrogen input in a semiarid meadow of China. Int J Environ Res Public Health 16(3):397

Michael K, Steffi N, Peter D (2011) The past, present, and future of biofuels—biobutanol as promising alternative. In: dos Santos MA (ed) Biofuel production—recent developments and prospects, pp 451–486

Mneney EE, Mantel SH, Mark B (2001) Use of random amplified polymorphic DNA markers to reveal genetic diversity within and between populations of cashew (Anacardium occidentale L). J Hortic Sci Biotech 77(4):375–383

Mtui G (2007) Trends in industrial and environmental biotechnology research in Tanzania. Afr J Biotechnol 6(25):2860–2867

Mtui SYG (2011) Involvement of biotechnology in climate change adaptation and mitigation: improving agricultural yield and food security. Int J Biotechnol Mol Biol Res 2(13):222–231

Ogunseitan OA (2003) Biotechnology and industrial ecology: new challenges for a changing global environment. Afr J Biotechnol 2(12):593–601

Osman AI (2020a) Catalytic hydrogen production from methane partial oxidation: mechanism and kinetic study. Chem Eng Technol 43:641–648. https://doi.org/10.1002/ceat.201900339

Osman AI (2020b) Upcycling brewer's spent grain waste into activated carbon and carbon nanotubes for energy and other applications via two-stage activation. J Chem Technol Biotechnol 95:183–195. https://doi.org/10.1002/jctb.6220

Peix A, Rivas-Boyero AA, Mateos PF, Rodriguez-Barrueco C, Martínez-Molina E, Velazquez E (2001) Growth promotion of chickpea and barley by a phosphate solubilizing strain of Mesorhizobium mediterraneum under growth chamber conditions. Soil Biol Biochem 33: 103–110

Pelletier N, Audsley E, Brodt S, Garnett T, Henrikkson P, Kendall A, Kramer K, Murphy D, Nemecek T, Troell M (2011) Energy intensity of agriculture and food systems. Annu Rev Env Resour 36:233–246

Perrott KW, Sarathchandra SU, Dow BW (1992) Seasonal and fertilizer effects on the organic cycle and microbial biomass in a hill country soil under pasture. Aust J Soil Res 30:383–394

Pimentel D (1980) Handbook of energy utilization in agriculture. CRC Press, Boca Raton, FL, p 475

Pimentel D, Hurd LE, Bellotti AC, Forster MJ, Oka IN, Sholes OD, Whitman RJ (1973) Food production and the energy crisis. Science 182:443–449

Purakayastha TJ (2019) A review on biochar modulated soil condition improvements and nutrient dynamics concerning crop yields: pathways to climate change mitigation and global food security. Chemosphere 227:345–365. https://doi.org/10.1016/j.chemosphere.2019.03.170

Qaim M (2009) The economics of genetically modified crops. Ann Rev Resources Econ 1:665–693

Qaim M, Zilberman D (2003) Yield effects of genetically modified crops in developing countries. Science 299:900–902

Rabalais NN, Turner RE, Wiseman WJ Jr, Dortch Q (1998) Consequences of the 1993 Mississippi River flood in the Gulf of Mexico. Regul Rivers: Res Manage 14:161–177

Saini S, Sharma M, Kumar A (2018) Global warming and climate change: next generation biofuels and role of biotechnology. Int J Life Sci Pharma Res 8(2):52–57

Saravanakumar D, Samiyappan R (2007) ACC deaminase from *Pseudomonas fluorescens* mediated saline resistance in groundnut (*Arachis hypogea*) plants. J Appl Microbiol 102(5):1283–1292

Savci S (2012) An agricultural pollutant: chemical fertilizer. Int J Environ Sci Dev 3:73

Schmidt HP (2019) Pyrogenic carbon capture and storage. GCB Bioenergy 11:573–591. https://doi.org/10.1111/gcbb.12553

Semida WM (2019) Biochar implications for sustainable agriculture and environment: a review. S Afr J Bot 127:333–347. https://doi.org/10.1016/j.sajb.2019.11.015

Sims R (2014) Transport. In: Edenhofer O et al (eds) Climate change 2014: mitigation of climate change contribution of working group III to the fifth assessment report of the intergovernmental panel on climate change. Cambridge University Press, Cambridge, UK, p 2. IEA. 2019 Renewables 2019. Paris. See https://www.iea.org/reports/renewables-2019

Soccol CR, Vandenberghe LPS, Costa B, Woiciechowski AL, de Carvalho JC, Medeiros ABP, Francisco AM, Bonomi LJ (2005) Brazilian biofuel program: an overview. J Sci Ind Res 64: 897–904

Steinshamn H, Thuen E, Bleken MA, Brenoe UT, Ekerholt G, Yri C (2004) Utilization of nitrogen (N) and phosphorus (P) in an organic dairy farming system in Norway. Agric Ecosyst Environ 104(3):509–522

Stringer LC, Dyer JC, Reed MS, Dougill AJ, Twyman C, Mkwambisi D (2009) Adaptation to climate change, drought and desertification: local insights to enhance policy in Southern Africa. Environ Sci Policy 12:748–765

Sunagawa S (2015) Structure and function of the global ocean microbiome. Science 348:1261359

Tawaraya K, Naito M, Wagatsuma T (2006) Solubilization of insoluble inorganic phosphate by hyphal exudates of arbuscular mycorrhizal fungi. J Plant Nutr 29:657–665

Terrer C, Vicca S, Hungate BA, Phillips RP, Prentice IC (2016) Mycorrhizal association as a primary control of carbon dioxide fertilization effect. Science 353(6294):72–74

Tesfahun W (2018) Climate change mitigation and adaptation through biotechnology approaches: a review. Cogent Food Agric 4:1512837

Treasury HM (2009) Green biotechnology and climate change. Eur Biol 12. Retrieved from http://www.docstoc.com/docs/15021072/GreenBiotechnology-andClimate-Change

Van Camp W (2005) Yield enhancing genes: seeds for growth. Curr Opin Biotechnol 16:147–153

Vessey JK (2003) Plant growth promoting Rhizobacteria as biofertilizers. Plant Soil 255:571–586. https://doi.org/10.1023/A:1026037216893

Wang W, Vinocur B, Altman A (2003) Plant responses to drought, salinity and extreme temperatures: towards genetic engineering for stress tolerance. Planta 218:1–14

Wang RK, Cao ZH, Hao YJ (2014) Overexpression of a R2R3 MYB gene MdSIMYB1 increases tolerance to multiple stresses in transgenic tobacco and apples. Physiol Plant 150:76–87

Weber CF, Zak DR, Hungate BA, Jackson RB, Vilgalys R, Evans RD (2011) Response of soil cellulolytic fungal communities to elevated atmospheric carbon dioxide are complex and variable across five ecosystem. Environ Microbiol 13(10):2778–2793

West TO, Post WM (2002) Soil organic carbon sequestration rates by tillage and crop rotation: a global analysis. Soil Sci Soc Am J 66:930–1046

Woods J, Williams A, Hughes JK, Black M, Murphy R (2010) Energy and the food system. Philos Trans Royal Soc B: Biol Sci 365(1554):2991–3006

Xiao L (2019) Biochar promotes methane production at high acetate concentrations in anaerobic soils. Environ Chem Lett 17:1347–1352. https://doi.org/10.1007/s10311-019-00863-3

Yang A, Dai X, Zhang WH (2012) A R2R3-type MYB gene, OsMYB2, is involved in salt, cold, and dehydration tolerance in rice. J Exp Bot 63:2541–2556

Yefetto L, Davis DJ, Money NP (2009) Biomechanics of invasive growth by Armillaria rhizomorphs. Fungal Genet Biol 46:688–694

Zhang G, Chen M, Li L, Xu Z, Chen X, Guo J, Ma Y (2009) Overexpression of the soybean GmERF3 gene, an AP2/ERF type transcription factor for increased tolerances to salt, drought, and diseases in transgenic tobacco. J Exp Bot 60:3781–3796

Zhang Z, Liu X, Wang X, Zhou M, Zhou X, Ye X, Wei X (2012) An R2R3 MYB transcription factor in wheat, Ta PIMP 1, mediates host resistance to Bipolaris sorokiniana and drought stresses through regulation of defense-and stress-related genes. New Phytol 196:1155–1170

Zhao C, Liu B, Piao S, Wang X, Lobell DB, Huang Y, Huang M, Yao Y, Bassu S, Ciais P (2017) Temperature increase reduces global yields of major crops in four independent estimates. Proc Natl Acad Sci 114:9326–9331

Zhe D, Mithcell PD (2011) Can conventional crop producers also benefit from Bt technology? Agric Appl Assoc Ser. Paper No. 103584

Zhu X, Qi L, Liu X, Cai S, Xu H, Huang R, Li J, Wei X, Zhang Z (2014) The wheat ethylene response factor transcription factor pathogen-induced ERF1 mediates host responses to both the necrotrophic pathogen *Rhizoctonia cerealis* and freezing stresses. Plant Physiol 16414:99–1514

Use of Synthetic Ecology Approach in Exploring Plant–Microbial Interactions Under Habitat-Imposed Stresses

Ahmad Mahmood, Hande Bardak, Duygu Bozdoğan, Taku Yamaguchi, Ali Hamed, Ryota Kataoka, and Oğuz Can Turgay

Abstract Soil microbes have received an attention due to their possible roles in plant growth promotion, avoidance of biotic and abiotic stress, soil remediation, and reclamation. However, climate change may affect microbial ecology and function, especially in marginal lands; therefore, we need to explore viable options of helping the soil microbes and bioinoculants perform better. In recent times, focus has been shifting toward harnessing groups of microorganisms or constructing consortia of selected microbes due to the possibilities of multiple roles, better adaptation, and ease in association with the hosts when compared with singular strains. Here, we reviewed role of this synthetic ecology approach in ameliorating biotic and abiotic stresses from plants with a focus on marginal lands, which are on the rise in scenarios of changing climate. In addition, possible mechanisms behind may include efficient and cooperative metabolism, closer interaction among microbes, and with the hosts, division of labor, and resilience in plant phenotype. Although further investigation in application strategies after successfully building the consortia would be required, utilization of well-matched consortia may help crop production particularly under changing climate scenarios, which have or may lead to adverse effect of functionality of applied single strains.

Keywords Synthetic ecology · Plant growth-promoting microorganisms · Consortium · Stress

A. Mahmood · T. Yamaguchi · R. Kataoka (✉)
Department of Environmental Sciences, University of Yamanashi, Yamanashi, Japan
e-mail: rkataoka@yamanashi.ac.jp

H. Bardak · D. Bozdoğan · O. C. Turgay
Department of Soil Science and Plant Nutrition, Ankara University, Ankara, Turkey

A. Hamed
Department of Agronomy, University of Agriculture, Faisalabad, Pakistan

© The Author(s), under exclusive license to Springer Nature Singapore Pte Ltd. 2022
A. Vaishnav et al. (eds.), *Plant Stress Mitigators*,
https://doi.org/10.1007/978-981-16-7759-5_18

Introduction

Approximately 41% of the world's surface area is arid or semi-arid, and one-third of the world's population is living in these areas (Fujiyama 2019). In addition, drylands take up 41.3% of the earth's land surface and up to 44% of all cultivated land is in the drylands (Hori et al. 2011). Arid lands play an important role in the world's food production; however, agriculture has always faced environmental risks such as drought stress, salinity, alkalinity, erosion, pollution, desertification, and land degradation. Environmental changes caused by climate change are occurring all over the world, and these changes are especially seen in marginal areas. Therefore, we need new and effective approaches and strategies to implement sustainable agriculture in these areas. In this sense, there is a possibility of using soil microorganisms in the development of low-cost and sustainable land use system in countries with such land degradation. Soil microbiology is progressing steadily, and various information has come to be obtained. In recent years, the high-throughput sequencing analysis has made it possible to comprehensively obtain microbial community data in soil. Moreover, functional analysis has also begun to understand what role these microbial communities play in soil or in other ecosystems. A recent study investigated soil microbiome in an experimental grassland system and observed significant support of nutrient cycling among other ecosystem functioning by the incident microbes (Wagg et al. 2019). The diversity of microbes has been estimated ranging from thousands to a million microbial species in a few grams of soil (Allison and Martiny 2008), but there is still unclear in our understanding of how such diverse and functionally complex microbiome affects different ecosystem functions (Wagg et al. 2019). In this chapter, we will deal with the soil–plant–microorganism triadic relations to overcome the problems that could occur in marginal areas, which are facing major influence of climate change.

Influence of Climate Change on Soil Microbes in Marginal Region

Climate change can be defined as "statistically significant changes in the average condition of the climate or its variability over tens or more years." This change is the natural occurrence of carbon dioxide, methane, nitrogen monoxide, ozone, chlorofluorocarbon, and water vapor released into the atmosphere as a result of natural events and predominantly human activities. Such changes in climate have led to or are projected to contribute to severe direct and indirect consequences, including rises in temperature, changes in weather patterns, irregular rainfalls, dry spells, rise of sea levels, heat-related fatalities, extinction risks for wildlife, vegetation shifts, and economic losses (Anderson and Bows 2011; Watts et al. 2015; IEA 2017; Pecl et al. 2017; Mcole 2018; Masson-Delmotte et al. 2018).

Although all spheres of life have been affected by the changing climate, agriculture is the most sensitive to this threat due to its dependence on weather and climate. Among the impacts of climate change on agriculture, it has been observed and anticipated that shifts in plant and animal adaptation, new pests, deleterious effects on plant physiology, soil degradation, and changes in weather patterns will have the major effects, and all of which will lead to a decrease in crop yield and animal production. Problems with water resources, declines in freshwater availability, decreases in water flow, and increases in evapotranspiration have already started showing their effects, and this matter is deemed the most important issue of this century.

Implications of climate change are foreseen as more severe on the marginal lands than that on other lands, and more areas would be vulnerable to becoming marginal. Under the scenarios of land degradation, desertification, and aridity, marginal land utilization has been seen as an approach toward addressing these soil loss issues. However, the changing climate may lead to nonutilization of such soils, which may consequently lead to loss of sequestered carbon from soils. Therefore, climate change is predicted to affect all dimensions of agriculture, but more importantly, the physical, chemical, and biological characteristics of soils. The field of soil biology, particularly that of microbiology, has gained more importance in recent times due to its role in ameliorating certain effects of climate change on plants (Das et al. 2019). Therefore, a variety of studies have not only focused on the effects of climate change on soil biology but also focused on interactions among soil–plant–microbes (Compant et al. 2010; Classen et al. 2015; de Vries and Griffiths 2018).

The effects of climate change on soil microbes can be divided into three major categories: those on biomass, functioning, and diversity. The response of soil microbes to climate change depends on the type of microorganisms present because microorganisms differ in their physiology, stress tolerance, and growth; therefore, quite differential responses are prevalent (Bahram et al. 2018). Generally, an increase in temperature will affect the structure and functioning of soil microbes and cause changes in available substrates. However, the microbial response to elevated temperature is directly linked with that to soil moisture (Classen et al. 2015), which fluctuates according to the warming conditions. Instances of deleterious effects of warming have been observed on fungal and bacterial growth rates (Pietikäinen et al. 2005). Similarly, a loss of microbial efficiency under warming has also been reported (Frey et al. 2013).

With regard to the second component of climate change, although an increase in CO_2 has been projected to enhance microbial activity and biomass (Lipson et al. 2005; Blagodatskaya et al. 2010), reports of microbial diversity inhibition (Ma et al. 2017), reduced growth, and shifts in edaphic microbial communities (Yang et al. 2019) have been observed. For instance, a multifactorial study has found that warming led to an increase in fungal abundance, while an increase in temperature and CO_2 levels led to an increase in bacterial richness (Castro et al. 2010). Shifts in bacterial and fungal communities in response to changes in precipitation have also been found in the study by Castro et al. (2010). Consequently, it can be concluded that climate change will affect soil microorganisms to different extents, depending on the type of microbes. Many studies conducted by different researchers (Gottfried et al. 2012; Langley and Hungate 2014; Zhou et al. 2015) have identified shifts in species distribution and the functioning of terrestrial ecosystems in the interactions of some species in response to climate change (De Vries et al. 2012). De Angelis et al. (2015) concluded that prolonged warming of forest soil leads to changes in microbial communities in temperate forest soils. Therefore, with the increasing temperature of 5 °C in temperate forests, it increased the bacterial–fungal ratio by changing the relative abundance of microbes such as bacteria in the soil (Bintanja 2018). A 10-year cross-biome study observed an increase in acidic soil bacteria alongside ecosystem-specific responses (Dunbar et al. 2012). In another study, it was observed that climate change caused shifts in population ranks in archaea, fungi, and some bacterial groups in Australian meadows (Hayden et al. 2012). Another study showed that the population of actinobacteria decreased with increasing temperature, which is due to higher temperatures, increased respiration (Luo 2007), and sensitivity to excess CO_2 in the environment. They reported that the group may negatively affect global warming (Goodfellow and Williams 1983).

Utilization of plant- and animal-associated microbes in enhancing productivity has been on the rise; therefore, the functionality of these microbes under changing climate conditions is important in agriculture. The function of plant-beneficial microbes was found to shift from positive to neutral under elevated temperatures (Heinze et al. 2016), which suggests that changing climate may lead to nonfunctionality of the existing microbiome, and we may need to explore more resilient biofertilizers. Similarly, commercial biofertilizers consist of a long history of exploring, characterization, and validation, and changes in climate may deter their performance. Such abruptions would therefore render the inoculum ineffective, with its need for the search for potential plant-beneficial microbes to commence.

Plant-Associated Microbiome

Food security under circumstances of growing populations, decreasing land areas, deteriorating soil resources, and climate change has been very challenging (Mahmood and Kataoka 2019). The estimated world population would reach 9.8 billion by the mid-century (United Nations 2017), which would require an exponential increase in food production. Following this, land and soil resources are continuously being exploited in many parts of the world where there is no legislation for these resources' conservation or where there are gaps in the laws and in their implementation. Along with the legislation issues are problems in the lack of awareness, mismanagement, and conventional crop husbandry practices that lead to land degradation. Economic extremes leading to noncultivation of land are also added to the deterioration of soil resources. In recent times, climate change has worsened due to frequent disasters, land degradation, and decreases in water availability. All these issues have led to reduced food production, which needs immediate remedy.

Fig. 1 Barley growth inoculated with *Piriformospora indica* ATCC204458 and endophytic fungal isolate. Barley seeds were surface-sterilized and then placed on the medium in which each fungus was grown to infect the roots of the seedlings. The control seedlings were solid medium without endophytes. After infection, the seedlings were transferred to soil

Intensive agricultural practices have been employed to enhance food production per unit area; however, excessive use of agricultural inputs has threatened soil and agricultural ecosystems. For instance, chemical fertilizers have been used to meet crop nutrition needs, but their incautious use has led to several environmental and economic issues (Mahmood and Kataoka 2020). Similarly, this incautious use has been seen with pesticides, which are extensively applied in intensive agriculture and have led to the contamination of soil and other environmental components, as well as the food. Buildup of pesticidal chemicals in soil, even at lower concentrations, can be deleterious to the edaphic microbial community, plant growth, and soil processes, and all of which lead to contaminated food production (Vryzas 2018). Therefore, continuous efforts toward finding sustainable solutions have been made to identify the potential areas in this domain.

Biological fertilizers or biofertilizers have emerged as an attractive alternative. Their multiple roles, sustainability, and other application prospects have made them the most preferred choice. Especially, there are many researches that figure out the plant endophytes. Some specific endophytes have an ability to enhance the plant growth (Fig. 1). Biofertilizers involve epiphytic or endophytic microorganisms, including bacteria, fungi, archaea, and cyanobacteria, among which bacteria and fungi have been employed more often (Mahmood and Kataoka 2019). Recent advances have enhanced our understanding of plant microbiome, which is a collective term for all the incident microbes; however, it is believed that we are still at the very surface of this newest "black box" (Stanley and van der Heijden 2017; Orozco Mosqueda et al. 2018; Compant et al. 2019; Arif et al. 2020). The plant microbiome as a broader term encompasses microbes living within the rhizosphere, endosphere, and phyllosphere. It is important that while there are chances of vertical transfer of microbes, horizontal transfer leads to the formation of endospheric and phyllospheric microbiomes, both of which remain a subset of the rhizosphere microbiome (Hodgson et al. 2014; Wiewióra et al. 2015; Frank et al. 2017; Shahzad et al. 2018). The rhizosphere microbes can also originate from bulk soil, which remains to be of least importance in the context of crop production.

Compared with the structure of the microbiome, their functions are much more complex and least explored (Stanley and van der Heijden 2017). Although many of the plant-associated microbiomes function like those of nitrogen fixation, the phosphorus and potassium solubilization, iron and other nutrients chelation, 1-aminocyclopropane-1-carboxylate (ACC) deaminase activity, plant hormone production, and biocontrol of pathogens have been well-established; however, new dimensions keep rising as we move forward (Berg et al. 2014; Stanley and van der Heijden 2017; Dubey et al. 2019; Jones et al. 2019). In recent trends, much focus has been on the efficient application of potential microbes, utilization of marginal soil, remediation of polluted soils utilizing microorganisms, and integrated use of organic and biofertilizers in application in applied research, while understanding the complete picture of the microbe–plant–soil interaction and utilizing advanced techniques have been studied in fundamental side. Recent studies have explored the role of plant-associated microbes in improving plant growth, health, productivity, and secondary metabolism (Korenblum and Aharoni 2019; Berg et al. 2020; Zuluaga et al. 2020), food quality (Hirt 2020), induced systemic resistance (Wang et al. 2020), pathogen resistance (Wei et al. 2019), bioremediation (Chaudhary and Shukla 2019; Rylott and Bruce 2019), utilization of contaminated soils (Ren et al. 2019), and facilitating phytoremediation (He et al. 2020).

All these and other functions will benefit the plants and soil if desired microbes are applied successfully and are adapted to performing their functions in a stress-free environment. Application of such plant-beneficial microorganisms has been well debated, and several approaches have been proposed for improved inoculation (Mahmood et al. 2016; Mahmood and Kataoka 2019). However, timing of application, sole application or consortium, quantity of inoculum to be applied, which organ of plant to be targeted for application, and hitting the association button are challenges that still need to be addressed.

Recent efforts in understanding the plant–microbe interaction have brought forward the need to understand the cumulative effect rather than that of singular factors. As host microbe, particularly, the plant–microbe interaction is a complex process; an understanding of the complete picture would better enable us to ensure efficient inoculation of applied microbes. Recent efforts based on contemporary understanding have led to multiple approaches, including those of rhizosphere engineering (Ahkami et al. 2017), biofilm application (Backer et al. 2018), use of different carriers such as biochar (Belcher et al. 2019), dual bioaugmentation (Roane et al. 2001), and use of synthetic ecology (Said and Or 2017). The synthetic ecology approach is a specialized construction for the application of microbial inoculum to plants for enhanced growth and soil remediation objectives. Utilization of the synthetic ecology approach offers multiple benefits, including resilient plant morphology (Finkel et al. 2017), better capture of resources and enhanced metabolite exchange (Said and Or 2017), division of labor (Roell et al. 2019), collective metabolism, stress avoidance, better microbe–microbe interaction, and possibilities of better performance. Therefore, microbial consortia, either isolated originally or prepared synthetically, may have the answers to our search for certain aims.

Plant–Microbe Interaction Under Drought and Salt Stress Conditions

The frequency of dry periods is increasing worldwide, which is accompanied by the consecutive occurrence of drought and salinity in the croplands. Soils are categorized as salt-affected when the total salt concentration (i.e., electrical conductivity) exceeds 20 mM or dS/m (Abrol et al. 1988). Soil salinization is one of the most prevalent reasons for soil degradation and is a continuous threat to agricultural production, especially in arid and semi-arid regions. It is estimated that because of inadequate irrigation schemes, at least 20% of all irrigated lands worldwide are salt-affected (Pitman and Läuchli 2002). Some higher estimates are also presented, stating that half of the irrigation systems worldwide are under the influence of salinization, alkalization, or waterlogging (Szabolcs 1994; Hu et al. 2005). In addition, according to the Food and Agriculture Organization of the United Nations (FAO 2002), soil salinization reduces arable lands that could be used for agriculture by 12% on an annual basis.

Under both drought and salinity conditions, soil water potential decreases, meaning that osmotic stress in plants increases, which has the greatest effect on plant growth among all single environmental factors. Under high osmotic pressure, root and stem length as well as dry weight are significantly reduced, and several growth stages of the plant, including germination, seedling, and flowering, are impaired. Saline conditions are characterized by extreme ratios of Na^+/Ca^{2+}, Na^+/K^+, Ca^{2+}/Mg^{2+}, and Cl^-/NO_3^-. At the rapid onset of salinity (osmotic phase), stomatal closure is followed by the inhibition of shoot elongation and new leaf formation. At an extended term (ionic phase), the accumulation of salt reaches toxic levels (Roy et al. 2014). In addition, under drought conditions, the rate of transpiration decreases and the active transport systems and membrane permeability are impaired. These conditions disrupt nutrient acquisition of plants, either by influencing nutrient uptake or translocation by reducing the osmotic potential of root media or by the reduction in nutrient availability due to competition with other ions. This competition is generally characterized by Na^+-induced Ca^{2+}/K^+ deficiencies and Ca^{2+}-induced Mg^{2+} deficiencies. It is also known that salinity reduces nitrogen (N) and phosphorus (P) uptake and causes deficiency symptoms (Alam 1999). In addition to these nutritional imbalances caused by salinity, excessive amounts of Na^+ and Cl^- cause specific ion toxicities; therefore, the concentration of these ions should be kept low in the cytoplasm for viable cellular processes (Grattan and Grieve 1999; Alam 1999). In high-order plants, various mechanisms exist to retrieve these adverse effects of salinity and drought such as (1) synthesis of compatible solutes, (2) nutrient transport and acquisition, (3) synthesis of antioxidants, and (4) hormone modulation.

In higher plants, low molecular weight organic solutes such as proline, glycine betaine, polyols, trehalose, sorbitol, and malate, which are compatible with enzymatic functions, are mainly used for turgor regulation. Besides regulating osmotic pressure, these osmolytes are also thought to stabilize protein complexes and membrane structures under drought and salinity stress (Murata et al. 1992; Cushman 2001). Being the most profound component of salinity in the soil, the transport and compartmentalization of Na^+ are important in alleviating salt stress. In order to maintain a low concentration of Na^+ in the cytoplasm, plants use transport proteins that mediate ion flux and maintain ion homeostasis. These proteins can be categorized as pumps, channels, and carriers. The H^+ ATPase and H^+ pyrophosphatase pumps mainly drive secondary ion and nutrient transport processes and maintain excess Na^+ in the vacuole through the direct utilization of metabolic energy. Meanwhile, carriers such as the Na^+/H^+ antiporter and K^+/H^+ symporter couple the uphill transport of a solute to the downhill transport of another in the same (symport) or opposite (antiport) direction.

At the cellular level, salt overly sensitive (SOS) pathway proteins phosphorylate the Na^+/H^+ antiporter, which results in the transfer of Na^+ from the cytoplast to the apoplast. The influx of Na^+ and K^+ ions is mediated by common proteins; however, since only K^+ is an essential cofactor for cellular functions, K^+ and Na^+ influx needs to be differentiated. Channels such as inward rectifier K^+ channels mediate the passive transport of ions and confer a salt-induced reduction in conductivity, which decreases the influx of Na^+ and efflux of K^+ ions to the cytoplasm under high salinity conditions. In addition, aquaporins (or water channels) are also noteworthy in maintaining water balance under osmotic stress. Aquaporins are members of the major intrinsic proteins (MIPs), which are abundant in the plasma membrane and tonoplast of plant cells (plasma intrinsic proteins (PIPs)). These proteins regulate root hydraulic conductance (L), and by decreasing the cell water potential, plants can continue to procure water from salinized environments (Niu et al. 1995; Jacoby 1999; Cushman 2001; Qin et al. 2016). Moreover, salt stress affects the electron transport chain (ETC), which ultimately results in the accumulation of reactive oxygen species (ROS). Reactive oxygen species triggers several toxic reactions to plants, including DNA damage, protein degradation, and membrane destabilization. Therefore, under stress conditions, several antioxidant enzymes such as peroxidases, glutathione reductase (GR), superoxide dismutase (SOD), and proline catalase (CAT) and nonenzymatic antioxidants are released to relieve this stress (Aghaei et al. 2009).

Phytohormones regulate plant growth under various abiotic stress conditions via conserved signal transduction pathways. Abscisic acid (ABA), ethylene, salicylic acid (SA), and jasmonic acid (JA) are regarded as stress response hormones; others including auxin, gibberellin (GA), cytokinins (CKs), brassinosteroids (BRs), and strigolactones (SLs) are classified as growth promotion hormones (Yu et al. 2020). Abscisic acid levels rapidly increase upon osmotic stress and trigger a signaling cascade, which results in the regulation of stomatal closure and in the production of sugar-derived osmolytes to alleviate salt-induced stress. Calcium also acts as a secondary messenger in the ABA signaling pathway. Similar to other stress response hormones, levels of ethylene increase under salt stress, and it regulates plant salt stress responses. Excessive ethylene is known to severely retard root development. Besides being toxic to plants, ethylene also inhibits mycorrhizal colonization and rhizobial nodulation. Ethylene production in plants is correlated with the endogenous ACC deaminase enzyme, which catalyzes the conversion of ACC (ethylene precursor) to ammonia and α-ketobutyrate and reduces ethylene levels (Mahajan and Tuteja 2005; Deepti et al. 2014). Under osmotic pressure, auxins such as indole acetic acid (IAA) regulate root formation through a complicated crosstalk pathway with ABA. Under mild salt stress, ABA activates auxin and triggers root formation, while as the NaCl concentrations increase, excessive amounts of ABA disrupt auxin functions and stop root formation. Under severe salt stress, levels of other growth-promoting hormones such as GA, CKs, BRs, and SLs decrease, and plant growth is inhibited, which is a profound adaptation mechanism of plants to salt stress (Yu et al. 2020).

The abovementioned mitigation strategies of plants in response to abiotic stress pave the way for practical implications aimed at increasing crop productivity under salinity and drought conditions. The conventional method deployed in this regard is the production of transgenic crops. Even though manipulation of the expression of several genes has been reported to increase salt tolerance of transgenic plants under greenhouse or laboratory conditions, successful applications in the field are still limited (Roy et al. 2014) Certain drawbacks exist regarding this situation—first, salt tolerance is mediated through complex signaling pathways. Therefore, mediation of a single or a few genes often results in undesirable alterations (Wang et al. 2003); second, salt stress in nature is generally accompanied by alkaline conditions. Therefore, the approach of engineering plants for abiotic stress is complex and daunting; third, the application of molecular techniques is limited because some important crop species such as *Brassica* and *Triticum* are hexaploid or tetraploid; fourth, constitutive expression of transcription factors can result in yield penalties (Roy et al. 2014); last, but not least, is the long debates regarding public concerns regarding the consumption of transgenic plants. Therefore, increasing efforts are put forth in finding alternative methods for increasing plant fitness under salt and drought stress. As an alternative technology to genetic modification, technology utilizing microorganisms is attracting attention.

Microbial communities coevolved with their plant hosts under several environmental conditions; therefore, it is assumed that they have crucial roles in increasing plant fitness under unfavorable environmental stress conditions. From an ecological perspective, it can be said that microbial genome acts as a "pangenome" and has the capacity for enhancing the abiotic stress tolerance of plants. Plant growth-promoting rhizobacteria (PGPR) and endophytic microorganisms are already known for their growth-promoting effects and have been used as biofertilizers for many years. Similarly, under abiotic stresses such as salinity and drought, microorganisms can induce some physiochemical changes in plants and increase their tolerance to stress factors, which is called induced systemic tolerance (IST). In addition to IST, studies have shown that PGPR can act as a biocontrol agent for several plant diseases, even under stress conditions, which is termed as induced systematic resistance (ISR) (Paul and Harshad 2014).

Several stress mitigation mechanisms triggered by microbial communities exist, some of which are provided below. Plant growth-promoting rhizobacteria are capable of secreting phytohormones such as IAA, ABA, GA, and cytokinins. Plant associations with these bacteria have been shown to promote plant growth under salinity and drought stress. Isolation of several rhizobacteria from four weeds thriving under high salinity showed that all strains were capable of producing the phytohormones indole-3-acetic acid (IAA), gibberellic acid (GA3), trans-zeatin-riboside (tZR), and ABA, and inoculation of these strains to soybean seedlings treated with or without 20 dS/m NaCl resulted in better growth and higher proline contents when compared with the control setting (Naz et al. 2009).

In a study performed by Deepti et al. (2014) on pea (*Pisum sativum*), the ACC deaminase producer rhizobacteria *Arthrobacter protophormiae*, PGPR *Rhizobium leguminosarum*, and arbuscular mycorrhizal fungus (AMF) *Glomus mosseae* acted synergistically and increased plant weight by 53% under high salinity. Enhanced nodulation and AMF colonization were observed along with reduced proline content, lipid peroxidation, and increased pigment content under 200 mM salt conditions. In another study, under drought stress, the fresh weights of the bacterially treated tomato (*Lycopersicon esculentum* Mill cv. F144) and pepper (*Capsicum annuum* L. cv. Maor) seedlings were approximately twice those of the control plants. The inoculated ACC deaminase producer strain *Achromobacter piechaudii* ARV8 also increased its P and K acquisition under high salinity conditions and improved water use efficiency (Mayak et al. 2004).

In a study performed with wheat (*Triticum aestivum* L.) at high salinity levels (15 dS/m), plant height, root length, plant biomass, and grain yield increased up to 37%, 70%, 116%, and 111%, respectively, compared with those of the control upon inoculation with a *Pseudomonas fluorescens* strain, which was reported to be positive for ACC deaminase, IAA, and phosphate solubilization (Nadeem et al. 2010). Other data showed that when modulated with the IAA-overproducing *Sinorhizobium meliloti* strain, the host plant (*Medicago truncatula*) showed reduced symptoms of senescence, lower ethylene levels, higher shoot dry weight, and better nitrogen-fixing capacity under salinity treatments of up to 0.3 M. The results were attributed to the remodulation of phytohormones, with a higher IAA content in the nodules and roots and a decreased IAA level in the shoots (Bianco and Defez 2009).

Salt-tolerant PGPR has also been reported to increase the nutrient uptake of plants in several studies. In a study conducted by Upadhyay and Singh (2015) with wheat in saline soils, the N, K, and P contents were increased in the leaves of plants treated with salt-tolerant PGPR strains. In addition, the dry weight and shoot biomass as well as the proline and total soluble sugar content were increased after inoculation with SU44 *Bacillus aquimaris* and SU8 *Bacillus aquimaris*. Plant growth-promoting rhizobacterium is also known to increase the K^+/Na^+ ratio of plants grown under elevated salt conditions. *Bacillus subtilis* strain GB03 was reported to decrease shoot and root Na^+ levels and therefore improve the K^+/Na^+ ratio in white clover (Han et al. 2014). The same strain also triggered the induction of a high-affinity K^+ transporter (*HKT1*) in the shoots of *Arabidopsis* (Brassicaceae) and increased the shoot to root Na^+ recirculation. With a dual function, expression of *HKT1* was downregulated in the roots of plants through the secretion of volatile organic compounds (VOCs) by PGPR, which resulted in a restricted flow of Na^+ in the roots (Zhang et al. 2008).

Another means of PGPR-conferred IST is through the regulation of aquaporins under salt stress conditions. In maize plants infected with *Pantoea agglomerans* (gammaproteobacteria) and *Bacillus megaterium*, gene expression was upregulated for both *PIP2* and *ZmPIP11* that encode plant aquaporins under salt stress, resulting in an increase in the root hydraulic conductance (Gond et al. 2015). In a similar study, inoculation with *Azospirillum brasilense* (alphaproteobacteria) triggered the transcription of *HvPIP21* in barley (Zawoznik et al. 2011). Plant growth-promoting rhizobacterium is also known to change the expression of ROS-scavenging genes. In a study performed by Sukweenadhi et al. (2015), *A. thaliana* plants inoculated with *Paenibacillus yonginensis* were found to be more resistant to salt stress, drought stress, and heavy metal (aluminum) stress than the control. The saline-responsive genes (*AtRSA1*, *AtVQ9*, and *AtWRKY8*) that were used as markers to check the responses of *A. thaliana* to drought and salt stress were the genes regulating the ROS detoxification and ion homeostasis.

In another study, two *Bacillus* isolates were demonstrated to enhance the mRNA levels of various ROS-scavenging enzymes and induced the tolerance to abiotic stresses in potato tubers subjected to salt, drought, and heavy metal stress. Proline content was also reported to increase in tubers treated with PGPR (Gururani et al. 2013). Likewise, in addition to the genes involved in antioxidative responses, it has been reported that PGPR can also induce the expression of genes encoding the synthesis of compatible solutes. Expression of proline biosynthetic genes (i.e., *P5CS1* and *P5CS2*) was found to be upregulated in tomato plants, and *Arabidopsis* inoculated with *Enterobacter* sp. EJ01 was also capable of producing ACC deaminase and IAA and improved the vegetative growth of host plants exposed to salt stress (Kim et al. 2014).

Transplantation of beneficial microorganisms into plants to increase the yield is an old technique that dates back to the nineteenth century (Bashan 1998). However, inoculant formulations need to be carefully designed because the plant microbiome is highly complex and is under the influence of strong habitat-filtering mechanisms (Qin et al. 2016). After the amendment of soil with associative, nonsymbiotic PGPR, the bacterial population tends to decrease progressively (van Elsas et al. 1986; Bashan and Levanony 1988). Unlike sterile soils, the inoculated microorganisms may not find an empty niche and may have to compete with native microflora that is already present in natural settings. In addition, inoculants need to compete with several stress conditions and need to survive to reach a sufficient population size for them to be able to provide intended benefits to plants. It should also be kept in mind that microorganisms in a pure culture behave differently in a microbial community (Bashan 1998; van Elsas et al. 1986). For these reasons, there is a shift from single-cell studies to community studies, and it is pointed out by several studies that using mixed inoculants with synergistic interactions enhance the plant fitness better than single strains (Baha and Bekki 2015; Janzen et al. 1992; Drozdowicz and Ferreira Santos 1987; Frommel et al. 1991; De Roy et al. 2014).

Therefore, a clear understanding is required to disentangle how biotic and abiotic factors shape the plant microbiome and how the members of a microbial community are linked together. A range of synthetic ecology approaches exist to reveal the functions of a microbial community in a diverse ecological niche. In that regard, both traditional techniques and modern techniques have their advantages and drawbacks. Isolation of microorganisms (culture-dependent methods) is necessary to carry out detailed studies on specific traits and to identify the genetic components behind these traits (Bakker et al. 2013). Meanwhile, only a small percentage of microorganisms are cultivable; therefore, omic technologies are required to characterize the functional and metabolic potential of the whole microbiome without the biases inherent in cultivation-based techniques.

The full shotgun metagenomic approach is used to determine which genes are enriched in a specific microhabitat. The metatranscriptomic method is used to identify the abundance of transcripts and link the function of a gene with a specific trait (Qin et al. 2016). Likewise, metaproteogenomics is used to study metabolic pathways and allows us to determine when and where the genes are being expressed (Guttman et al. 2014). In addition, DNA microarrays are being used to determine microbial function in soils (Mcgrath et al. 2010). Moreover, a software tool named phylogenetic investigation of communities by reconstruction of unobserved bacterial states (PICRUSt) has been developed, which uses marker genes and a database of reference genomes to elucidate the functional composition of the microbiome. This tool has a higher efficiency than whole-metagenome shotgun sequencing (Langille et al. 2013). Another similar tool is called Tax4Fun, which uses 16sRNA sequence data to predict the functional profiles of samples (Kathrin et al. 2015). These technological advances to predict microbial functionality are being supported by various experimental studies, which show that plants utilize more than a single microorganism to combat drought and salinity stress.

In a study investigating the potential of the whole soil microbiome to help *A. thaliana* to combat drought stress, it was shown that the microbiome isolated from a sympatric soil with a history of exposure to host plant in a natural setting increased the total plant biomass. Meanwhile, the nonsympatric soil microbiome derived from corn and pine soils had no such effect (Zolla et al. 2013). Likewise, it was shown that plant fitness increased in *Brassica rapa* inoculated with a microbiota that has adapted to multigeneration drought treatments (Lau and Lennon 2012). Rhizosphere engineered through artificial selection of the microbiome under water-deficit stress symptoms produced significantly more biomass and root system length, dry weight, and surface area than the negative control. At each round of this host-mediated selection, changes in the overall community composition were monitored (Jochum et al. 2019).

Another recent study also showed that at the respective salinities, changes in rhizosphere microbial communities induced by salinity promoted the germination and growth of *Hibiscus hamabo* (Yuan et al. 2019). Therefore, the introduction of a selected consortium has great potential for increasing plant fitness under stress conditions. However, it is known that multiple operational taxonomic units (OTUs) can perform the same function, and this type of functional redundancy is very common among microbiomes. In other words, only a portion of microorganisms are required to fulfill the required responses. Meanwhile, under habitat stress, these redundant species may also take over and sustain ecosystem functionality (Dopheide et al. 2015; Purahong et al. 2014; Souza et al. 2015). Therefore, the concepts of core microbiome and minimal microbiome are introduced for the rational design of synthetic microbial communities (SMCs).

The core microbiome (CM) is defined as a group of microorganisms that are persistent within the host microbiome and are essential in providing a potentially critical function within the habitat in which they are found. Meanwhile, there can still be functional redundancy within the CM, therefore, minimal microbiome (MM), which is the smallest possible and functionally indispensable subset of microbes within the total microbiome (de Vos 2013). Identifying the CM is critically important in defining a persistent community to achieve the desired functions and predict community responses to perturbations. Therefore, characterization of a CM also depends on the ecological question addressed. To determine the CM, five parameters are defined by Shade and Handelsman (2012). The first parameter is the membership that considers the shared taxa (or genes) that are present across two or more microbiomes in the defined environmental setting. Composition (or dominance) is the second parameter, which represents the abundance of each OTU. Contributions of rare, dominant, and common OTUs are unique and should not be overlooked. Phylogeny is the third parameter, which states that OTUs from the same lineage also have similar functional capabilities and show similar responses to environmental stimuli. The fourth parameter is persistence, which depends on temporal and spatial sampling of OTUs. For a persistent community, spatial organization is necessary to balance the antagonistic and synergetic interactions. Persistently detected microbiomes under each setting provide a quantitative framework for understanding CM. The last parameter is connectivity, and as the name implies, it includes information about OTU interactions. Species can interact when located in close proximity. Network analyses such as quantitative insights into microbial ecology (QIIME) or local similarity analysis (LSA) can be used to understand interacting species (Shade and Handelsman 2012). In other words, an understanding of the functional complementarity, synergistic effects, and microbial assembly is important in constructing a synthetic community for the desired functions (Qin et al. 2015).

Currently, the majority of SMC studies are conducted through selection of PGPR with multiple desired functions and combining them with an empirical methodology to assess the results (Baha and Bekki 2015; Santhanam et al. 2015; Thijs et al. 2014a, b). As mentioned earlier, in most of the studies, coculture systems have proven to be more effective in increasing plant fitness compared to inoculation with a single strain. Meanwhile, there are also reports stating that using combination of strains has no additional effect, which indicates that constructing compatible and effective SMCs has challenges (Rosier et al. 2016). Addressing this challenge requires the integration of several approaches. While constructing SMCs, it is rationally expected to maximize synergism and minimize antagonism among the consortia. These interactions can be unidirectional or bidirectional, and in order to understand them, one common method used is similarity-based network construction, where the cooccurrence of two species is detected over time series. In the case of complex relationships where one species affects several others, additional mathematical and statistical models such as the Lotka–Volterra framework are being used. This framework estimates the parameters governing species-to-species interactions.

Meanwhile, an inference model should consider the interactions simultaneously in a complex microbial network, where different algorithms are developed, such as a smooth response surface (SRS) or rule-based microbial network (RMN) (Tsai et al. 2015). In addition, forward engineering strategies are also being employed. In the study performed by Harcombe et al. (2014), dynamic flux balance analysis was used to perform time-dependent metabolic simulations of microbial ecosystems. Another computational framework was developed by Chiu et al. (2014), the Goldilocks principle, which is used to model microbial communities by predicting enhanced metabolic activity. It is proposed that species need to be neither too close nor too distant to observe high levels of emergent metabolic capacity. The predictive power of these models is uncertain until they are experimentally tested. In that sense, complementary approaches also exist where synthetic combinations of strains are screened for desired properties. For example, in the study of Friedman et al. (2017), survival of a small set of species in pairwise competition is used to predict survival in multispecies competitions.

Assaying microbial phenotypic combinations with high throughput is challenging because constructing strain combinations is very difficult due to limitations of combinatorial space; however, recent advances in microdroplet technology (kchip) ease these limitations and allow the determination of how biotic and abiotic factors drive metabolic decision-making and microbial interactions. This screening technology can also be used to suggest community design principles with improved accuracy (Kehe et al. 2019). Additionally, it is suggested that mimicking the natural environment should be considered in experimental settings while constructing SMCs. In a study performed by Yuan et al. (2016), SMCs were subjected to serial culturing under saline conditions, and upon inoculation, the SMCs were capable of efficiently improving the salt tolerance of plants (Qin et al. 2016; Yuan et al. 2016).

Plant–Microbe Interaction Under Heavy Metal Stress

Application of microbial consortia to improve growth and ameliorate biotic and abiotic stress in plants has been reported (Sarma et al. 2015; Woo and Pepe 2018; Compant et al. 2019; Hashmi et al. 2019). The application of consortia, particularly aimed at dealing with heavy metals, has been employed through two different approaches: (a) by enhancing or (b) decreasing the uptake of metals by the plants. In the first case, instances of increased accumulation, often termed as bioconcentration of metals by plants with the help of microbes in a consortium, have also been observed (Ullah et al. 2015; Ahemad 2019). However, recent times have seen a rise in heavy metal-polluted soils worldwide (Puschenreiter et al. 2005); therefore, our focus should be on the utilization of polluted soils for crop production with the minimum uptake of such metals by the plants. Multiple similar instances have been reported (Table 1), where reduced uptake of metals enhanced plant growth promotion and where improvements in plant physiology have been observed.

A variety of microbes, plants, and metals have been investigated in such studies. The microbes, however, were mainly bacteria (Table 1). Nevertheless, fungi associated with heavy metal-contaminated soils have also been observed (Hall 2002; Mishra et al. 2017), but their utilization in plant growth experiments is limited, although their role in detoxification of such metals is well-advocated (Cetin et al. 2011). On the other hand, certain plant species, including vegetables, fodders, and undomesticated plants, have been explored for their potential in growing on contaminated sites with the help of metal-tolerant or metal-resistant microbes. Lastly, multiple metals that are incidental in agricultural or other ecosystems have been explored where cadmium has been more common (Table 1).

It is noticeable that many of the microbial isolates employed in the abovementioned studies (Table 1) came from metal-polluted soils and showed metal tolerance or resistance. This indicates that contaminated soils can be a potential source of microbes, which could be helpful in addressing the issues of heavy metal contamination. The contamination, however, is more complicated in the field as compared to that tested under axenic conditions. Many of the sites are cocontaminated with multiple groups of pollutants, including heavy metals (Ye et al. 2017; Zhang et al. 2020). In such circumstances, many factors limit the functionality of microbes, which is a slower process than the continuous climatic and management transformations. In such limiting scenarios, synthetic ecology approaches have been termed successful because of the possibility of coworking of the applied consortia. The combined application of different microbial inoculums leads to combined metabolism, simultaneous degradation of different pollutants or different forms of pollutants, and induction of secondary metabolism, and it can sometimes lead to complete degradation or transformation, which is seldom the case with single strains. The synergy among the consortium members has also been postulated to help enhance the efficiency of the plural over that of the singular strains (Ghazali et al. 2004; He et al. 2004).

Table 1 Application of microbial consortia against heavy metals stress

Microbes in consortium	Isolated from/ characteristics of microbes	Plant to which consortium was applied	Plant growth-promoting traits	Incident heavy metals	Plant response to application of consortium	Reference
Pseudomonas sp. *Bacillus* sp.	Metal-contaminated soil/plant growth-promoting	Spinach	ACC deaminase activity, indole acetic acid and siderophore production, phosphate solubilization	Cadmium, lead, zinc	Reduced accumulation of metals and increased plant growth	Shilev et al. (2020)
Pseudomonas aeruginosa, *Burkholderia gladioli*	No info given/plant growth-promoting	*Lycopersicon esculentum*	–	Cadmium	Improved plant growth, photosynthesis, and accumulation of phenolic compounds and osmoprotectants	Khanna et al. (2019)
Proteus sp., *Pseudomonas* sp., *Ensifer meliloti*	Metal-polluted soil/ metal-tolerant	*Medicago sativa*	Phosphate solubilization, nitrogen fixation, siderophores, and indole acetic acid production	Copper, lead, zinc	Reduced accumulation of metals and increased plant growth	Raklami et al. (2019)
Bacillus megaterium, *Pantoea agglomerans*	*Lactuca sativa* and *Beta vulgaris* rhizospheric soils contaminated with aluminum/metal-resistant	*Vigna radiata*	IAA, ACC deaminase, siderophore production, phosphate solubilization	Aluminum	Improved plant growth and reduced metal uptake	Silambarasan et al. (2019)
Bacillus sp. and *Pseudomonas* sp.	Endophytes and rhizosphere bacteria associated with *Festuca rubra*, *Agrostis capillaris*, and *Arabidopsis thaliana*	*Brassica napus* *Festuca rubra*	ACC deaminase activity	Lead, cadmium, zinc	Improved plant growth under heavy metal conditions	Grobelak et al. (2018)

	growing in heavy metal-contaminated soil/plant growth-promoting					
Alcaligenes sp., *Bacillus* sp., *Curtobacterium* sp., *Microbacterium* sp.	Rhizosphere of *Prosopis laevigata* and *Sphaeralcea angustifolia* grown in heavy metal-contaminated soil/ metal-resistant	*Brassica nigra*	Selective siderophore and indole acetic acid production, phosphate solubilization	Zinc	Improved root development under heavy metal conditions	Roman-Ponce et al. (2017)
Endophytes consortium: *Kushneria* sp., *Micrococcus* sp., *Bacillus* sp., and *Halomonas* sp. Rhizosphere consortium: *Vibrio* sp., *Pseudoalteromonas* spp. (2 strains), *Staphylococcus* sp.	Endophytes and rhizosphere bacteria associated with *Arthrocnemum macrostachyum* growing in contaminated soil/plant growth-promoting and multiple metal-resistant	*Arthrocnemum macrostachyum*	Selective nitrogen fixation, indole acetic acid, siderophore production, and phosphate solubilization	Arsenic, cadmium, cobalt, copper, nickel, lead, zinc	Enhanced seed germination	Navarro-Torre et al. (2016)
Micrococcus yunnanensis, *Vibrio sagamiensis*, and *Salinicola peritrichatus*	Endophytes of *Spartina maritima* growing on contaminated area/ heavy metal-tolerant	*Spartina maritima*	Selective nitrogen fixation, phosphates solubilization, and production of indole acetic acid, siderophores, ACC deaminase activity	Arsenic, cadmium, cobalt, copper, iron, manganese, nickel, lead, zinc	Improved plant growth and photosynthetic activity, decreased metal uptake	Mesa et al. (2015)
Aeromonas aquariorum, *Pseudomonas composti*, *Bacillus* sp.	Rhizosphere of *Spartina densiflora*/copper resistant	*Spartina densiflora*	Selective phosphate solubilization and siderophores and indole acetic acid production	–	Increased seed germination and protection from fungal contamination	Andrades-Moreno et al. (2014)

(continued)

Table 1 (continued)

Microbes in consortium	Isolated from/ characteristics of microbes	Plant to which consortium was applied	Plant growth-promoting traits	Incident heavy metals	Plant response to application of consortium	Reference
Bradyrhizobium sp., *Pseudomonas* sp. and *Ochrobactrum cytisi*	Rhizosphere of legume plants growing at contaminated soil except for *Bradyrhizobium*/ metal-resistant plant growth-promoting	*Lupinus luteus*	–	Arsenic, cadmium, copper, lead, zinc	Improved plant biomass and decreased metal accumulation	Dary et al. (2010)

In relation to heavy metals, microbes can either facilitate the precipitation or transform the metallic characteristics, which leads to detoxification of the metals (Lee et al. 2008). Microbial strains have differential functional groups with an affinity toward metals; therefore, different strains in the consortium may lead to better adsorption. Similarly, the incidence of some metals or other pollutants inhibits the functionality and growth of some microorganisms. Under such circumstances, a greater the number of microbes in the inoculum would be better because the microbes can help one another in evading such limitations in a synergistic system. These mutualistic interactions among the microbes can be due to the reduced bioavailability (Roane et al. 2001), immobilization (Sprocati et al. 2006), and biosorption and genetic features such as horizontal gene transfer.

Plant–Microbe Interaction Under Miscellaneous Stresses

Multiple biotic and abiotic stresses have been ameliorated using microbial consortia (Table 2). *Bacillus*, *Pseudomonas*, and *Trichoderma* have been the most common genera in such studies, while a variety of plants have been employed. In contrast to heavy metal studies (Table 1), many of the microbes used here (Table 2) held plant growth-promoting characteristics. Similarly, there were many instances of enhanced uptake of pollutants like those of hydrocarbons with the help of microbial consortia (Escalante-Espinosa et al. 2005). However, in this chapter, the major focus was on the utilization of contaminated soils without compromising plant growth (Table 2). As mentioned above, polluted soils have been on the rise, which can be reclaimed through the utilization of microbes along with other approaches. Soil pollution in relation to crop production and animal husbandry has been of several types, but evolving issues, commonly termed as emerging pollutants, include those of organic pollutants and pesticides, microplastics, personal care products, steroids, anthelmintics, synthetic nanomaterials, cyanotoxins, endocrine disruptors, pharmaceuticals, antibiotic residues, antibiotic resistance genes, and hormones (Snow et al. 2019).

The incidence of such complex pollution scenarios has made crop production very difficult; consequently, food or feed grown in contaminated soils is not safe. Therefore, the urgent cleanup of soils, along with restriction of the release of such pollutants to agricultural and other allied ecosystems, is necessary. This cleanup must be quick, economical, environmentally friendly, efficient, in situ, and publicly acceptable if it needs to deal with the extent of pollution. Biological techniques, especially the use of microbes, have such potential; however, an issue with the use of individual microbes, as mentioned above, is a compromise of their functionality upon environmental stress. Therefore, many studies have looked for potential microorganisms in the target polluted soils (Tables 1 and 2). This exploration offers resilient and evolved microbes, which may lead to efficient soil remediation.

Besides looking for already-evolved microbes, researchers have been using different approaches such as bioaugmentation, biostimulation, genetic engineering, biofilm mediation, and enrichment in enhancing the efficiency of potential microbes.

Table 2 Application of microbial consortia against miscellaneous stresses

Microbes in consortium	Isolated from/characteristics of microbes	Plant to which consortium was applied	Plant growth-promoting traits	Incident stress	Plant response to application of consortium	Reference
Pseudomonas spp., *Serratia proteamaculans*, *Alcaligenes* sp., *Bacillus* sp.	Petroleum hydrocarbons enriched unplanted and rhizosphere soils of saline solid waste management unit/petroleum hydrocarbons and salinity-tolerant (pH range 4–9)	Saltgrass	Indole acetic acid and siderophores production, phosphate solubilization	Petroleum hydrocarbons (and salinity)	Enhanced plant biomass under contaminated soil conditions	Xia et al. (2020)
Bacillus cereus, *Bacillus subtilis*, *Serratia* sp.	Forest soil/plant growth-promoting	Sweet pepper	–	*Phytophthora* blight	Protection from disease, improvement of fruit quality, and soil properties	Zhang et al. (2019)
Bacillus cereus, *Bacillus altitudinis*, *Comamonas* sp., *Stenotrophomonas maltophilia*	Oil field/no info given	Maize	–	Oily sludge	Enhanced germination percentage, protein and proline content (bacterial application integrated with fertilizer)	Shahzad et al. (2016)
Bacillus cereus, *Bacillus subtilis* *Serratia* sp.	Forest soil/plant growth-promoting	*Lycopersicon esculentum*	–	Chilling stress	Protection from chilling stress	Wang et al. (2016)
Burkholderia, *Variovorax*, *Bacillus* sp., *Pseudomonas* sp., *Ralstonia* sp.	2,4-Dinitrotoluene contaminated and adjacent grassland soils/2,4-dinitrotoluene resistant	*Arabidopsis thaliana*	IAA, phosphate solubilization, siderophores among others	2,4-Dinitrotoluene	Enhanced root length	Thijs et al. (2014b)

Cupriavidus basilensis, *Stenotrophomonas chelatiphaga*, *Methylobacterium thiocyanatum*, *Sphingomonas panni*, *Pseudomonas* spp., *Variovorax ginsengisoli*, *Pseudomonas vranovensis*, *Pseudomonas azotoformans*	Rhizosphere and endosphere of Acer pseudoplatanus growing in trinitrotoluene (TNT)-contaminated soil/no info given	*Agrostis capillaris*	Selective IAA and siderophore production, ACC deaminase activity, phosphate solubilization among others	Trinitrotoluene (TNT)	Detoxification of TNT, improved plant growth, and health under TNT-contaminated conditions	Thijs et al. (2014a)
Pseudomonas aeruginosa, *Bacillus subtilis*, *Trichoderma harzianum*	Rhizosphere of *Pisum sativum* and agricultural soil/no info given	*Pisum sativum*	–	Oxidative stress	Increased plant protection activities	Jain et al. (2013)
Pseudomonas sp., *Trichoderma* sp., *Rhizobium* sp.	Rhizosphere and nodules of chickpea except for *Trichoderma*/no info given	Chickpea	–	*Sclerotium rolfsii*	Protection from disease	Singh et al. (2013)
Azotobacter spp., *Azospirillum* spp., *Pseudomonas fluorescens*	Rhizosphere of *Santalum album*, *Tamarindus indica*, and *Ailanthus excelsa*/plant growth-promoting	*Lycopersicon esculentum*	IAA, nitrogenase	*Fusarium oxysporum*	Enhanced plant growth and disease protection	Kannan and Sureendar (2009)

In the organic pollutants' domain, for example, Hesselsoe et al. (2005) collected the microbial consortia and grew them under methane conditions, which led to degradation of phthalic acid esters. Similarly, a bacterial consortium was employed to successfully remove 1,1,1-trichloro-2,2-*bis*(*p*-chlorophenyl)ethane (DDT) and benzo(a)pyrene when integrated with a static magnetic field. Similarly, for a hydrocarbon-contaminated soil, an integration of bacterial consortium with compost and biochar was employed to enhance phytoremediation by ryegrass (Hussain et al. 2018). Other approaches employed together with microbial consortia include the application of a consortium embedded in chitosan beads (Angelim et al. 2013), biochar as a carrier of the consortium (Wang et al. 2019), and widely used enrichment (Li et al. 2020). All these instances indicate that microbial consortia have the potential for soil remediation and can be integrated with other approaches for efficient application.

Challenges

The soil microbiome is highly diverse and accounts for up to a quarter of the diversity of the earth (Wagg et al. 2019). Many studies have focused on microbial community analysis; however, currently, the interest is shifting to linking microbiome composition and diversity to function (Graham et al. 2016; Wagg et al. 2019). The extremely high microbial diversity on small spatial scales has led to the hypothesis that these highly diverse microbiomes are functionally redundant (Allison and Martiny 2008). However, it remains unclear how such microbe-diverse and functionally complex microbiomes affect the functioning of ecosystems. Therefore, the synthetic ecology approach is still in its primitive stage as compared to that of sole application or commercial bioinoculants. A comprehensive review of such scenarios has been provided by Said and Or (2017), who suggest exploring the initiation, functionality, and handling strategies for the successful employment of consortia. In such circumstances, first, the consortium preparation should be focused, where it should be started, how the mixtures or consortia are to be prepared, and what should be the ratio of mixing that would not affect the functionality of the desired microbes. The functionality being the most important aspect needs investigation for its durability and actual performance in the applied circumstances. Another part is the joint use of different classes of microbes such as bacteria, fungi, archaea, and cyanobacteria. The employment of fungi (Migahed et al. 2017) and algae (Muñoz et al. 2006) in a bacterial consortium has already been investigated for the regulation of heavy metal concentrations in soils and plants. However, further fundamental research in this domain is required to understand the benefits and potential threats. In addition, better control of consortia through the manipulation of energy sources should be sought, which would take us closer to obtaining the desired microbial functions.

References

Abrol IP, Yadav JSP, Massoud FI (1988) Salt affected soils and their management. Food and Agriculture Organization of the United Nations, Soils Bull, vol 39. FAO, Rome, Italy

Aghaei K, Ehsanpour AA, Komatsu S (2009) Potato responds to salt stress by increased activity of antioxidant enzymes. J Integr Plant Biol 51:10951103

Ahemad M (2019) Remediation of metalliferous soils through the heavy metal resistant plant growth promoting bacteria: paradigms and prospects. Arab J Chem 12(7):13651377

Ahkami AH, Allen White R, Handakumbura PP, Jansson C (2017) Rhizosphere engineering: enhancing sustainable plant ecosystem productivity. Rhizosphere 3:233243

Alam SM (1999) Nutrient uptake by plants under stress conditions. In: Pessarakli M (ed) Handbook of plant and crop stress. Routledge, New York

Allison SD, Martiny JBH (2008) Resistance, resilience, and redundancy in microbial communities. Proc Natl Acad Sci U S A 105:11512–11519

Anderson K, Bows A (2011) Beyond 'dangerous' climate change: emission scenarios for a new world. Philos Trans R Soc A Math Phys Eng Sci 369(1934):20–44

Andrades-Moreno L, del Castillo I, Parra R, Doukkali B, Redondo-Gómez S, Pérez-Palacios P, Caviedes MA, Pajuelo E, Rodríguez-Llorente ID (2014) Prospecting metal-resistant plant-growth promoting rhizobacteria for rhizoremediation of metal contaminated estuaries using Spartina densiflora. Environ Sci Pollut Res 21(5):3713–3721

Angelim AL, Costa SP, Farias BCS, Aquino LF, Melo VMM (2013) An innovative bioremediation strategy using a bacterial consortium entrapped in chitosan beads. J Environ Manag 127:1017

Arif I, Batool M, Schenk PM (2020) Plant microbiome engineering: expected benefits for improved crop growth and resilience. Trends Biotechnol 38(12):1385–1396

Backer R, Rokem JS, Ilangumaran G, Lamont J, Praslickova D, Ricci E, Subramanian S, Smith DL (2018) Plant growth promoting Rhizobacteria: context, mechanisms of action, and roadmap to commercialization of biostimulants for sustainable agriculture [Review]. Front Plant Sci 9:1473. English

Baha N, Bekki A (2015) An approach of improving plant salt tolerance of lucerne (Medicago sativa) grown under salt stress: use of bioinoculants. J Plant Growth Regul 34:169–182

Bahram M, Hildebrand F, Forslund SK, Anderson JL, Soudzilovskaia NA, Bodegom PM, Bengtsson-Palme J, Anslan S, Coelho LP, Harend H et al (2018) Structure and function of the global topsoil microbiome. Nature 560(7717):233–237

Bakker PA, Berendsen RL, Doornbos RF, Wintermans PC, Pieterse CM (2013) The rhizosphere revisited: root microbiomics. Front Plant Sci 4:165

Bashan Y (1998) Inoculants of plant growth promoting bacteria for use in agriculture. Biotechnol Adv 16:729–770

Bashan Y, Levanony H (1988) Migration, colonization and adsorption of Azospirillum brasiliense to wheat roots. Lectins-Bio Biochem Clinical Biochem 6:69–83

Belcher RW, Kim HS, Buege B, Jarand ML, Cheiky MC, Sills RA (2019) Biochar as a microbial carrier. Google Patents

Berg G, Mahnert A, MoisslEichinger C (2014) Beneficial effects of plant associated microbes on indoor microbiomes and human health? Front Microbiol 5:1515

Berg S, Dennis PG, PaungfooLonhienne C, Anderson J, Robinson N, Brackin R, Royle A, DiBella L, Schmidt S (2020) Effects of commercial microbial biostimulants on soil and root microbial communities and sugarcane yield. Biol Fertil Soils 56(4):565580

Bianco C, Defez R (2009) Medicago truncatula improves salt tolerance when nodulated by an indole-3-acetic acid overproducing Sinorhizobium meliloti strain. J Exp Bot 60:3097–3107

Bintanja R (2018) The impact of Arctic warming on increased rainfall. Sci Rep. https://doi.org/10.1038/s41598-018-34450-3

Blagodatskaya E, Blagodatsky S, Dorodnikov M, Kuzyakov Y (2010) Elevated atmospheric CO_2 increases microbial growth rates in soil: results of three CO_2 enrichment experiments. Glob Chang Biol 16(2):836–848

Castro HF, Classen AT, Austin EE, Norby RJ, Schadt CW (2010) Soil microbial community responses to multiple experimental climate change drivers. Appl Environ Microbiol 76(4):999–1007

Cetin SC, Karaca A, Kizilkaya R, Turgay OC (2011) Role of plant growth promoting bacteria and fungi in heavy metal detoxification. In: Sherameti I, Varma A (eds) Detoxification of heavy metals. Springer, Berlin, pp 369–388

Chaudhary T, Shukla P (2019) Bioinoculants for bioremediation applications and disease resistance: innovative perspectives. Indian J Microbiol 59(2):129–136

Chiu HC, Levy R, Borenstein E (2014) Emergent biosynthetic capacity in simplemicrobial communities. PLoS Comput Biol 10:e1003695

Classen AT, Sundqvist MK, Henning JA, Newman GS, Moore JAM, Cregger MA, Moorhead LC, Patterson CM (2015) Direct and indirect effects of climate change on soil microbial and soil microbial-plant interactions: what lies ahead? Ecosphere 6(8):art130

Compant S, Van Der Heijden MGA, Sessitsch A (2010) Climate change effects on beneficial plant–microorganism interactions. FEMS Microbiol Ecol 73(2):197–214

Compant S, Samad A, Faist H, Sessitsch A (2019) A review on the plant microbiome: ecology, functions, and emerging trends in microbial application. J Adv Res 19:29–37

Cushman JC (2001) Osmoregulation in plants: implications for agriculture. Am Zool 41:758–769

Dary M, Chamber-Pérez MA, Palomares AJ, Pajuelo E (2010) "In situ" phytostabilisation of heavy metal polluted soils using Lupinus luteus inoculated with metal resistant plant-growth promoting rhizobacteria. J Hazard Mater 177(1):323–330

Das S, Ho A, Kim PJ (2019) Editorial: Role of microbes in climate smart agriculture [Editorial]. Front Microbiol 10:2756. English

De Angelis KM, Grace P, Topçuoğlu BD, van Diepen Linda TA, Varney RM, Blanchard JL, Melillo J, Frey SD (2015) Long-term forest soil warming alters microbial communities in temperate forest soils. Front Microbiol 6:104

De Roy K, Marzorati M, Pieter VDA, Tom VDW, Boon N (2014) Synthetic microbial ecosystems: an exciting tool to understand and apply microbial communities. Environ Microbiol 16:1472–1481

De Vos WM (2013) Fame and future of faecal transplantations–developing next generation therapies with synthetic microbiomes. Microb Biotechnol 6:316–325

De Vries FT, Griffiths RI (2018) Chapter 5—Impacts of climate change on soil microbial communities and their functioning. In: Horwath WR, Kuzyakov Y (eds) Developments in soil science. Elsevier, Amsterdam, pp 111–129

De Vries FT, Liiri ME, Bjørnlund L et al (2012) Land use alters the resistance and resilience of soil food webs to drought. Nat Clim Chang 2:276–280. https://doi.org/10.1038/nclimate1368

Deepti B, Nidhi B, Deepamala M, Chandan SC, Alok K (2014) ACC deaminase containing Arthrobacter protophormiae induces NaCl stress tolerance through reduced ACC oxidase activity and ethylene production resulting in improved nodulation and mycorrhization in Pisum sativum. J Plant Physiol 171:884–894

Dopheide A, Lear G, He Z, Zhou J, Lewis GD (2015) Functional gene composition, diversity and redundancy in microbial stream biofilm communities. PLoS One 10:e0123179

Drozdowicz A, Ferreira Santos GM (1987) Nitrogenase activity in mixed cultures of Azospirillum with other bacteria. Z Mikrobiol 142:487493

Dubey A, Malla MA, Khan F, Chowdhary K, Yadav S, Kumar A, Sharma S, Khare PK, Khan ML (2019) Soil microbiome: a key player for conservation of soil health under changing climate. Biodivers Conserv 28(8):2405–2429

Dunbar J et al (2012) Common bacterial responses in six ecosystems exposed to 10 years of elevated atmospheric carbon dioxide. Environ Microbiol 14:1145–1158

Escalante-Espinosa E, Gallegos-Martínez ME, Favela-Torres E, Gutiérrez-Rojas M (2005) Improvement of the hydrocarbon phytoremediation rate by Cyperus laxus Lam. inoculated with a microbial consortium in a model system. Chemosphere 59(3):405–413

Finkel OM, Castrillo G, Herrera Paredes S, Salas González I, Dangl JL (2017) Understanding and exploiting plant beneficial microbes. Curr Opin Plant Biol 38:155–163

Frank AC, Saldierna Guzmán JP, Shay JE (2017) Transmission of bacterial endophytes. Microorganisms 5(4):70

Frey SD, Lee J, Melillo JM, Six J (2013) The temperature response of soil microbial efficiency and its feedback to climate. Nat Clim Chang 3(4):395–398

Friedman J, Higgins LM, Gore J (2017) Community structure follows simple assembly rules in microbial microcosms. Nat Ecol Evol 1:109

Frommel MI, Pazos GS, Nowak J (1991) Plant growth stimulation and biocontrol of Fusarium wilt (Fusarium oxysporum f. sp. lycopersici) by coinoculation of tomato seeds with Serratia plymuthica and Pseudomonas sp. Fitopatologia 26:6673

Fujiyama H (2019) Toward sustainable dryland agriculture-preventing salt accumulation in soil and remediating salt-affected soil. Ghodo Publisher, pp 185–195

Ghazali FM, Rahman RNZA, Salleh AB, Basri M (2004) Biodegradation of hydrocarbons in soil by microbial consortium. Int Biodeterior Biodegradation 54(1):6167

Gond SK, Torres MS, Bergen MS, Helsel Z, White JF (2015) Induction of salt tolerance and upregulation of aquaporin genes in tropical corn by rhizobacterium Pantoea agglomerans. Lett Appl Microbiol 60:392–399

Goodfellow M, Williams ST (1983) Ecology of actinomycetes. Annu Rev Microbiol 37:18

Gottfried M, Pauli H, Futschik A, Diğerleri VE (2012) Dağ bitki örtüsünün iklim değişikliğine kıta çapında tepkisi. Nat Clim Chang 2:111–115

Graham EB, Knelman JE, Schindlbacher A, Siciliano S, Breulmann M, Yannarell A et al (2016) Microbes as engines of ecosystem function: when does community structure enhance predictions of ecosystem processes? Front Microbiol 7:214. https://doi.org/10.3389/fmicb.2016.00214

Grattan SR, Grieve CM (1999) Mineral nutrient acquisition and response by plants grown in saline environments. In: Pessarakli M (ed) Handbook of plant and crop stress. Marcel Dekker, New York

Grobelak A, Kokot P, Świątek J, Jaskulak M, Rorat A (2018) Bacterial ACC deaminase activity in promoting plant growth on areas contaminated with heavy metals. J Ecol Eng 19(5):150–157

Gururani MA, Upadhyaya CP, Baskar V, Venkatesh J, Nookaraju A, Park SW (2013) Plant growth promoting rhizobacteria enhance abiotic stress tolerance in Solanum tuberosum through inducing changes in the expression of ROS scavenging enzymes and improved photosynthetic performance. J Plant Growth Regul 32:245–258

Guttman DS, Mchardy AC, SchulzeLefert P (2014) Microbial genome enabled insights into plant microorganism interactions. Nat Rev Genet 15:797–813

Hall JL (2002) Cellular mechanisms for heavy metal detoxification and tolerance. J Exp Bot 53(366):111

Han QQ, Lü XP, Bai JP, Qiao Y, Paré PW, Wang SM et al (2014) Beneficial soil bacterium *Bacillus subtilis* (GB03) augments salt tolerance of white clover. Front Plant Sci 5:525

Harcombe WR, Riehl WJ, Dukovski I, Granger BR, Betts A, Lang AH, Bonilla G, Kar A, Leiby N, Mehta P, Marx CJ, Segre D (2014) Metabolic resource allocation in individual microbes determines ecosystem interactions and spatial dynamics. Cell Rep 7:1104–1115

Hashmi I, Paul C, AlDourobi A, Sandoz F, Deschamps P, Junier T, Junier P, Bindschedler S (2019) Comparison of the plant growth promotion performance of a consortium of Bacilli inoculated as endospores or as vegetative cells. FEMS Microbiol Ecol 95(11):fiz147

Hayden HL, Mele PM, Bougoure DS, Allan CY, Norng S, Piceno YM, Brodie EL, Desantis TZ, Andersen GL, Williams AL, Hovenden MJ (2012) Changes in the microbial community structure of bacteria, archaea and fungi in response to elevated CO_2 and warming in an Australian native grassland soil. Environ Microbiol 14:3081–3096

He F, Hu W, Li Y (2004) Biodegradation mechanisms and kinetics of azo dye 4BS by a microbial consortium. Chemosphere 57(4):293301

He W, Megharaj M, Wu CY, Subashchandrabose SR, Dai CC (2020) Endophyte assisted phytoremediation: mechanisms and current application strategies for soil mixed pollutants. Crit Rev Biotechnol 40(1):3145

Heinze J, Gensch S, Weber E, Joshi J (2016) Soil temperature modifies effects of soil biota on plant growth. J Plant Ecol 10(5):808–821

Hesselsoe M, Boysen S, Iversen N, Jørgensen L, Murrell JC, McDonald I, Radajewski S, Thestrup H, Roslev P (2005) Degradation of organic pollutants by methane grown microbial consortia. Biodegradation 16(5):435448

Hirt H (2020) Healthy soils for healthy plants for healthy humans: how beneficial microbes in the soil, food and gut are interconnected and how agriculture can contribute to human health. EMBO Rep 21(8):e51069

Hodgson S, de Cates C, Hodgson J, Morley NJ, Sutton BC, Gange AC (2014) Vertical transmission of fungal endophytes is widespread in forbs. Ecol Evol 4(8):1199–1208

Hori Y, Stuhlberger C, Simonett O (2011) Desertification: a visual synthesis, United Nations Convention to Combat Desertification (UNCCD), ISBN: 978-92-95043-49-7

Hu C, Delgado JA, Zhang X, Ma L (2005) Assessment of groundwater use by wheat (Triticum aestivum L.) in the Luancheng Xian Region and potential implications for water conservation in the Northwestern North China Plain. J Soil Water Conserv 60(2):80–88

Hussain F, Hussain I, Khan AHA, Muhammad YS, Iqbal M, Soja G, Reichenauer TG, Zeshan, Yousaf S (2018) Combined application of biochar, compost, and bacterial consortia with Italian ryegrass enhanced phytoremediation of petroleum hydrocarbon contaminated soil. Environ Exp Bot 153:8088

IEA (2017) Coal 2017: analysis and forecasts to 2022. IEA, Paris

Jacoby B (1999) Mechanisms involved in salt tolerance of plants. In: Handbook of plant and crop stress, 2nd edn. Marcel Dekker, New York. ISBN: 0824719484, p 108

Jain A, Singh A, Singh S, Singh HB (2013) Microbial consortium-induced changes in oxidative stress markers in pea plants challenged with Sclerotinia sclerotiorum. J Plant Growth Regul 32(2):388–398

Janzen RA, Rood SB, Dormaar JF, McGill WB (1992) Azospirillum brasilense produces gibberellin in pure culture on chemically defined medium and in coculture on straw. Soil Biol Biochem 24:1061–1064

Jochum MD, McWilliams KL, Pierson EA, Jo YK (2019) Host mediated microbiome engineering (HMME) of drought tolerance in the wheat rhizosphere. PLoS One. https://doi.org/10.1371/journal.pone.0225933

Jones P, Garcia BJ, Furches A, Tuskan GA, Jacobson D (2019) Plant host associated mechanisms for microbial selection [Review]. Front Plant Sci 10:862. English

Kannan V, Sureendar R (2009) Synergistic effect of beneficial rhizosphere microflora in biocontrol and plant growth promotion. J Basic Microbiol 49(2):158–164

Kathrin PA, Bernd W, Rolf D, Peter M (2015) Tax4Fun: predicting functional profiles from metagenomic 16S rRNA data. Bioinformatics 31:2882–2884

Kehe J, Kulesa A, Ortiz A, Ackerman CM, Thakku SG, Sellers D, Kuehn S, Gore J, Friedman J, Blainey PC (2019) Massively parallel screening of synthetic microbial communities. Proc Natl Acad Sci U S A 116(26):12804–12809

Khanna K, Kohli SK, Ohri P, Bhardwaj R, Al-Huqail AA, Siddiqui MH, Alosaimi GS, Ahmad P (2019) Microbial fortification improved photosynthetic efficiency and secondary metabolism in Lycopersicon esculentum plants under Cd stress. Biomol Ther 9(10):581

Kim K, Jang YJ, Lee SM, Oh BT, Chae JC, Lee KJ (2014) Alleviation of salt stress by Enterobacter sp. EJ01 in tomato and Arabidopsis is accompanied by upregulation of conserved salinity responsive factors in plants. Mol Cell Biol 37:109–117

Korenblum E, Aharoni A (2019) Phytobiome metabolism: beneficial soil microbes steer crop plants' secondary metabolism. Pest Manag Sci 75(9):2378–2384

Langille MG, Zaneveld J, Caporaso JG, Mcdonald D, Knights D, Reyes JA et al (2013) Predictive functional profiling of microbial communities using 16S rRNA marker gene sequences. Nat Biotechnol 31:814–821
Langley JA, Hungate BA (2014) Plant community feedbacks and long-term ecosystem responses to multifactored global change. AoB Plants 6:12
Lau JA, Lennon JT (2012) Rapid responses of soil microorganisms improve plant fitness in novel environments. Proc Natl Acad Sci U S A 109:14058–14062
Lee EY, Lim JS, Oh KH, Lee JY, Kim SK, Lee YK, Kim K (2008) Removal of heavy metals by an enriched consortium. J Microbiol 46(1):2328
Li J, Wu C, Chen S, Lu Q, Shim H, Huang X, Jia C, Wang S (2020) Enriching indigenous microbial consortia as a promising strategy for xenobiotics' cleanup. J Clean Prod 261:121234
Lipson DA, Wilson RF, Oechel WC (2005) Effects of elevated atmospheric CO_2 on soil microbial biomass, activity, and diversity in a chaparral ecosystem. Appl Environ Microbiol 71(12):8573–8580
Luo Y (2007) Terrestrial carbon cycle feedback to climate warming. Annu Rev Ecol Evol Syst 38:683–712
Ma J, Wang S, Xue L, Liu D, Zhu X, Ma J, Huang C, Yu H, Jiang S (2017) Research of the impact of elevated CO_2 on soil microbial diversity. Energy Procedia 114:3070–3076
Mahajan S, Tuteja N (2005) Cold, salinity and drought stresses: an overview. Arch Biochem Biophys 444:139–158
Mahmood A, Kataoka R (2019) Application of endophytes through seed priming. In: Hasanuzzaman M, Fotopoulos V (eds) Priming and pretreatment of seeds and seedlings: implication in plant stress tolerance and enhancing productivity in crop plants. Singapore, Springer, pp 509–521
Mahmood A, Kataoka R (2020) Metabolite profiling reveals a complex response of plants to application of plant growth promoting endophytic bacteria. Microbiol Res 234:126421
Mahmood A, Turgay OC, Farooq M, Hayat R (2016) Seed biopriming with plant growth promoting rhizobacteria: a review. FEMS Microbiol Ecol 92(8):fiw112
Masson-Delmotte V, Zhai P, Pörtner HO, Roberts D, Skea J, Shukla PR, Pirani A, Moufouma-Okia W, Péan C, Pidcock R, Connors S, Matthews JBR, Chen Y, Zhou X, Gomis MI, Lonnoy E, Maycock T, Tignor M, Waterfield T (2018) Global warming of 1.5°C. An IPCC Special Report on the impacts of global warming of 1.5°C above pre-industrial levels and related global greenhouse gas emission pathways, in the context of strengthening the global response to the threat of climate change. World Meteorological Organization Technical Document
Mayak S, Tirosh T, Glick BR (2004) Plant growth promoting bacteria confer resistance in tomato plants to salt stress. Plant Physiol Biochem 42:565–572
Mcgrath KC, Mondav R, Sintrajaya R, Slattery B, Schmidt S, Schenk PM (2010) Development of an environmental functional gene microarray for soil microbial communities. Appl Environ Microbiol 76:7161–7170
Mcole S (2018) Fourth warmest year in continued warming trend, according to NASA, NOAA
Mesa J, Mateos-Naranjo E, Caviedes MA, Redondo-Gómez S, Pajuelo E, Rodríguez-Llorente ID (2015) Endophytic cultivable bacteria of the metal bioaccumulator Spartina maritima improve plant growth but not metal uptake in polluted marshes soils. Front Microbiol 6:1450
Migahed F, Abdelrazak A, Fawzy G (2017) Batch and continuous removal of heavy metals from industrial effluents using microbial consortia. Int J Environ Sci Technol 14(6):1169–1180
Mishra J, Singh R, Arora NK (2017) Alleviation of heavy metal stress in plants and remediation of soil by rhizosphere microorganisms [Mini Review]. Front Microbiol 8:1706
Muñoz R, Alvarez MT, Muñoz A, Terrazas E, Guieysse B, Mattiasson B (2006) Sequential removal of heavy metals ions and organic pollutants using an algal bacterial consortium. Chemosphere 63(6):903911
Murata N, Mohanty PS, Hayashi H, Papageorgiou GC (1992) Glycine betaine stabilizes the association of extrinsic proteins with the photosynthetic oxygen evolving complex. FEBS Lett 296:187–189

Nadeem SM, Zahir ZA, Naveed M, Asghar HN, Arshad M (2010) Rhizobacteria capable of producing ACC-deaminase may mitigate salt stress in wheat. Soil Sci Soc Am J 74:533–542

Navarro-Torre S, Mateos-Naranjo E, Caviedes MA, Pajuelo E, Rodríguez-Llorente ID (2016) Isolation of plant-growth-promoting and metal-resistant cultivable bacteria from Arthrocnemum macrostachyum in the Odiel marshes with potential use in phytoremediation. Mar Pollut Bull 110(1):133–142

Naz I, Bano A, Ul Hassan T (2009) Isolation of phytohormones producing plant growth promoting rhizobacteria from weeds growing in Khewra salt range, Pakistan and their implication in providing salt tolerance to Glycine max L. Afr J Biotechnol 8(21)

Niu X, Bressan RA, Hasegawa PM, Pardo JM (1995) Ion homeostasis in NaCl stress environments. Plant Physiol 109:735–742

Orozco Mosqueda MC, RochaGranados MC, Glick BR, Santoyo G (2018) Microbiome engineering to improve biocontrol and plant growth promoting mechanisms. Microbiol Res 208:2531

Paul D, Harshad L (2014) Plant growth promoting rhizobacteria to improve crop growth in saline soils: a review. Agron Sustain Dev 34:737–752

Pecl GT, Araújo MB, Bell JD, Blanchard J, Bonebrake TC, Chen I-C, Clark TD, Colwell RK, Danielsen F, Evengård B (2017) Biodiversity redistribution under climate change: impacts on ecosystems and human well-being. Science 355(6332):eaai9214

Pietikäinen J, Pettersson M, Bååth E (2005) Comparison of temperature effects on soil respiration and bacterial and fungal growth rates. FEMS Microbiol Ecol 52(1):49–58

Pitman MG, Läuchli A (2002) Global impacts of salinity and agricultural ecosystem. In: Läuchli A, Lüttge U (eds) Salinity: environment plants molecules. Kluwer Academic, The Netherlands, pp 3–20, 203–229

Purahong W, Schloter M, Pecyna MJ, Kapturska D, Däumlich V, Mital S et al (2014) Uncoupling of microbial community structure and function in decomposing litter across beech forest ecosystems in central Europe. Sci Rep 4:7014

Puschenreiter M, Horak O, Friesl W, Hartl W (2005) Low cost agricultural measures to reduce heavy metal transfer into the food chain–a review. Plant Soil Environ 51(1):111

Qin Y, Druzhinina IS, Pan X, Yuan Z (2016) Microbially mediated plant salt tolerance and microbiome based solutions for saline agriculture. Biotechnol Adv 34:1245–1259

Qin S, Miao Q, Feng W, Wang Y, Zhu X, Xing K et al (2015) Biodiversity and plant growth promoting traits of culturable endophytic actinobacteria associated with Jatropha curcas L. growing in Panxi dry-hot valley soil. Appl Soil Ecol 93:47–55. https://doi.org/10.1016/j.apsoil.2015.04.004

Raklami A, Oufdou K, Tahiri A-I, Mateos-Naranjo E, Navarro-Torre S, Rodríguez-Llorente ID, Meddich A, Redondo-Gómez S, Pajuelo E (2019) Safe cultivation of Medicago sativa in metal-polluted soils from semi-arid regions assisted by heat- and metallo-resistant PGPR. Microorganisms 7(7):212

Ren XM, Guo SJ, Tian W, Chen Y, Han H, Chen E, Li BL, Li YY, Chen ZJ (2019) Effects of plant growth promoting bacteria (PGPB) inoculation on the growth, antioxidant activity, Cu uptake, and bacterial community structure of rape (Brassica napus L.) grown in Cu contaminated agricultural soil. Front Microbiol 10(1455) English

Roane TM, Josephson KL, Pepper IL (2001) Dual bioaugmentation strategy to enhance remediation of cocontaminated soil. Appl Environ Microbiol 67(7):3208–3215

Roell GW, Zha J, Carr RR, Koffas MA, Fong SS, Tang YJ (2019) Engineering microbial consortia by division of labor. Microb Cell Factories 18(1):35

Roman-Ponce B, Reza-Vázquez DM, Gutiérrez-Paredes S, De Haro-Cruz MJ, Maldonado-Hernández J, Bahena-Osorio Y, Estrada-De Los Santos P, Wang ET, Vásquez-Murrieta MS (2017) Plant growth-promoting traits in rhizobacteria of heavy metal-resistant plants and their effects on Brassica nigra seed germination. Pedosphere 27(3):511–526

Rosier A, Bishnoi U, Lakshmanan V, Sherrier DJ, Bais HP (2016) A perspective on interkingdom signaling in plant beneficial microbe interactions. Plant Mol Biol 90:537–548

Roy SJ, Negrão S, Tester M (2014) Salt resistant crop plant. Curr Opin Biotechnol 26:115–124

Rylott EL, Bruce NC (2019) Right on target: using plants and microbes to remediate explosives. Int J Phytoremediation 21(11):1051–1064

Said BS, Or D (2017) Synthetic microbial ecology: engineering habitats for modular consortia [hypothesis and theory]. Front Microbiol 8(1125) English

Santhanam R, Luu VT, Weinhold A, Goldberg J, Oh Y, Baldwin IT (2015) Native root associated bacteria rescue a plant from a sudden wilt disease that emerged during continuous cropping. Proc Natl Acad Sci U S A 112:5013–5020

Sarma BK, Yadav SK, Singh S, Singh HB (2015) Microbial consortium mediated plant defense against phytopathogens: readdressing for enhancing efficacy. Soil Biol Biochem 87:2533

Shade A, Handelsman J (2012) Beyond the Venn diagram: the hunt for a core microbiome. Environ Microbiol 14:4–12

Shahzad A, Saddiqui S, Bano A (2016) The response of maize (Zea mays L.) plant assisted with bacterial consortium and fertilizer under oily sludge. Int J Phytoremediation 18(5):521–526

Shahzad R, Khan AL, Bilal S, Asaf S, Lee IJ (2018) What is there in seeds? Vertically transmitted endophytic resources for sustainable improvement in plant growth. Front Plant Sci 9:2424. eng

Shilev S, Babrikova I, Babrikov T (2020) Consortium of plant growth-promoting bacteria improves spinach (Spinacia oleracea L.) growth under heavy metal stress conditions. J Chem Technol Biotechnol 95(4):932–939

Silambarasan S, Logeswari P, Cornejo P, Kannan VR (2019) Role of plant growth–promoting rhizobacterial consortium in improving the Vigna radiata growth and alleviation of aluminum and drought stresses. Environ Sci Pollut Res 26(27):27647–27659

Singh A, Sarma BK, Upadhyay RS, Singh HB (2013) Compatible rhizosphere microbes mediated alleviation of biotic stress in chickpea through enhanced antioxidant and phenylpropanoid activities. Microbiol Res 168(1):33–40

Snow DD, Cassada DA, Biswas S, Malakar A, D'Alessio M, Carter LJ, Johnson RD, Sallach JB (2019) Detection, occurrence, and fate of emerging contaminants in agricultural environments. Water Environ Res 91(10):1103–1113

Souza RC, Hungria M, Cantão ME, Vasconcelos ATR, Nogueira MA, Vicente VA (2015) Metagenomic analysis reveals microbial functional redundancies and specificities in a soil under different tillage and crop-management regimes. Appl Soil Ecol 86:106–112

Sprocati AR, Alisi C, Segre L, Tasso F, Galletti M, Cremisini C (2006) Investigating heavy metal resistance, bioaccumulation and metabolic profile of a metallophile microbial consortium native to an abandoned mine. Sci Total Environ 366(2):649–658

Stanley CE, van der Heijden MGA (2017) Microbiome-on-a-chip: new frontiers in plant–microbiota research. Trends Microbiol 25(8):610–613

Sukweenadhi J, Kim YJ, Choi ES, Koh SC, Lee SW, Kim YJ et al (2015) Paenibacillus yonginensis DCY84(T) induces changes in Arabidopsis thaliana gene expression against aluminum, drought, and salt stress. Microbiol Res 172:7–15

Szabolcs I (1994) Soils and salinisation. In: Pessarakali M (ed) Handbook of plant and crop stress. Marcel Dekker, New York, pp 3–11

The Food and Agriculture Organization of the United Nations (FAO) (2002) Crops and drops: making the best use of water for agriculture. FAO, Rome, Italy. http://www.fao.org/docrep/w5146e/w5146e0a.htm

Thijs S, Van Dillewijn P, Sillen W, Truyens S, Holtappels M, D'Haen J, Carleer R, Weyens N, Ameloot M, Ramos J-L et al (2014a) Exploring the rhizospheric and endophytic bacterial communities of Acer pseudoplatanus growing on a TNT-contaminated soil: towards the development of a rhizocompetent TNT-detoxifying plant growth promoting consortium. Plant Soil 385(1):15–36

Thijs S, Weyens N, Sillen W, Gkorezis P, Carleer R, Vangronsveld J (2014b) Potential for plant growth promotion by a consortium of stress-tolerant 2,4-dinitrotoluene-degrading bacteria: isolation and characterization of a military soil. Microb Biotechnol 7(4):294–306

Tsai KN, Lin SH, Liu WC, Wang D (2015) Inferring microbial interaction network from microbiome data using RMN algorithm. BMC Syst Biol 9:54

Ullah A, Heng S, Munis MFH, Fahad S, Yang X (2015) Phytoremediation of heavy metals assisted by plant growth promoting (PGP) bacteria: a review. Environ Exp Bot 117:2840

United Nations (2017) In: Department of Economics and Social Affairs PD (ed) World population prospects: the 2017 revision, key findings and advance tables. United Nations, New York

Upadhyay SK, Singh DP (2015) Effect of salt tolerant plant growth promoting rhizobacteria on wheat plants and soil health in a saline environment. Plant Biol 17:288–293

Van Elsas JD, Kijkstra AF, Govaert JM, van Veen JA (1986) Survival of Pseudomonas fluorescens and Bacillus subtilis introduced into two soils of different texture in field microplots. FEMS Microbial Ecol 38:151–160

Vryzas Z (2018) Pesticide fate in soil sediment water environment in relation to contamination preventing actions. Curr Opin Environ Sci Health 4:59

Wagg C, Schlaeppi K, Banerjee S, Kuramae EE, van der Heijden MGA (2019) Fungal-bacterial diversity and microbiome complexity predict ecosystem functioning. Nat Commun 10:4841

Wang RC, Okamoto M, Xing XJ, Crawford NM (2003) Microarray analysis of the nitrate response in Arabidopsis roots and shoots reveals over 1,000 rapidly responding genes and new linkages to glucose, trehalose6phosphate, iron, and sulfate metabolism. Plant Physiol 132:556–567

Wang C, Wang C, Gao Y-L, Wang Y-P, Guo J-H (2016) A consortium of three plant growth-promoting Rhizobacterium strains acclimates *Lycopersicon esculentum* and confers a better tolerance to chilling stress. J Plant Growth Regul 35(1):54–64

Wang C, Gu L, Ge S, Liu X, Zhang X, Chen X (2019) Remediation potential of immobilized bacterial consortium with biochar as carrier in pyrene Cr (VI) cocontaminated soil. Environ Technol 40(18):2345–2353

Wang KD, Borrego EJ, Kenerley CM, Kolomiets MV (2020) Oxylipins other than jasmonic acid are xylem resident signals regulating systemic resistance induced by *Trichoderma virens* in maize. Plant Cell 32(1):166–185

Watts N, Adger WN, Agnolucci P, Blackstock J, Byass P, Cai W, Chaytor S, Colbourn T, Collins M, Cooper A (2015) Health and climate change: policy responses to protect public health. Lancet 386(10006):1861–1914

Wei Z, Gu Y, Friman VP, Kowalchuk GA, Xu Y, Shen Q, Jousset A (2019) Initial soil microbiome composition and functioning predetermine future plant health. Sci Adv 5(9):eaaw0759

Wiewióra B, Żurek G, Pańka D (2015) Is the vertical transmission of Neotyphodium lolii in perennial ryegrass the only possible way to the spread of endophytes? PLoS One 10(2): e0117231

Woo SL, Pepe O (2018) Microbial consortia: promising probiotics as plant biostimulants for sustainable agriculture [opinion]. Front Plant Sci 9:1801

Xia M, Chakraborty R, Terry N, Singh RP, Fu D (2020) Promotion of saltgrass growth in a saline petroleum hydrocarbons contaminated soil using a plant growth promoting bacterial consortium. Int Biodeterior Biodegradation 146:104808

Yang S, Zheng Q, Yuan M, Shi Z, Chiariello NR, Docherty KM, Dong S, Field CB, Gu Y, Gutknecht J et al (2019) Long-term elevated CO_2 shifts composition of soil microbial communities in a Californian annual grassland, reducing growth and N utilization potentials. Sci Total Environ 652:1474–1481

Ye S, Zeng G, Wu H, Zhang C, Dai J, Liang J, Yu J, Ren X, Yi H, Cheng M, Zhang C (2017) Biological technologies for the remediation of cocontaminated soil. Crit Rev Biotechnol 37(8):1062–1076

Yu Z, Duan X, Luo L, Dai S, Ding Z, Xia G (2020) How plant hormones mediate salt stress responses. Trends Plant Sci. https://doi.org/10.1016/j.tplants.2020.06.008

Yuan Z, Druzhinina IS, Labbé J, Redman R, Qin Y, Rodriguez R, Zhang C, Tuskan GA, Lin F (2016) Specialized microbiome of a halophyte and its role in helping nonhost plants to withstand salinity. Sci Rep 6:32467

Yuan Y, Brunel C, van Kleunen M, Li J, Jin Z (2019) Salinity induced changes in the rhizosphere microbiome improve salt tolerance of Hibiscus hamabo. Plant Soil 443:525–537

Zawoznik MS, Ameneiros M, Benavides MP, Vázquez S, Groppa MD (2011) Response to saline stress and aquaporin expression in Azospirillum-inoculated barley seedlings. Appl Microbiol Biotechnol 90:1389–1397

Zhang H, Kim MS, Sun Y, Dowd SE, Shi H, Paré PW (2008) Soil bacteria confer plant salt tolerance by tissue specific regulation of the sodium transporter HKT1. Mol Plant-Microbe Interact 21:737–744

Zhang L-N, Wang D-C, Hu Q, Dai X-Q, Xie Y-S, Li Q, Liu H-M, Guo J-H (2019) Consortium of plant growth-promoting Rhizobacteria strains suppresses sweet pepper disease by altering the rhizosphere microbiota. Front Microbiol 10(1668)

Zhang H, Yuan X, Xiong T, Wang H, Jiang L (2020) Bioremediation of cocontaminated soil with heavy metals and pesticides: influence factors, mechanisms and evaluation methods. Chem Eng J 398:125657

Zhou J, He Z, Yang Y, Deng Y, Tringe SG, Alvarez-Cohen L (2015) High-throughput metagenomic technologies for complex microbial community analysis: open and closed formats. MBio 6(1):e02288–e02214

Zolla G, Badri DV, Bakker MG, Manter DK, Vivanco JM (2013) Soil microbiomes vary in their ability to confer drought tolerance to Arabidopsis. Appl Soil Ecol 68:1–9

Zuluaga MYA, Lima Milani KM, Azeredo Gonçalves LS, Martinez de Oliveira AL (2020) Diversity and plant growth promoting functions of diazotrophic/N scavenging bacteria isolated from the soils and rhizospheres of two species of Solanum. PLoS One 15(1):e0227422

Drought Tolerance Mechanisms in Crop Plants

Arash Hemati, Morteza Mofidi-Chelan, Arian Amirifar, Ebrahim Moghiseh, and Behnam Asgari Lajayer

Abstract The morphologic traits of plants that prevent drought are part of what is known as a xeromorphic structure or organization. Some of them save water, and others use water. Thus, only a few of these traits are likely to be found in water savers, while most of them are found among water consumers. These characteristics are more common in sun-exposed leaves than in the ones that are exposed to shade and in the lower ones. Plants respond to environmental stimuli making differences in behavior, morphology, anatomy, physiology, phenology, and reproductive organs. Related to stress resistance, adaptations that are of vital importance include aspects that prevent the destruction of vital vegetative tissues as well as the extensive production and proliferation of reproductive organs. There are two mechanisms for stress avoidance in plants: Tension avoidance and stress tolerance are types of stress resistance in which plants reach thermodynamic equilibrium without being damaged by stress. Plants are equipped with a "stress tolerance" method (interacting with stress) and are able to prevent, reduce, or repair stress-induced strain. Two characteristics that enable the plant to escape drought and produce remarkable performance are (1) phenologic development speed and (2) formability in terms of developmental stages.

Keywords Stress · Plant · Drought · Water

A. Hemati · B. Asgari Lajayer (✉)
Department of Soil Science, Faculty of Agriculture, University of Tabriz, Tabriz, Iran
e-mail: hasgari@tabrizu.ac.ir

M. Mofidi-Chelan
Department of Range and Watershed Management, Faculty of Natural Resources, Urmia University, Urmia, Iran

A. Amirifar
Department of Biology, Urmia Branch, Islamic Azad University, Urmia, Iran

E. Moghiseh
Nuclear Agriculture Research School, Nuclear Science and Technology Research Institute, Karaj, Iran

© The Author(s), under exclusive license to Springer Nature Singapore Pte Ltd. 2022
A. Vaishnav et al. (eds.), *Plant Stress Mitigators*,
https://doi.org/10.1007/978-981-16-7759-5_19

Introduction

The dynamic state of water in plants is mainly controlled by opposite processes such as evapotranspiration and water uptake (Heshmat et al. 2020). Whenever the intensity of evapotranspiration is more than the intensity of water uptake for a long time, the volume of the water (turgor pressure) decreases. This leads to decrease in cell turgor and further negation of water potential in cells, and reduction and hydration of protoplasm and cell walls. Eventually, it leads to growth suppression (De Swaef and Steppe 2010).

The cause of water stress in the plant is an increase in water loss or insufficient water uptake or a combination of both. Water deficiency in plants is made up of two parts: short-term stress and long-term stress. In short-term stress, water deficiency is caused and triggered by lag between evapotranspiration and water uptake (Moustafa-Farag et al. 2020). The reason for the delay in absorption in the middle of the day is related to the plant's resistance to water movement, rate of absorption, and transpiration that is being controlled by various factors. The amount of transpiration is determined by the structure and surface of the shoot (especially surface of the leaf), size of the stomata, and other various factors that affect the vapor pressure gradient between plant and the air. Lack of coordination between absorption and evapotranspiration makes one superior to another. If evapotranspiration is more than absorption, reduction in water content and loss of cell turgor of leaf cells are the first sign of transpiration that leads to wilting. In general, superiority of transpiration over absorption, especially on sunny days when there is a temporary shortage of water in the middle of the day, even when the soil is generally moist, leads to water shortage and temporary wilting. In such situation, if the water is present in soil and the atmospheric tension is reduced, the wilting is reversible and the growth becomes possible (Wilson and Greaves 1993).

In long-term stress, the amount of available water in the soil is responsible for water stress in the plant. If transpiration continues and soil water content keeps decreasing, access to the water content gets limited (Mao et al. 2020). Due to the gradual decrease in soil water potential over several days, its hydraulic conductivity is also rapidly reduced and more water is needed to move sufficient water from the soil to roots and to compensate for evapotranspiration at a slope of greater potential between the roots and the soil (Hayat et al. 2020).

In such case, the movement of water to the roots is so slow that it is not possible for plant to get to the normal state; hence, the plant gets wielded and damaged all the time (Wilson and Greaves 1993).

The morphological traits and characteristics of plant that prevent drought are part that is known as xeromorphic structure. Some of them save water, and others cause water consumption. Therefore, only a few of these traits are likely to be found in water savers. While most are found among water consumers (they may change from consumer to reservoir in critical situations), these traits may be genetic or may vary depending on the environment, so the plant is to likely develop these properties to a higher level under moderate drought conditions rather than high humidity conditions. These characteristics are usually more common in sun-exposed leaves than leaves that are grown in shade and upper leaves rather than the leaves that are in lower altitude. In *Zea mays*, *Sorghum bicolor* and *Nicotiana tabacum* in low LWP keep their stomata opener in the upper leaves in comparison with the lower ones (De Micco and Aronne 2012).

Lower leaves are more mesomorph than the upper ones because (a) they are closer to the water reservoir and (b) they are developed throughout the life of the plant when the water supply is present at its best. Because cell development depends on the tensile force of the turgor pressure, a smaller force results in less tensile strength, so the upper leaves when becoming mature are supposed to have traits like (1) cells and intercellular space are smaller than the lower leaves, (2) having the stomata and more conducted tissue, and (3) having thicker and waxier cell wall (Ludlow 1989; Aronson et al. 1992).

Under severe drought conditions, leaf growth may be very slow. In this case, excess photosynthetic production is mainly transferred to the roots, resulting in increased root growth and higher root-to-shoot ratio. Physiological differences are usually accompanied by morphological differences. Several morpho-physiological traits such as plant height, water content, sugar accumulation, concentration of cell sap, the rate of evapotranspiration, etc. not only affect stress tolerance to limited soil moisture but also reveal how adaptive genotypes cope with drought via morpho-physiological changes. These morpho-physiological traits of plant that grow under moderate drought conditions are all known as Xerophilia (Peña-Valdivia et al. 2010).

Water or Drought Stress

From a general point of view, according to the dictionary definition and context, the word drought means continuous dry climate. Drought is a meteorological term and generally refers to period without significant rainfall (Kumar et al. 1994). Due to the plant habitat in both soil and air, air dryness is due to low humidity and is often accompanied by hot and dry winds.

Air dryness occurs even in conditions where soil moisture is relatively high (Nayyar and Gupta 2006). Soil dryness occurs when the amount of moisture added to the soil is less than its losses by evapotranspiration. Drought can be divided into two general categories depending on available water sources—(1) meteorological drought displays the lack of water at the surface and groundwater resources; (2) agricultural drought indicates the situation that amount of rainfall and soil moisture is insufficient to meet the water needs of the plant to grow and reach the optimum state (Kumar et al. 1994; Nayyar and Gupta 2006).

Water stress may be due to a lack or excess of water in the plant habitat. The first word is water depletion or drought, which is called "water scarcity stress" (abbreviated form: water stress) or drought stress. The terms water stress and drought stress are often used interchangeably. The term drought stress has the advantage over water stress that is only related to water scarcity and has never been more pervasive (Levitt 1980).

Such condition is commonly referred to as "dehydration stress" when the plant is artificially induced to evapotranspiration and water loss. Levitt (1980) states that there is a mixture and inability to use and distinguish between cause and effect (stress and strain) because the plant is dehydrated in the strain (the result of stress), and thus, dehydration has the meaning of withering; it should only be used for deficiencies that push the plant to the edge of withering.

Hsiao (2000) suggests the term dehydration only when the plant loses more than half of its water content. In parallel with this definition, it can be said that whenever the amount of plant's water content is more than 50% but still inadequate, the plant suffers from dehydration resulting from excessive evaporation. The stress that can cause water loss in the liquid state is called osmotic stress (Black and ong 2000).

Drought Stress and Competition of Organs for Water Absorption

The growth stage of different parts of the plant is the main factor controlling the distribution of water in the plant. Therefore, meristems can compete favorably, because the synthesis of cellular materials such as proteins helps to create matrix (and osmotic potential), which leads to a reduction in water potential and a slope toward meristematic tissue. Also, some parts of the plants with high amount of soluble material such as photosynthetic leaves are able to compete optimally for water uptake (De Swaef and Steppe 2010). Older leaves provide the need for nutrients in young leaves and fruits, and if the plant is exposed to severe water stress, the old leaves will die first. Drought stress is very harmful to fruit and seed during the filling period (Moustafa-Farag et al. 2020).

In sugar beet, under drought stress, the supply of carbohydrates from the leaves to the roots is disrupted, which leads to reduced root growth in the plant. Inflammation reduction and soil's mechanical strength are also important in reducing the growth of sugar beetroots. Under these conditions, the growth of the shoot is more affected than the roots (Harvery and Dutton 1993).

Adaptation to Environmental Stresses

Plants respond to environmental stimuli, making differences in behavior, morphology, anatomy, physiology (phenology), and reproductive organs. Such changes that make it possible to withstand the pressures and problems of the environment are called adaptation. Adaptations that have vital importance include the inhibitory aspects of the destruction of vital vegetative tissues as well as the extensive production and proliferation of reproductive organs. Sometimes, depending on the various habitats, some temporary changes in plant characteristics, such as wilting and leaf curl, may occur during dry winds, which do not have genetic aspects. Plants produce other vegetative propagations in addition to seeds, tubers, or forms. All of these adaptations are genetic (Bohnert et al. 1995; Cortes and Sinclair 1986).

Plants can withstand stress survive, despite damage or parts of the plant (seeds, buds, dormant cells) may be stressed, while other parts of the plant, such as meristems and fleshy organs, may be more prone to damage caused by stress (Cortes and Sinclair 1986).

It has been reported that stress resistance is related to the plants' ability to tolerate adverse factors (Ghassemi et al. 2018, 2020; Khoshmanzar et al. 2019; Saghafi et al. 2019a, b). Cortes and Sinclair (1986) also defined stress tolerance as follows: Stress tolerance is the capacity of a plant to survive and grow, even if affected by an unsuitable environment and conditions.

In the case of stress tolerance, despite the fact that the plant is exposed to stress, it is done differently by reducing or removing the strain. Stress tolerance is a type of stress resistance in which plants reach thermodynamic equilibrium without being damaged by stress. The plant is equipped with the "stress tolerance" resistance method (interacting with stress) and is able to prevent, reduce, or repair stress–strain. The plant may use one or both methods against a stress. In the process of evolution, the selection has been toward stress avoidance mechanisms, which are more effective than tolerance mechanisms in order to resist stress (De Swaef et al. 2010). Ephemerals are drought-tolerant plants and, at least theoretically, do not require greater resistance to mesophytes. These plants complete their life cycle in a dry environment without water stress and survive drought as seeds in the dry season.

Drought Escape in Crop Plants

Two characteristics that make it possible for the plant to escape drought and produce remarkable performance are (Levitt 1980): (1) phenologic development speed and (2) formability in terms of developmental stages.

In cases where the probability of drought increases during the life cycle, the shorter the growth period, the higher the yield in comparison with the most of the years.

Too much early maturity can also limit yield. In addition, a crop with limited growth has a relationship between yield and ripening time, so the plants that have a shorter life span usually have lower yields.

In conclusion, plants that have a rapid phenologic period may be more prone to malfunction in rainy years. Formation in terms of developmental stages eliminates this defect. Most crops have some ductility if they are in favorable conditions. The formability of the developmental stages means the ability to flower and form seeds, in the presence of water and the absence of other constraints. In wheat, lack of drought compared to water shortage prolongs the pollination period and reaches physiological maturity. Under these conditions, the number of tillers produced by the cluster increases and the grain size increases. Early maturity has brought many benefits to the breed in terms of drought resistance, and because the plants of semi-arid and temperate regions are exposed to accidental drought, it is better to transfer some characteristics of drought resistance to them (Moustafa-Farag et al. 2020; Mao et al. 2020).

Drought Resistance in Crop Plants

Plants that cannot escape drought adapt to these conditions in the following ways: (1) drought avoidance and (2) drought tolerance.

Avoiding Drought

By definition, a drought-tolerant plant must maintain its water potential when faced with drought stress. According to another definition, "the ability of a plant to maintain a relatively high leaf water potential (close to zero) when exposed to dry air or soil conditions" (Levitt 1980).

There are two distinct types of land avoidance agents: (1) "water savers" who avoid drought by conserving water, and (2) "water consumers" by adequate and rapid absorption of water, and at the same time water consumption. They avoid drought by performing activities listed above. In both cases, the adaptation takes place with the continuation of inflammation and high water potential (De Micco et al. 2012; Kumar et al. 1994).

Water Conservation by Reservoirs

Water reservoirs are able to limit evapotranspiration for a long time before wilting occurs, so water storage in plants has two main effects: (1) It causes the plant to continue to develop in spite of the conditions that cause it to wilt and stop growing in non-water-storing plants; (2) if the water shortage intensifies, the probability of survival of the water-storing plant is higher, and this action is done with the elimination of growth. Fleshy plants are examples of water savers. Drought resistance in plants with moderate drought tolerance or even a number of high resistance mesophytic plants is often mainly due to water storage. Storage and conservation of water by stomata is affected by closure mechanisms, existence of cuticle barrier, and reduction in evapotranspiration level (complexity, folding, shading, and falling of leaves). Also, some plants have developed numerous adaptations to reduce water loss such as thick cuticles, small leaf size, leaf hairs, sunken stomata, root adaptations, water storage in different organs, crassulacean acid metabolism (CAM), and c4 photosynthetic pathways (Bohnert et al. 1995; Cortes and Sinclair 1986).

Rapid Water Uptake by Water Consumers

Compatibility through water storage causes plant survival to drought, but with the exception of some CAM plants, growth is significantly reduced. This decrease is due to the correlation between transpiration and photosynthesis, in other words the correlation between stomatal openness and photosynthesis.

Drought-resistant seedlings of *Eucalyptus viminalis* do not avoid drought but are capable of relatively high levels of evapotranspiration. This ability to maintain high evapotranspiration may be associated with the ability to continue photosynthesis and growth with low water potential. Water consumers are able to maintain a high water content compared to those who are less adaptable. Turgor pressure under drought conditions and continuation of growth need to discharge huge amount of waters. It is clear that the role of water consumers could change by evapotranspiration, rain, and also water uptake (De Swaef and Steppe 2010; Hsiao 2000).

Water flow resistance varies according to the different species. The main resistance is in the roots, but there are differences in current resistance in the stems and leaves. Decreased resistance to water flow can be achieved by increasing the diameter and number of xylems. In some xerophytes, the veins are closer to each other than in mesophytes, and the length of the vein network is longer per unit area of the leaf surface. The vascular system of an inbred line of maize had a high resistance to water flow compared to the drought-tolerant variety (Sharp et al. 1994).

Continuous water uptake requires the presence of deep and branched root and low resistance to water flow inside the plant. Undoubtedly, lack of water increases the ratio of root to shoot of the plant and increases root depth. Also, the root system of a plant with lower osmotic potential of water may continue to transfer and uptake water from the soil with low osmotic potential, which is not possible for other plants (Hsiao 2000).

Drought Tolerance (Low Water Condition of the Plant)

True drought tolerance is achieved when plants reduce their water potential due to drought. The practical definition of water tolerance is "the ability of a plant to continue to function at a certain level of plant water potential" (Levitt 1980; Khoshru et al. 2020). Maintaining cell turgor is very important in this case because many biochemical and physiological processes are sensitive to it. Keeping cell turgor in leaf water deficiency is completely or partially achieved by adjusting the osmotic state, increasing the elasticity, or reducing the cell size (Barlow et al. 1977).

Osmotic regulation of cell is achieved in two ways: (1) reduction in cell water due to the dryness of the environment and (2) accumulation of excess solutions due to the activation of mechanisms that increase the concentration of particles in the cell sap solution. In this case, despite the avoidance of wilting and sufficient turgor pressure for growth, a drop in the free energy of water is inevitable. Also, drought tolerance mechanisms can maintain tissue turgor because of the conditions that lead to a decrease in water potential and an increase in tissue elasticity. Small cells have more elasticity than large cells. Lack of water during leaf growth usually reduces cell size, thus increasing elasticity and maintaining turgor. In general, plants use the following mechanisms to withstand drought stress: (1) avoid injuries caused by limited growth, (2) tolerance against hunger, and (3) resistance to protein breakdown (De Micco et al. 2012; Kumar et al. 1994).

Conclusion

The cause of water stress in the plant is an increase in water loss or insufficient water absorption or a combination of both. Water shortage in the plant is divided into short-term stress and long-term stress. Plants can withstand stress and survive despite induced damage, or parts of the plant (seeds, buds, and dormant cells) may be stressed, while other parts of the plant, such as meristems and fleshy organs, may be resistant to stress. Plants choose two approaches to stress, drought escape or drought resistance. Two characteristics that enable the plant to escape drought and produce remarkable performance are (1) phenologic development speed and (2) formability in terms of developmental stages. Plants that cannot escape drought adapt to these conditions in the following two ways: (1) drought avoidance and (2) drought tolerance. The plant may use one or both methods against a stress. In the process of evolution, the selection has been toward stress avoidance mechanisms that are more effective than stress tolerance mechanisms.

References

Aronson J, Kigel J, Shmida A, Klein J (1992) Adaptive phenology of desert and Mediterranean populations of annual plants grown with and without water stress. Oecologia 89:17–26

Barlow E, Boersma L, Young L (1977) Photosynthesis, transpiration and leaf elongation in corn seedling at suboptimal soil temperatures. Agron J 69(1):95–100

Black C, Ong C (2000) Utilization of light and water in tropical agriculture. Agric For Meteorol 104:25–47

Bohnert HJ, Nelson DE, Jensen RG (1995) Adaptations to environmental stresses. Plant Cell 7(7): 1099–1111

Corte PM, Sinclair TR (1986) Water relations of field-grown soybean under drought. Crop Sci 26(5):993–998

De Micco V, Aronne G (2012) Morpho-anatomical traits for plant adaptation to drought. In: Aroca R (ed) Plant responses to drought stress. Springer, Berlin. https://doi.org/10.1007/978-3-642-32653-0_2

De Swaef T, Steppe K (2010) Linking stem diameter variations to sap flow, turgor and water potential in tomato. Funct Plant Biol 37:429

Ghassemi S, Farhangi-Abriz S, Faegi-Analou R, Ghorbanpour M, Asgari Lajayer B (2018) Monitoring cell energy, physiological functions and grain yield in field-grown mung bean exposed to exogenously applied polyamines under drought stress. J Soil Sci Plant Nutr 18(4): 1108–1125

Ghassemi S, Delangiz N, Asgari Lajayer B, Saghafi D, Magi F (2020) Review and future prospects on the mechanisms related to cold stress resistance and tolerance in medicinal plants. Acta Ecol Sin. https://doi.org/10.1016/j.chnaes.2020.09.006

Harvery CW, Dutton JV (1993) Root quality and processing. In: Cooke DA, Scott RK (eds) The sugar beet crops. Champan and Hall, London, pp 517–617

Hayat F, Ahmed MA, Zarebanadkouki M, Javaux M, Cai G, Carminati A (2020) Transpiration reduction in maize (Zea mays L) in response to soil drying. Front Plant Sci 10:1695. https://doi.org/10.3389/fpls.2019.01695

Heshmat K, Asgari Lajayer B, Shakiba MR, Astatkie T (2020) Assessment of physiological traits of Common bean cultivars in response to water stress and Molybdenum levels. J Plant Nutr. https://doi.org/10.1080/01904167.2020.1822395

Hsiao TC (2000) Leaf and root growth in relation to water status. HortScience 35:1051–1058

Khoshmanzar E, Aliasgharzad N, Neyshabouri MR, Khoshru B, Arzanlou M, Asgari Lajayer B (2019) Effects of Trichoderma isolates on tomato growth and inducing its tolerance to water deficit stress. Int J Environ Sci Technol 17:869–878

Khoshru B, Moharramnejad S, Hosseinzadeh Gharajeh N, Asgari Lajayer B, Ghorbanpour M (2020) Plant microbiome and its important in stressful agriculture. In: Varma A, Tripathi S, Prasad R (eds) Plant microbiome paradigm. Springer Nature, Switzerland. https://doi.org/10.1007/978-3-030-50395-6_2

Kumar A, Singh DP, Singh P (1994) Influence of water stress on photosynthesis, transpiration, water-use efficiency and yield of *Brassica juncea* L. Field Crop Res 37(2):95–101

Levitt J (1980) Response of plant environmental stress, vol 2. Academic, New York, pp 30–42

Ludlow MM (1989) Strategies of response to water stress. In: Kreeb KH, Richter H, Hinckley TM (eds) Structural and functional responses to environmental stresses: water shortage. SPB Academic Publishing, The Hague, pp 269–281

Mao W, Zhu Y, Wu J, Ye M, Yang J (2020) Modelling the salt accumulation and leaching processes in arid agricultural areas with a new mass balance model. J Hydrol 591:125329

Moustafa-Farag M, Mahmoud A, Arnao MB, Sheteiwy MS, Dafea M, Soltan M, Elkelish A, Hasanuzzaman M, Ai S (2020) Melatonin-induced water stress tolerance in plants: recent advances. Antioxidants 9:809

Nayyar H, Gupta D (2006) Differential sensitivity of C_3 and C_4 plants to water deficit stress: association with oxidative stress and antioxidants. Environ Exp Bot 58(1–3):106–113

Peña-Valdivia CB, Sánchez-Urdaneta AB, Rangel J, Muñoz J, García-Nava R, Velázquez R (2010) Anatomical root variations in response to water deficit: wild and domesticated common bean (*Phaseolus vulgaris* L.). Biol Res 43:417–427

Saghafi D, Delangiz N, Asgari Lajayer B, Ghorbanpour M (2019a) An overview on improvement of crop productivity in saline soils by halotolerant and halophilic PGPRs. 3Biotech 9:261. https://doi.org/10.1007/s13205-019-1799-0

Saghafi D, Ghorbanpour M, Shirafkan Ajirloo H, Asgari Lajayer B (2019b) Enhancement of growth and salt tolerance of canola (*Brassica napus* L.) seedlings by halotolerant Rhizobium strains containing ACC-deaminase activity. Plant Physiol Rep 24:225–235

Sharp R, Wu Y, Voetberg G, Saab I, Lenobel M (1994) Confirmation that ABA accumulation is required for maize primary root elongation at low water potentials. J Exp Bot 45:1743–1751

Wilson JM, Greaves JA (1993) Development of and water stress in crop plants. In: Adaptation of food crops to temperature and water stress. AVRDC, Shanhua, Taiwan, pp 389–398

Microbe-Mediated Amelioration of Salinity Stress in Crops

Monika, A. Kumar, N. Kumar, Mamta, N. Yadav, S. Kumar, and S.S. Arya

Abstract Salinity is the key constraint that affects the crop growth, metabolism, and yield because of high abundance of salts present in the soil. The area and agriculture crop affected by salinity stress are increasing day by day. Salinity stress disturbs many physiological, biochemical, and molecular parameters in crop plants. Therefore, there is urgent need of promising candidate, which helps to mitigate salinity stress, favors plant growth, and also has environment-friendly impact. The characterization and exploitation of soil microbes (especially mycorrhizal fungi, PGPR, endophytes such as *Piriformospora indica* and cyanobacteria) in agriculture open new alternatives to overcome salinity stress. Amelioration of salinity in plant occurs through plant growth-promoting rhizobacteria (PGPB) by applying different strategies such as they introduce synthesis of antioxidative enzyme to cope with reactive oxygen species, which is generated during salt stress in plants, and stimulates accumulation of osmolyte in plants, and plants inoculated with PGPB have high K^+/Na^+ ratio that favors salinity tolerance. Besides these, PGPB produces various hormones (auxin, cytokinins, and gibberellins) to mitigate salt stress. Arbuscular mycorrhizal fungi (AMF) equipped with fascinating mechanisms are useful in mitigating the adverse effect of salinity stress. AMF inoculation in plant increases the mineral nutrient uptake of K, Fe, Ca, Mg, Mn, and Zn, reduces the uptake of Na^+, and accumulation of proline and phenol increases, which also reduced the effect of salinity in plants. Several AMF species produced various antioxidative enzymes such as catalase (CAT), superoxide dismutase (SOD), glutathione reductase (GR), and peroxidase (POD) that help to minimize the effect of ROS produced at the time of salinity stress. An endophyte, *P. indica*, colonizes with a broad range of plant species. *P. indica* root colonization helps in amelioration of salt stress by manipulating hormone signaling pathways and enhanced root cell division by production of

Monika
Department of Microbiology, CCS Haryana Agricultural University, Hisar, Haryana, India

A. Kumar · N. Kumar · Mamta · N. Yadav · S.S. Arya (✉)
Department of Botany, Maharshi Dayanand University, Rohtak, Haryana, India

S. Kumar
Department of Environmental Science, Maharshi Dayanand University, Rohtak, Haryana, India

© The Author(s), under exclusive license to Springer Nature Singapore Pte Ltd. 2022
A. Vaishnav et al. (eds.), *Plant Stress Mitigators*,
https://doi.org/10.1007/978-981-16-7759-5_20

IAA hormone, which results in better absorption of nutrient and plant growth. Cyanobacteria can survive and live under extreme salinity and further utilized to increase the soil fertility.

Keywords Arbuscular mycorrhizal fungi (AMF) · Cyanobacteria · Plant growth-promoting rhizobacteria (PGPB) · Salinity stress · Plant growth-promoting rhizobacteria (PGPR)

Introduction

Salinization of soil is major constraint that constitutes severe loss and impairs crop production worldwide. Soil salinity is the major widespread environmental stress that decreased the plant growth and yield. Excess salinization in soil degrades the fertility of soil and declines the quality of water, which in turn leads to marginal lands. The quality of land and water is crucially for agriculture; for an instance of over-exploded population, there is urgency to maintain food security for enhancing crop production.

Salinity in agricultural land mentions the manifestation of large amount of salts (which are soluble) in the soil. Salinization is the process in environment that occurs with the addition of water-miscible salts in soil that contains sodium (Na^+), magnesium (Mg^{2+}), calcium (Ca^{2+}), potassium (K^+), chlorides, carbonate, and bicarbonate. Depending upon the soil, extracted solution differs in their electrical conductivity, and if electrical conductivity is more than 20 mM, then the soil can be categorized as saline soil. The salinization is mainly primary and secondary in nature. Salinization process is natural primarily, and secondary salinization process is carried out by anthropogenic activity. Soil originates from mineral weathering. Salinity is mainly due to the dissolved salts and deposition of them in the soil. The anthropogenic or secondary factors include crop irrigation with highly salted water, and other factors may include excessive use of inorganic fertilizers and soil amendments with different fertilizers such as gypsum composts and manures. Accumulation of excessive salt like Na^+ destroys soil structure, raises pH of soil, reduces aeration and filtration in soil, deteriorates soil physical and chemical properties, and increases water runoff and soil erosion. Moreover, Na^+ is mainly known for disruptive secondary clay minerals by dispersal in soil. Dispersion in soil occurs through replacement of sodium ions with other coagulators such as calcium and magnesium, which are absorbed on the surface or inner layer of soil. The high salt concentration due to ionic imbalance affects the growth of plants by poor water uptake and mineral nutrient by the roots. Salinity is one of the serious factors, which affect agriculture crops with inauspicious effect on germination, plant vigor, and crop yield. In majority of sanity-affected areas, the yield decreases at very low salinity percentage in soil. Some of the crops are very adversely affected by salt concentration such as rice, and 30% of worldwide rice fields are affected due to high salinity. Salinity in soil is reducing the agricultural area and adds 1–2% saline area every year worldwide basis, and problem is very critical in dry areas.

Emphasis on production of salinity stress-resistant crop has been the main objective with much less success. Successful remediation of salt-affected areas needs well-defined, easy, and inexpensive and environment-friendly approaches. Recently, there is great emphasis on use of sustainable agriculture through the use of beneficial microorganisms. An alternative strategy is use of salt-tolerant microbes to reduce effect of salinity. Rhizospheric soil microbes improved the growth of agricultural crops grown in salt-affected areas. Microorganism that lives in close association with crops grown in saline soil has unique properties to grow in salt-affected areas. Microbes are equipped with different mechanisms, which reduce the harmful effect of salinity on crops and improve the crop production. Salt-tolerant microbes improve plant and water relationship in soil through intricate mechanisms. Soil microbes have a unique property; they manipulate signaling of phytohormones and trigger other mechanisms, which collectively enhance stress tolerance in crops. The microbes isolated from arid regions and oceans were found to be helpful in amelioration of salinity stress and enhancement of agriculture output in salt-affected areas. Voluminous literature available reveals that the inoculation of salt-tolerant microbes in salt-affected soil reduces the effect of salinity stress and improves the crop production such as *Pseudomonas fluorescens* strain that was isolated from the rhizosphere of date palm, which enhanced root growth in corn seedling under high salt concentration (Zerrouk et al. 2016). Similarly, wheat inoculated with salt-tolerating microbe *Serratia* sp., isolated from a halophilic environment, improves salt tolerance and shoot biomass in *Triticum aestivum* (Fig. 1).

Plant Growth-Promoting Rhizobacteria (PGPR)

The PGPR help in mitigation of salt stress by regulates water potential and stomatal opening through their effect on hydraulic conductivity (properties of vascular plant that describes with which ease fluid moves through pore spaces) and transpiration process. Marulanda et al. (2010) reported improved hydraulic activity in maize plants inoculated with plant growth-promoting rhizobacteria as compared to uninoculated ones. The improved hydraulic conductivity was because of increased expression of plasma membrane aquaporin proteins. PGPR makes osmolyte accumulation and phytohormonal signaling that prevents plant from osmotic shock due to salinization. Chen et al. (2007) observed that osmolyte proline accumulation was increased in *Arabidopsis thaliana* on inoculation of *Bacillus subtilis*, which confirm plant salt tolerance. The expression of 14 genes was altered in *Oryza sativa* plant on inoculation of *Bacillus amyloliquefaciens* PGPR (Nautiyal et al. 2013). Under hydroponic state, four genes [SOS1, EREBP, NADP-malic enzyme, and SERK1] were upregulated, whereas two genes SNF1 (serine threonine protein kinase SAPK4) and GIG (glucose insensitive growth) were downregulated. In greenhouse condition, only one gene MAPK5 was upregulated rhizospheric microorganisms that stimulate carbohydrate metabolism and its transport, which improves photosynthesis

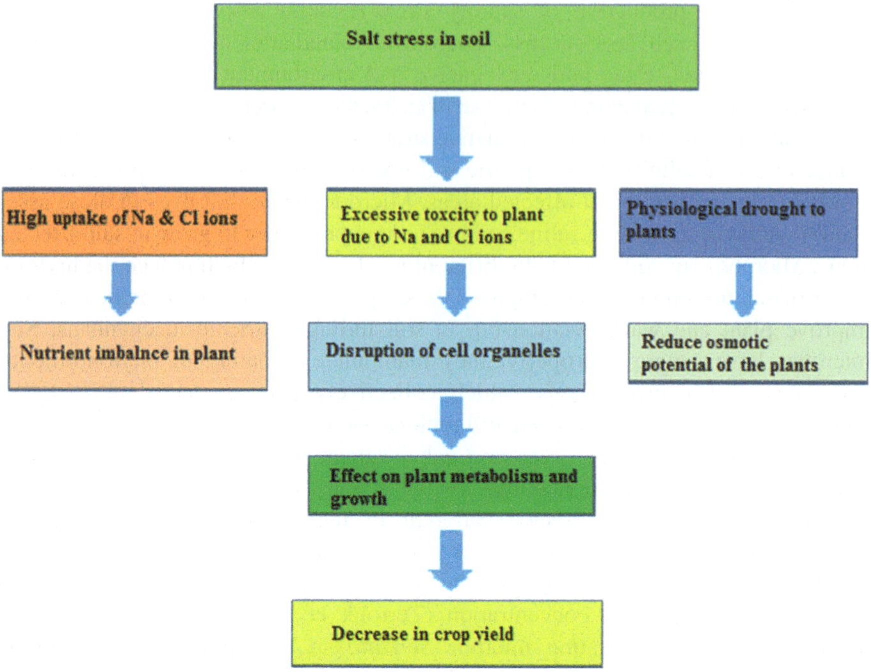

Fig. 1 Diagrammatic representation of salinity stress impacts on physiology and biochemistry of plant. Salt stress leads to drought, osmotic imbalance, and disruption of plant metabolism and cell organelles. This ultimately affects the plant growth and yield production

and growth rate. Upadhyay and Singh (2015) observed that plant inoculated with *B. aquimaris* strains has more soluble and reducing sugar in *Triticum aestivum* during salt stress in field conditions, which result in higher plant biomass and high NPK uptake (Upadhyay and Singh 2015). Similarly, on inoculation of *Azospirillum brasilense* and *Pantoea dispersa* rhizobacterial isolates in plants of *Capsicum annuum* showed high plant biomass under high salinity stress is an example of increased photosynthesis and stomatal conductance (del Amor and Cuadra-Crespo 2012). Microbes exposed to high salinity or pH fluctuations accumulate larger quantity of osmolyte in their cytosol. Biosynthesis of osmoprotectant like trehalose, proline, and glycine is more in PGPR as compared to associated host plant. These osmolytes were absorbed through root and prevent cellular oxidative damage and helpful in maintain osmotic balance during high salt stress.

A microarray analysis of bean nodules showed overexpression of trehalose-5-phosphate gene on coinoculation *Paenibacillus polymyxa* and *Rhizobium tropici* isolates. Overexpression of trehalose-6-phosphate gene acts as an osmoprotectant by improved nodulation efficiency and photosynthesis in plants. PGPR is capable of maintaining ionic balance in plant cells and by keeping low level of Na^+ and Cl^- and high level of K^+/Na^+ in shoot. Maize plants inoculated with Azotobacter strains improved the uptake of K^+ inside the plant tissue with the exclusion of Na^+ outside

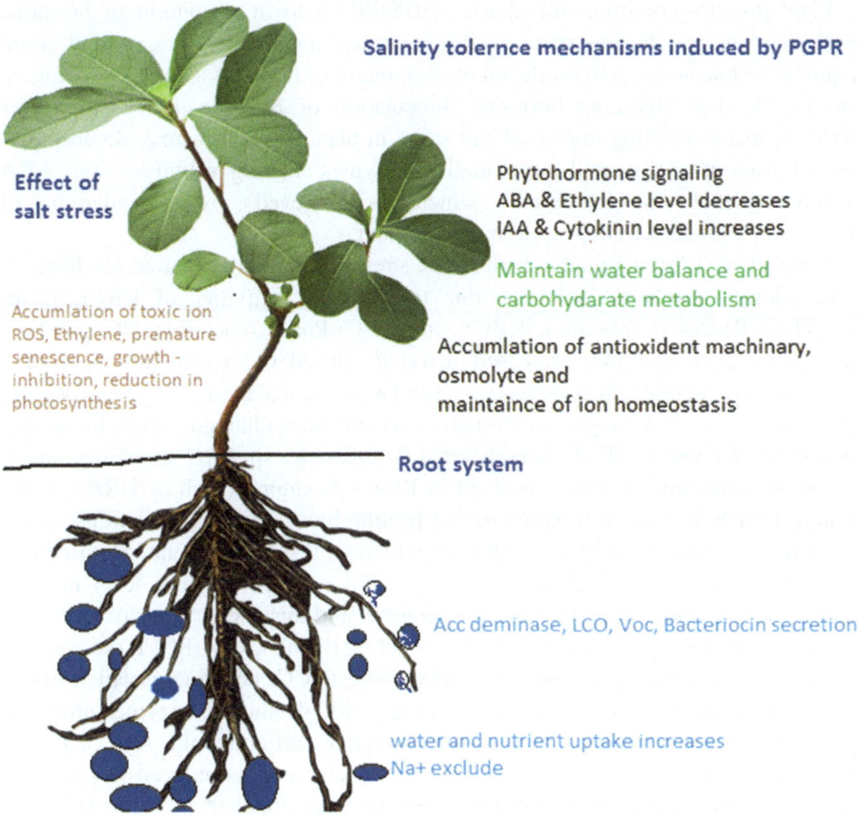

Fig. 2 Demonstration of salinity stress resistance mechanisms offered by plant growth-promoting rhizobacteria in inoculated plant. Rhizospheric microbes act as protective barrier against salt stress by enhanced nutrient acquisition and water absorption, stomatal conductance, stimulating osmolyte accumulation, and carbohydrate metabolism

the plant tissue during salt stress. Chlorophyll content is also increased on inoculation of PGPR. Osmolytes such as proline and polyphenol contents are also improved in leaves on inoculation of soil bacteria Azotobacter during salinity (Rojas-Tapias et al. 2012). Similarly, inoculation of PGPR *Burkholderia phytofirmans* in *Arabidopsis thaliana* plant is helpful in regulation effect of salinity and also induces tolerance to salt in plant. Pinedo et al. (2015) demonstrate that the expression pattern of genes involved in ion balance mechanism was modified after salinity stress and PsJN gene that governs salt tolerance also undergoes some molecular changes on inoculation of PGPR. Niu et al. (2016) reported that PGPR inoculation during salinity stress is helpful in mitigation of salinity by the regulation of genes involved in salt stress mechanism. On inoculation of PGPR, *Bacillus subtilis* in a halophyte grass *Puccinellia tenuiflora* manifest low sodium accumulation under high stress due to upregulation of PtHKT1 and PtSOS1 genes (Fig. 2).

Plant growth-promoting rhizobacteria (PGPR) help in regulation of hormone level in plant by altering secretion of hormones and metabolites as and when required. Indole acetic acid produced by soil microbes in plants is the most common and well-studied signaling hormone. Inoculation of *Bacillus amyloliquefaciens* SQR9 in maize seedling improved salt stress in plant through increased concentration of osmoprotectant and antioxidative enzymes. During salinity stress, ABA hormone level is increased, which is relieved by inoculation of *B. amyloliquefaciens* SQR9 strain in maize plant.

Chen et al. (2016) found out that salt stress mitigation in maize seedling on inoculation of bacterial isolate is due to increased activities of RBCL genes (encoding RuBisCo subunits), RBCS, and H(C)-Ppase (encoding HC pumping pyrophosphatase and the decreased activities of NCED expression (encoding 9-cisepoxycarotenoid dioxygenase). Similarly, inoculation of *Enterobacter* sp. *EJ01* strain in Arabidopsis and tomato plant enhanced plant growth by mitigating salty stress. Kim et al. (2014) demonstrated *Enterobacter* sp. *EJ01* strain inoculation caused upregulation of genes involved in ROS scavenging (such as DRE-binding protein DREB2b) and overexpression of proline biosynthesis gene and ascorbate peroxidase in salt-stressed plants. During salt stress, level of ethylene increased that help in regulation over high salt-stressed condition. High ethylene level in plant inhibits transcription of auxin biosynthesis gene, and this results in reduced growth and development of plant. Plant growth-promoting rhizobacteria (PGPR) secrete one enzyme, i.e., 1-aminocyclopropane-1-carboxylase (ACC) deaminase, which mover powers ethylene biosynthesis process in plant. ACC deaminase enzymes produced by PGPR, which convert ACC into α-ketobutyrate and ammonia. ACC acts as a precursor for ethylene biosynthesis. Hence, ethylene level is reduced in plant and plant now able to resume its normal growth and development process. There are many reports, which conform that ACC deaminase enzymes regulate salt stress in inoculated plant by modulating ethylene level (Glick et al. 2007). Nadeem et al. (2009) found that inoculation of *Pseudomonas fluorescens* and *Enterobacter* spp. in salt-affected maize seedling helps in regulation of stress by maintaining low ethylene level and higher K^+/Na^+ ratios. Maize seedling inoculated with PGPR produced higher yield due to increased NPK uptake. Similarly, okra (*Abelmoschus esculentus*) plants inoculated with UPMR18 strain of *Enterobacter* sp. increased the stress tolerance and are able to grow under 75 mM concentration of NaCl in pot house condition. It was supposedly due to the increased expression of antioxidant enzymes and upregulation of transcription of stress regulating gene (Habib et al. 2016).

There are comparatively few studies on the role of abscisic acid (ABA) hormone on plant–microbe interaction during salinity stress. But there are some reports, which indicated that the PGPR alter the signaling pathways and biosynthesis of ABA, which is helpful in improving growth of salt-stressed crop. Bharti et al. (2016) demonstrated that *Dietzia natronolimnaea* STR1, which is well known as halotolerant, introduces salinity stress mechanism in wheat crop by the modulation of ABA signaling cascade and it was done by upregulation of aABARE (ABA-responsive gene) and TaOPR1 (12-oxophytodienoate reductase 1) leading to TaWRKY and TaMYB stimulation, followed by expression of stress response genes including upregulation of TaST (a salt stress-induced gene). It also modulates the tissue-specific ion transporters and SOS pathway-related gene (Table 1). Yao et al. (2010) reported that seed inoculation of cotton (*G. hirsutum*) and *Pseudomonas putida* reduced ABA content and increased plant biomass. Similarly, Kang et al. (2014) recognized that role of salicylic acid and gibberellin in amelioration of salinity stress and it was proved that plants inoculated with PGPR show downregulation of ABA which increased the gibberellin (GA_4) and SA (salicylic acid) contents.

Arbuscular Mycorrhizal Fungi

Plants are inhibited various types of microorganisms on internal and external surfaces. Arbuscular mycorrhizal fungi (AMF) are most frequently colonized microbes among rhizospheric microorganisms. AMF are considered as most integral component of ecosystem and have appreciable role in mitigation of salt stress in plants. AMF inoculation in saline soil mitigates negative impact of salinity on crops by increasing total yield and enhances the activity of antioxidant enzymes including catalase, ascorbate peroxidase, superoxide dismutase, and glutathione reductase may result in lower level of lipid peroxidation and electrolyte leakage. Accumulation of osmolytes such as proline and phenol in AMF inoculated plant also causes significant reduction in harmful radicals indirectly responsible for cell damage due to high stress. Moreover, the level of important elements like salicylic acid and jasmonic acid is increased, which is again helpful in elimination of deleterious ions. Plant growth and biomass are adversely affected during salt stress. This may be due to the unavailability of major nutrients and expenditure of energy to cope with the toxic effect of different salt. However, colonization of mycorrhiza was found to mitigate the harmful effect of salt stress on plant by increasing plant growth, biomass, and better crop yield as compared to noninoculated plant. Giri et al. (2007) concluded that colonization of mycorrhizal in *Acacia nilotica* improves growth of plant during salt stress. Mycorrhizal-inoculated plant had higher root and shoot dry weight as compared to nonmycorrhizal plant. Similarly, Al-Karaki (2000) obtained more plant biomass and total yield in mycorrhizal-inoculated tomato plant as compared to nonmycorrhizal tomato plant. *Cucurbita pepo* plants colonized by *Glomus intraradices* have better plant yield, nutrient content, and quality of fruits even during high salt condition. AM-inoculated plant survives and grows better during high salt stress condition than non-AM plant.

Arbuscular mycorrhizal fungi (AMF) are beneficial symbiotic organism forms associated with many land plants. This symbiotic association improves nutrient availability in soil and enhances water absorption capacity of land plant, therefore enhancing plant growth in normal and harsh condition (Gamalero et al. 2010; Alqarawi et al. 2014a, b). AMF colonization in saline environment improves plant growth and vigor, and changes induced by AMF at nutritional, morphological, and

Table 1 Role of different microbes in amelioration of salinity stress in plant

Microbe	Plant	Contribution of microbes in mitigation of salinity stress in plant	References
Bacillus safensis, *Ochrobactrum pseudogregnonense* *Enterobacter cloacae*, *Pseudomonas putida*, *Serratia ficaria*, and *P. fluorescens*	*Triticum aestivum*	Production of volatile organic compounds, improvement in germination %, rate, index, nutrient status of the wheat plants	Chakraborty et al. (2013), Nadeem et al. (2012)
Alcaligenes faecalis, *Bacillus pumilus*, *Ochrobactrum* sp. *P. pseudoalcaligenes*, *B. pumilus*	*Oryza sativa*	Accumulation of osmolytes such as proline, glycine betaine	Bal et al. (2013), Jha et al. (2013)
Azospirillum sp.	(*Triticum aestivum*)	Stimulation of persistent exudation of flavonoids	Zarea et al. (2012)
Streptomyces sp.	*Triticum aestivum*	Improve nutrient availability	Sadeghi et al. (2012)
Pseudomonas sp., *Bacillus* sp., *Variovorax* sp.	*Persea gratissima*	Increase in K^+ level and decrease in Na^+ level	Nadeem et al. (2012)
Azotobacter chroococcum	*Z. mays*	High stomatal conductance and Photosynthesis	Rojas-Tapias et al. (2012)
P. pseudoalcaligenes, *P. putida*	*Cicer arietinum*	Increase in activity of ACC deaminase enzyme and enhancement in growth and nodulation of chick pea	Patel et al. (2012)
Brachybacterium saurashtrense, *Brevibacterium casei*, *Haererohalobacter* sp.	*Arachis hypogaea*	Higher Ca^{2+}, phosphorus, and nitrogen content, more K^+/Na^+ ratio	Shukla et al. (2012)
P. extremorientalis, *P. chlororaphis*	*Phaseolus vulgaris*	Upregulation of genes involved in stress Tolerance	Egamberdieva (2011)
Glomus intraradices	*Glycine max*	Accumulation of carbohydrates	Porcel and Ruiz-Lozano (2004)
Bacillus, *Burkholderia*, *Enterobacter*, *Microbacterium*, *Paenibacillus*	*T. aestivum*	Regulation of sodium transporter HKT1	Upadhyay et al. (2011)
P. fluorescens, *P. aeruginosa*, *P. stutzeri*	*Lycopersicon esculentum*	Reduction in Na^+ level and increase in K^+ level	Tank and Saraf (2010)
Pseudomonas sp.	*Solanum melongena*	Reduction in Na^+ level and increase in K^+ level	Fu et al. (2010)
Azospirillum sp.	*Triticum durum*	Stimulation of persistent exudation of flavonoids	Nabti et al. (2010a, b)
Glomus clarum, *Glomus etunicatum*	*Vigna radiata*, *Capsicum annuum*, *Triticum aestivum*	Decreased Na^+ in root and shoot and increased concentration of K^+ in root	Kaya et al. (2009)

(continued)

Table 1 (continued)

Microbe	Plant	Contribution of microbes in mitigation of salinity stress in plant	References
Piriformospora indica	*Oryza sativa*	High potassium (K^+)/ sodium (Na^+) ratio Improved salinity stress tolerance mechanism	Zhang et al. (2014)
Piriformospora indica	*Hordeum vulgare*	High K^+/Na^+ and Ca^{2+}/Na^+ ratios Improved salinity stress tolerance mechanism	Alikhani et al. (2013)
Bacillus amyloliquefaciens SN13	*Oryza sativa*	Upregulation of SOS1, EREBP, SERK1, NADP-Me2	Nautiyal et al. (2013)
Bacillus amyloliquefaciens SQR9	*Zea mays*	Upregulation of RBCS, RBCL, HKT1, NHX1, NHX2, and NHX3	Chen et al. (2016)
Bacillus megaterium	*Zea mays*	Improved expression of two ZmPIP isoforms	Marulanda et al. (2010)
Dietzia natronolimnaea	*Triticum aestivum*	Downregulation of ABA, upregulation of SOS gene, regulation over ion transporter system	Bharti et al. (2016)
Cyanobacteria and cyanobacterial extracts	*Oryza sativa, Triticum aestivum, Zea mays, Gossypium hirsutum*	Phytohormones as elicitor molecule	Singh et al. (2014)
Pseudomonas simiae AU	*Glycine max*	Decrease in root NaCl accumulation and increase in proline and chlorophyll content Upregulation of RuBisCo protein	Vaishnav et al. (2015)
Enterobacter sp. *UPMR18 (ACC deaminase)*	*Abelmoschus esculentus*	Increase antioxidant enzyme activities and upregulation of ROS pathway genes	Habib et al. (2016)
Enterobacter sp.	Rice	Reduction in ethylene production, increased activity of antioxidant enzymes	Sarkar et al. (2018)
Klebsiella sp.	Oat	Alteration in level of hormone More root and shoot growth	Sapre et al. (2018)
Rhizophagus irregularis *Chryseobacterium humi ECP37* *Ochrobactrum haemophilum ZR3-5*	Sunflower	High antioxidant response Less Na^+ content More biomass and nutritional content	Pereira et al. (2016)

(continued)

Table 1 (continued)

Microbe	Plant	Contribution of microbes in mitigation of salinity stress in plant	References
Hartmannibacter diazotrophicus	Barley	Increased ACC deaminase enzyme production More root and shoot growth	Suarez et al. (2015)
Thalassobacillus denorans (NCCP-58) *Oceanobacillus kapialis (NCCP-76)*	Rice	Increased germination ability root and shoot growth High protein, chlorophyll, and nutrient contents low Na^{+} level	Shah et al. (2017)
Bacillus subtilis	Wheat	Increased SA content Low proline and malondialdehyde content to induce systemic resistance	Lastochkina et al. (2017)
Halomonas maura *Ensifer meliloti*	Alfalfa	Increased root and shoot weight High water content Better yield	Martinez et al. (2015)
Enterobacter ludwigii	*Festuca arundinacea*	Regulation of transport protein More nitrogen fixation and yield	Kapoor et al. (2017)
Enterobacter cloacae HSNJ4	Canola	Modulation in hormone level Less ethylene and malondialdehyde More root shoot growth and yield	Li et al. (2017)
Arthrobacter scleromae SYE-3	Lettuce Radish Chinese cabbage	Significantly improves root, shoot, and plant length	Hong and Lee (2017)
Bacillus spp. *Alcaligenes* spp. *Proteus* spp. *Aneurinibacillus aneurinilyticus*	Chilli	Improves plant growth parameters	Patel et al. (2017)

physiological level improve resistance during salt stress condition (Tang et al. 2009; Hashem et al. 2015; Yang et al. 2015; Mo et al. 2016). For example, Hajiboland et al. (2010) concluded that high salt stress condition decreases the plant biomass of tomato plant at maturity, but the effect of salinity stress was not observed in mycorrhizal-inoculated plant. Shekoofeh et al. (2012) concluded that mycorrhizal inoculation of *Ocimum basilicum* protects against the salt stress in soil and improved

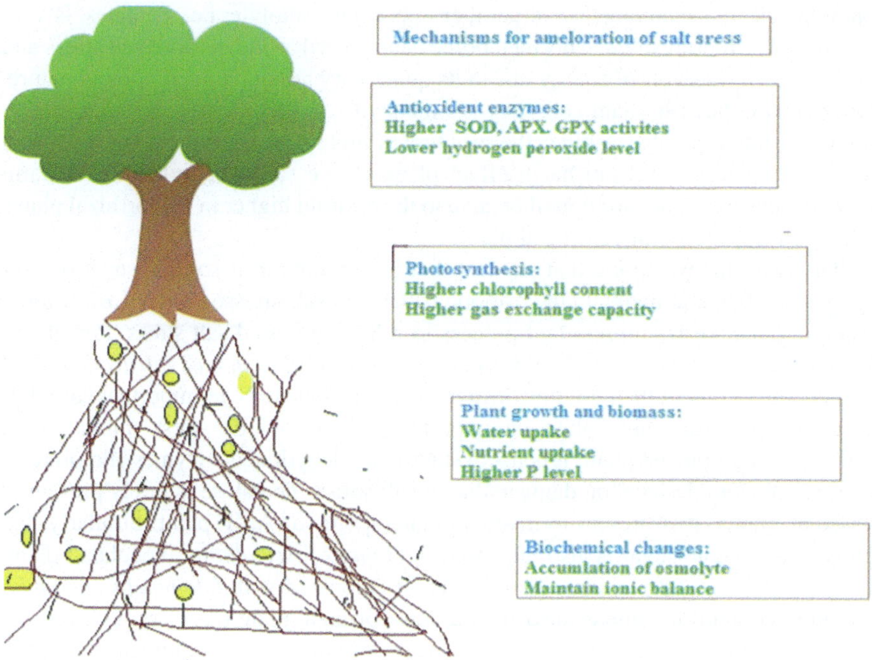

Fig. 3 Demonstration of salt tolerance mechanism opted by arbuscular mycorrhizal fungi (AMF) during inoculation in plant such as increase in water and nutrient uptake, especially P level, osmolyte accumulation, and production ROS scavenging gene

uptake of nutrients in plant, enhanced chlorophyll content, and boost up water-holding capacity. AMF-inoculated plant showed increase in calcium (Ca), phosphorus (P), nitrogen (N), and potassium (K) contents to improve plant biomass as noninoculated plant (Balliu et al. 2015). The mycorrhizal association is well known for better mineral water absorption, particularly phosphorus and potassium. The improvement in plant biomass during salinity stress is due to better phosphorus acquisition in AMF-inoculated plant compared to control.

It is well documented that AMF plays an important role in amelioration of salinity stress in crops. AMF uses different mechanisms to counteract the high salinity, i.e., accumulation of proline content, soluble sugars, glycine betaine, production of ROS scavenging enzymes, alteration in hormones level, increase in photosynthesis rate, and chlorophyll content, which measure the importance of AMF in deals with the salt stress deleterious effects (Fig. 3).

On AMF colonization, the osmolytes increased during salinity stress. Sharifi et al. (2007) observed that the proline accumulation is increased in AMF-inoculated plant during salinity stress. Under salinity stress, many plants accumulate that proline is nontoxic and protective osmolyte, which keeps ionic balance and water potential. Proline also serves as a substitute of energy for utilization under salt stress. Several authors claim that high proline concentration in AMF plant as compared to non-AM plant (Table 1). Accumulation of another important osmolyte, i.e., betaines, is very common during salt stress. This metabolite can be used as indicator of salt stress and also acts as a reservoir of energy due to its quaternary ammonium compound nature. Once accumulated in plant cell, they are metabolized very slowly. It is not merely nontoxic, but it can also stabilize the structure of protein and enzyme complexes, and also save membranes from harsh effect of excessive salt accumulation. Al-Garni (2006) measures concentration of betaine to the twofold higher in mycorrhizal plants than compared to nonmycorrhizal plant.

There are many reports that suggest that accumulation of soluble sugar acts as osmoprotectant and maintain the osmotic balance in salt-stressed plant. Lowering in osmotic potential is an important process to relieve imposed salt stress. Porcel and Ruiz-Lozano (2004) reported colonized soybean roots by *Glomus intraradices* had higher sugar content than nonmycorrhizal plant. Trehalose is an important carbohydrate in mycorrhiza that plays an important role in abiotic stress mitigation in mycorrhizal colonized plant. It helps in stabilizing dehydrated enzymes and protects membrane from desiccation damage due to salt stress. Trehalose is rarely present in vascular plants; it is present in mycelium and spores of AMF plant. Similarly, the content of soluble sugar is also increased in *Glomus fasciculatum*-inoculated plant. During salt stress, *Phragmites australis* plant colonized with *Glomus fasciculatum* had high content of soluble sugar than nonmycorrhizal plant (Al-Garni 2006).

Plant hormones play crucial role in plant growth and development and have remarking effect on quality and yield. The level of hormones is altered during salt stress. Level of ABA and ethylene is increased due to stress imposed by different salts on plant. AMF colonization in salinity-affected areas improves the plant growth and development. AMF colonization regulates hormone level in salt-affected plant, and for example, ABA plays an important role during salinity stress by checking plant development. It has been documented that mycorrhizal colonization in plant can alter ABA level in host plant. Jahromi et al. (2008) stated that the level of ABA is lower in lettuce plant as compared to mycorrhizal colonized plant, and therefore, mycorrhizal plants are less affected by salt stress. It was also mentioned that the effects of AMF inoculation on ABA content vary upon plant inoculated with AM. It is also reported that the level of stress hormone ethylene also lowers AMF-colonized plant as compared to non-AM plant.

Free polyamines such as spermidine (Spd), putrescine (put), and spermine (Spm) have major contribution in mitigation salinity stress by maintaining osmotic potential during stress. Free polyamines are small organic cations useful in growth and development of plants. Putrescine (Put) synthesized directly from ornithine decarboxylase pathway or indirectly by arginine decarboxylation pathway from its precursor ornithine. Putrescine (Put) also synthesized other two free polyamines spermine (Spm) and spermidine (Spd) through the successive addition of aminopropyl groups. These three cations play important role in plant responses to environmental stress such as high salinity stress, high osmolarity, and antioxidative stress.

These three free polyamines regulate root development in plant during saline condition. However, during saline situation the level of free polyamines is reduced.

To cope up with devasting effect of salinity of plant, it is necessary to revert back the concentration of free polyamines. This work is done by arbuscular mycorrhizal fungi on inoculation in saline conditions. Inoculation of AMF in host plant during saline environment increases concentration of free polyamines. These free polyamines maintain osmotic balance inside plant. Sannazzaro et al. (2006) speculated that the level of free polyamines higher in *Glomus intraradices* colonized Lotus plant. High salt concentration inside the plant tissue increases the activities of antioxidant enzymes like ascorbate peroxidase (APX), catalase (CAT), superoxide dismutase (SOD), and peroxidase (POD). AMF inoculation ameliorated the negative effect of salinity on growth parameter by enhancement of activity of antioxidant enzymes that favor accumulation of osmotic components, which in turn help in maintaining tissue water content. Different studies stated that there is remarkable increase in antioxidant enzyme activities in mycorrhizal-inoculated plant. Mycorrhizal-inoculated soybean plant has higher activities of antioxidant enzymes, i.e., SOD, APOX, and peroxidase than in nonmycorrhizal plant. However, catalase and polyphenol peroxidase enzyme activities remain unchanged in both the plants (Ghorbanli et al. 2004). It means the antioxidant activity depends on type of AMF species involved in mycorrhization. Similarly, Alguacil et al. (2003) also confirmed enhanced antioxidant enzyme activities in *Retama sphaerocarpa* and *Olea europaea*. The increased activity of superoxide dismutase activities detoxifies superoxide ion into hydrogen peroxide, which in turn is detoxified by CAT and peroxidase enzymes. It was also observed that the hydrogen peroxide is also detoxified by the activity of APOX enzymes. The upregulated glutathione reductase increased the availability of NADP, which in turn extracts electrons from oxygen to lowers down production of O_2. The content of micronutrient on iron, copper, manganese, and zinc in mycorrhizal plant was improved due to the total SOD activity. These studies explain that the mycorrhizal inoculation in plants activates many antioxidant enzymes that may be due to improved growth and absorption of macronutrients like nitrogen, phosphorus, and potassium PK in plants.

Salinity stress sometimes favors physiological drought in which Na^+ and Cl^- ions bind with H_2O that is essential to be mobilized in plant. Studies suggested the mycorrhization colonization can help plant in such situation. Many scientists investigated that AMF-inoculated plants had higher water content as compared to non-AM plant (Colla et al. 2008). The improved hydraulic root conductance is accompanied by altered root system and has longer and healthier root of AMF-colonized plant. All these parameters such as higher soluble sugar, higher stomatal conductance, and lower osmotic potential are due to mycorrhizal colonization validate colonized plants to use water efficiently and more coherently keeping low level of carbon dioxide inside the cells. Therefore, gas exchange capacity increased in AMF colonized plant.

Salinity stress has adverse effect on nitrogen fixation process, and nodule formed by bacteria is easy target for destruction. This may be due to premature nodule senescence caused by increased level of lytic activation, inefficient nodule formation, and loss of nitrogen fixation in saline environment. Inoculation of AMF balanced the harmful effects of salinity on nitrogen fixation and nodulation process.

Giri and Mukerji (2004) point out a positive effect of AMF on legume–rhizobium symbiosis and describe positive impact of this tripartite interaction on nodule formation process. Colonization of AMF in legume plants increased number of nodule in plant. Moreover, leghemoglobin content was also improved upon AMF colonization.

Cyanobacteria

The abundance of cyanobacteria is affected by pH of soil, and they are abundant in alkaline environment, while they are generally absent in acidic environments. There is close relationship between cyanobacteria (blue-green algae) density and soil reaction. Cyanobacteria *Spirulina* and *Arthrospira* very rich in protein can tolerate pH range of 8.0–11.0. However, cyanobacteria can tolerate high Na^+ ions and can grow on the soil surface. Cyanobacteria can tolerate high sodicity by accumulating various organic and inorganic metabolites, which act as a osmoregulator in high alkaline soil. The soil regions that have presence of cyanobacterial species from a long period had more fertile soil and making soil particles more arable by decreasing pH, exchangeable Na^+, increased content nitrogen, phosphorus, organic matter, and water-holding capacity of soil. All these improved osmolarities in soil by lowering the level of sodium ion, which is indicator of salinity stress in plant. It also improves the hydraulic conductivity of alkaline soil.

Cyanobacteria are able to survive in saline soil and able to tolerate harsh conditions like osmotic imbalance and low nutrient level imposed due to salinity in salt-affected soils. Colonization of cyanobacteria in soil is tremendously useful for plant to develop resistant against salt stress and grow better during high salt stress conditions. Different cyanobacteria secrete a variety of bioactive metabolites, which induce systematic resistance in plant and useful to fight during stress. Rodriguez et al. (2006) demonstrated that colonization of *Scytonema hofmanni* in rice seedling had positive effect on growth of plant. The product released by *Scytonema hofmanni* extracellularly alters the effect of high NaCl concentration on rice seedling. Besides this, extracelluar product of *Scytonema hofmanni* also induced some biochemical changes in rice seedling. This extracelluar product was found very similar to gibberellic acid because it improves shoot growth during salt stress, which is a characteristic function of gibberellic acid in plant. During salt stress, the level of abscisic acid rises and level of gibberellic acid becomes low. Abscisic acid is very important to induce physiological and biochemical changes in plant. It causes inhibition in growth during stress, which results in stunted growth of plant. Therefore, cyanobacterial colonization is helpful in reducing the level of abscisic acid and maintaining the level of gibberellic acid, IAA, and cytokine, so can be applied in the field to induce growth in salt-affected crops like rice (Table 1).

Arora et al. (2010) described that cyanobacteria can survive in salt-affected regions by increasing the production of exopolysaccharides to prevent harmful effects of salinity on plasma membrane. It also offers resistance to salinity by utilizing excess nitrogen into ammonia for nitrogen fixation process, which is further utilized by plants.

Sergeeva et al. (2002) purposed that cyanobacteria are able to produce phytohormone IAA, which is also supported by Ahmed et al. (2010). Salt stress in cyanobacteria is relieved by accumulating organic solutes by maintaining inorganic ion-level contribution of ion transport processes and different metabolic adjustments. Different scientists carried out research on stress-responsive proteins, and it was found that an inoculation of cyanobacteria of the expression of stress-responsive protein is increased (Malam et al. 2007).

Cyanobacteria adopted a number of strategies to mitigate salt stress. These strategies are very useful for amelioration of salt stress in plant. Nowadays, biotechnologist utilizes different strategies of cyanobacterium as a powerful tool to develop salt stress-resistant crops. For an instance, the expression of cyanobacterial flavodoxin is used to generate multiresistant crop, which is also resistant to salt stress (de la Pena et al. 2010). Tamoi et al. (2007) characterized a gene (Ssglc) from *Synechocystis* codes for α-β-1-4-glucanase-like protein. The amino acid sequence of Ssglc gene showed similarity with the amino acid sequence of glycoside hydrolase gene, which is important to cope up with high salt stress condition. The gene Ssglc found from *Synechocystis* found deals with salt stress and is used to cope with salt stress in different plants. Similarly, Gaber and El-Assal (2012) were able to clone slr1562 gene in *E. coli* characterized by *Synechocystis* cyanobacterium to generate multiresistant cell. The gene significantly enhanced the resistance of *E. coli* cells from various stresses like drought, oxidative, and salt stresses.

Application of cyanobacteria results in enrichment of soil by fixing nitrogen and improves soil structure and electrical conductivity. Inoculation of cyanobacteria improved the oxidation–reduction potential of soil and reduced electrical conductivity and enhanced nitrogenase and dehydrogenase activities (Swarnalakshmi et al. 2007). Blue-green algae decrease Na^+ in soil and increase the availability Ca^+. Moreover, polysaccharide excreted by cyanobacteria species improves soil structure by improving soil-binding capacity. Soil property improves with enhanced hydraulic conductivity. Nutrient-holding capacity of saline soil is also improved by the application of cyanobacteria in the form of organic fertilizers. Availability of major plant nutrient phosphorous and sulfur is also increased in application of cyanobacteria as organic fertilizer in saline soil. Hence, blue-green algae are not only grown in saline condition, but it also influences the physicochemical properties of salt-affected soil by enriching them with major micronutrients such as carbon, nitrogen, and phosphorus.

Piriformospora indica

P. indica is a root-colonizing and growth-promoting endophytic fungus found in Indian Thar Desert. *P. indica* is one of the potent root symbionts of plant and helps in growth promotion of various agriculture and medicinal plant. Among various

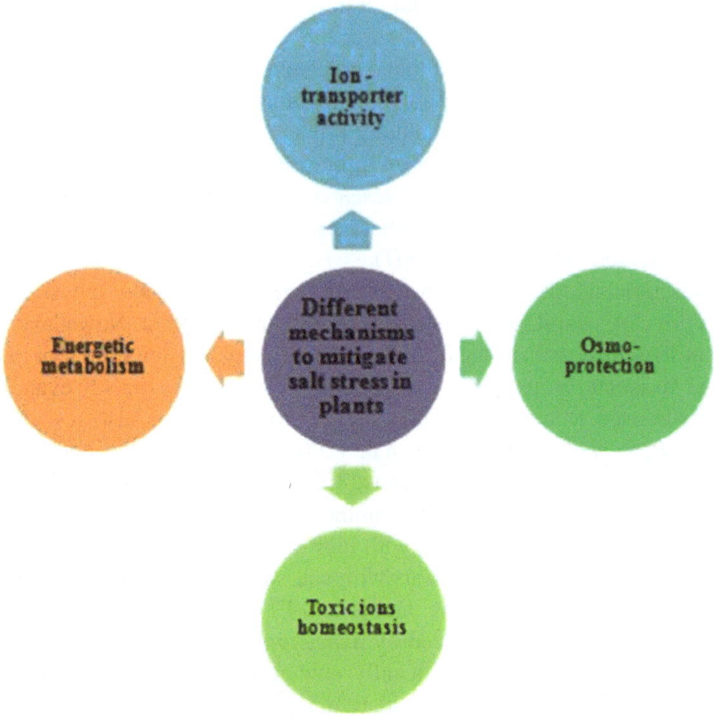

Fig. 4 Strategies employed by endophytic fungus *P. indica* in amelioration of salinity stress. It maintains osmotic concentration by exchange of ions, produced different ROS scavenging genes, and secretes some osmoprotectant like proline and betaine

endophytes, *P. indica* are able to maintain mutualistic association with most of the agricultural crops. It extends its symbiotic relationship with different bryophytes, pteridophytes, angiosperms, and gymnosperms plants. *P. indica* was found to be effective in mitigation of various biotic and abiotic stresses. Salinity stress is one of the major stresses mitigated on inoculation of *P. indica* in plant. *P. indica* use different strategies to cope with salt stress. It mitigates many harmful effect of salt stress by the production if various growth hormones maintain osmotic concentration by exchange of ions and produced different ROS scavenging genes and osmolyte metabolites (Fig. 4). Gill et al. (2016) revealed that *P. indica* inoculation in salt-stressed plant improves the plant resistance to salt stress by improving cellular osmolarity and turgor. Inoculation of *P. indica* also upregulated the production of many stress-responsive gene and proline synthesizing gene, i.e., P5C5 (Abo-Doma et al. 2016). Endophytic fungus produced proline, which helps in stabilization of membranous structure through osmoregulation. It can also be used as water reservoir for different metabolic activities. It also induced salt stress tolerance by reducing oxidative species produced vigorously during stressed conditions and causing damage to cellular structures. So, proline prevents cellular structure damage, and it can also use as energy substitute during nutrient scarcity. Sadique et al. (2018) observed that *P. indica* inoculation in rice fields improved the growth of rice seedling and mitigated the diminishing effect of osmotic stress. The endosymbiotic interaction between seedlings of rice and *P. indica* improves plant biomass and uptake of nutrients significantly. Further, it was also found that endophytic fungus inoculation in rice seedling helps in maintaining osmotic balance through increased production of proline. It also improves the chlorophyll number and antioxidant capacity of leaves. The interaction of endophytic fungus with rice plants also upregulated the activity of stress gene, i.e., pyrroline-5-carboxylate synthase (P5CS), which is essential for the production of proline osmolyte for maintaining ionic balance. Similarly, Tariq et al. (2017) emphasized that inoculation of *P. indica* increases the content of phosphorus through the improved nutrient uptake, which in turn reduced malondialdehyde level and enhanced the production of osmolytes. Ngwene et al. (2016) measure the content of zinc that is increasing on inoculation of *P. indica* in rice crop and it was also found that the influence of salinity stress is lower in rice upon inoculation. It is because zinc is part of antioxidant complexes (Cu/Zn-SOD) and it plays major role in scavenging of ROS generated during salinity stress in plant leaves. This prevents the damage to cell membrane which may impose due to generation of ROS species.

Conclusion

There is urgent need of ecologically and environment-friendly approaches for mitigation of salinity stress in crops. Application of beneficial microorganisms (salt-tolerant) is a major substitute to improved soil fertility of salt-affected regions, and nowadays, it is widely practiced by farmers. Microbes live in association or present in rhizosphere or phyllosphere or inside plant root offer enhanced plant growth and stress alleviation by different mechanisms. Commercialization of bioinoculants to increase crop productivity and yield in saline ecosystem is an implicit layout for saline soil agriculture. Studies from past give wide exposure to our understanding of multicomplex phenomena of establishing plant–microbe interaction in mitigation of salt stress in plants; still, there are many directions that need to be uncovered. Research on these issues in upcoming time will develop better understanding of the mechanisms and pathways used by microbes in salinity stress. Our knowledge about molecular aspects of salt stress mitigation process in plant is limited to few genes. The role of stress gene associated with each microbe should be addressed. Further identification of gene responsible for production of antioxidants, enzymes, and various osmoregulators provides in-depth molecular knowledge of mechanisms involved during salt stress resistance. The system biology deals with plant–microbe interactions that provide more comprehensive knowledge about regulatory mechanisms that altered by rhizosphere bacteria in crops during salt stress. Moreover, GC-LCMS techniques will more explore the ultrastructure details of bioactive metabolites secreted by cyanobacteria in salt-affected crops. The ultrastructure details of AMF colonized plants remain unexplored till now. The role of AMF in salt stress amelioration is commendable, but the detailed mechanisms, and biochemical and physiological changes imposed during salt stress in plant and in AMF are still very low. The various transporter systems and signaling molecule involved in plant-AMF need to be discovered. Various techniques like proton-induced X-ray emission should incorporate to reveal exact distribution of macro–micronutrition during plant–microbe interaction.

There are so many microbes, which are still unknown and can be used for amelioration of salt stress in crops. Various omic technologies like metagenomics, metabolomics, and proteomics provide more molecular, physiological, and biochemical details about known and unknown microbial species to mitigate salt stress. In future, we can hope that commercialized product of different microbes will come in market so that the salt-affected lands can be used to maximize agriculture production.

References

Abo-Doma A, Edrees S, Abdel-Aziz S (2016) The effect of mycorrhiza growth and expression of some genes in barley. Egypt J Genet Cytol 40(2):301–313

Ahmed A, Hasnain A, Akhtar S, Hussain A, Yasin AUG, Wahid A (2010) Antioxidant enzymes as bio-markers for copper tolerance in safflower (*Carthamus tinctorius* L.). Afr J Biotechnol 9(33): 5441–5544

Al-Garni SMS (2006) Increasing NaCl-salt tolerance of a halophytic plant Phragmites australis by mycorrhizal symbiosis. American-Eurasian J Agric Environ Sci 1:119–126

Alguacil MM, Herandez JA, Roldan A (2003) Antioxidant enzyme actives in shoots from three mycorrhizal shrub species afforested in a degraded semi arid soil. Physiol Plantarum 118:562–570

Alikhani M, Khatabi B, Sepehri M, Nekouei MK, Mardi M, Salekdeh GH (2013) A proteomics approach to study the molecular basis of enhanced salt tolerance in barley (*Hordeum vulgare* L.) conferred by the root mutualistic fungus *Piriformospora indica*. Mol BioSyst 9:1498–1510. https://doi.org/10.1039/c3mb70069k

Al-Karaki GN (2000) Growth of mycorrhizal tomato and mineral acquisition under salt stress. Mycorrhiza 10:51–54

Alqarawi AA, Abd Allah EF, Hashem A (2014a) Alleviation of salt-induced adverse impact via mycorrhizal fungi in *Ephedra aphylla* Forssk. J Plant Interact 9(1):802–810. https://doi.org/10.1080/17429145.2014.949886

Alqarawi AA, Hashem A, Abd Allah EF, Alshahrani TS, Huqail AA (2014b) Effect of salinity on moisture content, pigment system, and lipid composition in *Ephedra alata* Decne. Acta Biol Hung 65(1):61–71. https://doi.org/10.1556/ABiol.65.2014.1.6

Arora M, Kaushik A, Rani N, Kaushik CP (2010) Effect of cyanobacterial exopolysaccharides on salt stress alleviation and seed germination. J Environ Biol 31(5):701–704

Bal HB, Nayak L, Das S, Adhya TK (2013) Isolation of ACC deaminase producing PGPR from rice rhizosphere and evaluating their plant growth promoting activity under salt stress. Plant Soil 366(1–2):93–105. https://doi.org/10.1007/s11104-012-1402-5

Balliu A, Sallaku G, Rewald B (2015) AMF Inoculation enhances growth and improves the nutrient uptake rates of transplanted, salt-stressed tomato seedlings. Sustainability 7:15967–15981. https://doi.org/10.3390/su71215799

Bharti N, Pandey SS, Barnawal D, Patel VK, Kalra A (2016) Plant growth promoting rhizobacteria *Dietzia natronolimnaea* modulates the expression of stress responsive genes providing protection of wheat from salinity stress. Sci Rep 6:34768. https://doi.org/10.1038/srep34768

Chakraborty N, Ghosh R, Ghosh S, Narula K, Tayal R, Datta A, Chakraborty S (2013) Reduction of oxalate levels in tomato fruit and consequent metabolic remodeling following overexpression of a fungal oxalate decarboxylase1[W]. Plant Physiol 162(1):364–378

Chen M, Wei H, Cao J, Liu R, Wang Y, Zheng C (2007) Expression of *Bacillus subtilis* proBA genes and reduction of feedback inhibition of proline synthesis increases proline production and confers osmotolerance in transgenic Arabidopsis. J Biochem Mol Biol 40:396–403

Chen L, Liu Y, Wu G, Veronican NK, Shen Q, Zhang N (2016) Induced maize salt tolerance by rhizosphere inoculation of *Bacillus amyloliquefaciens* SQR9. Physiol Plant 158:34–44. https://doi.org/10.1111/ppl.12441

Colla G, Rouphael Y, Cardarelli M, Tullio M, Rivera CM, Rea E (2008) Alleviation of salt stress by arbuscular mycorrhizal in zucchini plants grown at low and high phosphorus concentration. Biol Fertil Soils 44:501–509

de la Pena TC, Redondo FJ, Manrique E, Lucas MM, Pueyo JJ (2010) Nitrogen fixation persists under conditions of salt stress in transgenic Medicago truncatula plants expressing a cyanobacterial flavodoxin. Plant Biotechnol J 8:954–965

del Amor FM, Cuadra-Crespo P (2012) Plant growth-promoting bacteria as a tool to improve salinity tolerance in sweet pepper. Funct Plant Biol 39:82–90. https://doi.org/10.1071/Fp11173

Egamberdieva D (2011) Survival of *Pseudomonas extremorientalis* TSAU20 and *P. Chlororaphis* TSAU13 in the rhizosphere of common bean (*Phaseolus vulgaris*) under saline conditions. Plant Soil Environ 57(3):122–127

Fu QL, Liu C, Ding NF, Lin YC, Guo B (2010) Ameliorative effects of inoculation with the plant growth-promoting *Rhizobacterium Pseudomonas sp.* DW1 on growth of eggplant (*Solanum melongena* L.) seedlings under salt stress. Agr Water Manag 97(12):1994–2000

Gaber A, El-Assal SE (2012) Cyanobacterium *Synechocystis sp.* PCC 6803 Glutaredoxin Gene (slr1562) protects *Escherichia coli* against abiotic stresses. Am J Agric Biol Sci 7:88–96

Gamalero E, Berta G, Massa N, Glick BR, Lingua G (2010) Interactions between Pseudomonas putida UW4 and *Gigaspora rosea* BEG9 and their consequences for the growth of cucumber under salt-stress conditions. J Appl Microbiol 108(1):236–245

Ghorbanli M, Ebrahimzadeh H, Sharifi M (2004) Effects of NaCl and mycorrhizal fungi on antioxidative enzymes in soybean. Bio Plant 48(4):575–581

Gill SS, Gill R, Trivedi DK, Anjum NA, Sharma KK, Ansari MW, Ansari AA, Johri AK, Prasad R, Pereira E (2016) *Piriformospora indica*: potential and significance in plant stress tolerance. Front Microbiol 7:1–20

Giri B, Mukerji KG (2004) Mycorrhizal inoculant alleviates salt stress in Sesbania aegyptiaca and Sesbania grandiflora under field conditions: evidence for reduced sodium and improved magnesium uptake. Mycorrhiza 14:304–312

Giri B, Kapoor R, Mukerji KG (2007) Improved tolerance of *Acacia nilotica* to salt stress by arbuscular mycorrhiza, *Glomus fasciculatum*, may be partly related to elevated K^+/Na^+ ratios in root and shoot tissues. Microbial Ecol 54:753–760

Glick BR, Cheng Z, Czarny J, Duan J (2007) Promotion of plant growth by ACC deaminase-producing soil bacteria. Eur J Plant Pathol 119:329–339. https://doi.org/10.1007/s10658-007-9162-4

Habib SH, Kausar H, Saud HM (2016) Plant growth-promoting rhizobacteria enhance salinity stress tolerance in okra through ROS-scavenging enzymes. Biomed Res Int 2016:6284547. https://doi.org/10.1155/2016/6284547

Hajiboland R, Aliasgharzadeh N, Laiegh SF, Poschenrieder C (2010) Colonization with arbuscular mycorrhizal fungi improves salinity tolerance of tomato (Solanum lycopersicum L.) plants. Plant Soil 331:313–327. https://doi.org/10.1007/s11104-009-0255-z

Hashem A, Abd Allah EF, Alqarawi AA, Aldubise A, Egamberdieva D (2015) Arbuscular mycorrhizal fungi enhances salinity tolerance of *Panicum turgidum* Forssk by altering photosynthetic and antioxidant pathways. J Plant Interact 10(1):230–242. https://doi.org/10.1080/17429145.2015.1052025

Hong SH, Lee EY (2017) Phytostabilization of salt accumulated soil using plant and biofertilizers: field application. Int Biodeterior Biodegrad 124:188–195

Jahromi F, Aroca R, Porcel R, Ruiz-Lozano JM (2008) Influence of salinity on the in vitro development of Glomus intraradices and on the in vivo physiological and molecular responses of mycorrhizal lettuce plants. Microb Ecol 55:45–53

Jha M, Chourasia S, Sinha S (2013) Microbial consortium for sustainable rice production. Agroecol Sustain Food Syst 37(3):340–362. https://doi.org/10.1080/10440046.2012.672376

Kang SM, Khan AL, Waqas M, You YH, Kim JH, Kim JG (2014) Plant growth-promoting rhizobacteria reduce adverse effects of salinity and osmotic stress by regulating phytohormones and antioxidants in *Cucumis sativus*. J Plant Interact 9:673–682. https://doi.org/10.1080/17429145.2014.894587

Kapoor R, Gupta MK, Kumar N, Kanwar SS (2017) Analysis of nhaA gene from salt tolerant and plant growth promoting *Enterobacter ludwigii*. Rhizosphere 4:62–69

Kaya C, Ashraf M, Sonmez O, Aydemir S, Tuna AL, Cullu MA (2009) The influence of arbuscular mycorrhizal colonisation on key growth parameters and fruit yield of pepper plants grown at high salinity. Sci Hortic 121:1–6. https://doi.org/10.1016/j.scienta.2009.01.001

Kim K, Jang YJ, Lee SM, Oh BT, Chae JC, Lee KJ (2014) Alleviation of salt stress by *Enterobacter sp*. EJ01 in tomato and Arabidopsis is accompanied by up-regulation of conserved salinity responsive factors in plants. Mol Cells 37:109–117. https://doi.org/10.14348/molcells.2014.2239

Lastochkina O, Pusenkova L, Yuldashev R, Babaev M, Garipova S, Blagova D, Khairullin R, Aliniaeifard S (2017) Effects of *Bacillus subtilis* on some physiological and biochemical parameters of *Triticum aestivum* L. (wheat) under salinity. Plant Physiol Biochem 121:80–88

Li H, Lei P, Pang X, Li S, Xu H, Xu Z, Feng X (2017) Enhanced tolerance to salt stress in canola (*Brassica napus* L.) seedlings inoculated with the halotolerant *Enterobacter cloacae* HSNJ4. Appl Soil Ecol 119:26–34

Malam IO et al (2007) Effects of the inoculation of cyanobacteria on the microstructure and the structural stability of a tropical soil. Plant Soil 290:209–219

Martinez R, Espejo A, Sierra M, Ortiz-Bernad I, Correa D, Bedmar E, Lopez-Jurado M, Porres JM (2015) Co-inoculation of *Halomonas maura* and *Ensifer meliloti* to improve alfalfa yield in saline soils. Appl Soil Ecol 87:81–86

Marulanda A, Azcon R, Chaumont F, Ruiz-Lozano JM, Aroca R (2010) Regulation of plasma membrane aquaporins by inoculation with a Bacillus megaterium strain in maize (*Zea mays* L.) plants under unstressed and salt-stressed conditions. Planta 232:533–543. https://doi.org/10.1007/s00425-010-1196-8

Mo Y, Wang Y, Yang R, Zheng J, Liu C, Li H, Ma J, Zhang Y, Wei C, Zhang X (2016) Regulation of plant growth, photosynthesis, antioxidation and osmosis by an arbuscular mycorrhizal fungus in watermelon seedlings under well-watered and drought conditions. Front Plant Sci 7:644. https://doi.org/10.3389/fpls.2016.00644

Nabti E, Sahnoune M, Ghoul M, Fischer D, Hofmann A, Rothballer M (2010a) Restoration of growth of durum wheat (Triticum durum var. waha) under saline conditions due to inoculation with the rhizosphere bacterium *Azospirillum brasilense* NH and extracts of nutrient uptake rates of transplanted, salt-stressed tomato seedlings. Sustainability 7:15967–15981. https://doi.org/10.3390/su71215799

Nabti E, Sahnoune M, Ghoul M, Fischer D, Hofmann A, Rothballer M, Schmid M, Hartmann A (2010b) Restoration of growth of durum wheat (Triticum durum var. waha) under saline conditions due to inoculation with the rhizosphere bacterium AzosThe marine alga *Ulva lactuca*. J Plant Growth Regul 29(1):6–22. https://doi.org/10.1007/s00344-009-9107-6

Nadeem SM, Zahir ZA, Naveed M, Arshad M (2009) Rhizobacteria containing ACC-deaminase confer salt tolerance in maize grown on salt-affected fields. Can J Microbiol 55:1302–1309. https://doi.org/10.1139/W09-092

Nadeem SM, Shaharoona B, Arshad M, Crowley DE (2012) Population density and functional diversity of plant growth promoting rhizobacteria associated with avocado trees in saline soils. Appl Soil Ecol 62:147–154. https://doi.org/10.1016/j.apsoil.2012.08.005

Nautiyal CS, Srivastava S, Chauhan PS, Seem K, Mishra A, Sopory SK (2013) Plant growth-promoting bacteria *Bacillus amyloliquefaciens* NBRISN13 modulates gene expression profile of leaf and rhizosphere community in rice during salt stress. Plant Physiol Biochem 66:1–9. https://doi.org/10.1016/j.plaphy.2013.01.020

Ngwene B, Boukail S, Söllner L, Franken P, Andrade-Linares D (2016) Phosphate utilization by the fungal root endophyte *Piriformospora indica*. Plant Soil 405(1–2):231–241

Niu SQ, Li HR, Pare PW, Aziz M, Wang SM, Shi HZ (2016) Induced growth promotion and higher salt tolerance in the halophyte grass *Puccinellia tenuiflora* by beneficial rhizobacteria. Plant Soil 407:217–230. https://doi.org/10.1007/s11104-015-2767-z

Patel D, Jha CK, Tan KN, Sara FM (2012) Growth enhancement of chickpea in saline soils using plant growth-promoting rhizobacteria. J Plant Growth Regul 31(1):53–62. https://doi.org/10.1007/s00344-011-9219-7

Patel S, Jinal HN, Amaresan N (2017) Isolation and characterization of drought resistance bacteria for plant growth promoting properties and their effect on chilli (*Capsicum annum*) seedling under salt stress. Biocat Agric Biotechnol 12:85–89

Pereira SIA, Moreira H, Argyres K, Castro PML, Marques APGC (2016) Promotion of sunflower growth under saline water irrigation by the inoculation of beneficial microorganisms. Appl Soil Ecol 105:36–47

Pinedo I, Ledger T, Greve M, Poupin MJ (2015) *Burkholderia phytofirmans* PsJN induces long-term metabolic and transcriptional changes involved in Arabidopsis thaliana salt tolerance. Front Plant Sci 6:466. https://doi.org/10.3389/fpls.2015.00466

Porcel R, Ruiz-Lozano JM (2004) Arbuscular mycorrhizal influence on leaf water potential, solute accumulation and oxidative stress in soybean plants subjected to drought stress. J Exp Bot 55: 1743–1750

Rodriguez AA, Stella AM, Storni MM, Zulpa G, Zaccaro MC (2006) Effects of cyanobacterial extracellular products and gibberellic acid on salinity tolerance in Oryza sativa L. Saline Syst 2: 7–18

Rojas-Tapias D, Moreno-Galvan A, Pardo-Diaz S, Obando M, Rivera D, Bonilla R (2012) Effect of inoculation with plant growth-promoting bacteria (PGPB) on amelioration of saline stress in maize (Zea mays). Appl Soil Ecol 61:264–272. https://doi.org/10.1016/j.apsoil.2012.01.006

Saddique AB, Ali Z, Khan AS, Rana AI, Shamsi MI (2018) Inoculation with the endophyte *Piriformospora indica* significantly affects mechanisms involved in osmotic stress in rice. Rice 11:34. https://doi.org/10.1186/s12284-018-0226-1

Sadeghi A, Karimi E, Dahaji PA, Javid MG, Dalvand Y, Askari H (2012) Plant growth promoting activity of an auxin and siderophore producing isolate of Streptomyces under saline soil conditions. World J Microb Biot 28(4):1503–1509. https://doi.org/10.1007/s11274-011-0952-7

Sannazzaro AI, Ruiz OA, Alberto EO, Menéndez AB (2006) Alleviation of salt stress in Lotus glaber by Glomus intraradices. Plant Soil 285:279–287

Sapre S, Gontia-Mishra I, Tiwari S (2018) *Klebsiella sp.* confers enhanced tolerance to salinity and plant growth promotion in oat seedlings (Avena sativa). Microbiol Res 206:25–32

Sarkar A, Ghosh PK, Pramanik K, Mitra S, Soren T, Pandey S, Mondal MH, Maiti TK (2018) A halotolerant *Enterobacter sp.* displaying ACC deaminase activity promotes rice seedling growth under salt stress. Res Microbiol 169:20–32

Sergeeva E, Liamir A, Begrman B (2002) Evidence for production of the Phytohormone indole—acetic acid by cyanobacteria. Planta 215:229–238

Shah G, Jan M, Afreen M, Anees M, Rehman S, Daud MK, Malook I, Jamil M (2017) Halophilic bacteria mediated phytoremediation of salt-affected soils cultivated with rice. J Geochem Expl 174:59–65

Sharifi M, Ghorbanli M, Ebrahimzadeh H (2007) Improved growth of salinity-stressed soybean after inoculation with salt pre-treated mycorrhizal fungi. J Plant Physiol 164(9):1144–1151

Shekoofeh E, Sepideh H, Roya R (2012) Role of mycorrhizal fungi and salicylic acid in salinity tolerance of *Ocimum basilicum* resistance to salinity. J Biotech 11(9):2223–2227

Shukla PS, Agarwal PK, Jha B (2012) Improved salinity tolerance of *Arachis hypogaea* (L.) by the interaction of halotolerant plant growth- promoting rhizobacteria. J Plant Growth Regul 31(2): 195–206. https://doi.org/10.1007/s00344-011-9231-y

Singh R, Singh P, Sharma R (2014) Microorganism as a tool of bioremediation technology for cleaning environment: a review. Proc Int Acad Ecol Environ Sci 4:1–6

Suarez C, Cardinale M, Ratering S, Steffens D, Jung S, Montoya ZMA, Geissler-Plaum R, Schnell S (2015) Plant growth-promoting effects of *Hartmannibacter diazotrophicus* on summer barley (*Hordeum vulgare* L.) under salt stress. Appl Soil Ecol 95:23–30

Swarnalakshmi K, Dhar D, Singh P (2007) Evaluation of blue-green algal inoculation on specific soil parameters. Acta Agron Hungarica 55:307–313

Tamoi M, Kurotaki H, Fukamizo T (2007) Beta-1,4-glucanase like protein from cyanobacterium synechocystis sp PCC 6803. Mol Cell Proteomics 13:3519–3532

Tang M, Chen H, Huang JC, Tian ZQ (2009) AM fungi effects on the growth and physiology of *Zea mays* seedlings under diesel stress. Soil Biol Biochem 41:936–940. https://doi.org/10.1016/j.soilbio.2008.11.007

Tank N, Saraf M (2010) Salinity-resistant plant growth promoting rhizobacteria ameliorates sodium chloride stress on tomato plants. J Plant Interact 5(1):51–58. https://doi.org/10.1080/17429140903125848

Tariq A, Pan K, Olatunji OA, Graciano C, Li Z, Sun F, Sun X, Song D, Chen W, Zhang A (2017) Phosphorous application improves drought tolerance of *Phoebe zhennan*. Front Plant Sci 8:1561

Upadhyay SK, Singh DP (2015) Effect of salt-tolerant plant growth promoting rhizobacteria on wheat plants and soil health in a saline environment. Plant Biol 17:288–293. https://doi.org/10.1111/plb.12173

Upadhyay SK, Singh JS, Singh DP (2011) Exopolysaccharide-producing plant growth-promoting rhizobacteria under salinity condition. Pedosphere 21(2):214–222. https://doi.org/10.1016/S1002-0160(11)60120-3

Vaishnav A, Kumari S, Jain S, Varma A, Choudhary DK (2015) Putative bacterial volatile-mediated growth in soybean (*Glycine max* L. Merrill) and expression of induced proteins under salt stress. J Appl Microbiol 119:539–551. https://doi.org/10.1111/jam.12866

Yang Y, Han X, Liang Y, Ghosh A, Chen J, Tang M (2015) The combined effects of Arbuscular Mycorrhizal fungi (AMF) and lead (Pb) stress on Pb accumulation, plant growth parameters, photosynthesis, and antioxidant enzymes in *Robinia pseudoacacia* L. PLoS One 10(12): e0145726. https://doi.org/10.1371/journal.pone.0145726

Yao LX, Wu ZS, Zheng YY, Kaleem I, Li C (2010) Growth promotion and protection against salt stress by *Pseudomonas putida* Rs-198 on cotton. Eur J Soil Biol 46:49–54. https://doi.org/10.1016/j.ejsobi.2009.11.002

Zarea MJ, Hajinia S, Karimi N, Goltapeh EM, Rejali F, Varma A (2012) Effect of Piriformospora indica and Azospirillum strains from saline or non-saline soil on mitigation of the effects of NaCl. Soil Biol Biochem 45:139–146. https://doi.org/10.1016/j.soilbio.2011.11.006

Zerrouk IZ, Benchabane M, Khelifi L, Yokawa K, Ludwig-Müller J, Baluska F (2016) A Pseudomonas strain isolated from date-palm rhizospheres improves root growth and promotes root formation in maize exposed to salt and aluminum stress. J Plant Physiol 191:111–119

Zhang WY, Wang AA, Hao RC, Yang T (2014) Endophytic fungus *Piriformospora indica* promotes growth and confers drought tolerance in sesame (*Sesamum indicum* L.). Chinese J Oil Crop Sci 1(11)

Emerging Technologies to Understand Plant–Microbe Responses on Climatic Change

P. Kiruthika Lakshmi and C. Usha

Abstract Plants and its rhizosphere region serve as a hostile habitat harboring various microorganisms, and these microbes interact with the host plants via antagonism or commensalism or symbiosis. Climatic factors have a significant role in modulation of the plant microbiota, which in turn influences the plant growth and yield. Microbial diversity in the rhizosphere has been explored to a greater extent, but research on the environmental factors controlling the plant–microbe interaction still remains scanty. Deciphering the complexity in plant–microbe relationship is very crucial to understand the nature of interaction and to develop strategies for crop improvement and productivity. Advancement in molecular techniques and various "omics" approaches, like metabolomics, transcriptomics, genomics, proteomics, and metagenomics, have emerged as a valuable tool for understanding the influence of climatic change in fine-tuning of plant–microbe interaction. Technological development in computational biology and bioinformatics accelerated the biological research addressing the intricate microbial interactions and provides a clear insight on fabricating several plant mitigation strategies mediated by microorganisms to overcome the abiotic stresses. This review focuses on various emerging technologies to understand plant–microbe responses on climatic change.

Keywords Plant · Microbe interaction · Stress response · Genomics · Transcriptomics · Proteomics · Metabolomics

Introduction

Ever-shifting climatic conditions made a greater impact on gene regulation, metabolism, and physiology of plants (Ramegowda and SenthilKumar 2015). Some plants have the inherent potential to modulate their gene expression for surviving in the

P. K. Lakshmi (✉)
Department of Microbiology, The Madura College, Madurai, Tamil Nadu, India

C. Usha
Department of Biotechnology, Lady Doak College, Madurai, Tamil Nadu, India

© The Author(s), under exclusive license to Springer Nature Singapore Pte Ltd. 2022
A. Vaishnav et al. (eds.), *Plant Stress Mitigators*,
https://doi.org/10.1007/978-981-16-7759-5_21

stressful environment, while most of the plants could not cope up with these abiotic stress factors (Tank and Saraf 2010). To meet the increasing nutritional demand, agricultural activities should be promoted amid of other undesirable climatic changes.

Nowadays, farmers are highly depending on agrochemicals for increasing the crop productivity, but it often poses a serious threat to the ecosystem (Singh et al. 2011). Hence, it creates an urge to focus on a sustainable and eco-friendly approach to enhance the crop yield. In this situation, plant–microbe interactions play a very crucial role in sustaining agriculture even under stressed conditions. But, the mechanism behind this interaction is very complex and interwoven. Molecular analysis on the plant stress response, signaling pathway, and gene regulation could reveal the tactics for using microbes as a catalyst to improve the productivity (Bhattacharyya et al. 2016). With this in mind, we have discussed various conventional methods and high-throughput "omics" techniques, which are being applied worldwide, to understand the effect of climatic change on plant–microbe association.

Impact of Climatic Change on Plant–Microbe Interaction

Microbial physiology in plants entirely depends on the environmental factors like pH, temperature, moisture, and nutrient availability. (Kheyrodin and Ghazvinian 2012) (Fig 1). For instance, the increase in carbon dioxide concentration favors the growth of mycorrhizal fungi such as *Pisolithus tinctorius* and *Hebeloma crustuliniforme* in *Pinus silvestris* L. (Ineichen et al. 1995; Fransson et al. 2005). Similarly, the colonization and establishment of plant growth-promoting bacteria are determined by the soil type and prevailing temperature conditions (Egamberdiyeva and Hoflich 2003). Root-nodulating *Rhizobium sp.* exhibited its temperature optima around 36 °C in the desert legume *Prosopis glandulosa*, whereas the increase in temperature (up to 30 °C) significantly reduced the establishment of *Burkholderia phytofirmans* in rhizosphere region of tomato plants (Waldon et al. 1989).

Further, the microbial colonization in plants like *Pinus oaxacana* and *Pinus muricata* was drastically suppressed by the drought stress (Compant et al. 2010). Sharma et al. (2014) also reported the negative impact of drought on establishment of mycorrhizal fungi. However, few reports are available on the positive impact of drought on colonization of rhizobacteria. Hence, for utilizing the microorganism as a biofertilizer or biopesticide, it becomes mandatory to explore the molecular level impact of interaction, mechanism of action, and its efficiency for the development of resistant varieties and crop improvement strategies (Sharma et al. 2014).

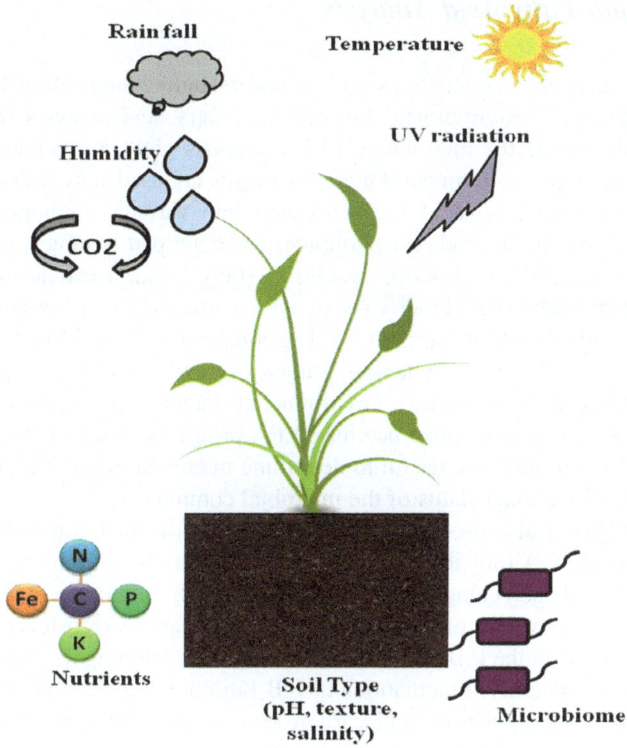

Fig. 1 Environmental factors influencing plant–microbe interaction

Conventional Methods for Understanding of Plant–Microbe Responses

Plants serve as a vital component in our environment. The microbes were considered as a major part in plant–microbe interactions (Vorholt 2012). The consequences of widespread studies recommend the use of microbes in an improved approach for boosting crop yield. Further, it has also been reported that the microbes will play a significant part in disease propagation and control (Reid 2011). Various conventional methods to study plant–microbe interactions are sequencing, chromatography, mass spectrometry, phospholipid fatty acid, microscopy, Fourier transform infrared spectroscopy (FTIR), isotope probing, nuclear magnetic resonance, and real-time PCR (Wu et al. 2009).

Phospholipid Fatty Acid Analysis

Microbiologists have done much research in understanding the relationship between microbiota and their environment. Phospholipid fatty acid analysis (PLFA) was reported to determine the microbiota. PLFA analysis gives a quantitative data of the microbiota in an environment. Organic solvents are used for extraction of fatty acids. The extracted fatty acid is fractionated into various components such as neutral lipids, glycolipids, and phospholipids. The latter part is subjected to alkaline methanolysis (mild) to produce FAMEs (fatty acid methyl esters). Gas chromatography–mass spectrometry is extensively used for the quantitative analysis of FAMEs (White and Ringelberg 1998). The composition of these PLFA profiles is revealed with details of chain length, saturation, and branching (Leckie 2005). Metabolic status of the organism, environmental factors, and interaction of toxic substances with microbiota influence the PLFA profile (Reinsch et al. 2014). Subsequently, PLFA profiles are useful to determine microbial biomass and to provide insight into the functional status of the microbial community.

PLFA analysis is also used in soil study to make a distinction between bacterial and fungal biomass. A fungal-to-bacterial (F:B) ratio can be determined by dividing the sum of the mole percentage values of the fungal fatty acid markers by the sum of the mole percentage values of the bacterial fatty acid markers (Frostegard and Baath 1996). An increase in the F:B ratio indicates the increase in fungal concentrations in the microbial community. Fluctuations in F:B ratio are related to nutrient cycling, organic matter decomposition, and carbon sequestration (Bragazza et al. 2015). Further, F:B ratio is used to evaluate the effect of agricultural practices on soil microbial communities (Zhang et al. 2012).

Terminal Restriction Fragment Length Polymorphism (TRFLP)

Terminal restriction fragment length polymorphism (T-RFLP) is a culture-independent technique with advanced resolution than cultivation-based methods (Nithya and Pandian 2012). This technique is used to reveal the composition and diversity of soil microbial communities in different ecological communities (Singh et al. 2006). Earlier, multiple enzyme digestion TRFLP is used to reveal the environmental bacterial community (Knauth et al. 2005). Single digestion TRFLP peaks illustrate operational TRFLP unit (OTUs) on the diversity of endophytic bacteria in leaves. Knauth et al. (2005) reported that 16S rRNA gene analysis has been performed to reveal the presence of rhizosphere and rhizoplane endophytic bacterial communities in potato and rice. Further, Ikeda et al. (2010) reported the presence of bacterial endophytes in phyllosphere of soybean, rice, and maize.

Denaturing Gel Gradient Electrophoresis (DGGE)

Recently, scientists have employed the use of primers in amplifying and analyzing the 16S rRNA genes. Denaturing gel gradient electrophoresis provides information to monitor the structure and dynamics of microbial populations under the influence of environmental changes. This approach has been employed to analyze microbiota such as proteobacteria (Henckel et al. 1999), actinomycetes (Heuer et al. 1997), ammonia-oxidizing bacteria (Kowalchuk et al. 1998), Archaea (Òvreas et al. 1997), and fungi (Smit et al. 1999).

DGGE was developed in the 1980s for the detection of point mutations and was first used for microbial analysis early 1990s (Muyzer et al. 1993). The technique separates DNA fragments of the same or similar length but of different sequences by electrophoresis in a gradient of a denaturant. In the denaturing gradient, DNA fragments of similar length are separated by electrophoresis. The DNA fragments migrate under the influence of an electric field and reach a site in the gradient, in which the strands separate, and thereby, the mobility of the fragment decreases rapidly. Finally, a picture composed of an array of bands with different intensities is obtained. Band intensities correspond to the frequency of individual PCR products in the reaction mixture.

Sakurai et al. (2007) reported that DGGE was also used for the analysis of functional genes. Depending on the intensity of the band, approximately one hundred of bands can be identified on a gel. However, the limitation of this method is intensity higher than 0.1–1% of the total intensity that can be technically assessed. Further, Qin et al. (2016) reported that characterization of bacterial populations is performed by using DGGE and Illumina MiSeq in *Triticum aestivum* L. Pei et al. (2017) identified that DGGE identified 16 genera in *Dendrobium officinale* (Pei et al. 2017).

Isotope Probing

The plant–microbe interaction is found to be more intense in the rhizosphere region because of the release of organic molecules by the plant roots. Microbes in turn facilitate the plant growth by releasing growth factors, by solubilizing nutrients, and by improving soil structure. This complex interaction between plants and the rhizosphere microbes has been studied conventionally by culture-based approach until the advent of isotope probing. This technique does not involves culturing of microbes but helps to characterize the microbes based on its ability to assimilate carbon molecules present in the root exudates and in the rhizosphere soil.

In this technique, generally a carbon substrate labeled with stable isotope is used to segregate metabolically active and inactive members present in the rhizobacterial community. Here, the organisms can be exposed to the isotope directly by adding ^{13}C-labeled carbon compounds to the rhizosphere region or by supplementing $^{13}CO_2$

during the cultivation. Later on, the carbon assimilating potential of the rhizobacteria will be screened by analyzing the RNA, DNA, and phospholipid fatty acids (Prosser et al. 2006).

Pulse labeling using $^{13}CO_2$ is considered to be a practically feasible approach to check the nutritional dependency of microorganisms on plant root exudates (Rangel-Castro et al. 2005). Pulse labeling involves cultivation of plants in a pot system, kept in a closed chamber supplemented with $^{13}CO_2$. The CO_2 got assimilated, fixed, and released in the root exudates for being up taken by the microbes in the rhizosphere (Ostle et al. 2000). Hence, understanding the pattern of carbon flow between the plants and the microorganisms, whether it may be a pathogen or synergistic one, is highly important for crop management under stressed conditions. It also serves as a basis for manipulating resistant varieties using plant genetic engineering approach.

However, specificity of the stable isotope probing (SIP) technique has to be increased to study more specific plant–microbe interactions. RNA-based SIP exhibited greater sensitivity even at low isotope concentration. It can be further increased by using highly specific gene probes (Griffiths et al. 2004). For example, ^{13}C-labeled mRNA for a specific functional gene can be utilized for detecting the expression of such gene under the influence of climatic stressors.

Advent of Omics Revolution for Understanding Plant–Microbe Responses

Omics enables a computational biology approach toward understanding the intricate interactions between genes, proteins, and metabolites (Table 1). This integrated approach relies on various disciplines of biology such as analytical methods, bioinformatics, and computational methods. The National Agricultural Biotechnology Council outlined its plan for developing a sustainable bio-based economy in its report "Vision for Agricultural Research and Development in the 21st Century." The council urged agricultural research and development programs to address sustainable bio-based industries (Emon 2016). Omics can facilitate the development of

Table 1 Techniques applied in various omics approaches to study plant–microbe interaction

Omics Approach	Techniques applied
Genomics: study of total cellular DNA	DNA microarray, genome sequencing, transcription factors, next-generation sequencing
Transcriptomics: study of total cellular mRNA	RNA sequencing, northern blotting, RT-PCR
Proteomics: study of total cellular proteins	Gene cloning, affinity chromatography, protein or peptide fractionation, gel electrophoresis, mass spectrometry (MS), ESI-MS, MALDI-TOF
Metabolomics: study of total cellular metabolites	HPLC, gas chromatography, NMR, FRIR, X-ray crystallography, LC-MS

agricultural research in food, health, energy, feedstock, and specialty chemicals while helping to protect, improve, and remediate the environment.

Genomics provide knowledge-based approaches for plant biotechnology, enabling accurate and controllable methods for molecular breeding and marker-assisted selection accelerating the development of new crop varieties. Omics era can be introduced into plants, such as the production of biopharmaceuticals and industrial compounds (Ahmad et al. 2012). Gene expression studies recognize functional gene products that provide the phenotypic information and can be used for crop improvements. By adding a specific gene/genes to a plant, or knocking down a gene with RNAi, the desirable phenotype can be created more quickly than through traditional plant breeding.

Gene-Sequencing Methods

A major progress related to the molecular ecology technique for analysis of microbial diversity of soil is based on the isolation and characterization of microbes. The complete microbiome of soil can be analyzed for the extraction of DNA/RNA and the biochemical markers. DNA exemplifies the phylogenetic relationship and the functional property of microbes. Molecular approach involves the process of DNA and RNA extraction from the soil. The combined microbial genome constituting isolation of DNA from microbial communities is referred to as a metagenome. Metagenomics involves isolation and cloning of fragments of DNA that comprises of a variety of operons and genes. The total DNA extraction from the environmental samples can be evaluated by cloning, amplification by PCR, high-throughput sequencing, and microarray hybridization (Barret et al. 2013).

Microbial diversity can be quantitatively determined using the PCR technique. Ribosomal subunit sequences act as molecular markers of microbes. Principally, a comparative gene analysis of 18S or 16S ribosomal rRNA is the most common molecular technique for recognizing microbes. Several microbial genomes have rRNA gene as a gene marker. Evaluation of 18S and 16S rRNA gene sequences is the foundation for comparing richness, composition, and assembly of the microbial groups. PCR products contribute analogous or indistinguishable variable regions and are known as operational taxonomic units (OTUs). A variety of the PCR products are defined by distinguished molecular processes that allow a molecular intuition for revealing the makeup of a specific microbial group. Diversity of the amplicon is determined by sequencing and cloning methods. Sequence evaluation of 16S or 18S rDNA amplicons can be performed using high-throughput next-generation sequencing.

Shotgun Metagenomics

Shotgun metagenomics is a high-throughput sequencing of the whole DNA of a sample. It has abundant advantages over metabarcoding, predominantly for general surveillance purpose. Evaluating shotgun metagenomics data requires numerous complex operations. The reads must be assembled in the contigs of the metagenome and are then combined to form the complete genome of the microorganisms by a process known as "binning." If the microorganisms of concern have low abundance or when other problems make assembly difficult, then it is also feasible to perform assembly-free taxonomic profiling. Shotgun metagenomic sequencing has a considerable improvement of detecting all microorganisms (including both fungi and bacteria) present in a sample. Moreover, the procedure has the prospective for reaching resolution at strain level. Furthermore, shotgun metagenomics can re-examine the entire genome of the detected pathogens, making it perfect for sequencing unculturable ones (Duan et al. 2009). However, the procedure presents few limitations too. Firstly, sequencing complex genomes, marker genes may only be recovered at low frequency. Thereby, taxonomic identification of all species is difficult toward species whose full genomes are there in databases. Secondly, the sequencing intensity required to capture all the species is much higher than in metabarcoding, causing much higher expenses. Binning, while giving a lot of information and potentially being able to characterize new pathogens, is not viable when there is high biodiversity (Knight et al. 2018), making it fundamental to use software such as BUSCO and CheckM (Parks et al. 2015) to test the wholeness and quality of assemblies.

Transcriptomics

The entire set of RNA, also known as the transcriptome, is edited and becomes mRNA. It carries information to the ribosome, the protein factory of the cell, translating the message into protein. Transcriptomics has been described as expression profiling, as it is a study of the expression levels of mRNAs in a given cell population. Unlike the genome, which is approximately fixed for a given cell line with the exception of mutations, the transcriptome is dynamic as it is essentially a reflection of the genes that are dynamically expressed at any given time under a variety of circumstances. Transcriptomics determines the changes in the pattern of gene expression and also evaluates the internal and external factors such as biotic and abiotic stress. Transcriptomics is an existing tool for understanding biological systems. Transcriptomic techniques such as next-generation sequencing (NGS) provide capability for further considering the functional essentials of the genome (Valdes et al. 2013).

Proteins are ubiquitous in plants and are accountable for numerous cell functions. Through proteomics, it can be determined whether expression of mRNA results in protein synthesis to further elucidate gene function. Proteomics can also enhance the understanding of mechanisms of resistance, mode of action, and biodegradation of pesticides, aiding in the detection of novel effective and safe pesticides. Translational plant proteomics is an extension of proteomics from expression to functional, structural, and finally the translation and manifestation of desired traits. Through translational proteomics, the outcomes of proteomics for food authenticity, food security, food safety, energy sustainability, human health, increased economic principles, and environmental stewardship can be applied. Metabolic profiling provides an instant illustration of what is happening in the cell, for example, during fruit ripening, identifying key compounds vital for imparting taste and aroma.

Monitoring changes in metabolite patterns can lead to quality improvements for nutrition and plant health (Dixon et al. 2006). Metabolic profiling by mass spectrometry (MS) and nuclear magnetic resonance (NMR) analyses has been used to determine metabolic responses to herbicides and examine metabolic regulation and metabolite changes associated with environmental conditions of light, temperature, humidity, soil type, salinity, fertilizers, pests, pesticides, and genetic disturbances (Dixon et al. 2006). The comprehensive study of metabolites may be supportive for emerging pesticides, decreasing pesticide usage, increasing nutritive values, or assisting with other key traits.

Proteomics

Proteins are the key molecule that determines the phenotypic expression of the plants under stressed environment.

Environmental conditions influence the expression of genes in the plant system and in microbes. Hence, proteomic-based research helps us to unravel the metabolism, physiology, plant-mediated signaling, and other mechanisms behind the plant–microbe interaction (Kosova et al. 2015). It also provides a clear insight on the regulation of stress response and facilitates the identification of stress-related protein molecules (Silva-Sanchez et al. 2015). Proteomics in combination with bioinformatics tools has been applied widely for analyzing the response of various abiotic and biotic stresses on plants (Wang et al. 2016; Xu et al. 2015; Kosova et al. 2015). This protein information will be useful for developing crop varieties resistant to various stress factors.

Microbes have proven their capability to protect the plants various stress factors. Instead of employing the whole microorganism, identification and incorporation of protein, which helps to alleviate the stress, are considered to be a practically convenient strategy. An et al. (2016) carried out a proteomic-based investigation on the mechanism of cold tolerance in Cassava plant and concluded that the increase in malondialdehyde synthesis, chlorophyll production, and leakage of electrolytes confer for the acclimatization of cassava plants to extremely cold conditions. Similar type of proteomic studies was carried out by Chen et al. (2015) and Balbuena et al. (2011) on cold-tolerant alfalfa and sunflower, respectively. Nowadays, protein profiles of halophilic microorganisms are being exploited to a greater extent to identify and incorporate the genes, which can facilitate the plants to thrive under saline environment. de Lorenzo et al. (2009) identified the role of Srlk enzyme, a receptor kinase molecule rich in leucine, present in *Medicago truncatula* in regulation of salinity stress. Protein product of Ds-26-16, a novel gene from halotolerant *Dunaliella salina,* revealed the function of Ds-26-16 in amelioration of salt tolerance in tobacco plants and *E. coli* (Wang et al. 2016). Ghabooli et al. (2013) analyzed the protein profile of barley plants and reported that the plant enhances the synthesis of antioxidants and photosynthetic rate in order to ameliorate them from drought stress.

Nutrient deficiency in the soil like nitrogen starvation suppresses the crop growth and productivity as well. Role of microorganisms in biological nitrogen fixation is an inevitable one. Mass spectrometry-based proteomic analysis by Lery et al. (2011) revealed that overexpression of SP70-1143 signal proteins in sugarcane may result in enhanced nitrogen fixation by its symbiont *Gluconacetobacter diazotrophicus.* In line with this, mass spectrometric characterization by Kandasamy et al. (2009) showed that the application of *Pseudomonas fluorescens* to rice plants stimulates the expression of proteins like glutathione S-transferase, ribulose bisphosphate carboxylase, P23 co-chaperone, and thioredoxin H and enhances plant growth. *Sinorhizobium meliloti* also exhibited its potential to enhance the synthesis of photosynthetic proteins in crops (Chi et al. 2010).

In addition to the abovementioned climatic stresses, the presence of xenobiotics in the environment also poses a serious threat to the plant kingdom. *Pseudomonas sp.* is considered to be a boon microbe with potential to degrade xenobiotic compounds. Moreover, it is well known for its ability to produce siderophores (Cunrath et al. 2015), plant growth-promoting capabilities (Sorty et al. 2016), and as a biocontrol agent (Wang et al. 2015). Hence, proteome profile of *Pseudomonas sp.* has been investigated widely by researches all over the world (Kim et al. 2007; Reardon and Kim 2002).

Generally, microorganisms exhibit a highly diverse metabolic pathway, which makes them to endure under extreme situations. Understanding the interaction of microbes and plants under varying climatic conditions may pave the way for agricultural sustainability. Proteomics acts as a valuable tool for achieving this. Basically, proteomic investigation involves peptide/protein separation, purification, identification, estimation, and characterization of single protein moiety. But, metaproteomics is a newly emerged branch of proteomics, which deals with protein profile of microbial communities found in the ecosystem simultaneously. It helps to determine the nature of association existing between various microorganisms and with the host plant in the selected ecosystem. Moreover, advancement in high-throughput protein sequencing makes proteome identification, a highly feasible one (Shevchenko et al. 2001). However, limitations in the availability of protein sequences, related to plant–microbe interaction, in the reference database have to be resolved.

Metabolomics

One of the latest approaches used to study the expression of various genes involved in plant–microbe association is metabolomics. Metabolomics deals with the entire metabolic profile of an organism that reflects the activation or inhibition of various biochemical pathways regulated by the environmental factors (Bundy et al. 2005). The level of metabolites produced by the plants, which influences the associated microbial communities, may vary with its growth and other physiological conditions. Even, metabolites found in the root were reported to be different from that of root nodules.

Zhalnina et al. (2018) demonstrated the influence of metabolites secreted from the grass roots on the succession of rhizobacterial community. Mass spectrometry-based metabolic profiling is being applied to identify the disease causing potential of phytopathogens such as *Phytophthora sojae* (Zhu et al. 2018) and *Plasmopara viticola* (Negrel et al. 2018) in plants like soybean and grapevine, respectively. Plants produce secondary metabolites in order to protect them from stress conditions. Schliemann et al. (2008) compared the metabolites present in non-mycorrhizal and mycorrhizal roots of *Medicago truncatula* and documented the increase in defensive molecules like fatty acids, amino acids, apocarotenoids, tyrosol, and isoflavonoids in the mycorrhizal roots. Similar type of investigation was carried out by Pedone-Bonfim et al. (2013) in *Anadenanthera colubrina* seedling, in which mycorrhizal seedling exhibited higher concentration of tannins, flavonoids, phenols, and other secondary metabolites. Interaction among *Lotus japonicus* roots and *Mesorhizobium loti* enhanced the production of phenolic acids, when compared with the control plants (Rispail et al. 2010).

Metabolome of the plant can provide a clear insight on various signaling molecule, which can stimulate various biochemical reactions in the interacting microbial community (Micallef et al. 2009). Inoculation of *Trichoderma sp.* in pea plants stimulates the production of auxin-like compounds and enhanced the plant growth during abiotic stresses (Vinale et al. 2008). Microorganisms can also produce various growth factors like gibberellins, IAA, siderophores, and cytokinins as a result of their cellular metabolism and promote the plant growth directly and indirectly (Robin et al. 2006). Several reports authenticated the influence of IAA produced by various rhizobacteria on the germination and growth of wheat seedlings under saline-stressed environment (Sorty et al. 2016). Mishra et al. (2016) documented the influence of phosphate solubilizing *Bacillus sp.* in enhancing the productivity of fennel plants grown in saline soil.

Similarly, siderophore production by microorganisms such as *Pseudomonas fluorescens* favored the plant growth by accumulating iron and helps the plant to ameliorate stress imposed by heavy metals (Diels et al. 2002). Accumulation of xenobiotics in the environment is a matter of serious concern as it affects the plant growth drastically. Metabolomic-based research on the ability of *Nocardia sp.* to degrade hydrocarbons and its potential to produce plant growth-promoting traits could be useful in developing the strategy to overcome the stress imposed by xenobiotics on several plants (Raymond et al. 1967).

Hence, understanding the influence of climatic factors in various metabolic pathways, operating simultaneously in plants and microbes, is necessary for increasing the yield amid of different abiotic stresses (Breitling et al. 2008). Although metabolomics serves as a promising approach for determining the fine-tuning among plant–microbe association, challenges related to the cost, instrumentation, and supportive database references have to be addressed to make it more applicable.

Challenges

Tremendous improvement in molecular-based technologies like DNA cloning, DNA and protein sequencing, terminal restriction fragment length polymorphism (TRFLP), stable isotope probing (SIP), denaturing gradient gel electrophoresis (DGGE), fluorescence in situ hybridization (FISH), next-generation sequencing (NGS), genomics, transcriptomics, proteomics, and metabolomics deepened our knowledge on plant–microbe interaction (Hao and Xiao 2017). All these techniques are focusing on the mechanism behind the interaction of a single microbe with the host plant and the factors influencing their association. But in case of crop field, plants will not found in isolation. It creates a nutrient-enriched environment, which supports colonization of wide variety of microbes forming a microbiome. Hence, findings based on single microbe perspective may not suffice to unravel the actual scenario.

Currently, metagenomics has evolved as a latest approach for integrated understanding of the plant microbiome. However, several hurdles in metagenomics-based research such as the presence of non-culturable phytopathogen in microbial consortia, problems in understanding its pathogenesis, and contamination of food by toxic variants have to be resolved. Moreover, the formulation of special media and strategies for culturing the non-culturable organisms has to be developed, to gain a clear idea on nature of each microbe found in the microbial consortia. This knowledge on microorganisms will enable the scientist to formulate microorganism-based mitigation strategy for plants from various stress factors (Sarhan et al. 2019).

In addition, lack of reference sequences and related information in nucleotide and protein databases also complicated further research in plant–microbe responses on climatic change. Cost of analysis, technical expertise, and availability of sophisticated instruments still remain as drawbacks in research investigations based on molecular techniques (Breitling et al. 2008). Yet, another challenge in applying microorganisms as biofertilizer or biopesticides is its sustainability under field conditions. In order to address this pitfall, either the microbe has to be engineered in such a way that it can resist any environmental fluctuations or the microorganism-derived compound, which has the potential to accomplish the task should be used.

Conclusion and Future Prospects

In the present trend of climate-smart farming, microorganism-based mitigation strategies are inevitable (Hamilton et al. 2016). To develop such strategies and for exploring microorganisms as biofertilizer, biocontrol agent, and stress ameliorator, a clear understanding about the plant microbiome is needed. Various omics approaches are widely applied to study the in-depth mechanism behind plant–microbe interaction like metabolic and biochemical pathways, production of metabolites, role of different proteins and enzymes, gene cascade, and gene regulation, during varying climatic conditions (Knief et al. 2012). Application of DNA sequencing technique provides information regarding the identification of colonizing microorganisms. On the other hand, genomics involves structural and functional analysis of various genes associated with plant–microbe interaction. Researches on level of gene expression and protein profiling are being carried out by applying high-throughput tools like transcriptomics and proteomics. Metabolomics deals with the identification and estimation of plant metabolites in response to various stress factors (Allwood et al. 2010).

Thus, the multi-omics technology can be utilized for genetic manipulation and development of novel PGPR strains, which can be used as an ecologically safe and sustainable approach for stress management in plants. Recently, Bhattacharyya et al. (2017) analyzed whole-genome sequence of *Bacillus aryabhattai* and documented the presence of unique genes involved in siderophore production, uptake of metal ion, chemotaxis, and phosphate solubilization. The study confirmed the ability of *Bacillus aryabhattai* to be used as an efficient PGPR as it can interact positively with the plants and promotes their growth even under stressed conditions.

Hence, the multi-omics approach occupied a unique position in research works pertaining to plant–microbe interaction. Studies on human beings revealed the linkage between human microbiome and various diseases. But in case of plants, the involvement of plant microbiome in disease outbreak is still in rudimentary stage. Hence, the nature of association needs to be explored further using more advanced technologies (Furnkranz et al. 2012). Further, greater efforts are necessary to identify, characterize, and utilize microorganisms isolated from various stressful environments for the mitigation of plants from climatic stresses (Olukolu et al. 2016). To conclude, the data and knowledge obtained from the modern "omics" techniques are highly useful in the field of plant biotechnology and plant breeding programs that aims to achieve sustainable agriculture.

References

Ahmad P, Ashraf M, Younis M, Hu X, Kumar A, Akram NA, Al-Qurainy F (2012) Role of transgenic plants in agriculture and biopharming. Biotechnol Adv 30:524–540

Allwood JW, Clarke A, Goodacre R, Mur LAJ (2010) Dual metabolomics: a novel approach to understanding plant-pathogen interactions. Phytochemistry 71:590–597

An F, Li G, Li QX, Li K, Carvalho LJ, Ou W et al (2016) The comparatively proteomic analysis in response to cold stress in cassava plantlets. Plant Mol Biol Report 34:1095–1110

Balbuena TS, Salas J, Martínez-Force E, Garcés R, Thelen JJ (2011) Proteome analysis of cold acclimation in sunflower. J Proteome Res 10:2330–2346

Barret M, Tan H, Egan F, Morrissey JP, Reen J, O'Gara F (2013) Exploiting new systems-based strategies to elucidate plant-bacterial interactions in the rhizosphere. In: de Bruijn FJ (ed) Molecular microbial ecology of the rhizosphere. Wiley Blackwell, Hoboken, New Jersey

Bhattacharyya PN, Goswami MP, Bhattacharyya LH (2016) Perspective of beneficial microbes in agriculture under changing climatic scenario: a review. J Phytology 8:26–41

Bhattacharyya C, Bakshi U, Mallick I, Mukherji S, Bera B, Ghosh A (2017) Genome-guided insights into the plant growth promotion capabilities of the physiologically versatile *Bacillus aryabhattai* strain AB21. Front Microbiol 8:411

Bragazza L, Bardgett RD, Mitchell EAD, Buttler A (2015) Linking soil microbial communities to vascular plant abundance along a climate gradient. New Phytol 205:1175–1182

Breitling R, Vitkup D, Barrett MP (2008) New surveyor tools for charting microbial metabolic maps. Nat Rev Microbiol 6:156–161

Bundy JG, Willey TL, Castell RS, Ellar DJ, Brindle KM (2005) Discrimination of pathogenic clinical isolates and laboratory strains of *Bacillus cereus* by NMR-based metabolomic profiling. FEMS Microbiol Lett 242:127–136

Chen J, Han G, Shang C, Li J, Zhang H, Liu F et al (2015) Proteomic analyses reveal differences in cold acclimation mechanisms in freezing-tolerant and freezing-sensitive cultivars of alfalfa. Front Plant Sci 6:105

Chi F, Yang P, Han F, Jing Y, Shen S (2010) Proteomic analysis of rice seedlings infected by *Sinorhizobium meliloti* 1021. Proteomics 10:1861–1874

Compant S, van der Heijden MG, Sessitsch A (2010) Climate change effects on beneficial plant-microorganism interactions. FEMS Microbiol Ecol 73:197–214

Cunrath O, Gasser V, Hoegy F, Reimmann C, Guillon L, Schalk IJ et al (2015) A cell biological view of the siderophore pyochelin iron uptake pathway in *Pseudomonas aeruginosa*. Environ Microbiol 17:171–185

Diels L, van der Lelie N, Bastiaens L (2002) New developments in treatment of heavy metal contaminated soils. Rev Env Sci Biotechnol 1:75–82

Dixon R, Gang D, Charlton A, Fiehn O, Kuiper H, Reynolds T, Tjeerdema R, Jeffery E, German JB, Ridley W, Seiber J (2006) Applications of metabolomics in agriculture. J Agric Food Chem 54: 8984–8994

Duan Y, Zhou L, Hall DG, Li W, Doddapaneni H, Lin H, Liu L, Vahling CM, Gabriel DW, Williams KP et al (2009) Complete genome sequence of citrus huanglongbing bacterium, "Candidatus Liberibacter asiaticus" obtained through metagenomics. Mol Plant Microbe Interact 22:1011–1020

Egamberdiyeva D, Hoflich G (2003) Influence of growth promoting bacteria on the growth of wheat in different soils and temperatures. Soil Biol Biochem 35:973–978

Emon JV (2016) The omics revolution in agricultural research. J Agric Food Chem 64:36–44

Fransson PM, Taylor AF, Finlay RD (2005) Mycelial production, spread and root colonisation by ectomycorrhizal fungi *Hebeloma crustuliniforme* and *Paxillus involutus* under elevated atmospheric CO_2. Mycorrhiza 15:25–31

Frostegard A, Baath E (1996) The use of phospholipid fatty acid analysis to estimate bacterial and fungal biomass in soil. Biol Fertil Soils 22:59–65

Furnkranz M, Lukesch B, Müller H, Huss H, Grube M, Berg G (2012) Microbial diversity inside pumpkins: microhabitat-specific communities display a high antagonistic potential against phytopathogens. Microb Ecol 63:418–428

Ghabooli M, Khatabi B, Ahmadi FS, Sepehri M, Mirzaei M, Amirkhani A et al (2013) Proteomics study reveals the molecular mechanisms underlying water stress tolerance induced by *Piriformospora indica* in barley. J Proteomics 94:289–301

Griffiths RI, Manefield M, Bailey MJ, Whiteley AS, Ostle N, McNamara N, O'Donnell AG (2004) $^{13}CO_2$ pulse labelling of plants in tandem with stable isotope probing: methodological considerations for examining microbial function in the rhizosphere. J Microbiol Methods 58:119–129

Hamilton CE, Bever JD, Labbe J, Yang X, Yin H (2016) Mitigating climate change through managing constructed microbial communities in agriculture. Agric Ecosyst Environ 216:304–308

Hao DC, Xiao P (2017) Rhizosphere microbiota and microbiome of medicinal plants: from molecular biology to omics approaches. Chin Herb Med 9:199–217

Henckel T, Friedrich M, Conrad R (1999) Molecular analyses of the methane-oxidizing microbial community in rice field soil by targeting the genes of the 16S rRNA, particulate methane monooxygenase, and methanol dehydrogenase. Appl Environ Microbiol 65:1980–1990

Heuer H, Krsek M, Baker P, Smalla K, Wellington EMH (1997) Analysis of actinomycete communities by specific amplification of genes encoding 16S rRNA and gel-electrophoretic separation in denaturing gradients. Appl Environ Microbiol 63:3233–3241

Ikeda S, Okubo T, Anda M, Nakashita H, Yasuda M, Sato S, Kaneko T, Tabata S, Eda S, Momiyama A et al (2010) Community- and genome-based views of plant-associated bacteria: plant-bacterial interactions in soybean and rice. Plant Cell Physiol 51:1398–1410

Ineichen K, Wiemken V, Wiemken A (1995) Shoots, roots and ectomycorrhiza formation of pine seedlings at elevated atmospheric carbon dioxide. Plant Cell Environ 18:703–707

Kandasamy S, Loganathan K, Muthuraj R, Duraisamy S, Seetharaman S, Thiruvengadam R, Ponnusamy B, Ramasamy S (2009) Understanding the molecular basis of plant growth promotional effect of Pseudomonas fluorescens on rice through protein profiling. Proteome Sci 7:47

Kheyrodin H, Ghazvinian K (2012) Effect of climate change on soil global microorganisms. J Biol Chem Res 29:310–319

Kim SI, Choi JS, Kahng HY (2007) A proteomics strategy for the analysis of bacterial biodegradation pathways. OMICS 11:280–294

Knauth S, Hurek T, Brar D, Reinhold-Hurek B (2005) Influence of different Oryza cultivars on expression of nifH gene pools in roots of rice. Environ Microbiol 7:1725–1733

Knief C, Delmotte N, Chaffron S, Stark M, Innerebner G, Wassmann R, von Mering C, Vorholt JA (2012) Metaproteogenomic analysis of microbial communities in the phyllosphere and rhizosphere of rice. ISME J 6:1378–1390

Knight R, Vrbanac A, Taylor BC, Aksenov A, Callewaert C, Debelius J, Gonzalez A, Kosciolek T, McCall LI, McDonald D et al (2018) Best practices for analysing microbiomes. Nat Rev Microbiol 16:410–422

Kosova K, Vitamvas P, Urban MO, Klima M, Roy A, Prasil IT (2015) Biological networks underlying abiotic stress tolerance in temperate crops—a proteomic perspective. Int J Mol Sci 16:20913–20942

Kowalchuk GA, Bodelier PLE, Heilig GHJ, Stephen JR, Laanbroek HJ (1998) Community analysis of ammonia-oxidising bacteria, in relation to oxygen availability in soils and root-oxygenated sediments, using PCR, DGGE and oligonucleotide probe hybridisation. FEMS Microbiol Ecol 27:339–350

Leckie SE (2005) Methods of microbial community profiling and their application to forest soils. For Ecol Manage 220:88–106

Lery LMS, Hemerly AS, Nogueira EM, von Krüger WMA, Bisch PM (2011) Quantitative proteomic analysis of the interaction between the endophytic plant-growth-promoting bacterium *Gluconacetobacter diazotrophicus* and sugarcane. Mol Plant Microbe Interact 24:562–576

de Lorenzo L, Merchan F, Laporte P, Thompson R, Clarke J, Sousa C et al (2009) A novel plant leucine-rich repeat receptor kinase regulates the response of *Medicago truncatula* roots to salt stress. Plant Cell 21:668–680

Micallef SA, Channer S, Shiaris MP, Colón-Carmona A (2009) Plant age and genotype impact the progression of bacterial community succession in the Arabidopsis rhizosphere. Plant Signal Behav 4:777–780

Mishra BK, Meena KK, Dubey PN, Aishwath OP, Kant K, Sorty AM et al (2016) Influence on yield and quality of fennel (*Foeniculum vulgare* Mill.) grown under semi-arid saline soil, due to application of native phosphate solubilizing rhizobacterial isolates. Ecol Eng 97:327–333

Muyzer G, De Wall EC, Uitterlinden AG (1993) Profiling of complex microbial populations by denaturing gradient gel electrophoresis analysis of polymerase chain reaction-amplified genes coding for 16S rRNA. Appl Environ Microbiol 59:695–700

Negrel L, Halter D, Wiedemann-Merdinoglu S, Rustenholz C, Merdinoglu D, Hugueney P, Baltenweck R (2018) Identification of lipid markers of *Plasmopara viticola* infection in grapevine using a non-targeted metabolomic approach. Front Plant Sci 9:360

Nithya C, Pandian SK (2012) Evaluation of bacterial diversity in Palk Bay sediments using terminal-restriction fragment length polymorphisms (T-RFLP). Appl Biochem Biotechnol 167:1763–1777

Olukolu BA, Tracy WF, Wisser R, De Vries B, Balint-Kurti PJ (2016) A genome-wide association study for partial resistance to maize common rust. Phytopathology 106:745–751

Ostle N, Benham D, Sleep D, Ineson P (2000) Carbon assimilation and turnover in grassland vegetation using an in situ $^{13}CO_2$ pulse labelling system. Rapid Commun Mass Spectrum 14: 1345–1350

Òvreas L, Forney L, Daae FL, Torsvik V (1997) Distribution of bacterioplankton in meromictic lake Saelenvannet, as determined by denaturing gradient gel electrophoresis of PCR-amplified gene fragments coding for 16S rRNA. Appl Environ Microbiol 63:3367–3373

Parks DH, Imelfort M, Skennerton CT, Hugenholtz P, Tyson GW, Check M (2015) Assessing the quality of microbial genomes recovered from isolates, single cells, and metagenomes. Genome Res 25:1043–1055

Pedone-Bonfim MV, Lins MA, Coelho IR, Santana AS, Silva FS, Maia LC (2013) Mycorrhizal technology and phosphorus in the production of primary and secondary metabolites in cebil (*Anadenanthera colubrina* (Vell.) Brenan) seedlings. J Sci Food Agric 93:1479–1484

Pei C, Mi C, Su L, Liu W, Li O, Hu X (2017) Diversity of endophytic bacteria of Dendrobium officinalis based on culture-dependent and culture-independent methods. Biotechnolo Biotechnol Equip 31:112–119

Prosser JI, Rangel-Castro JI, Killham K (2006) Studying plant-microbe interactions using stable isotope technologies. Curr Opin Biotechnol 17:98–102

Qin Y, Druzhinina IS, Pan X, Yuan Z (2016) Microbially mediated plant salt tolerance and microbiome-based solutions for saline agriculture. Biotechnol Adv 34:1245–1259

Ramegowda V, SenthilKumar M (2015) The interactive effects of simultaneous biotic and abiotic stresses on plants: mechanistic understanding from drought and pathogen combination. J Plant Physiol 176:47–54

Rangel-Castro JI, Prosser JI, Killham K, Nicol GW, Meharg A, Ostle N, Anderson IC, Scrimgeour CM, Ineson P (2005) Stable isotope probing analysis of the influence of liming on root exudates utilization by soil microorganisms. Environ Microbiol 7:828–838

Raymond RL, Jamison VW, Hudson JO (1967) Microbial hydrocarbon co-oxidation. I. Oxidation of mono- and dicyclic hydrocarbons by soil isolates of the genus *Nocardia*. Appl Microbiol 15: 857–865

Reardon KF, Kim KH (2002) Two-dimensional electrophoresis analysis of protein production during growth of *Pseudomonas putida* F1 on toluene, phenol, and their mixture. Electrophoresis 23:2233–2241

Reid A (2011) Microbes helping to improve crop productivity. Microbe 6:435

Reinsch S, Michelsen A, Sarossy Z, Egsgaard H, Schmidt IK, Jakobsen I, Ambus P (2014) Short-term utilization of carbon by the soil microbial community under future climatic conditions in a temperate heathland. Soil Biol Biochem 68:9–19

Rispail N, Hauck B, Bartholomew B, Watson AA, Nash RJ, Webb KJ (2010) Secondary metabolite profiling of the model legume *Lotus japonicus* during its symbiotic interaction with *Mesorhizobium loti*. Symbiosis 50:119–128

Robin A, Mougel C, Siblot S, Vansuyt G, Mazurier S, Lemanceau P (2006) Effect of ferritin overexpression in tobacco on the structure of bacterial and pseudomonad communities associated with the roots. FEMS Microbiol Ecol 58:492–502

Sakurai M, Suzuki K, Onodera M, Shinano T, Osaki M (2007) Analysis of bacterial communities in soil by PCR-DGGE targeting protease genes. Soil Biol Biochem 39:2777–2784

Sarhan MS, Hamza MA, Youssef HH, Patz S, Becker M, ElSawey H, Nemr R, Daanaa HA, Mourad EF, Morsi AT, Abdelfadeel MR, Abbas MT, Fayez M, Ruppel S, Hegazi NA (2019) Culturomics of the plant prokaryotic microbiome and the dawn of plant-based culture media—a review. J Adv Res 19:15–27

Schliemann W, Ammer C, Strack D (2008) Metabolite profiling of mycorrhizal roots of *Medicago truncatula*. Phytochemistry 69:112–146

Sharma AK, Bhattacharyya PN, Rajkhowa DJ, Jha DK (2014) Impact of global climate change on beneficial plant-microbe association. Ann Biol Res 5:36–37

Shevchenko A, Sunyaev S, Loboda A, Schevchenko A, Bork P, Ens W et al (2001) Charting the proteomes of organisms with unsequenced genomes by MALDI-quadrupole time of flight mass spectrometry and BLAST homology searching. Anal Chem 73:1917–1926

Silva-Sanchez C, Li H, Chen S (2015) Recent advances and challenges in plant phosphoproteomics. Proteomics 15:1127–1141

Singh BK, Nazaries L, Munro S, Anderson IC, Campbell CD (2006) Use of multiplex terminal restriction fragment length polymorphism for rapid and simultaneous analysis of different components of the soil microbial community. Appl Environ Microbiol 72:7278–7285

Singh JS, Pandey VC, Singh DP (2011) Efficient soil microorganisms: a new dimension for sustainable agriculture and environmental development. Agric Ecosyst Environ 140:339–353

Smit E, Leeang P, Glandorf B, van Elsas JD, Wernars K (1999) Analysis of fungal diversity in the wheat rhizosphere by sequencing of cloned PCR-amplified genes encoding 18S rRNA and temperature gradient gel electrophoresis. Appl Environ Microbiol 65:2614–2621

Sorty AM, Meena KK, Choudhary K, Bitla UM, Minhas PS, Krishnani KK (2016) Effect of plant growth promoting bacteria associated with halophytic weed (*Psoralea corylifolia* L.) on germination and seedling growth of wheat under saline conditions. Appl Biochem Biotechnol 180: 872–882

Tank N, Saraf M (2010) Salinity-resistant plant growth promoting rhizobacteria ameliorates sodium chloride stress on tomato plants. J Plant Interact 5:51–58

Valdes A, Ibanez C, Simó C, García-Cañas V (2013) Recent transcriptomics advances and emerging applications in food science. Trends Anal Chem 52:142–154

Vinale F, Sivasithamparam K, Ghisalberti EL, Marra R, Barbetti MJ, Li H et al (2008) A novel role for Trichoderma secondary metabolites in the interactions with plants. Physiol Mol Plant Pathol 72:80–86

Vorholt JA (2012) Microbial life in the phyllosphere. Nat Rev Microbiol 10:828–840

Waldon HB, Jenkins MB, Virginia RA, Harding EE (1989) Characteristics of woodland rhizobial populations from surface- and deep-soil environments of the Sonoran Desert. Appl Environ Microbiol 55:3058–3064

Wang LY, Xie YS, Cui YY, Xu J, He W, Chen HG et al (2015) Conjunctively screening of biocontrol agents (BCAs) against fusarium root rot and fusarium head blight caused by *Fusarium graminearum*. Microbiol Res 177:34–42

Wang Y, Hu B, Du S, Gao S, Chen X, Chen D (2016) Proteomic analyses reveal the mechanism of *Dunaliella salina* Ds-26-16 gene enhancing salt tolerance in *Escherichia coli*. PLoS One 11: e0153640

White DC, Ringelberg DB (1998) Signature lipid biomarker analysis. In: Burlage RS, Atlas R, Stahl D, Geesey G, Sayler G (eds) Techniques in microbial ecology, Techniques in microbial ecology. Oxford University Press, New York

Wu CH, Bernard SM, Andersen GL, Chen W (2009) Developing microbe–plant interactions for applications in plant-growth promotion and disease control, production of useful compounds, remediation and carbon sequestration. J Microbial Biotechnol 2:428–440

Xu J, Lan H, Fang H, Huang X, Zhang H, Huang J (2015) Quantitative proteomic analysis of the rice (Oryza sativa L.) salt response. PLoS One 10:e0120978

Zhalnina K, Louie KB, Hao Z, Mansoori N, da Rocha UN, Shi S, Cho H, Karaoz U, Loqué D, Bowen BP, Firestone MK, Northen TR, Brodie EL (2018) Dynamic root exudate chemistry and microbial substrate preferences drive patterns in rhizosphere microbial community assembly. Nat Microbiol 3:470–480

Zhang B, He H, Ding X, Zhang X, Zhang X, Yang X, Filley TR (2012) Soil microbial community dynamics over a maize (*Zea mays* L.) growing season under conventional- and no-tillage practices in a rainfed agroecosystem. Soil Tillage Res 124:153–160

Zhu L, Zhou Y, Li X, Zhao J, Guo N, Xing H (2018) Metabolomics analysis of soybean hypocotyls in response to *Phytophthora sojae* infection. Front Plant Sci 9:1530

Functional Diversity of Microbes in Rhizosphere: A Key Player for Soil Health Conservation Under Changing Climatic Conditions

Barkha Sharma and Kailash Chand Kumawat

Abstract Under changing climatic conditions, the negative impact on the soil leads to damage and destruction of soil and plant health. Here, plant microbiome (including both rhizospheric and phyllospheric) is crucial in plant growth promotion, recycling of nutrient, and reestablishing a safe and sustainable ecosystem. The microbes associated with plant growth-promoting (PGP) traits have surfaced as substantial and conducive sustainable agriculture tool. PGP microbes stimulate growth of plant either directly via releasing plant growth regulators, solubilizing phosphorus, potassium, and zinc, and biological nitrogen fixation or indirectly by the production of siderophore, ammonia, HCN, and other secondary metabolites, which are hostile to pathogenic microbes. The PGP microbes belonged to different genera such as *Achromobacter*, *Aspergillus*, *Azospirillum*, *Azotobacter*, *Bacillus*, *Erwinia*, *Methylobacterium*, *Paenibacillus*, *Pantoea*, *Penicillium*, *Planomonospora*, *Pseudomonas*, *Rhizobium*, *Serratia*, and *Streptomyces*. These PGP microbes might be utilized as biofertilizers/bioenhancers/bioprotectants/ bioinoculants as a green alternative of agrochemicals in sustainable agriculture. The use of PGPM/AMF thus constitutes a potential strategy for increasing the availability of essential nutrients and speeding up sustainable agriculture augmentation. Henceforth, PGPR can be considered as amenable, sustainable, and inexpensive tool for abiotic and biotic stress tolerance and productivity of crops/plants.

Keywords Nutrient solubilization · Rhizospheric microbes · Siderophore production · Sustainable agriculture

---

B. Sharma
Department of Microbiology, G. B. Pant University of Agriculture & Technology, Pantnagar, Uttarakhand, India

K. C. Kumawat (✉)
Department of Microbiology, Punjab Agricultural University, Ludhiana, Punjab, India

Department of Industrial Microbiology, Jacob Institute of Biotechnology and Bio-engineering Sam Higginbottom University of Agriculture, Technology and Sciences, Prayagraj, Uttar Pradesh, India
e-mail: kailash-cobsmb@pau.edu; kailash.kumawat@shiats.edu.in

© The Author(s), under exclusive license to Springer Nature Singapore Pte Ltd. 2022
A. Vaishnav et al. (eds.), *Plant Stress Mitigators*,
https://doi.org/10.1007/978-981-16-7759-5_22

Introduction

Worldwide, the exponential growth due to industrialization and urbanization has placed enormous constraints on the available cultivated land to increase agricultural production. In intensive cropping system, the exorbitant and inappropriate usage of agrochemicals, *i.e.*, pesticides, herbicides, inorganic/synthetic fertilizers, and various types of fungicides, have been adopted. However, there can be no ignorance toward the rising awareness and concern about its harmful impact on soil fecundity plus environmental excellence. The elevated expense of noxious agrochemicals, difficulty to satisfy their demand, and their detrimental environmental legacy have urged agricultural scientists to explore or originate a substitute to enhance crop productivity in a sustainable manners, with rhizospheric microbiome the central function (Vaxevanidou et al. 2015; Kumar and Gopal 2015). The seldomness of beneficial rhizospheric microbiome and their frequently surprising nature and biosynthetic abilities have made them potential candidates for resolving particularly challenges of bioscience and other agricultural fields as well.

Inherently attractive in getting social, economic, and environmental advantages, the inoculation of native rhizospheric microbiome also determines a magnificent advancement of research to offer effective means of protecting environment and its surveillance for sustainable agriculture system (Cai et al. 2013). Rhizospheric microbiome is crucial due to their various unique properties such as secondary metabolites and plant growth regulator production, biological nitrogen fixation, and essential nutrient solubilization ability (Kour et al. 2019a, 2019b; Rana et al. 2019). These microbial potentials encourage researchers to comprehend the intensely diversified undiscovered rhizospheric microbiome, not just for environment sustenance but also to protect this beneficial microbial diversity for humankind (Dwibedi and Saxena 2019). Hence, un-tapping and deeper knowledge of entire rhizospheric microbes are becoming increasingly essential to study the positive interactions for sustainable agriculture, notably during rhizospheric microbiome-based cropping approaches (Busby et al. 2017). This chapter attempts to elucidate the latest trend of roles of beneficial plant growth-promoting microorganisms (PGPMs) in addition to their cardinal underlying processes encouraging plant growth and agricultural productivity in the sustainable manners.

Characterization of Root Microbiome

The rhizosphere, a hotspot for diverse microorganisms and most complex ecosystem on earth, is a very narrow region surrounded and influenced by plant root (Hinsinger et al. 2009) (Table 1). Rhizospheric microbiome harbors thousands of organisms including bacteria, fungi, oomycetes, nematodes, protozoa, algae, viruses, archaea, and arthropods (Kapoor and Mukerji 2006; Kour et al. 2019a, 2019b). The structure and composition of microbial community affecting dynamics of rhizosphere depend on the following factors such as (i) root exudates: Root exudates or rhizodeposits

Table 1 Types of microorganisms present in the rhizosphere of different plants

Plant	Microorganism	References
Arabidopsis thaliana	*Bacillus subtilis*	Ryu et al. (2005)
Brassica napus	Pseudomonads	Mahaffee and Kloepper (1997)
Brassica oleracea var. italica	AM fungi	Tanwar et al. (2014)
Capsicum frutescens	*Trichoderma viride; P. fluorescens*	Ananthi et al. (2014)
Cicer arietinum	*P. fluorescens; Trichoderma; Rhizobium; Mesorhizobium*	Yadav et al. (2013); Tripathi et al. (2013); Verma et al. (2014)
Cucumis sativus	*Trichoderma*	Moharam and Negim (2012)
Fragaria × *ananassa* (Strawberry)	*Pseudomonas; Bacillus*	Esitken et al. (2010)
Oryza sativa	*Citrobacter freundii*	Nguyen et al. (2003)
Pennisetum glaucum	*P. fluorescens*	Arora et al. (2003)
Pisum sativum	*Pseudomonas fluorescence, P. putida, Rhizobium leguminosarum*	Ma et al. (2003)
Solanum lycopersicum	*B. amyloliquefaciens, Burkholderia cepacia*	Adesemoye et al. (2010); Recep et al. (2009);
Trifolium repens	*Rhizobium leguminosarum*	Lemanceau et al. (1995).
Triticum aestivum	*Aeromonas hydrophila, B. insolitus; Azospirillum brasilense; Alcaligenes faecalis; B. cereus; Enterobacter hormaechei*	Ashraf et al. (2004); Dobbelaere et al. (2003); Egamberdiyeva (2006);
Zea mays	*Enterobacter sakazakii; Azospirillum; Bradyrhizobium; Ideonella Klebsiella oxytoca*	Babalola et al. (2003); Roesch et al. (2007); Babalola and Odhiambo (2008)

(*e.g.*, exudates—dead fine roots, gases, low and high molecular weight organic compounds including amino acids, sugars, mucilage, and phenolics, border cells—microbial proteins, phytoalexins, antibiotics, extracellular enzyme, and growth factors) are primary drivers, which govern rhizospheric microbiome and their effects; hence, plants regulate the diversity of rhizospheric zone according to benefits in terms of growth and health, and it receives from the selective and unique microbiota (Cook et al. 1995); (ii) physical together with chemical composition of rhizosphere soil, e.g., requirement of water as plants in species-rich communities utilize more soil water compared to those with low diversity; and (iii) plant surface: Especially root surface zone is found to be the most active zone in terms of selective presence of diverse taxonomic range of microbial communities than rhizosphere soil or other parts of plants where microbial population appears stable (De la Fuente Cantó et al.

2020). Selective influence of wheat, clover, bentgrass, and ryegrass on rhizospheric microbial communities due to the secretion of carbohydrates, carboxylic acids, and amino acids by plant roots is solely responsible for the discrimination in the microbial diversity. This selection, *i.e.,* "rhizosphere effect," occurs at strain level instead of at higher taxonomic levels; (iv) plant genotype: The genotype of plant also influences the composition of rhizosphere microbiome. The rhizosphere of wild diploid with tetraploid wheat variety comprises greater proportion of bacterial and actinomycete populations than the hexaploid variety (White 2008). The diploid DD and VV chromosomes controlling the stimulation and secretion of exudation regulate these two groups. (v) Climate conditions: Amid growth season, the environment of soil ecosystem endures severe changes resulting in altered local climate and then fluctuated microbial communities and associated nutrient dynamics (Dos Santos et al. 2021).

Bacteria

Rhizospheric environment harbors about 10^9 bacterial species in one gram soil and their composition depends on zone of root, type, health and age of plant, root exudation, condition of soil and environment, and neighbor community (Rovira 1965). Depending on the technique used for isolation and characterization, the total number of bacterial operational taxonomic units (OTUs) fluctuates from <100 to more than 55,000. The dominant rhizobacterial phyla are Proteobacteria, Actinobacteria, Bacteroidetes, and Firmicutes together with Acidobacteria (Mendes et al. 2011). In rhizosphere, the most abundant microbial communities have great impact on the growth and development of plant through metabolic activities of plant growth-promoting rhizobacteria (PGPR), *e.g.*, *Bacillus, Rhizobium, Leclercia, Pseudomonas*, *Serratia*, *Proteus*, *Stenotrophomonas*, *Burkholderia*, *Comamonas*, *Sphingobacterium* sp., *Flavobacterium* sp., *Verrucomicrobium*, *Deinococcus sp.*, and *Acidobacterium.* In a report, 5619 OTUs dominated by *Acidobacteria* and *Proteobacteria* were detected in oak rhizosphere and *alpha-*, *beta-*, and *gammaproteobacteria* were detected in rhizosphere of *Xerocomus pruinatus* and *Scleroderma citrinum* using pyrosequencing (Uroz et al. 2010; Uroz et al. 2012).

Fungi

Among different microorganisms, fungi hold prominent position and are found associated with rhizosphere either as pathogens or as symbionts. The type of their interactions, which regulate the abundance of specific community, depends on several intrinsic and extrinsic phenomena such as type of soil, development of plant, and diverse root exudates (Bhaduri et al. 2015). They regulate the plant rhizosphere and soil ecosystem by controlling the nutrient cycling, environment sustainability, and plant development and growth. The prominent rhizospheric soil fungi majorly belong to Zygomycetes, Ascomycetes, Basidiomycetes, and Glomeromycetes, in which Zygomycetes and Deuteromycetes establish as saprophytes due to their fast growth *via* metabolizing simple sugars, *Gibberella* establish as symbionts, which stimulate plant growth, whereas *Pythium, Rhizoctonia, Aphanomyces,* and *Armillaria* (only during flooded condition) are plant pathogenic fungi (Hays and Watson 2019). Mycorrhizal fungi are unique and ubiquitous group of fungi, which fabricate mutualistic interaction with almost 80% of the terrestrial plants. They maintain the rhizosphere dynamics by not only providing an increasing the surface area for maximum nutrient and water uptake but also traversing through the small pores, which is impossible by the roots. Most of the time this association is beneficial, but there can be time when they become mildly detrimental to plant or plant started to feed on them (Tecon and Or 2017). Mycorrhizal fungi can be divided into two categories: ectomycorrhiza (composed of hyphal sheath/mantle covering the tip of roots and hartig net of hyphae that weave between the plant root cortex) and endomycorrhiza (fungus penetrate root cortical cells of vascular plant). *Boletus betulicola* is a small host ranged, whereas *Pisolithus tinctorius* is wide hot ranged (established association with 46 tree species and 8 genera as reported yet) ectomycorrhiza (Ajwa 2015). Arbuscular mycorrhizal fungi (AM fungi) are paramount mycorrhizal group as they make symbiotic association with most of the land plant, mangrove plant, and grasslands because their branched arbuscules grow within the cortex of roots to facilitate the mutual transfer of nutrients and carbon. Some AM fungi include *Glomus tenue*, *Rhizophagus irregularis*, and *Scutellospora* (Orlowska et al. 2002; Pierre et al. 2014).

Over and above 80% of crops such as wheat, soybean, and corn existing in the ancient and ubiquitous symbiotic association with AMF attain phosphorus (P), water, and other nutrients through the hypophosphoric microbes colonizing extraradical hyphae from soil. AMF lacks the ability to release phosphatases outside the hyphae for the solubilization of organic P present in soil; henceforth, they attract rhizospheric P-solubilizing bacteria (PSB), which are deficient of carbon source, by secreting hyphal exudates enriched with carbon. Colonization of PSB on AMF hyphal surface stimulates the release of alkaline phosphatase (ALP) and phytases to facilitate organic and phytate P mineralization (Zhang et al. 2018). The formation of arbuscules upregulates the P transporter genes and expression of the transporter proteins on periarbuscular membrane of the host plant. The increase in P mineralization also leads to alteration in native rhizospheric microbiota and recruitment of more ALP-producing bacterial communities (Kobae 2019).

Archaea

Archaea domain is unique and cosmopolitan, and found inhabiting soils, oceans, human skin, and even extremophilic condition (arid and semi-arid regions). They are important in the view of decomposition and nutrient (nitrogen, carbon, and sulfur) cycling, most prominently in ammonification in soil, nitrification in ocean, and methane production (Buee et al. 2009). Crenarchaeota and Euryarchaeota are dominant in rhizosphere soil. They coexist with bacteria as they fill different habitats in a shared common space. Archaeal communities are found associated with the major crop of rice (Van As 2012). Rice crop is known for the 10–25% global production of methane gas because its roots provide a healthy niche to methane emitting archaea such as members of Methanosarcinaceae, Methanosaetaceae, Methanomicrobiaceae, and Methanobacteriaceae families. A specific archaeal monooxygenase genes found in *Littorella uniflora* genome contribute in nitrification in freshwater sediments. The root exudates (such as oxygen and carbon) from the root of rice attract ammonia-oxidizing archaea, hence increasing their number in rhizospheric soil rather in bulk soil (Chin et al. 2004). As per previous studies, it can be concluded that archaea is common but not major rhizospheric microbiota, which is vital for sustainable ecological functions in plants.

Culturable Rhizomicrobiome

A traditional approach for the identification of composition of microorganisms by cultivation on defined media and then evaluating and characterizing the abundance of genes through molecular techniques such as next-generation sequencing technologies, e.g., Roche 454 and Illumina sequencing. This technique is suitable for cultivable aerobes, facultative anaerobes, denitrifying bacteria and archaea, and methanogens (Li et al. 2014). According to the rhizosphere microbial ecology, the availability of specific signaling molecules regulates the activity and composition of microbial communities in which proximal appearance of symbionts and fast-growing opportunistic species (*i.e.*, r-strategists than k-strategist) around the roots. *Pseudomonas*, *Burkholderia*, and *Rhizobium* are microbes that can be easily identified by culture-dependent approach (Winkler et al. 2017). Root-colonizing microorganisms exhibiting PGP traits such as phosphate solubilization, siderophore production, antimicrobial secretion, and nitrogen fixation are frequently higher in activity and composition than in bulk soil (Vaikuntapu et al. 2014). Garbeva et al. (2008) demonstrated the higher relative abundance of *Rhizoctonia solani* bacterial antagonists in comparison with the bulk soil in maize and *Lolium* sp.

Unculturable Rhizomicrobiome

Owing to the fact that culture-dependent techniques identify only 1% of microorganisms present in a natural environment and remaining 99% cannot be cultured on designed media, and unculturable methods allow the characterization of unknown microbes (VerBerkmoes et al. 2009). One of the high-throughput culture-independent strategies, *i.e.*, metagenomics, has higher probability of finding novel microbes having useful characteristic features and also helps in uncovering the hidden feature involved in plant growth and support. One striking example is the recently discovered novel bacterial group of Acidobacteria in the rhizosphere of oak by metagenomics as they are difficult to cultivate on media (Uroz et al. 2010). Other rhizosphere-dominant groups infer through uncultured techniques are Firmicutes, Actinobacteria, Bacteroidetes, and Proteobacteria (Lagos et al. 2015). *Massilia* sp., belonging to oxalobacteraceae family (β-proteobacteria), were principally isolated from clinical samples and assigned as novel genus not more than 15 years ago (Zhang et al. 2006). Through culture-independent techniques, it was detected in many environmental samples such as dust air and soil worldwide as well as placed in dominant root-colonizing category of microorganisms, which was underestimated in the course of use of culturable techniques (Blatny et al. 2011).

Climate change is the major concern globally. Most of the changes are detrimental to the living system and environment. Major regulators of climate destruction are misbalance in CO_2 and O_2 concentration, temperature, humidity, precipitation, and UV radiation (Ali 2012). As all the environment changes greatly exert impact on soil, rhizosphere also affected severely. Rhizosphere being the hotspot and most active zone for microorganisms reflects this alteration on soil and plant health. Elevated CO2 emission increases photosynthesis and water uptake efficiency in C_3 and C_4 cycling in plants, but these effects vary according to species of plant and all biotic and abiotic factors. It also alters the root exudates, hence alter the composition of microbial community (Augustine et al. 2011). There are limited data related to climate change effect on microbial community, but all reports suggest a long-term shift in microbiome.

Interactive Mechanisms Between Root and Microorganisms

Soil microorganisms modify the rhizosphere environment according to their needs also making it more favorable to live (Jouquet et al. 2006). Their contribution in maintaining the health and fitness of host plant is also significant. A myriad of interactions exist in rhizosphere neutral interactions, positive symbiotic interactions, and negative pathogenic interactions depending on the type of partners involved in the association (Somers et al. 2004). Neutral interactions involve no partnership or two species are neither benefitted nor harmed from each other in an ecological niche. This kind of interactions is extensively common and found in the rhizosphere of all plant. Saprophytes are one example of neutral interaction, as they are crucial for the organic matter decomposition, which is a major step of nutrient cycling, hence their impact (beneficial or detrimental) on plant is indirect (Brimecombe et al. 2007).

Positive microbial interactions involving benefits for the host plant can be categorized into three classes.

(i) Microbial partner is responsible for plant nutrients (either in generation of nutrients or in their uptake). This class of microbial partner may or may not in direct communication with plant but affects soil biotic or abiotic factors: (ii) In biocontrol agents, these microorganisms protect plant host from pathogens and (iii) plant growth-promoting microorganisms (PGPMs) directly associate with the plant production and regulation (Tellier et al. 2010).

Symbiosis or mutualisms is a positive interaction in which both partners are benefitted from each other either in terms of nutrients, space, or protection. This relationship can be permanent in which their association is necessary for their survival or they make association with each other according to their needs (Selim and Zayed 2017). In the association of leguminous plant species and nitrogen-fixing bacterial species (e.g., *Rhizobia*), the bacteria convert the atmospheric inert nitrogen into biologically useful ammonia, which is utilized by plants and plants provide carbon source and living space to bacteria. Symbiotic relationship is also known to exist between plant roots and fungi (Kiers and Heijden 2006). A brilliant example is mycorrhizal association, especially arbuscular mycorrhiza (AMs) described in section. *Laccaria Bicolor* provide nitrogen and other nutrients to *Populus* and *Arabidopsis* plants. Also, it also confers the coastal grasses to survive during heat and salt stress (Orłowska, 2002). A mutualistic *Bacillus subtilis* induce systemic resistance (ISR) in plant by secreting volatile compounds like butanediol by regulating plant's sodium ion transporter. The mutualist *B. subtilis* released butanediol, which elicit ISR by the modulation of transcription of sodium ion transporter 1 in plants (Lin et al. 2020). Commensalism is other positive interaction in which one partner is benefitted, while other remains unaffected. *Bacillus cereus* stimulates growth of *Cytophaga–Flavobacterium* group by releasing peptidoglycan in soybean rhizosphere (Peterson et al. 2006).

Detrimental associations occurring in rhizosphere include competition, parasitism, antagonism, and amensalism. Competition is actually a negative interaction in which both partners experience a reduced fitness owing to the interaction. Completion amid two species is either for nutrients (carbon and oxygen sources), space, or mates (Singh and Faull 2020). In rhizosphere, nutrient and living space are main reason for competition between soil-inhabiting microbes. Microorganisms use several strategies like antibiotic or poisonous compound production, to win over opponent. They also secrete some compound to signal the increase in surface area of root for increased area for colonization (Boddy 2016). Some microorganisms (e.g., *Pseudomonas*) have competitive advantage to utilize other metabolic strategies to sequester glucose effectively by converting it into gluconic acid and 2-ketogluconic acid (Nihorimbere et al. 2011). Completion for trace elements such as manganese, zinc, copper, and iron is also very significant between rhizospheric microbiome, because these elements are essential for microbial growth and their scarcity can lead them to death (Zia et al. 2020). Siderophore-producing bacteria sequester iron by solubilizing it into ferric ion under limiting conditions (Loper 1988). Plant pathogenesis is major concern issue, and these pathogens can cause mild diseases or destroy a million acre of crops and natural resources. Dominant plant pathogens are fungi. Parasitism involves a parasite, which harms its host and lives on it until the host dies. When the harm causes disease in host, the process is termed as pathogenesis. Generally, host tries to develop a defense mechanism, but when this mechanism fails and pathogen invades all pre-formed resistance barriers, the pathogenesis occurs. Fungal pathogens secrete plant cell wall-degrading substances. Common plant pathogens are *Erwinia carotovora, Ralstonia* spp., *Phytophthora infestans, and Pythium* sp. (Newton et al. 2010).

Antagonisms between two partners mean one partner put forth adverse effect on other while there is no impact on first partner. Some microbe releases enzymes like cellulose, laccase, chitinase, protein, and lipid to lyse the cell wall of their antagonistic partner. In amensalism, one of the partners leaves detrimental effects on other while it remains unaffected. The microbes involved in amensalism produce antimicrobial substances (antibiosis) that can affect either temporarily leading to damage of the host or permanently like completely killing the host. Some microorganism releases these compounds when confronted with the predators as defense mechanism, e.g., *Pseudomonas* sp. (Legaz et al. 2018). These chemicals can also negatively affect the plant growth-promoting microorganisms leading to loss of rhizospheric diversity. The antimicrobial compounds can be antibiotics, which are "low molecular weight" secondary metabolites that interfere or block the metabolic process of host. *Pseudomonas fluorescens* releases 2, 4-diacetylphloroglucinol (DAPG) to inhibit the growth of pathogen *Erwinia carotovora* (Vey et al. 2001). Other bacteria like *Bacillus*, *Streptomyces,* and *Pseudomonas* secrete bioactive lipopeptides (such as viscosin and massetolide), which efficiently destroy the cell integrity of *N. americana* resulting in its death. Other antimicrobial is rosmarinic acid (RA), released from *Ocimum basilicum* to inhibit *Phytophthora cinnamomi* (Abedini et al. 2013).

Factor Affecting the Root Microbiome Structure and Function

Rhizospheric microbiome, which affects plant structure and function, is known plant growth-promoting microorganisms (PGPMs). PGP rhizobacterial genera include *Bacillus subtilis*, *Pseudomonas*, *Azospirillum*, *Serratia*, and *Rhizobium*, and PGP fungi include *Trichoderma* and nonpathogenic strains of *Fusarium oxysporum*. They can be used directly as crop inoculants. The impact caused on the plant can be direct or indirect. PGPMs enhance plant growth by biofertilization, phytostimulation, and biocontrol (Rana et al. 2020).

Direct Impact of Climate Change on Root Microbiome

Direct PGPMs live in close vicinity with the roots where they grow in and around the plant tissue, thus provide mechanical support, better nutrient, and mineral supply

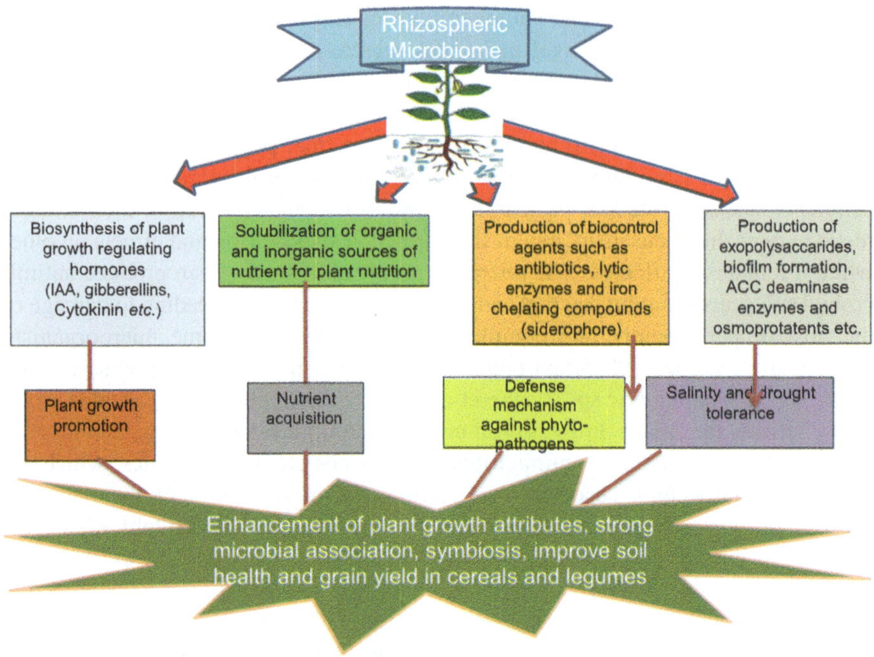

Fig. 1 Systematic diagram represented direct or indirect mechanism of plant growth promotion by rhizospheric microbiomes

(Kumar et al. 2019). Various direct mechanisms, which are regulated by root microbiome, are as follows and depicted in Fig. 1

Nitrogen Fixation

Nitrogen (N) is a fundamental plant essential nutrient, and a constrictive factor in agriculture due to its negligible amount of accessible form owing to emission or leaching; henceforth, bacteria with potential for making 78% inaccessible atmospheric N available for plants in the form of ammonia by nitrogenase complex play a critical role (Anas et al. 2020). Biologically, nitrogen-fixing (BNF) bacteria are ubiquitous. Biological N fixation can be performed by microbes: symbiotically and nonsymbiotically. Symbiotic N fixation is the chief mechanism for maximum atmospheric N fixation, but it is restricted to only leguminous plants and trees forming interactions such as mycorrhizal roots with *Frankia*. The most acknowledged symbiotic microbes are *Anabaena, Nostoc*, *Rhizobium*, *Bradyrhizobium*, *Sinorhizobium*, and *Mesorhizobium* (Roper and Gupta 2016). Contrarily, free-living diazotrophs perform the nonsymbiotic biological N fixation and they elicit the growth of nonleguminous plants. Several studies demonstrated that N-fixing bacteria, free-living together with *Rhizobium* strain, efficiently prompt the development parameters of nonleguminous plants, e.g., radish and rice, reducing dependency on N-based fertilizers. Nonsymbiotic N-fixing bacterial genera such as *Azoarcus*, *Azospirillum*, *Burkholderia*, *Gluconacetobacter*, and *Pseudomonas* were isolated from different rhizospheric soils (Kennedy et al. 2004).

Phosphate Solubilization

Phosphorus (P) is a second most vital nutrient for plants with poor attainability in numerous agricultural soils. Nowadays, several agricultural soils contain relatively elevated overall P content because of the prolonged use of P-containing fertilizers. Nevertheless, large proportion of this P is present in the mineral forms (monobasic (H2PO4) and the diabasic (H2PO4) ions) and is only slowly available to plants. Majority of the insoluble P forms compound with aluminum and iron in acidic soils, and calcium in basic soils (Schneider et al. 2019). The solubilization of insoluble organic P has been endowed to the efficacy of rhizospheric microbiota to lower pH by releasing organic acids (e.g., gluconate, citrate, succinate) and protons (amid NH^+ assimilation). In several previous researches, focusing P-solubilizing microbes (PSMs) including *Bacillus*, *Flavobacterium*, *Enterobacter*, *Aspergillus*, *Klebsiella*, *Kluyvera*, *Fusarium*, *Streptomyces*, *Pantoea*, *Sclerotium*, *Serratia*, and *Pseudomonas* genera were isolated from diverse rhizospheric soils. These microbes grow on media containing insoluble P (*e.g.*, tricalcium phosphate). They do not just assimilate P, but solubilize it in plethora more than their nutritional requirements as well, *ergo* making it accessible for plant uptake. Hence, PGPMs recorded to solubilize phosphorus exhibit a possible condition in field for better plant growth (Thakur et al. 2014).

Siderophore Production

Like N and P, iron (Fe) is also present on earth surface, yet unavailable to plant. PGPMs secrete a Fe-chelating molecule, siderophore, to sequester insoluble Fe^{3+} by reducing it into soluble Fe^{2+}. Microbial siderophores are low molecular weight small peptides molecules having side chains and functional groups, which offer a high-affinity set of ligands to coordinate ferric ions. They can also form complex with metal including Al, Cd, Cu, Ga, In, Pb, and Zn (Khan et al. 2019). During Fe limiting conditions, plants detect the bacterial ferric–siderophore complex through direct communication with bacteria, hence improving the Fe uptake. They also regulate enhanced plant Fe uptake when other metals such as nickel and cadmium are present. The process involves the following steps: binding of Fe^{+3} with microbial siderophore, translocation of siderophore-Fe (1:1) complex inside microbial cell, reduction of Fe^{+3} to Fe^{+2}, and release of Fe^{+2} outside through iron transporters. As they also make complex with other metals, henceforth, they can be used to alleviate metal toxicity in plants *e.g.*, arsenic toxicity (Saha et al. 2013).

Phytohormone Production

Phytohormones occurring in low concentrations are small signaling molecules that act as chemical messengers and know to be involved in regulation of growth and development of plants. These organic compounds regulate biochemical, physiological, and morphological mechanisms in plants and precisely control their synthesis. Myriad of fungal and bacterial species can secrete and modulate plant hormones and their level in plants (Pindi et al. 2014).

Indole-3-Acetic Acid (IAA)

Vast majority of the essential plant–microbial interactions focused on auxin production. IAA is a major plant auxin. 80% of rhizospheric microbiota synthesize and release auxin as secondary metabolite. The IAA is responsible for plethora of plant development processes including division, expansion, and differentiation of plant cells and tissues, stimulation lateral and adventitious root elongation, seed and tuber germination, synthesis of various metabolites, stress resistance, and many more (Spaepen et al. 2007). Rhizobacteria produce IAA in a variety of methods, although it may also be synthesized in smaller quantities by tryptophan-independent mechanisms (Goyal et al. 2019). Phytopathogens mostly manufacture the IAA *via* índole acetamide pathway to stimulate tumor production in plants. Utility of this pathway in beneficial bacteria is not obvious. Contrastingly, the acid indole pyruvic pathway is the primary route in PGP beneficial bacteria. Root elongation by IAA facilitates increased water and nutrient absorption by plants due to increased surface area. Rhizobacteria also release IAA to loosen the cell wall, which in turn releases more root exudates to the surroundings, hence attracting more beneficial microbiota (Martínez-Viveros et al. 2010).

Cytokinins are prevalent in bacteria, algae, and higher plants. Their major roles include cell division, elongation of roots, seed germination, flower and fruit germination, nutrient signaling, delayed senescence, and plant–pathogen interaction, and they are also regulated by auxins and control apical dominance in plants, which is also a major purpose for their agricultural utilization in yield and quality improvement of crops (Smolikova et al. 2018; De Salamone et al. 2005). Kinetin, a type of synthetic cytokinins, is produced by yeast. Several PGPR are known to express cytokinins genes, e.g., inoculation of *Bacillus subtilis* increases plant growth by increasing cytokinin concentration. Genetically modified *Sinorhizobium meliloti* overproduces cytokinin (5× than wild type) to uphold the alfalfa plant against drought stress (Liu et al. 2013).

Gibberellins (GAs) are chemicals produced naturally by plants but also found in bacteria and fungi. The GAs is tetracyclic diterpenoid carboxylic acids with C_{20} or C_{19} carbon molecules in which GA_1, GA_3, GA_4, and GA_{20} are produced by bacteria (Kundan et al. 2015). Bacterially produced GAs promote plant growth in exogenic association with the plant-producing GAs. GA stimulates and activates vital plant growth processes comprising stem elongation, root growth inhibition, seed germination, sex expression, flowering, fruit setting, dormancy, photosynthesis rate and cellular respiration, and senescence. They also act as transducers/chemical messenger of elicitor signals. The predominant GA-producing PGPB are *Achromobacter xylosoxidans*, *Gluconobacter diazotrophicus*, *Rhizobia*, *Azotobacter* sp., *Bacillus* sp., and *Azospirillum* sp. (Olanrewaju et al. 2017).

Ethylene

Ethylene is essential phytohormone, which promotes plant growth and development, but its impact on plant growth varies according to its concentration in root tissues. For example, its elevated concentrations are hazardous, due to the induction of defoliation and cellular processes, leading to stem and root growth inhibition in addition to premature senescence, ultimately resulting in reduction in crop functioning (Li et al. 2005). During the course of environmental stress including both biotic and abiotic, such as cold, draught, flooding, waterlogging, pathogenic infections, radiation, extreme temperatures, heavy metal and polyaromatic hydrocarbon toxicity, and salinity, plants start to synthesize 1-aminocyclopropane-1-carboxylate (ACC) (precursor for ethylene) (Tahir and Sarwar 2013). A fraction of this ACC is released out into the rhizosphere where it is reabsorbed by plant roots and then transformed into ethylene. This ethylene aggregation results in a downward spiral effect, which causes poor root growth due to diminished water and nutrient absorption, hence culminate a further stress. Some PGPR possess ACC deaminase enzyme, which helps in the degradation of ACC and breaks this downward cycle through decreasing ethylene level, thereby restoring healthy root system required to overcome environmental stress (Zarembinski and Theologis 1994). Rhizomicrobial genera exhibiting ACC deaminase include *Pseudomonas, Rhizobium, Acinetobacter, Bacillus, and Burkholderia* equilibrium in ethylene level that lead to better root and shoot development, mycorrhizal colonization in several plants, better rhizobial nodulation, enhanced N, P, and K uptake, *etc*. (Sapre et al. 2019).

Indirect Impact of Climate Change on Root Microbiome

Indirect effects caused by climate change also trigger entire microbial biomass of a habitat, which in turn stimulates several mechanisms of action involving the ability for antibiotic production, competing with phytopathogens for mineral and nutrients, exclusion of niche, activation of defense system *via* induced systemic resistance, and production of lytic enzymes, cyanide (HCN), ammonia, bacteriocin, and antimicrobial compounds. The microbes possessing these abilities for plant protection from pathogen are known as biocontrol agents (Cesco et al. 2012). Several root-colonizing microbes are known to produce antibiotics, *e.g*., oligomycin A, xanthobaccin, viscosinamide, pyoluteorin, tensin, and 2, 4-diacetyl phloroglucinol

(2, 4-DAPG). Antibiotics comprise a heterogeneous group of organic and low molecular weight compounds, which are detrimental to other microbes' growth or metabolic activities. Common antibiotic-producing microbes are *Bacillus*, *Pseudomonas*, *Burkholderia*, *etc*. One or more antibiotic production is the core mechanism linked with the PGP bacterial efficacy to function as antagonistic agents against phytopathogens (Raaijmakers and Mazzola 2012). Multiple antibiotics are isolated from both fungi and bacteria and their hyper-diversity interferes with the processes such as hindrance/inhibition of pathogen cell walls synthesis, influencing membrane structures of cells, and inhibition of the formation of initiation complexes on the ribosomal small subunit (Maksimov et al. 2011). *Pseudomonas* produces pyrrolnitrin to inhibit wide range of pathogens belonging to order Basidiomycota, Deuteromycota, and Ascomycota, especially *Rhizoctonia solani* and *Sclerotina sclerotiorum* (Beneduzi et al. 2012). Bacteriocins vary critically from standard antibiotics: They often have a comparatively narrow killing spectrum and are uniquely noxious to bacteria exhibiting close genetic linkage with the producer strain. Intriguingly enough, bacteriocins of *Bacillus* sp. are growing more noteworthy because of their occasionally broader inhibition spectra (including gram-negative bacteria, yeasts or fungi, some human/animal pathogenic gram-positive species) than that of most lactic bacterial bacteriocins (Gillor et al. 2008; Abriouel et al. 2011).

Competition for nutrients and habitat for colonization at the root surface is another one of indirect effects of root microbiome. PGPR often compete with disease-causing microbes for the nutrient available in trace amount and limit them. This can be clarified when there are plenty of nonpathogenic microbes in soil, which quickly colonize the plant surfaces and also consume available nutrients as well, thereby restricting the growth of pathogenic microbial agents (Backer et al. 2018). ISR is the phenomenon of enhanced resistance at specific plant sites where induction existed. ISR defense mechanism is triggered only during an attack of pathogenic agent. ISR is pathogen nonspecific but assists the plant in disease protection. ISR includes signaling of salicylic acid, jasmonate, and ethylene inside the plant, and these hormones trigger a set of plant genes, which protect the host plant against a broad spectrum of pathogens (Ramamoorthy et al. 2001). Lipopolysaccharides (LPS), flagella, siderophores, cyclic lipopeptides, 2, 4-diacetylphloroglucinol, homoserine lactones, and volatiles like acetoin and 2, 3-butanediol are microbial cell component known to induce ISR. *P. fluorescens* strain Pf1 has been reported to produce ISR against *Rhizoctonia solani*, causing sheath blight in rice (Rai et al. 2017).

Synthesis of hydrolytic enzymes, *e.g*., chitinases, glucanases, proteases, and lipases, can lyse pathogenic fungi. Their expression and secretion sometime only suppress plant pathogen directly. Chitinase enzyme secreted by *Serratia plymuthica* against *Bacillus cinerea* was reported to interfere in spore germination and germ tube elongation, while *S. marcescens'* chitinase is antagonistic against *Sclerotium rolfsii* (Oleńska et al. 2020). Rhizobacteria also act as biocontrol agents of weeds by colonizing on their surfaces and then suppressing their growth. Cyanide, a toxicant, is produced by most microorganisms including plants as survival mechanism by competing with their equivalents. In fact, inoculation of cyanide-releasing bacterial strains and biological weed controlling host-specific *Rhizobacteria* commonly do not have slightest bit of deleterious impact on the host plants (Naik et al. 2019). HCN, another toxicant, is synthesized as secondary metabolite by *Pseudomonas* and *Bacillus* species for the biocontrol of weeds. HCN is reported to cause inhibition of electron transport chain (ETC) and hindrance in energy supply to cell, resulting in cell death. It also appears that PGPR block smooth operation of enzymes, natural receptor reversion process, and action of cytochrome oxidase (Admassie et al. 2020).

Molecular Strategies for Shaping the Root Microbiome

Rhizosphere is an interlinked association of plant, microbe, and soil, in which microorganisms are the most facile component for the manipulation. A few microorganisms have a lot or important abilities that enable them to improve an ecosystem and also survive in unfavorable conditions such as various biotic and abiotic stresses, and nutrient-deficient condition, so the transfer of the ability to the other microbes can be a better alternative for shaping the root microbiome (Brenner et al. 2008). A healthy and sustainable rhizosphere is the base of agriculture and root exudates released by the native microbiome regulate it. Many genetic manipulation strategies revolve around the synthesis and release these bioactive compounds. The two technologies used in this era to manipulate and modify the rhizosphere microbiota are microbiome engineering and biotechnology (Meena et al. 2017).

Microbiome Engineering

Selective alteration of gene pool of a species that exist in host-mediated microbial interaction is the basis of microbiome engineering. In a rhizospheric ecosystem, engineering few microbes is enough to engineer whole microbiome. Microbiome engineering aims to improve the plant functions by synthetically selecting a host-specific microbe and implicating the change over several generations (Mueller and Sachs 2015). This can be done either by top-down strategy or bottom-up strategy (Table 2).

Top-Down Strategy

In top-down strategy, instead of choosing the organisms and their metabolic process to manipulate, the environmental variables and traits are added/deleted in/out of the system to function the biological process effectively and generate its influence on whole microbiome. "Top" in the term signifies the selective ecosystem where the process is functioning, and "top-down"' means methodology to improvise the

Table 2 Different techniques, advantages, and disadvantages

Methods	Advantages	Disadvantages	References
Top-down	Offers a framework for macroscale processes widely successful for wastewater treatment and bioremediation	Neglects the complex *in situ* metabolic networks driving microbial and linked chemical transformations Ignores processes that depend on intricate interactions between community members Molecular-scale microbiome processes are often ignored during design Limited system optimization through.	De Vrieze et al. (2017) Beites and Mendes (2015)
Bottom-up	Study biochemical phenomenon after isolation No size limitation	Inaccurate and/or incomplete metabolic network reconstructions Unknown functions of many genes, proteins, and metabolites are poorly understood Evolutionary pressures driving individual and community-level phenotypes Limited understanding of gene, metabolic, and ecosystem regulatory schemes	Göpfrich et al. (2018) Sivaloganathan and Brynildsen (2021)

desired physical, chemical, or biological functioning. A microbiome carrying the desired trait (fermenters, methanogens, nitrogen-fixers, *etc.*) is selected, and then, either phages or mobile genetic elements using horizontal transfer mechanisms are used to transfer these traits into the rhizosphere (Ke et al. 2020). Gene transfer using phages depends on the transduction process, a HGT strategy, used by various microbes. Phages are the viral particle whose replication and integration machinery help in introducing/deleting the desired gene into the host. In rhizosphere, they are known to add PGP traits effectively (Pratama et al. 2020). Mobile genetic elements (MGEs) include plasmids, integrons, and transposons. They follow HGT mechanisms such as transformation and conjugation to introduce desired PGP traits into the whole microbial community. Depending on the application, their reconstruction is achieved. Using *B. subtilis'* conjugation machinery, *nif* gene can be transiently delivered into nonmodel Firmicutes in soil (Brophy et al. 2018). All other PGP traits like phyto-hormone production can also be transferred into the whole rhizomicrobial community (Sanderson et al. 2019).

Bottom-Up Strategy

The bottom-up approach is based on the isolation of desired individual microbe from its native site, manipulation, and further inoculation of this modified organism into the host site. These modified microbes are called synthetic community (SynComs)

(Ke et al. 2020). "Bottom" refers to the natural genome of microorganism, and "bottom-up strategy" involves all the method of successful gene pool modification regulating the metabolic processes, biotic and abiotic resistance and resilience, *etc*. This technique not only works on model organisms including *Escherichia coli* and *Saccharomyces cerevisiae*) but also extends to the thousands of species of nonmodel organisms. This strategy also insures successful and efficient transfer of selective biosynthetic gene clusters (>50 kb DNA) and their maintenance generations after generations (Smolikova et al., 2018). The methodology uses three basic systems to modify the rhizomicrobial gene pool: phage integrase system (PIS), integrative and conjugative element (ICE) system, and chassis-independent recombinase-assisted genome engineering (CRAGE) system. PIS involves integration and replication of desired gene into phage genome at available attachment sites and transfer to the host. ICE system revolves around the use of MGEs where the cargo genes are attached to them and replicated into host chromosome using host machinery. tRNA genes and AT-rich regions are the ICE integration sites. Resistance to copper, arsenic, and cadmium in *Pseudomonas syringae* pv. *actinidiae* was accredited to the ICEs carrying resistance gene cluster (Yin et al. 2019). The CRAGE system is used to resolve the problem of integration, manipulation, and replication of large and complex gene clusters, hence easier to work on large host range of PGPMs. Cre recombinase of P1 phage flanks a landing pad (LP) with LoxP site where the reconstruction and assembling of cargo gene with the heterogeneous genes take place (Liu et al. 2020).

Rhizosphere microbiome engineering can also modify the habitat by regulating the root exudation. One method for manipulation is to interfere with biosynthetic metabolic pathway of exudates synthesis by regulating the expression of gene. In case of *P. aeruginosa,* transformation of citrate synthase gene (*csb*) results in quadruplex release of citrate from the seedling of tobacco plant (Ryan et al. 2009). Second method relies on controlling the concentration of root exudate *via* transporter proteins. Mutation in ABC transporter (*abcg30*) increased the amount of phenolics while decreasing some sugars and significantly enhanced the OTUs of PGPR in *Arabidopsis thaliana* (Badri et al. 2009). Numerous molecular tools have been utilized since a long time to modify the desired microbes as per our need. PCR-based fingerprinting techniques are electrophoresis-dependent, which are categorized into two groups: size-dependent and sequence-dependent. Size-dependent techniques include terminal restriction fragment length polymorphism (T–RFLP), single-strand conformation polymorphism (SSCP), automated ribosomal intergenic spacer analysis (ARISA/RISA), and random amplified polymorphic DNA (RAPD), whereas sequence-dependent technique included denaturing gradient gel electrophoresis (DGGE) and temperature gradient gel electrophoresis (TGGE). These techniques based on DNA fragment amplification and their utilization provide a keen insight of structure and function of rhizospheric microbiome (Bokulich and Mills 2012).

Effect of Rhizospheric Microbiome on Sustainable Agriculture

Indigenous rhizospheric microorganisms are a diverse community of innate microbial consortium, which occupy the rhizosphere and phyllosphere of entire living entities both inside and outside, having imperial role in biogeochemical cycle, elemental N fixation, soil fertility and soil health improvement, P/Zn/Fe/potash solubilization, and plant growth promotion (Kumar and Gopal 2015). Without this microbial diversity, living and survival on this vivacious planet will be miserable and melancholic for human race. Therefore, the central aim of sustainable agriculture system is environmental rehabilitation and protection by using indigenous rhizospheric microbes in order to transform the good-for-nothing and worthless waste into valuable bioresources. Rhizospheric microbial diversity also protects the crop from the invasion of severe disease-causing phyto-pathogens by competing for vital nutrients or receptors on host cells by synthesizing and releasing bacteriocins, and other inhibitory/suppressive moieties such as siderophore (Fe chelating compound), hence creating unfavorable environment for pathogen colonization (Sadi et al. 2006).

These beneficial microbes improve the nutrient accessibility to host plant and also enhance their water-holding capacity through biosynthesis of exopolysaccharides and biofilms, making the crop plants to have sufficient nutrient/water for plant growth promotion under climatic changing conditions. Han et al. (2006) demonstrated that the integrated application of *Bacillus megaterium var. phosphaticum* and *B. mucilaginosus* enhanced the phosphorous, potash, zinc, and iron availability in the soil, thereby enhancing the water and nutrient utilization efficiency in the pepper and cucumbers. Co-inoculation of fluorescent *Pseudomonas* strains raised *Vigna mungo* yield by threefold compared to the control treatment. Similarly, Gupta and Pandey (2019) reported that consortium of *Aneurinibacillus aneurinilyticus* and *Paenibacillus* sp. mitigated the adverse impact of salt stress in French bean and intensified root–shoot length, fresh weight, plant biomass, total chlorophyll content, and seedling nodulation. Therefore, indigenous micobiomes offer an ideal and favorable environment to rectify or conserve microflora and microfauna along with other microbial diversity that consequently enhances the quality of life of higher plants and animals including human (Bhattacharyya et al. 2016).

Conclusions and Future Prospective

Inappropriate use of agrochemicals has damaged the agroecosystem and impaired ecological cycling. Indigenous beneficial rhizospheric microbial communities with their uncountable beneficial attributes act as an effective means to sustain agriculture by limiting the utilization of agrochemicals and combating plant diseases to improve the crop productivity. Rhizospheric microbiome possesses ability to stimulate the plant growth, and induce stress tolerance and better plant and soil health. Various indigenous potential microbial inoculants were created to attain field prosperity by either fabricating smart microbial consortia or engineering the microbial communities having plant growth-promoting characters. To boost and support crop productivity, soil health, application of recent breeding practices, nano-science, and agribiotechnology science are explored for strain improvement and precision deliverance of agroinputs. This had favorable outcome on crop productivity, plant stress tolerance, balanced nutrient recycling, and soil health. The integration of modern scientific tools and techniques with multidisciplinary science can help in improved governance of rhizospheric microflora, rhizospheric biology, and increased crop productivity under sustainable manners.

References

Abedini A, Roumy V, Mahieux S, Biabiany M, Standaert-Vitse A, Rivière C, Hennebelle T (2013) Rosmarinic acid and its methyl ester as antimicrobial components of the hydro-methanolic extract of *Hyptis atrorubens* Poit. (Lamiaceae). Evid Based Complement Alternat Med 2013

Abriouel H, Franz CM, Omar NB, Gálvez A (2011) Diversity and applications of *Bacillus* bacteriocins. FEMS Microbiol Rev 35:201–232

Adesemoye AO, Torbert HA, Kloepper JW (2010) Increased plant uptake of nitrogen from 15N-depleted fertilizer using plant growth-promoting rhizobacteria. Appl Soil Ecol 46:54–58

Admassie M, Assefa F, Alemu T (2020) Different mechanism, application, methods and properties of plant growth promoting rhizosphere bacteria. Appl Microbiol 6:177

Ajwa AHA (2015) Evaluation the effect of ectomycorrhizal fungi on Prunus cerasifera x salicina growth compared with chemical and organic fertilizer.

Ali M (2012) Climate change impacts on plant biomass growth. Springer Science & Business Media

Ananthi M, Selvaraju P, Sundaralingam K (2014) Effect of bio-priming using bio-control agents on seed germination and seedling vigour in chilli (*Capsicum annuum* L.) 'PKM 1'. J Hortic Sci Biotech 89:564–568

Anas M, Liao F, Verma KK, Sarwar MA, Mahmood A, Chen ZL, Li YR (2020) Fate of nitrogen in agriculture and environment: agronomic, eco-physiological and molecular approaches to improve nitrogen use efficiency. Biol Res 53:1–20

Arora P, Sehgal S, Kawatra A (2003) Content and HCl-extractability of minerals as affected by acid treatment of pearl millet. Food Chem 80:141–144

Ashraf M, Hasnain S, Berge O, Mahmood T (2004) Inoculating wheat seedlings with exopolysaccharide-producing bacteria restricts sodium uptake and stimulates plant growth under salt stress. Biol Fertil Soils 40:157–162

Augustine DJ, Dijkstra FA, Hamilton EW, Morgan JA (2011) Rhizosphere interactions, carbon allocation, and nitrogen acquisition of two perennial North American grasses in response to defoliation and elevated atmospheric $^{CO}2$. Oecologia 165:755–770

Babalola OO, Odhiambo GD (2008) Effect of inoculation with *Klebsiella oxytoca* '10 mkr 7'on striga suicidal germination in *Zea mays*. World Appl Sci J 3:57–62

Babalola OO, Osir EO, Sanni AI, Odhiambo GD, Bulimo WD (2003) Amplification of 1-amino-cyclopropane-1- carboxylic (ACC) deaminase from plant growth promoting rhizobacteria in Striga-infested soil. Afr J Biotechnol 2:157–160

Backer R, Rokem JS, Ilangumaran G, Lamont J, Praslickova D, Ricci E, Smith DL (2018) Plant growth-promoting rhizobacteria: context, mechanisms of action, and roadmap to commercialization of bio-stimulants for sustainable agriculture. Front Plant Sci 9:1473

Badri DV, Quintana N, El Kassis EG, Kim HK, Choi YH, Sugiyama A. Vivanco JM (2009). An ABC transporter mutation alters root exudation of phyto-chemicals that provoke an overhaul of natural soil microbiota. Plant Physiol 151: 2006-2017.

Beites T, Mendes MV (2015) Chassis optimization as a cornerstone for the application of synthetic biology based strategies in microbial secondary metabolism. Front Microbiol 6:906

Beneduzi A, Ambrosini A, Passaglia LM (2012) Plant growth-promoting rhizobacteria (PGPR): their potential as antagonists and bio-control agents. Gene Mol Biol 35:1044–1051

Bhaduri D, Pal S, Purakayastha TJ, Chakraborty K, Yadav RS, Akhtar MS (2015) Soil quality and plant-microbe interactions in the rhizosphere. In: Sustainable agriculture reviews. Springer, Cham, pp 307–335

Bhattacharyya PN, Goswami MP, Bhattacharyya LH (2016) Perspective of beneficial microbes in agriculture under changing climatic scenario: a review. J Phytology 8:26–41

Blatny JM, Ho J, Skogan G, Fykse EM, Aarskaug T, Waagen V (2011) Airborne *Legionella* bacteria from pulp waste treatment plant: aerosol particles characterized as aggregates and their potential hazard. Aerobiologia 27:147–162

Boddy L (2016) Interactions between fungi and other microbes. In: The fungi. Academic Press, pp 337–360

Bokulich NA, Mills DA (2012) Next-generation approaches to the microbial ecology of food fermentations. BMB Rep 45:377–389

Brenner K, You L, Arnold FH (2008) Engineering microbial consortia: a new frontier in synthetic biology. Trends Biotechnol 26:483–489

Brimecombe MJ, De Leij FAM, Lynch JM (2007) Rhizo-deposition and microbial populations. In: The rhizosphere biochemistry and organic substances at the soil-plant interface. Taylor & Francis Group, New York, USA

Brophy JA, Triassi AJ, Adams BL, Renberg RL, Stratis-Cullum DN, Grossman AD, Voigt CA (2018) Engineered integrative and conjugative elements for efficient and inducible DNA transfer to un-domesticated bacteria. Nat Microbiol 3:1043–1053

Buee M, De Boer W, Martin F, Van Overbeek L, Jurkevitch E (2009) The rhizosphere zoo: an overview of plant- associated communities of microorganisms, including phages, bacteria, archaea, and fungi, and of some of their structuring factors. Plant and Soil 321:189–212

Busby PE, Soman C, Wagner MR, Friesen ML, Kremer J, Bennett A et al (2017) Research priorities for harnessing plant microbiomes in sustainable agriculture. PLoS Biol 15:e2001793

Cai M, Yao J, Yang H, Wang R, Masakorala K (2013) Aerobic biodegradation process of petroleum and pathway of main compounds in water flooding well of Dagang oil field. Bioresour Technol 144:100–106

Cesco S, Mimmo T, Tonon G, Tomasi N, Pinton R, Terzano R, Nannipieri P (2012) Plant-borne flavonoids released into the rhizosphere: impact on soil bio-activities related to plant nutrition. Biol Fertil Soils 48:123–149

Chin KJ, Lüders T, Friedrich MW, Klose M, Conrad R (2004) Archaeal community structure and pathway of methane formation on rice roots. Microb Ecol 47:59–67

Cook RJ, Thomashow LS, Weller DM, Fujimoto D, Mazzola M, Bangera G, Kim DS (1995) Molecular mechanisms of defense by rhizobacteria against root disease. Proc Natl Acad Sci U S A 92:4197–4201

De la Fuente Cantó C, Simonin M, King E, Moulin L, Bennett MJ, Castrillo G, Laplaze L (2020) An extended root phenotype: the rhizosphere, its formation and impacts on plant fitness. Plant J 103:951–964

De Salamone IG, Hynes RK, Nelson LM (2005) Role of cytokinins in plant growth promotion by rhizosphere bacteria. In: PGPR: biocontrol and biofertilization. Springer, Dordrecht, The Netherlands, pp 173–195

De Vrieze J, Christiaens ME, Verstraete W (2017) The microbiome as engineering tool: Manufacturing and trading between microorganisms. N Biotechnol 39:206–214
Dobbelaere S, Vanderleyden J, Okon Y (2003) Plant growth-promoting effects of diazotrophs in the rhizosphere. Crit Rev Plant Sci 22:107–149
Dos Santos DG, Coelho CCDS, Ferreira ABR, Freitas-Silva O (2021) Brazilian coffee production and the future microbiome and mycotoxin profile considering the climate change scenario. Microorganisms 9:858
Dwibedi V, Saxena S (2019) Diversity and phylogeny of resveratrol-producing culturable endophytic fungi from Vitis species in India. 3 Biotech 9(5):182
Egamberdiyeva D (2006) Comparative analysis of the dynamics and functions of rhizosphere soil microbial community in two ecosystems of the Chatkal Biosphere Reserve. United Nations educational, Scientific and Cultural Organization, Final Report Tashkent State University of Agriculture, Uzbekistan
Esitken A, Yildiz HE, Ercisli S, Donmez MF, Turan M, Gunes A (2010) Effects of plant growth promoting bacteria (PGPB) on yield, growth and nutrient contents of organically grown strawberry. Sci Hortic 124:62–66
Garbeva P, van Elsas JD, van Veen JA (2008) Rhizosphere microbial community and its response to plant species and soil history. Plant Soil 302:19–32
Gillor O, Etzion A, Riley MA (2008) The dual role of bacteriocins as anti-and probiotics. Appl Microbiol Biotechnol 81:591–606
Göpfrich K, Platzman I, Spatz JP (2018) Mastering complexity: towards bottom-up construction of multifunctional eukaryotic synthetic cells. Trends Biotechnol 36:938–951
Goyal D, Prakash O, Pandey J (2019) Rhizospheric microbial diversity: an important component for abiotic stress management in crop plants toward sustainable agriculture. In: New and future developments in microbial biotechnology and bioengineering. Elsevier, pp 115–134
Gupta S, Pandey S (2019) ACC deaminase producing bacteria with multifarious plant growth promoting traits alleviates salinity stress in French bean (*Phaseolus vulgaris*) plants. Front Microbiol 10:1506. https://doi.org/10.3389/fmicb.2019.01506
Han HS, Supanjani E, Lee KD (2006) Effect of co-inoculation with phosphate and potassium solubilizing bacteria on mineral uptake and growth of pepper and cucumber. Plant Soil Environ 52:130–136
Hays Z, Watson D (2019) Fungal ecology, diversity and metabolites. Scientific e-Resources
Hinsinger P, Bengough AG, Vetterlein D, Young IM (2009) Rhizosphere: biophysics, biogeochemistry and ecological relevance. Plant and Soil 321:117–152
Jouquet P, Dauber J, Lagerlöf J, Lavelle P, Lepage M (2006) Soil invertebrates as ecosystem engineers: intended and accidental effects on soil and feedback loops. Appl Soil Ecol 32:153–164
Kapoor R, Mukerji KG (2006) Rhizosphere microbial community dynamics. In: Microbial activity in the rhizosphere. Springer, Berlin, Heidelberg, pp 55–69
Ke J, Wang B, Yoshikuni Y (2020) Microbiome engineering: synthetic biology of plant-associated microbiomes in sustainable agriculture. Trends Biotechnol 39:244–261
Kennedy IR, Choudhury ATMA, Kecskés ML (2004) Non-symbiotic bacterial diazotrophs in crop-farming systems: can their potential for plant growth promotion be better exploited? Soil Biol Biochem 36:1229–1244
Khan A, Singh J, Upadhayay VK, Singh AV, Shah S (2019) Microbial bio-fortification: a green technology through plant growth promoting microorganisms. In: Sustainable green technologies for environmental management. Springer, Singapore, pp 255–269
Kiers ET, Heijden MGVD (2006) Mutualistic stability in the arbuscular mycorrhizal symbiosis: exploring hypotheses of evolutionary cooperation. Ecology 87:1627–1636
Kobae Y (2019) Dynamic phosphate uptake in arbuscular mycorrhizal roots under field conditions. Front Environ Sci 6:159
Kour D, Rana KL, Yadav N, Yadav AN, Kumar A, Meena VS, Saxena AK (2019a) Rhizospheric microbiomes: biodiversity, mechanisms of plant growth promotion, and biotechnological

applications for sustainable agriculture. In: Plant growth promoting rhizobacteria for agricultural sustainability. Springer, Singapore, pp 19–65

Kour D, Rana KL, Yadav N, Yadav AN, Singh J, Rastegari AA (2019b) Agriculturally and industrially important fungi: current developments and potential biotechnological applications. In: Yadav AN, Singh S, Mishra S, Gupta A (eds) Recent advancement in white biotechnology through fungi, volume 2: perspective for value-added products and environments. Springer International Publishing, Cham, pp 1–64

Kumar BL, Gopal DVRS (2015) Effective role of indigenous microorganisms for sustainable environment. 3 Biotech 5:867–876

Kumar A, Singh AK, Choudhary KK (eds) (2019) Role of plant growth promoting microorganisms in sustainable agriculture and nanotechnology. Woodhead Publishing

Kundan R, Pant G, Jadon N, Agrawal PK (2015) Plant growth promoting rhizobacteria: mechanism and current prospective. J Fertil Pestic 6:9

Lagos L, Maruyama F, Nannipieri P, Mora ML, Ogram A, Jorquera MA (2015) Current overview on the study of bacteria in the rhizosphere by modern molecular techniques: a mini–review. J Soil Sci Plant Nutr 15:504–523

Legaz ME, Sánchez-Elordi E, Santiago R, de Armas R, Fontaniella B, Millanes AM, Vicente C (2018) Metabolic responses of sugarcane plants upon different plant–pathogen interactions. In: Plant metabolites and regulation under environmental stress. Academic Press, pp 241–280

Lemanceau P, Corberand T, Gardan L, Latour X, Laguerre G, Boeufgras J, Alabouvette C (1995) Effect of two plant species, flax (*Linum usitatissimum* L.) and tomato (*Lycopersicon esculentum* Mill.), on the diversity of soilborne populations of fluorescent pseudomonads. Appl Environ Microbiol 61:1004–1012

Li JZ, Chapman B, Charlebois P, Hofmann O, Weiner B, Porter AJ, Jennings A (2014) Comparison of illumina and 454 deep sequencing in participants failing raltegravir-based antiretroviral therapy. PLoS One 9:e90485

Li J, Yang H, Peer WA, Richter G, Blakeslee J, Bandyopadhyay A, Titapiwantakun B, Undurraga S, Khodakovskaya M, Richards EL, Krizek B, Murphy AS, Gilroy S, Gaxiola R (2005) *Arabidopsis* H^+ – P Pase AVP1 regulates auxin-mediated organ development. Science 310:121–125

Lin CH, Lu CY, Tseng AT, Huang CJ, Lin YJ, Chen CY (2020) The ptsG gene encoding the major glucose transporter of *Bacillus cereus* C1L participates in root colonization and beneficial metabolite production to induce plant systemic disease resistance. Mole Plant-Microb Interac 33:256–271

Liu F, Xing S, Ma H, Du Z, Ma B (2013) Cytokinin-producing, plant growth-promoting rhizobacteria that confer resistance to drought stress in *Platycladus orientalis* container seedlings. Appl Microbiol Biotechnol 97:9155–9164

Liu H, Robinson DS, Wu ZY, Kuo R, Yoshikuni Y, Blaby IK, Cheng JF (2020) Bacterial genome editing by coupling Cre-lox and CRISPR-Cas9 systems. PLoS One 15:e0241867

Loper JE (1988) Role of fluorescent siderophore production in biological control of *Pythium ultimum* by a *Pseudomonas fluorescens* strain. Phytopathology 78:166–172

Ma W, Guinel FC, Glick BR (2003) *Rhizobium leguminosarum biovar viciae* 1-aminocyclopropane-1-carboxylate deaminase promotes nodulation of pea plants. Appl Environ Microbiol 69:4396–4402

Mahaffee WF, Kloepper JW (1997) Temporal changes in the bacterial communities of soil, rhizosphere and endorhiza associated with field-grown cucumber (*Cucumis sativus* L.). Microb Ecol 34:210–223

Maksimov IV, Abizgil'Dina RR, Pusenkova LI (2011) Plant growth promoting rhizobacteria as alternative to chemical crop protectors from pathogens. Appl Biochem Microbiol 47:333–345

Martínez-Viveros O, Jorquera MA, Crowley DE, Gajardo GMLM, Mora ML (2010) Mechanisms and practical considerations involved in plant growth promotion by rhizobacteria. J Soil Sci Plant Nutr 10:293–319

Meena VS, Meena SK, Verma JP, Kumar A, Aeron A, Mishra PK, Dotaniya ML (2017) Plant beneficial rhizospheric microorganism (PBRM) strategies to improve nutrients use efficiency: a review. Ecol Eng 107:8–32

Mendes R, Kruijt M, De Bruijn I, Dekkers E, van der Voort M, Schneider JH, Raaijmakers JM (2011) Deciphering the rhizosphere microbiome for disease-suppressive bacteria. Science 332: 1097–1100

Moharam MH, Negim OO (2012) Biocontrol of *Fusarium* wilt disease in cucumber with improvement of growth and mineral uptake using some antagonistic formulations. Commun Agric Appl Biol Sci 77:53–63

Mueller UG, Sachs JL (2015) Engineering microbiomes to improve plant and animal health. Trends Microbiol 23:606–617

Naik K, Mishra S, Srichandan H, Singh PK, Sarangi PK (2019) Plant growth promoting microbes: Potential link to sustainable agriculture and environment. Biocatal Agric Biotechnol 21:101326

Newton AC, Fitt BD, Atkins SD, Walters DR, Daniell TJ (2010) Pathogenesis, parasitism and mutualism in the trophic space of microbe–plant interactions. Trends Microbiol 18:365–373

Nguyen T, Sherratt PJ, Pickett CB (2003) Regulatory mechanisms controlling gene expression mediated by the antioxidant response element. Annu Rev Pharmacol Toxicol 43:233–260

Nihorimbere V, Ongena M, Smargiassi M, Thonart P (2011) Beneficial effect of the rhizosphere microbial community for plant growth and health. Biotechnol Agron Soc Environ 15:327–337

Olanrewaju OS, Glick BR, Babalola OO (2017) Mechanisms of action of plant growth promoting bacteria. World J Microbiol Biotechnol 33:1–16

Oleńska E, Małek W, Wójcik M, Swiecicka I, Thijs S, Vangronsveld J (2020) Beneficial features of plant growth- promoting rhizobacteria for improving plant growth and health in challenging conditions: a methodical review. Sci Total Environ 140682

Orlowska E, Zubek S, Jurkiewicz A, Szarek-Łukaszewska G, Turnau K (2002) Influence of restoration on arbuscular mycorrhiza of *Biscutella laevigata* L. (Brassicaceae) and *Plantago Lanceolata* L. (Plantaginaceae) from calamine spoil mounds. Mycorrhiza 12:153–159

Peterson SB, Dunn AK, Klimowicz AK, Handelsman J (2006) Peptidoglycan from *Bacillus cereus* mediates commensalism with rhizosphere bacteria from the *Cytophaga-Flavobacterium* group. Appl Environ Microbiol 72:5421–5427

Pierre MJ, Bhople BS, Kumar A, Erneste H, Emmanuel B, Singh YN (2014) Contribution of arbuscular mycorrhizal fungi (AM fungi) and *Rhizobium* inoculation on crop growth and chemical properties of rhizospheric soils in high plants. IOSR-JAVS 7:45–55

Pindi PK, Sultana T, Vootla PK (2014) Plant growth regulation of Bt-cotton through *Bacillus* species. 3 Biotech 4:305–315

Pratama AA, Terpstra J, de Oliveria ALM, Salles JF (2020) The role of rhizosphere bacteriophages in plant health. Trends Microbiol 28:709–718

Raaijmakers JM, Mazzola M (2012) Diversity and natural functions of antibiotics produced by beneficial and plant pathogenic bacteria. Annu Rev Phytopathol 50:403–424

Rai A, Rai PK, Singh S (2017) Exploiting beneficial traits of plant-associated fluorescent pseudomonads for plant health. Agro Environ Sustain:19–41

Ramamoorthy V, Viswanathan R, Raguchander T, Prakasam V, Samiyappan R (2001) Induction of systemic resistance by plant growth promoting rhizobacteria in crop plants against pests and diseases. Crop Prot 20:1–11

Rana KL, Kour D, Sheikh I, Dhiman A, Yadav N, Yadav AN, Rastegari AA, Singh K, Saxena AK (2019) Endophytic fungi: biodiversity, ecological significance and potential industrial applications. In: Yadav AN, Mishra S, Singh S, Gupta A (eds) Recent advancement in white Biotechnology through fungi, Diversity and enzymes perspectives, vol 1. Springer, Switzerland, pp 1–62

Rana KL, Kour D, Yadav AN, Yadav N, Saxena AK (2020) Agriculturally important microbial biofilms: biodiversity, ecological significances, and biotechnological applications. In: New and future developments in microbial biotechnology and bioengineering: Microbial biofilms. Elsevier, pp 221–265

Recep K, Fikrettin S, Erkol D, Cafer E (2009) Biological control of the potato dry rot caused by *Fusarium* species using PGPR strains. Biol Control 50:194–198

Roesch LFW, de Quadros PD, Camargo FA, Triplett EW (2007) Screening of diazotrophic bacteria *Azospirillum* spp. for nitrogen fixation and auxin production in multiple field sites in southern Brazil. World J Microbiol Biotechnol 23:1377–1383

Roper MM, Gupta VVSR (2016) Enhancing non-symbiotic N2 fixation in agriculture. Open Agric J 10(1)

Rovira AD (1965) Interactions between plant roots and soil microorganisms. Annu Rev Microbiol 19:241–266

Ryan PR, Dessaux Y, Thomashow LS, Weller DM (2009) Rhizosphere engineering and management for sustainable agriculture. Plant and Soil 321:363–383

Ryu CM, Hu CH, Locy RD, Kloepper JW (2005) Study of mechanisms for plant growth promotion elicited by rhizobacteria in *Arabidopsis thaliana.* Plant and Soil 268:285–292

Sadi T, Jeffey LSH, Rahim N, Rashdi AA, Nejis NA, Hassan R (2006) Bio prospecting and management of microorganisms. National conference on agro biodiversity conservation and sustainable utilization, pp 129–130

Saha R, Saha N, Donofrio RS, Bestervelt LL (2013) Microbial siderophores: a mini review. *J Basic Microbiol 53*:303–317

Sanderson H, Brown RS, Hania P, McAllister TA, Majury A, Liss SN (2019) Antimicrobial resistant genes and organisms as environmental contaminants of emerging concern: addressing global public health risks. In: Management of emerging public health issues and risks. Academic Press, pp 147–187

Sapre S, Gontia-Mishra I, Tiwari S (2019) ACC deaminase-producing bacteria: a key player in alleviating abiotic stresses in plants. In: Plant growth promoting rhizobacteria for agricultural sustainability. Springer, Singapore, pp 267–291

Schneider KD, Thiessen Martens JR, Zvomuya F, Reid DK, Fraser TD, Lynch DH, Wilson HF (2019) Options for improved phosphorus cycling and use in agriculture at the field and regional scales. J Environ Qual 48:1247–1264

Selim SM, Zayed MS (2017) Microbial interactions and plant growth. In: Plant-microbe interactions in agro-ecological perspectives. Springer, Singapore, pp 1–15

Singh J, Faull JL (2020) Antagonism and biological control. In: Mukerji KG, Garg KL (eds) Biocontrol of plant diseases. CRC Press, pp 167–177

Sivaloganathan DM, Brynildsen MP (2021) Phagosome bacteria interactions from the bottom up. Ann Rev Chem Biomol Eng 12:309–331

Smolikova G, Lukasheva E, Mavropolo-Stolyarenko G, Bilova T, Soboleva A, Tsarev A, Frolov A (2018) Discovering seed proteome: challenges and perspectives. doi: https://doi.org/10.20944/preprints201812.0129.v1

Somers E, Vanderleyden J, Srinivasan M (2004) Rhizosphere bacterial signalling: a love parade beneath our feet. Crit Rev Microbiol 30:205–240

Spaepen S, Vanderleyden J, Remans R (2007) Indole-3-acetic acid in microbial and microorganism-plant signaling. FEMS Microbiol Rev 31:425–448

Tahir M, Sarwar MA (2013) Plant growth promoting rhizobacteria (PGPR): a budding complement of synthetic fertilizers for improving crop production. Group 19:79–87

Tanwar A, Aggarwal A, Parkash V (2014) Effect of bio-inoculants and superphosphate fertilizer on the growth and yield of broccoli (*Brassica oleracea L. var. italica* Plenck). New Zeal J Crop Hortic Sci 42:288–302

Tecon R, Or D (2017) Biophysical processes supporting the diversity of microbial life in soil. FEMS Microbiol Rev 41:599–623

Tellier A, de Vienne DM, Giraud T, Hood M, Refrégier G (2010). Theory and examples of reciprocal influence between hosts and pathogens, from short-term to long term interactions: co-evolution, co-speciation and pathogen speciation following host shifts.

Thakur D, Kaushal R, Shyam V (2014) Phosphate solubilizing microorganisms: role in phosphorus nutrition of crop plants-a review. Agric Rev 35:159–171

Tripathi P, Singh PC, Mishra A, Chaudhry V, Mishra S, Tripathi RD, Nautiyal CS (2013) *Trichoderma* inoculation ameliorates arsenic induced phytotoxic changes in gene expression and stem anatomy of chickpea (Cicer arietinum). Ecotoxicol Environ Saf 89:8–14

Uroz S, Buée M, Murat C, Frey-Klett P, Martin F (2010) Pyro-sequencing reveals a contrasted bacterial diversity between oak rhizosphere and surrounding soil. Environ Microbiol Rep 2: 281–288

Uroz S, Oger P, Morin E, Frey-Klett P (2012) Distinct ecto-mycorrhizospheres share similar bacterial communities as revealed by pyro-sequencing-based analysis of 16S rRNA genes. Appl Environ Microbiol 78:3020–3024

Vaikuntapu PR, Dutta S, Samudrala RB, Rao VR, Kalam S, Podile AR (2014) Preferential promotion of *Lycopersicon esculentum* (Tomato) growth by plant growth promoting bacteria associated with tomato. Ind J Microbiol 54:403–412

Van As J (2012) The story of life & the environment. Penguin Random House South Africa

Vaxevanidou K, Christou C, Kremmydas GF, Georgakopoulos DG, Papassiopi N (2015) Role of indigenous arsenate and iron (III) Respiring microorganisms in controlling the mobilization of arsenic in a contaminated soil sample. Bull Environ Contam Toxicol 94:282–288

VerBerkmoes NC, Denef VJ, Hettich RL, Banfield JF (2009) Functional analysis of natural microbial consortia using community proteomics. Nat Rev Microbiol 7:196–205

Verma JP, Yadav J, Tiwari KN, Jaiswal DK (2014) Evaluation of plant growth promoting activities of microbial strains and their effect on growth and yield of chickpea (*Cicer arietinum* L.) in India. Soil Biol Biochem 70:33–37

Vey A, Hoagland RE, Butt TM (2001) Toxic metabolites of fungal biocontrol agents. In: Butt TM et al (eds) Fungi as biocontrol agents: progress, problems and potential, vol 1. CABI, Wallingford, pp 311–346

White GJ (2008) The effect of conditionally dispensable chromosomes of *Nectria haematococca* MPVI on rhizosphere colonization and the identification of a gene cluster for homoserine utilization. The University of Arizona

Winkler MK, Boets P, Hahne B, Goethals P, Volcke EI (2017) Effect of the dilution rate on microbial competition: r- strategist can win over k-strategist at low substrate concentration. PLoS One 12:e0172785

Yadav SK, Dave A, Sarkar A, Singh HB, Sarma BK (2013) Co-inoculated bio-priming with *Trichoderma, Pseudomonas* and *Rhizobium* improves crop growth in *Cicer arietinum* and *Phaseolus vulgaris*. Int J Agric Environ Biotechnol 6:255–259

Yin Y, Ni PE, Deng B, Wang S, Xu W, Wang D (2019) Isolation and characterization of phages against *Pseudomonas syringae* pv. *actinidiae*. Acta Agric Scandi 69:199–208

Zarembinski TI, Theologis A (1994) Ethylene biosynthesis and action: a case of conservation. Plant Mol Biol 26:1579–1597

Zhang YQ, Li WJ, Zhang KY, Tian XP, Jiang Y, Xu LH, Lai R (2006) Massilia dura sp. nov., *Massilia albidiflava* sp. nov., *Massilia plicata* sp. nov. and *Massilia lutea* sp. nov., isolated from soils in China. Int J Syst Evol Microbiol 56:459–463

Zhang L, Shi N, Fan J, Wang F, George TS, Feng G (2018) Arbuscular mycorrhizal fungi stimulate organic phosphate mobilization associated with changing bacterial community structure under field conditions. Environ Microbiol 20:2639–2651

Zia R, Nawaz MS, Siddique MJ, Hakim S, Imran A (2020) Plant survival under drought stress: implications, adaptive responses, and integrated rhizosphere management strategy for stress mitigation. Microbiol Res 242:12,662

The Role of Polyamines in Drought Tolerance in Plants

Shiva Aliyari Rad, Arash Rostami Azar, Zahra Dehghanian, Khosro Balilashaki, Elaheh Kavusi, and Behnam Asgari Lajayer

Abstract Polyamines are compact protonates that include amine groups that are present in most living beings (PAs). Putrescine, spermidine, spermine, and its isomer thermospermine are the most common polyamines in plants. Those compounds can be present in hydroxycinnamic acids in free or conjugated forms. Via regulation of PA biosynthesis, conjugation, and transport, their homeostasis is closely regulated. In recent years, genetic methods have unraveled the main roles of PAs in stress defense. Indeed, in reaction to various abiotic and biotic pressures, the levels of different PAs rise. The PA pathway is activated by drought, which results in an ABA-dependent putrescine. The variations between drought-resistant and reactive cultivars are likely to connect with the plants' capacity to collect many PA over a minimal level. Consequently, PA defense molecular mechanisms seem to be complicated and responsive to stress. In this chapter, we discuss the current advances on the PA metabolism regarding the tolerance to the drought stress and its influence on plants resistant to drought.

Keywords Drought stress · Metabolism · Polyamines

S. Aliyari Rad · Z. Dehghanian · E. Kavusi
Department of Biotechnology, Faculty of Agriculture, Azarbaijan Shahid Madani University, Tabriz, Iran

A. Rostami Azar
Department of Plant Breeding and Biotechnology, Faculty of Agriculture, University of Kurdistan, Sanandaj, Iran

K. Balilashaki
Department of Horticultural Science, Faculty of Agriculture, University of Guilan, Rasht, Iran

B. Asgari Lajayer (✉)
Department of Soil Science, Faculty of Agriculture, University of Tabriz, Tabriz, Iran

© The Author(s), under exclusive license to Springer Nature Singapore Pte Ltd. 2022

A. Vaishnav et al. (eds.), *Plant Stress Mitigators*,
https://doi.org/10.1007/978-981-16-7759-5_23

Introduction

Plants, as sessile organisms, are often affected by an assortment of unfavorable biotic or abiotic ecological factors. As plants could not flee from adverse conditions unlike animals, severe stresses are main limiting factor to growth and seriously restricting the productivity of high-value agricultural products. As such, reactions to environmental stress, depending on its intensity and period, cause considerable variation in potential and crop yields. Abiotic stresses including drought, floods, excessive temperatures, salinity, environmental toxins, and material deficiencies are considered as the primary reasons of crop failure, and in the world. More than half of decrement in the production of main yearly and persistent produce is related to these agents (Wang et al. 2003; Saghafi et al. 2019a, b, 2020; Khoshmanzar et al. 2019). So, understanding how plants conform to and withstand abiotic stress is essential to successful use of genetic resources to enhance stress toleration as well as to increase yields. PAs (polyamines) are aliphatic nitrogenous bases having low molecular weight and more than two amino groups that have strong biological process (Xu et al. 2009; Vuosku et al. 2018).

In eukaryotic and prokaryotic cells, they are broadly distributed (Liu et al. 2017; Mustafavi et al. 2018). PAs are accumulated within live organisms in three distinct kinds. Others are free (F-PA), some are covalently connected (CC-PA), and some are not covalent (NCC-PAs) (Gholami et al. 2013). There are two separate groups of CC-PAs including soluble (PSCC-PAs) and insoluble (PISCC-PAs) based on their solubility in perchloric acid.

Free form of PAs is more common in higher plants than the other forms. Some of the key plant PAs such as Put (putrescine), Spd (spermidine), and Spm (spermine) are playing roles in controlling physiological functions like flower growth and embryonic development (Xu 2015), senility of cells, and fruit ripening and growth (Xu et al. 2014; Mustafavi et al. 2018). They are also part of the stress response process (de Oliveira et al. 2016; Mustafavi et al. 2018).

Covalent bond between F-PAs and a molecular compound by amid bond will lead to formation of binding PAs (PSCC-PAs). This compound can be a phenolic combination and a derived therefrom, like hydroxycinnamic acid or ferulic acid (Lue et al. 2009; Martin-Tanguy 2010). The largest PAs of plants are made in this way (Bassard et al. 2010).

According to investigations, PSCC-PAs are involved in the allergic response to external pest as secondary metabolites (Kumar et al. 1997), and also in plant morphology (De Oliveira et al. 2018; Mustafavi et al. 2018). F-PAs are binding to biomacromolecules via ionic bonds and form bound PAs or PISCC-PAs. Nucleic acids, uronic acids, proteins, or lignin are some of these macromolecules.

In the pH spectrum of physiology, F-PAs can take protons and positive charge. Electrostatistically, interaction between these compounds with positive biomacromolecules such as acidic proteins and membrane phospholipids causes formation of NCC-PAs in the organism (Igarashi and Kashiwagi 2015). These forms of PAs are involved in various processes. They are controlling enzyme activity, are a part of DNA replication and gene transcription, are involved in cell division and make stability of membrane, and play multiple roles in plant growth and biological development.

Dispensation and PA Metabolism in Plant

PAs are present in different types of eukaryotic cells, as well as prokaryotic ones (Liu et al. 2016a, b, 2017). This is also true in plant, and they are found even in plant tumors. There are several types of PA. PA is mostly available in its free form in higher plants. Put, SPD, Spm, and thermospermine are most prevalent PAs in higher plants (t-Spm) (Sobieszczuk-Nowicka 2017; Takahashi et al. 2017) and cadaverine (Cad) (Regla-Márquez et al. 2015; Nahar et al. 2016). Additional PAs are only available within level and circumstances of plants. Polyamines illustrate tissue- and organ-specific templates of dispensation in plants. For the highest abundance of PAs in plant, leaves are related to Put (up to three times more than the rest), while the most level of PAs in other organs is related to Spd (Takahashi et al. 2017).

In a study on carrot, Put was observed in the cytoplasm of cells, while cell walls contain Spm (Cai et al. 2006). Various kinds of PAs also exhibit various localization templates in cells. PAs' distribution trends can be attributed to their special characteristics. Higher PA biosynthesis and content are correlated with more severe plant growth and metabolism in total (Zhao and Qin 2004; Cai et al. 2006). Polyamines have been shown to be implicated in several mechanisms in plant, like embryogenic competence (Silveira et al. 2006), cellular suicide (Kim et al. 2013), fruit maturing (Gomez-Jimenez et al. 2010), specification of xylem (Tisi et al. 2011), and formation of biofilm (Lee et al. 2009).

Recent studies show the involvement of PA in adaptations to different difficult conditions including such stresses. This has been proven by recording changes in the amount of PAs under stress circumstances. This issue has been researched in various plant species. The first research was done subject to potassium deficiency (Richards and Coleman 1952). Some other abiotic stresses, including temperature changes, and reduction in nutrients and drought, were investigated in other studies (Liu et al. 2007).

It has been found that sometimes after abiotic stress, the level of all three key PAs increases (Yang et al. 2007). Nevertheless, in certain instances, only one type of the three PAs displays a substantial enhancement. For example, the levels of Put in apple callus increased under the influence of salinity treatment, whereas Spd and Spm experienced just slight changes (Liu et al. 2006). In return, salt stress increased Spd content in sweet orange callus (Wang and Liu 2009), and Spd and Spm in grape (*Vitis vinifera*) plants (Ikbal et al. 2014). In a study, 18 rice varieties were reported to show significant changes in the levels of Spm in a long period of drought treatment (Do et al. 2014).

These results show that the aggregation of PA is impacted by multiple agents, such as the species of plant, tolerance potential against stresses, forms and circumstances of stress, and the physiological state of the organs studied. Various outcomes of studies can be explained by the presence of complex PA interactions under abiotic stress. The vastness of the PA tank may be compared with potential for stress tolerance and confirms the role of PAs in responding to stresses.

Commonly, resistant genotypes have higher levels of PAs in contrast to the sensitive ones (Hatmi et al. 2015); nevertheless, under certain abiotic stresses, genotypes with different stress tolerance may have various contents of PA accumulation. Spd and Spm levels mainly were more in tolerant genotypes, while their level was higher in the more sensitive genotypes of one species (Liu et al. 2004).

Biosynthesis of Polyamine

The production of putrescine results from PA biosynthesis. It has two amine groups and is a precurrent to the production of Spd and Spm (Xu et al. 2009). The first route involves removing the eighth carbon atom from arginine (Arg) that makes agmatine (Agm) and CO_2; the second nitrogen atom, N-carbamoyl Put (NCPA) and NH_3, is removed from the Agm. NCH is hydrolyzed after that by the removal of its carbamoyl group, NCA, CO_2, and the NH_3 form. After that, NCAH is hydrolyzed Put and NH_3 form.

Arginase hydrolyzes Arg and produces ornithine (Orn) in the second pathway and eliminates the carbon atom no. 1 from Put and CO_2 group by ortho-ornithine decarboxylase (ODC) (Docimo et al. 2012; Pegg 2016). In plants, this is basic pathway for the synthesis of Put (Pegg 2016).

Transforming Arg into citrulline (Cit) is first step in the third path. Then, it is decarboxylated to form Put using citrulline decarboxylase (CDC) enzymatic activity (Ouyang et al. 2017; De Oliveira et al. 2018). Arabidopsis thaliana and some Brassicaceae species lack the ODC gene. (Hanfrey et al. 2010). It means the presence of ornithine pathway is not necessary for natural development. Up to now, only sesame was discovered in Cit pathway, which makes plants abound in other pathways.

The activity of the ADC and ODC is suppressed, respectively, by DFMA (difluoromethyl arginine) and difluoromethyl ornithine (DFMO) as irreversible competitive inhibitors (Grossi et al. 2016). Spd and Spm are produced from Put and aminopropyl residues, which are provided by methionine (Vuosku et al. 2018).

Catabolism of Polyamine

Polyamines (PAs) are aliphatic polycations with low molecular weight, and all organisms, including plants, have PAs (Hussain et al. 2011; Ghassemi et al. 2018). Plants have three main PAs, which are Put (putrescine), Spm (spermine), and Spd (spermidine), while other forms of them can be found, such as cadaverine.

Many studies have been performed in plants on the biosynthetic pathway of PAs (Pegg and Casero 2011; Gupta et al. 2013). The difference between animals and plants is in the number of putrescine precursors. In plants, 2 precursors including l-ornithine and l-arginine form Put; but in animals, this task is performed by l-ornithine. The activities of arginine decarboxylase (ADC, EC 4.1.1.19) and ornithine decarboxylase (ODC, EC 4.1.1.17) form Put in three steps. Then, with the addition of decarboxylated S-adenosylmethionine (dcSAM) moiety donated aminopropyl (dcSAM), Spd synthase converts Put into Spd. S-adenosylmethionine decarboxylase (SAMDC, EC 4.1.1.50) and methionine adenosyltransferase (EC 2.5.1.6) are used to catalyze methionine and synthesize dcSAM. Spd is subsequently transformed into spmor thermospermine utilizing synthase of thermospermine (ACL5, EC 2.5.1.79) and Spm synthase (SPMS, EC2.5.1.22). It is necessary to remember that no genes are found in Arabidopsis thaliana to encode ODC (Hanfrey et al. 2001), which means this plant uses ADC pathway to generate Put. Eventually, different organs of an organism may differ in PA synthesis; for example, Spd and Spm are predominant in tobacco shoot apical meristem (Nicotiana tabacum), whereas Put is abundant in roots (Moschou et al. 2008).

Catalysis of PAs in their novo synthesis pathway is done by two enzyme groups: hydrolase and catalase. The amino groups are mainly forming ammonia and 4-aminobutanal, peroxide (H_2O_2) at the first step, by catalyzing Put and cadaverine (Cad) oxidation using CuAOs (Moschou et al. 2012). These protein groups often respond with Put and Cad in comparison with Spd and Spm (Moschou et al. 2012); however, their reaction with Spd has also been observed in some species like A. thaliana (Planas-Portell et al. 2013).

CuAO protein is abundant in dicot plants and particularly in the Fabaceae family such as peas, chickpeas, lentils, and soybean seedlings (Cona et al. 2006).

Until today, a low number of plants such as A. thaliana and chickpea have been reported to contain CuAO genes (Møller and McPherson 1998; Rea et al. 1998; Planas-Portell et al. 2013). Four CuAO genes have been reported in A. thaliana ($AtAO_1$, $AtCuAO_1$, $AtCuAO_2$, and $AtCuAO_3$), and absolutely, they are more than this (Planas-Portell et al. 2013). PAOs are opposed to CuAOs. Monocots possess higher PAO content, and Spd, Spm, and their derivatives are favored (Marcé et al. 1995). In terminal metabolism, PAO shows simpler PAO, while in back conversion, PAO is involved.

Circles show the relevant cis-acting elements in the ADC promoter. The sequence that codes ADC is shown in box, and the horizontal line indicates the promoter. X regulates gene expression as a transcription factor. Arrows signify the promotion or activation of the associated mechanisms, while blunted arrows imply inhibition. Spd and Spm terminal catabolism is carried out in the first group, which produces 1,3-diaminopropane (DAP), H_2O_2, and N-(3-aminopropyl)-4-aminobutane (Spm catabolism) or 4-aminobutane (Spd catabolism) (Angelini et al. 2010; Moschou et al. 2012).

The second category converts PA to Spd and then Spd in Put (Moschou et al. 2012; Mo et al. 2015). Many plant species have PAO genes, which can be mentioned: A. thaliana (Fincato et al. 2011), poplar (Tuskan et al. 2006), rice (Ono et al.

2012), barley (Cervelli et al. 2001), tobacco (Yoda et al. 2006), maize (Cervelli et al. 2000), cotton (Mo et al. 2015), apple (Kitashiba et al. 2006), and sweet orange (*Citrus sinensis*) (Wang and Liu 2015).

By studying the sequenced genomes, it has been determined that small gene families encode plant PAO proteins. A. thaliana, rice, and sweet orange are containing 5, 7, and 6 PAO genes, respectively. Nevertheless, to date just a few PAOs have been found related to the first group. ZmPAO in maize and $OsPAO_7$ in rice are PAOs engaged in PA terminal catabolism (Liu et al. 2014a).

The second group of PAO genes are more abundant. For instance, the five PAO genes ($AtPAO_1$-$AtPAO_5$) have been tested for four rice genes ($OsPAO_1$, $OsPAO_3$, $OsPAO_4$, and $OsPAO_5$) and one in sweet orange ($CsPAO_3$) is playing roles in polyamine back transformation (Tavladoraki et al. 2006; Moschou et al. 2008; Kamada-Nobusada et al. 2008; Ono et al. 2012; Liu et al. 2014a; Wang and Liu 2015). H_2O_2 increases due to the PA catabolism, and when its level is high, it acts as a low-level signaling molecule or as a toxic compound. The PA catabolism proportion to biosynthesis might be viewed as a significant element-generating tolerance or death to abiotic stress plant cell death (Moschou et al. 2008). This indicates that, depending on its cellular levels, H_2O_2 that derived from PA can play a significant role in maintaining ROS homeostasis.

Transporting Polyamine

The discovery of PA transport mechanisms at the molecular level in plant and animal cells has still to be clearly identified. Findings from Put and Spd uptake studies in carrot cells showed that an electrical transmembrane gradient guides PAs to enter into the cells. An anti-port process among external and internal PAs makes this possible (Pistocchi et al. 1987). Some studies have shown that the transferring of Put through the plasmalemma in maize roots is similar to that of animals and is done by carrier-mediated procedure. (DiTomaso et al. 1992). This procedure needs energy and has been reported in mammalian cells (Casero and Marton 2007) and Saccharomyces cerevisiae (Nozaki et al. 1996). Poulin et al. have stated that in yeast, a supposed serine/threonine protein kinase, NRP_1, involves Spd uptake activation and was presumably needed to reactivate numerous nitrogen permeases (Kaouass et al. 1997, 1998).

PA transport in mammalian cells depends on antizymes (Mitchell et al. 1994). E. coli has four PA transport systems: (1) Spd preferential uptake system (PotABCD), (2) Put-specific uptake system (PotFGHI), (3) Put transport system (PotE), and (4) cadaverine transport system (CadB) (Igarashi and Kashiwagi 2006; Igarashi and Kashiwagi 2010).

PA transport was discovered to be present in carrot and mitochondria vacuoles and protoplastics of helianthus (Pistocchi et al. 1987; Bagni and Tassoni 2001; Tassoni et al. 2008, 2010).

Conjugates of Polyamine

Plant PAs are hydroxycinnamic acid conjugates and hydroxycinnamic amides. As a part of certain leguminous seeds, caffeoylputrescine (paucine) was first discovered in 1893 (Tiburcio et al. 1990). Conjugates of PA have already been identified in a large number of plants such as coumaroylputrescine, coumaroyl agmatine, dicoumaroylspermidine, diferuloyl spermine, and feruloyltyramine (Martin-Tanguy 1997). Recently, it has been determined that the Arabidopsis flower buds contain a few Spd hydroxycinnamic acid conjugates, which have roles in floral growth (Fellenberg et al. 2009).

Spd hydroxycinnamoyltransferase (SHT) gene, possibly involved in forming tricoumaroyl, tricaffeoyl, and triferuloyl Spd, was found in the anther tapetum of Arabidopsis (Grienenberger et al. 2009). In addition, it has been found that in Arabidopsis breeds, two novel acyltransferase genes are controlling Spd conjugate aggregation (disinapoyl-Spd and sinapoyl-(glucose)-Spd) (Luo et al. 2009). After determining the abundance of PA conjugate aggregation in the seeds, Facchini et al. (2002) introduced them as N reserves for germination of seeds. Arabidopsis has not yet documented the discovery of transferases that acylate other PAs (Alcázar et al. 2010b).

In the physiological and the molecular activities of plants, such as growth cycle, cell division, flora, cellular wall cross-linking and the reaction of stresses, and PA conjugates such PAs (Put, Spd, Spm), are directly and indirectly effective (Luo et al. 2009; Bassard et al. 2010; Moschou et al. 2012). PA conjugates have also been considered as final products. They also can intermediate contributing to the complexities of phenolic metabolism and their cross-talk with the N metabolism (Bassard et al. 2010).

Phenolamide functions are investigated in plants (Bassard et al. 2010). A phenolic movement combines PA or deaminated aromatic amino acids, and forms a secondary significant metabolite of phenolamides. Throughout the plant kingdom, they have been registered, typically as major phenolic constituents of reproductive organs and seeds. The much more definitive recent discoveries point to their importance in stress responses. PA conjugate induces protection to the cell wall and toxicity to predators and pathogens, by translocation and interconversion between free and conjugated forms (Bassard et al. 2010; Moschou et al. 2012). Because the catabolism/reverse conversion issue of free PA is extensive, little is known about the turnover of conjugated PA. Take into account the activity of $AtPAO_1$ and $AtPAO_5$ in tapetum as enzymes in the bias of various PA conjugates (Fellenberg et al. 2009; Grienenberger et al. 2009; Fincato et al. 2012), it can be said that these PAOs have roles in conjugated PA homeostasis (Moschou et al. 2012).

Polyamines and Drought Stress in Plants

Via various mechanisms and in related agronomic species such as species of Poaceae family, PAs make resistance against many stresses, especially drought stress (Agudelo-Romero et al. 2014; Rouphael et al. 2016). This may be owing to its aggregation in the cell, which provides direct stress support. Also when PA degrades, its products are signals to initiate protection against stress.

Plant adaptation to drought includes a variety of sophisticated physiological methods and modifications that may be made in the short or long term (e.g., pore closure reduces carbon absorption) (Chaves et al. 2002). When plants have been exposed to drought stress, they must organize processes to rescue cell hyperosmolarity, oxidative stress, and osmotic imbalance (Rangan et al. 2014); so, probably physiological changes will occur at multiple stages of metabolism. Application of PA increases osmoprotective molecules and reduces oxidative harm caused by the stress, and these molecules mainly act in other essential processes of plant tolerance.

In addition, PAs have a relation with the molecules, which are related to the stress-like reactive nitrogen and sulfur. The glyoxalase mechanism improving the expression and stress control of major antioxidant enzymes (SOD, CAT) improving Spm treatment in mung bean has decreased the compound toxicity of methylglyoxal. The result of these processes was plant immunity to drought/heat stress and salt stress (Nahar et al. 2016).

A similar pattern has also been recorded for tomatoes (Sánchez-Rodríguez et al. 2016). In addition, recent white clover studies have shown that exogenous Spd application increases the amount of soluble carbohydrates (WSCs) like sucrose and sorbitol and this increases the resistance to drought (Li et al. 2015). PA biosynthesis inhibition decreased the function of cytosolic Ca^{2+} and antioxidant enzymes, whereas the additional application of Spd mitigated these results. In addition, Spd increases NO in the cell by strengthening the activation of NR and NOS. Both NO and Spd are activating signal cascades by activating the signaling cascade in a NO-mediated process to raise the levels to enzymatic antioxidative machinery, which improves the resistance to drought stress (Peng et al. 2016).

By the way, in Arabidopsis during the severe drought, NO has also been reported recently, demonstrating that the stomach closure not only depends on ABA but also depends on NO (Wang et al. 2015).

Remarkably, after H_2S therapy, the PA content increased, thereby providing substantial proof that PAs and reactive species have interactions with each other; here, this interaction is done by regulating the level of PA under situations of drought. Most importantly, H_2S was shown to increase the sugar-related genes and PA biosynthesis, which increase Spinacia oleracea's resistance to drought stress (Chen et al. 2016).

It has also been proven lately that the ABA signaling pathway to address drought with H_2S in wheat is required (Ma et al. 2016), which means H_2S plays important role in physiological and metabolic processes of plant (Xie et al. 2014; García-Mata and Lamattina 2010). In addition, several studies have been conducted to investigate the roles of PA in species- or tissue-specific activities under dry circumstances Triticale is a synthetic hybrid of wheat and rye. It has been found that when this plant is responding to the drought stress, cell wall-PA binding is increasing and this makes a decrease in free PAs at late stages of development, which shows that probably PAs have a role for cell wall reinforcement in abiotic stress situations (Hura et al. 2015).

With respect to wheat, higher PAs under drought situations have been shown to be implicated not only in stress resistance, but also in seed filling (Liu et al. 2016a, b), which is strongly interdicted during water depletion. Most importantly, it was shown that ABA and zeatin levels were significantly increased and the rate of evolution of ethylene was reduced with Spd and Spm, particularly in grains, promoting wheat grain filling in water deficiency, indicating that PA affects this mechanism, which is regulated by interactions between hormones (Liu et al. 2016a, b).

Polyamines, ABA, and NO

Polyamines, NO, and ABA are involved in a complex network of multiple physiological and stress responses as small molecules, and when plant is responding to stresses, they have cross talk between each other (Wimalasekera et al. 2011a). In abiotic stress circumstances, mainly ABA activates the expression of downstream genes and other responses, which are related to physiology of plant (Klingler et al. 2010). It has been found that ABA makes a pathway for oxidation of polyamines with increasing deposition of them in grape, which makes secondary defensive impacts like stomata closure (Toumi et al. 2010). Put and ABA are stimulating biosynthesis of each other, in response to stresses (Alcázar et al. 2010a). It is rational to conclude that by causing closure and decreasing aperture, polyamines control stomatal responses. ABA and NO are interfering in parts of this process (Klingler et al. 2010; Alcázar et al. 2010a; Wimalasekera et al. 2011a). In plants, PAs quickly make the NO to burst, which means NO is a possible intermediate in PA-mediated signaling (Wimalasekera et al. 2011a, b; Hussain et al. 2011).

The copper amino oxidase 1 enzyme of Arabidopsis ($CuAO_1$) is one possible contributor to ABA-induced NO development, which causes Put to degrade (Wimalasekera et al., 2011a). They also suggested that most of the responses to stresses mediated by polyamines are described by synthesis of NO. CuAO and PAO are producing NO in the metabolism of polyamine (Wimalasekera et al. 2011a, b). Polyamine regulates activity of nitrate reductase (NR) (Rosales-Corral et al. 2012). It inhibits NR by enhancing NO and the interaction of 14-3-3 protein with NR; the H–ATPase regulates the 14-3-3 protein association and enables the NR function to be activated.

The altered aggregation of arginine by the regulation of the metabolic pathway of polyamine had a major impact on the degree of NO. In addition, NO acts as a stress signal to alter structure and behavior of protein, as well as protein–protein interaction, mainly by protein post-translational modification (Tanou et al. 2014). In Arabidopsis, PAs induce NO and NO-contributed S-nitrosylated proteins (Wimalasekera et al. 2011b). 271 S-nitrosyl proteins were discovered as commonly regulated proteins in Put, Spd, and Spm in citrus in the salinity stress, suggesting that polyamine and stress relation is dependent on NO (Tanou et al. 2014). First, NOs modulated by polyamines are contributing to stress responses mediated by polyamine in both NOS and NR pathways, and NO regulation triggered by polyamine and polyamine-dependent nitrosoproteome shifts.

Filippou et al. (2013) offered fresh insight into the impact of the addition of sodium nitroprusside (SNP) as donor of NO on biosynthesis of PA and proline. They observed that the expression of most of the genes, which are part of the PA metabolism and the activities of underlying enzyme, is modulating by SNP, thereby SNP is controlling polyamine and proline levels in the overview of Medicago truncatula. The cross talks between ABA and NO, between polyamines and ABA, and among polyamines and NO formed a dynamic integration between PA, ABA, and NO, which is responding to numerous stresses.

Polyamine Catabolism in Plants Growth

Increasing findings indicate that the catabolism of PA is especially involved in the growth of plant. A range of evidence indicates that PA oxidation is playing role in PCD (programmed cell death) or cell suicide and differentiation of xylems in apoplast, which is associated with ROS (reactive oxygen species) (Podlesakova et al. 2019; Corpas et al. 2019).

Møller and McPherson (1998) discovered as early as 1998 that in xylem tissues of Arabidopsis roots, CuAO localization precedes with lignin synthesis, and H_2O_2 of apoplast was produced by PAO increases leaf blade length in Zea mays (Rodriguez et al. 2009). Furthermore, by overexpressing the ZmPAO gene and reducing the S-adenosyl methionine decarboxylase (SAMDC) accompanied by RNA, the disruption of PA catabolism in tobacco promotes the differentiation of vascular cells and starts cell suicide in root tips (Moschou et al. 2008; Tisi et al. 2011).

Alabdallah et al. (2017) suggested that $AtPAO_5$ is involved in the interactions of auxins with cytokinins, which is essential for good differentiation of xylem and regulates the growth of Arabidopsis by oxidation activity of t-Spm (Kim et al. 2014). PAs regulate ion channeling in both stress and normal situations in plant cells. They are influencing the transport along with ROS or acting like messengers (Pottosin et al. 2014; Pegg 2014).

H_2O_2 produced by Spd oxidase has been reported to regulate Ca^{2+} canals, which are activated by hyperpolarization in pollen plasma membrane and growth of pollen tubes (Wu et al. 2010). All genes of AtPAO family have different expression patterns in Arabidopsis thaliana. The space between meristems and the elongation regions of root, and also anther tapetum are the primary locations of $AtPAO_1$, and $AtPAO_2$ is more abundant in the quiescent center, pollen, and primary cells of columella, while columella, guard cells, and pollen are containing $AtPAO_3$. Moreover, in the root vascular system and in hypocotyls, $AtPAO_5$ was specifically expressed (Fincato et al. 2012).

In addition, $AtPAO_5$'s gene structure was very distinct from the other AtPAOs (Fincato et al. 2011). Flowers and largely sepals showed its highest expression during different growth phases (Takahashi et al. 2010). $AtPAO_5$ is called proteasomal cytosomal controlled protein oxidase/dehydrogenase (Ahou et al. 2014), which regulates the development of Arabidopsis through oxidation activity of t-Spm (Liu et al. 2014b; Kim et al. 2014). In contrast, $OsPAO_1$ rice is functionally considered as an ortholog of $AtPAO_5$ (Liu et al. 2014b) and $OsPAO_7$ is playing role in lignin synthesis.

Polyamines and Plant Senescence

Drought tolerance is the issue of many researches on the interaction of PAs with water stress (Ebeed et al. 2017), and only a few of them are concentrating on tolerance to waterlogging. Polyamines are controlling the extent and time of closing and opening of the potassium channel and apertures in the membrane of guard cells. In plants, PAs regulate water content using this property (Liu et al. 2000). It has been confirmed that the sufficient amount of foliar application of Put will activate processes related to the physiology of plant and starts the synthesis of materials, like free amino acids, soluble sugars, and proline, which are involved in osmotic regulation. This will retrieve the detrimental impacts on plant biomass of drought stress and boost the state and content of some bioactive materials (Sánchez-Rodríguez et al. 2016; Mohammadi et al. 2018). Put treatments increased growth indexes and seed germination in alfalfa in drought condition induced by various polyethylene glycol density (PEG 4000), in both laboratory and in pot experiments (Zeid and Shedeed 2006) (hypocotyl duration, root and shoot fresh and dry mass). The acl5/Spms mutant of the Arabidopsis is hypersensitive to high salt and drought and is unable to develop Spm. Pretreatment of this mutant with Spm (but not with Put and Spd) healed this phenotype, which means that Spm deficiency is the reason of the drought hypersensitivity (Yamaguchi et al. 2007). Mycorrhizal masson pine has a great level of Spm and a high ratio of (Spd+Spm)/Put, and shows high tolerance against drought stress (Xu et al. 2009). Spm was most closely linked to drought tolerance apples among the three major endogenous PAs (Liu et al. 2010). Identical findings in cherry tomatoes have been obtained (Montesinos-Pereira et al. 2015). It has been observed in wheat that the high levels of Spd and Spm decreased inhibition impacts of drought stress and increased grain filling and tolerance to drought, while Put showed the reverse effect (Yang et al. 2016). All these findings suggest that plants' reaction to PAs in both osmotic and water stresses is dependent on the species of plants and even in some cases it differs in various parts of the one plant (Sen et al. 2018).

Drought Adaptation of PA Levels Through Genetic Modification

In response to drought, the ADC_2 expression is strongly upregulated and converted into Put aggregation. In an experiment, A. thaliana plants that had homologous ADC_2 genes were transformed under the CaMV 35s promoter to determine role of this gene in drought tolerance. The multiple examined lines (Thuiller et al. 2005; Porter 2005; Seki et al. 2007) showed that ADC_2 has different degrees of expression and also there are differences in Put concentration (Alcázar et al. 2005). Complete transgenic line content of the place was 12- to 2-fold greater than that of the wild type (Put the material line 2.1 $>$ 3.6 $>$ 7.2 $>$ wild type) (Alcázar et al. 2005; Alcázar et al. 2010b). These lines and wild forms have been subjected to conditions of drought stress by withdrawing water for 14 days and evaluating their survival rates (Alcázar et al. 2010b). Seven days after rewatering, plants that restored growth have been counted and the level of resistance was scored. It is important to note that plants accumulating more content of Put were more resistant against drought stress (Alcázar et al. 2010b). Therefore, line 2.1 that had 12 times more Put showed 75% of rescue relative to the wild form (12 percent) (Alcázar et al. 2010b). The increasing tolerance to the drought circumstance is alongside with reduction in stomata aperture and transpiration rate (Alcázar et al. 2010b). This information is confirming that the PAs are controlling the aperture of stomata by ROS and NO signaling modulation (Alcázar et al. 2010b). Our detection shows that the modulation of the PA pathway will achieve increased drought resistance in plants.

Concluding Remarks and Future Perspectives

The identification of PAs involved in increasing drought resistance has only recently begun. Evidence points to ROS signaling intervention, likely by PA-recycling loops involving back conversion of PA, and also cross talks with ABA, which is the most important stress-related hormone. We hope new insights for crop protection against environmental change will be generated by determining the roles of PA and increasing natural variety for PA material control.

References

Agudelo-Romero P, Ali K, Choi YH, Sousa L, Verpoorte R, Tiburcio AF, Fortes AM (2014) Perturbation of polyamine catabolism affects grape ripening of Vitis vinifera cv. Trincadeira. Plant Physiol Biochem 74:141–155

Ahou A, Martignago D, Alabdallah O, Tavazza R, Stano P, Macone A et al (2014) A plant spermine oxidase/dehydrogenase regulated by the proteasome and polyamines. J Exp Bot 65:1585–1603

Alabdallah O, Ahou A, Mancuso N, Pompili V, Macone A, Pashkoulov D et al (2017) The Arabidopsis polyamine oxidase/dehydrogenase 5 interferes with cytokinin and auxin signaling pathways to control xylem differentiation. J Exp Bot 68:997–1012
Alcázar R, García-Martínez JL, Cuevas JC, Tiburcio AF, Altabella T (2005) Overexpression of ADC2 in Arabidopsis induces dwarfism and late-flowering through GA deficiency. Plant J 43(3):425–436
Alcázar R, Altabella T, Marco F, Bortolotti C, Reymond M, Koncz C, Carrasco P (2010a) Polyamines: molecules with regulatory functions in plant abiotic stress tolerance. Planta 231(6): 1237–1249
Alcázar R, Planas J, Saxena T, Zarza X, Bortolotti C, Cuevas J, Bitrián M et al (2010b) Putrescine accumulation confers drought tolerance in transgenic Arabidopsis plants over-expressing the homologous Arginine decarboxylase 2 gene. Plant Physiol Biochem 48(7):547–552
Angelini R, Cona A, Federico R, Fincato P, Tavladoraki P, Tisi A (2010) Plant amine oxidases "on the move": an update. Plant Physiol Biochem 48:560–564
Bagni N, Tassoni A (2001) Biosynthesis, oxidation and conjugation of aliphatic polyamines in higher plants. Amino Acids 20(3):301–317
Bassard JE, Ullmann P, Bernier F, Al E (2010) Phenolamides: bridging polyamines to the phenolic metabolism. Phytochemistry 71:1808–1824
Cai Q, Zhang J, Guo C, Al E (2006) Reviews of the physiological roles and molecular biology of polyamines in higher plants. J Fujian Educ Coll 7:118–124
Casero RA, Marton LJ (2007) Targeting polyamine metabolism and function in cancer and other hyperproliferative diseases. Nat Rev Drug Discov 6(5):373–390
Cervelli M, Tavladoraki P, Agostino SD, Angelini R, Federico R, Mariottini P (2000) Isolation and characterization of three polyamine oxidase genes from Zea mays. Plant Physiol Biochem 38: 667–677
Cervelli M, Cona A, Angelini R, Polticelli F, Federico R, Mariottini P (2001) A barley polyamine oxidase isoform with distinct structural features and subcellular localization. Eur J Biochem 268:3816–3830
Chaves MM, Pereira JS, Maroco J, Rodrigues ML, Ricardo CPP, Osório ML, Carvalho I et al (2002) How plants cope with water stress in the field? Photosynthesis and growth. Ann Bot 89(7):907–916
Chen J, Shang YT, Wang WH, Chen XY, He EM, Zheng HL, Shangguan Z (2016) Hydrogen sulfide-mediated polyamines and sugar changes are involved in hydrogen sulfide-induced drought tolerance in Spinacia oleracea seedlings. Front Plant Sci 7:1173
Cona A, Rea G, Angelini R, Federico R, Tavladoraki P (2006) Functions of amine oxidases in plant development and defence. Trends Plant Sci 11:80–88
Corpas FJ, Del Rio LA, Palma JM (2019) Plant peroxisomes at the crossroad of NO and H2O2 metabolism. J Integr Plant Biol
De Oliveira LF, Navarro BV, Cerruti G, Al E (2018) Polyamines and amino acid related metabolism: the roles of arginine and ornithine are associated with the embryogenic potential. Plant Cell Physiol 59:1084–1098
DiTomaso JM, Hart JJ, Kochian LV (1992) Transport kinetics and metabolism of exogenously applied putrescine in roots of intact maize seedlings. Plant Physiol 98(2):611–620
Do PT, Drechsel O, Heyer AG, Hincha DK, Zuther E (2014) Changes in free polyamine levels, expression of polyamine biosynthesis genes, and performance of rice cultivars under salt stress: a comparison with responses to drought. Front Plant Sci 5:182
Docimo T, Reichelt M, Schneider B, Al E (2012) The first step in the biosynthesis of cocaine in Erythroxylum coca: the characterization of arginine and ornithine decarboxylases. Plant Mol Biol 78:599–615
Ebeed HT, Hassan NM, Aljarani AM (2017) Exogenous applications of Polyamines modulate drought responses in wheat through osmolytes accumulation, increasing free polyamine levels and regulation of polyamine biosynthetic genes. Plant Physiol Biochem 118:438–448

Facchini PJ, Hagel J, Zulak KG (2002) Hydroxycinnamic acid amide metabolism: physiology and biochemistry. Can J Bot 80(6):577–589

Fellenberg C, Böttcher C, Vogt T (2009) Phenylpropanoid polyamine conjugate biosynthesis in Arabidopsis thaliana flower buds. Phytochemistry 70(11-12):1392–1400

Fincato P, Moschou PN, Spedaletti V, Tavazza R, Angelini R, Federico R et al (2011) Functional diversity inside the Arabidopsis polyamine oxidase gene family. J Exp Bot 62:1155–1168

Fincato P, Moschou PN, Ahou A, Angelini R, Roubelakis-Angelakis KA, Federico R et al (2012) The members of Arabidopsis thaliana PAO gene family exhibit distinct tissue- and organ-specific expression pattern during seedling growth and flower development. Amino Acids 42: 831–841

Filippou P, Antoniou C, Fotopoulos V (2013) The nitric oxide donor sodium nitroprusside regulates polyamine and proline metabolism in leaves of Medicago truncatula plants. Free Radic Biol Med 1(56):172–183

García-Mata C, Lamattina L (2010) Hydrogen sulphide, a novel gasotransmitter involved in guard cell signalling. New Phytol 188(4):977–984

Ghassemi S, Farhangi-Abriz S, Faegi-Analou R, Ghorbanpour M, Asgari Lajayer B (2018) Monitoring cell energy, physiological functions and grain yield in field-grown mung bean exposed to exogenously applied polyamines under drought stress. J Soil Sci Plant Nutr 18(4): 1108–1125

Gholami M, Fakhari AR, Ghanati F (2013) Selective regulation of nicotine and polyamines biosynthesis in tobacco cells by enantiomers of ornithine. Chirality 25:22–27

Gomez-Jimenez MC, Paredes MA, Gallardo M, Sanchez-Calle IM (2010) Mature fruit abscission is associated with up-regulation of polyamine metabolism in the olive abscission zone. J Plant Physiol 167:1432–1441

Grienenberger E, Besseau S, Geoffroy P, Debayle D, Heintz D, Lapierre C, Pollet B et al (2009) A BAHD acyltransferase is expressed in the tapetum of Arabidopsis anthers and is involved in the synthesis of hydroxycinnamoyl spermidines. Plant J 58(2):246–259

Grossi M, Phanstiel O, Rippe C, Al E (2016) Inhibition of polyamine uptake potentiates the anti-proliferative effect of polyamine synthesis inhibition and preserves the contractile phenotype of vascular smooth muscle cells. J Cell Physiol 231:1334–1342

Gupta KJ, Brotman Y, Segu S, Zeier T, Zeier J, Persijn ST et al (2013) The form of nitrogen nutrition affects resistance against Pseudomonas syringae pv. phaseolicola in tobacco. J Exp Bot 64:553–568

Hanfrey C, Sommer S, Mayer MJ, Burtin D, Michael AJ (2001) Arabidopsis polyamine biosynthesis: absence of ornithine decarboxylase and the mechanism of arginine decarboxylase activity. Plant J 27:551–560

Hanfrey C, Sommer S, Mayer MJ, Al E (2010) Arabidopsis polyamine biosynthesis: absence of ornithine decarboxylase and the mechanism of arginine decarboxylase activity. Plant J 27:551–560

Hatmi S, Gruau C, Trotel-Aziz P, Villaume S, Rabenoelina F, Baillieul F et al (2015) Drought stress tolerance in grapevine involves activation of polyamine oxidation contributing to improved immune response and low susceptibility to Botrytis cinerea. J Exp Bot 66:775–787

Hura T, Dziurka M, Hura K, Ostrowska A, Dziurka K (2015) Free and cell wall-bound polyamines under long-term water stress applied at different growth stages of× Triticosecale Wittm. PLoS One 10(8):e0135002

Hussain SS, Ali M, Ahmad M, Siddique KHM (2011) Polyamines: natural and engineered abiotic and biotic stress tolerance in plants. Biotechnol Adv 29:300–311

Igarashi K, Kashiwagi K (2006) Bacterial and eukaryotic transport systems. In: Polyamine cell signaling. Springer, pp 433–448

Igarashi K, Kashiwagi K (2010) Characteristics of cellular polyamine transport in prokaryotes and eukaryotes. Plant Physiol Biochem 48(7):506–512

Igarashi K, Kashiwagi K (2015) Modulation of protein synthesis by polyamines. IUBMB Life 67: 160–169

Ikbal FE, Hernández JA, Barba-Espín G, Koussa T, Aziz A, Faize M et al (2014) Enhanced salt-induced antioxidative responses involve a contribution of polyamine biosynthesis in grapevine plants. J Plant Physiol 171:779–788

Kamada-Nobusada T, Hayashi M, Fukazawa M, Sakakibara H, Nishimura M (2008) A putative peroxisomal polyamine oxidase, AtPAO4, is involved in polyamine catabolism in Arabidopsis thaliana. Plant Cell Physiol 49:1272–1282

Kaouass M, Audette M, Ramotar D, Verma S, De Montigny D, Gamache I, Torossian K et al (1997) The STK2 gene, which encodes a putative Ser/Thr protein kinase, is required for high-affinity spermidine transport in Saccharomyces cerevisiae. Mol Cell Biol 17(6):2994–3004

Kaouass M, Gamache I, Ramotar D, Audette M, Poulin R (1998) The spermidine transport system is regulated by ligand inactivation, endocytosis, and by the Npr1p Ser/Thr protein kinase in Saccharomyces cerevisiae. J Biol Chem 273(4):2109–2117

Khoshmanzar E, Aliasgharzad N, Neyshabouri MR, Khoshru B, Arzanlou M, Asgari Lajayer B (2019) Effects of Trichoderma isolates on tomato growth and inducing its tolerance to water deficit stress. Int J Environ Sci Technol 17:869–878

Kim NH, Kim BS, Hwang BK (2013) Pepper arginine decarboxylase is required for polyamine and gamma-aminobutyric acid signaling in cell death and defense response. Plant Physiol 162: 2067–2083

Kim DW, Watanabe K, Murayama C, Izawa S, Niitsu M, Michael AJ et al (2014) Polyamine oxidase5 regulates arabidopsis growth through thermospermine oxidase activity. Plant Physiol 165:1575–1590

Kitashiba H, Honda C, Moriguchi T (2006) Identification of polyamine oxidase genes from apple and expression analysis during fruit development and cell growth. Plant Biotechnol J 23:425–429

Klingler JP, Batelli G, Zhu JK (2010) ABA receptors: the START of a new paradigm in phytohormone signalling. J Exp Bot 61(12):3199–3210

Kumar A, Taylor M, Altabella T, Al E (1997) Recent advances in polyamine research. Trends Plant Sci 2:124–130

Lee J, Sperandio V, Frantz DE, Longgood J, Camilli A, Phillips MA et al (2009) An alternative polyamine biosynthetic pathway is widespread in bacteria and essential for biofilm formation in Vibrio cholerae. J Biol Chem 284:9899–9907

Li Z, Jing W, Peng Y, Zhang XQ, Ma X, Huang LK, Yan Y (2015) Spermine alleviates drought stress in white clover with different resistance by influencing carbohydrate metabolism and dehydrins synthesis. PLoS One 10(4):e0120708

Liu K, Fu HH, Bei QX, Luan S (2000) Inward potassium channel in guard cells as a target for polyamine regulation of stomatal movements. Plant Physiol 124:1315–1325

Liu HP, Dong BH, Zhang YY, Liu ZP, Liu YL (2004) Relationship between osmotic stress and the levels of free, conjugated, and bound polyamines in leaves of wheat seedlings. Plant Sci 166: 1261–1267

Liu JH, Kitashiba H, Wang J, Ban Y, Moriguchi T (2007) Polyamines and their ability to provide environmental stress tolerance to plants. Plant Biotechnol J 24:117–126

Liu Y, Zuo Z, Hu J (2010) Effects of exogenous polyamines on growth and drought resistance of Apple seedlings. J Northwest Forestry Univ 25:000039–000042

Liu T, Kim DW, Niitsu M, Berberich T, Kusano T (2014a) Oryza sativa polyamine oxidase 1 back-converts tetraamines, spermine and thermospermine, to spermidine. Plant Cell Rep 33:143–151

Liu T, Wook Kim D, Niitsu M, Berberich T, Kusano T (2014b) POLYAMINE OXIDASE 1 from rice (Oryza sativa) is a functional ortholog of Arabidopsis POLYAMINE OXIDASE 5. Plant Signal Behav 9:e29773

Liu Y, Liang H, Lv X, Liu D, Wen X, Liao Y (2016a) Effect of polyamines on the grain filling of wheat under drought stress. Plant Physiol Biochem 100:113–129

Liu Q, Nishibori N, Imai I, Al E (2016b) Response of polyamine pools in marine phytoplankton to nutrient limitation and variation in temperature and salinity. Mar Ecol Prog Ser 544:93–105

Liu W, Tan M, Zhang C, Al E (2017) Functional characterization of murB-potABCD operon for polyamine uptake and peptidoglycan synthesis in Streptococcus suis. Microbiol Res 207:177–187

Luo J, Fuell C, Parr A, Hill L, Bailey P, Elliott K, Fairhurst SA et al (2009) A novel polyamine acyltransferase responsible for the accumulation of spermidine conjugates in Arabidopsis seed. Plant Cell 21(1):318–333

Ma D, Ding H, Wang C, Qin H, Han Q, Hou J, Lu H et al (2016) Alleviation of drought stress by hydrogen sulfide is partially related to the abscisic acid signaling pathway in wheat. PLoS One 11(9):e0163082

Marcé M, Brown DS, Capell T, Figueras X, Tiburcio AF (1995) Rapid high-performance liquid chromatographic method for the quantitation of polyamines as their dansyl derivatives: application to plant and animal tissues. J Chromatogr B Biomed Appl 666(2):329–335

Martin-Tanguy J (1997) Conjugated polyamines and reproductive development: biochemical, molecular and physiological approaches. Physiol Plant 100(3):675–688

Martin-Tanguy J (2010) Conjugated polyamines and reproductive development: biochemical, molecular and physiological approaches. Physiol Plant 100:675–688

Mitchell J, Judd G, Bareyal-Leyser A, Ling S (1994) Feedback repression of polyamine transport is mediated by antizyme in mammalian tissue-culture cells. Biochem J 299(1):19–22

Mo H, Wang ZY, Zhang G, Zhang JF, Ma ZY (2015) Cotton polyamine oxidase is required for spermine and camalexin signalling in the defence response to Verticillium dahliae. Plant J 83: 962–975

Mohammadi H, Ghorbanpour M, Brestic M (2018) Exogenous putrescine changes redox regulations and essential oil constituents in field-grown Thymus vulgaris L. under well-watered and drought stress conditions. Ind Crop Prod 122:119–132

Møller SG, Mcpherson MJ (1998) Developmental expression and biochemical analysis of the Arabidopsis atao1 gene encoding an H2O2-generating diamine oxidase. Plant J 13:781–791

Montesinos-Pereira D, Barrameda-Medina Y, Romero L, Ruiz JM, Sánchez-Rodríguez E (2015) Genotype differences in the metabolism of proline and polyamines under moderate drought in tomato plants. Plant Biol 16:1050–1057

Moschou PN, Delis ID, Paschalidis KA, Roubelakis-Angelakis KA (2008) Transgenic tobacco plants overexpressing polyamine oxidase are not able to cope with oxidative burst generated by abiotic factors. Physiol Plant 133:140–156

Moschou P, Wu J, Cona A, Tavladoraki P, Angelini R, Roubelakis-Angelakis K (2012) The polyamines and their catabolic products are significant players in the turnover of nitrogenous molecules in plants. J Exp Bot 63(14):5003–5015

Mustafavi SH, Badi HN, Sekara A, Al E (2018) Polyamines and their possible mechanisms involved in plant physiological processes and elicitation of secondary metabolites. Acta Physiol Plant 40:102

Nahar K, Hasanuzzaman M, Rahman A, Alam M, Mahmud JA, Suzuki T, Fujita M (2016) Polyamines confer salt tolerance in mung bean (Vigna radiata L.) by reducing sodium uptake, improving nutrient homeostasis, antioxidant defense, and methylglyoxal detoxification systems. Frontiers in. Plant Sci 7:1104

Nozaki T, Nishimura K, Michael AJ, Maruyama T, Kakinuma Y, Igarashi K (1996) A second gene encoding a putative serine/threonine protein kinase which enhances spermine uptake in Saccharomyces cerevisiae. Biochem Biophys Res Commun 228(2):452–458

Ono Y, Kim DW, Watanabe K, Sasaki A, Niitsu M, Berberich T et al (2012) Constitutively and highly expressed Oryza sativa polyamine oxidases localize in peroxisomes and catalyze polyamine back conversion. Amino Acids 42:867–876

Ouyang J, Song C, Chen D (2017) Research progress on heat-tolerance mechanism and transports of polyamines in plant. Mol Plant Breeding 15:3286–3294

Pegg AE (2014) The function of spermine. IUBMB Life 66:8–18

Pegg AE (2016) Functions of polyamines in mammals. J Biol Chem 291:14904–14912

Pegg AE, Casero RA (2011) Current status of the polyamine research field. Methods Mol Biol 720: 3–35

Peng D, Wang X, Li Z, Zhang Y, Peng Y, Li Y, He X et al (2016) NO is involved in spermidine-induced drought tolerance in white clover via activation of antioxidant enzymes and genes. Protoplasma 253(5):1243–1254

Pistocchi R, Bagni N, Creus JA (1987) Polyamine uptake in carrot cell cultures. Plant Physiol 84(2): 374–380

Planas-Portell J, Gallart M, Tiburcio AF, Altabella T (2013) Copper-containing amine oxidases contribute to terminal polyamine oxidation in peroxisomes and apoplast of Arabidopsis thaliana. BMC Plant Biol 13:109

Podlesakova K, Ugena L, Spichal L, Dolezal K, De Diego N (2019) Phytohormones and polyamines regulate plant stress responses by altering GABA pathway. N Biotechnol 48:53–65

Porter JR (2005) Rising temperatures are likely to reduce crop yields. Nature 436(7048):174–174

Pottosin I, Velarde-Buendia AM, Bose J, Zepeda-Jazo I, Shabala S, Dobrovinskaya O (2014) Cross-talk between reactive oxygen species and polyamines in regulation of ion transport across the plasma membrane: implications for plant adaptive responses. J Exp Bot 65:1271–1283

Rangan P, Subramani R, Kumar R, Singh AK, Singh R (2014, 2014) Recent advances in polyamine metabolism and abiotic stress tolerance. Biomed Res Int

Rea G, Laurenzi M, Tranquilli E, D'ovidio R, Federico R, Angelini R (1998) Developmentally and wound-regulated expression of the gene encoding a cell wall copper amine oxidase in chickpea seedlings. FEBS Lett 437:177–182

Regla-Márquez CF, Canto-Flick A, Avilés-Viñas SA, Al E (2015) Cadaverine: a common polyamine in zygotic embryos and somatic embryos of the species Capsicum chinense Jacq. PCTOC 124:253–264

Richards FJ, Coleman RG (1952) Occurrence of putrescine in potassium deficient barley. Nature 170:47,981

Rodriguez AA, Maiale SJ, Menendez AB, Ruiz OA (2009) Polyamine oxidase activity contributes to sustain maize leaf elongation under saline stress. J Exp Bot 60:4249–4262

Rouphael Y, Colla G, Bernardo L, Kane D, Trevisan M, Lucini L (2016) Zinc excess triggered polyamines accumulation in lettuce root metabolome, as compared to osmotic stress under high salinity. Front Plant Sci 7:842

Rosales-Corral SA, Acuña-Castroviejo D, Coto-Montes A, Boga JA, Manchester LC, Fuentes-Broto L, Korkmaz A, Ma S, Tan DX, Reiter RJ (2012) Alzheimer's disease: pathological mechanisms and the beneficial role of melatonin. J Pineal Res 52(2):167–202

Saghafi D, Delangiz N, Asgari Lajayer B, Ghorbanpour M (2019a) An overview on improvement of crop productivity in saline soils by halotolerant and halophilic PGPRs. 3Biotech 9:261. https://doi.org/10.1007/s13205-019-1799-0

Saghafi D, Ghorbanpour M, Shirafkan Ajirloo H, Asgari Lajayer B (2019b) Enhancement of growth and salt tolerance of canola (*Brassica napus* L.) seedlings by halotolerant Rhizobium strains containing ACC-deaminase activity. Plant Physiol Rep 24:225–235

Saghafi D, Asgari Lajayer B, Ghorbanpour M (2020) Engineering bacterial ACC deaminase for improving plant productivity under stressful conditions. In: Sharma V, Salwan R, Al-Ani LKT (eds) Molecular aspects of plant beneficial microbes in agriculture. Elsevier Inc. https://doi.org/10.1016/B978-0-12-818469-1.00022-5

Sánchez-Rodríguez E, Romero L, Ruiz J (2016) Accumulation of free polyamines enhances the antioxidant response in fruits of grafted tomato plants under water stress. J Plant Physiol 190:72–78

Seki M, Umezawa T, Urano K, Shinozaki K (2007) Regulatory metabolic networks in drought stress responses. Curr Opin Plant Biol 10(3):296–230

Sen S, Ghosh D, Mohapatra S (2018) Modulation of polyamine biosynthesis in Arabidopsis thaliana by a drought mitigating Pseudomonas putida strain. Plant Physiol Biochem 129:180–188

Silveira V, Santa-Catarina C, Tun NN, Scherer FE, Handro W, Guerra MP, Floh IS (2006) Polyamine effects on the endogenous polyamine contents, nitric oxide release, growth and differentiation of embryogenic suspension cultures of *Araucaria angustifolia* (Bert.) O. Ktze. Plant Sci 71:91–98

Sobieszczuk-Nowicka E (2017) Polyamine catabolism adds fuel to leaf senescence. Amino Acids 49:49–56

Takahashi Y, Cong R, Sagor GH, Niitsu M, Berberich T, Kusano T (2010) Characterization of five polyamine oxidase isoforms in Arabidopsis thaliana. Plant Cell Rep 29:955–965

Takahashi Y, Tahara M, Yamada Y, Al E (2017) Characterization of the polyamine biosynthetic pathways and salt stress response in Brachypodium distachyon. J Plant Growth Regul 37:625–634

Tanou G, Ziogas V, Belghazi M, Christou A, Filippou P, Job D, Fotopoulos V, Molassiotis A (2014) Polyamines reprogram oxidative and nitrosative status and the proteome of citrus plants exposed to salinity stress. Plant Cell Environ 37(4):864–885

Tassoni A, Franceschetti M, Bagni N (2008) Polyamines and salt stress response and tolerance in Arabidopsis thaliana flowers. Plant Physiol Biochem 46(5-6):607–613

Tassoni A, Bagni N, Ferri M, Franceschetti M, Khomutov A, Marques MP, Fiuza SM et al (2010) Helianthus tuberosus and polyamine research: past and recent applications of a classical growth model. Plant Physiol Biochem 48(7):496–505

Tavladoraki P, Rossi MN, Saccuti G, Perez-Amador MA, Polticelli F, Angelini R et al (2006) Heterologous expression and biochemical characterization of a polyamine oxidase from Arabidopsis involved in polyamine back conversion. Plant Physiol 141:1519–1532

Thuiller W, Lavorel S, Araújo MB, Sykes MT, Prentice IC (2005) Climate change threats to plant diversity in Europe. Proc Natl Acad Sci 102(23):8245–8250

Tiburcio A, Kaur-Sawhney R, Galston A (1990) Polyamine metabolism. Biochem Plants 16:283–325

Tisi A, Federico R, Moreno S, Lucretti S, Moschou PN, Roubelakis-Angelakis KA et al (2011) Perturbation of polyamine catabolism can strongly affect root development and xylem differentiation. Plant Physiol 157:200–215

Toumi I, Moschou PN, Paschalidis KA, Bouamama B, Salem-Fnayou AB, Ghorbel AW, Mliki A, Roubelakis-Angelakis KA (2010) Abscisic acid signals reorientation of polyamine metabolism to orchestrate stress responses via the polyamine exodus pathway in grapevine. J Plant Physiol 167(7):519–525

Tuskan GA, Difazio S, Jansson S, Bohlmann J, Grigoriev I, Hellsten U et al (2006) The genome of black cottonwood, Populus trichocarpa (Torr. & Gray). Science 313:1596–1604

Vuosku J, Karppinen K, Muilu-Mäkelä R, Kusano T, Sagor GHM, Avia K (2018) Scots pine aminopropyl transferases shed new light on evolution of the polyamine biosynthesis pathway in seed plants. Ann Bot 121:1243–1256

Wang J, Liu JH (2009) Change in free polyamine contents and expression profiles of two polyamine biosynthetic genes in citrus embryogenic callus under abiotic stresses. Biotechnol Biotechnol Equip 29:1289–1293

Wang W, Liu JH (2015) Genome-wide identification and expression analysis of the polyamine oxidase gene family in sweet orange (Citrus sinensis). Gene 555:421–429

Wang W, Vinocur B, Altman A (2003) Plant responses to drought, salinity and extreme temperatures: towards genetic engineering for stress tolerance. Planta 28:1–14

Wang P, Du Y, Hou YJ, Zhao Y, Hsu CC, Yuan F, Zhu X et al (2015) Nitric oxide negatively regulates abscisic acid signaling in guard cells by S-nitrosylation of OST1. Proc Natl Acad Sci 112(2):613–618

Wimalasekera R, Tebartz F, Scherer GF (2011a) Polyamines, polyamine oxidases and nitric oxide in development, abiotic and biotic stresses. Plant Sci 181(5):593–603

Wimalasekera R, Villar C, Begum T, Scherer GF (2011b) COPPER AMINE OXIDASE1 (CuAO1) of Arabidopsis thaliana contributes to abscisic acid-and polyamine-induced nitric oxide biosynthesis and abscisic acid signal transduction. Mol Plant 4(4):663–678

Wu J, Qu H, Shang Z, Jiang X, Moschou PN, Roubelakis-Angelakis KA et al (2010) Spermidine oxidase-derived H2O2 activates downstream Ca2+ channels which signal pollen tube growth. Plant J 63:1042–1053

Xie Y, Mao Y, Zhang W, Lai D, Wang Q, Shen W (2014) Reactive oxygen species-dependent nitric oxide production contributes to hydrogen-promoted stomatal closure in Arabidopsis. Plant Physiol 165(2):759–773

Xu L (2015) The effect of polyamine on flower bud differentiation and bud germination of chrysanthemum. Shandong Agricultural University, pp 31–36

Xu C, Wu X, Zhang H (2009) Impact of D-Arg on drought resistance and endogenous polyamines in mycorrhizal Pinus massoniana. J Nanjing Forestry Univ 33:019–023

Xu L, Xing ST, Sun X (2014) Effects of polyamines on hormones contents and the relationship with the flower bud differentiation in chrysanthemum. J Plant Physiol 50:1195–1202

Yamaguchi K, Takahashi Y, Berberich T, Al E (2007) A protective role for the polyamine spermine against drought stress in Arabidopsis. Biochem Biophys Res Commun 352:486–490

Yang JC, Zhang JH, Liu K, Wang ZQ, Liu LJ (2007) Involvement of polyamines in the drought resistance of rice. J Exp Bot 58:1545–1555

Yang L, Liang H, Lv X, Liu D, Wen X, Liao Y (2016) Effect of polyamines on the grain filling of wheat under drought stress. Plant Physiol Biochem 100:113–129

Yoda H, Hiroi Y, Sano H (2006) Polyamine oxidase is one of the key elements for oxidative burst to induce programmed cell death in tobacco cultured cells. Plant Physiol 142:193–206

Zeid IM, Shedeed ZA (2006) Response of alfalfa to putrescine treatment under drought stress. Biol Plant 50:635–640

Zhao FG, Qin P (2004) Protective effect of exogenous polyamines on root tonoplast functions against salt stress in barley seedlings. Plant Growth Regul 42:97–103

Portfolio of Drought Stress Response and Genetic Enhancement Strategies for Development of Future Drought-Tolerant Crop

Hemangini Parmar, Ajay Kumar Venkatapuram, Afreen Rashid, K. Sangeetha, Sahil Mehta, Malireddy K. Reddy, Anjana Goel, and V. Mohan M. Achary

Abstract Global climate irregularities and the freshwater shortage have become a serious ecological problem posing threats to global food security. Drought stress threatens every stage of plant development and causes huge crop loss during panicle development and anthesis. Crop productivity and quality are important aspects to ensure food security. Hence, it is essential to understand the knowledge about the response of plants towards the drought stress for improving the existing varieties for high yield under the water-limited conditions to fulfill the growing food demand from an increasing population. Even if conventional and mutation breeding were successfully used in the development in the past for developing desired varieties, due to technological limitations, these methods cannot fulfill the predicted 2050 future global food demands. Further, genetically modified crops are usually not considered owing to regulatory concerns. Due to the complex nature of the drought resistance mechanism which is governed by many QTLs, genes, transporters, stress proteins, transcription factors, hormones, metabolites, and microRNAs, a multipronged

Malireddy K. Reddy, Anjana Goel, and V. Mohan M. Achary have contributed equally to this work.

H. Parmar
Crop Improvement Group, International Centre for Genetic Engineering and Biotechnology, Aruna Asaf Ali Marg, New Delhi, India

Department of Biotechnology, GLA University, Mathura, Uttar Pradesh, India

A. K. Venkatapuram · A. Rashid · K. Sangeetha · M. K. Reddy · V. M. M. Achary (✉)
Crop Improvement Group, International Centre for Genetic Engineering and Biotechnology, Aruna Asaf Ali Marg, New Delhi, India

S. Mehta
Crop Improvement Group, International Centre for Genetic Engineering and Biotechnology, Aruna Asaf Ali Marg, New Delhi, India

School of Agricultural Sciences, K.R. Mangalam University, Sohna, Haryana, India

A. Goel
Department of Biotechnology, GLA University, Mathura, Uttar Pradesh, India

© The Author(s), under exclusive license to Springer Nature Singapore Pte Ltd. 2022

A. Vaishnav et al. (eds.), *Plant Stress Mitigators*,
https://doi.org/10.1007/978-981-16-7759-5_24

approach is required to develop new stress resilience varieties. High-throughput genomics, phenomics, and employing new breeding techniques such as genome editing offer a rapid targeted improvement of the genotype of agriculturally important traits. In the current book chapter, we have summarized the biochemical and genetic aspects of drought tolerance in plants. Furthermore, we also focused on the current technical limitations and future application of the genome-editing systems for crop improvement for sustainable agriculture under water-limited conditions.

Keywords Plant · Productivity · Drought · MAS · Genome editing · CRISPR/Cas9

Introduction

The changing global climatic conditions aggravate disaster events like drought, salinity, emerging of new pathotypes, etc. and affect the crop yield and ultimately threaten food security. Rain-fed agriculture covers 80% of the world's total food production, including almost 60% of global cereals production regions (FAO 2008; UNESCO 2009). Among all biological stresses, drought stress is the single largest abiotic factor that negatively influences plant growth and fertility by altering water and nutrient relations and inhibiting water-use efficiency which ultimately limits agriculture productivity (Takahashi et al. 2020). Drought stress on the plant productivity depends on the development stage and duration of water deficit in the soil. For instance, among various stress, drought alone contributes 23–60% yield loss in maize during the vegetative stage, 66–88% in the reproductive stage (Farooq et al. 2009; Kamara et al. 2003), 65–91% in rice during reproductive and grain filling stages (Bouman et al. 2005; Serraj et al. 2011), approximately 46–72% in soybean (Samarah et al. 2006), 49–57% during seed filling stage in barley (Samarah 2005), etc. (Table 1).

Upon dehydration stress, the plant operates morphological, physiological, and molecular adaptation mechanisms including reduction of water loss from the leaf surface, increase in root surface area, enrichment of antioxidant defense system, production of osmolytes, modulation of the hormonal pathway, transcriptional regulation, and activation of stress-responsive genes, as to mitigate the drought stress (Hund et al. 2009; Shinwari et al. 2020). Reactive oxygen species (ROS) generation is a general consequence of plant response to drought stress that causes damage to the plant cell membrane, biomolecules, and photosynthesis apparatus. Plant neutralizes excess ROS production through enzymatic and non-enzymatic systems and maintains redox balance in the cellular environment (Gill and Tuteja 2010). Plants exhibit drought tolerance strategies by maintaining the osmotic balance and protecting plant cells and biomolecules to function under water-deficient stress (Slama et al. 2015). Osmolytes include amino acid, sugars, proline, sucrose, trehalose, glycine betaine, and polyhydric alcohols are act as osmoprotectants and increase during drought conditions. Under stress, these compounds help to maintain membrane integrity, scavenge ROS, protect enzymes and maintain their biological functions (Per et al. 2017).

Table 1 Grain yield losses in some selected cereal crop on drought stress

Crops	Development stages	Yield loss	References
Wheat	Pre-anthesis	18–53%	Majid et al. (2007)
	Anthesis	11–39%	Jatoi et al. (2011)
	Booting to maturity	37%	Shamsi et al. (2010)
	Grain filling to maturity	31%	Shamsi et al. (2010)
	Grain filling to maturity	35%	Shamsi and Kobraee (2011)
Rice	Lowland severe reproductive stage	65%–91%	Vikram et al. (2011) Ghimire et al. (2012) Dixit et al. (2012, 2014)
	Upland severe reproductive stage	80%–97%	Bernier et al. (2007) Dixit et al. (2012, 2014)
Pearl Millet	Pre-anthesis and anthesis	65%	Winkel et al. (1997)
Barley	Severe reproductive stage	73%–87%	Samarah et al. (2009)
Maize	Vegetative	23–60	Farooq et al. (2009)
	Reproductive	63–88	Kamara et al. (2003)
	Grain filling	79–81	Farooq et al. (2009)

The phytohormone abscisic acid (ABA) plays an important role to relay drought stress responses by regulating stomatal closure and expression of stress-responsive genes (Cutler et al. 2010). During drought, ABA accumulation is enhanced in the leaf vasculature tissue, further spreads to all other tissues, regulates stomatal movement, slows photosynthesis rate, and modulates stress resistance genes related to drought (Fig. 1). Plant root systems recognize water-deficit conditions in the soil through ABA accumulation and transmit the signal to the leaves to counter drought stress conditions. Plants respond to drought stress mediated through ABA-dependent regulatory systems and ABA-independent as well (Takahashi et al. 2018).

Drought tolerance is highly complex as the functions of the molecular mechanism in response to drought stress are coordinately controlled by many genes, transcription factors (MYB, NAC, bZIP, AP2/ERF, and AREB/ABF), protein kinases that regulate stomatal movement, and stress-responsive genes involved in the drought-tolerance mechanisms (Kim et al. 2010; Joshi et al. 2016; Wang et al. 2016a, b). Peptides function as mobile signaling molecules involved in long-distance organ-to-organ communication under water-deficit stress has been recently reported (Takahashi et al. 2019). Based on conventional and modern breeding methods, many QTLs for agronomic important traits associated with drought resistance have been identified in several crops including rice, maize, wheat pearl millet, foxtail millet, sorghum, soybean, etc. (Khan et al. 2016).

Drought tolerance mechanism is complex in nature and requires an integrated and comprehensive approach to develop new varieties with sustainable grain yield under drought conditions. Several important drought-tolerant QTLs, genes, TFs, and small RNAs have been characterized in many plant species through various genetic engineering methods. Many key genes associated with drought-tolerant QTLs are not fully characterized. Hence, reverse genetics approaches are quite advantageous for accurate identification and characterization of these stress-related genes under

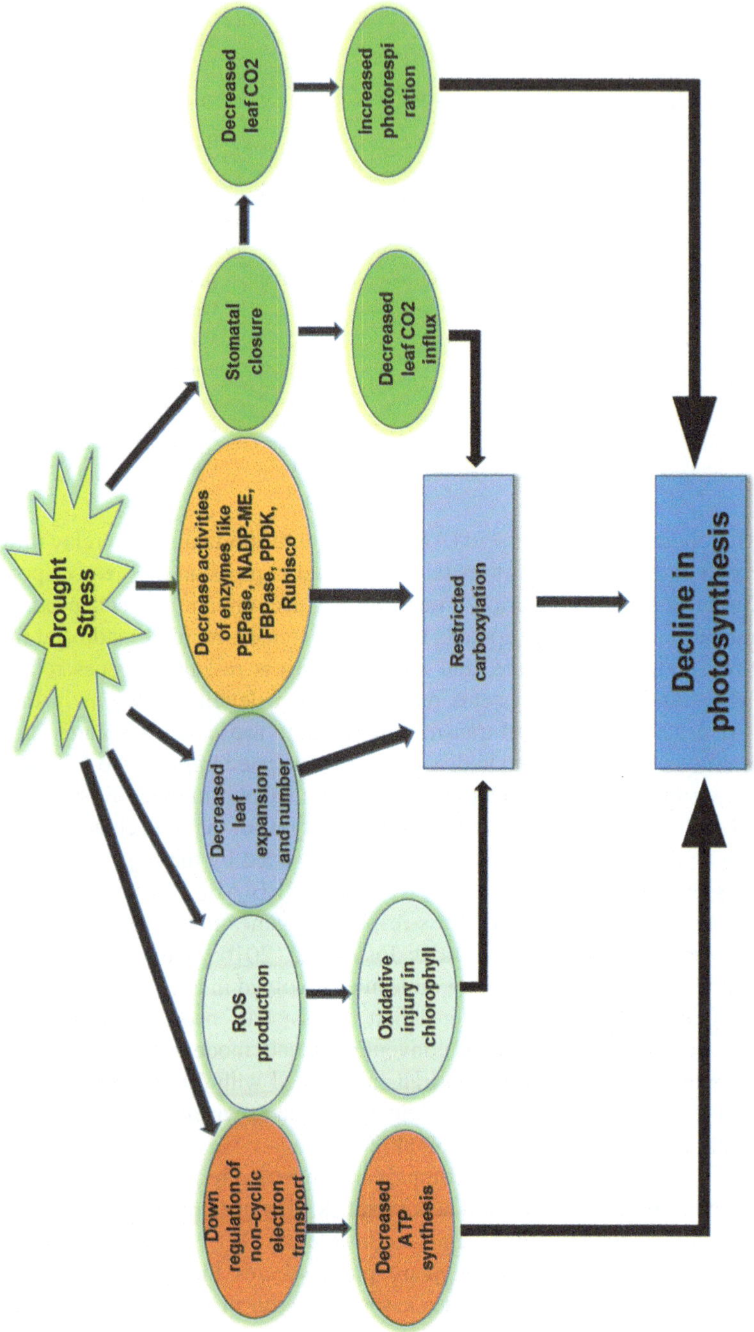

Fig. 1 Impact of drought stress on the photosynthesis in plants

water stress conditions (Billah et al. 2021). Genome editing is a recent addition to the new breeding technology (NBT), a reliable, target-specific, fast, less laborious method to manipulate the genome of existing elite cultivars in various ways. Recently, CRISPR/Cas9 technologies have been usefully used in many crops for improving various agronomically important traits including drought tolerance. In this review, we discussed various aspects of drought stress tolerance, including tissue/organ-specific response to drought, and innovative applications, and their utility to answer biological questions related to drought for future food security.

Genetic Basis of Drought Tolerance

Being stalkless by nature, the plants are particularly sensitive to the drought conditions throughout their life cycle, especially in the reproductive stage where it leads to sterility and decline in plant biomass production, quality, and energy (Widawsky and O'Toole 1990; Xu et al. 2005; Kalefetoğlu Macar and Ekmekci 2009; Seleiman et al. 2021). This is because of the reason that drought tolerance in crops is highly complex at the genetic as well as physiological levels (Blum 2005). Over the several decades, conventional breeding is the most effective way and successfully used to enhance the yield potential of many crops under drought conditions. Drought stress is governed by various minor genes (polygenes) having additional effects in their expression (Zhao et al. 2008; Mohammadi et al. 2005; Serba and Yadav 2016). Therefore, the chromosomal loci encompassing such types of genes are denoted as quantitative trait loci (QTL). Marker-assisted selection (MAS) includes molecular markers that map close to definite genes or QTLs, which can be used to select individuals with the target traits that have been established (Varshney et al. 2013).

At the genetic level, multiple QTLs putatively contribute to the drought tolerance up to a great extent. In this regard, developing robust drought-tolerant crop varieties is one of the topmost priorities of breeders in the last few decades. To overcome this problem, several drought tolerance traits such as root characteristics, grain yield, osmotic adjustment, relative water contents, etc., have been genetically dissected first through QTL mapping followed by practicing QTL breeding (Li et al. 2000; Shen et al. 2001; Xu et al. 2005). For example, QTLs linked with drought tolerance have been identified in various crops, i.e., *Oryza sativa* (Tripathy et al. 2000; Zhao et al. 2008; Gomez et al. 2010; Mohd Ikmal et al. 2019, 2021), *Triticum aestivum* (Pinto et al. 2010; Bennett et al. 2012; Kumar et al. 2020a, b), *Zea mays* (Agrama and Moussa 1996; Prasanna et al. 2009; Trachsel et al. 2016; Zhou et al. 2019; Hu et al. 2021), *Setaria italica* (Jia et al. 2013), *Pennisetum glaucum* (Bidinger et al. 2007), *Glycine max* (Yang et al. 2014; Khan et al. 2018, 2020), Sorghum bicolor (Sintayehu et al. 2018; Kamal et al. 2021), and other economically important crops (Table 2).

Yet, several drought-associated QTLs characterized are not stable in diverse environments. Further, a QTL can possess positive or negative additive effects that depend on the drought condition due to strong genotype by environment interaction (G × E) (Collins et al. 2008). Also, certainty and accuracy in QTL recognition are complicated. Drought tolerance is a complex mechanism that is prompted by various

Table 2 Successful reports of QTL breeding for enhancing drought tolerance in economically important crops

S. No.	Crop	QTL	Variety used	Effect	Reference
1	Rice	*qtl12.1*	24 progenies with contrasting genotype at qtl12.1	Improved grain yield under drought stress condition through slight improvement in water uptake in genotypes with Way Rarem derived allele	Bernier et al. (2009)
		qDTY 2.2, qDTY 3.1, and qDTY 12.1	MR219	Improved yield of introgressed lines over MR219 under drought	Shamsudin et al. (2016)
		$qDTY_{3.1}$, $qDTY_{6.1}$, and $qDTY_{6.2}$	TDK1	Lines with combination of $qDTY_{3.1}$ and $qDTY_{6.1}$ showed higher drought tolerance	Dixit et al. (2017)
		qDTY12.1 and *qDTY2.3*	FUNAABOR-2	Introgression lines with both the QTLs showed more yield potential than lines with single or no QTL under drought stress	Anyaoha et al. (2019)
		qDTY1.1	Pusa Basmati 1	Seven NILs performed better in terms of agronomic and overall grain quality under reproductive stage drought stress	Dhawan et al. (2021)
		qDTY2.1 and qDTY3.1	Pusa 44	Fourteen NILs showed improved yield, grain quality and higher drought tolerance potential under drought stress condition	Dwivedi et al. (2021)
2	Chickpea	"*QTL-hotspot*" containing QTLs for root and drought tolerance traits	JG 11	Introgressed lines showed higher rooting depth, root length density and higher root dry weight	Varshney et al. (2013)
		"*QTL-hotspot*" region with QTLs for drought tolerance traits	Pusa 372, Pusa 362, and DCP 92-3	BGM 10216 showed 16% yield improvement as compared to its respective recurrent parent Pusa 372	Bharadwaj et al. (2021)
3	Wheat	QTLs from wild emmer wheat	Uzan, Bar Nir and Zahir	Improved grain yield and biomass in NILs under drought stress	Merchuk-Ovnat et al. (2016)
		QTLs linked with drought tolerance	GW322	Eighteen homozygous were phenotypically	Todkar et al. (2020)

(continued)

Table 2 (continued)

S. No.	Crop	QTL	Variety used	Effect	Reference
				more superior than the recurrent parent	
		Qyld.csdh.7AL	HUW234, HUW468, K307 and DBW17	Lines with introgression gave higher yield under rainfed and irrigated conditions	Gautam et al. (2021)
4	Barley	40 QTLs	Wild barley introgression lines (S42ILs)	S42IL-121 showed 17% increase in thousand grain weight under terminal drought stress	Honsdorf et al. (2017)
5	Sorghum	Stg QTLs	Meko, Teshale and Gambella 1107	Stg introgression lines Teshale/E36-1, Meko/B35-selection 120, and Teshale/B35-selection 2 performed better under drought stress in terms of growth and physiological traits	Sintayehu et al. (2018)
		Stg1–Stg4	Tabat and Wad Ahmed	Grain yield of most of the introgression lines was higher as compared to their recurrent parents	Kamal et al. (2021)

alterations in plant phenology and controlled by numerous quantitative trait loci (QTLs) (Fleury et al. 2010).

Understanding these complex plant responses is a quite challenging task without learning the physiological and genetic bases. Unless identifying the molecular mechanism associated with the stability of grain yield, it is a very tedious task to develop improved crop varieties against drought stress either via modern genetics or traditional breeding approaches (Sinclair 2011). New insights in drought tolerance for crop improvement through developments in plant physiology, phenotyping, and systematic plant genomics lead to enable crop breeders to enhance the crop yield by employing the latest gene network information and approaches for plant improvement (Tuberosa 2012). Improved knowledge about the physiology of the plant led to developing an understanding of complex drought tolerance-related traits which have been identified via molecular markers; therefore, the associated loci enable to choose the genotypes that ameliorate the crop yield in response to drought conditions. For instance, in maize, the yield is adversely correlated with the "anthesis-silking interval" which is raised in case of water scarcity (Duvick 2005). Genotypes involved in post-flowering under drought stress in sorghum denoted as stay-green phenotypes show the favorable effect on yield under critical drought; therefore major and minor QTLs were categorized as *Stg2*, *Stg3*, and *Stg4* which help to regulate the expression of the stay-green trait (Harris et al. 2007). In the case of rice, various QTLs related to drought tolerance have been reported, whereas fewer QTLs are

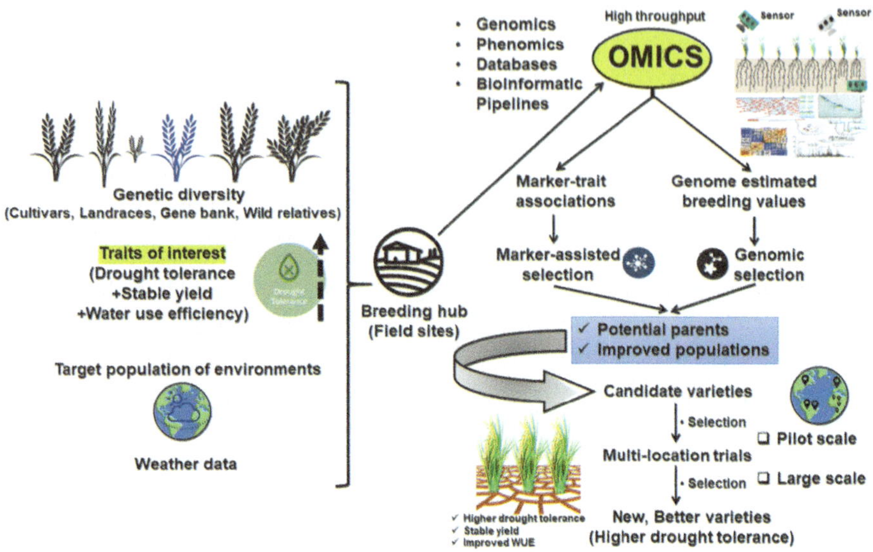

Fig. 2 Multipronged breeding programs employ marker-assisted selection (MAS) and genomic selection (GS) to develop the new varieties with enhanced drought tolerance and stable yield under stressed conditions

identified for grain quality and yield under drought stress. For instance, QTL responsive for an additive effect at the reproductive stage on grain yield under drought stress has been identified on rice chromosome no. 1 and flanked by RM431, RM11943 in the three populations N22/MTU1010, N22/Swarna, and N22/IR64 (Vikram et al. 2011). In upland cultivating rice under drought, a QTL with a significant effect on grain yield has been linked to advanced root architecture (Bernier et al. 2007). Based on the literature, the overall focus for sustainable crop product should be on selecting those QTLs that are highly heritable and affect the grain yield and water use efficiency directly under drought conditions only (Tardieu et al. 2018; Ding et al. 2021; Seleiman et al. 2021).

In addition to marker-assisted breeding (MAB), recently, genomic selection (GS) has also surfaced as the alternative approach for improving drought tolerance in crops (Shikha et al. 2017; Crossa et al. 2017; Krishnappa et al. 2021). In the GS method, the progress of enhancing tolerance depends entirely on data of genetic markers present in the whole genome (Fig. 2). Additionally, the effectiveness of GS programs also depends on the selection of genome estimated breeding values (GEBVs) in breeding populations via phenotyping (Seleiman et al. 2021). Interestingly, there is no prerequisite for knowledge of QTLs, unlike QTL breeding (Nakaya and Isobe 2012). Till the date, the GS for improving drought tolerance has been applied in only a few crops such as *Triticum aestivum*, *Zea mays*, *Cicer arietinum*, and *Pisum sativum* (Shikha et al. 2017; Annicchiarico et al. 2017; Cerrudo et al. 2018; Li et al. 2018; Badu-Apraku et al. 2019; Juliana et al. 2020; Kumar et al. 2021).

Molecular Basis of Drought Tolerance

To design effective improvement strategies for drought stress tolerance it is important to understand the physiological adaptions of plants. Several candidate genes and regulatory proteins have been identified via reverse and genetic approaches in the past and their functions are well characterized under drought stress. Numerous kinases such as CBL (calcineurin B-like) interacting protein kinase (CIPK), mitogen-activated protein kinases (MAPKs), calcium-dependent protein kinases (CDPKs), and sucrose non fermenting protein (SNF1)-related kinase 2 (SnRK2) are involved in drought stress response. Phytohormones are associated with signaling pathways; among them ABA mediate tolerance in the drought stress. Therefore, coordination of the three classes of proteins, including (a) the Pyrabactin Resistance 1 (PYR1) and/or PYR1-like protein (PYL) and/or Regulatory component of the ABA receptor (RCAR) (herewith referred as PYLs), (b) Protein phosphatase 2C (PP2C), and (c) SnRK2s regulates ABA mediated drought tolerance in plants (Joshi et al. 2016). Under normal condition, PP2Cs are usually associated with SnRK2s kinases and remains inactive (dephosphorylated state) in the absence of ABA. In water-deficit conditions, the PP2C phosphatase activity is inhibited due to the binding of ABA with PYLs. Consequently, SnRK2s undergo autophosphorylation which in turn phosphorylates downstream effector molecules and provides drought tolerance (Fig. 3). Thus, improved plant growth and drought resistance can be achieved by altering some key enzymes of ABA biosynthetic pathways (Park et al. 2008).

For instance, ABA-related genes govern various roles in metabolism, signaling, and localization (Osakabe et al. 2014). ABA signaling pathway regulates stomatal closure, for example, the *CPK10* gene mediates stomatal movement through Ca^{2+} and ABA signaling networks in rice and *Arabidopsis* (Zou et al. 2010). Similarly, the *OsCDPK7* gene acts as a positive regulator of drought and salt stress in rice (Saijo et al. 2000). Previous studies revealed the *OsCIPK23* gene is induced via abiotic stress conditions and the knockdown mutant lines show sensitivity under drought conditions (Yang et al. 2008). 9-cis-epoxycarotenoid dioxygenase 3 (NCED3) catalyzes an important step in the ABA biosynthesis pathway and its overexpression gives rise to drought tolerance (Iuchi et al. 2001; Endo et al. 2008). The membrane-localized ABC transporters ABCG25 and ABCG40 are responsible for the accurate localization of ABA require for its proper function. Similarly among three ABA transport systems, one of the members belonging to the nitrate transporter family (AIT1/NRT1.2/NPF4.6) has been identified, involved in ABA efflux and influx in the plant (Kang et al. 2010; Kanno et al. 2012). In the case of MAP kinases such as *Os*MPK5 and an RAF-like MAPKKK gene *DSM1* have been reported which are associated with drought resistance in rice (Ning et al. 2010; Xiong and Yang 2003; Sinha et al. 2011). In *Arabidopsis*, the *SnRK2* gene (*SnRK2C*) confers drought tolerance by controlling stress-responsive genes expression (Umezawa et al. 2004).

Transcription factors are an essential group of regulatory proteins which play a critical role in downstream gene expression and modulate signaling cascades at the

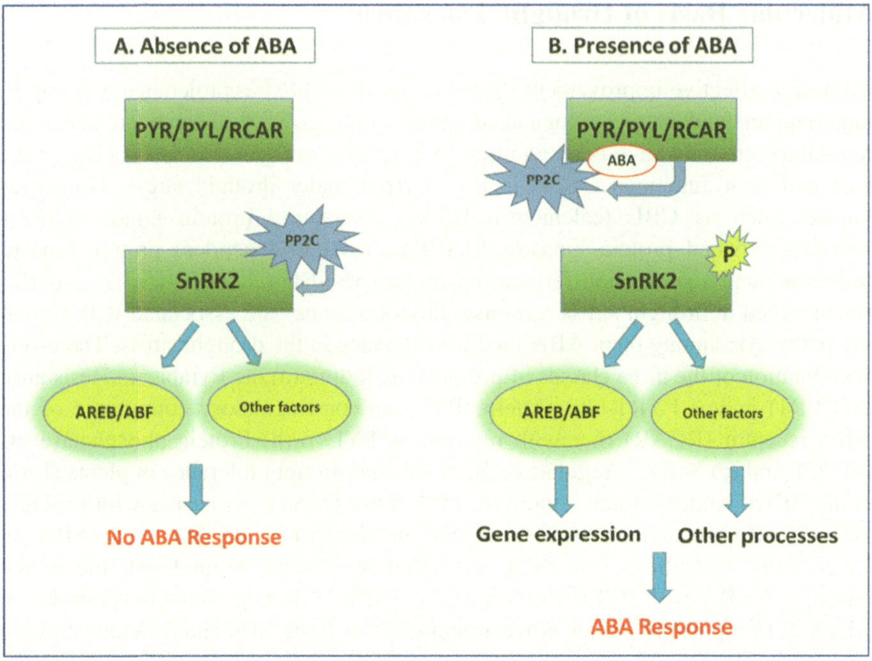

Fig. 3 Model of ABA perception and signaling in plants under (**A**) normal and (**B**) stress conditions

transcriptional level in response to drought (Osakabe et al. 2014). Various transcription factor families members have been reported to be involved in drought tolerance; e.g., APETALA2/Ethylene-responsive element binding protein (AP2/EREBP), ABA-responsive elements (ABRE), C-repeat/drought-responsive/low-temperature-responsive elements (CRT/DRE/LTRE), NAM-ATAF1/2-CUC2 (NAC), zinc finger, basic leucine zipper (bZIP), Myeloblastosis (MYB), and Myelocytomatosis (MYC) are regulatory elements present in the promoter region of the genes and encode for stress-inducible dehydrins. The ABRE binding factors (ABFs or AREBs) and CBF4/DREB1D binds to ABRE/CRT/DRE/LTRE, MYBFs, and MYCFs regulatory elements respectively through the ABA-dependent signaling pathway. Similarly, DREB2A and DREB2B bind to CRT/DRE/LTRE via ABA independent signaling pathway (Zhu 2002; Yamaguchi-Shinozaki and Shinozaki 2005). In *Arabidopsis*, the AP2/EREBP domain transcription factor SHN activates wax biosynthesis which modifies cuticle properties as a result of confers drought tolerance (Aharoni et al. 2004).

In rice, dehydration responsive element-binding factors (DREB) consist of the AP2 domain; *Os*DREBs and ARAG1 involve in drought stress (Chen et al. 2008; Zhou et al. 2010). The bZIP transcription factors subfamily consisting of ABA-responsive element-binding proteins/ factors (AREBs/ABFs) plays a key role in drought conditions and is mediated by the ABRE-dependent ABA signaling network (Fig. 4). In *Arabidopsis* ABA-mediated positive regulation of drought tolerance is facilitated by the coordination of AREB1, AREB2, and AREB3 (Yoshida et al. 2010).

In addition, NAC transcription factors are involved in imparting drought tolerance apart from contributing to total plant growth and development. Therefore, stress-responsive NAC gene *SNAC1* of rice and drought-inducible nuclear transcription factor (NFYA5) participate in stomatal regulation (Li et al. 2008; Hu et al. 2006). *OsNAC10* gene is regulated by a root-specific promoter RCc3 which increases the root diameter in overexpression lines and is mainly responsible for drought tolerance at the reproductive stage and thus results in improved grain yield in natural and drought conditions (Jeong et al. 2010).

Similarly, bZIP transcription factors in rice such as *OsbZIP23* and overexpression of OsbZIP46CA1 (a constitutively active form of *OsbZIP46* with a deletion of domain D) show a significant increase in drought and osmotic stresses tolerance in rice (Xiang et al. 2008; Tang et al. 2012). In petunia, the dehydration stress tolerance is achieved by the constitutive overexpression of a Cys2/His2 (C_2H_2)-type zinc finger protein that encodes for the ZPT2-3 gene (Sugano et al. 2003). The zinc finger protein DST acts as a negative regulator of drought and salt tolerance. Thed st knockout mutant lines showed increased stomatal closure and a decreased leaf stomatal density in rice (Huang et al. 2009). Overexpression of zinc-finger transcription factors ZAT10 and ZAT12 induces cytosolic ascorbate peroxidase encoding genes (*APXs*) which perform as ROS scavengers for chloroplast proteins under drought stress (Karpinski et al. 1997; Davletova et al. 2005). Under stress conditions many genes are involved in ROS homeostasis and metabolism, including manganese superoxide dismutase (MnSOD) driven by oxidative stress-inducible promoter SWPA2 in rice (Wang et al. 2005).

An amino acid proline is derived from glutamate biosynthesis; therefore, osmoprotectant 1-pyrroline-5-carboxylate synthetase (P5CS) production is the rate-limiting step in this biosynthesis pathway. *AtP5CS1* and *AtP5CS2* genes are differentially regulated by *ABA1*, *ABI1*, and *AXR2* and thus they encode for P5CS in *Arabidopsis* (Verbruggen et al. 1993). Proline synthesis could be occurred by ornithine-delta aminotransferase (OAT), producing glutamate semialdehyde (GSA) and pyrroline-5-carboxylate (P5C) which further converts into proline (Delauney and Verma 1993; Roosens et al. 1998). The betaine aldehyde dehydrogenase (*BADH*) is involved in the biosynthesis of betaine osmoprotectant in plants (Zhang et al. 2011). On the other hand, the trehalose is synthesized through trehalose-6-phosphate synthase (*TPS1*) (Yeo et al. 2000). Various retrograde signaling cascades have been identified in response to drought stress. The chloroplast stress signal 3′-phosphoadenosine-5′-phosphate (PAP) accumulate in the plastids, confers drought tolerance to plants and SAL1 regulates its levels by dephosphorylating PAP to AMP. Similarly, methylerythritol cycloid phosphate (MEcPP), a precursor of isoprenoids synthesized in the chloroplastic in methylerythritol phosphate (MEP) pathway. Thus, both retrograde pathways are associated with the modification of nuclear gene expression under drought conditions (Wilson et al. 2009, Estavillo et al.

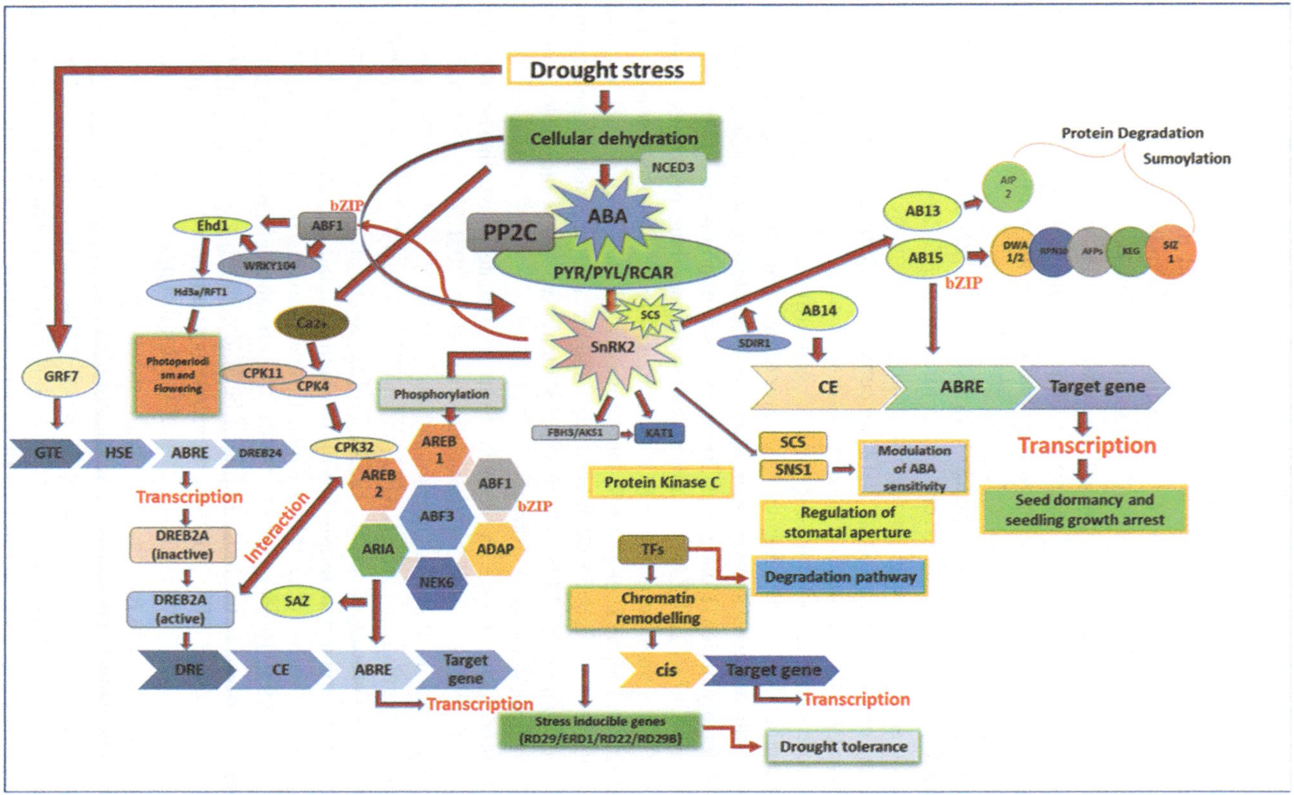

Fig. 4 Schematic representation of signaling cascades involving various transcription factors under drought conditions

2011, Xiao et al. 2012). It is quite difficult to explain the whole genetics behind complex drought stress mechanisms in plants here, as drought stress is modulated by unique signaling cascades and genes.

Engineering Drought Tolerance Through Genome Editing

Till now traditional breeding has successfully been used to improve drought tolerance in various important crops including wheat, rice, maize, and soybean (Ashraf 2010). But, many of these promising lines are unable to meet the optimum yield under drought stress conditions. Drought stress tolerance is multigenic and is complex by the number of physiological traits, underlying genes, and pathways. Hence identification of individual effects and introgressing them into elite cultivars is quite challenging for the development of modern cultivars. Also, the methods are time-consuming, laborious, and cost-intensive. Mutagenesis techniques and conventional breeding can introduce undesirable characters along with the targeted traits. Furthermore, the crossing is possible between the same plant species that limit the introduction of new genes and traits. Also, the availability of genetic variability in the same germplasm is the biggest bottleneck (Tabassum et al. 2021). Hence, there is an urgent need to utilize novel genomic approaches to improve existing cultivars further to enhance yield potential under dehydration stress.

Genome editing methods are recent addition to the new breeding technology (NBT) are capable of transferring any plant species with the agronomic useful trait and offer great potential for speeding up the crop breeding program. The popular genome editing technologies including zinc finger nucleases (ZNFs), transcription activator-like effector nucleases (TALENS), and CRISPR/Cas9 have been introduced in the past decade and are widely used as gene-editing tools in various crop plants. CRISPR is clustered regularly interspaced short palindromic repeats and the CRISPR-associated protein (Cas9) system is the latest genome editing tool that is capable of site-specific modification of the genome of any organism in several ways including insertion, deletion, base substitution, and replacement. This technology is also used to study gene repression, activation, and demethylation for functional genomic study. Various Cas9 endonucleases such as Cas9, Cas12a, Cpf1, FnCpf1, LbCpf1, Cas12b, Cas13a, Cas13b, and modified Cas9 forms including dCas9 and nCas9 have been characterized and utilized for genome modification in the various organisms (Cebrian-Serrano and Davies 2017; Naeem et al. 2020).

Among the above CRISPR technologies, the CRISPR/Cas9 system is cost-effective, reliable, efficient, and widely used for improving different traits including grain yield, herbicide resistance, disease resistance, stress resilience, and nutritional quality in several crop plants, such as wheat, rice, barley, maize, citrus, cucumber, soybean, tomato, and tobacco (Hussain et al. 2018a, b; Tabassum et al. 2021). Additionally, base editing and prime editing tools are widely used to create single to multiple bases polymorphism in the genome (Komor et al. 2016; Hua et al. 2018; Hua et al. 2020a, b; Shimatani et al. 2017). Using the above genome editing technology many important traits such as improved grain yield, herbicide tolerance, disease resistance, and also abiotic stress tolerance have been developed in various crops (Lin et al. 2020; Xu et al. 2020; Tang et al. 2020; Li et al. 2020; Hua et al. 2020a, b). CRISPR/Cas9 system has been successfully used to improve abiotic stress tolerance in many crops including maize, wheat, rice, tomato, *Arabidopsis*, and *Physcomitrella patens* (Ahmed et al. 2021; Hussain et al. 2018a, b).

Abscisic acid (ABA) is the key phytohormone that plays an important role in stress response, seed germination, and plant development. Thus, manipulating the ABA signaling pathway is one of the crucial and key targets for the development of drought tolerance (Budak et al. 2015). CRISPR/Cas9 genome editing tool has been used to simultaneously edit three ABA receptor genes *PYL1*, *PYL4*, and *PYL6* (Miao et al. 2018). The edited rice lines increased the number of grains by 31% than wild-type plants under drought conditions. Similarly, the CRISPR/Cas9 knockout ABA receptor Pyrabactin Resistance 9 (*OsPYL9*) showed enhanced drought tolerance and improved grain yield under normal and limited water conditions in rice (Usman et al. 2020). Further, the *OsPYL9* mutant rice plant exhibit reduced stomata number, stomatal conductance, malondialdehyde content, and transpiration rate. They showed enhanced cuticle wax, panicle number, ABA content and antioxidants (CAT and SOD), and survival rate compared to wild-type rice plants (Usman et al. 2020). Improved drought-tolerant rice plants are also generated through Cas9-based targeting mutation of Enhanced Response to ABA1 (*ERA1*). The tolerance mechanism of the *ERA1* mutant plant is exhibited by regulating the stomatal conductance (Ogata et al. 2020). Plants respond to drought consequences through ABA-dependent or -independent signaling cascades. CRISPR/Cas9 generated *osmotic stress/ABA–activated protein kinase2* knockout showed that *OsSAPK2* is involved in ABA-mediated seed dormancy, salinity, drought, and osmotic stress tolerance in rice mediated through the ABA-cascade pathway.

Further, the mutant showed decreased reactive oxygen species, stomatal closure through the accumulation of compatible solutes that leads to stomatal closure, downregulation of anion channels (*OsSLAC1* and *OsSLAC7*), and upregulation of stress-related genes (*OsLEA3*, *OsbZIP23*, *OsRab16b*, *OsRab21*, and *OsOREB1*). The *OsSAPK2* mutant lines were vulnerable to stresses; thus the study highlights the critical role of *OsSAPK2* in the ABA signaling cascade (Lou et al. 2017). CRISPR/Cas9 system has been utilized to develop knockout mutant of *Drought and salt tolerance* (*OsDST*) gene, a zinc finger TF in rice. The mutant rice lines showed enhanced drought tolerance through reduced stomatal density which eventually improve the water retention in the leaf (Kumar et al. 2020b). Similarly, the mutation in the *OsmiR535* showed improved drought tolerance, insensitivity to ABA, improved lateral root system (73% more), increased shoot length (30% more), and improved primary root length. The *OsmiR535* knockout mutant lines also exhibited improved tolerance towards salinity, osmotic stress and shoot length increased 86.8% more, increased lateral roots (514% as compared with line overexpressing MIR535) and primary root length (Yue et al. 2020). Reduced transpiration rate through leaves is an important factor for drought tolerance. Hence, changing leaf morphology is a key approach to improve crop yield under drought stress conditions.

Multiplexed CRISPR/Cas9 genome editing approach used to develop double mutant of *SEMI-ROLLED LEAF 1* and *SEMI-ROLLED LEAF 2* (*OsSRL1* and *OsSRL2*) in rice resulted in drought tolerance. The mutants showed curled leaves, reduced stomata number, stomatal conductance, transpiration rate, and malondialdehyde (MDA) content and mutants exhibit higher survival percentage with improved antioxidant enzymes catalase and superoxide dismutase (Liao et al. 2019). OPEN STOMATA 2 (*OST2*) is a prominent plasma membrane H^+ ATPase responsible for stomatal response in *Arabidopsis*. The *ost2* mutant lines had a significantly high degree of stomatal closure with low transcriptional water loss compared to wild type (Osakabe et al. 2016).

The gene family AUXIN REGULATED GENE INVOLVED IN ORGAN SIZE (ARGOS) negatively regulates ethylene signaling and provides enhanced drought tolerance and improved grain yield under water-limited conditions (Shi et al. 2017). The endogenous transcript level of ARGOS8 is relatively low in maize. Shi and coworkers used CRISPR/Cas9 approach following the HDR pathway to insert the maize native *GOS2* gene promoter into the 5′ untranslated region of ARGOS8. The CRISPR inserted ARGOS8 variants of maize lines confer better tolerance against drought stress. Furthermore, the novel mutant line showed enhanced grain yield and no yield loss under water-limited conditions. Thus, the study provides the potential use of CRISPR/Cas9 for the creation of novel allelic variation in the crop genome for the breeding program. MAPK kinases cascades phosphorylate target protein and perceive extracellular signals and regulate various cellular physiological and biochemical responses in plants including drought response (Sinha et al. 2011).

The report suggests that *Os*MSRMK2, *Os*MPK5, *At*MPK3, and *Zm*MPK3 genes showed a significant response to drought stress. Hence, these genes could act as a future target to understanding drought response in plants. NPR1 is crucial for plant defense mechanisms. The study suggests that reduced expression of *Md*NPR1 has been reported in drought-responsive apple trees (Bassett et al. 2014). Further, transgenic overexpression of *At*NPR1 in *Oryza sativa* showed a hypersensitive response against drought stress (Quilis et al. 2008). CRISPR/Cas9 genome editing system has been applied to develop edited lines of non-expresser of pathogenesis-related gene 1 (*NPR1*) gene to study drought stress tolerance in tomato plants (Li et al. 2019). The loss of function of *slnpr1* mutants resulted in sensitivity to drought stress, wider stomatal aperture, higher electrolytic leakage, decreased levels of antioxidant enzymes, and increased level of malondialdehyde (MDA) and H_2O_2 compared to wild-type tomato plants. Further, the drought-responsive genes such as *SlGST*, *SlDHN*, and *SlDREB* are downregulated in *slnpr1* lines to drought stress (Li et al. 2019). CRISPR/Cas9 knockout approach has been demonstrated to develop null mutant of *SlMAPK3* gene to understand drought response in tomato plants (Wang et al. 2017a, b). The *slmapk3* mutant lines showed serious wilting symptoms, accumulate high H_2O_2, low levels of antioxidant enzymes, and severe membrane damage under drought conditions compared to wild-type plants. Also, the *slmapk3* knockout mutants resulted in significant differential expression of drought-responsive genes including *SlLOX*, *SlGST,* and *SlDREB* (Wang et al. 2017a, b). The results highlight the crucial role of SlMAPK3 in response to drought stress in tomatoes. Hence, CRISPR/Cas tool can be used to develop a mutant variant of the drought-responsive pathway targeting downstream negative players of the drought tolerance mechanism.

Concluding Remarks

Response of plant to water scarcity is highly spontaneous and dynamic and is a complex process and poses a serious threat to global food security. Understanding the plant response to drought is crucial for improving crop varieties with stable high yields under limited water conditions. In the last several decades, genetic breeding methods have made tremendous contributions to the development of stress resilience cultivars. Plant responses to drought stress in a complex manner involving three levels of protection including biochemicals, physiological, and molecular. Plant response mechanisms are myriads signaling pathways involving many genes; thus untangling the responses for practical application in crops requires a multipronged approach. The breeding program requires sufficient genetic diversity to find the right allelic combinations. Also, the existing breading methods are not sufficient to pyramid multiple alleles in an effective way and within a short period. New advancements in biotechnology can create new sources which can be potentially used to make the breeding process fast and forward direction. The advent of genome editing technology including CRISPR/Cas9 offers precise genetic modification has overcome the above drawbacks and transformed agriculture science to create genetic variation and enrich the gene pool of elite crops in a short time.

Drought stress tolerance mechanism involves multigenic traits associated with complex metabolic pathways genes. Multiplex CRISPR/Cas9 genome editing technology with independent sgRNAs pyramided into a single CRISPR/Cas9 expression vector has made it possible to simultaneously edit many alleles to develop robust drought-tolerant genotype through targeted alterations. In addition, CRISPR/Cas-based cis-regulatory element sequence alternation to change the expression of genes that are very important in drought tolerance mechanism hold a great promise for future development of stress adapted plant. Single-nucleotide polymorphism (SNPs) and microRNA (mRNA) both play a very important role in gene expression and their functions in a plant which are closely associated with many agronomic traits. The latest base editing and prime editing tools offer a wider application of CRISPR technologies to create SNPs in the genome and alteration of microRNA binding genomic regions thus alter the function of gene and regulatory region which can be implemented to create drought-tolerant crop varieties to mitigate drought and other abiotic stress conditions. In conclusion, CRISPR/Cas-based technologies indubitably accelerate the breeding programs and thus implication of technology promises towards the development of climate-resilient crops for future food security.

Acknowledgments Hemangini Parmar thank to CSIR for awarding direct-SRF (09/512(0244)/2019-EMR-1).

Conflict of Interests All the authors mutually declare that there is no conflict of interest. Furthermore, this article does not contain any studies with human participants or animals performed by any of the authors during preparation.

References

Agrama HA, Moussa ME (1996) Mapping QTLs in breeding for drought tolerance in maize (Zea mays L.). Euphytica 91(1):89–97

Aharoni A, Dixit S, Jetter R, Thoenes E, Van Arkel G, Pereira A (2004) The SHINE clade of AP2 domain transcription factors activates wax biosynthesis, alters cuticle properties, and confers drought tolerance when overexpressed in Arabidopsis. Plant Cell 16(9):2463–2480

Ahmed T, Noman M, Shahid M, Muhammad S, Tahir ul Qamar M, Ali MA, Maqsood A, Hafeez R, Ogunyemi SO, Li B. (2021) Potential application of CRISPR/Cas9 system to engineer abiotic stress tolerance in plants. Protein Pept Lett 28(8):861–877

Annicchiarico P, Nazzicari N, Pecetti L, Romani M, Ferrari B, Wei Y, Brummer EC (2017) GBS-based genomic selection for pea grain yield under severe terminal drought. Plant Genome 10(2)

Anyaoha CO, Fofana M, Gracen V, Tongoona P, Mande S (2019) Introgression of two drought QTLs into FUNAABOR-2 early generation backcross progenies under drought stress at reproductive stage. Rice Sci 26(1):32–41

Ashraf M (2010) Inducing drought tolerance in plants: recent advances. Biotechnol Adv 28(1): 169–183

Badu-Apraku B, Talabi AO, Fakorede MA, Fasanmade Y, Gedil M, Magorokosho C, Asiedu R (2019) Yield gains and associated changes in an early yellow bi-parental maize population following genomic selection for Striga resistance and drought tolerance. BMC Plant Biol 19 (1):1–7

Bassett CL, Baldo AM, Moore JT, Jenkins RM, Soffe DS, Wisniewski ME, Norelli JL, Farrell RE (2014) Genes responding to water deficit in apple (Malus × domestica Borkh.) roots. BMC Plant Biol 14(1):1–2

Bennett D, Reynolds M, Mullan D, Izanloo A, Kuchel H, Langridge P, Schnurbusch T (2012) Detection of two major grain yield QTL in bread wheat (Triticum aestivum L.) under heat, drought and high yield potential environments. Theor Appl Genet 125(7):1473–1485

Bernier J, Kumar A, Ramaiah V, Spaner D, Atlin G (2007) A large-effect QTL for grain yield under reproductive-stage drought stress in upland rice. Crop Sci 47(2):507–516

Bernier J, Serraj R, Kumar A, Venuprasad R, Impa S, RP VG, Oane R, Spaner D, Atlin G. The large-effect drought-resistance QTL qtl12. 1 increases water uptake in upland rice. Field Crops Res 2009;110(2):139-146.

Bharadwaj C, Tripathi S, Soren KR, Thudi M, Singh RK, Sheoran S, Roorkiwal M, Patil BS, Chitikineni A, Palakurthi R, Vemula A (2021) Introgression of "QTL-hotspot" region enhances drought tolerance and grain yield in three elite chickpea cultivars. Plant Genome 14(1):e20076

Bidinger FR, Nepolean T, Hash CT, Yadav RS, Howarth CJ (2007) Quantitative trait loci for grain yield in pearl millet under variable postflowering moisture conditions. Crop Sci 47(3):969–980

Billah M, Aktar S, Brestic M, Zivcak M, Khaldun AB, Uddin M, Bagum SA, Yang X, Skalicky M, Mehari TG, Maitra S (2021) Progressive genomic approaches to explore drought-and salt-induced oxidative stress responses in plants under changing climate. Plan Theory 10(9):1910

Blum A (2005) Drought resistance, water-use efficiency, and yield potential—are they compatible, dissonant, or mutually exclusive? Aust J Agric Res 56(11):1159–1168

Bouman BA, Peng S, Castaneda AR, Visperas RM (2005) Yield and water use of irrigated tropical aerobic rice systems. Agric Water Manag 74(2):87–105

Budak H, Hussain B, Khan Z, Ozturk NZ, Ullah N (2015) From genetics to functional genomics: improvement in drought signaling and tolerance in wheat. Front Plant Sci 19(6):1012

Cebrian-Serrano A, Davies B (2017) CRISPR-Cas orthologues and variants: optimizing the repertoire, specificity and delivery of genome engineering tools. Mamm Genome 28(7):247–261

Cerrudo D, Cao S, Yuan Y, Martinez C, Suarez EA, Babu R, Zhang X, Trachsel S (2018) Genomic selection outperforms marker assisted selection for grain yield and physiological traits in a maize doubled haploid population across water treatments. Front Plant Sci 9:366

Chen JQ, Meng XP, Zhang Y, Xia M, Wang XP (2008) Over-expression of OsDREB genes lead to enhanced drought tolerance in rice. Biotechnol Lett 30(12):2191–2198

Collins NC, Tardieu F, Tuberosa R (2008) Quantitative trait loci and crop performance under abiotic stress: where dowe stand? Plant Physiol 147(2):469–486

Crossa J, Pérez-Rodríguez P, Cuevas J, Montesinos-López O, Jarquín D, De Los CG, Burgueño J, González-Camacho JM, Pérez-Elizalde S, Beyene Y, Dreisigacker S (2017) Genomic selection in plant breeding: methods, models, and perspectives. Trends Plant Sci 22(11):961–975

Cutler SR, Rodriguez PL, Finkelstein RR, Abrams SR (2010) Abscisic acid: emergence of a core signaling network. Annu Rev Plant Biol 61:651–679

Davletova S, Schlauch K, Coutu J, Mittler R (2005) The zinc-finger protein Zat12 plays a central role in reactive oxygen and abiotic stress signaling in Arabidopsis. Plant Physiol 139(2):847–856

Delauney AJ, Verma DP (1993) Proline biosynthesis and osmoregulation in plants. Plant J 4(2):215–223

Dhawan G, Kumar A, Dwivedi P, Gopala Krishnan S, Pal M, Vinod KK, Nagarajan M, Bhowmick PK, Bollinedi H, Ellur RK, Ravikiran KT (2021) Introgression of qDTY1. 1 Governing reproductive stage drought tolerance into an elite basmati rice variety "pusa basmati 1" through marker assisted backcross breeding. Agronomy 11(2):202

Ding Z, Ali EF, Elmahdy AM, Ragab KE, Seleiman MF, Kheir AM (2021) Modeling the combined impacts of deficit irrigation, rising temperature and compost application on wheat yield and water productivity. Agric Water Manag 244:106,626

Dixit S, Swamy BM, Vikram P, Ahmed HU, Cruz MS, Amante M, Atri D, Leung H, Kumar A (2012) Fine mapping of QTLs for rice grain yield under drought reveals sub-QTLs conferring a response to variable drought severities. Theor Appl Genetics 125(1):155–169

Dixit S, Singh A, Cruz MT, Maturan PT, Amante M, Kumar A (2014) Multiple major QTL lead to stable yield performance of rice cultivars across varying drought intensities. BMC Genetics 15(1):1–3

Dixit S, Singh A, Sandhu N, Bhandari A, Vikram P, Kumar A (2017) Combining drought and submergence tolerance in rice: marker-assisted breeding and QTL combination effects. Mol Breed 37(12):1–2

Duvick DN (2005) The contribution of breeding to yield advances in maize (Zea mays L.). Adv Agron 86:83–145

Dwivedi P, Ramawat N, Dhawan G, Gopala Krishnan S, Vinod KK, Singh MP, Nagarajan M, Bhowmick PK, Mandal NP, Perraju P, Bollinedi H (2021) Drought tolerant near isogenic lines (NILs) of Pusa 44 developed through marker assisted introgression of qDTY2. 1 and qDTY3. 1 Enhances yield under reproductive stage drought stress. Agriculture 11(1):64

Endo A, Sawada Y, Takahashi H, Okamoto M, Ikegami K, Koiwai H, Seo M, Toyomasu T, Mitsuhashi W, Shinozaki K, Nakazono M (2008) Drought induction of Arabidopsis 9-cis-epoxycarotenoid dioxygenase occurs in vascular parenchyma cells. Plant Physiol 147(4):1984–1993

Estavillo GM, Crisp PA, Pornsiriwong W, Wirtz M, Collinge D, Carrie C, Giraud E, Whelan J, David P, Javot H, Brearley C (2011) Evidence for a SAL1-PAP chloroplast retrograde pathway that functions in drought and high light signaling in Arabidopsis. Plant Cell 23(11):3992–4012

Farooq M, Aziz T, Wahid A, Lee DJ, Siddique KH (2009) Chilling tolerance in maize: agronomic and physiological approaches. Crop Pasture Sci 60(6):501–516

Fleury D, Jefferies S, Kuchel H, Langridge P (2010) Genetic and genomic tools to improve drought tolerance in wheat. J Exp Bot 61(12):3211–3222

Food and Agriculture Organization of the United Nations (2008) Food outlook: global market analysis. FAO, Rome

Gautam T, Saripalli G, Kumar A, Gahlaut V, Gadekar DA, Oak M, Sharma PK, Balyan HS, Gupta PK (2021) Introgression of a drought insensitive grain yield QTL for improvement of four Indian bread wheat cultivars using marker assisted breeding without background selection. J Plant Biochem Biotechnol 30(1):172–183

Ghimire KH, Quiatchon LA, Vikram P, Swamy BM, Dixit S, Ahmed H, Hernandez JE, Borromeo TH, Kumar A (2012) Identification and mapping of a QTL (qDTY1. 1) with a consistent effect on grain yield under drought. Field Crops Res 131:88–96

Gill SS, Tuteja N (2010) Reactive oxygen species and antioxidant machinery in abiotic stress tolerance in crop plants. Plant Physiol Biochem 48(12):909–930

Gomez SM, Boopathi NM, Kumar SS, Ramasubramanian T, Chengsong Z, Jeyaprakash P, Senthil A, Babu RC (2010) Molecular mapping and location of QTLs for drought-resistance traits in indica rice (Oryza sativa L.) lines adapted to target environments. Acta Physiol Plant 32(2):355–364

Harris K, Subudhi PK, Borrell A, Jordan D, Rosenow DH, Klein P, Klein R, Mullet J (2007) Sorghum stay-green QTL individually reduce post-flowering drought-induced leaf senescence. J Exp Bot 58(2):327–338

Honsdorf N, March TJ, Pillen K (2017) QTL controlling grain filling under terminal drought stress in a set of wild barley introgression lines. PLoS One 12(10):e0185983

Hu H, Dai M, Yao J, Xiao B, Li X, Zhang Q, Xiong L (2006) Overexpressing a NAM, ATAF, and CUC (NAC) transcription factor enhances drought resistance and salt tolerance in rice. Proc Natl Acad Sci 103(35):12,987–12,992

Hu W, Zhang J, Yan K, Zhou Z, Zhao W, Zhang X, Pu Y, Yu R (2021) Beneficial effects of abscisic acid and melatonin in overcoming drought stress in cotton (Gossypium hirsutum L.). Physiol Plant 173(4):2041–2054

Hua K, Tao X, Yuan F, Wang D, Zhu J-K (2018) Precise A-T to G-C base editing in the rice genome. Mol Plant 11:627–630

Hua K, Jiang Y, Tao X, Zhu JK (2020a) Precision genome engineering in rice using prime editing system. Plant Biotechnol J 18(11):2167

Hua K, Tao X, Liang W, Zhang Z, Gou R, Zhu J (2020b) Simplified adenine base editors improve adenine base editing efficiency in rice. Plant Biotechnol J 18:770–778

Huang XY, Chao DY, Gao JP, Zhu MZ, Shi M, Lin HX (2009) A previously unknown zinc finger protein, DST, regulates drought and salt tolerance in rice via stomatal aperture control. Genes Dev 23(15):1805–1817

Hund A, Ruta N, Liedgens M (2009) Rooting depth and water use efficiency of tropical maize inbred lines, differing in drought tolerance. Plant Soil 318(1):311–325

Hussain M, Farooq M, Sattar A, Ijaz M, Sher A, Ul-Allah S (2018a) Mitigating the adverse effects of drought stress through seed priming and seed quality on wheat (Triticum aestivum L.) productivity. Pak J Agric Sci 55(2)

Hussain B, Lucas SJ, Budak H (2018b) CRISPR/Cas9 in plants: at play in the genome and at work for crop improvement. Brief Funct Genomics 17(5):319–328

Iuchi S, Kobayashi M, Taji T, Naramoto M, Seki M, Kato T, Tabata S, Kakubari Y, Yamaguchi-Shinozaki K, Shinozaki K (2001) Regulation of drought tolerance by gene manipulation of 9-cis-epoxycarotenoid dioxygenase, a key enzyme in abscisic acid biosynthesis in Arabidopsis. Plant J 27(4):325–333

Jatoi WA, Baloch MJ, Kumbhar MB, Khan NU, Kerio MI (2011) Effect of water stress on physiological and yield parameters at anthesis stage in elite spring wheat cultivars. Sarhad J Agric 27(1):59–65

Jeong JS, Kim YS, Baek KH, Jung H, Ha SH, Do Choi Y, Kim M, Reuzeau C, Kim JK (2010) Root-specific expression of OsNAC10 improves drought tolerance and grain yield in rice under field drought conditions. Plant Physiol 153(1):185–197

Jia H, Wan H, Yang S, Zhang Z, Kong Z, Xue S, Zhang L, Ma Z (2013) Genetic dissection of yield-related traits in a recombinant inbred line population created using a key breeding parent in China's wheat breeding. Theor Appl Genet 126(8):2123–2139

Joshi R, Wani SH, Singh B, Bohra A, Dar ZA, Lone AA, Pareek A, Singla-Pareek SL (2016) Transcription factors and plants response to drought stress: current understanding and future directions. Front Plant Sci 7:1029

Juliana P, He X, Kabir MR, Roy KK, Anwar MB, Marza F, Poland J, Shrestha S, Singh RP, Singh PK (2020) Genome-wide association mapping for wheat blast resistance in CIMMYT's international screening nurseries evaluated in Bolivia and Bangladesh. Sci Rep 10(1):1–4

Kalefetoğlu Macar T, Ekmekci Y (2009) Alterations in photochemical and physiological activities of chickpea (Cicer arietinum L.) cultivars under drought stress. J Agron Crop Sci 195(5):335–346

Kamal NM, Gorafi YS, Abdeltwab H, Abdalla I, Tsujimoto H, Ghanim AM (2021) A new breeding strategy towards introgression and characterization of stay-green QTL for drought tolerance in Sorghum. Agriculture 11(7):598

Kamara AY, Menkir A, Badu-Apraku B, Ibikunle O (2003) The influence of drought stress on growth, yield and yield components of selected maize genotypes. J Agric Sci 141:43–50

Kang J, Hwang JU, Lee M, Kim YY, Assmann SM, Martinoia E, Lee Y (2010) PDR-type ABC transporter mediates cellular uptake of the phytohormone abscisic acid. Proc Natl Acad Sci 107(5):2355–2360

Kanno Y, Hanada A, Chiba Y, Ichikawa T, Nakazawa M, Matsui M, Koshiba T, Kamiya Y, Seo M (2012) Identification of an abscisic acid transporter by functional screening using the receptor complex as a sensor. Proc Natl Acad Sci 109(24):9653–9658

Karpinski S, Escobar C, Karpinska B, Creissen G, Mullineaux PM (1997) Photosynthetic electron transport regulates the expression of cytosolic ascorbate peroxidase genes in Arabidopsis during excess light stress. Plant Cell 9(4):627–640

Khan A, Sovero V, Gemenet D (2016) Genome-assisted breeding for drought resistance. Curr Genomics 17(4):330–342

Khan MA, Tong F, Wang W, He J, Zhao T, Gai J (2018) Analysis of QTL–allele system conferring drought tolerance at seedling stage in a nested association mapping population of soybean [G lycine max (L.) Merr.] using a novel GWAS procedure. Planta 248(4):947–962

Khan MA, Tong F, Wang W, He J, Zhao T, Gai J (2020) Molecular characterization of QTL-allele system for drought tolerance at seedling stage and optimal genotype design using multi-locus multi-allele genome-wide association analysis in a half-sib population of soybean (Glycine max (L.) Merr.). Plant Gen Resour 18(5):295–306

Kim TH, Böhmer M, Hu H, Nishimura N, Schroeder JI (2010) Guard cell signal transduction network: advances in understanding abscisic acid, CO2, and Ca2+ signaling. Annu Rev Plant Biol 61:561–591

Komor AC, Kim YB, Packer MS, Zuris JA, Liu DR (2016) Programmable editing of a target base in genomic DNA without double-stranded DNA cleavage. Nature 533(7603):420–424

Krishnappa G, Rathan ND, Sehgal D, Ahlawat AK, Singh SK, Singh SK, Shukla RB, Jaiswal JP, Solanki IS, Singh GP, Singh AM (2021) Identification of novel genomic regions for biofortification traits using an SNP marker-enriched linkage map in wheat (Triticum aestivum L.). Front Nutr 8

Kumar A, Saripalli G, Jan I, Kumar K, Sharma PK, Balyan HS, Gupta PK (2020a) Meta-QTL analysis and identification of candidate genes for drought tolerance in bread wheat (Triticum aestivum L.). Physiol Mol Biol Plants 26(8):1713–1725

Kumar VS, Verma RK, Yadav SK, Yadav P, Watts A, Rao MV, Chinnusamy V (2020b) CRISPR-Cas9 mediated genome editing of drought and salt tolerance (OsDST) gene in indica mega rice cultivar MTU1010. Physiol Mol Biol Plants 26(6):1099

Kumar S, Singh VP, Saini DK, Sharma H, Saripalli G, Kumar S, Balyan HS, Gupta PK (2021) Meta-QTLs, ortho- MQTLs, and candidate genes for thermotolerance in wheat (Triticum aestivum L.). Mol Breed 41(11):1–22

Li WX, Oono Y, Zhu J, He XJ, Wu JM, Iida K, Lu XY, Cui X, Jin H, Zhu JK (2008) The Arabidopsis NFYA5 transcription factor is regulated transcriptionally and posttranscriptionally to promote drought resistance. Plant Cell 20(8):2238–2251

Li Z, Shen LS, Courtois B, Lafitte R (2000) Development of near isogenic introgression line (NIIL) sets for QTLs associated with drought tolerance in Rice. Molecular approaches for the genetic improvement of cereals for stable production in water-limited environments, p 103

Li Y, Ruperao P, Batley J, Edwards D, Khan T, Colmer TD, Pang J, Siddique KH, Sutton T (2018) Investigating drought tolerance in chickpea using genome-wide association mapping and genomic selection based on whole-genome resequencing data. Front Plant Sci 9:190

Li R, Liu C, Zhao R, Wang L, Chen L, Yu W, Zhang S, Sheng J, Shen L (2019) CRISPR/Cas9-mediated SlNPR1 mutagenesis reduces tomato plant drought tolerance. BMC Plant Biol 19(1): 1–3

Li H, Li J, Chen J, Yan L, Xia L (2020) Precise modifications of both exogenous and endogenous genes in rice by prime editing. Mol Plant 13(5):671–674

Liao S, Qin X, Luo L, Han Y, Wang X, Usman B, Nawaz G, Zhao N, Liu Y, Li R (2019) CRISPR/Cas9-induced mutagenesis of semi-rolled Leaf1, 2 confers curled leaf phenotype and drought tolerance by influencing protein expression patterns and ROS scavenging in Rice (Oryza sativa L.). Agronomy 9(11):728

Lin Q, Zong Y, Xue C, Wang S, Jin S, Zhu Z, Wang Y, Anzalone AV, Raguram A, Doman JL, Liu DR (2020) Prime genome editing in rice and wheat. Nat Biotechnol 38(5):582–585

Lou D, Wang H, Liang G, Yu D (2017) OsSAPK2 confers abscisic acid sensitivity and tolerance to drought stress in rice. Front Plant Sci 8:993

Majid SA, Asghar R, Murtaza G (2007) Yield stability analysis conferring adaptation of wheat to pre-and post-anthesis drought conditions. Pak J Bot 39(5)

Merchuk-Ovnat L, Barak V, Fahima T, Ordon F, Lidzbarsky GA, Krugman T, Saranga Y (2016) Ancestral QTL alleles from wild emmer wheat improve drought resistance and productivity in modern wheat cultivars. Front Plant Sci 7:452

Miao C, Xiao L, Hua K, Zou C, Zhao Y, Bressan RA, Zhu JK (2018) Mutations in a subfamily of abscisic acid receptor genes promote rice growth and productivity. Proc Natl Acad Sci 115(23): 6058–6063

Mohammadi M, Taleei A, Zeinali H, Naghavi MR, Ceccarelli S, Grando S, Baum M (2005) QTL analysis for phenologic traits in doubled haploid population of barley. Int J Agric Biol 7(5):820–823

Mohd Ikmal A, Nurasyikin Z, Riana TNA, Puteri Dinie Ellina Z, Wickneswari R, Noraziyah AA (2019) Drought yield QTL (qDTY) with consistent effects on morphological and agronomical traits of two populations of new rice (Oryza sativa) lines. Plants 8(6):186

Mohd Ikmal A, Noraziyah AA, Wickneswari R (2021) Incorporating drought and submergence tolerance QTL in Rice (Oryza sativa L.)—the effects under reproductive stage drought and vegetative stage submergence stresses. Plan Theory 10(2):225

Naeem M, Majeed S, Hoque MZ, Ahmad I (2020) Latest developed strategies to minimize the off-target effects in CRISPR-Cas-mediated genome editing. Cells. 9(7):1608

Nakaya A, Isobe SN (2012) Will genomic selection be a practical method for plant breeding? Ann Bot 110(6):1303–1316

Ning J, Li X, Hicks LM, Xiong L (2010) A Raf-like MAPKKK gene DSM1 mediates drought resistance through reactive oxygen species scavenging in rice. Plant Physiol 152(2):876–890

Ogata T, Ishizaki T, Fujita M, Fujita Y (2020) CRISPR/Cas9-targeted mutagenesis of OsERA1 confers enhanced responses to abscisic acid and drought stress and increased primary root growth under nonstressed conditions in rice. PloS One 15(12):e0243376

Osakabe Y, Osakabe K, Shinozaki K, Tran LS (2014) Response of plants to water stress. Front Plant Sci 5:86

Osakabe Y, Watanabe T, Sugano SS, Ueta R, Ishihara R, Shinozaki K, Osakabe K (2016) Optimization of CRISPR/Cas9 genome editing to modify abiotic stress responses in plants. Sci Rep 6:26,685

Park HY, Seok HY, Park BK, Kim SH, Goh CH, Lee BH, Lee CH, Moon YH (2008) Overexpression of Arabidopsis ZEP enhances tolerance to osmotic stress. Biochem Biophys Res Commun 375(1):80–85

Per TS, Khan NA, Reddy PS, Masood A, Hasanuzzaman M, Khan MI, Anjum NA (2017) Approaches in modulating proline metabolism in plants for salt and drought stress tolerance: phytohormones, mineral nutrients and transgenics. Plant Physiol Biochem 115:126–140

Pinto RS, Reynolds MP, Mathews KL, McIntyre CL, Olivares-Villegas JJ, Chapman SC (2010) Heat and drought adaptive QTL in a wheat population designed to minimize confounding agronomic effects. Theor Appl Genet 121(6):1001–1021

Prasanna BM, Beiki AH, Sekhar JC, Srinivas A, Ribaut JM (2009) Mapping QTLs for component traits influencing drought stress tolerance of maize (Zea mays L) in India. J Plant Biochem Biotechnol 18(2):151–160

Quilis J, Peñas G, Messeguer J, Brugidou C, Segundo BS (2008) The Arabidopsis AtNPR1 inversely modulates defense responses against fungal, bacterial, or viral pathogens while conferring hypersensitivity to abiotic stresses in transgenic rice. Mol Plant-Microbe Interact 21(9):1215–1231

Roosens NH, Thu TT, Iskandar HM, Jacobs M (1998) Isolation of the ornithine-δ-aminotransferase cDNA and effect of salt stress on its expression in Arabidopsis thaliana. Plant Physiol 117(1): 263–271

Saijo Y, Hata S, Kyozuka J, Shimamoto K, Izui K (2000) Over-expression of a single Ca2+-dependent protein kinase confers both cold and salt/drought tolerance on rice plants. Plant J 23(3):319–327

Samarah NH (2005) Effects of drought stress on growth and yield of barley. Agron Sustain Dev 25(1):145–149

Samarah NH, Mullen RE, Cianzio SR, Scott P (2006) Dehydrin-Like Proteins in soybean seeds in response to drought stress during seed filling. Crop Sci 46(5):2141–2150

Samarah NH, Alqudah AM, Amayreh JA, McAndrews GM (2009) The effect of late-terminal drought stress on yield components of four barley cultivars. J Agron Crop Sci 195(6):427–441

Seleiman MF, Al-Suhaibani N, Ali N, Akmal M, Alotaibi M, Refay Y, Dindaroglu T, Abdul-Wajid HH, Battaglia ML (2021) Drought stress impacts on plants and different approaches to alleviate its adverse effects. Plants 10(2):259

Serba DD, Yadav RS (2016) Genomic tools in pearl millet breeding for drought tolerance: status and prospects. Front Plant Sci 7:1724

Serraj R, McNally KL, Slamet-Loedin I, Kohli A, Haefele SM, Atlin G, Kumar A (2011) Drought resistance improvement in rice: an integrated genetic and resource management strategy. Plant Prod Sci 14(1):1–4

Shamsi K, Kobraee S (2011) Bread wheat production under drought stress conditions. Ann Biol Res 2(3):352–358

Shamsi K, Petrosyan M, Noor-Mohammadi G, Haghparast R (2010) The role of water deficit stress and water use efficiency on bread wheat cultivars. J Appl Biosci 35:2325–2331

Shamsudin NA, Swamy BM, Ratnam W, Cruz MT, Raman A, Kumar A (2016) Marker assisted pyramiding of drought yield QTLs into a popular Malaysian rice cultivar, MR219. BMC Genet 17(1):1–4

Shen L, Courtois B, McNally KL, Robin S, Li Z (2001) Evaluation of near-isogenic lines of rice introgressed with QTLs for root depth through marker-aided selection. Theor Appl Genet 103 (1):75–83

Shi J, Gao H, Wang H, Lafitte HR, Archibald RL, Yang M, Hakimi SM, Mo H, Habben JE (2017) ARGOS 8 variants generated by CRISPR-Cas9 improve maize grain yield under field drought stress conditions. Plant Biotechnol J 15(2):207–216

Shikha M, Kanika A, Rao AR, Mallikarjuna MG, Gupta HS, Nepolean T (2017) Genomic selection for drought tolerance using genome-wide SNPs in maize. Front Plant Sci 8:550

Shimatani Z, Kashojiya S, Takayama M, Terada R, Arazoe T, Ishii H, Teramura H, Yamamoto T, Komatsu H, Miura K, Ezura H (2017) Targeted base editing in rice and tomato using a CRISPR-Cas9 cytidine deaminase fusion. Nature Biotechnol 35(5):441–443

Shinwari ZK, Jan SA, Nakashima K, Yamaguchi-Shinozaki K (2020) Genetic engineering approaches to understanding drought tolerance in plants. Plant Biotechnol Rep 14(2):151–162

Sinclair TR (2011) Challenges in breeding for yield increase for drought. Trends Plant Sci 16(6): 289–293

Sinha AK, Jaggi M, Raghuram B, Tuteja N (2011) Mitogen-activated protein kinase signaling in plants under abiotic stress. Plant Signal Behav 6(2):196–203

Sintayehu S, Adugna A, Fetene M, Tirfessa A, Ayalew K (2018) Study of growth and physiological characters in stay-green QTL introgression Sorghum bicolor (L.) lines under post-flowering drought stress. Cereal Res Commun 46(1):54–66

Slama I, Abdelly C, Bouchereau A, Flowers T, Savouré A (2015) Diversity, distribution and roles of osmoprotective compounds accumulated in halophytes under abiotic stress. Ann Bot 115(3): 433–447

Sugano S, Kaminaka H, Rybka Z, Catala R, Salinas J, Matsui K, Ohme-Takagi M, Takatsuji H (2003) Stress-responsive zinc finger gene ZPT2-3 plays a role in drought tolerance in petunia. Plant J 36(6):830–841

Tabassum J, Ahmad S, Hussain B, Mawia AM, Zeb A, Ju L (2021) Applications and potential of genome-editing systems in rice improvement: current and future perspectives. Agronomy 11(7): 1359

Takahashi F, Kuromori T, Sato H, Shinozaki K (2018) Regulatory gene networks in drought stress responses and resistance in plants. Surv Strategies Extreme Cold Desiccation:189–214

Takahashi F, Hanada K, Kondo T, Shinozaki K (2019) Hormone-like peptides and small coding genes in plant stress signaling and development. Curr Opin Plant Biol 51:88–95

Takahashi F, Kuromori T, Urano K, Yamaguchi-Shinozaki K, Shinozaki K (2020) Drought stress responses and resistance in plants: From cellular responses to long-distance intercellular communication. Front Plant Sci 11:1407

Tang N, Zhang H, Li X, Xiao J, Xiong L (2012) Constitutive activation of transcription factor OsbZIP46 improves drought tolerance in rice. Plant Physiol 158(4):1755–1768

Tang X, Sretenovic S, Ren Q, Jia X, Li M, Fan T, Yin D, Xiang S, Guo Y, Liu L, Zheng X (2020) Plant prime editors enable precise gene editing in rice cells. Mol Plant 13(5):667–670

Tardieu F, Simonneau T, Muller B (2018) The physiological basis of drought tolerance in crop plants: a scenario-dependent probabilistic approach. Annu Rev Plant Biol 69:733–759

Todkar L, Harikrishna GP, Jain N, Singh PK, Prabhu KV (2020) Introgression of drought tolerance QTLs through marker assisted backcross breeding in wheat (Triticum aestivum L.). Indian J Genet 80(2):209–212

Trachsel S, Sun D, SanVicente FM, Zheng H, Atlin GN, Suarez EA, Babu R, Zhang X (2016) Identification of QTL for early vigor and stay-green conferring tolerance to drought in two connected advanced backcross populations in tropical maize (Zea mays L.). PloS one 11(3): e0149636

Tripathy JN, Zhang J, Robin S, Nguyen TT, Nguyen HT (2000) QTLs for cell-membrane stability mapped in rice (Oryza sativa L.) under drought stress. Theor Appl Genet 100(8):1197–1202

Tuberosa R (2012) Phenotyping for drought tolerance of crops in the genomics era. Front Physiol 3: 347

Umezawa T, Yoshida R, Maruyama K, Yamaguchi-Shinozaki K, Shinozaki K (2004) SRK2C, a SNF1-related protein kinase 2, improves drought tolerance by controlling stress-responsive gene expression in Arabidopsis thaliana. Proc Natl Acad Sci 101(49):17,306–17,311

UNESCO (2009) World water assessment programme (United Nations), UN-water. Water in a changing world

Usman B, Nawaz G, Zhao N, Liao S, Liu Y, Li R (2020) Precise editing of the ospyl9 gene by rna-guided cas9 nuclease confers enhanced drought tolerance and grain yield in rice (Oryza sativa l.) by regulating circadian rhythm and abiotic stress responsive proteins. Int J Mol Sci 21(21):7854

Varshney RK, Gaur PM, Chamarthi SK, Krishnamurthy L, Tripathi S, Kashiwagi J, Samineni S, Singh VK, Thudi M, Jaganathan D (2013) Fast-track introgression of "QTL-hotspot" for root traits and other drought tolerance traits in JG 11, an elite and leading variety of chickpea. Plant Genome 6(3).: Plantgenome, 2013-07

Verbruggen N, Villarroel R, Van Montagu M (1993) Osmoregulation of a pyrroline-5-carboxylate reductase gene in Arabidopsis thaliana. Plant Physiol 103(3):771–781

Vikram P, Swamy BM, Dixit S, Ahmed HU, Cruz MT, Singh AK, Kumar A (2011) qDTY 1.1, a major QTL for rice grain yield under reproductive-stage drought stress with a consistent effect in multiple elite genetic backgrounds. BMC Genetics 12(1):1–5

Wang FZ, Wang QB, Kwon SY, Kwak SS, Su WA (2005) Enhanced drought tolerance of transgenic rice plants expressing a pea manganese superoxide dismutase. J Plant Physiol 162(4):465–472

Wang C, Lu W, He X, Wang F, Zhou Y, Guo X, Guo X (2016a) The cotton mitogen-activated protein kinase kinase 3 functions in drought tolerance by regulating stomatal responses and root growth. Plant Cell Physiol 57(8):1629–1642

Wang Y, Sun T, Li T, Wang M, Yang G, He G (2016b) A CBL-interacting protein kinase TaCIPK2 confers drought tolerance in transgenic tobacco plants through regulating the stomatal movement. PLoS One 11(12):e0167962

Wang L, Chen L, Li R, Zhao R, Yang M, Sheng J, Shen L (2017a) Reduced drought tolerance by CRISPR/Cas9-mediated SlMAPK3 mutagenesis in tomato plants. J Agric Food Chem 65(39): 8674–8682

Wang M, Mao Y, Lu Y, Tao X, Zhu JK (2017b) Multiplex gene editing in rice using the CRISPR-Cpf1 system. Mol Plant 10(7):1011–1013

Widawsky DA, O'Toole JC (1990) Prioritizing the rice biotechnology research agenda for eastern India. Rockefeller Foundation, New York

Wilson PB, Estavillo GM, Field KJ, Pornsiriwong W, Carroll AJ, Howell KA, Woo NS, Lake JA, Smith SM, Harvey Millar A, Von Caemmerer S (2009) The nucleotidase/phosphatase SAL1 is a negative regulator of drought tolerance in Arabidopsis. Plant J 58(2):299–317

Winkel T, Renno JF, Payne WA (1997) Effect of the timing of water deficit on growth, phenology and yield of pearl millet (Pennisetum glaucum (L.) R. Br.) grown in Sahelian conditions. J Exp Bot 48(5):1001–1009

Xiang Y, Tang N, Du H, Ye H, Xiong L (2008) Characterization of OsbZIP23 as a key player of the basic leucine zipper transcription factor family for conferring abscisic acid sensitivity and salinity and drought tolerance in rice. Plant Physiol 148(4):1938–1952

Xiao Y, Savchenko T, Baidoo EE, Chehab WE, Hayden DM, Tolstikov V, Corwin JA, Kliebenstein DJ, Keasling JD, Dehesh K (2012) Retrograde signaling by the plastidial metabolite MEcPP regulates expression of nuclear stress-response genes. Cell 149(7):1525–1535

Xiong L, Yang Y (2003) Disease resistance and abiotic stress tolerance in rice are inversely modulated by an abscisic acid–inducible mitogen-activated protein kinase. Plant Cell 15(3): 745–759

Xu JL, Lafitte HR, Gao YM, Fu BY, Torres R, Li ZK (2005) QTLs for drought escape and tolerance identified in a set of random introgression lines of rice. Theor Appl Genet 111(8):1642–1650

Xu W, Zhang C, Yang Y, Zhao S, Kang G, He X, Song J, Yang J (2020) Versatile nucleotides substitution in plant using an improved prime editing system. Mol Plant 13(5):675–678

Yamaguchi-Shinozaki K, Shinozaki K (2005) Organization of cis-acting regulatory elements in osmotic-and cold-stress-responsive promoters. Trends Plant Sci 10(2):88–94

Yang W, Kong Z, Omo-Ikerodah E, Xu W, Li Q, Xue Y (2008) Calcineurin B-like interacting protein kinase OsCIPK23 functions in pollination and drought stress responses in rice (Oryza sativa L.). J Genetics Genomics 35(9):531–5S2

Yang W, Wang M, Yue A, Wu J, Li S, Li G, Du W (2014) QTLs and epistasis for drought-tolerant physiological index in soybean (Glycine max L.) across different environments. Caryologia 67 (1):72–78

Yeo ET, Kwon HB, Han SE, Lee JT, Ryu JC, Byu MO (2000) Genetic engineering of drought resistant potato plants by introduction of the trehalose-6-phosphate synthase (TPS1) gene from Saccharomyces cerevisiae. Mol Cells 10(3):263–268

Yoshida T, Fujita Y, Sayama H, Kidokoro S, Maruyama K, Mizoi J, Shinozaki K, Yamaguchi-Shinozaki K (2010) AREB1, AREB2, and ABF3 are master transcription factors that cooperatively regulate ABRE-dependent ABA signaling involved in drought stress tolerance and require ABA for full activation. Plant J 61(4):672–685

Yue E, Cao H, Liu B (2020) OsmiR535, a potential genetic editing target for drought and salinity stress tolerance in Oryza sativa. Plants 9(10):1337

Zhang F, Chen S, Chen F, Fang W, Chen Y, Li F (2011) SRAP-based mapping and QTL detection for inflorescencerelated traits in chrysanthemum (Dendranthema morifolium). Mol Breed 27 (1):11–23

Zhao CX, Guo LY, Jaleel CA, Shao HB, Yang HB (2008) Prospective for applying molecular and genetic methodology to improve wheat cultivars in drought environments. C R Biol 331:579–586

Zhou ML, Ma JT, Pang JF, Zhang ZL, Tang YX, Wu YM (2010) Regulation of plant stress response by dehydration responsive element binding (DREB) transcription factors. Afr J Biotechnol 9(54):9255–9269

Zhu JK (2002) Salt and drought stress signal transduction in plants. Annu Rev Plant Biol 53(1): 247–273

Zou JJ, Wei FJ, Wang C, Wu JJ, Ratnasekera D, Liu WX, Wu WH (2010) Arabidopsis calcium-dependent protein kinase CPK10 functions in abscisic acid-and Ca2+-mediated stomatal regulation in response to drought stress. Plant Physiol 154(3):1232–1243

Zhou R, Kan X, Chen J, Hua H, Li Y, Ren J, Feng K, Liu H, Deng D, Yin Z (2019) Drought-induced changes in photosynthetic electron transport in maize probed by prompt fluorescence, delayed fluorescence, P700 and cyclic electron flow signals. Environ Exp Bot 158:51–62

Alleviating Abiotic Stress in Plants Using Nanoparticles

Yogita Sharma and Rachna Bhateria

Abstract Nanoparticles have distinct characteristics and various uses. In plants, excellent results have been shown by nanoparticles for abiotic stress alleviation. Abiotic stress is significant problem around the globe, in particular salinity, heavy metal, and drought stress. Such stresses can lead to several problems, viz. production of reactive oxygen species, damage of membranes, photosynthesis inhibition, etc. which can reduce the development and growth by changing physiological, biochemical, and molecular reactions consequently affecting plant productivity. Several ecosystems have witnessed the changes due to extreme climatic conditions, frequent droughts, areas damaged by salinity, higher temperatures, and unusual rainfall. The productivity and yield of crop have also declined worldwide as a consequence of changes in ecosystem. Many studies have been done on metallic nanoparticles and tolerance related to abiotic stress in plants. Use of nanoparticles can prove to an effective solution to overcome impact of abiotic stress and to support plant growth and development. Nanoparticles boost up the antioxidant defense mechanisms. Extensive research is needed to analyze the behavior of NPs on crops under these stress conditions. Plants response under abiotic stress needs to be understood. This chapter focuses on significant role of nanoparticles to reduce abiotic stress specifically salinity, drought, and heavy metals. Mechanism behind the reduction of abiotic stress in plants due to application of nanoparticles is discussed.

Keywords Salt stress · Heavy metal stress · Drought stress · Nanoparticles

Introduction

In ecosystem, plants encounter a range of abiotic stresses, like drought, heat, salt, cold temperatures, etc. The prevalence of the stresses has been increased worldwide (Khan et al. 2017). Plants are unable to physically migrate away from their current

Y. Sharma · R. Bhateria (✉)
Department of Environmental Sciences, Maharshi Dayanand University, Rohtak, Haryana, India

© The Author(s), under exclusive license to Springer Nature Singapore Pte Ltd. 2022
A. Vaishnav et al. (eds.), *Plant Stress Mitigators*,
https://doi.org/10.1007/978-981-16-7759-5_25

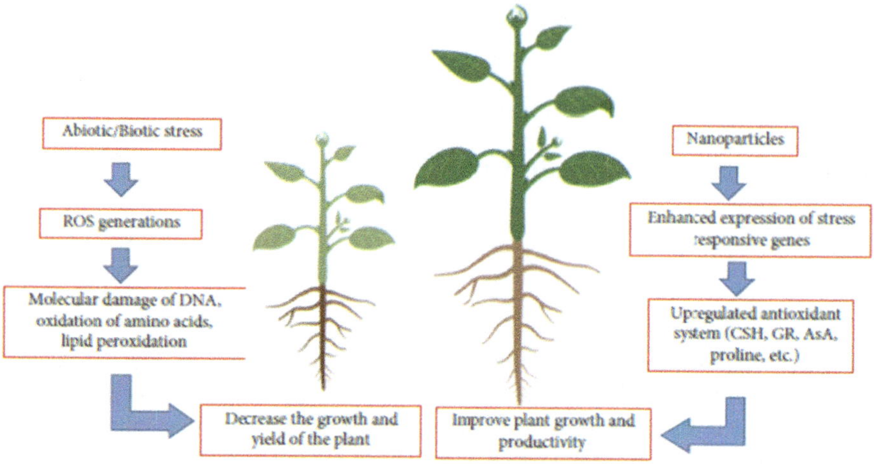

Fig. 1 Mechanism of abiotic stress reduction in plant (Ali et al. 2021)

position in order to avoid the effects of environmental stress, such as abiotic stress. Heavy metals (HMs) pollution, salinity, and drought stress are among the abiotic stresses that have been helpful to lower agricultural production by many orders of magnitude (Haider et al. 2021; Godoy et al. 2021). Plant metabolism is disrupted as a result of these alterations under abiotic stress, allowing rearrangement of the metabolic network to maintain the key metabolic activities (Rajput et al. 2021; Verma et al. 2020; Minkina et al. 2020; Ayup et al. 2015).

Drought, salinity, and heat stress are all circumstances that plants may adapt to. Plants' molecular and cellular responses to abiotic stress have been studied extensively (Choudhury et al. 2017). Against the abiotic stress, a primary response of plant involves the transitory provocation of cytoplasmic calcium reactive oxygen species (ROS), abscisic acid, and increased mitogen-activated protein kinase (MAPK) pathways (Zhang et al. 2020). However, modulating proteins involved in cellular damage protection and regulating stress-specific gene expression come under a response for advanced stage.

The adaptation of plants to a specific abiotic stress situation needs a unique response suited to the different environmental states of the plant. As a result, a combination of environmental factors may be different in biochemical, molecular, and physiological processes activated by one source of stress than those that arise from a little different combination. The following points are shown by the transcriptome profiling of plant exposures to different abiotic conditions: A slight overlap between the plant's reactions to stress circumstances like heat, drought, cooling, salt, bright light or mechanical stress caused a slightly unique reaction in each stress condition examined (Rizhsky et al. 2004). Although a range of abiotic and biotic stresses are associated with ROS, the ROS gene network from *Arabidopsis* was reported to respond to various stress treatment options (Fig. 1). The studies showed that every abiotic stress demands a distinct adapted response.

Depending upon the plant's requirement, it may give distinct response based on combination of different stresses.

Situations of abiotic stress generate major losses to world agricultural yields. Intensive study has been conducted on individual stress situations such as drought, salinity, etc. (Cushman and Bohnert 2000). However, different plants and crops frequently undergo a mix of several abiotic stresses in fields. In drought stress-affected areas, crops are exposed to a combination of stresses like salinity and heat (Moffat 2002). As revealed by studies done, also the metabolic as well as molecular responses of plant under stress are unique, i.e., mixture of stresses like heat and drought gives an exclusive response. This could not be directly extrapolated from the response of plants to each of these different stresses applied individually (Suzuki et al. 2005).

In recent years, significant technological improvements and innovations in agriculture have been created to encounter the mounting difficulties of sustainable agricultural output and food safety (Kou et al. 2018; Dwivedi et al. 2016). By 2050, the globe will need to produce 50 percent more food to fulfill the demands of 9 billion people. Because land and water supplies are limited and the aim to enhance the productivity can only be fulfilled by technical interventions. The paucity of cultivable land, dependency on conventional crops, and irrigation are the main factors which prompted the scientists to investigate some innovative and modern techniques. Because of nanomaterials' obvious exceptional and extraordinary capabilities, it does not sound much astonishing that efforts are being done for the betterment of the agricultural sector utilizing ideas and application of nanotechnology (Grillo et al. 2015; Handford et al. 2014).

Nanotechnology

Nanotechnology is the most dynamic subject of material science study, and nanoparticles (NPs) manufacturing is growing quickly worldwide. Consequently, some features, e.g., size (in a range from 1 to 100 nm), structure, and shape of NPs show totally new or better properties (Taran et al. 2017; Nejatzadeh 2021). Nanoparticles may be inorganic and organic NPs. Inorganic NPs include metallic NPs (such as gold (Au), silver (Ag), copper (Cu), aluminum (Al)), magnetic NPs (such as cobalt, iron, nickel), and semiconductor NPs (such as zinc oxide (ZnO), zinc sulfide (ZnS), cadmium sulfide (CdS)) while organic NPs include carbon NPs (such as carbon nanotubes and quantum dots) (Chouhan 2018; Taran et al. 2017).

As NPs have different characteristics, it is possible to employ inorganic NPs to produce sustainable crops (Parisi et al. 2015; Nejatzadeh 2021). Agricultural practices integrate different products as well as inventions synthesize and raised from engineered NPs, for e.g., nanofertilizers, nanosensors, and also nanopesticides were founded in the last 10 years in order to improve the quality of agricultural system requiring small quantities of output as well as waste in comparison to conventional techniques (Servin et al. 2015; Liu and Lal 2015).

Role of Nanoparticles in Abiotic Stress Reduction

NPs have several extremely substantial characteristics that, due to their very tiny size, distinguish them from their bulk analogs. NPs have a larger surface area and are more reactive than bulk materials and are more soluble. The aim of sustainable agriculture, with a possible impact on decreasing the negative impact of abiotic stress was thus accomplished worldwide (Alabdallah and Hasan 2021). NPs are very important in agriculture in reducing abiotic stress in plants. The nanotechnology intervention has shown that it has not only been successful in removing nonbiodegradable metals, but in detoxifying slowly decaying pollutants (Sebastian et al. 2018).

NPs are used as sensing materials (Chaudhary et al. 2018; Duhan et al. 2017), herbicides, pesticides, and nanofertilizers (Liu and Lal 2015). They can also be applied for improvisation of plant nutrients. Additionally, to release the agrochemicals in controlled concentration, NPs can act as best carriers (Medina-Pérez et al. 2019; Shang et al. 2019; Lv et al. 2019). However, some NPs with distinct physiochemical characteristics naturally enhance plant's development and tolerance to stress instead of behaving as nanocarriers. Biofunction of NPs determines their method of application, physicochemical characteristics (foliar, hydroponics, soil), and also it depends on quantity (Kim et al. 2017).

Abiotic stress is a serious problem for the environment. The principal abiotic stresses impacting plant development and productivity are salinity, drought, and extreme hot as well as low temperatures (Hasan et al. 2020a, b; Jahan et al. 2021; Alharbi et al. 2021). Several of metallic NPs such silver NPs (AgNPs), gold NPs (AuNPs), copper NPs (CuNPs), iron NPs (FeNPs), titanium oxide NPs (TiO_2NPs), zinc NPs (ZnNPs), zinc oxide NPs (ZnONPs) were applied on different crop species for stress tolerance, seed germination, and plant growth (Latef et al. 2017; Taran et al. 2017).

Salinity Stress in Plants

Salt stress affects the plants drastically and hence productivity decreases (Zafar et al. 2020). Alteration of ion balance, mineral nutrition, photosynthetic efficacy, water status, and stomatal activity comprises physiological mechanisms that impact salinity stress. Salinity influences the development of plants by altering their physiology through ionic and osmotic stress both at cellular and at whole plant level. High salt content in the root zone also impacts the availability of water or the potential of water since the plants have extremely little water (Sonam et al. 2014). Reduced water in the plant causes the plants to suffer from osmotic stress. It also produces imbalances in the absorption of beneficial ions such as potassium (K^+), calcium (Ca^{2+}), and magnesium (Mg^{2+}), and substitutes it with the absorption of harmful ions such as sodium ion (Na^+) and chlorine (Cl^{-1}).

Various salt stress physiological mechanisms involve membrane instability from K^+ and Ca^{2+} replaced by Na^+, membrane permeability, rise in rate of respiration and mineral distribution, ion toxicity, plant growth variations, and decrease in photosynthetic efficiency. Reduced photosynthesis causes stomata closure, reduced CO_2 assimilation inside the plant cells, reduced green pigments, and leaf area (Shahid et al. 2011). Salinity stress results in the production of ROS, for e.g., H_2O_2, O_2, and also OH, which affects deoxyribonucleic acid (DNA), ribonucleic acid (RNA), and proteins, destructs chlorophyll, and damages meristem activity also (Habib et al. 2012). New methods are thus continuously needed to reduce the harmful effects of these stresses on plants.

ROS is produced in many plant cell organelles like peroxisomes, plasma membranes, mitochondria, and chloroplasts under normal and stress circumstances. In plants, excessively ROS production is related to oxidative damage and affected at developmental level, genotype, and salinity stress. Plant reacts to the harmful effect of salts by enhancing their antioxidant defense (Alharbi et al. 2021). Many antioxidant enzymes function in such antioxidant systems such as catalase (CAT), superoxide dismutase (SOD), dehydro ascorbate reductase, ascorbate peroxidase, glutathione reductase, monodehydroascorbate reductase, proline, anthocyanin, and glycine betaine.

Role of Nanoparticles to Reduce Salinity Stress in Plants

Due to the environment friendly applications in agriculture sector, numerous metal NPs have gained substantial attention in recent years (Mahakham et al. 2017; Chouhan 2018). Silver Nanoparticles (AgNPs) are becoming more popular in agriculture because of their influence on stress tolerance. Although study has revealed that AgNPs decreased salt stress efficiently (Almutairi 2016; Abou-Zeid and Ismail 2018). AgNPs have improved stress tolerance in crops by controlling nutrient shortages, raising enzyme reactions, and also assist in adherence of abiotic stress plant growth-promoting bacteria to plant roots.

The quantities or concentration of different salts in plants species encountered AgNPs considerably raised the levels of chloride, potassium, sodium, and also osmolality. Changes in the salinity of aquatic bodies control AgNPs stability and in low-salinity environments, AgNPs have been found to be more stable. Notably higher content of salts can prove to be harmful for growth and development of plant (Sagghatol-Islami 2010). Efforts have been done by researchers for plant propagation and germination because novel transgenic plant types have become more significant in development, management, and production. Priming of a seed is a better idea way to encourage plant germination (Alabdallah and Hasan 2021).

Like AgNPs, iron sulfate ($FeSO_4$) NPs foliar spray showed a favorable effect on salinity stress resistance in sunflower cultivars. The application of $FeSO_4$ NPs increased the net assimilation rate of CO_2, leaf area, chlorophyll concentration, dry weight of the shoot, CO_2 concentration, the maximum photochemical potential of photosystem II, and the iron (Fe) content, also reduced the substantial amount of Na^+ ions (Torabian et al. 2018).

Drought Stress in Plants

During development and growth, plants encounter different environmental stresses in natural and agricultural environment. Out of the other stresses, drought counts the much severe one and harms the productivity of the plants. Water has a significant role in plant growth, metabolism, and other physiological reactions as a plant body is comprised of 80–95% water (Brodersen et al. 2019; Abbasi and Abbasi 2010). For better production of agricultural crops, optimum water requirement should be fulfilled for osmoregulation, short- to long-distance transport, and single-cell expansion through cell membranes (Iwuala et al. 2020). The flow of the water may get affected during drought in plants and can be managed through opening of membrane channels known as water-permeability aquaporins.

The abiotic stress in plants activates the changes in its metabolic processes, which facilitate the restructuring of metabolic system so that the important reactions can be maintained. Plants in drought-like conditions close stomata, stop photosynthesis, lessen the leaf area, decrease growth and biomass, reduce the water potential, raise osmolytes quantity as well as stimulate ROS generation (Fig. 2) (Ibrahim 2016). As the plants are motionless, it leads to beginning of abiotic stress which further results into less production. In order to maintain food security, drought stress must be reduced and drought-tolerant crops should be developed (Hasan et al. 2020a).

Role of Nanoparticles in Drought Stress Reduction

In plants, the cellular organelles are targeted, and certain contents are released through the nanoparticle target (Seleiman et al. 2020; Cunningham et al. 2018). The function of antioxidants enzymes, i.e., CAT, SOD, and peroxidase (POD) were regulated and enhanced (Table 1) by the application of nanoparticles. For example, the activity of SOD in plants was amplified by applying TIO_2 NPs (Ghasemlou et al. 2019). In agriculture, different trace elements and their oxides of NPs were used for intensifying drought stress resistance in different plants (Table 1). The negative results of abiotic stress like drought, chilling stress, salinity, and HM toxicity were mitigated through silicon nanoparticles (SiNPs) application (Seleiman et al. 2020; Siddiqui et al. 2020). Growth and physio- and biochemical traits such as carbohydrates, proline, chlorophyll, carotenoids, and relative water contents were significantly improved in different plant species, when NPs were applied such as silica and ZnO nanoparticles (Seleiman et al. 2020). SiNPs also enhanced the drought resistance in wheat plants (Rizwan et al. 2019; Khan et al. 2019). Similarly, the salinity and drought stress in plants were also mitigated by ZnO nanoparticles application

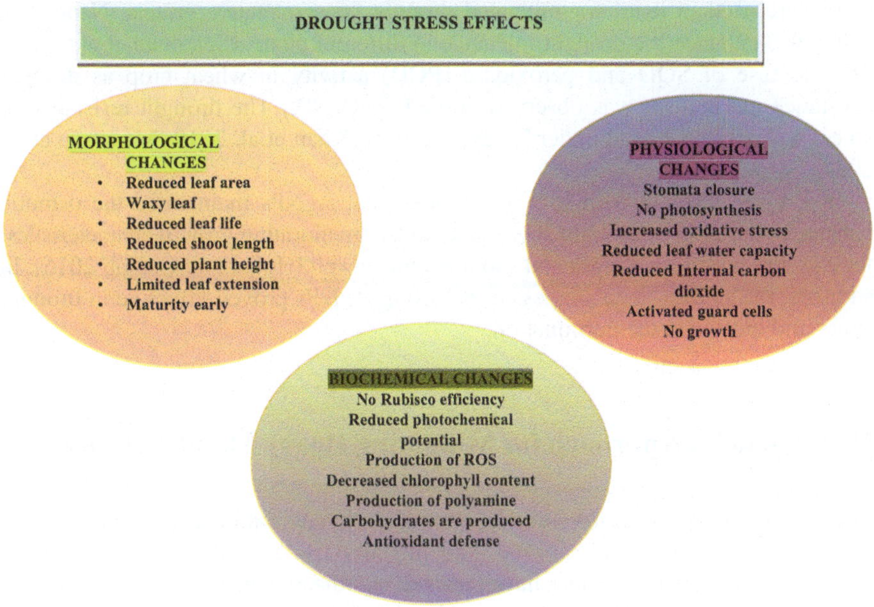

Fig. 2 Drought stress effects on plant

Table 1 Nanoparticles application to enhance drought stress tolerance

Nanoparticles	Mechanism	References
Silica nanoparticles	Enhanced tolerance for drought stress	Ashkavand et al. (2015)
Iron nanoparticles	Foliar spray was used	Davar et al. (2014)
Thiol-gated mesoporous silica nanoparticles	To release encapsulated ABA in controlled amount which enhanced AtGALK2 gene	Sun et al. (2018)
Titanium nanoparticles	Foliar spray improved contents of starch and also seed gluten	Jaberzadeh et al. (2013)
Zinc and copper nanoparticles	Decreased accumulation of MDA and increased RWC antioxidant enzymes	Taran et al. (2017)
Zinc oxide nanoparticles	Improved percentage and germination rate of soybean	Sedghi et al. (2013)

(Seleiman et al. 2020). During the early stage of growth, the application of ZnO NPs stimulated the seed reservoirs for sapling and enhanced the drought resistance in plants (Seydmohammadi et al. 2020).

Ferrous in combination with Zn were also reported to have a beneficial effect on plant resistance to drought stress. Plants grown under drought stress were mitigated through TIO_2 nanoparticles, consequently activated different compounds and ameliorated the adverse effects of water deficit (Seleiman et al. 2020; Movafeghi et al. 2018). To improve drought stress in plants, other NPs such as silver (Ag) and copper

(Cu) were used in lentil for mitigated drought stress negative effects. Nanosilica could also enhance the drought tolerance in different plants (Seleiman et al. 2021). The increase of SOD and peroxidase (POD) activity in wheat crop as drought resistance mechanism was observed through ZnO NPs. The drought resistance in wheat was also enhanced under Zn and Cu NPs (Khan et al. 2019; Maswada et al. 2020).

In a study done by Hojjat et al. reported that AgNPs maintained the osmotic balance in duration of drought stress in lentil by ameliorating growth characteristics, for e.g., dry and fresh weight and shoot length as well (Hojjat and Ganjali 2016). In addition, the studies done suggested that using AgNPs proved effective in drought stress and enhanced the germination.

Nanoparticles Approach for Mitigating Heavy Metal Toxicity

The heavy metals concerns with humans and its environment are copper, zinc, cadmium, chromium, lead, arsenic, and mercury (Raja et al. 2021; Zhao et al. 2021). The last several decades have received significant contributions in managing solar soils impacted by HMs pollution, using NPs which improve plant development and soil properties (Singh et al. 2021). The hazards of HMs contamination in cultivated and aquatic regions are severely affected by indiscriminate addition of different agrofertilizers (Zamora-Ledezma et al. 2021). The quantity of HMs within a particular matrix exceeds specific limitations, causes toxicity and genotoxic, carcinogenic, and mutagenic activity (Zamora-Ledezma et al. 2021; Kontaş and Bostancı 2020; Sall et al. 2020).

NPs intervention also enhances plant physiological and biochemical characteristics such as increasing defense enzyme production SOD, catalase (CAT), ascorbate peroxidase (APX), etc. improved nutrient uptake, reducing electrolytes loss, improved soluble proteins and pigments, lowered peroxidation, and increased phytochelatins glutathione, and proline. The above-mentioned characteristics and properties of specific crop make them or are responsible for their tolerance. Plants make use of two strategies for mitigation HM toxicity and organs' protection, i.e., limiting the HM accumulation and uptake using tolerance mechanism.

Noman et al. (2020) proposed the contribution to countering chromium (Cr)-induced toxicity in *Triticum aestivum* L. using bacterially generated copper nanoparticles (Cu NPs). The Cu NPs improved antioxidant pool, biomass, and growth, by their amendment in soil (25 and 50 mg kg^{-1}). The NPs have been applied to reduce production of ROS and Cr transfer to the plant by significant soil immobilization during Cr stress. Similar findings were observed on Cu-based microbially synthesized NPs aiding Cr-induced toxicity reduction and ensuing increase in nutrient absorption and biomass, also showed decreased translocation of plant components given by metal immobilization (Noman et al. 2020b).

The foliar application of titanium oxide (TiO_2) was used against the cadmium toxicity in Zea mays (Lian et al. 2020). Treatment using different-sized silicon NPs

(SiNPs) has resulted in a decrease in growth inhibition induced by mercury (Hg) (Li et al. 2020). Changes or additions done to improvisation in SiNPs resulted into reduced mercury accumulation in plant parts like shoot and roots. This was verified by X-ray fluorescence. This methodology has been proposed to make volatile HM like Hg unreactive.

Lowered levels of cadmium have been observed in *O. sativa* L. plantations treated with Au NPs. Significant reduction levels have been noticed in leaves and roots, i.e., 46.2% and 33%, respectively. Supplementing the nanoparticles with media resulted in an improved antioxidant defense enzyme system which makes it easier to reduce the oxidative stress triggered by cadmium in *O. sativa* L. Additionally, the Au NPs restricted the gene expression (related with transport of metals beyond the cells) (Jiang et al. 2021).

Fe_2O_3 NPs exhibited alleviation of increasing Cadmium (Cd) toxicity in plants for example, *Oryza sativa* (*O. sativa*) L. (Ahmed et al. 2021). This is because the applied Fe_2O_3NPs to plants reduced the ROS production or generation and also gene expression which used to support the cadmium transport in cells. For example, in T. *aestivum* L., the application of Fe_2O_3 NPs decreased the cadmium percentage to 70% (Figs. 3 and 4) (Manzoor et al. 2021).

Mechanism Behind Alleviation of Heavy Metal Stress in Plants Using NPs

Heavy metal stress affects physiology, morphology as well as biochemistry of plants. Many strategies have been adopted by plants to reduce the stress exerted by heavy metals (Fig. 5). For example, the quantity of HMs in soil is reduced (Moharem et al. 2019), HMs transport gene regulation (Cao et al. 2020), plants improve the physiological functions and its antioxidant systems (Wang et al. 2020), and also plants produce the protective agents like phytochelatins (Lian et al. 2020; Cao et al. 2020).

Nanoparticles absorb and also transform HMs in soil, which reduces their bio-availability and mobility in soil (Sebastian et al. 2019). Mercapto SiNPs transformed the Cd into its more stable form as reported by Wang et al. (2020). NPs accumulated on cell wall make HM unavailable for plants by binding with them and make a complex which gets adsorbed on the surface of the cell (Wang et al. 2021; Cui et al. 2018). Consequently, the biological activity of HM reduced and also their migration. Other than this, NPs also enhance protective agent production. For example, SiNPs encouraged the organic acid synthesis and damage to plant due to cadmium was reduced (Zhou et al. 2021). Apoplastic barriers control flow of ions, oxygen, and water besides protecting the plants with their physiological function (Yang et al. 2013). However, the NPs can influence the apoplastic barriers and reduce the HM concentration in roots (Rossi et al. 2017).

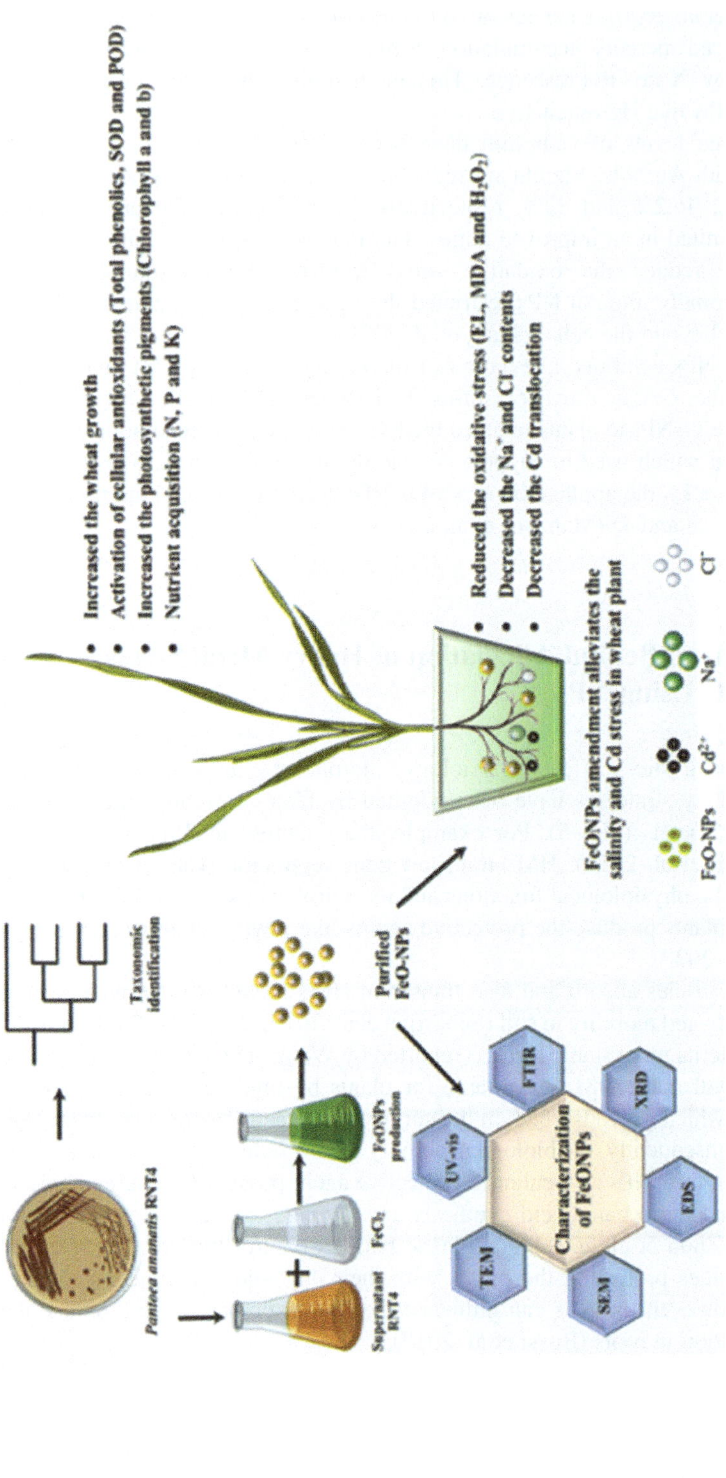

Fig. 3 Synthesis of FeO NPs and their use for alleviating both cadmium and salinity stress (Manzoor et al. 2021)

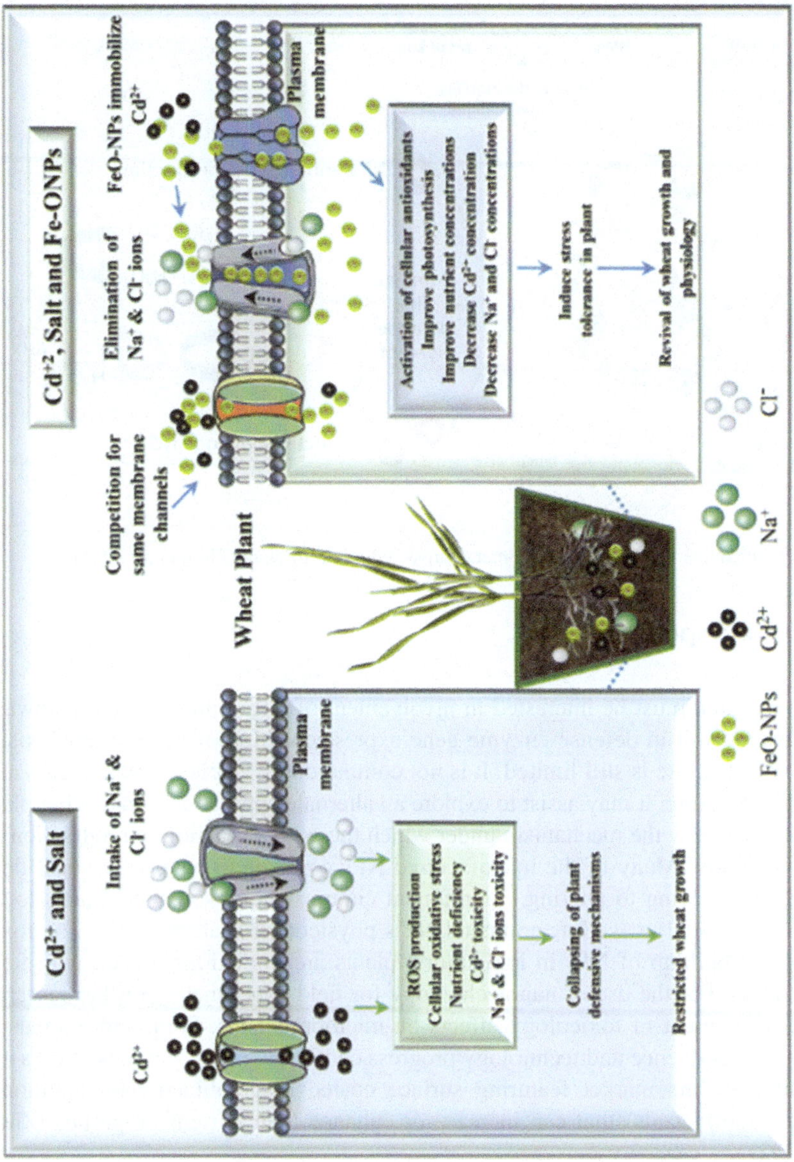

Fig. 4 Coalleviation of cadmium and salinity stress using biogenic FeO NPs (Manzoor et al. 2021)

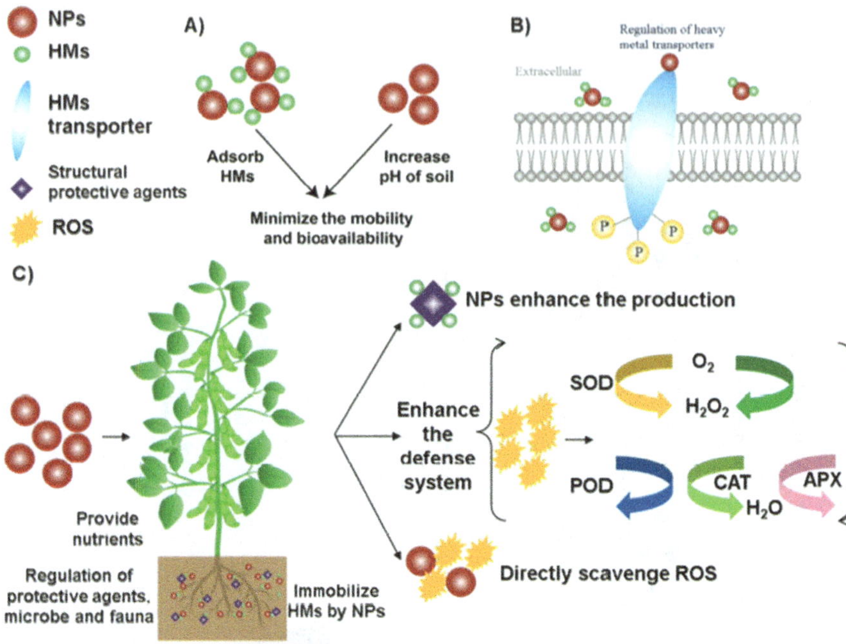

Fig. 5 Mechanism involved in heavy metal stress reduction in plant (Zhou et al. 2021)

Future Perspectives

Future research must be interested in involvement of NPs with signaling pathways that cause a shift in defense enzyme gene expression. Moreover, the use of costly NPs for agriculture is still limited. It is not commercially viable to treat AuNPs and AgNPs. Therefore, it may assist to explore an alternate approach of improving plant tolerance to know the mechanism under which these NPs function in modulation of plant immunity. Many of the manufactured NPs continue to have very small field applicability, owing to altering environment circumstances, soil type, plants to be treated, and metal as well as nonmetal NP's physicochemical character. The toxic effects and buildup of NPs in agricultural plants are limitations related with field applications. For the use of nanotechnology for field experiments, further research on the assessment of toxicology effects on microorganisms, flora, and animals is important. As science and technology progress continues, many varieties of NPs are accessible on the market featuring surface coated or combined with additional elements (surfactants) that can increase or enhance their current properties (Zhou et al. 2021).

Conclusion

Because of recent climate change, abiotic stresses represent a major danger to global food safety and sustainability in agriculture. Abiotic stresses have harmful effect on growth, production capacity, and development of plant. Thus, less yield and biomass threaten food security worldwide. Abiotic stress like drought affects life cycle of the plants. Also, it hampers productivity of different plants affecting their metabolic, physiological, and biochemical reactions. However, to solve the stress-related effects on plants, some mechanism has been adopted by plants for abiotic stress tolerance. So, some characteristics should be discovered that can be adopted by plants to tolerate the stress effects and their incorporation to the genotypes so that their productivity does not get reduced or affected. Some strategies can be followed by plants to avoid drought stress such as structural dynamics and growth pattern, less water loss during transpiration, leaf rolling, root to shoot ratio dynamics, root length increment, compatible solute accumulation, increased transpiration capacity, hormonal and osmotic regulation, and late senescence. Also, several researches have considered the useful role of NPs in salinity improvement, heavy metal, and drought stresses. Furthermore, their uses can protect plants against adverse effects caused by various abiotic stresses, since NPs show a relatively wide range of activities. NPs are shown to function, notably in terms of germination, development, and production of diverse plants, as a possible stimulant. Metal NPs in many experiments have demonstrated an improvement in stress tolerance but the mechanism of stress-resistance characteristics is still unclear. Such an understanding can assist to develop smart NPs in future, which can assist mitigate abiotic stress and help sustain better performance in agriculture. However, additional research is required to establish clear conclusions regarding the synergistic effect of the different therapies. Further study is desirable at many levels, including molecular and subcellular ones, to determine how nanoparticles prevent plant stress. Before making use of their potential advantages, care is required when assessing the toxicity impact of NPs on various species and ecosystems.

References

Abbasi T, Abbasi SA (2010) Biomass energy and the environmental impacts associated with its production and utilization. Renew Sustain Energy Rev 14(3):919–937

Abou-Zeid H, Ismail G (2018) The role of priming with biosynthesized silver nanoparticles in the response of Triticum aestivum L to salt stress. Egypt J Bot 58(1):73–85

Ahmed T, Noman M, Manzoor N, Shahid M, Abdullah M, Ali L et al (2021) Nanoparticle-based amelioration of drought stress and cadmium toxicity in rice via triggering the stress responsive genetic mechanisms and nutrient acquisition. Ecotoxicol Environ Saf 209:111,829

Alabdallah NM, Hasan MM (2021) Plant-based green synthesis of silver nanoparticles and its effective role in abiotic stress tolerance in crop plants. Saudi J Biol Sci

Alharbi BM, Elhakem AH, Alnusairi GS, Soliman MH, Hakeem KR, Hasan MM, Abdelhamid MT (2021) Exogenous application of melatonin alleviates salt stress-induced decline in growth and

photosynthesis in Glycine max (L.) seedlings by improving mineral uptake, antioxidant and glyoxalase system. Plant Soil Environ 67(4):208–220
Ali EF, El-Shehawi AM, Ibrahim OHM, Abdul-Hafeez EY, Moussa MM, Hassan FAS (2021) A vital role of chitosan nanoparticles in improvisation the drought stress tolerance in Catharanthus roseus (L.) through biochemical and gene expression modulation. Plant Physiol Biochem 161: 166–175
Almutairi ZM (2016) Influence of silver nano-particles on the salt resistance of tomato (Solanum lycopersicum) during germination. Int J Agric Biol 18(2):449–457
Ashkavand P, Tabari M, Zarafshar M, Tomásková I, Struve D (2015) Effect of SiO2 nanoparticles on drought resistance in hawthorn seedlings. Leśne Prace Badawcze 76(4)
Ayup M, Chen YN, Nyongesah MJ, Zhang YM, Rajput VD, Zhu CG (2015) Xylem anatomy and hydraulic traits of two co-occurring riparian desert plants. Iawa J 36(1):69–83
Brodersen CR, Roddy AB, Wason JW, McElrone AJ (2019) Functional status of xylem through time. Annu Rev Plant Biol 70:407–433
Cao F, Dai H, Hao PF, Wu F (2020) Silicon regulates the expression of vacuolar H+-pyrophosphatase 1 and decreases cadmium accumulation in rice (Oryza sativa L.). Chemosphere 240:124907
Chaudhary S, Umar A, Bhasin KK, Baskoutas S (2018) Chemical sensing applications of ZnO nanomaterials. Materials 11(2):287
Choudhury FK, Rivero RM, Blumwald E, Mittler R (2017) Reactive oxygen species, abiotic stress and stress combination. Plant J 90(5):856–867
Chouhan N (2018) Silver nanoparticles: Synthesis, characterization and applications. IntechOpen
Cui H, Shi Y, Zhou J, Chu H, Cang L, Zhou D (2018) Effect of different grain sizes of hydroxyapatite on soil heavy metal bioavailability and microbial community composition. Agr Ecosyst Environ 267:165–173
Cunningham FJ, Goh NS, Demirer GS, Matos JL, Landry MP (2018) Nanoparticle-mediated delivery towards advancing plant genetic engineering. Trends Biotechnol 36(9):882–897
Cushman JC, Bohnert HJ (2000) Genomic approaches to plant stress tolerance. Curr Opin Plant Biol 3(2):117–124
Davar ZF, Roozbahani A, Hosnamidi A (2014) Evaluation the effect of water stress and foliar application of Fe nanoparticles on yield, yield components and oil percentage of safflower (Carthamus tinctorious L.). Int J Adv Biol Biomed Res 2:1150–1159
Duhan JS, Kumar R, Kumar N, Kaur P, Nehra K, Duhan S (2017) Nanotechnology: the new perspective in precision agriculture. Biotechnol Rep 15:11–23
Dwivedi S, Saquib Q, Al-Khedhairy AA, Musarrat J (2016) Understanding the role of nanomaterials in agriculture. In: Microbial inoculants in sustainable agricultural productivity. Springer, New Delhi, pp 271–288
Ghasemlou F, Amiri H, Karamian R, Mirzaie-asl A (2019) Alleviation of the effects of on drought stress Verbascum nudicuale by methyl jasmonate and titanium dioxide nanoparticles. Iran J Plant Physiol 9(4):2911–2920
Godoy F, Olivos-Hernández K, Stange C, Handford M (2021) Abiotic stress in crop species: improving tolerance by applying plant metabolites. Plants 10(2):186
Grillo R, Rosa AH, Fraceto LF (2015) Engineered nanoparticles and organic matter: a review of the state-of-the-art. Chemosphere 119:608–619
Habib N, Ashraf M, Ali Q, Perveen R (2012) Response of salt-stressed okra (Abelmoschus esculentus Moench) plants to foliar-applied glycine betaine and glycine betaine-containing sugarbeet extract. South Afr J Bot 83:151–158
Haider FU, Liqun C, Coulter JA, Cheema SA, Wu J, Zhang R et al (2021) Cadmium toxicity in plants: Impacts and remediation strategies. Ecotoxicol Environ Saf 211:111,887
Handford CE, Dean M, Henchion M, Spence M, Elliott CT, Campbell K (2014) Implications of nanotechnology for the agri-food industry: opportunities, benefits and risks. Trends Food Sci Technol 40(2):226–241

Hasan MM, Ali MA, Soliman MH, Alqarawi AA, Abd Allah EF, Fang XW (2020a) Insights into 28-homobrassinolide (HBR)-mediated redox homeostasis, AsA–GSH cycle, and methylglyoxal detoxification in soybean under drought-induced oxidative stress. J Plant Interact 15(1): 371–385

Hasan MM, Alharby HF, Uddin MN, Ali MA, Anwar Y, Fang XW et al (2020b) Magnetized water confers drought stress tolerance in Moringa biotype via modulation of growth, gas exchange, lipid peroxidation and antioxidant activity. Pol J Environ Stud 1:29

Hojjat SS, Ganjali A (2016) The effect of silver nanoparticle on lentil seed germination under drought stress. Int J Farm Allied Sci 5(3):208–212

Ibrahim EA (2016) Seed priming to alleviate salinity stress in germinating seeds. J Plant Physiol 192:38–46

Iwuala E, Odjegba V, Sharma V, Alam A (2020) Drought stress modulates expression of aquaporin gene and photosynthetic efficiency in Pennisetum glaucum (L.) R. Br. genotypes. Curr Plant Biol 21:100,131

Jaberzadeh A, Moaveni P, Moghadam HRT, Zahedi H (2013) Influence of bulk and nanoparticles titanium foliar application on some agronomic traits, seed gluten and starch contents of wheat subjected to water deficit stress. Notulae Botanicae Horti Agrobotanici Cluj-Napoca 41(1): 201–207

Jahan MS, Shu S, Wang Y, Hasan M, El-Yazied AA, Alabdallah NM et al (2021) Melatonin pretreatment confers heat tolerance and repression of heat-induced senescence in tomato through the modulation of ABA- and GA-mediated pathways. Front Plant Sci 12:381

Jiang M, Dai S, Wang B, Xie Z, Li J, Wang L, Huang J (2021) Gold nanoparticles synthesized using melatonin suppress cadmium uptake and alleviate its toxicity in rice. Environ Sci Nano 8(4): 1042–1056

Jogawat A, Yadav B, Lakra N, Singh AK, Narayan OP (2021) Crosstalk between phytohormones and secondary metabolites in the drought stress tolerance of crop plants: a review. Physiol Plant 172(2):1106–1132

Khan A, Anwar Y, Hasan M, Iqbal A, Ali M, Alharby HF et al (2017) Attenuation of drought stress in Brassica seedlings with exogenous application of Ca2+ and H2O2. Plan Theory 6(2):20

Khan ZS, Rizwan M, Hafeez M, Ali S, Javed MR, Adrees M (2019) The accumulation of cadmium in wheat (Triticum aestivum) as influenced by zinc oxide nanoparticles and soil moisture conditions. Environ Sci Pollut Res 26(19):19,859–19,870

Kim DH, Gopal J, Sivanesan I (2017) Nanomaterials in plant tissue culture: the disclosed and undisclosed. RSC Adv 7(58):36,492–36,505

Kontaş S, Bostancı D (2020) Genotoxic effects of environmental pollutant heavy metals on Alburnus chalcoides (Pisces: Cyprinidae) inhabiting lower Melet river (Ordu, Turkey). Bull Environ Contam Toxicol 104(6):763–769

Kou TJ, Yu WW, Lam SK, Chen DL, Hou YP, Li ZY (2018) Differential root responses in two cultivars of winter wheat (Triticum aestivum L.) to elevated ozone concentration under fully open-air field conditions. J Agron Crop Sci 204(3):325–332

Latef AAHA, Alhmad MFA, Abdelfattah KE (2017) The possible roles of priming with ZnO nanoparticles in mitigation of salinity stress in lupine (Lupinus termis) plants. J Plant Growth Regul 36(1):60–70

Li Y, Zhu N, Liang X, Bai X, Zheng L, Zhao J et al (2020) Silica nanoparticles alleviate mercury toxicity via immobilization and inactivation of Hg (ii) in soybean (Glycine max). Environ Sci Nano 7(6):1807–1817

Lian J, Zhao L, Wu J, Xiong H, Bao Y, Zeb A et al (2020) Foliar spray of TiO2 nanoparticles prevails over root application in reducing Cd accumulation and mitigating Cd-induced phytotoxicity in maize (Zea mays L.). Chemosphere 239:124794

Liu R, Lal R (2015) Potentials of engineered nanoparticles as fertilizers for increasing agronomic productions. Sci Total Environ 514:131–139

Lv J, Christie P, Zhang S (2019) Uptake, translocation, and transformation of metal-based nanoparticles in plants: recent advances and methodological challenges. Environ Sci Nano 6(1):41–59

Mahakham W, Sarmah AK, Maensiri S, Theerakulpisut P (2017) Nanopriming technology for enhancing germination and starch metabolism of aged rice seeds using phytosynthesized silver nanoparticles. Sci Rep 7(1):1–21

Manzoor N, Ahmed T, Noman M, Shahid M, Nazir MM, Ali L et al (2021) Iron oxide nanoparticles ameliorated the cadmium and salinity stresses in wheat plants, facilitating photosynthetic pigments and restricting cadmium uptake. Sci Total Environ 769:145221

Maswada HF, Mazrou YS, Elzaawely AA, Eldein SMA (2020) Nanomaterials. Effective tools for field and horticultural crops to cope with drought stress: a review. Spanish J Agric Res 18(2):15

Medina-Pérez G, Fernández-Luqueño F, Campos-Montiel RG, Sánchez-López KB, Afanador-Barajas LN, Prince L (2019) Nanotechnology in crop protection: status and future trends. In: Nano-biopesticides today and future perspectives. Academic Press, pp 17–45

Minkina T, Rajput V, Fedorenko G, Fedorenko A, Mandzhieva S, Sushkova S et al (2020) Anatomical and ultrastructural responses of Hordeum sativum to the soil spiked by copper. Environ Geochem Health 42(1):45–58

Moffat AS (2002) Finding new ways to protect drought-stricken plants. Science 296:1226–1229

Moharem M, Elkhatib E, Mesalem M (2019) Remediation of chromium and mercury polluted calcareous soils using nanoparticles: Sorption–desorption kinetics, speciation and fractionation. Environ Res 170:366–373

Movafeghi A, Khataee A, Abedi M, Tarrahi R, Dadpour M, Vafaei F (2018) Effects of TiO2 nanoparticles on the aquatic plant Spirodela polyrrhiza: Evaluation of growth parameters, pigment contents and antioxidant enzyme activities. J Environ Sci 64:130–138

Nejatzadeh F (2021) Effect of silver nanoparticles on salt tolerance of Satureja hortensis l. During in vitro and in vivo germination tests. Heliyon 7(2):e05981

Noman M, Shahid M, Ahmed T, Tahir M, Naqqash T, Muhammad S, Aslam Z (2020) Green copper nanoparticles from a native Klebsiella pneumoniae strain alleviated oxidative stress impairment of wheat plants by reducing the chromium bioavailability and increasing the growth. Ecotoxicol Environ Saf 192:110303

Noman M, Ahmed T, Hussain S, Niazi MBK, Shahid M, Song F (2020b) Biogenic copper nanoparticles synthesized by using a copper-resistant strain Shigella flexneri SNT22 reduced the translocation of cadmium from soil to wheat plants. J Hazard Mater 398:1231

Parisi C, Vigani M, Rodríguez-Cerezo E (2015) Agricultural nanotechnologies: what are the current possibilities? Nano Today 10(2):124–127

Raja V, Lakshmi RV, Sekar CP, Chidambaram S, Neelakantan MA (2021) Health risk assessment of heavy metals in groundwater of industrial township Virudhunagar, Tamil Nadu, India. Arch Environ Contam Toxicol 80(1):144–163

Rajput VD, Singh RK, Verma KK, Sharma L, Quiroz-Figueroa FR, Meena M et al (2021) Recent developments in enzymatic antioxidant defence mechanism in plants with special reference to abiotic stress. Biology 10(4):267

Rizhsky L, Liang H, Shuman J, Shulaev V, Davletova S, Mittler R (2004) When defense pathways collide. The response of Arabidopsis to a combination of drought and heat stress. Plant Physiol 134(4):1683–1696

Rizwan M, Ali S, Ali B, Adrees M, Arshad M, Hussain A et al (2019) Zinc and iron oxide nanoparticles improved the plant growth and reduced the oxidative stress and cadmium concentration in wheat. Chemosphere 214:269–277

Rossi L, Zhang W, Schwab AP, Ma X (2017) Uptake, accumulation, and in planta distribution of coexisting cerium oxide nanoparticles and cadmium in Glycine max (L.) Merr. Environ Sci Technol 51(21):12,815–12,824

Sagghatol-Islami MC (2010) Effect of salinity on germination of three species of medicinal herbs, Chicory and Artichoke. Iran J Agric Res 8:818–823

Sall ML, Diaw AKD, Gningue-Sall D, Efremova Aaron S, Aaron JJ (2020) Toxic heavy metals: impact on the environment and human health, and treatment with conducting organic polymers, a review. Environ Sci Pollut Res 27:29,927–29,942

Sebastian A, Nangia A, Prasad MNV (2018) A green synthetic route to phenolics-fabricated magnetite nanoparticles from coconut husk extract: implications to treat metal contaminated water and heavy metal stress in Oryza sativa L. J Clean Prod 174:355–366

Sebastian A, Nangia A, Prasad MNV (2019) Cadmium and sodium adsorption properties of magnetite nanoparticles synthesized from Hevea brasiliensis Muell. Arg. bark: Relevance in amelioration of metal stress in rice. J Hazard Mater 371:261–272

Sedghi M, Hadi M, Toluie SG (2013) Effect of nano zinc oxide on the germination parameters of soybean seeds under drought stress. Annal West Univ Timisoara Ser Biol 16(2):73

Seleiman MF, Alotaibi MA, Alhammad BA, Alharbi BM, Refay Y, Badawy SA (2020) Effects of ZnO nanoparticles and biochar of rice straw and cow manure on characteristics of contaminated soil and sunflower productivity, oil quality, and heavy metals uptake. Agronomy 10(6):790

Seleiman MF, Almutairi KF, Alotaibi M, Shami A, Alhammad BA, Battaglia ML (2021) Nano-fertilization as an emerging fertilization technique: why can modern agriculture benefit from its use? Plan Theory 10(1):2

Servin A, Elmer W, Mukherjee A, De la Torre-Roche R, Hamdi H, White JC et al (2015) A review of the use of engineered nanomaterials to suppress plant disease and enhance crop yield. J Nanopart Res 17(2):1–21

Seydmohammadi Z, Roein Z, Rezvanipour S (2020) Accelerating the growth and flowering of Eustoma grandiflorum by foliar application of nano-ZnO and nano-CaCO 3. Plant Physiol Rep 25(1):140–148

Shahid MA, Pervez MA, Balal RM, Ahmad R, Ayyub CM, Abbas T, Akhtar N (2011) Salt stress effects on some morphological and physiological characteristics of okra (Abelmoschus esculentus L.). Soil Environ 30(1)

Shang Y, Hasan M, Ahammed GJ, Li M, Yin H, Zhou J (2019) Applications of nanotechnology in plant growth and crop protection: a review. Molecules 24(14):2558

Siddiqui H, Ahmed KBM, Sami F, Hayat S (2020) Silicon nanoparticles and plants: current knowledge and future perspectives. Sustain Agric Rev 41:129–142

Singh D, Sillu D, Kumar A, Agnihotri S (2021) Dual nanozyme characteristics of iron oxide nanoparticles alleviate salinity stress and promote the growth of an agroforestry tree, Eucalyptus tereticornis Sm. Environ Sci Nano 8(5):1308–1325

Sonam S, Anirudha R, Subhash C (2014) Effect of short-term salt stress on chlorophyll content, protein and activities of catalase and ascorbate peroxidase enzymes in pearl millet. Am J Plant Physiol 9(1):32–37

Sun D, Hussain HI, Yi Z, Rookes JE, Kong L, Cahill DM (2018) Delivery of abscisic acid to plants using glutathione responsive mesoporous silica nanoparticles. J Nanosci Nanotechnol 18(3): 1615–1625

Suzuki N, Rizhsky L, Liang H, Shuman J, Shulaev V, Mittler R (2005) Enhanced tolerance to environmental stress in transgenic plants expressing the transcriptional coactivator multiprotein bridging factor 1c. Plant Physiol 139(3):1313–1322

Taran N, Storozhenko V, Svietlova N, Batsmanova L, Shvartau V, Kovalenko M (2017) Effect of zinc and copper nanoparticles on drought resistance of wheat seedlings. Nanoscale Res Lett 12(1):1–6

Torabian S, Farhangi-Abriz S, Zahedi M (2018) Efficacy of FeSO4 nano formulations on osmolytes and antioxidative enzymes of sunflower under salt stress. Indian J Plant Physiol 23(2):305–315

Verma KK, Anas M, Chen Z, Rajput VD, Malviya MK, Verma CL et al (2020) Silicon supply improves leaf gas exchange, antioxidant defense system and growth in Saccharum officinarum responsive to water limitation. Plan Theory 9(8):1032

Wang Z, Yue L, Dhankher OP, Xing B (2020) Nano-enabled improvements of growth and nutritional quality in food plants driven by rhizosphere processes. Environ Int 142:105,831

Wang K, Wang Y, Wan Y, Mi Z, Wang Q, Wang Q, Li H (2021) The fate of arsenic in rice plants (Oryza sativa L.): influence of different forms of selenium. Chemosphere 264:128417

Yang C, Zhang X, Liu G, Zhang J, Bao M, Zhou Z (2013) Progress on the structure and physiological functions of apoplastic barriers in root. Bull Bot Res 33(1):114–119

Zafar S, Akhtar M, Perveen S, Hasnain Z, Khalil A (2020) Attenuating the adverse aspects of water stress on wheat genotypes by foliar spray of melatonin and indole-3-acetic acid. Physiol Mol Biol Plants 26(9):1751–1762

Zamora-Ledezma C, Negrete-Bolagay D, Figueroa F, Zamora-Ledezma E, Ni M, Alexis F, Guerrero VH (2021) Heavy metal water pollution: a fresh look about hazards, novel and conventional remediation methods. Environ Technol Innov:101504

Zhang Z, Ali S, Zhang T, Wang W, Xie L (2020) Identification, evolutionary and expression analysis of PYL-PP2C-SnRK2s gene families in soybean. Plan Theory 9(10):1356

Zhao Q, Bai J, Gao Y, Zhang G, Lu Q, Jia J (2021) Heavy metal contamination in soils from freshwater wetlands to salt marshes in the Yellow River Estuary, China. Sci Total Environ 774: 145,072

Zhou P, Adeel M, Shakoor N, Guo M, Hao Y, Azeem I et al (2021) Application of nanoparticles alleviates heavy metals stress and promotes plant growth: an overview. Nanomaterials 11(1):26

Climate Change Mitigation Through Agroforestry: Socioeconomic and Environmental Consequences

Jitendra Kumar and Ranjana Yadav

Abstract Regardless of the overall achievement of concentrated cultivating frameworks advanced in many pieces of the world, land pressure and climate change are harming farming systems in developing countries, both of which constitute a threat to food production. Planting sampling of poplar and eucalyptus in tropical places like India is a well-managed and profitable enterprise. Many farm lands in South Asia now include rapidly growing poplar as a major component of woodlots and shelterbelts. The study of Eucalyptus and Poplar as important plant species in agroforestry systems to counteract climate change and improve social, economic and environmental status is renewed in this chapter.

Keywords Agroforestry · Climate change · Eucalyptus · Poplar

Introduction

Climate change has evolved over the last two decades from a dispute about whether the earth is indeed warming to a greater focus on how to reduce and adapt to its effects. Environmental change is characterised as an adjustment of climate that is ascribed either straightforwardly or in a roundabout way to human movement. Natural climate variation recorded across similar time periods, as well as changes in the structure of the worldwide environment. Because of environmental change, societal structures, economic sectors and ecosystems are all at risk.

Agriculture is the human-driven business that is most vulnerable in response of climate change, and however, agroforestry can help small-holder agriculturalists adjust to environmental change, it also contributes to lowering atmospheric greenhouse gas (GHG) accretion (Smith et al. 2008). Regardless of the overall achievement of concentrated cultivating frameworks advanced in many pieces of the world,

J. Kumar
Department of Biosciences, UIBT, Chandigarh University, Mohali, Punjab, India

R. Yadav (✉)
Indian Plywood Industries Research & Training Institute, Centre, Mohali, Punjab, India

© The Author(s), under exclusive license to Springer Nature Singapore Pte Ltd. 2022

A. Vaishnav et al. (eds.), *Plant Stress Mitigators*,
https://doi.org/10.1007/978-981-16-7759-5_26

land pressure and climate change are significantly impacting agricultural systems in emerging countries, both of which constitute a threat to food supply (Singh and Pandey 2011).

Agriculture and forestry technology are combined in agroforestry to generate more interwoven, diversified, creative, economical, clean and durable habitat regimes. Because of its ability to trap carbon, agroforestry has been identified as an environmental change protector. It has numerous advantages to agriculturalist, assisting in climate change adaptation.

Agroforestry improves farmers' adaptive capacity to counteract climate change effects by developing more resilient agricultural systems and diversifying income streams (Udawatta and Godsey, 2010). Agroforestry contributes to food security by supplying farmers with a number of products and benefits, such as food, fodder and shade for livestock, as well as lumber and renewable wood energy (Young et al. 1998, Casarim et al. 2010). Agroforestry adds essentially to carbon storage, gives a scope of manging environment benefits and further boosts biodiversity when contrasted with customery agriculture (Kay et al. 2019, Pandey 2007, Nair 2004, Nair et al. 2009a, b, Murthy et al. 2013a, b, Newaj and Dhyani 2008).

Planting crops like poplar and eucalyptus in tropical regions like India is a well-managed and profitable activity. Many farm lands in south Asia now include rapidly growing poplar as a major component of woodlots and shelterbelts. Eucalyptus plantations provide farmers with more net revenue per hectare per year than about 60–70% of agriculture crops, and can help to boost farm profitability in the future (Agarwal and Saxena 2017). The most frequent tree species in government plantations, community programmes and private woodlots is Eucalyptus. This tree species thrives on poor soil and develops more quickly than most native tree species. After 1980, the *Populus deltoides* (poplar) was effectively incorporated into agroforestry and was widely planted in farmlands of Uttar Pradesh (UP), Haryana and Punjab. Poplars are fast-growing trees that recycle nutrients quickly because they lose a large number of leaves that degrade quickly. Poplar trees were used in Agri-silviculture systems, in which an agriculture crop is cultivated within rows of trees and on the field boundary (Ajit et al. 2011, Chauhan et al. 2010).

Environmental Effects of Eucalyptus

The main benefit of Eucalyptus is that it might potentially replace indigenous species as a source of fuelwood, preventing further damage of native forests. Eucalyptus planting affects land use in temperate and tropical areas, modifying microclimate, soil nutrients, drawing in seed dispersers and reducing rival grasslands (Lemenih and Teketay 2005). Farming is a different type of habitat for indigenous species of plants to regrow (Senbeta et al. 2002). Eucalyptus plantings tend to operate as foster ecosystems, fostering native woody species recruitment, development and progression (Senbeta and Teketay 2000, Yirdaw and Luukkanen 2003). Eucalyptus plantations that are well-managed can help to prevent soil erosion, but their effectiveness is dependent on a variety of conditions, including rain force, soil conditions, slant point and length, as well as the presence of plant and litter cover.

Eucalyptus is also noted for its coppicing abilities; when let to re-grow, it normally develops a dense and productive stand (Kidanu 2004). It may grow in a variety of conditions, including deforestation, wetlands, infertile and fatigued soil and desert area (Pohjonen and Pukkala 1990). Around 900 species of Eucalyptus can be observed everywhere in all over the world. More than 300 species of eucalyptus incorporate explosive oils, which can be located in different pieces of the plant relying on the species; however, are most plentiful in the leaves (Pino et al. 2002, Getahun 2002). Eucalyptus oil has various applications: medicinal, industrial and perfumery/flavouring.

Agroforestry is a sustainable land management technique that involves incorporating trees and shrubs into crop and livestock farming schemes. Eucalyptus plantation could be employed in agroforestry schemes, such as estates, field fences, agricultural trees and roadside plantings. Plantation gives a variety of benefits, including fuel wood and construction materials, and hence contributes significantly to better rural living (Palma et al. 2007, Singh 1993). Eucalyptus trees are being planted in field boundaries in India. Intercropping with *Eucalyptus camaldulensis* strips boosted cotton productivity under irrigation in Pakistan. This is because it can aid in the reduction of salinity. Single *E. camaldulensis* trees were planted in farm fences in Colombia for the wood product business. *Eucalyptus* trees were planted on farm fences in Ethiopia's highland Vertosols for a variety of reasons including fuelwood, lumber, as a commercial harvest, edge delineation, railing, soil and water management and enhancing harvest productivity (Kidanu 2004).

Climate change is one of the world's most pressing issues today. Climate change, mostly as a result of human activity, is caused by an increase in the concentration of carbon dioxide and other greenhouse gases in the atmosphere. Climate change will have significant implications for the environment, society, economy and politics. One of the most essential ways to preventing climate change is to plant trees. Eucalyptus is one of the greatest agroforestry systems due to rapid growth and fixing of more CO_2 through photosynthesis. Eucalyptus trees produce higher biomass in comparison to other tree species. It is well-established carbon sequestration proportional to biomass production.

Some researchers have shown that interplanting Eucalyptus species with nitrogen-fixing species such as Acacia and Albizia species can help with carbon sequestration. *Eucalyptus* plantation can aid in the mitigation of climate change while simultaneously generating revenue from carbon trading. Farmers, who are living in a variety of soil and climatic conditions, can cultivate eucalyptus (Hailemicael 2012).

Farmers prioritise planting eucalyptus on their land for a spread of reasons, together with the provision of seeds regionally and also the inconvertible fact that they are doing not any special treatment to stay for a long duration, its coppicing ability and rapid secondary growth (Zerfu 2000, Jose 2009), wide range adaptability in degraded land, swampy areas, unfertile and exhausted soil and dry areas (Pohjonen and Pukkala 1990). Due to rapid growth, farmers prefer eucalyptus to fulfil the increasing demand for wood-based industry, increasing revenue and deal with the ever-increasing shortage of wood commodities. Eucalyptus has a social value because of its widespread use, which offers work possibilities (Nair 2011, Buck et al. 1999).

Ecological Impacts of Eucalyptus

The problem is low and absent in regions where eucalyptus trees are intercropped with other species, whereas soil erosion and gully formation are more prevalent in areas where exclusively eucalyptus trees are grown (Zerga 2015, Jose and Gordon 2008, Murthy et al. 2016). The majority of eucalypts are poor erosion control trees. In a eucalyptus plantation region, ground cover growth and debris design were insufficient to prevent run-off. It is a tree with a dense canopy and little debris that grows quickly and has a heavy crown (Molua 2005, Stern 2007).

Eucalyptus Plantation's Effects on Water Resources

The primary criticisms levelled at Eucalyptus plantings in this regard are that they drain water resources and do not manage water flow as well as the native plants that they occasionally replace on slope catchments (FAO 1985). Eucalyptus plantations can, in theory, have an impact on any of these by changing the quantities. The goal of the plantings, as well as a balancing of the numerous costs and advantages in each case, determines whether these adjustments are useful or not (FAO 1985). The effects of Eucalyptus plantation on water have got a lot of press. Eucalyptus trees, more than any other tree, are known to absorb more water from the soil. According to the findings in Kenya, Eucalyptus species require more water than Pinus species (Dye and Bosch 2000), especially during early stage. Eucalyptus planting is commonly criticised for decreasing the ground water level and disrupting the hydrological cycle. However, there are limited empirical studies on eucalyptus water use and its direct impact on close agricultural productivity. Ground plants of eucalyptus species are limited in dry locations (FAO 1985) because of root competition and possibly allelopathic effects. The majority of people believe that eucalyptus consumes significantly more water than any other tree or agricultural crop. According to the studies, eucalyptus is highly effective water user. Davidson (1989) claims that eucalyptus generates greater biomass for the same amount of water used, which is economically viable and acceptable. In comparison to cotton/coffee/banana (3200), sunflower (2400), field pea (2000), cow pea (1667), soybeans (1430), potato (1000), sorghum (1000) and maize (1000) litres of water per kg of biomass produced, most eucalyptus species require 785 L of water/kg biomass produced, demonstrating that eucalyptus species are efficient water users. Large eucalyptus wood, on the hand, can greatly increase a Catchment's water supply while also raising water tables in flat areas downstream. Eucalyptus trees have a water-use advantage over other plants due to their high-water requirements and deep root systems, which can be problematic if eucalyptus trees are planted in arid places (Janger and Pender 2003).

One of the criticisms levelled about Eucalyptus plantations, is that they may change the local atmosphere temperature. This is owing to their high rate of evapotranspiration, which could result in a decrease in the water table. This rapid loss of soil water is supposed to have a negative impact on local rainfall levels, potentially leading to desertification. They include, among other things, temperature decrease, CO_2 fixation and shading. Climatic shifts happened when Eucalyptus plantations were established in Acacia forest regions in Senegal (FAO 2011). The impact of eucalyptus trees on microclimate is well-documented (Bernhard-Reversat 1988). The magnitude of these effects is proportional to the amount of leaf surface carried by the trees in ratio to the area of the ground covered. Average air temperatures are minimised, air and surface soil temperature extremes are also minimised and surface air humidity is higher in regions with tress than in place without trees (FAO 2011).

Poplar

One of the most common and favoured agroforestry species is poplar (*Populus deltoides*) and has been extensively planted in farmlands in Uttar Pradesh (UP), Haryana and Punjab after 1980. Poplar trees are planted in blocks, rows and on field bunds, while intercrops like wheat, paddy, sugarcane, maize, mustard, pulses and other crops are produced alongside the trees. Poplar is only grown on irrigated terrain since it has a high-water requirement due to its rapid growth. Though the farmlands vary in size, all values have been reported on a per hectare basis for the purposes of computations.

Carbon Sequestration Potential

Carbon capture and storage is the act of removing extra carbon dioxide from the atmosphere and storing it in another ‘reservoirs’. Agroforestry for sequestering carbon is attractive because:

(i) It sequesters carbon in plants and soils depending on the preconversion soil C.
(ii) Because more land is used actively for agricultural production, it reduces the need of shifting cultivation.
(iii) Agroforestry wood products can be used as a substitute for similar goods that are not collected sustainably from forest resources.
(iv) Because high-nutritive-value fodder species can serve to strengthen the diets of methane-producing ruminants while also sequestering carbon, agroforestry

practices may have dual mitigation advantages (Albrecht and Kandji 2003, Kumar and Nair 2011, Pandey 2002).

When wood products are not included, the interventions on poplar block and bund plantings have a sequestration potential of 1.33 ton Carbon per ha per year and 1.05 ton Carbon per ha per year, correspondingly, and 2.41 ton Carbon per ha per year and 1.80 ton Carbon per ha per year, respectively, when wood products are included (Shibu and Sougata 2012, Swamy and Puri 2005). CO_2FIX was utilised in India to assess the ability of agroforestry systems in the Indo-Gangetic plains to store and sequester carbon (Ajit et al. 2011), including sal (*Shorea robusta*), eucalyptus (*Eucalyptus tereticornis*), poplar (*Populus deltoides*) and teak (*Tectona grandis*) (Kaul et al. 2010, Chauhan et al. 2009). Simulated tests also suggest that retaining the rotation age at 9 years and continuing poplar cultivation for a long period will improve the soil C store and produce more biomass than a 6-year rotation with current management practices (Panwar et al. 2017).

Poplar-based agroforestry in Saharanpur (UP) and Yamuna Nagar (Haryana) retains 27–32 tons per hectare C in the overall implementation and 66–83 tons per hectare C in the Farming scheme more than a 7-year rotation period. According to Punjabi studies, the carbon content of poplar lumber throughout a seven-year cycle could be 23.57 tons per hectare, with roots, leaves, even tree bark all contributing equally (Rizvi et al. 2011).

Adaptation and Mitigation in Context of Agroforestry

Agroforestry holds a lot of promise when it comes to climate change adaptation and mitigation. We frequently discuss agroforestry's adaptation and mitigation measures in relation to global warming, but we must first define these terms. Agroforestry offers a unique chance to reconcile climate change reduction and adaptation goals (Sathaye and Ravindranath 1998, Kumar et al. 2019, Basu 2014).

As climate change and variability become a science, the job of agroforestry in lowering the susceptibility of farming schemes and the countryside populations who rely on them for their maintenance to climate change and inconsistency must be addressed (UNFCCC 2013).

Adaptation strategies of agroforestry systems are below mentioned:

1. Increased moisture and reduced warmth improve microclimate.
2. Shade effect contributes to reducing evaporation.
3. High species diversity reduces pest problems.
4. Preventing droughts in farming crops.

As a result, agroforestry is primarily composed of perennials plants with a high ability to withstand rough atmospheric conditions and, via their coping mechanisms, make the overall system sustainable. However, several agroforestry systems' adaptive techniques are briefly discussed.

Climate change adaptation is now unavoidable. Agroforestry research as an environment change adaptation and a barrier against environment inconsistency is still in its early stages. Most common strategy for agroforestry to qualify as a climate change adaptation is to broadening workflows and to increase the viability of smallholder agricultural arrangements (Rao et al. 2007, Pratap and Abhishek 2018). The function of agroforestry in lowering agroecosystems' vulnerability to climate change and variability and the people who rely on them needs to be better understood (Schoreneberger et al. 2012, Anwar et al. 2013, Kumar et al. 2011).

Mitigation is a term used to describe a remedy for a bad effect that has occurred in a restricted way, such as an increase in percentage of CO_2. Mitigation is the process of reducing GHG emissions by allowing carbon sequestration to occur (Rizvi et al. 2019). In layman's terms, adaptation can be done locally, while mitigation must be done internationally. Mitigation measures improve soil organic matter, which enhance soil fertility and quality, increasing crop output in agroforestry systems while also improving soil adaptive ability, making it a 'win-win' option (Smit and Skinner 2002, Verma et al. 2008, Yirefu et al. 2019).

Conclusion

Trees along with plantation crops have a lot of promise for sequestering carbon in the atmosphere and reducing global warming. Our agricultural systems, residences, institutions, marketplaces, parks and other public spaces all need to be effectively integrated with trees and plantations. Trees fill in as extra kind revenue at the hour of harvest disappointments. Agroforestry provides benefits to local people in multiple aspects (Fig. 1).

As a result, agroforestry systems exist in different forms in various regions of the world, and they exist as main function in combating negative effects of climate change by enhancing different tree crops, which enhance the carbon storage capacity in comparison to solely cultivating agricultural crops. Agroforestry systems can also help to reduce CO_2 emissions by eliminating the burning of forest-based fuelwood

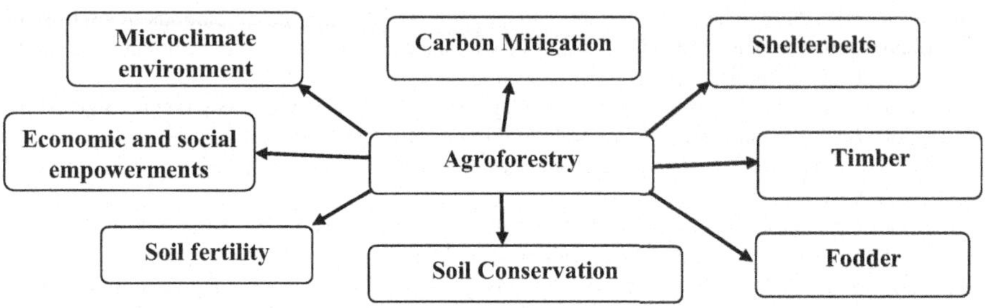

Fig. 1 Benefit of agroforestry (poplar and eucalyptus)

and preserving lands. The best way of promoting mitigation and adaptation interaction in the face of environment change is agroforestry.

References

Agarwal S, Saxena AK (2017) The puzzle of forest productivity: are forest development corporations solving it right? Centre for Science and Environment, New Delhi, p 47

Ajit DDK, Chaturvedi OP, Jabeem M, Dhyani SK (2011) Predictive models for dry weight estimation of above and below ground biomass components of Populus deltoids in India: Development and comparative diagnosis. Biomass Bioenergy 35:1145–1152

Albrecht A, Kandji ST (2003) Carbon sequestration in tropical agroforestry systems. Agric Ecosyst Environ 99:15–27

Anwar MR, Liu DE, Macadam I, Kelly G (2013) Adapting agriculture to climate change: a review. Theor Appl Climatol 113:225–245

Basu JP (2014) Agroforestry, climate change mitigation and livelihood security in India. N Z J For Sci 44(Suppl 1):S11

Bernhard-Reversat F (1988) Soil nitrogen mineralization under a Eucalyptus plantation and a natural *Acacia* Forest in Senegal. For Ecol Manag 23:233–244

Buck L, Lassoie J, Fernandes E (1999) Agro forestry in sustainable agricultural systems. CRC Press, Boca Raton

Casarim FM, Harris NL, Brown S (2010) USAID forest C calculator: data and equations for the agro forestry tool. Win Rock Int

Chauhan SK, Gupta N, Ritu, Yadav S, Chauhan R (2009) Biomass and carbon allocation in different parts of agroforestry tree species. Indian For 135(7):981–993

Chauhan SK, Sharma SC, Chauhan R, Gupta R (2010) Accounting poplar and wheat productivity for carbon sequestration Agri-silvicultural system. Indian Forester 136:1174–1182

Davidson J (1989) The eucalypt dilemma: argument for and against eucalypt planting in Ethiopia. The Forestry Research Center, Addis Ababa. African Sun Publishing, Oakland

Dye PJ, Bosch JM (2000) Sustained water yield in afforested catchments-the south African experience. In: Von Gadow K, Pukkala T, Tome M (eds) Sustainable forest management. Kluwer Academic Publishers, Dordrecht, pp 99–120

FAO (1985) The ecological effects of eucalyptus. FAO forestry paper. FAO, Rome

FAO (2011) Eucalyptus in East Africa: socio-economic and environmental issues. Planted Forests and Trees Working Paper, Forest Management Team, Forest Management Division. FAO, Rome

Getahun A (2002) Eucalyptus farming in Ethiopia: the case for eucalyptus woodlots in the Amhara region. Bahir Dar Conference Proceedings, Ethiopian Society of Soil Science, Bahir Dar, pp 137–153

Hailemicael BK (2012) The contribution of Eucalyptus woodlots to the livelihoods of small-scale farmers in tropical and subtropical countries with special reference to the Ethiopian highlands. Master Thesis. Swedish University of Agricultural Sciences, 43 pp

Janger P, Pender J (2003) The role of trees for sustainable management of less favoured lands: the case of Eucalyptus in Ethiopia. Forest Policy Economics 5:83–95

Jose S (2009) Agroforestry for ecosystem services and environmental benefits: an overview. Agrofor Syst 76(1):1–10

Jose S, Gordon AM (2008) Ecological knowledge and agro forestry design: an introduction. In: toward agro forestry design. Adv Agro Forestry 4:3–9

Kaul M, Mohren GMJ, Dadhwal VK (2010) Carbon storage and sequestration potential of selected tree species in India. Mitigat Adaptat Strategy Global Change 15:489–510

Kay S, Rega C, Moreno G, den Herder M, Palma JH, Borek R, Crous-Duran J, Freese D, Giannitsopoulos M, Graves A, Jäger M, Lamersdorf N, Memedemin D, Mosquera-Losada R, Pantera A, Paracchini ML, Paris P, Roces-Díaz JV, Rolo V, Rosati A, Sandor M, Smith J, Szerencsits E, Varga A, Viaud V, Wawer R, Burgess PJ, Herzog F (2019) Agroforestry creates carbon sinks whilst enhancing the environment in agricultural landscapes in Europe. Land Use Policy 83:581–593

Kidanu S (2004) Using eucalyptus for soil and water conservation in the highland vertisols of Ethiopia. Ph. D. Thesis. Wageningen University, Wageningen, Netherlands, 198 pp

Kumar BM, Nair PKR (2011) Carbon sequestration potential of agroforestry systems: opportunities and challenges. Advances in agroforestry. Springer Science, The Netherlands, 307 p

Kumar M, Ramachandran B, Nair PK (2011) Carbon sequestration potential of agroforestry systems: opportunities and challenges, vol 8. Springer Science & Business Media

Kumar R, Pandey A, Rana R, Yadav A (2019) Climate change and mitigation through agroforestry. Int J Curr Microbiol App Sci 8(6):1662–1667

Lemenih M, Teketay D (2005) Effect of prior land use on the recolonization of native woody species under plantation forests in the highlands of Ethiopia. For Ecol Manag 218:60–73

Molua EL (2005) The economics of tropical agroforestry systems: the case of agroforestry farms in Cameroon. Forest Policy Econ 7:199–211

Murthy IK, Gupta M, Tomar S, Munsi M, Tiwari R, Hedge GT, Ravindranath NH (2013a) C sequestration potential of agro forestry systems in India. J Earth Sci Climate Change 4:131

Murthy IK, Gupta M, Tomar S, Munsi M, Tiwari R, Hegde GT, Ravindranath NH (2013b) Carbon sequestration potential of agroforestry systems in India. J Earth Sci Climate Change 4(1):1–7

Murthy IK, Dutta S, Varghese V, Joshi PP, Kumar P (2016) Impact of agroforestry systems on ecological and socio-economic systems: a review. Global J Sci Front Res 16(5)

Nair PKR (2004) Agro forestry: trees in support of sustainable agriculture. In: Hillel H, Rosenzweig C, Powlson D (eds) Encyclopedia of soils in the environment. Elsevier, London, pp 35–44

Nair PKR (2011) Agro forestry systems and environmental quality: introduction. J Enviro Qual 40(3):784–790

Nair PKR, Nair VD, Kumar BM, Haile SG (2009a) Soil C sequestration in tropical agro forestry systems: a Swamy SL, Puri feasibility appraisal. Environ Sci Policy 12(8):1099–1111

Nair PKR, Kumar B, Nair VD (2009b) Agro forestry as a strategy for C sequestration. J Plant Nutr Soil Sci 172:10–23

Newaj R, Dhyani SK (2008) Agro forestry for C sequestration: scope and present status. Indian J Agro Forestry 10:1–9

Palma JHN, Graves AR, Bunce RGH, Burgess P, de Filippi R, Keesman KJ, Van Keulen, Herman, Liagre F, Mayus M, Moreno G, Reisner Y, Herzog F. Modeling environmental benefits of silvoarable agro forestry in Europe. Agric Ecosyst Environ, 2007; 119(3–4):320–334

Pandey DN (2002) Carbon sequestration in agroforestry systems. Clim Pol 2:367–377

Pandey DN (2007) Multifunctional agroforestry systems in India. Curr Sci 92(4):455–463

Panwar P, Chauhan S, Kaushal R, Das DK, Arora AG, Chaturvedi OP, Jain AK, Chaturvedi S (2017) Carbon sequestration potential of poplar-based agroforestry using the CO_2 FIX model in the indo-Gangetic region of India. Trop Ecol 58(2):439–447

Pino JA, Marbot R, Quert R, Garcia H (2002) Study of essential oils of Eucalyptus resinifera, E. tereticornis and Corymbia maculata (Hook.) grown in Cuba. Flavour Frag J 17:1–14

Pohjonen V, Pukkala T (1990) Eucalyptus globules in Ethiopia forestry. For Ecol Manag 36:19–31

Pratap T, Abhishek R (2018) Role of agroforestry in climate change mitigation. J Pharmacognosy Phytochem 7(2):241–243

Rao KPC, Verchot LV, Laarman J (2007) Adaptation to climate change through sustainable management and development of agroforestry systems. J SAT Agric Res 4(1):1–30

Rizvi RH, Dhyani SK, Yadav RS, Singh R (2011) Biomass production and carbon storage potential of poplar agroforestry systems in Yamuna Nagar and Saharanpur districts of North-Western India. Curr Sci 100:736–742

Rizvi RH, Newaj R, Chaturvedi OP, Prasad R, Handa AK, Alam B (2019) Carbon sequestration and CO_2 absorption by agroforestry systems: an assessment for central plateau and hill region of India. J Earth Syst Sci 128:56

Sathaye JA, Ravindranath NH (1998) Climate change mitigation in the energy and forestry sectors of developing countries. Annu Rev Energy Environ 23:387–437

Schoreneberger M, Bendrub G, Gooijer H, Soolanayakanahal R, Sauer T, Brandle J, Zhan X, Cuer D (2012) Branching out: agroforestry as a climate change mitigation and adaptation for agriculture. J Soil Water Conserv 67(5):128–136

Senbeta F, Teketay D (2000) Regeneration of indigenous woody species under the canopies of tree plantations in Central Ethiopia. Trop Ecol 42:175–185

Senbeta F, Teketay D, Naslund AB (2002) Native woody species regeneration in exotic tree plantations at Munessa-Shashemene Forest, southern Ethiopia. N Fores 24:131–145

Shibu J, Sougata B (2012) Agroforestry for biomass production and carbon sequestration: an overview. Agrofor Syst 86(2):105–111

Singh GB (1993) Role of agroforestry in improving the environment. Indian Fmg 33(3):15–19

Singh VS, Pandey DN (2011) Multifunctional agroforestry systems in India: Science-based policy options. RSPCB Occasional Paper No 4/2011, 2–35

Smit B, Skinner MW (2002) Adaptation options in agriculture to climate change: a typology. Mitig Adapt Strateg Glob Chang 7:85–114

Smith P, Martino D, Cai Z, Gwary D, Janzen HH, Kumar P, McCarl B, Ogle S, O'Mara F, Rice C, Scholes RJ, Sirotenko O, Howden M, McAllister T, Pan G, Romanenkov V, Schneider U, Towprayoon S, Wattenbach M, Smith JU (2008) Greenhouse gas mitigation in agriculture. Philos Trans Roy Soc B Biol Sci 363:789–813

Stern N (2007) The economics of climate change: the Stern review. Cambridge University Press, New York

Swamy SL, Puri S (2005) Biomass production and C-sequestration of Gmelina arborea in plantation and agro forestry system in India. Agrofor Syst 64:181–195

Udawatta RP, Godsey LD (2010) Agro forestry comes of age: putting science into practice. Agro Forest Syst 79:1–4

UNFCCC (2013) The nineteenth session of the Conference of the Parties (COP 19). PGE Narodowy

Verma KS, Kumar S, Bhardwaj DR (2008) Soil organic carbon stocks and carbon sequestration potential of agroforestry systems in H.P. Himalaya region of India. J Tree Sci 27(1):14–27

Yirdaw E, Luukkanen O (2003) Indegenous woody species diversity in Eucalyptus globules (Labill.) plantations in the Ethiopian highlands. Biod Conserv 23:567–582

Yirefu T, Yenenesh H, Zemedkun S (2019) Potential of agroforestry for climate change mitigation through carbon sequestration: review paper. Agri Res Tech Open Access J 22(3):556,196

Young A, Menz KM, Muraya P et al (1998) SCUAF—Version 4: a model to estimate soil changes under agriculture, agroforestry and forestry. Australian Centre for International Agricultural Research (ACIAR), Canberra, Australia, p 49

Zerfu H 2000 Ecological impact evaluation of Eucalyptus plantation in comparison with agricultural and grazing land-use types in the highlands of Ethiopia. Ph. D. Thesis. Institute of Forest Ecology, Vienna University of Agricultural Science, Vienna

Zerga B (2015) Ecological impacts of Eucalyptus plantation in Eza Wereda, Ethiopia. Int Inv J Agric Soil Sci 3(4):47–51

An Overview on Soybean Mosaic Virus and Its Management

Mansi Srivastava, Uma Bhardwaj, Nisha Choudhary, Rajarshi Kumar Gaur, and Rakesh Kumar Verma

Abstract Viruses have become one of the major concerns regarding the agricultural loss globally. One such virus is *Soybean mosaic virus* (SMV) of genus Potyvirus that infects the soybean all around the world. Agriculture plays an important role in Indian economy and also 70% of rural households depend on it in India. It also contributes employment to over 60% population, thus agricultural loss due to viral infection is a matter of concern. The genetic structure and evolution of virus are based on: Recombination, mutation, genetic drift, natural selection, and migration. So, it has become important to control the virus, and research about the genetic variation of viruses will help to prepare the diagnostic tool, learn about the epidemiology, spread, and elimination method of virus. In this chapter, we summarize the major factors responsible for virus evolution and control and the global and Indian status of soybean mosaic virus.

Keywords Potyvirus · Genome complexity · Mosaic virus · Disease management

Introduction

Virus diseases are of utmost concern in regard with the futile agriculture globally (Mumford et al. 2016). Agribusiness, human populace, and global climate change are some of the major reasons for the outburst of the harmful viruses. It was estimated that plant viral diseases effect the worldwide capital by >US$30 billion annually (Jones and Naidu 2019). Plant disease came into the knowledge of the man in the BC era as written down by various countries in their ceremonies and proposals made to farmers. The Japanese Empress, Kokan is supposed to be the first to record

M. Srivastava · U. Bhardwaj · N. Choudhary · R. K. Verma (✉)
Dept of Biosciences, School of Liberal Arts and Sciences, Mody University of Science and Technology, Lakshmangarh, Sikar, Rajasthan, India

R. K. Gaur
Department of Biotechnology, D.D.U. Gorakhpur University, Gorakhpur, Uttar Pradesh, India

© The Author(s), under exclusive license to Springer Nature Singapore Pte Ltd. 2022
A. Vaishnav et al. (eds.), *Plant Stress Mitigators*,
https://doi.org/10.1007/978-981-16-7759-5_27

the plant virus disease by mentioning about the yellowing leaves of *Eupatorium lindleyanum* in a poem noted down in 752 AD (Nayudu 2008).

Viruses are holoparasite that depends on the host to complete their life cycle. They have their genetic material either DNA or RNA enclosed in an outer covering made up of protein. Most of the plant viruses have RNA as their genetic material where majority lie with the ssRNA genome with positive polarity. Some plant viruses also have DNA as their genetic material (Gergerich and Dolja 2006). Virus-infected plant may or may not lack symptom on the basis of virus-host interaction and the surrounding conditions. Some infection can also lead to plant death. The symptoms in the leaf of the infected plant can be mosaic leaf pattern, vein banding, leaf necrosis, yellowing of leaves, vein clearing along with curled and rolled leaves. Flowers can show discoloration and distortion. Stems can get infected with tumors. Change in color, mosaic formation, and chlorotic ringspots can be seen in the eatable parts (Gergerich and Dolja 2006). Viruses exhibit diverse genetic variability and evolution because of their variable genomic forms and replication methods. The word quasispecies is used to describe the greatly diversed population of the RNA virus where the frequentness in the mutation of the virus population is decided by the fitness and the variation rate (Sanjuán and Domingo-Calap 2019). The main fundamental features of the RNA viruses are their fast mutation rate in their great population size, i.e., upto 10^{12} in an organism, their nature of natural selection, absence of proof-reading activity, and very small genome size, between 3 kb and −30 kb (Moya et al. 2004). The Genetic Structure and Evolution of RNA viruses are based on five criterions: Recombination, Migration, Mutation, Genetic drift, and Natural Selection (Moya et al. 2004). Recombination is a permeating phenomenon that produces diversity among various viruses. It gives viruses a chance to adjust in different surroundings with different hosts by making them bind with other viral genome and allowing the interchange of their genomes. It plays a major role in the evolution of viruses as they have role in the dawn of the viruses, their host adaptation, their pathogenicity and virulence, and in the variation of the transmission vectors (Pérez-Losada et al. 2015). Mutation allows alternations in the nucleotide sequences of the viruses leading to the formation of new variants. So, it is not a recombination that is responsible for mutations but it allows the amalgamations of the previous ones (Pérez-Losada et al. 2015). Migration (also known as gene flow) allows the conveyance of the viruses from the inoculation lot to different tissues (Moya et al. 2004). Genetic drift is caused due to the great population size of the virus leading to Bottle-neck phenomenon majorly during the inter-host transference. The instant result of genetic drift is the deprivation in the genetic diversity as only a little population contributes to the next generation (Sanjuán and Domingo-Calap 2019). Natural Selection is the capability of traits to decrease or increase in frequentness in the population depending on the progenitive achievements of those expressing them. Diversification is approved by the balancing selection, positive selection results in the increase in the frequentness of the favored alleles while negative selection helps to get rid of harmful alleles (Karlsson et al. 2014). The evolution seen in the RNA viruses is due to their great population size as well as their fast mutation rate as positive mutations help them in escaping from the dangers from the harmful ones (Pérez-Losada et al. 2015).

The Potyviridae Family

The Potyviridae with the (+) ve sense and ssRNA genome belongs to one of the largest plant- infecting RNA virus family (Verma et al. 2015). They have 680–900 nm large and 11–20 nm broad pliable filamentous particle. Their genome size is about 9.7 kb with monopartite RNA along with the exception of genus *Bymovirus* with bipartite genome. The VPg protein is linked covalently at the 5′ end and a poly(A) tail is attached at the 3′ end of the genome (Wylie et al. 2017). A polyprotein of 350 kDa is encoded with an open reading frame that has certain proteins: P1, HCPro, P3, P3N-PIPO, 6K1, CI, 6K2, NIa Pro, VPg, Nib, and CP (Valli et al. 2015; Verma et al. 2020). They have lone polyphyletic origin, at the minimum, share three of their genes: the coat protein, helicase region of CI protein, and the RdRp region of NIb protein with the other unlike viruses (Gibbs et al. 2020). The Arthropods or the plasmodiophorids are responsible for the parallel and the seeds for the upright transmission of the viruses (Wylie et al. 2017). The virus in the potyviridae family can be distinguished on the basis of pinwheel-shaped cytoplasmic inclusion bodies or common genomic organization (Valli et al. 2015). The family Potyviridae includes 12 genera and 235 species. The genera of the family potyviridae are: *Arepavirus*, *Bevemovirus*, *Brambyvirus*, *Bymovirus*, *Celavirus*, *Ipomovirus*, *Macluravirus*, *Poacevirus*, *Potyvirus*, *Roymovirus*, *Rymovirus*, and *Tritimovirus* according to ICTV report of virus taxonomy 2020.

Potyvirus

The genus Potyvirus is considered as one of the largest group of the plant virus among 34 different plant virus groups. The potyvirus genus, name taken from potato virus Y is one of the largest plant-infecting virus groups (Ward and Shukla 1991). It consists of more than 180 definite viruses that cause serious infections in floriculture, pasture, decorative, and agricultural crops (Ward and Shukla 1991). A study suggests that the potyviruses emerged because of the initial radiation that took place around 6600 years ago during the origin of agriculture (Gibbs et al. 2008). The genome arrangement and the working of proteins are highly preserved in potyviruses (Revers and García 2015; Gibbs and Ohshima 2010). They form cylindrical inclusion bodies that play a major role in providing the shape which characterizes them in potyvirus group. These viruses have flexible particle and are 680–900 nm long (Riechmann et al. 1992).

Transmission of Potyviruses

These viruses can be transmitted in a nonpersistent manner by aphids mainly by Aphidinae species (Gibbs et al. 2008) and in some cases by seeds (Johansen et al. 1994). Potyviruses have ample number of host range where both di- and monocotyledonous along with cultivated and uncultivated plants are affected (Nigam et al. 2019). Aphids play a major role as a stylet borne, noncirculative transmission vector for potyviruses. It has been observed that less than three aphid vectors or no familiar vectors have led to the transmission of more than 176 potyviruses. *Myzus persicae*, *Aphis gossypii*, and *Aphis craccivora* are some of the aphid species seen transmitting the potyviruses (Gadhave et al. 2020). The interactivity between the HC-Pro and the CP proteins helps in transmission of the virus (Valli et al. 2015). HC-Pro acts as a bridge between the mouthparts of the aphids and the viruses (Valli et al. 2015). The KLSC (KITC of potyvirus) motif of the N-terminus of the HC-Pro is important for the binding of the stylets of the aphids (Blanc et al. 1998; Jossey et al. 2013). The DAG motif of the CP is found to interact with the PTK motif of the C-terminus of the HC-Pro (Huet et al. 1994; Peng et al. 1998). The feeding character of the aphid is one of the major factors for the transmission work of the viruses (Gadhave et al. 2020).

Soybean Mosaic Virus

Soybean mosaic virus (SMV) is one of the major plant-infecting potyviruses that harm the important crops globally (Adams et al. 2005). SMV is one of the viruses included in the Bean common mosaic virus (BCMV) ancestory of potyvirus that were seen emerging in South and East Asia (Gibbs et al. 2008). Yield loss due to Soybean mosaic virus varies from 8% to 35%, although yield reduction up to 94% has been observed (Ross 1977). The SMV transmission usually takes place with the help of vectors such as aphids but sometimes seed transmission can also be seen (Abney et al. 1976; Bowers Jr and Goodman 1979). SMV can infect various hosts such as *Vigna angularis*, *Passiflora* spp., *Senna occidentalis*, *Pinellia ternata*, *Glycine max* (L), *Chenopodium amaranticolor*, *Dolichos lablab*, *Vigna unguiculata (L)*, *Datura stramonium*, *Cucumis sativus* where *G. max* showed perfect transmission (Nandakishor et al. 2017; Almeida et al. 2002; Yoon et al. 2018; Chen et al. 2004; Benscher et al. 1996).

The diverse nature of the SMV strain came into knowledge for the first time when few healthy soybean lines were affected by the isolated SMV strains (Ross 1969; Ross 1975). The variation among the isolated strains was decided on the basis of transmission, symptom development, their host range, and their structure (Hajimorad et al. 2018). The G2 and G7 strains of the SMV were the first whose complete genome sequence was discovered (Jain et al. 1992; Jayaram et al. 1992; Jayaram et al. 1991). Later on, many complete genome sequences were obtained from various countries such as South Korea, China, North America, Japan, and Iran that helped to learn about virus evolution (Hajimorad et al. 2018).

Genome Organization of SMV

SMV has particles that are filamentous in shape and are about 7500 Å in length and 120 Å in diameter with a hole in center of about 15 Å (Hajimorad et al. 2018). The RNA genome of SMV is (+) ve sense and single stranded with 9588 nucleotides having a large open reading frame (ORF) (Jayaram et al. 1992). The VPg is bound at the 5′ end and poly(A) tail at the 3′ end of the viral genome (Fig. 1). Out of 11, 10 proteins are obtained from polyprotein precursor (Riechmann et al. 1992). Proteolytic action takes place for the formation of 10 mature proteins and one fusion protein. The proteins are P1 (Potyvirus 1), HC-Pro (Helper component proteinase), P3 (Potyvirus 3), 6 K1 (6 kinase 1), CI (Cytoplasmic inclusion), 6 K2 (6 Kinase 2), VPg (genome linked protein), NIa (Nuclear inclusion a), NIb (Nuclear inclusion b), CP (Coat protein), and the overlapping gene PIPO at P3 protein forming P3N-PIPO (Nigam et al. 2019; Revers and García 2015; Chung et al. 2008). SMV was used to learn about the structure of potyvirus with the help of electron microscopy using records from fiber diffraction (Kendall et al. 2008).

Functions of SMV Proteins

P1: C-terminal site of this protein has serine protease domain for self-cleavage (Verchot et al. 1991); Host adaptation and symptom development can be caused by P1 (SMV)-Rieske Fe/S (Host) protein interaction (Shi et al. 2007); Helps in phylogenetic relationship studies (Yoshida et al. 2012), used for genome amplification (Verchot and Carrington 1995); P1 protein enhances symptoms and suppresses RNA Silencing (Young et al. 2012).

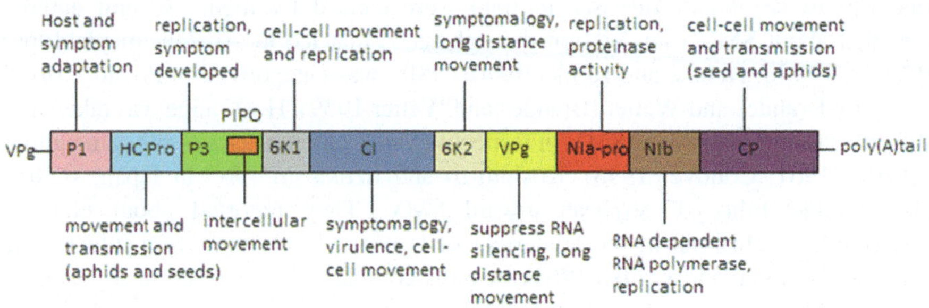

Fig. 1 Genome organization of Soybean Mosaic Virus (SMV)

HC-Pro: Cell-Cell movement (Rojas et al. 1997); long-distance movement (Cronin et al. 1995); aphid and seed transmission (Jossey et al. 2013); suppresses gene silencing (Kasschau and Carrington 1998).

P3: Avirulent and Symptom Determinant (Jenner et al. 2003); Feasible in Replication (Rodriguez-Cerezo et al. 1993).

P3N-PIPO: Present at plasmodesmata and allows intercellular movement of the virus (Wei et al. 2010); Golgi SNARE 12 (host) and P3N-PIPO interaction can attack plasmodesmata (Song et al. 2016).

6 K1: Cell-Cell movement (Hong et al. 2007); replication of virus and attacks the viral replication complex at primary level of infection (Cui and Wang 2016).

CI: RNA helicase for genome replication; cell to cell movement (Carrington et al. 1998); Symptomatology, virulence (Zhang et al. 2009); pathogenicity (Seo et al. 2009).

6 K2: Develops symptoms and allows long-distance movement (Spetz and Valkonen 2004).

VPg: Suppresses RNA silencing (Rajamäki et al. 2014); long-distance movement; replication and translation (Nigam et al. 2019).

NIa-Pro: Cellular localization; RNA-binding activity; proteinase activity and viral replication (Urcuqui-Inchima et al. 2001).

Nib (Replicase Gene): RNA-dependent RNA polymerase; viral replication (Hong and Hunt 1996).

CP (Coat Protein): Cell-Cell movement (Rojas et al. 1997); aphid and seed transmission (Jossey et al. 2013).

Global Status of SMV

Clinton (1916) was the first to report about the disease and symptoms caused by SMV in United States. In 1921, the virus nature was reported by Gardner and Kendrick (Gardner and Kendrick 1921). Then Heinze and Koehler revealed the physical properties of virus in 1940 (Heinze and Kohler 1940). The detailed SMV symptom was reported in 1948 by Conover (Conover 1948). SMV transmission through various aphid species was shown and it was thought (Wei et al. 2010) that majority of secondary diseases in field were caused by them. Several authors revealed about SMV transmission through seeds that led to 30–75% of yield loss (Conover 1948; Heinze and Kohler 1940). SMV was categorized in Potato virus Y group by Brandes and Wetter (Brandes and Wetter 1959). Host range was taken as a differentiating property for SMV in respect with the other viruses of the same group (Quantz 1961; Conover 1948). Koshimizu and Iizuka, in 1963 in Japan, studied about vulnerability of soybean toward SMV. They reported about globular intercellular inclusions in the epidermal cells of soybeans that were infected by SMV (Koshimizu and Lizuka 1963). A virus complex was reported in soybean by Walteks. He confirmed about four natural occurring virus in soybean in which one virus was SMV (Walteks 1958).

Indian Status of SMV

SMV was first observed in India by Nariani and Pingale in New Delhi (Nariani and Pingaley 1960), then it was found infecting soybean in Coimbatore of Tamil Nadu in 1970 (Usman et al. 1973), after that in Kanpur of Uttar Pradesh (Singh et al. 1976) and then in Karnataka (Naik and Murthy 1992). The First molecular proof of soybean-infecting SMV was found in mid-hill condition of Meghalaya which was confirmed by using methods such as RT-PCR, symptomatology, partial characterization of CI, NIB-CP protein, and transmission electron microscopy (Banerjee et al. 2014).

Management of Soybean Mosaic Virus

There have been recent technological developments to learn about the plant viruses concerning food safety globally. One of the methods to detect plant virus is through ELISA (Enzyme-Linked Immunosorbent Assay), a serological test. Here, the selected virus is made to bind with antibodies having specific enzymes in them which are then observed using color-producing enzymatic substrate. It is helpful in detecting filamentous virus (Clark and Adams 1977); other method is by nucleic acid hybridization where the viral genome is attacked by probe genome (Hull and Al-Hakim 1988). The use of PCR (Polymerase chain reaction) and RT-PCR (Real-time PCR) is also seen to detect the viruses. This method uses fluorophores, nucleotide-labeling chemistries, and developed oligoprobe hybridization. Its cycling time is also short and hence used to identify and do the quantitative assessment of the viral targets (Mackay et al. 2002).

It is very important to control the viruses infecting the various plant species. Virus isolation and recognition is the major and the unavoidable step required for the control of virus. Use of vector control agents such as insecticides or pesticides must be promoted against the virus-transmitting vectors (Verma et al. 2016). It is also good to use verified virus-secured seeds to avoid infection. There is also a need to improve the planting method. Plants with natural resistance must be used and if not, resistant genes should be introduced in the plants by conventional method; here genetic engineering can be used for the introduction of genes. Preadjustment of the RNA silencing protecting system of the plants can also help to control or prevent the virus infection. Although genetic engineering has provided various chances to produce plant having resistance from viruses, but there is still a need to learn more about this mechanism in order to avoid possible threats associated with it (Gergerich and Dolja 2006).

The research in the field of virus evolution and genetic variability will help to know about the exposure, epidemiology of the viruses (Grenfell et al. 2004; Mishra et al. 2020), design methods to eliminate diseases (Acosta-Leal et al. 2011) and prepare diagnostical tools (García-Arenal and McDonald 2003) that will benefit in farming. The ancestral record of an organism can be known by their fossil records. Since virus do not produce fossils, phylogenetic analysis plays a major role in learning about the evolutionary history of the viruses (Gergerich and Dolja 2006). The soybean mosaic virus proteins seize certain host factors to complete their life cycle. The characterization of target plant proteins through systems biology approach will lead to a better understanding of this competition between the plant proteins and potyviruses and would also help in designing certain strategies for management against the viral pathogens. Agriculture plays an important role in Indian economy and also 70% of rural households depend on it in India. It also contributes employment to over 60% population, thus agricultural loss due to viral infection is a matter of concern. Therefore, the strategies for virus infection management must be from seedling to harvesting of crops and should consider simultaneously cultural practice, chemical control, genetic resistance of host for sustainable crop production.

References

Abney TS, Sillings JO, Richards TL, Broersma DB (1976) Aphids and other insects as vectors of soybean mosaic virus. J Econ Entomol 69(2):254–256

Acosta-Leal R, Duffy S, Xiong Z, Hammond RW, Elena SF (2011) Advances in plant virus evolution: translating evolutionary insights into better disease management. Phytopathology 101(10):1136–1148

Adams MJ, Antoniw JF, Fauquet CM (2005) Molecular criteria for genus and species discrimination within the family Potyviridae. Arch Virol 150(3):459–479

Almeida ÁM, Sakai J, Souto ER, Kitajima EW, Fukuji TS, Hanada K (2002) Mosaic in Senna occidentalis in southern Brazil induced by a new strain of soybean mosaic virus. Fitopatol Bras 27(2):151–156

Banerjee A, Chandra S, Swer EKP, Sharma SK, Ngachan SV (2014) First molecular evidence of soybean mosaic virus (SMV) infection in soybean from India. Aust Plant Dis Notes 9(1):1–4

Benscher D, Pappu SS, Niblett CL, Varón de Agudelo F, Morales F, Hodson E, Lee RF (1996) A strain of soybean mosaic virus infecting Passiflora spp. in Colombia. Plant Dis 80(3):258–262

Blanc S, Ammar ED, Garcia-Lampasona S, Dolja VV, Llave C, Baker J, Pirone TP (1998) Mutations in the potyvirus helper component protein: effects on interactions with virions and aphid stylets. J Gen Virol 79(12):3119–3122

Bowers GR Jr, Goodman RM (1979) Soybean mosaic virus: infection of soybean seed parts and seed transmission. Phytopathology 69(6):569–572

Brandes J, Wetter C (1959) Classification of elongated plant viruses on the basis of particle morphology. Virology 8(1):99–115

Carrington JC, Jensen PE, Schaad MC (1998) Genetic evidence for an essential role for potyvirus CI protein in cell-to-cell movement. Plant J 14(4):393–400

Chen J, Zheng HY, Lin L, Adams MJ, Antoniw JF, Zhao MF, Chen JP (2004) A virus related to soybean mosaic virus from Pinellia ternata in China and its comparison with local soybean SMV isolates. Arch Virol 149(2):349–363

Chung BYW, Miller WA, Atkins JF, Firth AE (2008) An overlapping essential gene in the Potyviridae. Proc Natl Acad Sci 105(15):5897–5902

Clark MF, Adams AN (1977) Characteristics of the microplate method of enzyme-linked immunosorbent assay for the detection of plant viruses. J Gen Virol 34(3):475–483

Clinton GP (1916) Reports of the botanist for 1915: Soybeans. Annual Report, Connect.

Conover RA (1948) Studies of two viruses causing mosaic diseases in soybean. Phytopathology 38(9)
Cronin S, Verchot J, Haldeman-Cahill R, Schaad MC, Carrington JC (1995) Long-distance movement factor: a transport function of the potyvirus helper component proteinase. Plant Cell 7(5):549–559
Cui H, Wang A (2016) Plum pox virus 6K1 protein is required for viral replication and targets the viral replication complex at the early stage of infection. J Virol 90(10):5119–5131
Gadhave KR, Gautam S, Rasmussen DA, Srinivasan R (2020) Aphid transmission of Potyvirus: the largest plant-infecting RNA virus genus. Viruses 12(7):773
García-Arenal F, McDonald BA (2003) An analysis of the durability of resistance to plant viruses. Phytopathology 93(8):941–952
Gardner NW, Kendrick JB (1921) Soybean mosaic. J Agric Res 22:111–114
Gergerich RC, Dolja VV (2006) Introduction to plant viruses, the invisible foe. Plant Health Instructor 4
Gibbs A, Ohshima K (2010) Potyviruses and the digital revolution. Annu Rev Phytopathol 48:205–223
Gibbs AJ, Trueman JWH, Gibbs MJ (2008) The bean common mosaic virus lineage of potyviruses: where did it arise and when? Arch Virol 153(12):2177–2187
Gibbs AJ, Hajizadeh M, Ohshima K, Jones RA (2020) The potyviruses: an evolutionary synthesis is emerging. Viruses 12(2):132
Grenfell BT, Pybus OG, Gog JR, Wood JL, Daly JM, Mumford JA, Holmes EC (2004) Unifying the epidemiological and evolutionary dynamics of pathogens. Science 303(5656):327–332
Hajimorad MR, Domier LL, Tolin SA, Whitham SA, Saghai Maroof MA (2018) Soybean mosaic virus: a successful potyvirus with a wide distribution but restricted natural host range. Mol Plant Pathol 19(7):1563–1579
Heinze K, Kohler E (1940) Die Mosaikkrankheit der Sojabohne und ihre ubertragung durch insekten. Phytopath Z 13:207–242
Hong Y, Hunt AG (1996) RNA polymerase activity catalyzed by a potyvirus-encoded RNA-dependent RNA polymerase. Virology 226(1):146–151
Hong XY, Chen J, Shi YH, Chen JP (2007) The '6K1' protein of a strain of soybean mosaic virus localizes to the cell periphery. Arch Virol 152(8):1547–1551
Huet H, Gal-On A, Meir E, Lecoq H, Raccah B (1994) Mutations in the helper component protease gene of zucchini yellow mosaic virus affect its ability to mediate aphid transmissibility. J Gen Virol 75(6):1407–1414
Hull R, Al-Hakim A (1988) Nucleic acid hybridization in plant virus diagnosis and characterization. Trends Biotechnol 6(9):213–218
Jain RK, McKern NM, Tolin SA, Hill JH, Barnett OW, Tosic M, Shukla DD (1992) Confirmation that fourteen potyvirus isolates from soybean are strains of one virus by comparing coat protein peptide profiles. Phytopathology 82(3):294–299
Jayaram C, Hill JH, Miller WA (1991) Nucleotide sequences of the coat protein genes of two aphid-transmissible strains of soybean mosaic virus. J Gen Virol 72(4):1001–1003
Jayaram CH, Hill JH, Miller WA (1992) Complete nucleotide sequences of two soybean mosaic virus strains differentiated by response of soybean containing the Rsv resistance gene. J Gen Virol 73(8):2067–2077
Jenner CE, Wang X, Tomimura K, Ohshima K, Ponz F, Walsh JA (2003) The dual role of the potyvirus P3 protein of turnip mosaic virus as a symptom and avirulence determinant in brassicas. Mol Plant-Microbe Interact 16(9):777–784
Johansen E, Edwards MC, Hampton RO (1994) Seed transmission of viruses: current perspectives. Annu Rev Phytopathol 32(1):363–386
Jones RA, Naidu RA (2019) Global dimensions of plant virus diseases: current status and future perspectives. Annu Rev Virol 6:387–409
Jossey S, Hobbs HA, Domier LL (2013) Role of soybean mosaic virus–encoded proteins in seed and aphid transmission in soybean. Phytopathology 103(9):941–948

Karlsson EK, Kwiatkowski DP, Sabeti PC (2014) Natural selection and infectious disease in human populations. Nat Rev Genet 15(6):379–393
Kasschau KD, Carrington JC (1998) A counterdefensive strategy of plant viruses: suppression of post-transcriptional gene silencing. Cell 95(4):461–470
Kendall A, McDonald M, Bian W, Bowles T, Baumgarten SC, Shi J, Stubbs G (2008) Structure of flexible filamentous plant viruses. J Virol 82(19):9546–9554
Koshimizu Y, Lizuka N (1963) Studies on soybean virus diseases in Japan. Tokyo Nat Agric Exp Stn Bull 27:1–103
Mackay IM, Arden KE, Nitsche A (2002) Real-time PCR in virology. Nucleic Acids Res 30(6): 1292–1305
Mishra M, Verma RK, Marwal A, Sharma P, Gaur RK (2020) Biology and interaction of the natural occurrence of distinct Monopartite Begomoviruses associated with satellites in Capsicum annum from India. Front Microbiol 11:512957. https://doi.org/10.3389/fmicb.2020.512957
Moya A, Holmes EC, González-Candelas F (2004) The population genetics and evolutionary epidemiology of RNA viruses. Nat Rev Microbiol 2(4):279–288
Mumford RA, Macarthur R, Boonham N (2016) The role and challenges of new diagnostic technology in plant biosecurity. Food Security 8(1):103–109
Naik RG, Murthy KVK (1992) Characterization of a virus causing mosaic on soybean (Glycine max (L.) Merr) from Karnataka. Legume Res (India) 15:19–23
Nandakishor HV, Kumaraswamy B, Mane SS, Amrutha Veena G (2017) Host range studies of soybean mosaic virus. Int J Curr Microbiol App Sci 6(7):304–308
Nariani TK, Pingaley KV (1960) A mosaic disease of soybean Glycine max (L.) Merr. Indian Phytopathology 13(2)
Nayudu MV (2008) Plant viruses. Tata McGraw-Hill Education
Nigam D, LaTourrette K, Souza PF, Garcia-Ruiz H (2019) Genome-wide variation in potyviruses. Front Plant Sci 10:1439
Peng Y, Kadoury D, Gal-On A, Huet H, Wang Y, Raccah B (1998) Mutations in the HC-pro gene of zucchini yellow mosaic potyvirus: effects on aphid transmission and binding to purified virions. J Gen Virol 79(4):897–904
Pérez-Losada M, Arenas M, Galán JC, Palero F, González-Candelas F (2015) Recombination in viruses: mechanisms, methods of study, and evolutionary consequences. Infect Genet Evol 30: 296–307
Quantz L (1961) Untersuchungen über das Gewöhnliche Bohnenmosaik virus und das Sojamosaikvirus. J Phytopathol 43(1):79–101
Rajamäki ML, Streng J, Valkonen JP (2014) Silencing suppressor protein VPg of a potyvirus interacts with the plant silencing-related protein SGS3. Mol Plant-Microbe Interact 27(11): 1199–1210
Revers F, García JA (2015) Molecular biology of potyviruses. Adv Virus Res 92:101–199
Riechmann JL, Lain S, García JA (1992) Highlights and prospects of potyvirus molecular biology. J Gen Virol 73(1):1–16
Rodriguez-Cerezo E, Ammar ED, Pirone TP, Shaw JG (1993) Association of the non-structural P3 viral protein with cylindrical inclusions in potyvirus-infected cells. J Gen Virol 74(9): 1945–1949
Rojas MR, Zerbini FM, Allison RF, Gilbertson RL, Lucas WJ (1997) Capsid protein and helper component-proteinase function as potyvirus cell-to-cell movement proteins. Virology 237(2): 283–295
Ross JP (1969) Pathogenic variation among isolates of soybean mosaic virus. Phytopathology 59(6):829–832
Ross JP (1975) A newly recognized strain of soybean mosaic virus. Plant Dis Reporter 59(10): 806–808
Ross JP (1977) Effect of aphid-transmitted soybean mosaic virus on yields of closely related resistant and susceptible soybean lines 1. Crop Sci 17(6):869–872

Sanjuán R, Domingo-Calap P (2019) Genetic diversity and evolution of viral populations. Reference Module in Life Sciences

Seo JK, Lee SH, Kim KH (2009) Strain-specific cylindrical inclusion protein of soybean mosaic virus elicits extreme resistance and a lethal systemic hypersensitive response in two resistant soybean cultivars. Mol Plant-Microbe Interact 22(9):1151–1159

Shi Y, Chen J, Hong X, Chen J, Adams MJ (2007) A potyvirus P1 protein interacts with the Rieske Fe/S protein of its host. Mol Plant Pathol 8(6):785–790

Singh BR, Singh OK, Saxena HK (1976) A Mosaic disease of soybean at Kanpur. India Sci Culture 42:53–54

Song P, Zhi H, Wu B, Cui X, Chen X (2016) Soybean Golgi SNARE 12 protein interacts with soybean mosaic virus encoded P3N-PIPO protein. Biochem Biophys Res Commun 478(4): 1503–1508

Spetz C, Valkonen JP (2004) Potyviral 6K2 protein long-distance movement and symptom-induction functions are independent and host-specific. Mol Plant-Microbe Interact 17(5): 502–510

Urcuqui-Inchima S, Haenni AL, Bernardi F (2001) Potyvirus proteins: a wealth of functions. Virus Res 74(1/2):157–175

Usman KM, Ranganathan K, Kandaswamy TK, Damodaran AS, Ayyavoo R (1973) Studies on a mosaic disease of soybean. Madras Agric J 60(7):472–474

Valli A, García JA, López-Moya JJ (2015) Potyviridae. eLS:1–10

Verchot J, Carrington JC (1995) Evidence that the potyvirus P1 proteinase functions in trans as an accessory factor for genome amplification. J Virol 69(6):3668–3674

Verchot J, Koonin EV, Carrington JC (1991) The 35-kDa protein from the N-terminus of the potyviral polyprotein functions as a third virus-encoded proteinase. Virology 185(2):527–535

Verma RK, Mishra R, Petrov N et al (2015) Molecular characterization and recombination analysis of Indian isolate of onion yellow dwarf virus. Eur J Plant Pathol 143(3):437–445

Verma RK, Mishra R, Gaur RK (2016) Genetic variability of the replicase (NIb) gene of papaya ringspot virus in Northern India indicates common ancestry with isolates from China and Taiwan. J Plant Pathol 98(1):105–110

Verma RK, Mishra M et al (2020) Identification, genetic diversity and recombination analysis of watermelon mosaic virus isolates. 3 Biotech 10:257

Walteks HJ (1958) A virus disease complex in soybeans in Arkansas. Phytopathology 48(6)

Ward CW, Shukla DD (1991) Taxonomy of potyviruses: current problems and some solutions. Intervirology 32(5):269–296

Wei T, Zhang C, Hong J, Xiong R, Kasschau KD, Zhou X, Wang A (2010) Formation of complexes at plasmodesmata for potyvirus intercellular movement is mediated by the viral protein P3N-PIPO. PLoS Pathog 6(6):e1000962

Wylie SJ, Adams M, Chalam C, Kreuze J, López-Moya JJ, Ohshima K, Consortium IR (2017) ICTV virus taxonomy profile: Potyviridae. J Gen Virol 98(3):352

Yoon Y, Lim S, Jang YW, Kim BS, Bae DH, Maharjan R, Lee SH (2018) First report of soybean mosaic virus and soybean yellow mottle mosaic virus in Vigna angularis. Plant Dis 102(3): 689–689

Yoshida N, Shimura H, Yamashita K, Suzuki M, Masuta C (2012) Variability in the P1 gene helps to refine phylogenetic relationships among leek yellow stripe virus isolates from garlic. Arch Virol 157(1):147–153

Young BA, Stenger DC, Qu F, Morris TJ, Tatineni S, French R (2012) Tritimovirus P1 functions as a suppressor of RNA silencing and an enhancer of disease symptoms. Virus Res 163(2): 672–677

Zhang C, Hajimorad MR, Eggenberger AL, Tsang S, Whitham SA, Hill JH (2009) Cytoplasmic inclusion cistron of soybean mosaic virus serves as a virulence determinant on Rsv3-genotype soybean and a symptom determinant. Virology 391(2):240–248

An Overview of Microbial-Mediated Alleviation of Abiotic Stress Response in Plant

Ayushi Sharma, Usha, Saurabh Gupta, Kundan Kumar Chaubey, and Shoor Vir Singh

Abstract Plants adapt up to the quick variances and affliction of natural conditions on account of their inherent metabolic abilities. Plant microbiome offers basic help to the plants in getting supplements, opposing against infections, and enduring abiotic stresses. Work on plant-microorganism collaborations at biochemical, physiological, and subatomic levels set up those microbial affiliations to a great extent direct plant reactions toward stresses. In this chapter, we target summing up the ramifications of abiotic and biotic pressure and plant reactions created from that point as far as biochemical instruments followed by the organism interceded pressure alleviation measure. We further depict the job multiomics that assist plants with improving burdens.

Keywords Plant-microbe interaction · Salinity stress · Drought stress · Heavy metal stress

Introduction

Unfavorable natural conditions because of environmental change, joined with declining soil richness, undermine food security. Current agribusiness is confronting a squeezing circumstance where novel methodologies should be created for reasonable food creation and security. The plants are ceaselessly presented to biotic (e.g., nuisance and microorganism assaults) and abiotic stresses (e.g., dry spell, outrageous temperatures, saltiness, substantial metals) which lead to visit change and redesigning of the plant safeguard apparatus, just as including reconfiguration of the plant digestion (Stael et al. 2011; Nomura et al. 2012). Developmentally, plants have fostered a complex and dynamic safeguard framework that renders them as adjust as creatures in reacting adequately to steadily changing conditions (Balmer et al. 2018). Water shortfall (dry spell) has influenced 64.0% of the worldwide land

A. Sharma · Usha · S. Gupta (✉) · K. K. Chaubey · S. V. Singh
Department of Biotechnology, GLA University, Mathura, Uttar Pradesh, India
e-mail: saurabh.gupta@gla.ac.in

© The Author(s), under exclusive license to Springer Nature Singapore Pte Ltd. 2022
A. Vaishnav et al. (eds.), *Plant Stress Mitigators*,
https://doi.org/10.1007/978-981-16-7759-5_28

region, saltiness 6%, mineral insufficiency 9%, and cold 57% (Cramer et al. 2011). Out of the world's 5.2 billion ha of dryland horticulture, 3.6 billion ha is influenced by the issues of disintegration, soil debasement, and saltiness (Riadh et al. 2010). Many occasions plants get worked with in diminishing the weight of ecological anxieties with the help of the microbiome they occupy (Ngumbi and Kloepper 2014). Here, we report an overview of plant-microbe interaction for alleviation of different types of stresses in plants.

Influence of Drought Stress

Quickly changing environment elements convey dry season a genuine intimidation to the manageability of food creation frameworks all through the world (Kogan et al. 2019). Drought influences plant-water potential and turgor, enough to meddle with ordinary capacities (Hsiao 2000) changing physiological and morphological qualities in plants (Rahdari and Hoseini 2012). Growth decrease under dry spell pressure has been contemplated in a few harvests like grain (Samarah 2005), rice (Lafitte et al. 2007), wheat (Rampino et al. 2006), and maize (Kamara et al. 2003). Dry season pressure weakens germination by restricting water imbibition, and diminishes seedling power (Farooq et al. 2009; Kaya et al. 2006). Under dry spell pressure, the reduction in chlorophyll content was indication of photooxidation (Rahdari et al. 2012; Anjum et al. 2017). Diminishing of chlorophyll content in *Carthamus tinctorius* (Siddiqi et al. 2009), bean (Beinsan et al. 2003), and Paulownia imperialis (Astorga and Melendez 2010) was seen under dry spell pressure. Dry spell additionally influences biochemical exercises like nitrate reductase (NR), because of lower take-up of nitrate from the dirt (Caravaca et al. 2005). Drought pressure emphatically impacts plant development and yield arrangement (Bartlett et al. 2019; Tardieu et al. 2017). Drought-instigated decline in yield and yield parts of sugarcane (Vasantha et al. 2005), maize (Kamara et al. 2003), sunflower (Tahir et al. 2002), peanut (Furlan et al. 2012), wheat (Barnabás et al. 2008), cotton (Pettigrew 2004) has likewise been reported. It has been assessed that dry season pressure might diminish the worldwide chickpea creation by 33.0% (Kashiwagi et al. 2015). In soybean, dry spell pressure altogether diminished the quantity of branches and absolute seed yield (Frederick et al. 2001). Drought pressure lessens photosynthesis by restraining leaf region and diminishing the pace of photosynthesis per unit leaf region (Wahid and Rasul 2005). Dry spell pressure dials back carbon obsession rate by straightforwardly confining digestion or by restricting the CO section into the leaf (Farooq et al. 2009). During dry season pressure, restricted intercellular CO_2 fixation prompts the amassing of decreased photosynthetic electron transport parts that might possibly lessen the subatomic oxygen, bringing about the creation of ROS, which are harmful to photosynthetic mechanical assembly (Basu et al. 2016).

Plants adjust in different ways because of dry spell pressure, for example, modifications in development design, plant morphology, and safeguard components (Zandalinas et al. 2018). Various plant species immunized with IAA-creating microbes (Azospirillum) expanded root development or potentially upgraded arrangement of horizontal roots and roots hairs (Dimkpa et al. 2009), water and supplement take-up (Mantelin and Touraine 2004), assisting plants to adapt to water shortfall (Egamberdieva and Kucharova 2009). Azospirillum-wheat affiliation initiated decline in leaf water potential and expansion in leaf water content, which was ascribed to the creation of plant chemicals, for example, IAA by the microbes that upgraded root development and arrangement of horizontal roots in this way expanding take-up of water and supplements under dry season pressure (Arzanesh et al. 2011). Thiourea (exogenous) application has been displayed to further develop plant guard frameworks to a huge degree, which further develops phloem movement of photosynthates in crop plants and subsequently prompts dry spell and saltiness resilience in grains, heartbeats, and oilseeds (Singh and Singh 2017; Bhunia et al. 2015). Mineral supplements likewise assume significant part in further developing plant resistance against stress conditions (Waraich et al. 2012). Nitrogen as nitric oxide (NO) is an exceptionally receptive that secures plant against pressure conditions by going about as forager to ROS (Waraich et al. 2012). Under dry season pressure, phosphorus (P) helps plants to maintain leaf water possible which, thusly, upgrades photosynthetic rates and conductivity of stomata (Waraich et al. 2011).

Influence of Salinity in Plants

A significant test toward world agribusiness includes creation of 70.0% more food crop for an extra 2.3 billion individuals by 2050 overall (FAO 2009). Saltiness is a significant pressure restricting the expansion in the interest for food crops. More than 20% of developed land around the world (~ around 45 hectares) is influenced by salt pressure and the sum is expanding day by day. Plants based on versatile advancement can be grouped generally into two significant sorts: the halophytes (that can withstand saltiness) and the glycophytes (that cannot withstand saltiness and in the end die). Majority of major crop species have a place with this subsequent class. Accordingly saltiness is one of the most fierce natural burdens that hamper crop efficiency overall (Flowers 2004; Munns and Tester 2008).

Saltiness stress causes changes in several physiological and metabolic parameters, depending on the severity and duration of the stress, and subsequently reduces crop production. Soil salinity, in the form of osmotic pressure, is known to impede plant development, followed by particle harmfulness (Rahnama et al. 2010; James et al. 2011). Because of the osmotic pressure of excessive salt accumulating in soil and plants throughout the underlying times of saltiness stress, the water absorption limit of root frameworks decreases and water misfortune from leaves accelerates, saltiness stress is also referred to as hyperosmotic stress (Munns and Tester 2008).

Osmotic pressure in the underlying phase of saltiness stress causes different physiological changes, like interference of layers, supplement lopsidedness, impedes the capacity to detoxify receptive oxygen species (ROS), contrasts in the cell reinforcement proteins and diminished photosynthetic movement, and abatement in stomatal opening (Munns and Tester 2008; James et al. 2011). Saltiness stress is likewise considered as a hyperionic stress. One of the most hindering impacts of saltiness stress is the gathering of Na^+ and Cl^- particles in tissues of plants uncovered to soils with high NaCl focuses. Section of both Na^+ what's more, Cl^- into the cells causes serious particle awkwardness and overabundance take-up might cause critical physiological disorder(s). High Na^+ fixation represses take-up of K^+ particles, which is a fundamental component for development and improvement that results into lower usefulness and may even prompt demise (James et al. 2011; Ahmad 2010). In light of saltiness stress, the creation of ROS, like singlet oxygen, superoxide, hydroxyl extremist, and hydrogen peroxide, is improved (Ahmad 2010). Saltiness-induced ROS development can prompt oxidative harms in different cell parts like proteins, lipids, and DNA, hindering imperative cell elements of plants.

Hereditary varieties in salt resilience exist, and the level of salt resilience changes with plant species and assortments inside a species. Among major crops, grain (Hordeum vulgare) shows a more noteworthy level of salt resilience than wheat (*Triticum aestivum*) and rice (*Oryza sativa*). The level of variety is even more articulated on account of dicotyledons going from Arabidopsis thaliana, which is extremely delicate toward saltiness, to halophytes (Munns and Tester 2008). Over the most recent 20 years, lavish measure of examination has been done to comprehend the system of salt resilience in model plant Arabidopsis. Hereditary varieties and differential reactions to saltiness stress in plants contrasting in pressure resistance empower plant scholars to distinguish physiological systems, sets of qualities, and quality items that are associated with expanding pressure resilience and to join them in reasonable species to yield salt open-minded assortments. The fundamental point of this audit is to talk about research progresses on the complex physiological and subatomic instruments that are associated with plant saltiness resistance.

Influence of Temperature

Yields experience times of outrageous low temperatures in numerous areas of the world (Wang et al. 2016; Ruelland et al. 2009), and are presented to restricted water accessibility inferable from one or the other dry season or, on the other hand, upset water development and take-up under low encompassing temperatures (Beck et al. 2007). Plants that are exposed to both cold and a dry spell at the same time are hampered in their development and, as a result, are inconvenient to utilize (Tommasini et al. 2008). Although studies demonstrate that chilling and dry spells have a similar influence on stomatal advancement and leaf growth, the mechanisms of dry season-induced alterations in specific physiological cycles are substantially different from those triggered by chilling (Deng et al. 2012). The combined influence of freezing and dry season weights on plants has received little attention, and whether plant reactions to these are unique or common is also unclear. Plants may exhibit normal subatomic and physiological processes in response to openness cold and dry season (shared reaction), but others may respond specifically to a certain pressure component (Sewelam et al. 2014; Atkinson et al. 2013). Chilling pressure, like metabolic cycles in plants, thermodynamically reduces the energy of a few physiological processes (Hussain et al. 2016; Ruelland et al. 2009). It reduces germination rate and consistency, reduces seedling survival, and delays ontogenic plant development (Oliver et al. 2007), resulting in severe harvest production losses (Ruelland et al. 2009). Dry season pressure lessens the plant development by affecting different physiological just as biochemical capacities such as photosynthesis, chlorophyll combination, supplement digestion, particle take-up and movement, breath, and carbs digestion (Li et al. 2011; Jaleel et al. 2008). In any case, plants experience water shortage not just during dry spell; however, low temperature may likewise cause turgor stress at cell level (Yadav 2010) inferable from helpless root pressure-driven conductance and decreased root movement (Aroca et al. 2003). In plants, oxidative pressure is critical in wounds caused by cold and drought (Hussain et al. 2016). The production of reactive oxygen species (ROS) in plant cells is accelerated during the dry season and under cold conditions. In plants, excessive production of ROS produces oxidative damage at the cell level, disrupts cell films, and leads to catalyst deactivation, protein debasement, and ionic irregularity (Tarchoune et al. 2010). Plants have an exceptionally effective and modern antioxidative guard framework to control the overproduction of reactive oxygen species (Hussain et al. 2016). The ROS-induced harms and interruption of cell homeostasis are mitigated by the activity of various enzymatic (like catalase, glutathione peroxidise, SOD, peroxidase, GPX) and nonenzymatic (like carotenoids, a-tocopherol, ascorbic corrosive, and glutathione content) cell reinforcements (Chen et al. 2016; Gill and Tuteja 2010; Hussain et al. 2016). The degree of ROS production, as well as the degree of their detoxification by enzymatic or potentially nonenzymatic cancer prevention agent frameworks, is regulated in plants. The production of ROS and its control by strong antioxidative limits has been linked to plant resistance to abiotic stressors (Gill and Tuteja 2010). For a long period, studies have been stored that include cold and dry spells as explicit pressure factors. In general, insufficient attention has been paid to their combined influence and plant response to both of these. Some new investigations contemplate that the physiological and atomic reaction of the harvests to blend of two distinct natural anxieties is exceptional so that such reactions cannot be straightforwardly extrapolated when examined the setting of any of these pressure applied separately (Mittler 2006). In view of existing research on the effects of various stressors on crop plants, an effort is made to comprehend the present understanding of chilling and dry season weight on crop plants. We sought to provide an overview of common and unique responses of yield plants to dry spell and chilling stressors, as well as some potential moderation techniques for adapting to these stresses and minimizing their negative effects on crop plants.

Impact of Nutritional Stress

The supplement take-up conduct of plants is likewise impeded under stress. Excess salt (Na^+ and Cl^-) in the dirt influences accessibility of supplements by forcing contest during take-up, movement or allotting inside the plant (Rabie 2005). As a result, elevated Na^+ and Cl^- concentrations in the soil may suppress supplement-related activities and result in unfavorable Na^+:K^+, Na^+:Ca^{2+}, and Ca^{2+}:Mg^{2+} ratios (Abdel-Fattah and Asrar 2012). Such a marvel might cause an imbalance in the plant's ionic production, affecting the plant's physiological characteristics (Munns et al. 2006).

Temperature has an effect on the physio-substance and microbiological cycles in soils, which may alter the plant-supplement relationships (Yan et al. 2012).

Dry season pressure additionally diminishes the accessibility, take-up, move, and digestion of supplements (Farooq et al. 2009). Exposure of dry spell pressure in plants by and large abatements both the take-up of supplements by roots and movement from roots to shoots (Hu and Schmidhalter 2005). Drought pressure causes the distinction in dynamic vehicle and film porousness of cations (K^+, Ca^{2+}, and Mg^{2+}), in this manner bringing about diminished assimilation of these cations through roots (Farooq et al. 2009). Dry season pressure restricts the exercises of compounds engaged with supplement absorption. For example, the nitrate reductase exercises in the leaves and knobs of dhaincha (Sesbania aculeata L.) and normal bean (*Phaseolus vulgaris L.*) were fundamentally diminished under dry spell pressure (Ashraf and Iram 2005). Apart from full-scale supplement, dry season can instigate the insufficiencies of some miniature supplements, i.e., Mn, Fe, and Mo (Hu and Schmidhalter 2005).

A satisfactory stock of mineral components in the development medium is needed for plants to get by under natural burdens. AMF (Arbuscular mycorrhizal fungi) colonization has been displayed to work on supplement take-up and keep up with ionic homeostasis in plants filled in saline soils (Evelin et al. 2019). AMF colonization likewise impacts the fixation and profile of natural acids and polyamines in plants (Talaat and Shawky 2013). Natural acids assume significant part in bringing down soil electrical conductivity and expanding the accessibility of N, P, and K in soil. Polyamines help in holding particle homeostasis in plant cells by upgrading the take-up of supplements and water (Pang et al. 2007).

To mitigate the unfriendly impacts of abiotic stresses, nitrogen (N) treatment has been accounted for to play huge part in pressure easing (Waraich et al. 2011). Nitrogen as nitric oxide (NO) is an exceptionally responsive that ensures plant against pressure conditions by going about as scrounger to ROS (Waraich et al. 2012). Phosphorus (P) helps plants maintain leaf water throughout the dry season, which improves stomatal conductivity and photosynthetic rates (Waraich et al. 2011). Supplementation of potassium (K) doses and its uptake in plant parts may be beneficial for obtaining reasonable yields during dry spells (Valadabadi and Farahani 2010). Magnesium (Mg) additionally further develops root morphological characters which assist with expanding take-up of water and supplements by means of roots, while sustenance of boron increases sugar transport in the plant body and promotes seed germination and grain arrangement (Waraich et al. 2012). It was discovered that adding Se to a plant's diet increased its resilience to freezing (Hawrylak-Nowak et al. 2010). Under low temperature stress, calcium has been shown to improve the convergences of polyamines (spermidine and putrescine), amino acids, and chlorophyll content in red tidy (Schaberg et al. 2011). The utilization of plant mineral supplements is a potential choice to accomplish better harvest development and efficiency and to reduce the detrimental impacts of stresses in a feasible manner.

Effects of Oxidative Stress

Unevenness among age and safe detoxification of ROS address metabolic express that is alluded to as oxidative stress (Baier et al. 2005). ROS comprises of a gathering of synthetically receptive oxygen particles like hydroxyl extremist (OHH), H_2O_2,. O22, and O22. ROS are created as a result of metabolic routes in plant digestion, which result in the interchange of high-energy electrons, resulting in the production of atomic oxygen (Gill and Tuteja 2010). Excessive ROS aggregation leads to protein oxidation, film lipid peroxidation, DNA and RNA damage, and cell death (Apel and Hirt 2004). Salinity increases the degree of lipid peroxidation (Zhang et al. 2018; Evelin et al. 2012; Pedranzani et al. 2016), resulting in increased layer penetrability and particle loss from phones (Zhang et al. 2018; Evelin et al. 2012; Pedranzani et al. 2016; Felicia et al. 2017; Estrada et al. 2013).

Overproduction of ROS under chilling pressure causes genuine cell harms by quickly responding with fundamental cell structures (Sattler et al. 2000). Various investigations have detailed the higher ROS collection and oxidative harm under dry season pressure (Cruz de Carvalho 2008; Blokhina et al. 2003), owing fundamentally to restricted carbon dioxide obsession and higher photorespiration (Cruz de Carvalho 2008).

Influence of Biotic Stress

In their regular territories, plants coincide with exceptionally powerful microbial networks, some of which are destructive to establish health. Thus, for ideal development and advancement, plants should have a defensive safe and safeguard system. Several models have been proposed for depicting plant invulnerable reactions to biotic burdens, all of which have one thing in common: the natural-resistant framework is dependent on the detection and impression of nonself, harmed self, and modified self, all of which are collectively referred to as "risk" signals (Gust et al. 2017; Sanabria et al. 2010). In general, prepared physical and compound obstacles such as cutin and waxes (fingernail skin), cell divisions, antimicrobial catalysts, and auxiliary metabolites are used by plants in response to biotic worries to prevent or lessen the entry of various microorganisms (Sanabria et al. 2010; Kant et al. 2015). For a situation of a fruitful microorganism passage or plant modification and additionally harm, the second line of safeguard is dispatched, set off by the acknowledgment of peril flags through basically different transmembrane or intracellular example acknowledgment receptors (PRRs). This acknowledgment of risk signals initiates a few degrees of incited safeguard reactions through an intricate organization of sign transduction and intensification, both locally at the site of contamination and foundationally in far off tissues (Gust et al. 2017; Adam et al. 2018).

Microbe-Mediated Mitigation

In any abiotic habitat, microbial collaborations with agricultural plants are critical for the variety and durability of both partners. The phrase "induced systemic tolerance" (IST) refers to the recruitment of abiotic stress responses by organisms (Meena et al. 2017). Microbes contribute to the reduction of abiotic stressors in plants via their probable intrinsic metabolic and genetic capacities (Gopalakrishnan et al. 2015). The choice, screening, and utilization of stress open-minded microorganisms, in this manner, could be suitable choices to assist with conquering usefulness impediments of yield plants in pressure-inclined region. Bacteria that promote plant development use a variety of direct and indirect ways to achieve their goals (Braud et al. 2009; Hayat et al. 2010). Direct processes include the creation of plant growth regulators and the manufacture of bacterial chemicals that aid in the absorption of critical nutrients and micronutrients from the soil. Indirect processes, on the other hand, include antagonistic action against plant pathogenic organisms, the synthesis of HCN and antifungal chemicals, and abiotic stress tolerance. Pseudomonas sp., as well as Acinetobacter sp., was found in grain and oats to improve the production of deaminases for ACC and IAA in salt-affected soil (Chang et al. 2014).

Other than microorganisms, parasites especially the mycorrhiza are additionally significant plant development advertisers. These are chiefly partitioned into mycorrhizal organisms and VAM (vesicular-arbuscular mycorrhizal fungi). These growths may form endosymbiotic affiliations or stay distantly linked to the host plant (ectomycorrhizae) (VAM). Piriformospora indica, a root parasite endophyte, promotes salt resilience in grain (Baltruschat et al. 2008) and dry season resistance in Chinese cabbage (Sun et al. 2010) by increasing cell reinforcements and operating from a variety of angles (Franken 2012).

Effects of Heavy Metals

Plants generally rely upon soil answer for procure supplements for their development and formative cycle. Continued industrialization, serious agrarian practices, and anthropogenic exercises lead to substantial metals defilement in soil. The ongoing expansion in pollution of arable grounds with weighty metals is one of the main sources of misfortune in crop efficiency (Proshad et al. 2018). Heavy metals are metallic components that have a higher thickness than 4 g/cm^3, nondegradable, and furthermore toxic at low fixation (Duruibe et al. 2007; Ma et al. 2016a, b). Heavy metals are characterized into superfluous components (Cd, Pd, Hg, Cr, As, and Ag) being conceivably harmful to plants and fundamental micronutrients (Cu, Zn, Fe, Mn, Mo, Ni, and Co) which are significant for sound development and improvement of plants (Kalaivanan and Ganeshamurthy 2016). A few (conceivably harmful) substantial metals, like Cu, Zn, Ni, Co, Se, and Fe, are additionally fundamental components needed for the ideal exhibition of plants and become poisonous when amassed in overabundance in soil arrangement (Narendrula-Kotha et al. 2019; Khan et al. 2018). Soil defilement with substantial metals causes collection of these poisonous metals in plant parts, bringing about diminished harvest efficiency and expanded danger to creature and human well-being (Couto et al. 2018). Cd pressure lessens plant development, as apparent from plant hindering, decreased leaf region, and diminished shoot and root dry matter yield (Shah et al. 2019). Arsenic (As) is a cancer-causing component present worldwide that is hurtful to each type of life (Ghosh and Mukhopadhyay 2019). Normal harmful impacts of substantial metals incorporate restraint of development and photosynthesis, chlorosis, low biomass amassing, adjusted supplement osmosis, senescence, and water balance which at last can cause plant passing (Ali et al. 2011).

Phytoremediation is savvy and supportable methodology for expulsion of weighty metals (Ma et al. 2016a, b; Chirakkara et al. 2016). Further, on another hand, utilization of microorganisms upgraded the effectiveness of phytoremediation. Microorganisms are more delicate than other living creature and might be a decent marker of weighty metal pressure (Broos et al. 2004; Chen et al. 2014). It has been accounted for that Proteobacteria, Firmicutes, Actinobacteria, possibly eliminated higher grouping of Mn, Pb, and As from metal-dirtied soil (Zhang et al. 2015). AM parasites ease injurious impact of cadmium stress by lessening malonaldehyde and hydrogen peroxide (Hashem et al. 2016).

The utilization of silicon has added to the relief of weighty metals pressure in plants (Wu et al. 2013; Sahebi et al. 2015). Silicon acts by limiting metal and metalloid poison levels through the coprecipitation or complex formation of harmful metalloids in plant tissue and soil, just as incitement of cell reinforcement frameworks in plants (Savvas and Ntatsi 2015). In plants, Si influences the movement and appropriation of metals in different plant parts and permits them to get by under higher metal pressure (Zhang et al. 2008). Decreased Al content with Si application was seen in the stem, roots, and leaves of nut and rice seedlings (Singh et al. 2011; Shen et al. 2014). Heavy metal pressure actuates an abundance arrangement of responsive oxygen species (ROS), which brings about a few metabolic problems in crop plants (Adrees et al. 2015; Ahmad et al. 2019). In this specific situation, the enzymatic and nonenzymatic cancer prevention agent framework animated by Si assists with bringing down oxidative pressure by diminishing the creation of ROS.

Future Prospects and Conclusion

For taking apart more profound association components and interfacing the progressions at atomic levels with the resistance reactions against stresses, natural information dependent on the multiomics approaches should be created. Multiomics approaches are comprehensive and coordinated scientific techniques for the analyzation of perhaps the most perplexing and dynamic living arrangement of microbial connections with plant and adjusting the results created in the plants to assist them with conquering stresses.

References

Abdel-Fattah GM, Asrar AWA (2012) Arbuscular mycorrhizal fungal application to improve growth and tolerance of wheat (Triticum aestivum L.) plants grown in saline soil. Acta Physiol Plant 34:267–277

Adam AL, Nagy Z, Kátay G, Mergenthaler E, Viczián O (2018) Signals of systemic immunity in plants: progress and open questions. Int J Mol Sci 19:1146

Adrees M, Ali S, Rizwan M, Zia-ur-Rehman M, Ibrahim M, Abbas F, Farid M, Qayyum MF, Irshad MK (2015) Mechanisms of silicon-mediated alleviation of heavy metal toxicity in plants: a review. Ecotoxicol Environ Saf 119:186–197

Ahmad (2010, 2010) Growth and antioxidant responses in mustard (*Brassica juncea L.*) plants subjected to combined effect of gibberellic acid and salinity. Arch Agron Soil Sci 56

Ahmad P, Tripathi DK, Deshmukh R, Singh VP, Corpas FJ (2019) Revisiting the role of ROS and RNS in plants under changing environment. Environ Exp Bot

Ali S, Bai P, Zeng F, Cai S, Shamsi IH, Qiu B, Wu F, Zhang G (2011) The ecotoxicological and interactive effects of chromium and aluminum on growth, oxidative damage and antioxidant enzymes on two barley genotypes differing in Al tolerance. Environ Exp Bot 70:185–191

Anjum SA, Ashraf U, Tanveer M, Khan I, Hussain S, Shahzad B et al (2017) Drought-induced changes in growth, osmolyte accumulation and antioxidant metabolism of three maize hybrids. Front Plant Sci 8:69. https://doi.org/10.3389/fpls.2017.0006

Apel K, Hirt H (2004) Reactive oxygen species: metabolism, oxidative stress, and signal transduction. Annu Rev Plant Biol 55:373–399

Aroca R, Vernieri P, Irigoyen JJ, Sancher-Diaz M, Tognoni F, Pardossi A (2003) Involvement of abscisic acid in leaf and root of maize (Zea mays L.) in avoiding chilling-induced water stress. Plant Sci 165:671–679. https://doi.org/10.1016/S0168-9452(03)00257-7

Arzanesh MH, Alikhani HA, Khavazi K, Rahimian HA, Miransari M (2011) Wheat (Triticum aestivum L.) growth enhancement by Azospirillum sp. under drought stress. World J. Microbiol Biotechnol 27:197–205

Ashraf M, Iram A (2005) Drought stress induced changes in some organic substances in nodules and other plant parts of two potential legumes differing in salt tolerance. Flora 200:535–546

Astorga GI, Melendez LA (2010) Salinity effects on protein content, lipid peroxidation, pigments and proline in Paulownia imperialis and Paulowina fortune grown in vitro. Electron J Biotechnol 13:115

Atkinson NJ, Lilley CJ, Urwin PE (2013) Identification of genes involved in the response to simultaneous biotic and abiotic stress. Plant Physiol 162:2028–2041. https://doi.org/10.1104/pp.113.222372

Baier M, Kandlbinder A, Golldack D, Dietz K (2005) Oxidative stress andozone: perception, signalling and response. Plant Cell Environ 28:1012–1020

Balmer A, Pastor V, Glauser G, Mauch-Mani B (2018) Tricarboxylates induce defense priming against bacteriain Arab. Thaliana Front Plant Sci 9:1–15

Baltruschat H, Fodor J, Harrach BD, Niemczyk E, Barna B, Gullner G et al (2008) Salt tolerance of barley induced by the root endophyte Piriformospora indica is associated with a strong increase in anti-oxidants. New Phytol 180:501–510. https://doi.org/10.1111/j.1469-8137.2008.02583.x

Barnabás B, Jäger K, Fehér A (2008) The effect of drought and heat stress on reproductive processes in cereals. Plant Cell Environ 31:11–38

Bartlett MK, Detto M, Pacala SW (2019) Predicting shifts in the functional composition of tropical forests under increased drought and CO2 from trade-offs among plant hydraulic traits. Ecol Lett 22:67–77. https://doi.org/10.1111/ele.13168

Basu S, Ramegowda V, Kumar A, Pereira A (2016) Plant adaptation to drought stress. F1000 Res 5: 1554. https://doi.org/10.12688/f1000research.7678.1

Beck EH, Fettig S, Knake C, Hartig K, Bhattarai T (2007) Specific and unspecific responses of plants to cold and drought stress. J Biosci 32:501–510. https://doi.org/10.1007/s12038-007-0049-5

Beinsan C, Camen D, Sumalan R, Babau M (2003) Study concerning salt stress effect on leaf area dynamics and chlorophyll content in four bean local landraces from Banat areas. Fac Hortic 119:416–419

Bhunia SR, Verma IM, Sahu MP, Sharma NC, Balai K (2015) Effect of drip irrigation and bioregulators on yield, economics and water use of fenugreek (Trigonella foenum-graecum). J Spices Aromat Crop 24:102–105

Blokhina O, Virolainen E, Fagerstedt KV (2003) Antioxidants, oxidative damage and oxygen deprivation stress: a review. Ann Bot 91:179–194. https://doi.org/10.1093/aob/mcf118

Braud A, Jezequel K, Bazot S, Lebeau T (2009) Enhanced phytoextraction of an agricultural Cr-, Hg- and Pb contaminated soil by bioaugmentation with siderophore-producing bacteria. Chemosphere 74:280–286. https://doi.org/10.1016/j.chemosphere.2008.09.013

Broos K et al (2004) A survey of symbiotic nitrogen fixation by white clover grown on metal-contaminated soils. Soil Biol Biochem 36:633–640

Caravaca F, Alguacil MM, Herníandez JA, Roldaín A (2005) Involvement of antioxidant enzyme and nitrate reductase activities during water stress and recovery of mycorrhizal Myrtus communis and Phillyrea angustifolia plants. Plant Sci 169:191–197

Chang P, Gerhardt KE, Huang X-D, Yu X-M, Glick BR, Gerwing PD et al (2014) Plant growth-promoting bacteria facilitate the growth of barley and oats in salt impacted soil: implications for phytoremediation of saline soils. Int J Phytoremediation 16:1133–1147. https://doi.org/10.1080/15226514.2013.821447

Chen L et al (2014) Interaction of cd-hyperaccumulator Solanum nigrum L. and functional endophyte Pseudomonas sp. Lk9 on soil heavy metals uptake. Soil Biol Biochem 68:300–308

Chen W, Guo C, Hussain S, Zhu B, Deng F, Xue Y et al (2016) Role of xylo-oligosaccharides in protection against salinity-induced adversities in Chinese cabbage. Environ Sci Pollut Res 23: 1254–1264. https://doi.org/10.1007/s11356-015-5361-2

Chirakkara RA, Cameselle C, Reddy KR (2016) Assessing the applicability of phytoremediation of soils with mixed organic and heavy metal contaminants. Rev Environ Sci Biotechnol 15:299–326

Couto R, Comin JJ, Souza M, Ricachenevsky FK, Lana M, Gatiboni L et al (2018) Should heavy metals be monitored in foods derived from soils fertilized with animal waste? Front Plant Sci 9: 732. https://doi.org/10.3389/fpls.2018.00732

Cramer GR, Urano K, Delrot S, Pezzotti M, Shinozaki K (2011) Effectsof abiotic stress on plants: a systems biology perspective. BMC Plant Biol 11:163

Cruz de Carvalho MH (2008) Drought stress and reactive oxygen species: production, scavenging and signaling. Plant Signal Behav 3:156–165. https://doi.org/10.4161/psb.3.3.5536

Deng B, Du W, Liu C, Sun W, Tian S, Dong H (2012) Antioxidantresponse to drought, cold and nutrient stress in two ploidy levels of tobaccoplants: low resource requirement confers polytolerance in polyploids? Plant Growth Regul 66:37–47

Dimkpa C, Weinand T, Asch F (2009) Plant-rhizobacteria interactions alleviate abiotic stress conditions. Plant Cell Environ 32:1682–1694

Duruibe JO, Ogwuegbu MOC, Egwurugwu JN (2007) Heavy metal pollution and human bio toxic effects. Int J Phys Sci 2:112–118

Egamberdieva D, Kucharova Z (2009) Selection for root colonizing bacteria stimulating wheat growth in saline soils. Biol Fert Soil 45:561–573

Estrada B, Aroca R, Maathuis FJ, Barea JM, Ruiz-Lozano JM (2013) Arbuscular mycorrhizal fungi native from a Mediterranean saline area enhance maize tolerance to salinity through improved ion homeostasis. Plant Cell Environ 36:1771–1782. https://doi.org/10.1111/pce.12082

Evelin H, Giri B, Kapoor R (2012) Contribution of Glomus intraradices inoculation to nutrient acquisition and mitigation of ionic imbalance in NaCl-stressed Trigonella foenum-graecum. Mycorrhiza 22:203–217. https://doi.org/10.1007/s00572-011-0392-0

Evelin H, Devi TS, Gupta S, Kapoor R (2019) Mitigation of salinity stress in plants by arbuscular mycorrhizal Symbiosis: current understanding and new challenges. Front Plant Sci 10:470

FAO (2009) *High level expert forum—how to feed the world in 2050*, economic and social development. Food and Agricultural Organization of the United Nations, Rome, Italy

Farooq M, Wahid A, Kobayashi N, Fujita D, Basra SMA (2009) Plant drought stress: effects, mechanisms and management. Agron Sustain Dev 29:185–212. https://doi.org/10.1051/agro:2008021

Felicia V, Russia P, Ingraffia R, Giambalvo D, Frenda AS, Martinelli F (2017) Arbuscular mycorrhizal symbiosis mitigates the negative effects of salinity on durum wheat. PLoS One 12:e0184158. https://doi.org/10.1371/journal.pone.0184158

Flowers TJ (2004) Improving crop salt tolerance. J Exp Bot 55:307–319

Franken P (2012) The plant strengthening root endophyte Piriformospora indica: potential application and the biology behind. Appl Microbiol Biotechnol 96:1455–1464. https://doi.org/10.1007/s00253-012-4506-1

Frederick JR, Camp CR, Bauer PJ (2001) Drought-stress effects on branch and main stem seed yield and yield components of determinate soybean. Crop Sci 41:759–763. https://doi.org/10.2135/cropsci2001.413759x

Furlan A, Llanes A, Luna V, Castro S (2012) Physiological and biochemical responses to drought stress and subsequent rehydration in the symbiotic association peanut Bradyrhizobium sp. ISRN Agron 2012:1–8. https://doi.org/10.5402/2012/318083

Ghosh G, Mukhopadhyay DK (2019) Human health hazards due to arsenic and fluoride contamination in drinking water and food chain. In: Groundwater development and management. Springer, New Delhi, pp 351–369. https://doi.org/10.1007/978-3-319-75115-3_15

Gill SS, Tuteja N (2010) Reactive oxygen species and antioxidant machinery in abiotic stress tolerance in crop plants. Plant Physiol Biochem 48:909–930. https://doi.org/10.1016/j.plaphy.2010.08.016

Gopalakrishnan S, Sathya A, Vijayabharathi R, Varshney RK, Gowda CL, Krishnamurthy L (2015) Plant growth-promoting rhizobia: challenges and opportunities. 3Biotech 5:355–377

Gust AA, Pruitt R, Nürnberger T (2017) Sensing danger: key to activating plant immunity. Trends Plant Sci 22:779–791

Hashem A et al (2016) Alleviation of cadmium stress in Solanum lycopersicum L. by arbuscular mycorrhizal fungi via induction of acquired systemic tolerance. Saudi J. Soil Sci 23:272–281

Hawrylak-Nowak B, Matraszek R, Szymanska M (2010) Selenium modifies the effect of short-term chilling stress on cucumber plants. Biol Trace Elem Res 138:307–315

Hayat R, Ali S, Amara U, Khalid R, Ahmed I (2010) Soil beneficial bacteria and their role in plant growth promotion: are view. Ann Microbiol 60:579–598. https://doi.org/10.1007/s13213-010-0117-1

Hsiao A (2000) Effect of water deficit on morphological and physiological characterizes in Rice (Oryza sativa). J Agric For 3:93–97

Hu Y, Schmidhalter U (2005) Drought and salinity: a comparison of their effects on mineral nutrition of plants. J Plant Nutr Soil Sci 168:541–549

Hussain S, Khan F, Cao W, Wu L, Geng M (2016) Seed priming alters the production and detoxification of reactive oxygen intermediates in rice seedlings grown under sub-optimal temperature and nutrient supply. Front Plant Sci 7:439. https://doi.org/10.3389/fpls.2016.00439

Jaleel CA, Gopi R, Sankar B, Gomathinayagam M, Panneerselvam R (2008) Differential responses in water use efficiency in two varieties of Catharanthus roseus under drought stress. Comp Rend Biol 331:42–47. https://doi.org/10.1016/j.crvi.2007.11.003

James RA, Blake C, Byrt CS, Munns R (2011) Major genes for Na+ exclusion, Nax1 and Nax2 (wheat HKT1;4 and HKT1;5), decrease Na+ accumulation in bread wheat leaves under saline and waterlogged conditions. J Exp Bot 62:2939–2947

Kalaivanan D, Ganeshamurthy AN (2016) Mechanisms of heavy metal toxicity in plants. In: Abiotic stress physiology of horticultural crops. Springer, Berlin, Germany, pp 85–102

Kamara AY, Menkir A, Badu-Apraku B, Ibikunle O (2003) The influence of drought stress on growth, yield and yield components of selected maize genotypes. J Agric Sci 141:43–50. https://doi.org/10.1017/S0021859603003423

Kant MR, Jonckheere W, Knegt B, Lemos F, Liu J, Schimmel BCJ, Villarroel CA, Ataide LMS, Dermauw W, Glas JJ et al (2015) Mechanisms and ecological consequences of plant defence induction and suppression in herbivore communities. Ann Bot 115:1015–1051

Kashiwagi J, Krishnamurthy L, Purushothaman R, Upadhyaya HD, Gaur PM, Gowda CLL et al (2015) Scope for improvement of yield under drought through the root traits in chickpea (Cicer arietinum L.). Field Crop Res 170:47–54. https://doi.org/10.1016/j.fcr.2014.10.003

Kaya MD, Okçub G, Ataka M, Çıkılıc Y, Kolsarıcıa Ö (2006) Seed treatments to overcome salt and drought stress during germination in sunflower (Helianthus annuus L.). Eur J Agron 24:291–295. https://doi.org/10.1016/j.eja.2005.08.001

Khan ZI, Ugulu I, Sahira S, Ahmad K, Ashfaq A, Mehmood N et al (2018) Determination of toxic metals in fruits of Abelmoschus esculentus grown in contaminated soils with different irrigation sources by spectroscopic method. Int J Environ Res 12:503–511

Kogan F, Guo W, Yang W (2019) Drought and food security prediction from NOAA new generation of operational satellites. Geomatics Nat Hazards Risk 10:651–666. https://doi.org/10.1080/19475705.2018.1541257

Lafitte HR, Yongsheng G, Yan S, Lil ZK (2007) Whole plant responses, key processes, and adaptation to drought stress: the case of rice. J Exp Bot 58:169–175

Li C, Jiang D, Wollenweber B, Li Y, Dai T, Cao W (2011) Waterlogging pretreatment during vegetative growth improves tolerance to waterlogging after anthesis in wheat. Plant Sci 180: 672–678. https://doi.org/10.1016/j.plantsci.2011.01.009

Ma Y et al (2016a) Inoculation of Brassica oxyrrhina with plant growth-promoting bacteria for the improvement of heavy metal phytoremediation under drought conditions. J Hazard Mater 320: 36–44

Ma Y et al (2016b) The beneficial role of bacterial endophytes in heavy metal phytoremediation. J Environ Manag 174:14–25

Mantelin S, Touraine B (2004) Plant growth-promoting rhizobacteria and nitrate availability: impacts on root development and nitrate uptake. J Expt Bot 55:27–34

Meena KK, Sorty AM, Bitla UM, Choudhary K, Gupta P, Pareek A, Singh DP, Prabha R, Sahu PK, Gupta VK, Singh HB, Krishanani KK, Minhas PS (2017) Abiotic stress responses and microbe-mediated mitigation in plants: the Omics strategies. Front Plant Sci 8:172. https://doi.org/10.3389/fpls.2017.00172

Mittler R (2006) Abiotic stress, the field environment and stress combination. Trends Plant Sci 11: 15–19. https://doi.org/10.1016/j.tplants.2005.11.002

Munns R, James RA, Läuchli A (2006) Approaches to increasing the salt tolerance of wheat and other cereals. J Exp Bot 57:1025–1043

Munns R, Tester M (2008) Mechanisms of salinity tolerance. Annu Rev Plant Biol 59:651–681

Narendrula-Kotha R, Theriault G, Mehes-Smith M, Kalubi K, Nkongolo K (2019) Metal toxicity and resistance in plants and microorganisms in terrestrial ecosystems. In: Reviews of environmental contamination and toxicology, vol 249. Springer, Cham, Switzerland, pp 1–27. https://doi.org/10.1007/398_2018_22

Ngumbi E, Kloepper J (2014) Bacterial-mediated drought tolerance: currentand future prospects. Appl Soil Ecol 105:109–125

Nomura H, Komori T, Uemura S, Kanda Y, Shimotani K, Nakai K, Furuichi T, Takebayashi K, Sugimoto T, Sano S (2012) Chloroplast-mediated activation of plant immune signalling in Arabidopsis. Nat Commun 3

Oliver SN, Dennis ES, Dolferus R (2007) ABA regulates apoplastic sugar transport and is a potential signal for cold-induced pollen sterility in rice. Plant Cell Physiol 48:1319–1330. https://doi.org/10.1093/pcp/pcm100

Pang XM, Zhang ZY, Wen XP, Ban Y, Moriguchi T (2007) Polyamines, all-purpose players in response to environment stresses in plants. Plant Stress 1:173–188

Pedranzani H, Rodríguez-Rivera M, Gutiérrez M, Porcel R, Hause B, Ruiz-Lozano JM (2016) Arbuscular mycorrhizal symbiosis regulates physiology and performance of Digitaria eriantha plants subjected to abiotic stresses by modulating antioxidant and jasmonate levels. Mycorrhiza 26:141–152. https://doi.org/10.1007/s00572-015-0653-4

Pettigrew WT (2004) Physiological consequences of moisture deficit stress in cotton. Crop Sci 44: 1265–1272. https://doi.org/10.2135/cropsci2004.1265

Proshad R, Kormoker T, Mursheed N, Islam MM, Bhuyan MI, Islam MS, Mithu TN (2018) Heavy metal toxicity in agricultural soil due to rapid industrialization in Bangladesh: a review. Int J Adv Geosci 6(1):83–88

Rabie GH (2005) Influence of arbuscular mycorrhizal fungi and kinetin on the response of mungbean plants to irrigation with seawater. Mycorrhiza 15:225–230

Rahdari P, Hoseini SM (2012) Drought stress, a review. Int J Agron Plant Prod 3:443–446

Rahdari P, Hoseini SM, Tavakoli S (2012) The studying effect of drought stress on germination, proline, sugar, lipid, protein and chlorophyll content in Purslane (Portulaca oleraceae L.) leaves. J Med Plants Res 6:1539–1547

Rahnama A, James RA, Poustini K, Munns R (2010) Stomatal conductance as a screen for osmotic stress tolerance in durum wheat growing in saline soil. Funct Plant Biol 37:255–263

Rampino P, Pataleo S, Gerardi C, Perotta C (2006) Drought stress responses in wheat: physiological and molecular analysis of resistant and sensitive genotypes. Plant Cell Environ 29:2143–2152

Riadh K, Wided M, Hans-Werner K, Chedly A (2010) Responses of halophytes to environmental stresses with special emphasis to salinity. Adv Bot Res 53:117–145

Ruelland E, Vaultier MN, Zachowski A, Hurry V (2009) Cold signalling and cold acclimation in plants. Adv Bot Res 49:35–150. https://doi.org/10.1016/S0065-2296(08)00602-2

Sahebi M, Hanafi MM, Siti Nor Akmar A, Rafii MY, Azizi P, Tengoua F, Nurul Mayzaitul Azwa J, Shabanimofrad M (2015) Importance of silicon and mechanisms of biosilica formation in plants. BioMed Res Int 2015

Samarah NH (2005) Effects of drought stress on growth and yield of barley. AgronSustain Dev 25: 145–149

Sanabria NM, Huang JC, Dubery IA (2010) Self/nonself perception in plants in innate immunity and defense. Self/Nonself-Immune Recognit Signal 1:40–54

Sattler U, Calsou P, Boiteux S, Salles B (2000) Detection of oxidative base dna damage by a new biochemical assay. Arch Biochem Biophys 376:26–33. https://doi.org/10.1006/abbi.2000.1701

Savvas D, Ntatsi G (2015) Biostimulant activity of silicon in horticulture. Sci Hortic 196:66–81

Schaberg PG, Minocha R, Long S, Halman JM, Hawley GJ, Eagar C (2011) Calcium addition at the Hubbard brook experimental forest increases the capacity for stress tolerance and carbon capture in red spruce (Picea rubens) trees during the cold season. Trees 25:1053–1061

Sewelam N, Oshima Y, Mitsuda N, Ohme-Takagi M (2014) A step towards understanding plant responses to multiple environmental stresses: a genome-wide study. Plant Cell Environ 37: 2024–2035. https://doi.org/10.1111/pce.12274

Shah K, Nahakpam S, Chaturvedi V, Singh P (2019) Cadmium-induced anatomical abnormalities in plants. In: Cadmium toxicity and tolerance in plants. Academic Press Elsevier, pp 111–139. https://doi.org/10.1016/B978-0-12-814864-8.00005

Shen X, Xiao X, Dong Z, Chen Y (2014) Silicon effects on antioxidative enzymes and lipid peroxidation in leaves and roots of peanut under aluminum stress. Acta Physiol Plant 36: 3063–3069

Siddiqi EH, Ashraf M, Hussain M, Jamil A (2009) Assessment of intercultivar variation for salt tolerance in safflower (*Carthamustinctorius L.*) using gas exchange characteristics as selection criteria. Pak J Bot 41:2251–2259

Singh RP, Singh D (2017) Response of lentil to thiourea application under rainfed conditions of Central India. Int J Curr Microbiol App Sci 6:2556–2560. https://doi.org/10.20546/ijcmas.2017.603.289

Singh VP, Tripathi DK, Kumar D, Chauhan DK (2011) Influence of exogenous silicon addition on aluminium tolerance in rice seedlings. Biol Trace Elem Res 144:1260–1274

Stael, S, Rocha, AG, Robinson, AJ, Kmiecik, P, Vothknecht, UC, Teige (2011) M Arabidopsis calcium-bindingmitochondrial carrier proteins as potential facilitators of mitochondrial ATP-import and plastid SAM-import. FEBS Lett 585:3935–3940

Sun C, Johnson J, Cai D, Sherameti I, Oelmüeller R, Lou B (2010) Piriformospora indica confers drought tolerance in Chinese cabbage leaves by stimulating antioxidant enzymes, the expression of drought-related genes and the plastid-localized CAS protein. J Plant Physiol 167:1009–1017. https://doi.org/10.1016/j.jplph.2010.02.013

Tahir MHN, Imran M, Hussain MK (2002) Evaluation of sunflower (Helianthus annuus L.) inbred lines for drought tolerance. Int J Agric Biol 3:398–400

Talaat NB, Shawky BT (2013) Modulation of nutrient acquisition and polyamine pool in salt-stressed wheat (Triticum aestivum L.) plants inoculated with arbuscular mycorrhizal fungi. Acta Physiol Plant 35:2601–2610

Tarchoune I, Sgherric C, Izzo R, Lachaal M, Ouerghi Z, Navari-Izzo F (2010) Antioxidative response of Ocimum basilicum to sodium chloride and sodium sulphate salinization. Plant Physiol Biochem 48:772–777. https://doi.org/10.1016/j.plaphy.2010.05.006

Tardieu F, Draye X, Javaux M (2017) Root water uptake and ideotypes of the root system: whole-plant controls matter. Vadose Zo J 16:1–10. https://doi.org/10.2136/vzj2017.05.0107

Tommasini L, Svensson JT, Rodriguez EM, Wahid A, Malatrasi M, Kato K et al (2008) Dehydrin gene expression provides an indicator of low temperature and drought stress: transcriptome-based analysis of barley (Hordeum vulgare L.). Funct Integr Genomics 8:387–405. https://doi.org/10.1007/s10142-008-0081-z

Valadabadi SA, Farahani HA (2010) Studying the interactive effect of potassium application and individual field crops on root penetration under drought condition. Asian J Agric Sci 1:19–24

Vasantha S, Alarmelu S, Hemaprabha G, Shanthi RM (2005) Evaluation of promising sugarcane genotypes for drought. Sugar Technol 7:82–83

Wahid A, Rasul E (2005) Photosynthesis in leaf, stem, flower and fruit. In: Pessarakali M (ed) Handbook of photosynthesis, 2nd edn. CRC Press), Boca Raton, pp 479–497

Wang W, Chen Q, Hussain S, Mei J, Dong H, Peng S et al (2016) Pre-sowing seed treatments in direct-seeded early rice: consequences for emergence, seedling growth and associated metabolic events under chilling stress. Sci Rep 6:19,637

Waraich EA, Ahmad R, Ashraf MY (2011) Role of mineral nutrition in alleviation of drought stress in plants. Aust J Crop Sci 5:764–777
Waraich EA, Ahmad R, Halim A, Aziz T (2012) Alleviation of temperature stress by nutrient management in crop plants: a review. J Soil Sci Plant Nutr 12:221–244
Wu J, Shi Y, Zhu Y, Wang Y, Gong H (2013) Mechanisms of enhanced heavy metal tolerance in plants by silicon: a review. Pedosphere 23:815–825
Yadav SK (2010) Cold stress tolerance mechanisms in plants. a review. Agron Sustain Dev 30:515–527
Yan Q, Duan Z, Jingdong M, Xun L, Fei D (2012) Effects of rootzone temperature and N, P, and K supplies on nutrient uptake of cucumber (Cucumis sativus L.) seedlings in hydroponics. Soil Sci Plant Nutr 58:707–717
Zandalinas SI, Mittler R, Balfagón D, Arbona V, Gómez-Cadenas A (2018) Plant adaptations to the combination of drought and high temperatures. Physiol Plant 162:2–12
Zhang C, Wang L, Nie Q, Zhang W, Zhang F (2008) Long-term effects of exogenous silicon on cadmium translocation and toxicity in rice (Oryza sativa L.). environ. Exp Bot 62:300–307
Zhang J et al (2015) Health risk to residents and stimulation to inherent bacteria of various heavy metals in soil. Sci. Total Environ 508:29–36
Zhang W, Wang C, Lu T, Zheng Y (2018) Cooperation between arbuscular mycorrhizal fungi and earthworms promotes the physiological adaptation of maize under a high salt stress. Plant Soil 423:125–140. https://doi.org/10.1007/s11104-017-3481-9

The manufacturer's authorised representative in the EU is Springer Nature Customer Service Centre GmbH, Europaplatz 3, 69115 Heidelberg, Germany. If you have any concerns regarding our products, please contact ProductSafety@springernature.com

Printed and bound by CPI Group (UK) Ltd, Croydon, CR0 4YY
30/09/2026
02225301-0016